PHYSICAL GEOGRAPHY

A LANDSCAPE APPRECIATION

FIFTH EDITION

Tom L. McKnight

University of California, Los Angeles

Prentice Hall
Upper Saddle River, New Jersey 07458

Library of Congress Cataloging-in-Publication Data

McKnight, Tom L. (Tom Lee)
     Physical Geography : a landscape appreciation / Tom L. McKnight. —
5th ed.
        p.   cm.
     Includes bibliographical references and index.
     ISBN 0-13-440215-4
     1. Physical geography.   I. Title.
GB54.5.M39   1996
910'.02—dc20   95-34179
                                                                   CIP

Acquisition Editor: Ray Henderson
Development Editor: Ray Mullaney
Production Editor: Debra Wechsler
Director, Production & Manufacturing: David W. Riccardi
Editor-in-Chief: Paul Corey
Managing Editor, Production: Kathleen Schiaparelli
Marketing Manager: Kelly MacDonald
Copy Editor: Bert Zelman
Art Director: Heather Scott
Cover Designer: CopperLeaf Design
Creative Director: Paula Maylahn
Buyer: Trudi Pisciotti
Page Layout: Michael Bertrand
Photo Research: Tobi Zausner
Photo Editor: Lori Morris-Nantz
Illustrator: Maryland Cartographics
Art Coordinator: Patrice Van Acker, Rhoda Sidney

Cover Photograph Credits: Tony Stone Images, FPG International,
The Image Bank, Minden Pictures, Inc., Starlight, Inc.,
National Oceanic and Atmospheric Administration

 © 1996, 1993, 1990, 1987, 1984 by Prentice-Hall, Inc.
Simon & Schuster/A Viacom Company
Upper Saddle River, New Jersey 07458

Printed in the United States of America
10 9 8 7 6 5 4 3 2 1

ISBN 0-13-440215-4

Prentice-Hall International (UK) Limited, *London*
Prentice-Hall of Australia Pty. Limited, *Sydney*
Prentice-Hall Canada Inc., *Toronto*
Prentice-Hall Hispanoamerica, S.A., *Mexico*
Prentice-Hall of India Private Limited, *New Delhi*
Prentice-Hall of Japan, Inc., *Tokyo*
Simon & Schuster Asia Pte, Ltd., *Singapore*
Editora Prentice-Hall do Brasil, Ltda., *Rio de Janeiro*

For Clint and Jill,
who make me smile with pride

CONTENTS

5

Atmospheric Pressure and Wind   *106*

6

Atmospheric Moisture   *138*

7

Transient Atmospheric Flows and Disturbances   *170*

8

Climatic Zones and Types   *202*

9

The Hydrosphere   *242*

10

Cycles and Patterns in the Biosphere   *268*

11

Terrestrial Flora and Fauna   *288*

12

Soils   *328*

13

Introduction to Landform Study *364*

14

The Internal Processes *386*

15

Preliminaries to Erosion: Weathering and Mass Wasting *424*

16

The Fluvial Processes *446*

17

The Topography of Arid Lands *480*

18

Coastal Processes and Terrain   *508*

19

Solution Processes and Karst Topography   *530*

20

Glacial Modification of Terrain   *544*

PREFACE

The United States of America possesses many singular characteristics, most of them good, but some of them not so good. In the latter category is the unfortunate distinction that our citizens probably are the world's most geographically ignorant people. Despite our highly literate and educated society, on average we know relatively little about the geography of our own country, not to mention that of the rest of the world. In almost all other countries, geography is a basic field of study in both primary and secondary schools, as well as being a firmly established university subject. Over much of the world, schoolchildren are exposed to geographic training for most of their school years. This applies not only to such developed countries as England, Sweden, and New Zealand, but also to such less-developed lands as India, Tanzania, and Ecuador. Not so in the United States, where a pupil last hears the word *geography* in about the third grade, and the rare introduction of geographic course content at any higher level is usually muffled under the heading of "social studies."

Fortunately, this sad situation is now in a process of dynamic change. During the late 1980s our collective geographical ignorance became a matter of widespread discussion and concern, and some significant actions were taken to introduce or upgrade geographic education at various levels. The national education objectives of **America 2000** stipulate Geography as one of eight subjects that should comprise basic education in primary and secondary schools throughout the nation, and for which each state will be expected to develop functional standards. Thus in the mid-1990s there is a swelling enthusiasm for geographic training, although there is much lost ground to recover. Indeed, most American students continue to be surprised when they discover that geography courses are offered in colleges and universities. Nevertheless, geography is a well-established discipline in most of our institutions of higher learning, and its significance is sure to grow as more geographic content is introduced into the K–12 (kindergarten–twelfth grade) curriculum.

The writer believes that a useful definition of geography is "landscape appreciation", and has prepared this book with that theme in mind. "Landscape" is considered to include everything that one senses by sight, sound, and smell when looking out a window. "Appreciation" is used in the sense of understanding. Any proper exposition of geography should serve to heighten one's understanding of all that is seen, heard, and smelled at the window, whether an actual experience at a nearby window or a vicarious experience on the other side of the world. Thus it is the purpose of this book to make the environmental landscape of the world more understandable to the reader, at least at an introductory level.

What do you see when you cross the Mohave Desert from Los Angeles to Las Vegas? Three hundred miles of Not Much? A geographer sees 300 miles of Quite A Lot. It is hoped that this book will help the reader expand her or his capacity for landscape appreciation from the former to the latter.

FEATURES OF THE NEW EDITION

New to this revision are the following features:

- A completely revamped cartographic and illustration program, including:
- Over 240 new maps, with shaded relief where appropriate
- New climographs throughout
- Completely new line illustrations with numerous multi-part illustrations that capture sequence and evolution
- Over 150 new photographs
- Major photos are paired with locator maps to heighten basic geographic literacy
- New coverage of analemma is provided in Chapter 1. An updated discussion of *Geography as a Field of Learning* and career opportunities for geographers are also highlighted in this chapter.
- A new focus box on understanding topographic maps opens Chapter 2. New coverage of mapping provides a thematic overview on the impact of new technologies including automated cartography, GIS, and GPS.

- A completely revised discussion of atmospheric pressure is featured in Chapter 5. Twenty academic reviewers helped to completely reconceive the illustrations for greater pedagogical effectiveness. This chapter also contains a new focus box on wind energy.
- Chapter 7 has been updated to include coverage of Hurricane Andrew and new El Niño data from TOGA/COARE.
- The confusing terminology of *climate zone, climate type,* and *climate subtype* have been clarified in Chapter 8. This chapter also contains a new focus box on *An Arizona Monsoon.*
- An update of Chapter 9 includes 1990 soil taxonomy data and new coverage of Andisols. Soil distribution maps have been carefully redrawn from the latest data, and now feature companion maps for US and World distributions so students better understand differences in the local, national, and global distribution of soils.
- A more systematic Chapter 10 has been revised around the concept of energy flows. Photosynthesis has been recast in similar form.
- New sections on cooperation among species, mutualism, and commensalism have been added to Chapter 11. A new focus box on *Alien Animals of Australia* has been added to the chapter.
- Vulcanization now appears before deformation of Earth's crust in Chapter 14. A new section on the benefits of volcanic activity has been added.
- Chapter 18 has been significantly modernized. Greater emphasis is given to the issues of sea level fluctuations and global warming. This question is examined from an objective scientific viewpoint.

ACKNOWLEDGMENTS

Several dozen colleagues, students, and friends were helpful in the preparation of the original version of this book and its three succeeding editions. Their assistance has been gratefully acknowledged previously. A number of people were instrumental in the development of this particular revision, and I am delighted to recognize their contributions

Stephen Stadler of Oklahoma State University and **William R. Buckler** of Youngstown State University provided specialized expertise to the sections on weather and soils, respectively. **Darrel Hess** of City College of San Francisco was my right-hand man in terms of both ideas and details for the entire book.

In addition, useful reviews and suggestions for improvement were provided by:

Gary Anderson, *Santa Rosa Junior College*
Joseph Ashley, *Montana State University*
Daniel Balogh, *Cabrillo College*
Scott Brady, *Southeastern Louisiana University*
Richard Crooker, *Kutztown University*
David deLaubenfels, *Syracuse University*
Salvatore DiMaria, *University of New Mexico*
Mike DeVivo, *Bloomsburg University of Pennsylvania*
John Dixon, *University of Arkansas*
Vernon Domingo, *Bridgewater State College*
James Dyer, *University of North Dakota*
Orville Gab, *South Dakota State University*
Don Gary, *Nichols State University*
Frederic Glaser, *University of Texas-Pan American*
Jerry Green, *Miami (Ohio) University*
Perry Hardin, *Brigham Young University*
John Hayes, *Salem State College*
C. M. Head, *University of Central Florida*
Robert Hordon, *Rutgers University*
Hugh Hornstein, *Muskegon Community College*
Solomon Isiorho, *Indiana University-Purdue University, Fort Wayne*
Stephen Justham, *Kutztown University*
Ralph Lewis, *Eastern Oregon State College*
Robin Lyons, *Leeward Community College*
Andrew Marcus, *Montana State University*
Joann Mossa, *University of Florida*
Stanley Norsworthy, *California State University, Fresno*
J. L. Pasztor, *Delta College*
James Penn, *Southeastern Louisiana University*
Robert Picker, *San Francisco State University*
Joyce Quinn, *California State University, Fresno*
Bill Renwick, *Miami (Ohio) University*
G. L. Reynolds, *University of Central Arkansas*
Roger Sandness, *South Dakota State University*
Brent Skeeter, *Salisbury State University*
Dale Stevens, *Brigham Young University*
Tom Wikle, *Oklahoma State University*
David Williams, *Delta College*
Mark Williams, *University of Colorado-Boulder*
Deborah Woodcock, *University of Hawaii*
Craig ZumBrunnen, *University of Washington*

Prentice Hall personnel were notably helpful and efficient on the project. I am grateful to *Ray Mullaney* for marshaling his developmental forces; to *Debra Wechsler* for her cheerful and dogged professionalism in the face of a bonecrushing schedule; to *Ray Henderson* for his thoughtful supervision and unflagging support; and especially to *Irene Nunes* for her energetic, imaginative, and comprehensive editing.

1

INTRODUCTION TO EARTH

EARTH IS THE ANCESTRAL, AND THUS FAR THE ONLY, home of humankind. People live on the surface of Earth in a physical environment that is extraordinarily complex, extremely diverse, infinitely renewing, and yet ultimately fragile.

This habitable environment exists over almost the entire face of Earth, which means that its horizontal dimensions are vast [Figure 1-1(a)]. Its vertical extent, however, is very limited, as Figure 1-1(b) shows; the great majority of all earthly life inhabits a zone less than 3 miles (5 kilometers) thick, and the total vertical extent of the life zone is less than 20 miles (32 kilometers).

GEOGRAPHY AS A FIELD OF LEARNING

It is within this shallow life zone that geographers focus their interest and do their work. In fact, the word *geography* comes from the Greek word for "earth description." Geography has always been (and remains) a generalized—as opposed to a specialized—discipline. Its viewpoint is one of broad understanding. Two thousand years ago many scholars were more truly "earth describers" than anything else. However, during the first centuries of the Christian era, there was a trend away from generalized Earth description and toward more specific scholarly specializations. This narrowing of focus led to a variety of more specialized disciplines—such as geology, meteorology, economics, and botany—along with a concomitant eclipsing of geography. If "mother geography gave birth to many offshoot sciences," as some geographers have said, it is clear that these developing disciplines soon became better known than their progenitor.

It was not until the 1600s that there began a rekindling of interest in geography in the European world, and it was another two centuries before there was a strong impetus given to the discipline by geographers from various countries. A prominent geographic theme that has persisted through the centuries is that geographers study how things differ from place to place—in other words, geography is the areal differentiation of Earth's surface. There is a multiplicity of "things" to be found on Earth, and nearly all of them are distributed unevenly, thus providing a spatial, or geographic, aspect to the planet.

Table 1-1 lists the kinds of "things" to which we are referring. They are divided into two columns signify-

TABLE 1-1
The Elements of Geography

Physical elements	Cultural elements
Rocks	Population
Landforms	Settlements
Soil	Economic activities
Flora	Transportation
Fauna	Recreation activities
Climate	Languages
Water	Religion
Minerals	Political systems
	Traditions
	And many others

ing the two principal classifications in geography. The elements of **physical geography** are natural in origin, and for this reason physical geography is sometimes called environmental geography. The elements of **cultural geography** are those of human endeavor, so this branch is sometimes referred to as human geography. The list of physical geography elements is essentially complete as a broad tabulation. The list of cultural geography elements, on the other hand, is merely suggestive of a much longer inventory of both material and nonmaterial features.

All the items listed in Table 1-1 are familiar to us, and this familiarity highlights another basic characteristic of geography as a field of learning: Geography has no peculiar body of facts or objects of study that it can call wholly its own. The particular focus of geology is rocks, the attention of economics is fastened on economic systems, demography examines human population statistics, and so on. Geography, however, is a very broad field of inquiry and "borrows" its objects of study from related disciplines. Geographers, too, are interested in rocks and economic systems and population statistics, but only in certain aspects of these elements.

In simplest terms, geographers are concerned with the spatial, or distributional, aspects of the elements listed in Table 1-1. Thus geography is neither a physical science nor a social science; rather, it combines

A satellite image of Earth, centered on Japan. The picture is a composite created from thousands of separate images recorded by the Tiron-N series of meteorological satellites of NOAA. Most of Asia is shown on the left side of the image; Australia is at the bottom; the north polar ice cap shows clearly at the top, with the icy island of Greenland beyond it; a portion of North America is seen at upper right. (*Tom Van Sant/Geosphere Project, Santa Monica/Science Photo Library/Photo Researchers, Inc.*)

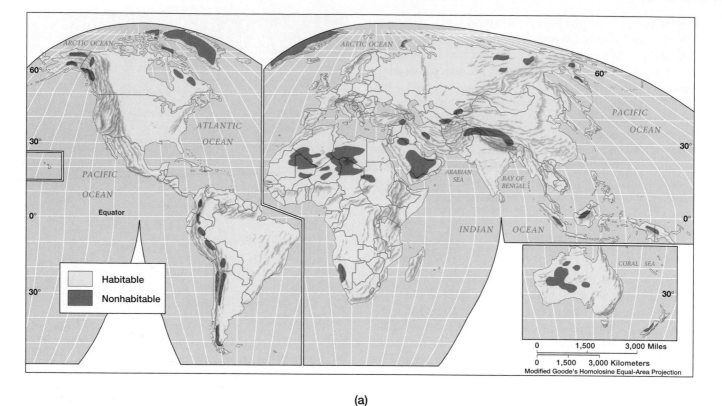

(a)

Figure 1-1 **(a)** Most of Earth's land surface is habitable, as indicated by this world map. The nonhabitable areas are either too hot, too cold, too wet, too dry, or too rugged to support much human life. These nonhabitable areas include parts of the Arctic, most of Greenland, the Antarctic, various mountainous regions, and several deserts.
(b) Vertically, the habitable zone surrounding Earth is extremely restricted. The blanket of air surrounding the planet is relatively thin: it extends only to an altitude of about 20 miles (32 kilometers) and then fades into the emptiness of space. To understand the relative thinness of the atmosphere, consider that Earth's radius is 4000 miles (6400 kilometers) while the "living zone" of the air is a layer only 3 miles (5 kilometers) high. If Earth were a peach, this living zone would be thinner than its fuzz.

(b)

characteristics of both and can be conceptualized as bridging the gap between the two. A geographer is not interested in specializing in the environment as a field of study as an ecologist does, nor in concentrating on human social relationships as a sociologist does. Instead, geography is concerned with the environment as it provides a home for humankind and with the way humans utilize that environmental home.

Another basic characteristic of geography is its interest in interrelationships. One cannot understand the distribution of soils, for example, without knowing something of the rocks from which the soils were derived, the slopes on which the soils developed, and the climate and vegetation under which they developed.

Similarly, it is impossible to comprehend the distribution of agriculture without an understanding of climate, topography, soil, drainage, population, economic conditions, technology, historical development, and a host of other factors, both physical and cultural. Thus the elements in Table 1-1 are enmeshed in an intricate web of interrelationships, all of which are encompassed in the discipline we call geography.

In short, the fundamental questions of geographic inquiry are, Why is What Where? and So What?

IN THIS BOOK

The subject matter of this book is physical geography. We shall examine the various components of the natural environment, the nature and characteristics of the physical elements, the processes involved in their development, their distribution over Earth, and their basic interrelationships. We shall largely ignore interrelationships with the various elements of cultural geography except where they help to explain the development or contemporary distribution patterns of the physical elements.

Although there are billions of celestial bodies in the universe, the environment of our planet is different from that of all the others, insofar as we know. This unique environment has produced a unique landscape: a combination of solids, liquids, and gases; organic and inorganic matter; a bewildering variety of life forms; and an extraordinary diversity of physical features that is not even approximated on any other known planet. In this book our study of geography is built around an attempt to "appreciate" (i.e., understand) that unique landscape. As we proceed from chapter to chapter, the notion of landscape development and landscape modification will serve as our central theme.

In this chapter our attention is focused on a few broad concepts—the place of Earth in the solar system, the basic physical characteristics of this planet, and the functional relationship between Earth and its prime source of energy, the sun. Understanding these concepts will help us comprehend the remarkable landscape we inhabit.

THE ENVIRONMENTAL SPHERES

From the standpoint of physical geography, the surface of Earth is a complex interface where the four principal components of the environment meet and to some degree overlap and interact (Figure 1-2). The solid, inorganic portion of Earth is the **lithosphere** (*litho* is Greek for "stone"), comprising the rocks of Earth's

Figure 1-2 The physical landscape of Earth's surface is composed of four overlapping and interacting systems called "spheres," all visible in this drawing. The atmosphere is the air we and all other living things breathe. The hydrosphere is the water of rivers, lakes, and oceans as well as the moisture present in soil and air. The biosphere is us plus all other living things, both plant and animal, as well as viral. The lithosphere is the soil and bedrock that cover Earth's surface.

crust as well as the broken and unconsolidated particles of mineral matter that overlie the solid bedrock. The lithosphere's surface is shaped into an almost infinite variety of landforms, both on the seafloors beneath the oceans and on the surfaces of the continents and islands.

The gaseous envelope of air that surrounds Earth is the **atmosphere** (*atmo* is Greek for "air"). It contains the complex mixture of gases needed to sustain life. Most of the atmosphere adheres closely to Earth's surface, being densest at sea level and rapidly thinning with increased altitude. It is a very dynamic sphere, kept in almost constant motion by solar energy and Earth's rotation.

The **hydrosphere** (*hydro* is Greek for "water") comprises water in all its forms. The oceans contain the vast majority of the water found on Earth and are the moisture source for most precipitation.

The **biosphere** (*bio* is Greek for "life") encompasses all the parts of Earth where living organisms can exist; in its broadest and loosest sense, the term also includes the vast variety of earthly life forms (properly referred to as "biota").

These four spheres are not four discrete and separated entities but rather are considerably intermingled. This intermingling is readily apparent when considering an ocean, a body that is clearly a major component of the hydrosphere and yet may contain a vast quantity of fish and other organic life that are part of the biosphere. An even better example is soil, which is composed largely of bits of mineral matter (lithosphere) but also contains life forms (biosphere), soil moisture (hydrosphere), and air (atmosphere) in pore spaces.

The four spheres can serve as important organizing concepts for the systematic study of Earth's physical geography and will be used that way in this book.

Before focusing our attention on the spheres, however, we set the stage by considering the planetary characteristics of Earth and noting the most important relationships between our planet and its basic source of energy, the sun.

THE SOLAR SYSTEM

We live on an extensive (to us), rotating, complicated mass of mostly solid material that orbits the enormous ball of superheated gases we call the sun.

The geographer's concern with spatial relationships properly begins with the relative location of this spaceship Earth in the universe.

Earth is one of the nine *planets* of the **solar system**, which also contains at least 44 *moons* revolving around the planets, scores of *comets* (composed of frozen gases that hold together small pieces of rock and metallic minerals), more than 50,000 *asteroids* (objects made of rock and/or metal that are mostly very small, but with a few as large as several hundred miles in diameter), and millions of *meteors* (most of them the size of sand grains).

The medium-sized star we call our sun is the central point of the solar system and makes up more than 99 percent of its total mass. The solar system is part of the Milky Way galaxy, which consists of perhaps 100,000,000,000 stars arranged in a disk-shaped cloud that is 100,000 light-years (1 light-year equals nearly 6 trillion miles) in diameter and 10,000 light-years thick at the center (Figure 1-3). The Milky Way galaxy is only one of at least a billion galaxies in the universe!

The vastness of the universe is clearly beyond normal comprehension. To begin to develop a feel for astronomical distances, we might consider a reduced-scale model of the universe. If the distance between Earth and the sun, which is 93,000,000 miles (150,000,000 kilometers), is taken to be 1 inch (2.5 centimeters), then the distance from Earth to the nearest star would be 4.5 miles (7.2 kilometers) and the distance from Earth to the next galaxy beyond the Milky Way would become about 150,000 miles (240,000 kilometers).

The origin of Earth, and indeed of the universe, is only partially understood since any attempt to trace billions of years of history with only tidbits of evidence is a task fraught with uncertainty. Most astronomers

Figure 1-3   A drawing of our galaxy, the Milky Way, seen in profile. The Milky Way is a spiral galaxy shaped like a thin disc with a central bulge. (J. Baum & N. Henbest/Science Photo Library/Photo Researchers, Inc.)

FOCUS

THE MOON

Because of its proximity, Earth's only natural satellite–the moon–became the first celestial body to be visited by humans (in 1969). It appears to be large because it is close (239,000 miles, or 385,000 kilometers, away from Earth), but the moon has a volume that is only 2 percent of Earth's volume and is only a tiny fraction of the size of the sun. However, it is one of the largest satellites in the solar system.

No life of any kind exists on the moon. Nor does it have an atmosphere, wind, or water. Its sky is black all the time, and stars are continuously visible. At night the moon becomes colder than any place on Earth, and during the day its temperature is slightly higher than that of boiling water at sea level.

The surface of the moon consists of both solid rock and broken bits of rock and natural glass. There are various lowlands and highlands, but the most numerous features of the moon's surface are craters. It is estimated that there are 30,000,000,000,000 craters that are at least 1 foot (30 centimeters) wide. It appears that the surface of the moon has changed little (apart from the formation of craters) since there were great outpourings of lava between 3 and 4 billion years ago.

The moon revolves around Earth once about every 27 days, which is also the time required for one rotation of the moon on its axis. It moves from west to east in the sky but appears to move in the opposite direction because Earth's rotation rate is much faster than the rate at which the moon revolves around Earth. The moon also travels with Earth as the latter revolves around the sun and is held in its orbit by the gravitational pull of Earth.

The moon is the brightest object in our night sky, but it gives off no light of its own. All moonlight is light reflected from the sun. The moon appears to change its size and shape from night to night, but this is simply because varying amounts of the lighted side of the moon are visible from Earth.

During the 27 days the moon takes to orbit Earth, we see it change from a skinny crescent to a full globe and back again. These apparent changes are referred to as the *phases* of the moon. The sequence begins when the moon is between the sun and Earth, at which time the sunlit side is hidden from our view and we see only a dim shape called the *new* moon. With each passing night, more and more of the moon's sunlit side is revealed as the waxing crescent gets larger and larger ("to wax" is an archaic term meaning "to increase"), and after about a week we see half the moon; this phase is called the first quarter. After about another seven nights, the moon moves to a point where Earth is between it and the sun, and we see its entire sunlit side; this is called the full moon. After another week the moon shape diminishes to half, called the last quarter phase. After one more week the new moon phase is reached again.

The moon is an exciting part of our skyscape. Moreover, it has physical effects on Earth, most notably in the generation of oceanic tides (Chapter 9). The moon's lack of atmosphere helps us understand many principles of physical geography, such as the black sky that results when there is no atmosphere to scatter blue light (Chapter 3), the temperature extremes that accompany an airless environment (Chapter 4), and the relatively changeless land surface that occurs when there is no running water to erode and deposit (Chapter 16).

---

believe that the celestial patterns in the universe began to develop some 15 billion years ago with a cosmic explosion now called the "big bang," after which all the material of the exploding mass began dispersing at enormous speeds.

The solar system apparently originated between 4.5 and 5 billion years ago when a *nebula*, a huge cold, diffuse cloud of gas and dust propelled through space by the big bang, began to contract inward owing to its own gravitational collapse, forming a hot, dense protostar (Figure 1-4). This hot center–our sun–was surrounded by a cold, revolving disk of gas and dust that formed the planets.

The solar system also contains at least 44 moons revolving around the planets. These planetary satellites may have been formed elsewhere in the nebula. Their orbits around the sun then brought them close enough to the planets to be captured by mutual gravitational influence.

Table 1-2 compares various features of the nine planets of the solar system.

All the planets move around the sun in elliptical orbits, with the sun located at one focus (Figure 1-5a). With the exception of Pluto, whose orbit is somewhat askew, all the planetary orbits are in nearly the same plane, as Figures 1-4 and 1-5(b) show, perhaps

Figure 1-4 The birth of the solar system. (1) Diffuse gas cloud, or nebula, begins to contract inward. (2) Cloud flattens into nebular disk as it spins around a central axis. (3) Particles in the outer parts of the disk collide with each other to form protoplanets. (4) Protoplanets coalesce into planets and settle into orbits around the hot center. (5) The final product: a central sun surrounded by nine orbiting planets.

revealing their relationship to the formative nebular disk. The four inner planets generally are smaller, denser, less oblate (that is, more nearly spherical) than the five outer planets and rotate more slowly on their axes. Also, the inner planets are composed principally of mineral matter and, except for airless Mercury, have diverse but relatively shallow atmospheres. By contrast, the outer planets tend to be much larger, at least 100 times more massive, less dense, and much more oblate, and they rotate more rapidly. They seem to be composed entirely of gases, principally hydrogen, which become liquid and then frozen toward the interior of the planets. Their chemical composition is similar to that of the sun, indicating that they were formed in a similar fashion. The outer planets generally have atmospheres that are dense and turbulent.

The sun rotates from west to east, and each of the planets revolves around it from west to east. Moreover, most of them rotate from west to east on their own axes. Generally, planets move more slowly and have a lower temperature as their distance from the sun increases.

THE SIZE AND SHAPE OF EARTH

Is Earth large or small? The answer to this question depends on one's frame of reference. If the frame of reference is the universe, Earth is almost infinitely small. The radius of our planet is only about 4000 miles (6400 kilometers), a miniscule distance on the scale of the universe, where, for instance, the moon is 239,000 miles (385,000 kilometers) from Earth, the sun is 93,000,000 miles (150,000,000 kilometers) away, and the nearest star is 25,000,000,000,000 miles (40,000,000,000,000 kilometers) distant.

In a human frame of reference, however, Earth is a very impressive mass. Its highest point is more than 29,000 feet (8800 meters) above sea level, and the deepest spot in the oceans is almost 36,000 feet (11,000 meters) below sea level (Figure 1-6). The vertical distance between these two extremes is about 12 miles (20 kilometers), which is minuscule when contrasted with the 8000-mile (12,875-kilometer) diameter of Earth. To illustrate this relationship, if we had a globe with a diameter the height of a 40-story building, the difference in elevation between the highest point of the conti-

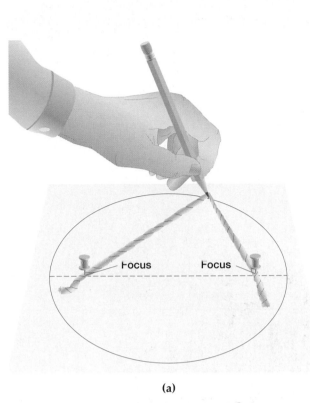

(a)

Figure 1-5 **(a)** Drawing an ellipse with the help of two tacks and a string. When held taut against the string, the pencil traces out an ellipse. The closer together the tacks are placed, the less ovate (or egg shaped) the ellipse. (When the tacks are so close together that they are in the same spot, the result is a circle.) Each tack represents one focus of the ellipse. **(b)** The planets of the solar system in their elliptical orbits around the sun.

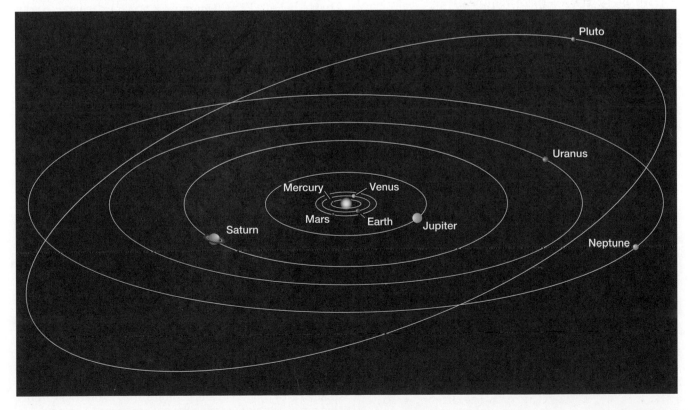

(b)

TABLE 1-2
Some Planetary Comparisons

Planet	Distance from sun Millions of miles	Distance from sun Millions of kilometers	Rotation time (days)	Revolution time (years)	Diameter Miles	Diameter Kilometers	Relative gravity force (Earth = 1)	Number of satellites
Mercury	36.8	59.2	59	0.2	3000	4830	0.38	0
Venus	66.9	108	243	0.6	7500	12100	0.89	0
Earth	92.6	149	1	1	8000	12900	1	1
Mars	144	214	1	1.9	4200	6760	0.38	2
Jupiter	482	776	0.4	12	88000	142000	2.64	16
Saturn	883	1420	0.4	30	75000	121000	1.17	18
Uranus	1777	2859	1	84	29300	47100	1.03	15
Neptune	2787	4484	1	165	27700	44600	1.50	8
Pluto	3658	5886	6	248	1400	2250	?	1

nents and the lowest point in the ocean basins would be represented by the width of a single brick on top of the building.

From time to time, various people have maintained that Earth has the shape of a flat disk. (A Flat Earth Society still exists in the United States today.) However, as early as the sixth century B.C., some Greek scholars correctly believed Earth to have a spherical shape. About 200 years before Christ, Eratosthenes, the director of the Greek library at Alexandria, calculated the circumference of Earth trigonometrically. He determined the angle of the noon sun rays at Alexandria and at another city, Syene, 600 miles (960 kilometers) away, measured the distance between the two localities, and from these angular and linear distances was able to estimate an Earth circumference of almost 26,700 miles (43,000 kilometers), which is reasonably close to the actual figure of 24,900 miles (40,000 kilometers). Some historians believe that Eratosthenes made errors in his calculations, errors that partially canceled one another and made his result more accurate than it deserved to be. Be that as it may, several other Greek scholars made independent calculations of Earth's circumference during this period, and all their results were close to reality, which indicates the antiquity of accurate knowledge of the size and shape of Earth.

In actuality Earth is almost, but not quite, spherical. The cross section revealed by a cut from pole to pole

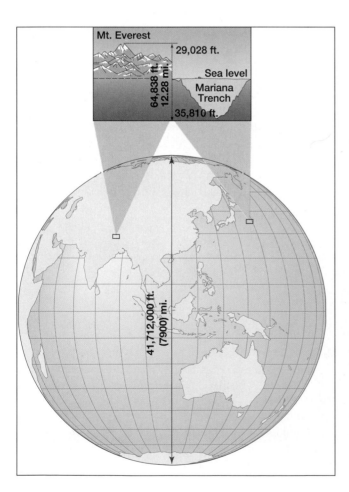

Figure 1-6 Earth's maximum relief (which means the difference in elevation between the highest and lowest points) is 64,838 feet (about 12 miles or 20 kilometers), from the top of Mount Everest to the bottom of the Mariana Trench in the Pacific Ocean.

FOCUS

N U M B E R S   A N D   M E A S U R E M E N T

The United States is in the process of shifting to the International System of Units (abbreviated SI from the French *Système International*) for the following categories of measurement:

Category	SI Base Unit
Length	Meter
Mass	Kilogram
Temperature	Degree Celsius
Electric current	Ampere
Luminous intensity	Candela
Amount of substance	Mole

The impetus for the changeover comes from businesses that seek SI standardization in order to increase their economies of scale when competing in foreign markets, where SI units are standard.

The International System came into being in 1960. Since its precursor, the metric system, was legalized for both interstate and intrastate trade in 1866, the current "metrication" of the United States presumably represents the phaseout of the traditional (or English) system as one of two alternative legal measuring systems for commerce. You have undoubtedly noticed that this book gives measurements in both systems.

The metric system was devised by French scientists in the midst of the French Revolution (1790s), responding to a directive by their government to "deduce an invariable standard for all the measures and all the weights" that would be both simple and scientific. The designers chose a standard unit of length from which could be derived measures for volume and mass, thereby relating these units to one another and to nature. The larger and smaller components of each unit could be obtained simply by moving the decimal point (i.e., multiplying or dividing by multiples of 10). The basic length unit chosen was the *metre* (which we spell *meter*), a word derived from the Greek *metron*, meaning "a measure." A meter was defined as one ten-millionth of the distance from the North Pole to the equator. The basic unit of mass was the *gram*, defined to be the mass of one cubic centimeter of water at its temperature of maximum density (4° C). The *liter*, defined as the volume of a cubic decimeter, was the basic unit of volume.

The Gregorian calendar, which is the one we use, and the measurement of time, now fairly standard throughout the world, represent quite a potpourri of numerical equivalencies. For instance, there are 60 seconds in a minute, 60 minutes in an hour, 24 hours in a day, and 7 days in a week. A year has 365.242199 days, or 52.177455 weeks, or 12 months, with each month having, variously, 28, 29, 30, or 31 days. Similarly, a circle has 360 degrees, each degree has 60 minutes, and each minute has 60 seconds. Why are these measurements not also being revised to a base-10 (decimal) standard? Although such a revision of our units of time and angular measurement would be costly and psychologically unsettling, it would be no more expensive or troublesome than the changes we are attempting to make in all the other SI categories.

Most other nondecimal units of measurement are expressed in multiples of 60 or 12. What is the origin of this tendency toward 12s and 60s? As early as 4500 B.C., traders carried Egyptian ideas of numeration systems, including base 10, to Babylonia. The Babylonians reorganized the Egyptian system so that written numbers would have a positional, or place, value. As far as we know, the Babylonians were the first people to develop this system, which is basic to all modern numeration. Also, since the Babylonians concentrated on studying astronomy, they were justifiably impressed by the increased number of even divisors in a system based on 12. For example, 10 has three divisors: 2, 5, and 10; whereas 12 has five divisors: 2, 3, 4, 6, and 12. Moreover, although 100 (and any other base-10 number ending in two zeros) is divisible by 2, 4, 5, 10, 25, 50, and 100, the number 144 (the "100" of a base-12 system) is evenly divisible by twice as many divisors: 2, 3, 4, 6, 8, 9, 12, 16, 18, 24, 36, 48, 72, and 144. For this reason, the Babylonians first chose 12 as the base for their number system. Later, they combined the advantages of both the base-10 and the base-12 systems by changing the base of their system to 60 (12 × 5).

Because the Greeks began their study of geometry with knowledge from the Babylonians, they adopted the Babylonian base 60 for measuring angles. Through time, most of the world's cultures have developed or returned to a numeration system based on 10, probably because we have 10 fingers on our hands and we first learn to count by using these fingers. However, most of the world's cultures have retained the dividing properties of 12 in measuring time and angles, as well as in an array of everyday commercial measurements such as

Dozen	12
Gross	144 (12 × 12)
Ream	20 quire = 480 sheets (now often 500)
Karat	1/24 gold (pure gold = 24 karats)

A table of equivalents between English and SI units appears in Appendix V.

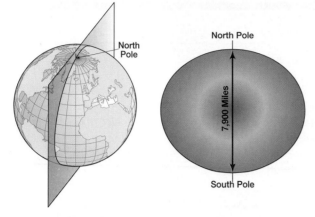

(a) A diameter through the poles   Resulting cross section

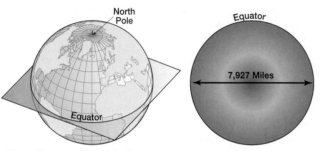

(b) A diameter through the equator   Resulting cross section

Figure 1-7   Earth is not quite a perfect sphere because its surface flattens a bit at the North Pole and the South Pole. Thus a cross section through the poles, shown in **(a)**, has a diameter slightly less than the diameter of a cross section through the equator, shown in **(b)**. The polar flattening causes bulging all around the equator, but because this bulging is evenly distributed all around the globe, it has no effect on the circular shape of the equatorial cross section.

would be an ellipse rather than a circle, but a similar cut through the equator would be circular (Figure 1-7). Any rotating body has a tendency to bulge around its equator and flatten at the polar ends of its rotational axis. Although the rocks of Earth may seem quite rigid and immovable to us, they are sufficiently pliable to allow Earth to develop a bulge in its midriff. The slightly flattened polar diameter of Earth is 7900 miles (12,714 kilometers), whereas the slightly bulging equatorial diameter is 7927 miles (12,757 kilometers), a difference of only about 0.3 percent. Thus our planet is properly described as an *oblate spheroid* rather than a true sphere.

In addition, the shape of Earth has obvious deviations from true sphericity that are the result of topographic irregularities on its surface. This surface varies in elevation from the highest mountain peak, Mount Everest at 29,028 feet (8850 meters) above sea level, to the presumed deepest ocean trench, the Mariana

Trench of the Pacific Ocean, at 35,810 feet (10,915 meters) below sea level, a total difference in elevation of 64,838 feet (19,800 meters). Although prominent on a human scale of perception, this difference is minor on a planetary scale, as Figure 1-6 illustrates. If Earth were the size of a basketball, Mount Everest would be an imperceptible pimple no greater than 0.00069 inch (about seven ten-thousandths of an inch, or 0.010 millimeter) high. Similarly, the Mariana Trench would be a tiny crease only 0.0009 inch (about one thousandth of an inch, or 0.023 millimeter) deep. This represents a depression smaller than the thickness of a sheet of paper.

Our perception of the relative size of topographic irregularities has been further confused by three-dimensional wall maps and globes that emphasize such landforms. To portray any noticeable appearance of topographic variation, the vertical distances on such maps usually are exaggerated 8 to 20 times their actual proportional dimensions.

The variation from true sphericity, then, is exceedingly minute. For most purposes, Earth may be properly considered a sphere.

THE GEOGRAPHIC GRID

Any understanding of the distribution of geographic features over Earth's surface requires some system of accurate location. Such a system should be capable of pinpointing with mathematical precision the position of any spot on the surface. The simplest technique for

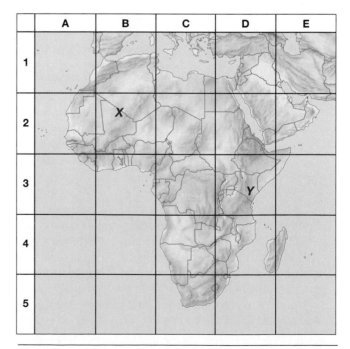

Figure 1-8   An example of a grid system. The location of the point X can be described as 2B or as B2; the location of Y is 3D or D3.

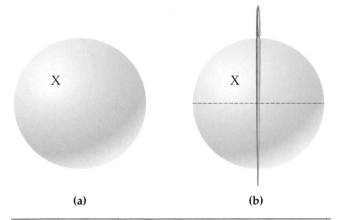

Figure 1-9 **(a)** On a perfectly round, perfectly clean Ping-Pong ball, it is impossible to describe the position of the point marked X. **(b)** If we stick a darning needle through the ball along any diameter and draw a line around the ball midway between the two points where the needle punctures the ball, we now have some reference points for describing where a point lies. Our X, for instance, is in the top left quadrant in the perspective shown here.

Figure 1-11 An imaginary plane bisecting Earth midway between the two poles defines the equator.

achieving this precision is to design two unvarying sets of lines that intersect at right angles, thus permitting the location of any point on the surface to be described (mathematically) by the appropriate intersection. Such a network of intersecting lines, referred to as a *grid system,* is shown in Figure 1-8. For purposes of location on Earth's surface, a geographic grid consisting of east-west lines and north-south lines has been devised.

If our planet were a nonrotating body, the problem of describing precise surface locations would be more formidable than it is. Imagine the difficulty of trying to describe the location of a particular point on a perfectly round, perfectly clean Ping-Pong ball. Because every point on the sphere is exactly like every other point, it is impossible to distinguish one point from another [Figure 1-9(a)]. If we stick a darning needle through the ball, however, and draw a circle around the ball midway between the two points where the needle pierces the surface, as in Figure 1-9(b), we produce "poles" and an "equator" and thus provide arbitrary reference points for a locational system. Fortunately, Earth does rotate, and we can use its rotation axis to describe locations on Earth just the way we use the needle to describe points on the Ping-Pong ball. Earth's rotation *axis* is a diameter line that connects the points of maximum flattening on Earth's surface (Figure 1-10). These points are called the *North Pole* and the *South Pole*. If we visualize an imaginary plane passing through Earth halfway between the poles and perpendicular to the axis of rotation, as in Figure 1-11, we have another valuable reference feature: *the plane of the equator.* Where this plane intersects Earth's surface is the imaginary midline of Earth, called simply the **equator**. We use the North Pole, South Pole, rotational axis, and equatorial plane as natural reference points for measuring and describing locations on Earth's surface. Without these

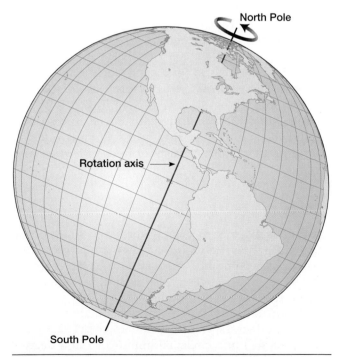

Figure 1-10 Earth rotates around its rotation axis, an imaginary line that passes through the points of maximum flattening described in Figure 1-7. These two points are what we call the North Pole and the South Pole.

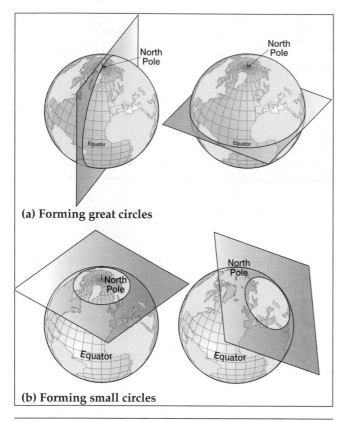

(a) Forming great circles

(b) Forming small circles

Figure 1-12   Comparison of great and small circles. **(a)** A great circle results from the intersection of Earth's surface with any plane that passes through Earth's center. **(b)** A small circle results from the intersection of Earth's surface with any plane that does not pass through Earth's center.

four points, a set of reference points would have to be chosen arbitrarily to establish an accurate locational system.

Any plane that is passed through the center of a sphere bisects that sphere (divides it into two equal halves) and creates what is called a **great circle** where it intersects the surface of the sphere [Figure 1-12(a)]. Planes passing through any other part of the sphere produce what are called **small circles** where they intersect the surface [Figure 1-12(b)]. Great circles have two principal properties of special interest:

1. A great circle is the largest circle that can be drawn on a sphere; it represents the circumference of that sphere and divides its surface into two equal halves, or hemispheres. For example, the sun illuminates one-half of Earth at any given moment. The edge of the sunlit hemisphere, called the **circle of illumination,** is a great circle that divides Earth between a light half and a dark half.
2. Only one great circle can be constructed to include any two given points (not diametrically opposite) on Earth's surface . The segment, or arc, of the great

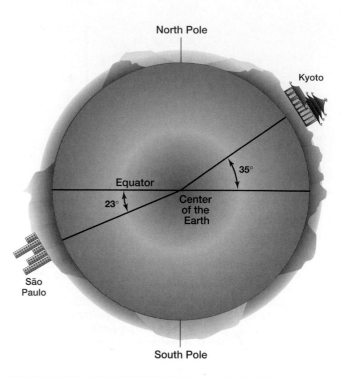

Figure 1-13   Measuring latitude. An imaginary line from Kyoto, Japan, to Earth's center makes an angle of 35° with the equator. Therefore Kyoto's latitude is 35° N. An imaginary line from São Paulo, Brazil, to Earth's center makes an angle of 23°, giving this South American city a latitude of 23° S.

circle connecting those two points is always the shortest route between the points. Such routes are known as great circle routes. They appear as curved lines on most maps.

The geographic grid used as the locational system for Earth is based on the principles just discussed. Furthermore, the system is closely linked with the various positions assumed by Earth in its orbit around the sun. Our earthly grid system is referred to as a **graticule** and consists of lines of *latitude* and *longitude*.

Latitude

Latitude is distance measured north and south of the equator. As shown in Figure 1-13, we can project a line from a given point on Earth's surface to the center of Earth. The angle between this line and the equatorial plane is the latitude of the point.

Like any other angular measurement, latitude is expressed in degrees, minutes, and seconds. There are 360 degrees (°) in a circle; 60 minutes (′) in a degree; and 60 seconds (″) in a minute. Latitude varies from 0° at the equator to 90° at either pole. Any position north

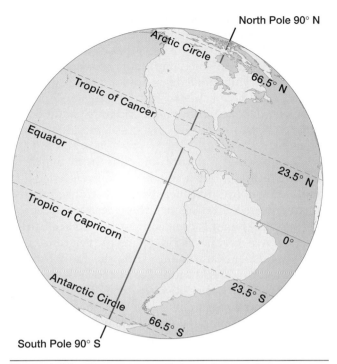

North Pole 90° N

Arctic Circle

66.5° N

Tropic of Cancer

Equator

23.5° N

0°

Tropic of Capricorn

Antarctic Circle

23.5° S

66.5° S

South Pole 90° S

Figure 1-14 The seven parallels used most often in geography. As discussed later, these latitudes represent special points where rays from the sun strike Earth's surface.

of the equator is north latitude, and any position south of the equator is south latitude.

A line connecting all points of the same latitude is called a **parallel** because it is parallel to all other lines of latitude. The equator is the parallel of 0° latitude, and it, alone of all parallels, constitutes a great circle. All other parallels are small circles, and all are aligned in true east-west directions on Earth's surface. Because parallels are imaginary lines, there can be an infinite number of them. There can be a parallel for each degree of latitude, for each minute, for each second, or for any fraction of a second. However many latitude lines one visualizes, their common property is that they are all parallel to each other.

Although it is possible to either construct or visualize an unlimited number of parallels, the seven shown in Figure 1-14 are of particular significance in a general study of Earth:

1. Equator, 0°
2. North Pole, 90° N
3. South Pole, 90° S
4. Tropic of Cancer, 23.5° N
5. Tropic of Capricorn, 23.5° S (Figure 1-15)
6. Arctic Circle, 66.5° N
7. Antarctic Circle, 66.5° S

Items 2 and 3 are of course points rather than lines but can be thought of as infinitely small parallels.

Each degree of latitude has a north-south length of about 69 miles (111 kilometers), as the second and third columns of Table 1-3 show. The length varies slightly with latitude because of the polar flattening of Earth.

LONGITUDE

Latitude comprises one-half of Earth's grid system, the north-south component. The other half is represented by **longitude**, which is distance measured east and west on Earth's surface. Again, this is angular distance, and so longitude is also measured in degrees, minutes, and seconds.

Longitude is represented by imaginary lines extending from pole to pole and crossing all parallels at right angles. These lines, called **meridians**, are not parallel to one another except where they cross the equator. They are farthest apart at the equator, becoming increasingly close together northward and southward and finally converging at the poles. A meridian, then, is a great semicircle that extends from one pole to the other, crossing all parallels of latitude perpendicularly and being aligned in a true north-south direction.

TABLE 1-3				
Lengths of Degrees of Latitude and Longitude				
	Length of 1° of latitude measured along a meridian		Length of 1° of longitude measured along a parallel	
Latitude (degrees)	Miles	Kilometers	Miles	Kilometers
0	68.703	110.567	69.172	111.321
10	68.726	110.605	68.129	109.641
20	68.789	110.705	65.026	104.649
30	68.883	110.857	59.956	96.488
40	68.998	111.042	53.063	85.396
50	69.121	111.239	44.552	71.698
60	69.235	111.423	34.674	55.802
70	69.328	111.572	23.729	38.188
80	69.387	111.668	12.051	19.394
90	69.407	111.699	0	0

Figure 1-15 The Tropic of Capricorn, like all other parallels of latitude, is an imaginary line. As a significant parallel, however, its location is often commemorated by a sign. This scene is near Alice Springs in the center of Australia. (TLM photo)

Figure 1-16 The prime meridian of the world; longitude 0°0′0″, at Greenwich, England. (James Lemass/Liaison International)

The distance between meridians varies significantly but predictably. At the equator, the surface length of one degree is about the same as that of one degree of latitude. However, since meridians converge toward the poles, the length of a degree of longitude decreases poleward, as the fourth and fifth columns of Table 1-3 show. The distance from one meridian to another diminishes to zero at the poles, where all meridians meet at a point.

The equator is a natural baseline from which to measure latitude, but no such natural reference line exists for longitudinal measurement. Consequently, for most of recorded history there was no accepted longitudinal baseline; each country would arbitrarily select its own "prime meridian" as the reference line for east-west measurement. Thus the French measured from the meridian of Paris, the Italians from the meridian of Rome, and so forth. There were at least 13 "prime meridians" in use in the 1880s. Not until about a century ago was standardization finally achieved; U.S. and Canadian railway executives adopted a standard time system for all North American railroads in 1883, and the following year an international conference was convened in Washington, D.C., to achieve the same goal on a global scale. After weeks of debate, the delegates (by a vote of 21 to 1, with 2 abstentions) chose the meridian passing through the Royal Observatory at Greenwich, England, just east of London, as the **prime meridian** for all longitudinal measurement (Figure 1-16). The principal argument for adopting the Greenwich Meridian as the prime meridian was a practical one: more than two-thirds of the world's shipping lines already used the Greenwich Meridian as a navigational base.

Thus an imaginary north-south plane passing through Greenwich and through Earth's axis of rotation represents the plane of the prime meridian. The angle between this plane and a plane passed through any other point and the axis of Earth is a measure of longitude. For example, the angle between the Greenwich plane and a plane passing through the city of Freetown (in the western African country of Sierra Leone) is 13 degrees, 15 minutes, and 12 seconds. Since the angle is formed *west* of the prime meridian, the longitude of Freetown is written 13°15′12″W (Figure 1-17).

Longitude is measured both east and west of the prime meridian to a maximum of 180° in each direction. Exactly halfway around the globe from the prime meridian, in the middle of the Pacific Ocean, is the 180th meridian, which is 180° of longitude removed from the prime meridian (Figure 1-18). All places on Earth, then, have a location that is either east longitude or west longitude, except for points exactly on

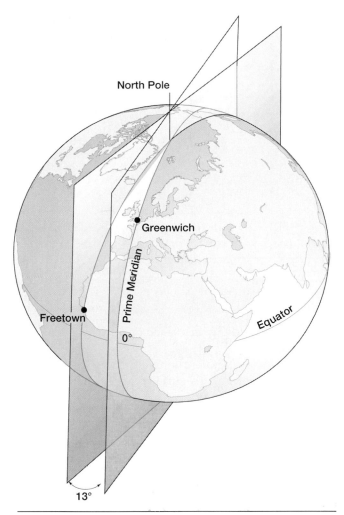

Figure 1-17 The meridians that mark longitude are defined by intersecting imaginary planes passing through the poles. Here are shown the planes for the prime meridian through Greenwich, England, and the meridian through Freetown, Sierra Leone.

the prime meridian (described simply as 0° longitude) or exactly on the 180th meridian (described as 180° longitude).

The network of intersecting parallels and meridians creates a *geographic grid* over the entire surface of Earth (Figure 1-19). The location of any place on Earth's surface can be described with great precision by reference to detailed latitude and longitude data. For example, at the 1964 World's Fair in New York City, a time capsule (a container filled with records and memorabilia of contemporary life) was buried. For reference purposes, the U.S. Coast and Geodetic Survey determined that the capsule was located at 40°28'34.089" north latitude and 73°43'16.412" west longitude. At some time in the future, if a hole were to be dug at the spot indicated by those coordinates, it would be within 6 inches (15 cm) of the capsule.

EARTH MOVEMENTS

Life on Earth is dependent on solar energy, and therefore the functional relationship between Earth and the sun is of vital importance. This relationship is not static because the perpetual motions of Earth continually change the geometric perspective between the two bodies. Two basic Earth movements—its daily rotation on its axis and its annual revolution around the sun—change its position with respect to the sun and thus are of special interest to geographers.

EARTH'S ROTATION ON ITS AXIS

Earth rotates toward the east on its axis (Figure 1-20), a complete rotation requiring 24 hours. The apparent "motion" of the sun, the moon, and the stars is, of course, just the opposite of the true direction of earthly rotation. All these celestial bodies appear to rise in the east and set in the west, which is an illusion created by the steady easterly spin of Earth.

Rotation causes all parts of Earth's surface except the poles to move in a circle around Earth's axis. Table

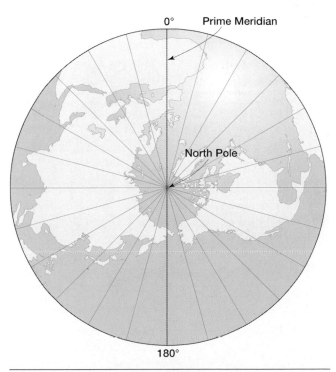

Figure 1-18 An overhead view of meridians radiating from the North Pole. Think of each line as the top edge of an imaginary plane passing through both poles. All the planes are perpendicular to the plane of the page.

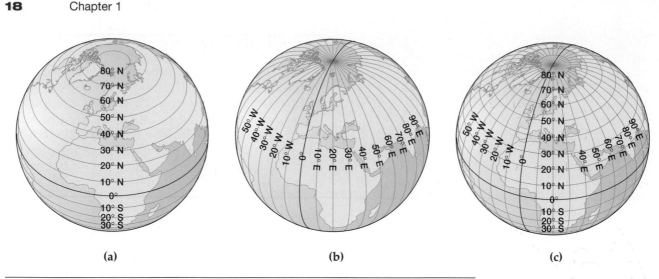

Figure 1-19   Development of the geographic grid. **(a)** Parallels of latitude.
(b) Meridians of longitude. **(c)** The two combined to form a complete grid system.

1-4 shows the speed of this motion at various latitudes. Although the speed of rotation varies from place to place, it is constant at any given place. This is the reason that we experience no sense of motion. Often one can get the same impression in a modern jetliner, where a smooth flight at cruising speed is much like sitting in a comfortable living room. Only during takeoff and landing is the sense of motion quite apparent. Similarly, the motion and speed of earthly rotation would become apparent to us only if that rotation rate suddenly increased or decreased—a very unlikely event.

Rotation has several striking effects on the physical characteristics of Earth's surface. Most important are the following:

1. The constancy of Earth's rotation in the same direction causes apparent deflection in the flow path of both air and water. The deflection is invariably toward the right in the Northern Hemisphere and toward the left in the Southern Hemisphere. This phenomenon is called the *Coriolis effect* and is discussed in detail in Chapter 5.

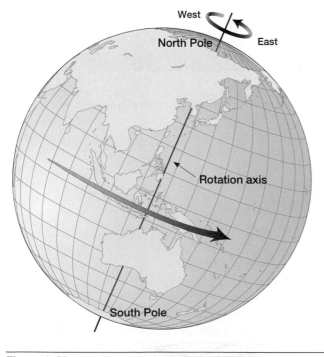

Figure 1-20   Earth rotates from west to east on a rotation axis that is tilted from the vertical.

TABLE 1-4		
Speed of Rotation of Earth's Surface at Selected Latitudes		
Latitude	*Miles per hour*	*Kilometers per hour*
0	1037.6	1669.9
10	1021.9	1642.0
20	975.4	1569.7
30	899.3	1447.3
40	795.9	1280.9
50	668.3	1075.5
60	520.1	837.0
70	355.9	572.8
80	180.8	291.0
90	0	0

2. The rotation of Earth brings any point on the surface through the increasing and then decreasing gravitational pull of the moon and the sun. Although the land areas of Earth are too rigid to be noticeably moved by these oscillating gravitational attractions, oceanic waters move onshore and then recede in a rhythmic pattern as a result of the interplay of earthly rotation with these gravitational forces. The rise and fall of water level constitutes the *tides*, which are discussed further in Chapter 9.

3. Undoubtedly the most important effect of earthly rotation is the diurnal (daily) alternation of light and darkness, as portions of Earth's surface are turned first toward and then away from the sun. This variation in exposure to sunlight greatly influences local temperature, humidity, and wind movements. Except for the organisms that live either in caves or in the ocean deeps, all forms of life have adapted to this sequential pattern of light and darkness. We human beings fare poorly when our circadian (daily) rhythms are misaligned as the result of high-speed air travel that significantly interrupts the normal sequence of daylight and darkness. We are left with a sense of fatigue and psychological distress known as jet lag, which can include unpleasant changes in our usual patterns of appetite and sleep.

EARTH'S REVOLUTION AROUND THE SUN

Another significant Earth motion is its *revolution around the sun*. Each revolution takes 365 days, 5 hours, 48 minutes, and 46 seconds, or 365.242199 days. This is known officially as the *tropical year* and for practical purposes is usually simplified to 365.25 days. (Astronomers define the year in other ways as well, but the duration is very close to that of the tropical year and need not concern us here.)

As mentioned earlier, the path followed by Earth in its journey around the sun is not a true circle but an ellipse (Figure 1-21). Because of this elliptical orbit, the Earth–sun distance is not constant; rather it varies from 91,445,000 miles (147,166,480 kilometers) at the **perihelion** (*peri* is from the Greek and means "about" or "around") position on January 3 to 94,555,000 miles (152,171,500 kilometers) at the **aphelion** (*ap* is from the Greek and means "away") position on July 4. The average Earth–sun distance is defined as one astronomical unit (1 AU) and is 92,955,806 miles (149,597,892 kilometers). Earth is 3.3 percent closer to the sun during the Northern Hemisphere winter than during the Northern Hemisphere summer, an indication that the varying distance between Earth and the sun is not an important determinant of seasonal temperature fluctuations.

THE ANNUAL MARCH OF THE SEASONS

The plane that passes through the sun and through every point of Earth's orbit around the sun is called the **plane of the ecliptic** (Figure 1-22). Our sense of orderly geometric relationships tends to make us think that the plane of the equator should coincide with the plane of the ecliptic. However, as we can readily see in Figure 1-22, this line of reasoning incorrectly assumes that Earth's rotation axis is perpendicular to the plane of the ecliptic. Such a positional relationship between Earth and the sun would create a very different world from the one in which we live. If these two planes did coincide, noontime sunlight would always strike Earth at the same angle at any one place. Daylight and darkness, caused by the rotation of Earth on its axis, would be the only temporal variations in the amount of energy reaching Earth. The result would be a world without seasons.

The plane of the equator is not coincident with the plane of the ecliptic, however, because Earth's rotation axis is not perpendicular to the ecliptic plane. Rather, the axis is tilted about 23.5° away from the perpendicular (Figure 1-23). This tilt is referred to as the **inclination** of the rotation axis. Moreover, the axis always points toward Polaris, the North Star, no matter where Earth is in its orbit around the sun. In other words, at any time during the year, Earth's rotation axis is parallel to its orientation at all other times. This is called the **polarity**, or *parallelism*, of the axis.

The combined effect of rotation, revolution, inclination, and polarity is such that the angle at which the sun's rays strike Earth changes throughout the year. This is a critical determinant of the amount of solar energy delivered to any spot on Earth. The basic generalization is that the more direct the angle at which the sunlight strikes Earth, the more effective is the resultant heating. Where the sun's rays strike Earth perpendicularly (at a 90° angle), solar energy is concentrated over the smallest possible surface area (Figure 1-24). Where the sun's rays strike Earth obliquely (at an angle smaller than 90°), the same amount of energy is spread over a much larger surface area. Thus the amount of energy reaching a particular surface area is significantly smaller. The angle at which the sun's rays strike Earth is fundamental in determining the amount of incoming solar radiation—**insolation**—reaching any given point on Earth.

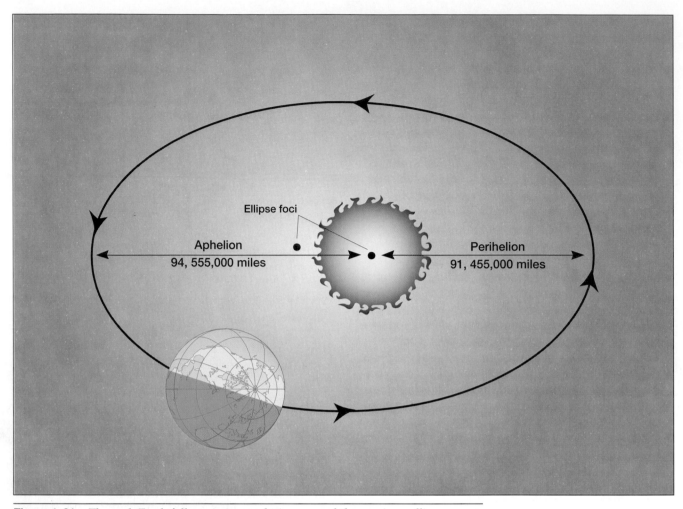

Figure 1-21   The path Earth follows in its revolution around the sun is an ellipse having the sun at one focus. The line marking the shortest distance between the sun and Earth's orbit is called the perihelion and is 91,455,000 miles (147,166,480 kilometers) long. The line marking the greatest distance between the sun and Earth's orbit is called the aphelion and is 94,555,000 miles (152,171,500 kilometers) long.

SOLSTICES

At all times, the sun illuminates one-half of Earth (Figure 1-25). On or about June 21 (the exact date varies slightly from year to year), the noon rays from the sun are perpendicular at the latitude line lying 23.5° north of the equator (Figure 1-26). This parallel, the **Tropic of Cancer**, marks the northernmost location reached by perpendicular rays in the annual cycle of Earth's revolution. At this time, because Earth is tilted on its axis the *circle of illumination* reaches 23.5° to the far side of the North Pole and stops short 23.5° to the near side of the South Pole, as Figure 1-26 shows. As Earth rotates on its axis, all points lying north of 66.5° N (in other words, within 23.5° of the North Pole) remain continuously within the circle of illumination, thus experiencing 24 continuous hours of daylight. By contrast, all points south of 66.5° S are always outside the

circle of illumination and thus have 24 continuous hours of darkness. These special parallels defining the 24 hours of light and dark on the solstice dates are the *polar circles*. The northern polar circle, at 66.5° N, is the **Arctic Circle;** the southern polar circle, at 66.5° S, is the **Antarctic Circle.**

On or about December 21 (slightly variable from year to year), the perpendicular rays of the sun strike the parallel of 23.5° S, the **Tropic of Capricorn.** Once again, the circle of illumination reaches to the far side of one pole and falls short on the near side of the other pole, but this time the illuminated hemispheres are reversed. Areas north of the Arctic Circle now always lie outside the circle of illumination and thus are in continuous darkness, whereas areas within the Antarctic Circle are continuously within the circle and thus in daylight for 24 hours.

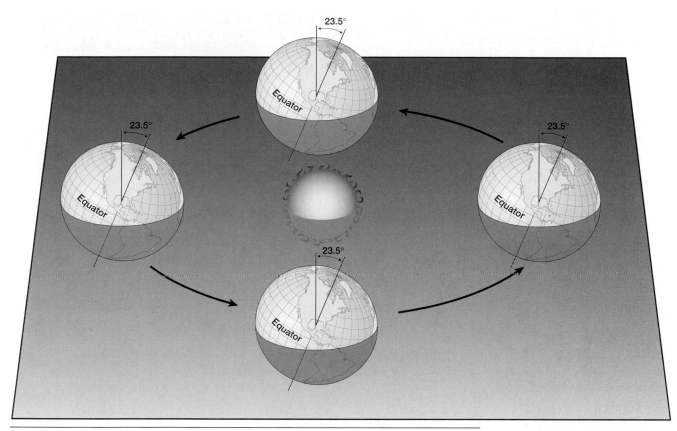

Figure 1-22   The plane of the ecliptic is the imaginary plane that passes through the sun as well as through every point on Earth's orbit around the sun. Because Earth's rotation axis is tilted, the plane of the ecliptic and the equatorial plane do not coincide.

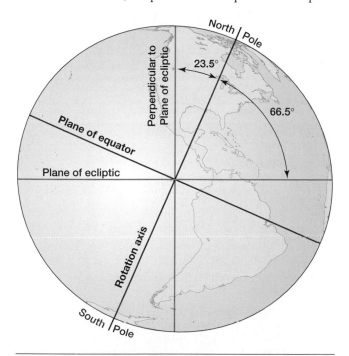

Figure 1-23   Earth's rotation axis forms an angle of 66.5° with the plane of the ecliptic, and this tilt is called the inclination of the axis. It is commonly described not in terms of this angle, however, but rather as being 23.5° from a line perpendicular to the plane of the ecliptic.

Although the angle of incoming perpendicular rays shifts 47° from June 21 to December 21, the relationships between Earth and the sun are very similar on those days, which are called the **solstices.** In the Northern Hemisphere, December 21 is called the *winter solstice* and June 21 is referred to as the *summer solstice.* (In the Southern Hemisphere, the seasonal designations are reversed.) Table 1-5 summarizes the conditions present during the winter and summer solstices. The enormous variations in length of daylight at different latitudes at the time of a solstice are shown in Table 1-6.

EQUINOXES

Figure 1-27 shows the circle of illumination on approximately March 20 and September 22 (as with solstice dates, these dates also vary slightly from year to year), which are the **equinoxes** (from the Latin, meaning "the time of equal days and nights"). March 20 is the *vernal* (spring) *equinox* in the Northern Hemisphere, and September 22 is the *autumnal equinox*; the seasonal terms are reversed in the Southern Hemisphere. The positional relationships of Earth and the sun are virtually identical on these two dates, and the following characteristics prevail:

(a)

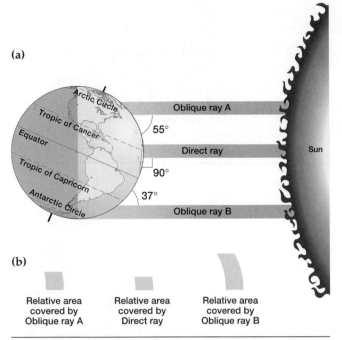

(b)

Relative area covered by Oblique ray A

Relative area covered by Direct ray

Relative area covered by Oblique ray B

Figure 1-24   The amount of solar radiation warming a given area on Earth's surface depends on the angle at which the ray strikes the surface. **(a)** A **direct ray** strikes the surface at a 90° angle, and an **oblique ray** strikes at an angle smaller than 90°. **(b)** The relative areas covered by the three rays shown in (a). Because all rays contain the same amount of heat energy, the three areas shown all receive the same amount of heat. The smallest area, that heated by the direct ray, is warmed most, therefore, and the largest, that heated by oblique ray B, is warmed least.

TABLE 1-6
Day Length at Time of June Solstice

Latitude	Day length	Noon sun angle (degrees)
90° N	24 h	23.5
80° N	24 h	33.5
70° N	24 h	43.5
60° N	18 h 53 min	53.5
50° N	16 h 23 min	63.5
40° N	15 h 01 min	73.5
30° N	14 h 05 min	83.5
20° N	13 h 21 min	86.5
10° N	12 h 43 min	76.5
0°	12 h 07 min	66.5
10° S	11 h 32 min	56.5
20° S	10 h 55 min	46.5
30° S	10 h 12 min	36.5
40° S	09 h 20 min	26.5
50° S	08 h 04 min	16.5
60° S	05 h 52 min	6.5
70° S	0	0
80° S	0	0
90° S	0	0

Source: After Robert J. List, *Smithsonian Meteorological Tables*, 6th rev. ed. Washington, DC: Smithsonian Institution, 1963, Table 171.

TABLE 1-5
Conditions at the Solstices

Winter Solstice, December 21	Summer Solstice, June 21
Vertical rays at Tropic of Capricorn	Vertical rays at Tropic of Cancer
Higher sun angle in Southern Hemisphere, as compared with same latitude in Northern Hemisphere	Higher sun angle in Northern Hemisphere, as compared with same latitude in Southern Hemisphere
Day length increases with increasing latitude in Southern Hemisphere	Day length increases with increasing latitude in Northern Hemisphere
Day length decreases with increasing latitude in Northern Hemisphere	Day length decreases with increasing latitude in Southern Hemisphere
Equator receives 12 hours of daylight	Equator receives 12 hours of daylight
Circle of illumination touches Arctic Circle on near side of Earth	Circle of illumination touches Arctic Circle on far side of Earth
Circle of illumination touches Antarctic Circle on far side of Earth	Circle of illumination touches Antarctic Circle on near side of Earth
24 hours of daylight from Antarctic Circle to South Pole	24 hours of daylight from Arctic Circle to North Pole
24 hours of darkness from Arctic Circle to North Pole	24 hours of darkness from Antarctic Circle to South Pole

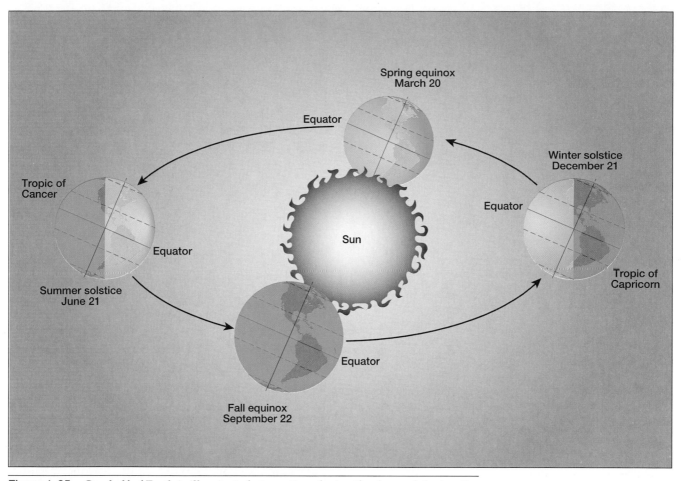

Figure 1-25   One half of Earth is illuminated at any time during the day and during the year. Here light blue is the illuminated half and dark blue is the unilluminated half. As noted earlier, the line between the two halves is called the *circle of illumination*. Note its position relative to the polar circles on June 21, date of the summer solstice, and December 21, date of the winter solstice.

1. The perpendicular rays of the sun strike the equator.
2. The circle of illumination just touches both poles.
3. The periods of daylight and darkness are each 12 hours long all over Earth, a situation that occurs only on these two dates.
4. The equinoxes represent the midpoints in the shifting of direct rays of the sun between the Tropic of Cancer and the Tropic of Capricorn.

Changes in Daylight and Darkness

From the March equinox until the June solstice, the period of daylight gradually increases everywhere north of the equator. Conversely, during this period day length diminishes for all points south of the equator.

From June 21 until September 22, these changes are reversed, with the days getting shorter in the Northern Hemisphere and longer in the Southern Hemisphere until September 22, when day and night are again equal all over Earth.

After the September equinox, the days of the Northern Hemisphere continue to become shorter and those of the Southern Hemisphere continue to lengthen. By the time of the December solstice, the days are shortened progressively northward, reaching a minimum of no daylight north of the Arctic Circle. During this same interval, day length gets longer south of the equator, reaching a maximum of 24 hours of continuous daylight south of the Antarctic Circle. Again, this pattern is reversed following the December solstice, and periods of daylight and darkness become equalized over the entire world at the March equinox.

Both day length and the angle at which the sun's rays strike Earth are principal determinants of the amount of insolation received at any particular latitude. Thus the tropical latitudes are always warm/hot because they always have high sun angles and consistent day lengths

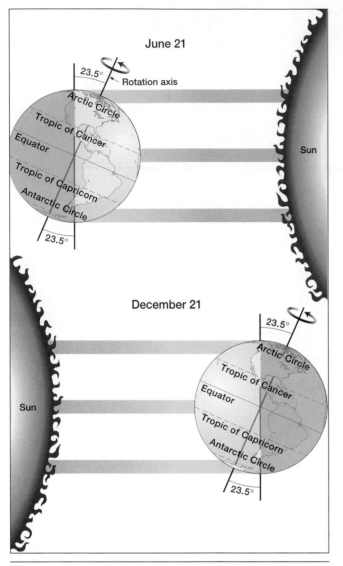

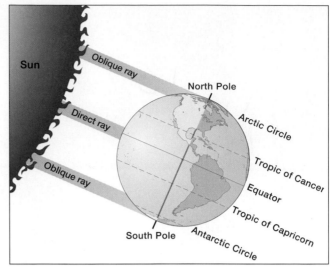

Figure 1-27 Earth-sun configuration during the spring and fall equinoxes, which usually occur on March 20 and September 22, respectively, in the Northern Hemisphere. On these two days the direct noon rays of the sun fall on the equator.

Figure 1-26 Earth-sun configuration during the winter and summer solstices. At these times, the sun's noon rays strike directly at 23.5° of latitude. During the June solstice (Northern Hemisphere summer), sunlight is concentrated in the Northern Hemisphere. During the December solstice (Northern Hemisphere winter), sunlight is concentrated in the Southern Hemisphere. Direct rays never get any farther south than 23.5° south latitude or any farther north than 23.5° north latitude. After June 21 they begin striking farther and farther south of the Tropic of Cancer, and after December 21 they begin striking farther and farther north of the Tropic of Capricorn.

that are close to 12 hours long. Indeed, the region between the Tropic lines is the only part of Earth that experiences 90° sun angles sometime during the year. Conversely the polar regions are consistently cold because they always have low sun angles in spite of 24-hour days in "summer." Only the areas poleward of 66.5° have 24-hour days or nights. Seasonal temperature differences are large in the midlatitudes because of sizable seasonal variations in sun angles and length of day.

TELLING TIME

> There are three natural units of time: the tropical year, marked by the return of the seasons; the lunar month, marked by the return of the new moon; and the day, marked by passage of the sun. All other units, such as the hour, minute and second, are man-made to meet the needs of society.
>
> (Malcolm Thomson, a Canadian authority on the physics of time)

In prehistoric times, the rising and setting of the sun probably were the principal means of telling time. As civilizations developed, however, more precise time-keeping was required. Early agricultural civilizations in Egypt, Mesopotamia, India, China, and even England, as well as the Aztec and Mayan civilizations in the New World, observed the sun and the stars in order to tell time and to keep accurate calendars.

Local solar noon can be determined by watching for the moment when objects cast their shortest shadows. The Romans used sundials to tell time (Figure 1-28) and gave great importance to the noon position, which they called the **meridian**—the sun's highest (*meri*) point of the day (*diem*). Our use of A.M. (*ante meridian*: before noon) and P.M. (*post meridian*: after noon) was derived from the Roman world.

When nearly all transportation was by foot, horse, or sailing vessel, it was difficult to compare time at different localities. In those days, each community set its

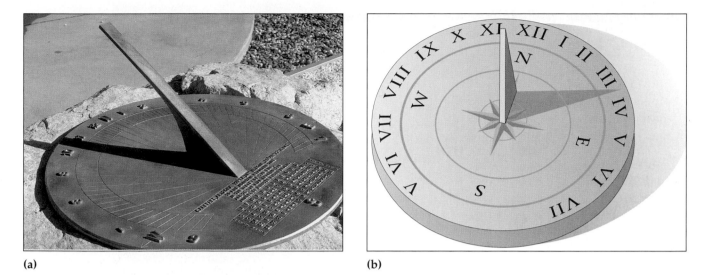

(a)   (b)

Figure 1-28   (a) *A* typical sundial. (Richard Kolar/Earth Scenes) (b) A sundial is the oldest known device for measuring time, having been used in Babylon at least as early as 2000 B.C. The main idea behind how it works is that the shadow of an object moves from one side of the object to the other as the sun appears to move across the sky during the course of a day. The two main parts of any sundial are a horizontal face and a vertical piece of metal—called a gnomon—in its center. The upper edge of the gnomon must slant upward from the dial face at an angle equal to the latitude at the place where the sundial is installed. The gnomom points toward the North Pole in the Northern Hemisphere and toward the South Pole in the Southern Hemisphere. When the sun shines on it, the gnomon casts a shadow that tells the time. The time shown in this drawing is 3:30 PM.

own time by correcting its clocks to high noon at the moment of the shortest shadow. A central public building, such as a temple in India or a county courthouse in Kansas, usually had a large clock or loud bells to toll the hour. Periodically, this time was checked against the shortest shadow.

STANDARD TIME

As the telegraph and railroad began to speed words and passengers between cities, the use of local solar time created increasing problems. A cross-country rail traveler in the United States in the 1870s might have experienced as many as 24 different times between the Atlantic and Pacific coasts. Eventually, the railroads stimulated the development of a standardized time system. At the previously mentioned 1884 International Prime Meridian Conference in Washington, the major nations agreed to divide the world into 24 standard time zones, each extending over 15° of longitude. The local solar time of the Greenwich (prime) meridian was chosen as the standard for the entire system. The prime meridian became the center of a time zone that extends 7.5° of longitude to the west and 7.5° to the east of the prime meridian. Similarly, the meridians that are multiples of 15, both east and west of the prime meridian, were set

as the *central meridians* for the 23 other time zones (Figure 1-29).

Although Greenwich mean time (GMT) is now referred to as **Universal Time Coordinated** (UTC), the prime meridian is still the reference for standard time. Since it is always the same number of minutes after the hour in all standard time zones (keeping in mind that a few countries, such as Saudi Arabia, do not adhere to standard time zones), we usually need to know only the correct time at Greenwich and the number of hours that our local time zone is later or earlier than the Greenwich meridian in order to know exact local time. Figure 1-29 shows the number of hours "fast" (later) or "slow" (earlier) for UTC in each time zone of the world.

Most of the nations of the world are sufficiently small in their east-west direction so as to lie totally within a single time zone. However, large countries may encompass several zones: Russia occupies nine time zones; including Alaska and Hawaii, the United States spreads over six (Figure 1-30); Canada, six; and Australia, three. In international waters, time zones are defined to be exactly 7°30' to the east and 7°30' to the west of the central meridians. Over land areas, however, zone boundaries vary to coincide with appropriate political and economic boundaries. For example, continental Europe from Portugal to Poland

FOCUS

THE ANALEMMA

An **analemma** is a scale shaped like the figure 8 that shows the latitude of direct overhead noon sunshine for every day of the year (Figure 1-3-A). Knowing this latitude, one can determine the angle of the noon sun for any place on Earth for any day of the year. The procedure is to compare the latitude of the place in question with the latitude that receives noon sun directly overhead. If the place in question is in the same hemisphere as the latitude that receives noon sun directly overhead, subtract the smaller from the larger number and then subtract that answer from 90°. If the place in question is in the opposite hemisphere, add the two numbers and then subtract that answer from 90°.

An analemma also shows the equation of time for each day of the year, which is the number of minutes earlier or later than noon that the sun arrives over the local meridian. Owing to minor variations in rotation and revolution, there are small but predictable fluctuations in the arrival of the overhead sun at different times of the year. As shown by the analemma, the sun sometimes arrives overhead at a meridian as much as 14 minutes after noon (in February) and as much as 16 minutes before noon (in November).

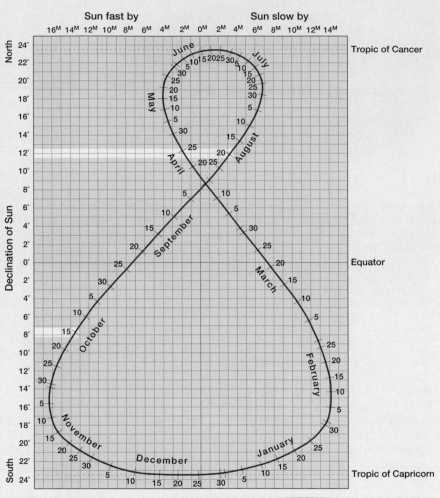

Figure 1-3-A   An analemma tells us the sun angle at any place on any day. To see how it works, let us first arbitrarily choose August 20 in Cairo, the latitude of which is 30° N. On this date, the noon sun is directly overhead at 12° N, and so our calculation is 30 - 12 = 18; 90 - 18 = 72. On August 20, therefore, the noon sun shining on Cairo makes an angle of 72° with the line tangent to Earth's surface at Cairo. The October 15 calculation for this same city shows us how to proceed when we have to deal with two hemispheres instead of one: 8 + 30 = 38; 90 - 38 = 52. The October 15 noon sun hits Cairo at an angle of 52°.

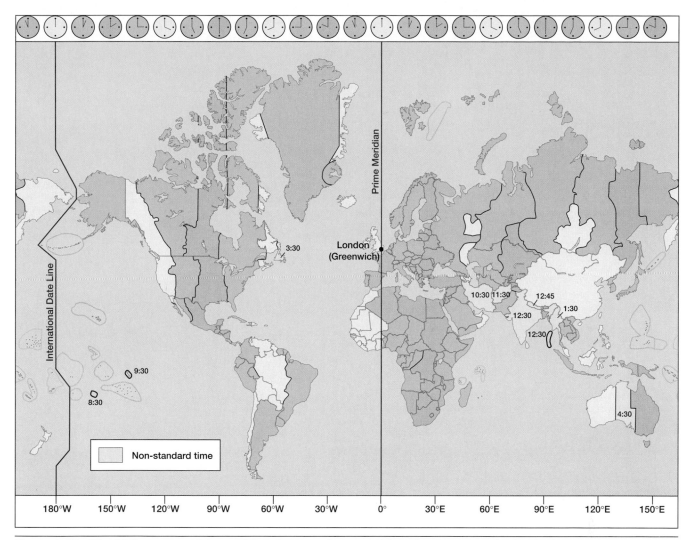

Figure 1-29 The 24 time zones of the world, each centered on a meridian that is some multiple of 15˚.

shares one time zone, although longitudinally covering about 30º. At the extreme, China extends across four 15˚ zones, but the entire nation, at least officially, observes the time of the 120th east meridian, which is the one closest to Beijing.

In each time zone there is a controlling meridian along which clock time is the same as sun time (i.e., the noon sun is directly overhead). On either side of that meridian, of course, clock time does not coincide with sun time. The deviation between the two is shown for one U.S. zone in Figure 1-31.

From the map of time zones of the United States (Figure 1-30), we can recognize a great deal of manipulation of the time zone boundaries for economic and political convenience. For example, the Central Standard Time Zone, centered on 90˚ W, extends all the way to 105˚ W (which is the central meridian of the

Mountain Standard Time Zone) in Texas in order to keep most of that state within the same zone. By contrast, El Paso, Texas, is officially within the Mountain Standard Time Zone, in accord with its role as a major market center for southern New Mexico, which observes Mountain Standard Time. In the same vein, northwestern Indiana is in the Central Standard Time Zone with Chicago.

THE INTERNATIONAL DATE LINE

In 1519 Ferdinand Magellan set out westward from Spain, sailing for East Asia with 241 men in five ships. Three years later the remnants of his crew (18 men in one ship) successfully completed the first circumnavigation of the globe. Although a careful log had been kept, the crew found that their calendar was one day

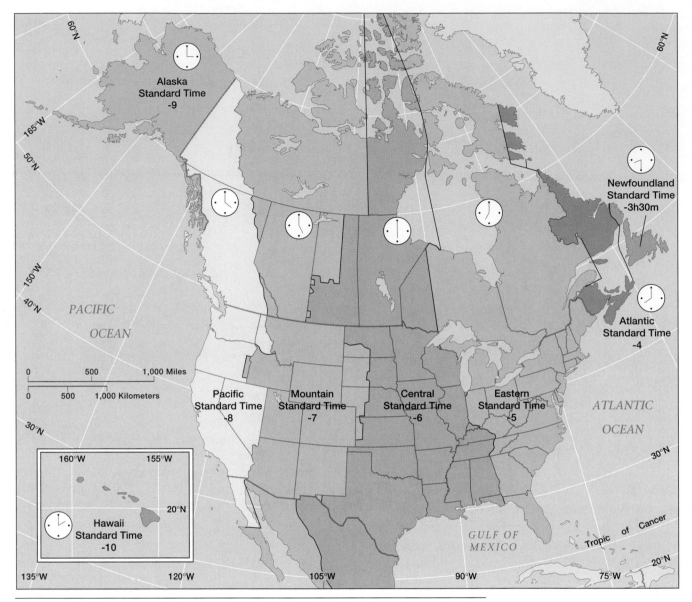

Figure 1-30 Times zones for Canada and the United States. Zone boundaries often do not follow meridians.

short of the correct date. This was the first human experience with time change on a global scale, the realization of which eventually led to the establishment of the international date line.

One advantage of establishing the Greenwich meridian as the prime meridian is that its opposite arc is in the Pacific Ocean. The 180th meridian, transiting the sparsely populated mid-Pacific, was chosen as the meridian at which new days begin and old days exit from the surface of Earth. The **international date line** deviates from the 180th meridian in the Bering Sea to include all of the Aleutian Islands of Alaska within the same day and again in the South Pacific to keep islands

of the same group (Fiji, Tonga) within the same day (Figure 1-32).

The new day first appears on Earth at midnight at the international date line (Figure 1-33). For the next 24 hours, the new day advances westward around the world, finally covering the entire surface for an hour at the end of this period. For the next 24 hours, this day leaves Earth, 1 hour at a time, making its final exit 48 hours after its first appearance. Except at Greenwich noon, two days exist on Earth: the more recent one (e.g., January 2) extending from the international date line westward to the current position of midnight, and the older one (e.g., January 1) extending the rest of the

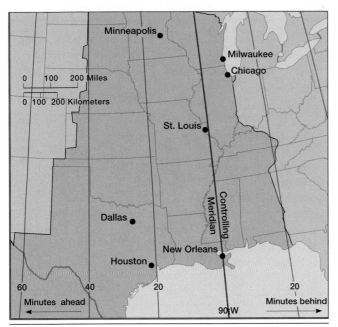

Figure 1-31   Standard clock time versus sun time. The noon sun is directly overhead in St. Louis and New Orleans because these two cities lie on the controlling meridian. For places not on the controlling meridian, the sun is directly overhead at some time other than standard time noon. In Chicago, for instance, the sun is directly overhead at 11:50 A. M., and in Dallas it is directly overhead at 12:28 P.M.

way to the date line. Thus when you cross the international date line going from west to east, the day becomes "earlier" (e.g., from January 2 to January 1); when you move across the line from east to west, the day becomes "later" (e.g., from January 1 to January 2).

The international date line is in the middle of a time zone. Consequently, there is no time zone (i.e., hourly) change at that point; there is a change only on the calendar, not on the clock.

DAYLIGHT SAVING TIME

To conserve energy during World War I, Germany ordered all clocks set forward by an hour. This practice allowed the citizenry to "save" an hour of daylight by extending the daylight period into the usual evening hours, thus reducing the consumption of electricity for lighting. The United States began a similar policy in 1918, but many localities declined to observe "summer time" until the Uniform Time Act made the practice mandatory in all states that had not deliberately exempted themselves. Arizona, Hawaii, and part of Indiana have exempted themselves from observance of daylight-saving time under this act.

Russia has adopted permanent daylight-saving time (and double daylight-saving time—2 hours ahead of sun time–in the summer). In recent years, Canada,

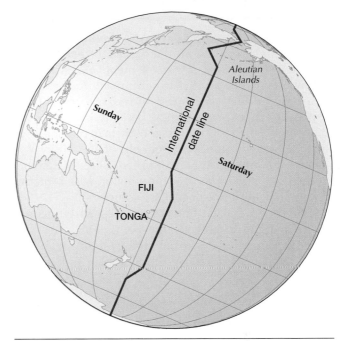

Figure 1-32   The international date line generally follows the 180th meridian, but it zigzags around various island groups.

Australia, New Zealand, and most of the nations of western Europe have also adopted daylight-saving time. In the Northern Hemisphere, many nations, like the United States, begin daylight-savings time early in April and resume standard time late in October. In the tropics, the lengths of day and night change little seasonally, and there is not much twilight. Consequently, daylight-saving time would offer little or no "savings" for tropical areas.

Figure 1-33   The masthead of *The Fiji Times*, a newspaper published in Suva, Fiji, proclaims that it is the first newspaper in the world each day. Since Suva is located just west of the international date line, this claim cannot be faulted. (Photo by Tobi Zausner)

CHAPTER SUMMARY

Geography is the study of the distribution of natural and cultural elements on the surface of Earth and of the interrelationships of these elements. Physical geography is concerned with the natural environment, and can conveniently be subdivided into four basic "spheres": lithosphere, atmosphere, hydrosphere, and biosphere. Earth is one of nine planets in our solar system, which is a minor component of the Milky Way galaxy. It is an almost perfect sphere, slightly flattened at the poles and bulging at the equator, with a circumference of 24,900 miles (40,000 kilometers).

Longitude and latitude lines are used by geographers to accurately locate any spot on Earth's surface. The imaginary geographic grid system formed by these lines is anchored by the the poles and equator,

which are determined by Earth's slight variance from a perfectly spherical shape and its rotation pattern.

The geometric relationship between Earth and the sun varies continuously as both move through interplanetary space. Moreover, Earth has perpetual motions of its own—rotation on its axis every 24 hours and revolution about the sun every 365.25 days—that change its position with respect to the sun and determine both the annual march of the seasons and the alternation of day and night. Standard time zones have been established over the world based on the 24 hours needed for one complete earthly rotation. The framework of time zones is anchored by the prime meridian on one side of the globe and the international date line on the other.

KEY TERMS

analemma	great circle	physical geography
Antarctic Circle	hydrosphere	plane of the ecliptic
aphelion	inclination	polarity
Arctic Circle	insolation	prime meridian
atmosphere	international date line	small circle
biosphere	latitude	solar system
circle of illumination	lithosphere	solstice
cultural geography	longitude	Tropic of Cancer
direct ray	meridian	Tropic of Capricorn
equator	oblique ray	Universal Time Coordinated
equinox	parallel	
graticule	perihelion	

REVIEW QUESTIONS

1. Identify and briefly describe the four environmental "spheres."
2. In what ways do the inner and outer planets of our solar system differ from each other?
3. Is Earth large or small? Explain.
4. The sphericity of Earth is not perfect. Is this fact important in our study of physical geography?
5. What are the major differences between *parallels* and *meridians*?
6. What is the relationship among a *great circle*, the *equatorial plane*, and the *circle of illumination*?
7. Explain why the *plane of the ecliptic* does not coincide with the *equatorial plane*.
8. What would be the effect on the annual march of the seasons if Earth's axis did not maintain *parallelism* as the planet circles the sun?

9. Why are vertical rays of the sun never experienced poleward of the tropic lines?
10. On June 21st, at 15° south latitude, what is the angle of the sun at noon?
11. If it is 3:00 P.M. Friday in Hong Kong (114° east longitude), what is the time and day in Los Angeles (118° west longitude)?
12. Towns A and B are on the same meridian; A is at latitude 10° north, and B is at latitude 10° south. How many miles apart are they?
13. Towns C and D are on the same parallel: C is at 10° east, and D is at 10° west. How many miles apart are they? (*Note*: This is a trick question.)
14. Points E and F are both on the equator: E is at 175° west, and F is at 179° east. How many miles apart are they?

SOME USEFUL REFERENCES

ALLABY, AILSA, AND MICHAEL ALLABY, eds. *The Concise Oxford Dictionary of Earth Sciences*. London and New York: Oxford University Press, 1990.

CLOUD, P., *Cosmos, Earth, and Man*. New Haven, CT: Yale University Press, 1978.

HOWSE, DEREK, *Greenwich Time and the Discovery of the Longitude*. London and New York: Oxford University Press, 1980.

RIDLEY, B. K., *The Physical Environment*. Chichester, England: Ellis Horwood Limited, 1979.

RINGWOOD, A. E., *Origin of the Earth and Moon*. New York: Springer-Verlag, 1979.

SCIENTIFIC AMERICAN EDITORS, *The Solar System: A Scientific American Book*. San Francisco: W. H. Freeman & Company, Publishers, 1975.

WHITTOW, J. B., *The Penguin Dictionary of Physical Geography*. Harmondsworth, England: Penguin, 1984.

2

PORTRAYING EARTH

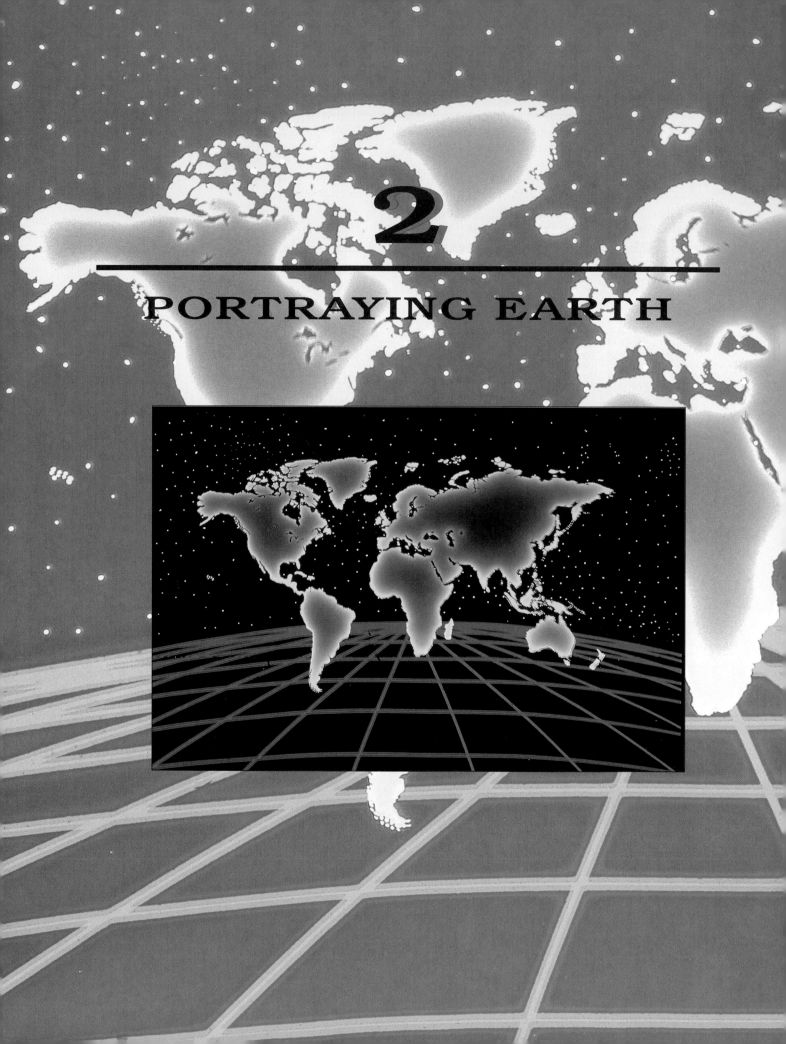

THE SURFACE OF EARTH IS THE FOCUS OF THE GEOGRApher's interest. The enormity and complexity of this surface would be difficult to comprehend and analyze without tools and equipment to aid in systematizing and organizing the varied data. Although many kinds of tools are used in geographic studies, the most important and most universal are maps because the mapping of any geographic feature is normally essential to understanding the spatial distributions and relationships of that feature.

Our concern in this chapter is with the usefulness of maps to geographers. In some cases, geographers deal with maps as an end in themselves, but more often than not maps serve geographers as a means to some end. This book is a case in point. It contains numerous maps of various kinds, each inserted in the book to further your understanding of some fact, concept, or relationship.

This chapter outlines the positive and negative attributes of maps and other tools for portraying any portion of Earth. Our purpose in the chapter is twofold:

1. To describe the basic characteristics of maps, including their capabilities and limitations.
2. To describe the various ways a landscape can be portrayed–through map projections, globes, photographs, and remotely sensed imagery.

THE NATURE OF MAPS

A **map** is a two-dimensional representation of the spatial distribution of selected phenomena–normally components of a landscape. In essence, a map is a scaled drawing of a portion of a landscape, representing the area at a reduced scale and showing only selected data. A map serves as a surrogate (a substitute) for any surface we wish to portray or study. Although any surface can be mapped—the lunar surface, for instance, or that of Mars—all the maps we are concerned with in this book portray portions of Earth's surface.

The basic attribute of maps is their ability to show distance, direction, size, and shape in their horizontal (that is to say, two-dimensional) spatial relationships. In addition to these fundamental graphic data, most maps show other kinds of information as well. Maps nearly always have a special purpose, and that purpose is usually to show the distribution of one or more phenomena (Figure 2-1). Thus a map may be designed to show street patterns, or the distribution of Tasmanians, or the ratio of sunshine to cloud, or the number of earthworms per cubic meter of soil, or any of an infinite number of other facts or combinations of facts. Because they depict graphically *what* is *where* and because they are often helpful in providing clues as to *why* such a distribution occurs, maps are indispensable tools for geographers.

Even so, it is important to realize that maps have their faults. First of all, no map is perfectly accurate. Most people understand that not everything written in a book is necessarily valid—the printed word is not infallible simply because it exists in black and white. These same people, however, may uncritically accept all information portrayed on a map as being true. Most people have insufficient experience with maps to transfer their skepticism about the written word to a similar suspicion of the data depicted on maps. Nevertheless, the inaccuracy of maps is ubiquitous, simply because a map is an attempt to portray an impossible geometrical relationship: a curved surface drawn on a flat piece of paper.

THE MATTER OF SCALE

A map is always smaller than the portion of Earth's surface it represents, and so any user who is to understand the areal relationships (distances or relative sizes, for example) depicted on that map must know how to use a map scale. The **scale** of a map gives the relationship between length measured on the map and the corresponding distance on the ground. Knowing the scale of a map makes it possible to measure distance, determine area, and compare sizes.

Scale can never be represented with perfect accuracy on a map because of the impossibility of rendering the curve of a sphere on the flatness of a sheet of paper without distortion. Therefore, a map scale cannot be constant (in other words, the same) all over the map. If the map is of a small area, however, the scale is so nearly perfect that it can be accepted uncritically throughout the map. If the map represents either a large portion or all of Earth's surface, however, there are enormous scale variations because of the significant distortions involved. Thus it is important for us to understand the capabilities and recognize the limitations of different kinds of maps and maps at different scales.

SCALE TYPES

There are several ways to portray scale on a map, but only three are widely used: the graphic method, the word method, and the fractional method (Figure 2-2).

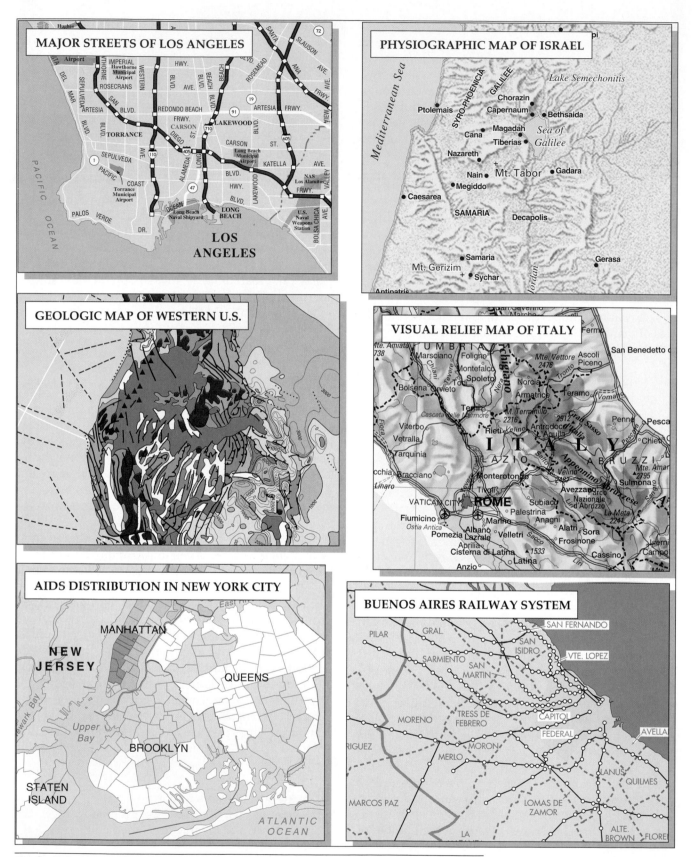

Figure 2-1 A sampler of different kinds of maps.

FOCUS

MAP ESSENTIALS

Maps come in an infinite variety of sizes and styles and serve a limitless diversity of purposes. Some are general-reference maps—a map of the world, say, or one of the western coast of Africa. Others are thematic maps, which means that they show the location or distribution of particular phenomena, as shown in Figure 2-1-A. Regardless of type, however, every map must contain a few basic components (discussed below) to facilitate their use. Omission of any of these essential components decreases the clarity of the map and makes it more difficult to interpret.

Title

This ought to be a brief summary of the map's content or purpose. It should identify the area covered and provide some indication of content, as "Road Map of Kenya," "River Discharge in Northern Europe," or "Seattle: Shopping Centers and Transit Lines."

Date

This should indicate the time span over which the information was collected. In addition, some maps also give the date of publication of the map. Most maps depict conditions or patterns that are temporary or even momentary. In order for a map to be meaningful, therefore, the reader must be informed when the data were gathered, as this information indicates how timely or out of date the map is.

Legend

Most maps use symbols, colors, shadings, or other devices to represent features or the amount, degree, or proportion of some quantity. Some symbols are self-explanatory, but it is

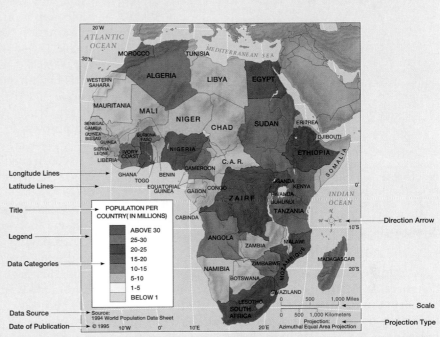

Figure 2-1-A   A typical thematic map, containing all the essentials.

usually necessary to include a legend box in a corner of the map to explain the symbolization.

Scale

Any map that serves as more than a pictogram must be drawn to scale, at least approximately. A graphic, verbal, or fractional scale is therefore necessary.

Direction

Direction is normally shown on a map by means of the geographic grid. Meridians are supposed to extend north-south, and parallels are east-west lines. If no grid is shown, direction may be indicated by a straight arrow pointing northward, which is called a north arrow. A north arrow is aligned with the meridians and thus points toward the north geographic pole.

Location

As we learned in Chapter 1, the standard system for locating places

on a map is a geographic grid showing latitude and longitude by means of parallels and meridians. Other grid systems are sometimes used for specifying locations because the latitude/longitude system, with its angular subdivisions, is cumbersome to use. These alternative systems are devised like the x and y coordinates of a graph. Some maps display more than one coordinate system.

Data Source

For most thematic maps, it is useful to indicate the source of the data.

Projection Type

On many maps, particularly small-scale ones, the type of projection is indicated.

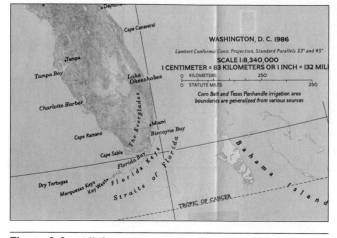

Figure 2-2   All three expressions of scale are shown on this map. Included are a fractional scale (1: 8,340,000), a word scale (1 centimeter = 83 kilometers or 1 inch = 132 miles), and a graphic scale (shown in both miles and kilometers). (Courtesy of National Geographic Society)

A **graphic scale** uses a line marked off in graduated distances. To determine the distance between two points on Earth's surface, one measures the distance between the points on the map and compares that length with a measurement along the line of the graphic scale belonging to the map. The advantage of a graphic scale is its simplicity: you determine approximate distances on the surface of Earth by measuring them directly on the map. Moreover, a graphic scale remains correct when a map is reproduced in a larger or smaller size because the length of the graphic scale line is changed precisely as the map size is changed.

A **word**, or **verbal, scale** states in words the ratio of the map scale length to the distance on Earth's surface, such as "one inch to one mile" or "five centimeters to ten kilometers."

A **fractional scale** compares map distance with ground distance by proportional numbers expressed as a fraction or ratio called a **representative fraction**. For example, a common fractional scale uses the representative fraction 1/63,360, usually expressed as the ratio 1:63,360. This notation means that 1 unit of distance on the map represents 63,360 of the same units on the ground. (We use this particular number because it is the number of inches in one mile. As noted below, scales are often given in a mixture of inches and miles, and so the 63,360 is a convenient standard scale number.) Thus 1 inch on the map represents 63,360 inches on the ground, 1 millimeter on the map represents 63,360 millimeters on the ground, and so forth. When no units are given in a fractional scale or ratio, the units are the same for both numbers. When the units are different for the two numbers, they must be expressed as, for instance, in the scale 1 inch = 1 mile.

LARGE AND SMALL SCALES

The adjectives "large" and "small" are comparative rather than absolute. In other words, scales are large or small only in comparison with other scales. A **large-scale map** is one that has a relatively large representative fraction, which means that the denominator is small. Thus, 1/10,000 is a larger fraction than, say, 1/1,000,000, and so a scale of 1:10,000 is large in comparison with one of 1:1,000,000; consequently, a map at a scale of 1:10,000 is called a large-scale map. Such a map portrays only a small portion of Earth's surface but portrays it in considerable detail. For example, if this page were covered with a map having a scale of 1:10,000, it would be able to show just a small part of a single county but that part would be rendered in great detail.

A **small-scale map** has a small representative fraction—in other words, one having a large denominator. A map having a scale of 1:10,000,000 is classified as a small-scale map. If it were covered with a map of that scale, this page would be able to portray about one-third of the United States but only in limited detail.

Figure 2-3 compares various scales and gives some idea of what is visible with each.

THE ROLE OF GLOBES

For properly portraying the sphericity of Earth, there is no substitute for a model globe (Figure 2-4). If manufactured carefully, a globe can be an accurate representation of the shape of our planet. The only thing changed in the transition from the immensity of Earth to the manageable proportions of a model globe is size. A globe is capable of maintaining the correct geometric relationships of meridian to parallel, of equator to pole, of continents to oceans. It can show comparative distances, comparative sizes, and accurate directions. It can represent, essentially without distortion, the spatial relationships of the features of Earth's surface.

A globe is not without disadvantages, however. For one thing, only half of it can be viewed at one time, and the periphery of the visible half is not easy to see (Figure 2-5). Moreover, almost any globe must be constructed at a very small scale, which means that it is incapable of portraying much detail. The principal problem with a globe, however, is that it is too cumbersome for almost any use other than classroom study or quiet contemplation. Because maps are much more portable and versatile than globes, therefore, there are literally billions of maps in use over the world, whereas globes are extremely limited both in number and in variety.

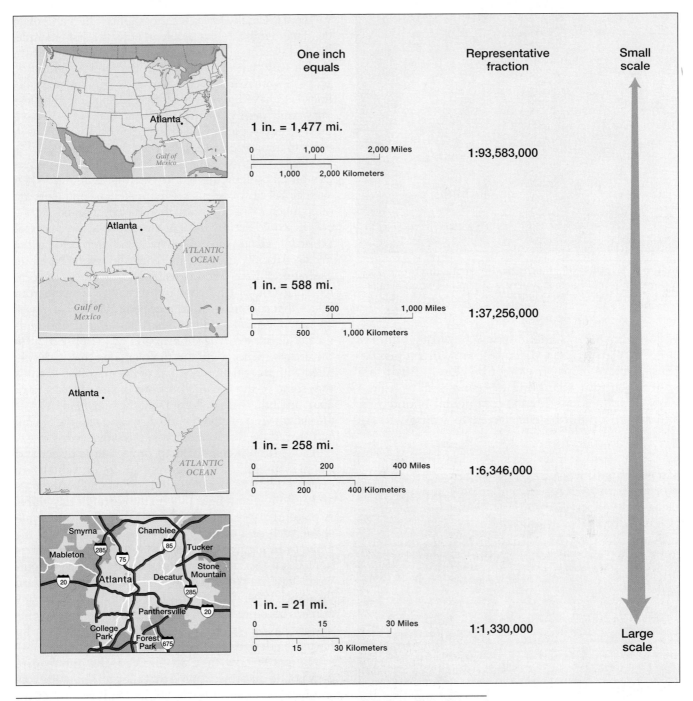

	One inch equals	Representative fraction	Small scale
	1 in. = 1,477 mi.	1:93,583,000	
	1 in. = 588 mi.	1:37,256,000	
	1 in. = 258 mi.	1:6,346,000	
	1 in. = 21 mi.	1:1,330,000	Large scale

Figure 2-3   Comparisons of distance and area at various map scales. A small-scale map portrays a large part of Earth's surface but depicts only the most salient features, whereas a large-scale map shows only a small part of the surface but in considerably more detail.

MAP PROJECTIONS

The challenge to the cartographer (mapmaker) is to combine the geometric exactness of a globe with the convenience of a flat map. This melding has been attempted for many centuries, and further refinements continue to be made. The fundamental problem is always the same: how to transfer data from a spherical surface to a flat piece of paper with a minimum of distortion?

A **map projection** is a system whereby the rounded surface of Earth is transformed in order to display it on a flat surface. No matter how the transformation

Figure 2-4   A model globe provides a splendid broad representation of Earth at a very small scale, but no details can be portrayed. (1989 David Sailors/The Stock Market)

is done, however, something will be wrong with the shapes, relative sizes, distances, and directions. Because a piece of paper cannot be closely fitted to a sphere without wrinkling or tearing, data from a globe (parallels, meridians, continental boundaries, and so on) cannot be transferred to a map without distortion.

Figure 2-5   Earth is indeed a globe. This is a composite of thousands of images made by the NOAA series of weather satellites. (Tom Van Sant/Geosphere Project, Santa Monica/Science Photo Library/Photo Researchers, Inc.)

There are many ways a cartographer can manipulate the data to mitigate the problem, however. For example, the grid system can be arranged on the flat map surface in such a way that one or more of the globe's geometric properties are retained. Another way to minimize distortions is to have the most distorted areas fall in the "less important" parts of the map. Still another way is to have the map interrupted by blank spaces in oceanic regions in order to minimize distortion of the continents (Figure 2-6). This last method is the most striking and is very effective on maps showing worldwide distributions. When such global distributions are mapped, the continents are often much more important than the oceans, and yet the oceans would occupy most of the map space in a normal projection. Hence, the projection can be interrupted in the Pacific, Atlantic, and Indian oceans and can be based on **central meridians** (meridians that pass through the center of each major landmass and serve as a baseline from which that continent can be mapped). With no land area far from a central meridian, shape and size distortion is greatly decreased. The interruption of the projection in the oceans creates a void in the map, one that is simply filled with information not part of the map. The result is that some of the oceanic portions of the map are torn apart and otherwise distorted, but the major landmasses are shown with relatively little distortion and the overall accuracy of the distribution pattern is enhanced.

The basic principle of map projection is direct and simple. Imagine a transparent globe on which are drawn meridians, parallels, and continental boundaries and also imagine a light bulb in the center of this globe. A piece of paper, either held flat or rolled into some shape such as a cylinder or a cone, is placed over the globe as in Figures 2-7, 2-8, and 2-9. When the bulb is lighted, all the lines on the globe are "projected" outward onto the paper. These lines are then sketched on the paper. When the paper is laid out flat, a map projection has been produced. (In actuality, very few map projections have ever been constructed this way. Nearly all have been derived by mathematical computation.)

No matter how they are derived, the one feature common to all projections is that they all show the correct location of latitude and longitude lines on Earth's surface. In other words, each projection consists of an orderly rearrangement of the geographic grid transposed from the globe to the map. The arrangement of the grid, however, varies from projection to projection. Indeed, the difference from one projection to another is just this difference in grid layout.

Because there is no possible way to avoid distortion completely, no map projection is perfect. Each of the many hundreds of different projections has been designed as a compromise to achieve some purpose. Each projection has some advantage over the others, but it also has its own particular limitations.

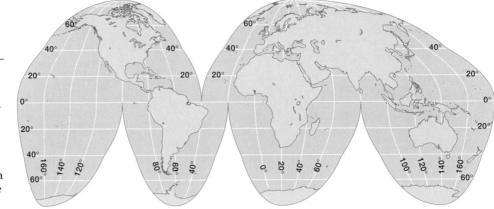

Figure 2-6 An interrupted projection of the world. The purpose of the interruptions is to portray certain areas (usually continents) more accurately, at the expense of portions of the map (usually oceans) that are not important to the map's theme. The map shown here is a Goode's interrupted homolosine projection.

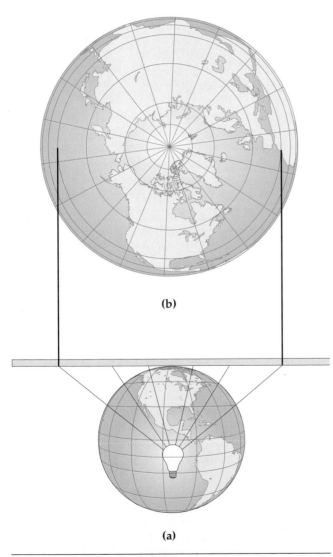

Figure 2-7 **(a)** The projection theory, as illustrated by a globe with a light in its center, projecting images onto an adjacent plane. **(b)** The resultant map goes by various names: azimuthal projection, plane projection, or zenithal projection.

THE MAJOR DILEMMA: EQUIVALENCE VS. CONFORMALITY

The cartographer wants any map to portray distances and directions accurately so that the sizes and shapes shown on the map are correct. However, such perfection is impossible, and so a compromise must be struck. Which to emphasize: size or shape? Which to sacrifice: shape or size? This is the central problem both in constructing and in choosing a map projection. The projection properties involved are called equivalence and conformality.

Equivalence   In an **equivalent projection**, the size ratio of any area on the map to the corresponding area on the ground is the same all over the map. To illustrate, suppose you have a world map before you and you place four dimes at different places (perhaps one on Brazil, one on Australia, one on Siberia, and one on South Africa). Calculate the area covered by each coin. If it is the same in all four cases, there is a good chance that the map is an equivalent projection—in other words, that there are equal areal relationships all over it.

Equivalent projections are very desirable because with them misleading impressions of size are avoided. (The world maps in this book are mostly equivalent projections because they are so useful in portraying distributions of the various geographic features we shall be studying.) They are by no means perfect, however. Equivalence is difficult to achieve on small-scale maps because shapes must be sacrificed to maintain proper areal relationships. Most equivalent world maps, which are small-scale maps, therefore display disfigured shapes. For example, as Figure 2-10b shows, Greenland and Alaska are usually shown as more squatty than they actually are and New Zealand as more stretched out in the north/south direction than it actually is.

Conformality   A **conformal projection** is one in which proper angular relationships are maintained so that the shape of something on the map is the same as

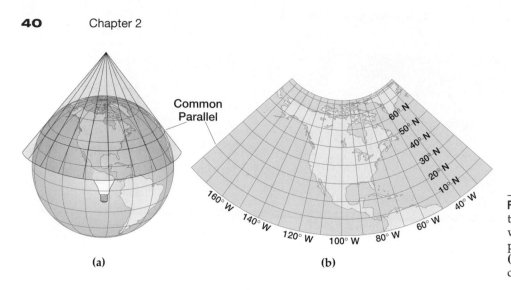
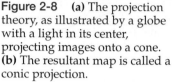

(a)   (b)

Figure 2-8 **(a)** The projection theory, as illustrated by a globe with a light in its center, projecting images onto a cone. **(b)** The resultant map is called a conic projection.

its shape on Earth. It is impossible to depict true shapes for large areas such as a continent, but they can be approximated, and for small areas the true shape can be shown on a conformal map. All conformal projections have meridians and parallels crossing each other at right angles, just as they do on a globe.

The outstanding problem with conformal projections is that the size of an area often must be considerably distorted in order to depict the proper shape. Thus the scale necessarily changes from one region to another. For example, a conformal map of the world normally greatly enlarges sizes in the higher latitudes. Figure 2-10a shows the conformal projection known as a Mercator projection, the type most familiar to schoolchildren.

Except for maps of very small areas (in other words, large-scale maps), where both can be closely approximated, conformality and equivalence cannot be maintained on the same projection, and thus the art of mapmaking, like politics, is an art of compromise. Figure 2-11 shows a Robinson projection, which is a compromise between equivalence and conformality that shows reasonably accurate shapes and has a reasonably constant scale, especially in the middle and lower latitudes.

As a rule of thumb, it can be stated that some projections are purely conformal, some are purely equivalent, none are both conformal and equivalent, and most are neither purely conformal nor purely equivalent but rather a compromise between the two.

More than a thousand types of map projections have been devised for one purpose or another. Most of them can be grouped into just a few families, as Appendix I shows. Projections in the same family generally have similar properties and related distortion characteristics.

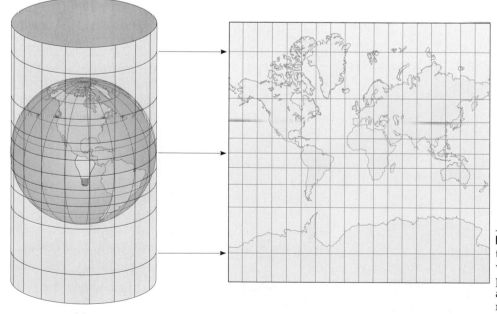

(a)   (b)

Figure 2-9 **(a)** The projection theory, as illustrated by a globe with a light in its center, projecting images onto an adjacent cylinder. **(b)** The resultant map is called a cylindrical projection.

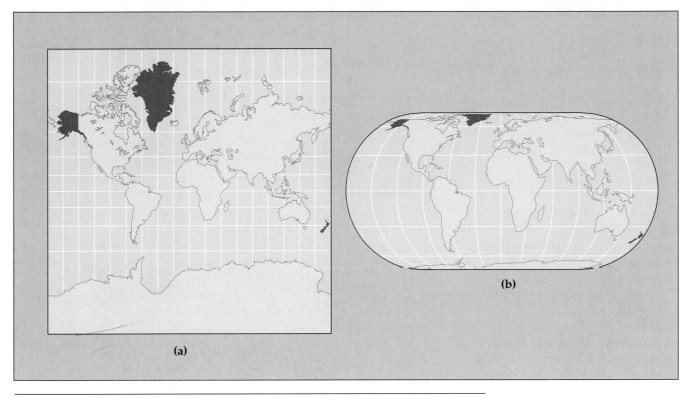

Figure 2-10 It is particularly difficult to portray the whole world accurately on a map. Compare the sizes and shapes of Alaska, Greenland, and New Zealand in these examples. **(a)** A conformal projection depicts accurate shapes, but the sizes are severely exaggerated. **(b)** An equivalent projection is accurate with regard to size, but shapes are badly distorted.

AUTOMATED CARTOGRAPHY

Cartographers have been at work since the days of the early Egyptians, but it is only in the last half century, with the introduction of computers in the 1950s, that their technology has advanced beyond simple manual drawing on a piece of paper. Computers provided incredible improvements in speed and data-handling ability, and automated cartography has become a prominent high-technology field.

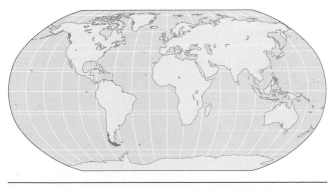

Figure 2-11 Many world maps are neither purely conformal nor purely equivalent, but a compromise between the two. One of the most popular compromises is the Robinson projection shown here.

One of the first widely used software packages for automated cartography was SYMAP, developed in 1965. It allowed the cartographer to input map boundaries along with data values. The computer then interpolated values of the mapped variable and produced a distribution map.

Since then there have been remarkable improvements in desktop and mainframe computers, making possible sophisticated, high-quality maps produced via relatively inexpensive and widely available software. Not only does the computer greatly reduce the time involved in map production, it allows the cartographer to examine alternative map layouts simply by changing details on the monitor. As an example of how computer-generated maps have improved since the days of SYMAP, just look at any of the maps in this book; all of them were made with desktop computers.

ISOLINES

Geographers employ a variety of cartographic devices to display data on maps. One of the most widespread devices for portraying the spatial distribution of some

FOCUS

MERCATOR: THE MOST FAMOUS PROJECTION

Although some map projections were devised centuries ago, there has been a continuing refining of projection techniques right up to the present day. Thus it is remarkable that the most famous of all projections, "invented" more than 400 years ago, is still in common usage today without significant modification. This is the **Mercator projection,** originated in 1569 by a Flemish geographer and cartographer.

Gerhardus Mercator produced some of the best maps and globes of his time. His place in history, however, is based largely on the fact that he developed a special-purpose projection that became inordinately popular for general-purpose use. The Mercator projection is essentially a navigational chart of the world, designed to facilitate oceanic navigation. Mercator's instructions accompanying the map stated clearly that its proper use would guide the mariner by simple compass direction to a destination but that the route plotted with the map would not necessarily be the quickest way to get to that destination.

The prime advantage of a Mercator map is that it shows loxodromes as straight lines. A **loxodrome,** also called a **rhumb line,** is a curve on the surface of a sphere that crosses all meridians at the same angle. A navigator, whether on a ship or on a plane, plots the shortest distance between origin and destination (a great circle) on some projection in which great circles are shown as straight lines, as shown in Figure 2-2-A, and then transfers that route to a Mercator projection by marking spots on the meridians that the great circle crosses. These

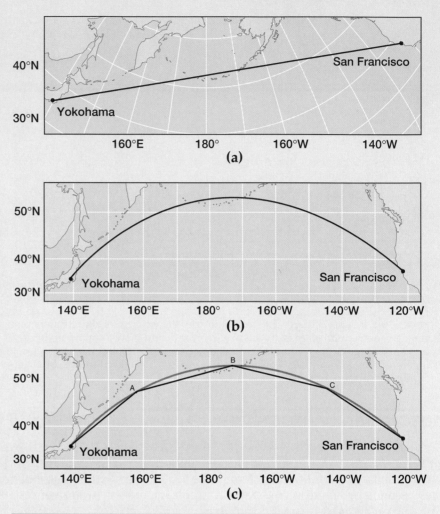

Figure 2-2-A The prime virtue of the Mercator projection lies in its usefullness for straight-line navigation. **(a)** The shortest distance between two locations—here San Francisco and Yokohama—can be plotted on any of several projections in which great circles are shown as straight lines. **(b)** The route plotted in **(a)** can be transferred to a Mercator projection with mathematical precision. **(c)** Then, still on the Mercator projection, straight-line loxodromes are substituted for the curved great circle obtained in **(b).** On the Mercator projection, straight-line loxodromes can then be substituted for the curved great circle. The loxodromes allow the navigator to maintain constant compass headings over small distances while still from point to point, approximating the curve of the great circle. The navigator can plot a straight-line course from Yokohama to A, then a new straight-line course from A to B, and so on across the Pacific to San Francisco.

spots are then connected by straight-line loxodromes, which approximate arcs of the great circle but consist of constant compass headings. This procedure allows the navigator to chart an approximately shortest distance route between origin and destination by making periodic changes in the compass course of the airplane or ship as it generally follows a great circle. Today, of course, these calculations are all done by computer.

Using our image of a transparent globe with a lightbulb at its center, you can envision making a Mercator projection by wrapping the globe with a cylinder of paper in such a way that the paper touches the globe only at the globe's equator, as in Figure 2-9. (We say that paper positioned this way is *tangent* to the globe at the equator, and the equator is called the *circle of tangency*.) The curved parallels and meridians of the globe then form a perfectly rectangular grid on the map, as Figure 2-2-B shows. A Mercator map is accurate at the equator and relatively undistorted in the low latitudes, but, because this is a conformal projection, size distortion increases rapidly in the middle and high latitudes. Because the projection method causes the meridi-

ans to appear as parallel lines rather than lines that converge at the poles, there is extreme east-west distortion in the higher latitudes.

In order to maintain conformality, Mercator compensated for the east-west stretching by spacing the parallels of latitude increasingly farther apart so that north-south stretching occurs at the same rate. This procedure allowed shapes to be approximated with reasonable accuracy, but at great expense to proper size relationships. Area is distorted by 4 times at the 60th parallel of latitude and by 36 times at the 80th parallel. If the North Pole were shown on a Mercator projection, it would be a line as long as the equator rather than a single point.

It is clear then that the Mercator projection is excellent for straight-line navigation but not for most other uses. Despite the obvious flaws associated with areal distortion in the high latitudes, however, Mercator projections have been widely used in American classrooms and atlases. Indeed, several generations of American students have passed through school with their principal view of the world provided by a Mercator map. This has created many misconceptions, not the least of which is confusion about the relative sizes of high-latitude landmasses. For example, on a Mercator projection, the island of Greenland appears to be as large as or larger than Africa, Australia, and South America. In actuality, however, Africa is 14 times larger than Greenland, South America is 9 times larger, and Australia is 3.5 times larger.

The Mercator projection, then, was devised several centuries ago for a specific purpose, and it still serves that purpose well. Its fame, however, is significantly due to its misuse.

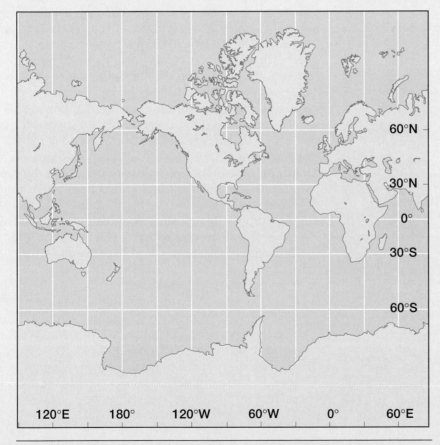

Figure 2-2-B The Mercator projection, in all its simplicity and exaggeration. Although the meridians are spaced equally, note that the parallels have uneven spacing, becoming farther apart as they approach the poles.

phenomenon is the *isoline* (from the Greek *isos*, "equal"), which is also called by a variety of related terms, such as isarithm, isogram, isopleth, and isometric line, all of which can be considered as synonymous for our purposes. The word **isoline** is a generic term that refers to any line that joins points of equal value of something. More than one hundred kinds of isolines have been identified by name, ranging from *isoamplitude* (used to describe radio waves) to *isovapor* (water vapor content of the air).

Some isolines represent tangible surfaces, such as the lines on a contour map that represent elevations on the land (Figure 2-12). Most, however, signify such intangible features as temperature and precipitation, and some express relative values, such as ratios or proportions (Figure 2-13). Only a few types of isolines are important in an introductory physical geography course:

Contour line–a line joining points of equal elevation

Isobar–a line joining points of equal atmospheric pressure

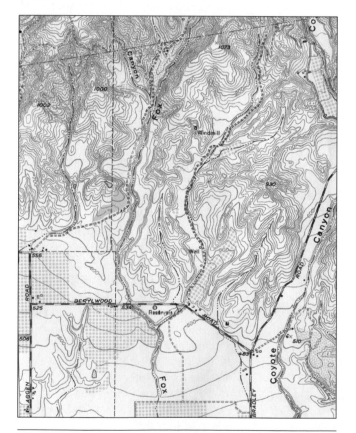

Figure 2-12 This portion of a typical United States Geological Survey quadrangle illustrates the use of contour lines, which are the brown lines joining parts of equal elevation. Where the contours are close together, the slope is steep; where the lines are far apart, the slope is gentle. This is a section of the Santa Paula, California quadrangle.

Isogonic line–a line joining points of equal magnetic declination

Isohyet–a line joining points of equal quantities of precipitation (*hyeto* is from the Greek, meaning "rain")

Isotherm–a line joining points of equal temperature

To construct an isoline, it is always necessary to estimate values that are not available. As a simple example, Figure 2-14 illustrates the basic steps in constructing an isoline map. Each dot in Figure 2-14(a) represents a data-collection station, and the number alongside each dot is the value recorded at that station. (These data could be values of any variable–temperatures, pressures, feet above sea level, or anything else. That specific detail does not concern us here.) Suppose we want to draw isolines that connect all the points having a value of 140. How we do this is shown in Figure 2-14(b), where **X**s have been placed at points where the cartographer estimates the values are 140. These points are located by interpolating between two recorded values. For example, at the center top are points labeled 138 and 154; the **X** between them is at a point estimated to be 140. A bit to the left and below that same 138 is 142, and the **X** between the 138 and the 142 marks the estimated 140 point between them. This process is continued for all the pairs that bracket 140 (which means we don't bother with, for instance, the pair 137/128 because those two numbers do not bracket 140). Once all the 140-bracketing pairs have been identified and their 140 points marked, the **X**s are joined as shown in Figure 2-14(c). The lines connecting the **X**s are the 140 isolines. In Figure 2-14(d), the interpolation process is repeated for other isolines, and in Figure 2-14(e) shading is added to clarify the pattern.

The basic characteristics of isolines are as follows:

1. They are always closed lines; that is, they have no ends. On a map, of course, an isoline is likely to extend beyond the edge, and when that happens, the closure is not seen. (This absence of closure is seen with all the isolines in Figure 2-14.)
2. Because they represent gradations in quantity, isolines can never touch or cross one another except under rare and unusual circumstances.
3. The numerical difference between one isoline and the next is called the **interval**. Although intervals can be varied according to the wishes of the mapmaker, it is normally more useful to maintain a constant interval all over a given map.
4. Isolines close together indicate a steep gradient (in other words, a rapid change); isolines far apart indicate a gentle gradient (Figure 2-15).

Edmund Halley (1656–1742), an English astronomer and cartographer, was not the first person to use iso-

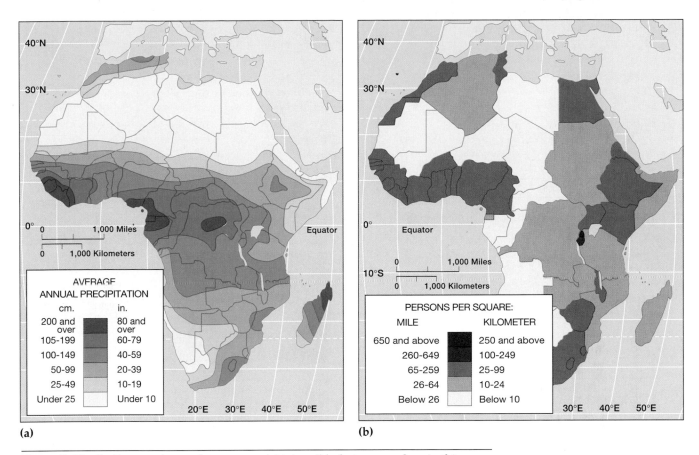

Figure 2-13 Isolines can be used to express: **(a)** intangible features, such as in this map that shows average annual precipitation for the continent of Africa; **(b)** ratios or proportions, such as in this map that displays the population density of Africa in terms of persons per square mile or square kilometer.

lines, but he published a map in 1700 that apparently was the first time that isolines appeared in print. This map showed isogonic lines in the Atlantic Ocean. Isoline maps are now commonplace and are very useful to geographers even though an isoline is an artificial construct; that is, it does not occur in nature. For instance, an isoline map can reveal spatial relationships that might otherwise go undetected. Patterns that are too large, too abstract, or too detailed for ordinary comprehension are often significantly clarified by the use of isolines.

THE GLOBAL POSITIONING SYSTEM

The **global positioning system (GPS)** is a satellite-based system for determining accurate positions on or near Earth's surface. It was developed in the 1970s and 1980s by the U.S. Defense Department to aid in guiding missiles, navigating aircraft, and controlling ground troops.

The system is based on a network of 24 high-altitude satellites configured so that a minimum of four are in view of any position on Earth. Each satellite continuously transmits both identification and positioning information that can be picked up by receivers on Earth (Figure 2-16 on page 48). The distance between a given receiver and each member in a group of four (or more) satellites is calculated by comparing clocks stored in both units. Because four satellites are used, it is possible to calculate the three-dimensional coordinates of the receiver's position. The system already has an accuracy greater than that of the best base maps. Commercial and private users can calculate position within 330 feet (100 meters); military users receive a more accurate signal, allowing position calculation within about 30 feet (10 meters).

The first receivers were the size of a file cabinet, but continued technological improvement has reduced them to the size of a paperback book and they are still shrinking (Figure 2-17). The cost is also diminishing rapidly; at the time of this writing a good receiver sells for less than $100.

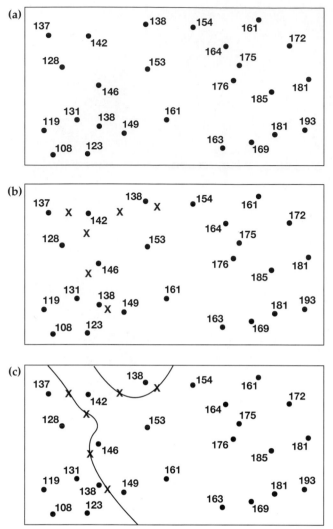

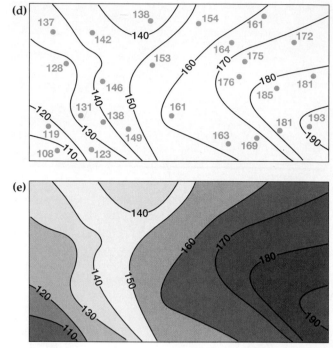

Figure 2-14 Drawing isolines. **(a)** Each dot represents a measuring station, and the number next to each dot is the quantity measured or calculated at that station. **(b)** The approximate location of one or more critical isolines (in this case the value of 140) is interpolated and marked by **X**s. **(c)** The isolines are drawn. **(d)** The other isolines at 10-unit intervals are interpolated and drawn. **(e)** Shading is added for clarity.

In 1983 President Ronald Reagan made access to the system free to the public, and astounding commercial growth has resulted. It is anticipated that eventually practically everything that moves in our society—airplane, truck, train, car, bus, tractor, bulldoze—will be equipped with a GPS receiver. Meanwhile, GPS had been employed in earthquake prediction, ocean floor mapping, volcano monitoring, and a variety of mapping projects. For example, recognizing that GPS is a relatively inexpensive way of collecting data, the Federal Emergency Management Agency (FEMA) has used the system for damage assessment following such natural disasters as floods and hurricanes.

GEOGRAPHIC INFORMATION SYSTEMS

Computer systems capable of integrating geographic data are called **Geographic Information Systems (GIS)**. To put it another way, a GIS is essentially a data-

base management system that facilitates the collection and analysis of geographic data (Figure 2-18).

These systems have grown out of a diverse group of technologies familiar to geographers: surveying, photogrammetry, computer cartography, spatial statistics, and remote sensing. Over the last quarter century or so, these technologies have followed parallel paths in developing hardware and software for collecting, storing, retrieving, manipulating, analyzing, and displaying geographic data from the real world. Geographic information systems have emerged as a by-product of the linking of these parallel developments.

A GIS is used mainly in overlay analysis, where two or more layers of data are superimposed, or integrated. A GIS treats each spatially distributed variable as a particular layer in a sequence of overlays. Each layer can consist of remotely sensed data, information gathered from published maps, or observations made in field study. As shown in Figure 2-19, input layers bring together such diverse elements as topography, vegetation, land use, land ownership, and land survey.

(a)   1

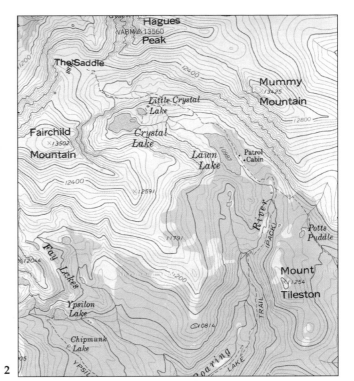

2

(b)   1

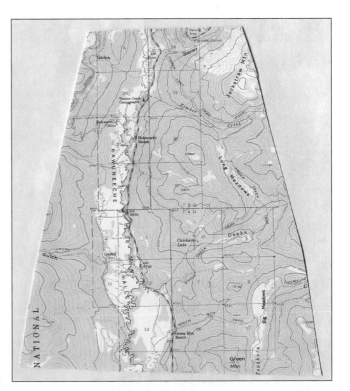

2

Figure 2-15   **(a)** Portrayal of steep slopes: (1) a photo of Hagues Peak and Mummy Mountain in Colorado's Rocky Mountain National Park. (2) A standard topographic map of the same area. **(b)** Portrayal of gentle slopes: (1) a photo of the upper Colorado River valley near Grand Lake in north central Colorado. (2) A standard topographic map of the same area. (TLM photos)

Figure 2-16   Global positioning system (GPS) satellites circling 11,000 miles (17,700 kilometers) above Earth broadcast signals that are picked up by the receiver in an ambulance and used to pinpoint the location of the ambulance at any moment. A transmitter in the ambulance then sends this location information to a dispatch center, and all other ambulances in the rescue system likewise receive and transmit *their* location information. Knowing the location of all ambulances at any given moment, the dispatcher is able to route the closest available vehicle to each emergency and then direct that vehicle to the nearest appropriate health facility.

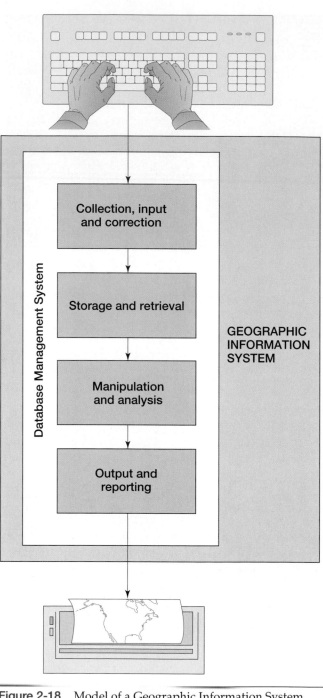

Figure 2-18   Model of a Geographic Information System (GIS).

Figure 2-17   A handheld satellite navigation receiver. It receives signals sent by the network of GPS satellites. The receiver reads data from up to four satellites and instantly calculates its position anywhere in the world to within 100 yards (100 meters). (David Parker/Science Photo Library/Photo Researchers, Inc.)

FOCUS

USGS TOPOGRAPHIC MAPS

The U.S. Geological Survey (USGS) is one of the world's largest mapping agencies and is primarily responsible for the country's National Mapping Program. The USGS produces a broad assortment of maps, but its topographic "quadrangles" are by far the ones most widely used by geographers. The quadrangles come in a variety of sizes and scales, but all are rectangles bordered by parallels and meridians rather than political boundaries. For many years the quadrangles were produced by surveys on the ground; today, however, they are created from aerial photographs, satellite imagery, and computer rectification.

The quadrangles use contour lines to portray the relief of the land–its shape, slope, and elevation (Figure 2-3-A). The quadrangles also portray a variety of other features. Six standard colors and shades distinguish various features:

◆ *Brown*–contour lines
◆ *Blue*–hydrographic (water) features
◆ *Black*–features constructed or designated by humans, such as buildings, roads, boundary lines, and names
◆ *Green*–woodlands, forests, orchards, vineyards
◆ *Red*–lines of the public land survey system and important roads
◆ *Purple*–features added from aerial photos during map revision

The standard symbols used on all USGS quadrangles are shown in Appendix II.

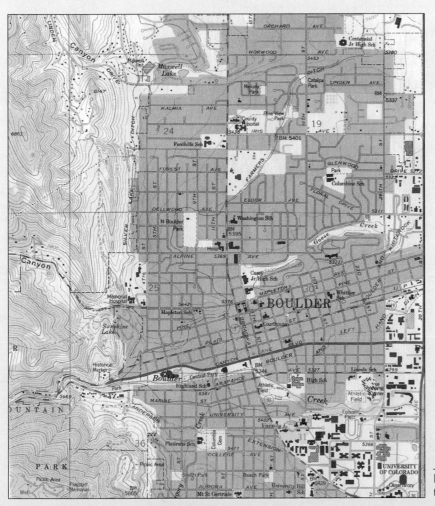

Figure 2-3-A   A typical USGS topographic map.

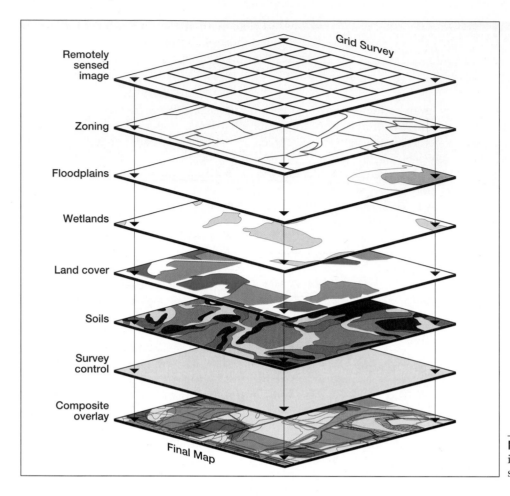

Remotely sensed image

Grid Survey

Zoning

Floodplains

Wetlands

Land cover

Soils

Survey control

Composite overlay

Final Map

Figure 2-19   Much GIS work involves layers of spatial data superimposed upon one another.

Details of these various components are converted to digital data and synthesized onto a reference map or data set.

Geographic information systems are used in a diverse array of applications concerned with geographic location. Because they provide impressive output maps and a powerful methodology for analytical studies, GIS systems can bring a new and more complete perspective to resource management, environmental monitoring, and environmental site assessment.

REMOTE SENSING

Throughout most of history, maps have been the only tools available to depict anything more than a tiny portion of Earth's surface with any degree of accuracy. However, the sophisticated technology developed in recent years permits precision recording instruments to operate from high-altitude vantage points, providing a remarkable new set of tools for the study of Earth. **Remote sensing**, broadly considered, is any measurement or acquisition of information by a recording device that is not in physical contact with the object under study–in this case, Earth's surface.

AERIAL PHOTOGRAPHS

Aerial photography was almost the only form of remote sensing used for geographic purposes until the last few decades. An **aerial photograph** is one taken from an elevated platform, such as a balloon, airplane, rocket, or satellite. The earliest aerial photographs were taken from balloons in France in 1858 and in the United States in 1860. A major problem with photographs taken from balloons–the lack of control of the platform–was overcome in the early twentieth century by the development of airplanes. By the time of World War I (1914–1918), systematic aerial photographic coverage was possible.

Depending on the camera angle, aerial photographs are classified as being either oblique or vertical (Figure 2-20). The advantage of oblique photographs, where the camera angle is less than 90°, is that features are seen from a more or less familiar point of view. The disadvantage is that, because of perspective, measurement is more difficult than on vertical photographs, which are taken with the optical axis of the camera approximately perpendicular to the surface of Earth.

Precise measurement is possible on vertical aerial photographs, and **photogrammetry** is the science of

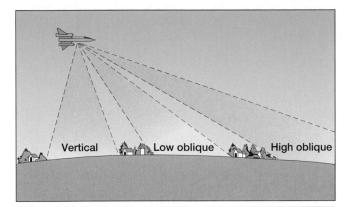

Figure 2-20 On the basis of the angle between the camera and Earth's surface, aerial photographs are classified as vertical, low oblique, or high oblique.

obtaining reliable measurements from photographs and, by extension, the science of mapping from aerial photographs. The workhorse of photogrammetry has been black-and-white vertical aerial photographs taken automatically at regular spatial intervals, where each position from which a photograph is taken is called either an *air* station or a *camera station*. The stations are usually close enough together to allow a considerable overlap in neighboring photographs. For instance, Figure 2-21(a) is a diagram of two vertical aerial photographs being taken sequentially at camera stations 1 and 2. Because of the distance between these two points, the resulting photographs overlap by 60 percent in the direction of flight. Figure 2-21(b) shows a sample of photographs taken in this way. When properly aligned, this overlap produces a three-dimensional appearance.

With a stereoscope (a binocular optical instrument) and two overlapping photographs, an observer can view an object simultaneously from two perspectives to obtain the mental impression of a three-dimensional model. Vertical distance (height) can be measured from the model, and contours can be plotted. When this information is then combined with two-dimensional measurements obtained with an interpreter's ruler of known scale, the length, breadth, and height of features can readily be ascertained.

Color and Color Infrared Sensing

Color photogrammetry was developed slowly in the 1940s and 1950s, with much of the improvements coming as a result of the importance of color aerial photographs during World War II (1939–1945). In contrast to black-and-white images, where only a few tones can be recognized, a large number of hues can be discriminated on color photographs.

The word *color* without any modifier refers to the visible-light region of the electromagnetic spectrum (Figure 2-22) and means any of the colors seen in what we call a rainbow. Another development of World War II, however, was **color infrared (color IR)** film, which is film sensitive to radiation in the infrared region of the spectrum. First known as camouflage-detection film because of its ability to discriminate living vegetation from the withering vegetation used to hide objects during the war, color IR film has become one of the most versatile tools available to the interpreter.

Because it can sense radiation beyond the visible-light region of the spectrum, color IR film is more versatile than black-and-white and (visible) color films. As shown in Figure 2-22, the infrared region of the electromagnetic spectrum is divided into near, middle, and far sections relative to the visible-light region. With color IR film photographing in the near infrared, blue is filtered out and photographic infrared added. The images produced in this way, even though they are false-color images (for example, living vegetation appears red), are still extremely valuable. Thus on photographs taken with conventional black-and-white or color film, it might be difficult to distinguish between synthetic turf and grass, but on those taken with color IR film, the synthetic surface comes out blue or green and the grass pink or red (Figure 2-23).

One of the major uses of color IR imagery is to evaluate the health of crops and trees (Figure 2-24).

Much of the usable portion of the near infrared cannot be detected by photographic emulsions. One result of this limitation has been the development of various optical-mechanical scanner systems that can sense far into the infrared and at the same time simulate color IR photographic imagery. The most widely used such system is **Landsat**, a series of satellites that orbit Earth at an altitude of 570 miles (915 kilometers) and can image all parts of the planet except the polar regions every nine days.

Thermal Infrared Sensing

None of the middle or far infrared part of the electromagnetic spectrum, called the **thermal infrared (thermal IR),** can be sensed with film, and as a result special supercooled scanners are needed. Thermal scanning senses the radiant temperature of objects and may be carried out either day or night. The photographlike images produced in this process are particularly useful for showing diurnal temperature differences between land and water and between bedrock and alluvium, for studying thermal water pollution, and for detecting forest fires.

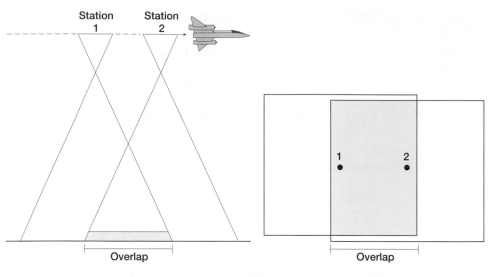

Overlap

Overlap

(a)

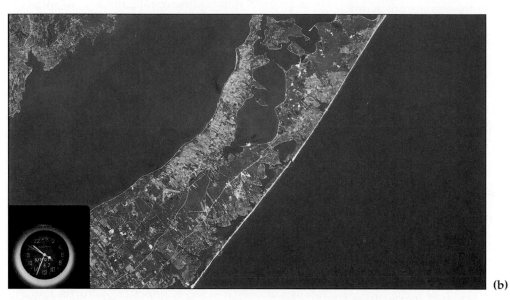

(b)

Figure 2-21 Overlapping aerial vertical photographs. **(a)** One photograph is taken when the plane is at the position labeled station 1, and a second is taken when the plane is at the position labeled station 2. Because of the distance between these two stations, the photographs overlap by about 60 percent. **(b)** Two overlapping vertical images of the same two photographs that might have been taken by the plane in **(a)**. (Photri)

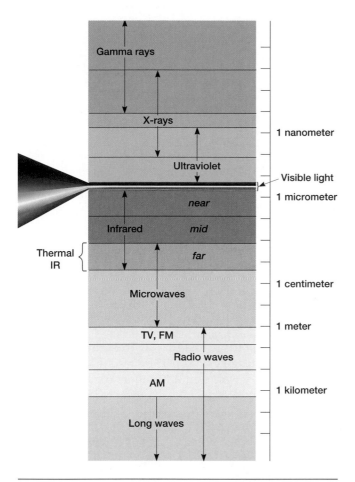

Figure 2-22 The electromagnetic spectrum. All we humans can sense is radiation from the visible-light region. Conventional photography can use only a small portion of the total spectrum. Various specialized remote-sensing scanners, such as color IR film and thermal scanners, are capable of "seeing" radiation from other parts of the spectrum.

By far the greatest use of thermal IR scanning systems to date has been on meteorological satellites. Although the spatial resolution (the size of the smallest feature that can be identified) is only on the order of a few kilometers, it is more than sufficient to provide details that allow weather forecasting that is far more accurate and complete than has ever before been possible.

MICROWAVE SENSING

Systems that sense wavelengths even longer than infrared ones are used in earth sciences, including microwave radiometry, which senses radiation in the 100-micrometer to 1-meter range. Although such systems have low spatial resolution, they are particularly useful for showing subsurface characteristics such as moisture.

RADAR AND SONAR SENSING

All the systems mentioned so far work by sensing the natural radiation emitted by or reflected from an object and are therefore characterized as passive systems. Another type of system, called an active system, has its own source of electromagnetic radiation. The most important active sensing system used in the earth sciences is radar, the acronym for **ra**dio **d**etection **a**nd **r**anging. Radar senses wavelengths longer than 1 millimeter, using the principle that the time it takes for an emitted signal to reach a target and then return to the sender can be converted to distance information.

Initially, radar images were viewed only on a screen, but they are now available in photographlike form (Figure 2-25). In common with some other sensors, radar is capable of operating by day or night, but it is unique in its ability to penetrate atmospheric moisture. Thus some wet tropical areas that could never be sensed by other systems now have been imaged by radar. Radar imagery is particularly useful for terrain analysis.

Another active remote sensing system, sonar (**so**und **na**vigation **r**anging), permits underwater imaging so that at last scientists can determine the form of that part of Earth's crust hidden by the world ocean.

Figure 2-23 A false-color IR aerial photograph of a portion of the UCLA campus in Los Angeles. The football practice field near the center of the photo is vegetated mostly with grass, which shows up red in this false-color imagery. Near the upper right corner of the football field is a small rectangle of synthetic turf, which takes on a dark green color in the image. (Courtesy of Norman Thrower/NASA)

Figure 2-24 A color infrared photo from *Apollo 9*, orbiting 110 miles (176 kilometers) above Earth's surface. The tiny red squares are irrigated fields. The large farming area in the center of the image is California's Imperial Valley. Near the bottom of the photo, the color of the Imperial Valley changes abruptly as the international border is reached. On the Mexican side of the boundary, irrigation is much less intensive and the crop plants show up less prominently. (NASA)

MULTISPECTRAL REMOTE SENSING

Although it is still common for remote sensing systems to gather data from only one part of the electromagnetic spectrum at a time, sophisticated sensors are increasingly multispectral, or multiband (the latter name because the various regions of the electromagnetic spectrum are sometimes called "bands"). These instruments image more than one region of the electromagnetic spectrum simultaneously from precisely the same location. Thus while traditional black-and-white photographic film is sensitive to only a narrow band of visible radiation, a Landsat satellite equipped with a multiband instrument images the surface of Earth in several spectrum regions at once, each designated for a unique application.

Landsat Sensor Systems The early NASA space missions (Mercury, Gemini, and Apollo) used multiband photography obtained through multicamera arrays. These imaging experiments were so successful that NASA then developed what was initially called the earth resources satellite series (ERTS) and later renamed Landsat. The 1970s and 1980s saw the launch of five Landsat satellites carrying a variety of sensor systems. Except for a low spatial resolution—260 feet (79 meters)—the multiband Landsat system with its continuous observation capability approaches the ideal in imagery production.

Figure 2-25 A false-color perspective view of the area around Mammoth Mountain, California. The image is derived from radar data combined with a digital elevation model. The large dark blue area is Lake Crowley. (NASA/Science Photo Library/Photo Researchers, Inc.)

The basic Landsat imaging instrument is a **multispectral scanning system (MSS)**, a four-band system that gathers a set of digital numbers for each picture element (*pixel*) collected. One Landsat MSS image cov-

Figure 2-26 A mosaic of Landsat images showing the Iberian peninsula and surrounding area. (Geospace/Science Photo Library/Photo Researchers, Inc.)

ers an area of 110 by 110 miles (185 by 185 kilometers) and includes more than 30 million pieces of data. Figure 2-26 is a typical Landsat MSS image. The first three Landsat satellites also carried a second sensor called the return beam vidicon (RBV) camera. The RBV system comprises three cameras aimed at the same 110- by 110-mile section of Earth and produces a more accurately measured image than do MSS scanners.

Landsat 4 and *Landsat* 5, launched in the early 1980s, carried an improved multispectral scanner known as the **thematic mapper**. Unlike the four general bands of the MSS, the seven bands of the thematic mapper are more narrowly defined, a refinement that provides improved resolution and greater imaging flexibility. A description of the primary applications for the various bands is provided in Table 2-1.

Unlike aerial photographs, which are produced on film, a Landsat image is digital and is displayed as a matrix of numbers, each number representing a single value for the specific pixel and band. These data are stored in the satellite, eventually transmitted to an Earth receiving station, numerically manipulated by a computer, and produced as a set of gray values and/or colors on a screen and/or hard-copy device (Figure 2-27).

SPOT IMAGERY

The newest sensor system, SPOT, an acronym for Système pour l'Observation de la Terre, was launched in 1986 by the French Centre National d'Études Spatiales. It has a resolution of either 33 by 33 feet (10 by 10 meters) or 66 by 66 feet (20 by 20 meters). The SPOT satel-

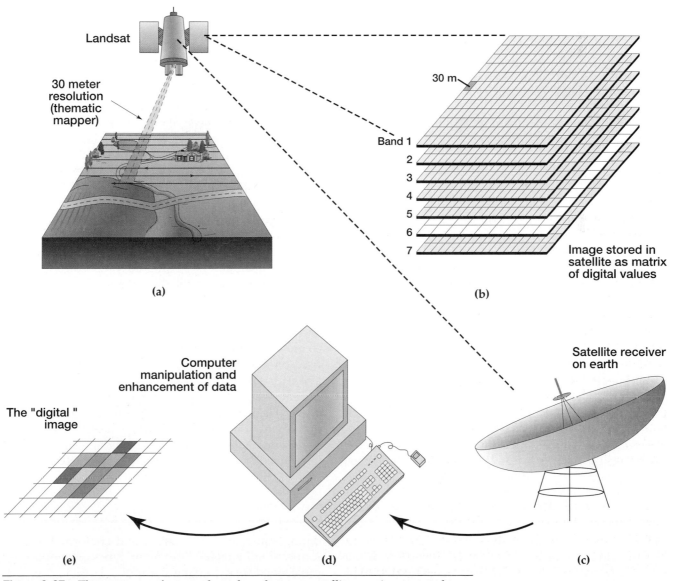

Figure 2-27 The sequence of events that takes place as a satellite scan is converted to a digital image. (Courtesy of Robert McMaster)

TABLE 2-1 Bands of the Landsat Thematic Mapper			
Band number	Bandwidth (micrometers)	Name of spectral region	Applications
1	0.45–0.52	Blue	Spots water penetration and vegetation
2	0.52–0.60	Green	Spots vegetation
3	0.63–0.69	Red	Spots vegetation (is sensitive to red pigments in plants)
4	0.76–0.90	Near IR	Measures biomass, analyzes soil
5	1.55–1.75	Middle IR	Detects soil moisture; excellent for hydrologic work
6	10.4–12.5	Thermal	Detects geothermal resources and vegetation stress
7	2.08–2.35	Middle IR	Geologic analysis

lite is similar to Landsat in orbit characteristics but carries a different type of sensor known as a high-resolution-visible (HRV) sensing system. In addition to significantly improved resolution, the SPOT satellite is capable of stereoscopic imaging, although coverage is somewhat variable with latitude.

THE ROLE OF THE GEOGRAPHER

A vast array of imagery of various types and of different areas is available to the interpreter, ranging from black-and-white photographs to the output of exotic sensing systems. Many foreign governments place restrictions on coverage of their territory, but the United States has made imagery easily accessible to its own citizens and to others. For example, all images from the U.S. civilian space program (including Landsat) are unclassified and are obtainable at a modest cost.

In using these images and equipment, the geographer should never lose sight of the major objective: to understand Earth better. The new imagery technology available to us is an adjunct to field study, geographic description, and maps but is not a substitute for any of these.

Certain types of imagery are useful for particular purposes; no single sensing system has universal ap-

plicability for all problems. Accordingly, each interpreter must select and obtain the best type of imagery for his/her special needs. For providing an overview of the lithosphere, for example, high-altitude space imagery has been of particular value and has led to important discoveries. This type of imagery might have limited value in detailed terrain studies, however, where large-scale color or black-and-white photographs would be more appropriate.

For studying the hydrosphere, to cite another example, images of different scales have proved useful. Satellite images of an entire hemisphere can tell us much about the water content in clouds, air masses, glaciers, and snowfields at a given time, but detailed conventional color photographs might be better for discriminating a complicated shoreline because of scale and the penetrating ability of the film. On the other hand, the biosphere and especially its vegetation often are best appreciated on color IR imagery of several scales–overall vegetation patterns on satellite images and detailed imagery down to 1:5000 for crop and forest inventory studies.

Features of human creation are generally not evident on very-high-altitude imagery, but they become increasingly clear as one approaches Earth. Thus survey patterns, transportation lines, rural settlements, and cities are best interpreted on imagery of intermediate or large scale.

CHAPTER SUMMARY

Many kinds of map projections have been devised in an effort to minimize the problem of distortion when data are transferred from a spherical surface (Earth) to a flat surface (map). None has been completely satisfactory because of the insoluble problem of accommo-

dating equivalence (true size) and conformality (true shape) simultaneously, but many types of compromise projections are exceedingly useful.

In recent years there has been a phenomenal expansion of high technology systems—automated cartography, the global positioning system (GPS), geographic information systems (GIS), and remote sensing—to make possible an increasingly graphic and accurate portrayal of Earth.

Remote sensing involves precision instruments operating from high-altitude vantage points and recording information from Earth's surface. Aerial photography has been used since the middle 1800s, but most types of remote sensing—color infrared, thermal infrared, microwave, radar, sonar—are of more recent vintage. In multispectral remote sensing, several bands of the electromagnetic spectrum are imaged simultaneously by a single instrument.

KEY TERMS

aerial photograph
central meridian
color infrared (color IR)
conformal projection
contour line
equivalent projection
fractional scale
geographic information system (GIS)
global positioning system (GPS)
graphic scale
interval

isobar
isogonic line
isohyet
isoline
isotherm
Landsat
large-scale map
loxodrome
map
map projection
Mercator projection
multispectral scanning system (MSS)

photogrammetry
remote sensing
representative fraction
rhumb line
scale
small-scale map
thermal infrared (thermal IR)
thematic mapper
verbal scale
word scale

REVIEW QUESTIONS

1. Explain the implications of the statement "No map is totally accurate."
2. Explain the difference between large-scale and small-scale maps.
3. Which is more useful, a graphic scale or a fractional scale? Why?
4. A globe can portray Earth's surface more accurately than a map, but globes are rarely used. Why?
5. Explain the concept of *equivalence*.
6. Explain the concept of *conformality*.
7. Why are there so many types of map projections?
8. A cylindrical map of the world always has considerable distortion, yet cylindrical projections are widely used. Why?
9. Distinguish between GPS and GIS.
10. Compare the advantages and disadvantages of vertical and oblique aerial photographs.
11. What are the advantages of color IR imagery?
12. What are the advantages of thermal IR imagery?

SOME USEFUL REFERENCES

DENT, BORDEN D., *Principles of Thematic Map Design*. Reading, MA: Addison-Wesley, 1984.

DURY, S.A., *A Guide to Remote Sensing*. Melbourne: Oxford University Press, 1990.

HOLZ, ROBERT K., *The Surveillant Science: Remote Sensing of the Environment*. New York: John Wiley & Sons, 1985.

JENSEN, JOHN R., *Introductory Digital Image Processing*. Englewood Cliffs, NJ: Prentice Hall, 1986.

MAGUIRE, D.J., M.F. GOODCHILD, AND D.W. RIND, eds., *Geographical Information Systems: Volume 1, Principles, and Volume 2, Applications*. Harlow, UK: Longman, 1991.

MONMONIER, MARK, *How to Lie with Maps*. Chicago: University of Chicago Press, 1991.

MONMONIER, M., AND G. SCHNELL, *Map Appreciation*. Englewood Cliffs, NJ: Prentice Hall, 1988.

MUEHRCKE, P.C., AND J.O. MUEHRCKE, *Map Use: Reading, Analysis, and Interpretation*, 3rd ed. Madison, WI: JP Publications, 1992.

RAPER, J., "GEOGRAPHICAL INFORMATION SYSTEMS," *Progress in Physical Geography*, vol. 17, 1993, pp. 493-502.

SABINS, F.J., JR., *Remote Sensing: Principles and Interpretation*, 2nd ed. San Francisco: W.H. Freeman, 1987.

STAR, JEFFREY, AND JOHN ESTES, *Geographic Information Systems: An Introduction*. Englewood Cliffs, NJ: Prentice Hall, 1990.

WOOD, DENIS, *The Power of Maps*. New York: Guilford Press, 1992.

3

INTRODUCTION TO THE ATMOSPHERE

EARTH IS DIFFERENT FROM ALL OTHER KNOWN PLANETS in a variety of ways. One of the most notable differences is the presence around our planet of an atmosphere distinctive from other planetary atmospheres. It is our atmosphere that makes life possible on Earth. The atmosphere supplies most of the oxygen that animals must have to survive, as well as the carbon dioxide needed by plants. It helps maintain a water supply, which is essential to all living things. It insulates Earth's surface against temperature extremes and thus provides a livable environment over most of the planet. It also shields Earth from much of the sun's ultraviolet radiation, which otherwise would be fatal to most life forms.

Air, generally used as a synonym for *atmosphere*, is not a specific gas but rather a mixture of gases, mainly nitrogen and oxygen. In addition, most air also contains minor but varying quantities of solid and liquid particles that can be thought of as impurities. The individual particles are mostly submicroscopic and therefore held in suspension in the air. Most air also contains some gaseous impurities.

Pure air is invisible because the gases in it are colorless, odorless, and tasteless. Gaseous impurities, on the other hand, can often by smelled, and the air may even become visible if enough submicroscopic solid and liquid impurities coalesce (stick together) to form particles large enough to either reflect or scatter sunlight. Clouds, by far the most conspicuous visible features of the atmosphere, represent such a coalescing of impurities, primarily water vapor.

The atmosphere completely surrounds Earth and can be thought of as a vast ocean of air with Earth at its bottom (Figure 3-1). It is held to Earth by gravitational attraction and therefore accompanies our planet in all its celestial motions. The attachment of Earth and atmosphere is a loose one, however, and the latter can therefore move on its own, doing things that the solid Earth cannot do.

Although the atmosphere extends outward at least 6000 miles (10,000 kilometers), most of its mass is concentrated at very low altitudes, as the gradations of color in Figure 3-1(a) indicate. More than half of the mass of the atmosphere lies below the elevation of North America's highest peak, Mount McKinley in Alaska, elevation 3.8 miles (6.2 kilometers), and more than 98 percent of it lies within 16 miles (26 kilometers) of sea level (Figure 3-2). Therefore, relative to Earth, the diameter of which is about 8000 miles (13,000 kilometers), the "ocean of air" we live in is a very shallow one.

In addition to reaching upward above Earth's surface, the atmosphere also extends slightly downward. Because air expands to fill empty spaces, it penetrates into caves and into crevices in rocks and soil. Moreover, it is dissolved in the waters of Earth and in the bloodstreams of organisms.

The atmosphere interacts with other components of the earthly environment, and it is instrumental in providing a hospitable milieu for life. Whereas we often speak of human beings as creatures of Earth, it is perhaps more accurate to consider ourselves creatures of the atmosphere. As surely as a crab crawling on the sea bottom is a resident of the ocean, a person living at the bottom of the ocean of air is a resident of the atmosphere.

COMPOSITION OF THE ATMOSPHERE

The chemical composition of pure, dry air at lower elevations (by which we mean lower than about 50 miles, or 80 kilometers) is simple and uniform, and the concentrations of the major components are basically unvarying over time. Certain minor gases and nongaseous particles vary markedly from place to place and from time to time, however, as does the amount of moisture in the air.

THE GASES

As Figure 3-3 and Table 3-1 show, nitrogen and oxygen are the two most abundant gases in the atmosphere. Nitrogen makes up more than 78 percent of the total, and oxygen makes up nearly 21 percent. Nitrogen is added to the air by the decay and burning of organic matter, by volcanic eruptions, and by the chemical breakdown of certain rocks, and it is removed by certain biological processes and by being washed away in rain or snow. Overall, the addition and removal of nitrogen gas are balanced, and consequently the quantity present in the air remains constant over time. Oxygen is produced by vegetation and is removed by a variety of organic and inorganic processes; its total quantity also apparently remains stable. The remaining 1 percent of the atmosphere's volume consists mostly of the inert gas *argon*. These three principal atmospheric gases—nitrogen, oxygen, argon—have a minimal effect on weather and climate and therefore need no further consideration here. The trace gases neon, helium, methane, krypton, and hydrogen also have little effect on weather and climate.

Tropical cloudscapes are often spectacular. This is a sunset scene on Rarotonga in the Cook Islands of the South Pacific. *(TLM photo)*

(a)

(b)

Figure 3-1 **(a)** Artist's rendering of the atmosphere, a "sea of air" blanketing Earth. **(b)** Earth's atmosphere is dense only as a thin layer surrounding our planet. (Joe Towers/ The Stock Market)

Several other gases occur in sparse but highly variable quantities in the atmosphere, and their influence on weather and climate is significant. The amount of *water vapor* present determines the humidity of the atmosphere. It is most abundant in air overlying warm, moist surface areas, such as tropical oceans, where water vapor may amount to as much as 4 percent of total volume. Over deserts and in polar regions, the amount of water vapor is but a tiny fraction of 1 percent. (For the atmosphere as a whole, the total amount of water vapor remains virtually constant. Thus the listing of it as a variable gas in Table 3-1 means variable in location, not variable in time.) Water vapor has a significant effect on weather and climate in that it is the source of all clouds and precipitation and is intimately involved in the storage, movement, and release of heat energy.

Atmospheric *carbon dioxide* also has a significant influence on climate, primarily because of its potent ability to absorb infrared radiation, which is the type of radiation that keeps the lower atmosphere warm. Carbon dioxide is distributed fairly uniformly in the lower layers of the atmosphere, but its concentration has been increasing for the last century or so, and the rate of accumulation has been accelerating, presumably because of the increased burning of fossil fuels. The long-range effect of increasing amounts of carbon dioxide in the atmosphere is debatable, but many scientists believe that the higher levels will cause the lower atmosphere to warm up enough to produce major, and still unpredictable, global climatic changes. The proportion of carbon dioxide in the atmosphere has been increasing at a rate of about 0.0002 percent (2 parts per million) per year, and at present is about 360 parts per million.

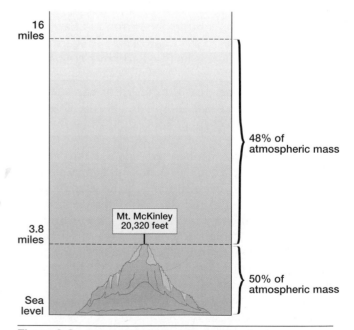

Figure 3-2 Most of the atmospheric mass is close to Earth's surface. More than half of the mass is below the elevation of Mount McKinley, North America's highest peak.

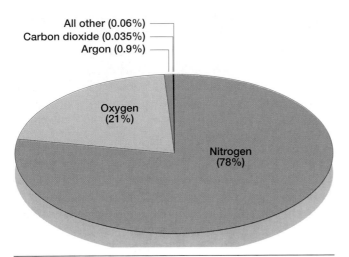

All other (0.06%)
Carbon dioxide (0.035%)
Argon (0.9%)

Oxygen
(21%)

Nitrogen
(78%)

Figure 3-3 Proportional volume of the gaseous components of the atmosphere. Nitrogen and oxygen are the dominant components.

Another minor but vital gas in the atmosphere is **ozone,** which is a molecule made up of three oxygen atoms joined together (O_3). For the most part, ozone is concentrated in a layer of the atmosphere called the **ozone layer,** which lies between 9 and 30 miles (15 and 48 kilometers) above Earth's surface (Figure 3-4). Ozone is an excellent absorber of ultraviolet solar radiation; it filters out enough of these rays to protect life forms from potentially deadly effects.

The other variable gases listed in Table 3-1—carbon monoxide, sulfur dioxide, nitrogen oxides, and various hydrocarbons—are increasingly being introduced into the atmosphere by emission from factories and automobiles. All of them are hazardous to life and may possibly have some effect on climate.

THE PARTICLES

The larger nongaseous particles in the atmosphere are mainly water and ice, which form clouds, rain, snow, sleet, and hail. There are also dust particles large

TABLE 3-1 **Principal Gases of Earth's Atmosphere**		
Gas	*Percent of volume of dry air*	*Concentration in parts per million parts of air*
Gases whose concentration does not change over time:		
Nitrogen (N_2)	78.084	
Oxygen (O_2)	20.948	
Argon (Ar)	0.934	
Neon (Ne)	0.00182	18.2
Helium (He)	0.00052	5.2
Methane (CH_4)	0.00015	1.5
Krypton (Kr)	0.00011	1.1
Hydrogen (H_2)	0.00005	0.5
Important variable-amount gases:		
Water vapor (H_2O)	0–4	
Carbon dioxide (CO_2)	0.036	360
Carbon monoxide (CO)		less than 100
Ozone (O_3)		less than 2
Sulfur dioxide (SO_2)		less than 1
Nitrogen dioxide (NO_2)		less than 0.2

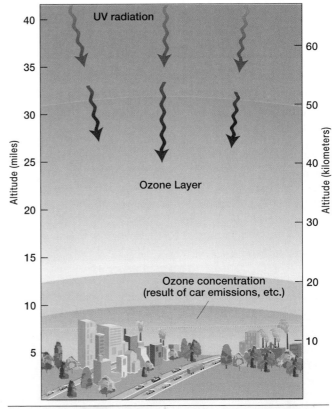

Figure 3-4 The ozone layer, which lies between 9 and 30 miles (15 and 48 kilometers) above Earth's surface, filters out much of the ultraviolet (UV) part of the sun's radiation. Closer to the surface, the effect of this gas is not so benign. The ozone resulting from automobile emissions and other sources is one of a group of gases called the "greenhouse gases" (discussed in Chapter 4). These gases are believed to be the cause of global warming.

enough to be visible, which are sometimes kept aloft in the turbulent atmosphere in sufficient quantity to cloud the sky (Figure 3-5), but they are too heavy to remain long in the air. Smaller particles, invisible to the naked eye, may remain suspended in the atmosphere for months or even years.

The solid and liquid particles found in the atmosphere are collectively called **particulates.** They have innumerable sources, some natural and some the result of human activities. Volcanic ash, wind-blown soil and pollen grains, meteor debris, smoke from wildfires, and salt spray from breaking waves are examples of particulates from natural sources. Particulates coming from human sources mostly consist of industrial and automotive emissions and smoke and soot from fires of human origin.

These tiny particles are most numerous near their places of origin—above cities, seacoasts, and active volcanoes. They may be carried great distances, however, both horizontally and vertically, by the restless atmosphere. They affect weather and climate in two major ways:

1. Many are hygroscopic (which means they absorb water), and water vapor condenses around them as they float about. This accumulation of water vapor molecules is a critical step in cloud formation, as we shall see in Chapter 6.
2. Some either absorb or reflect sunlight, thus decreasing the amount of solar energy that reaches the Earth's surface.

VERTICAL STRUCTURE OF THE ATMOSPHERE

The next five chapters deal with atmospheric processes and their influence on climatic patterns. Our attention in these chapters will be devoted primarily to the lower portion of the atmosphere, which is the zone in which most weather phenomena occur. Even though the upper layers of the atmosphere affect the environment of Earth's surface only minimally, it is still useful to have some understanding of the total atmosphere. Therefore we now note some general characteristics of the atmosphere in its total vertical extent, emphasizing vertical patterns of temperature, pressure, and composition.

TEMPERATURE

Most of us have had some personal experience with temperature variables associated with altitude. As we climb a mountain, for instance, we sense a decrease in temperature. Until about a century ago it was generally assumed that temperature decreased with increasing altitude throughout the whole atmosphere, but now we know that such is not the case.

As shown in Figure 3-6, the vertical pattern of temperature is complex, consisting of a series of layers[1] in

---

[1] A given layer (altitudinal zone) of the atmosphere has different names, depending on the characteristic or feature under discussion. In this section we are discussing thermal layers; later we shall be introduced to other names for the same or overlapping layers.

Figure 3-5 Dust particles sometimes cloud the sky for a short time over a limited part of Earth's surface. On some occasions, as in this scene from the central part of New South Wales, Australia, the term "dust storm" is very appropriate, and the visual effect is imposing, if not menacing. (TLM photo)

AUSTRALIA

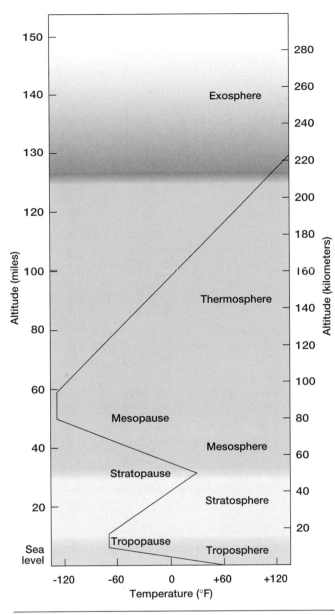

Figure 3-6 Thermal structure of the atmosphere. Air temperature decreases with increasing altitude in the troposphere and mesosphere and increases with increasing altitude in the stratosphere and thermosphere.

which temperature alternately decreases and increases. From the bottom up, these **thermal layers** are called the **troposphere, stratosphere, mesosphere, thermosphere** and **exosphere**. In addition to these five principal names, we also have a special name for the top of the first three layers: **tropopause, stratopause,** and **mesopause**. We use the -sphere name when talking about an entire layer and the -pause name when our interest is either in the upper portion of a layer or in the boundary between two layers.

The names troposphere and tropopause are derived from the Greek word *tropos* ("turn") and imply an overturning of the air in this zone as a result of vertical mixing and turbulence. The depth of the troposphere varies in both time and place (Figure 3-7). It is deepest over tropical regions and shallowest over the poles, deeper in summer than in winter, and varies with the passage of warm and cold air masses. On the average, the top of the troposphere (including the tropopause) is about 11 miles (18 kilometers) above sea level at the equator and about 5 miles (8 kilometers) above sea level over the poles.

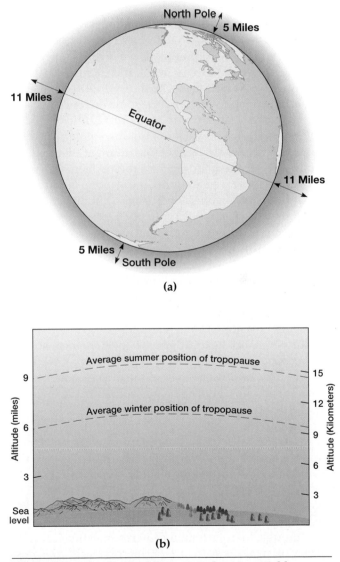

(a)

(b)

Figure 3-7 The depth of the troposphere is variable. **(a)** This thermal layer is deepest over the equator and shallowest over the poles. **(b)** It is deeper in summer than in winter.

The names stratosphere and stratopause come from the Latin *stratum* ("a cover"), implying a layered, or stratified, condition without vertical mixing. The stratosphere extends from an altitude of 11 miles above sea level to about 30 miles (48 kilometers). The names mesosphere and mesopause are from the Greek *meso* ("middle"). The mesosphere begins at 30 miles and extends to about 50 miles (80 kilometers) above sea level. Above the mesopause is the thermosphere (from the Greek *therm*, meaning "heat"), which begins at an altitude of 50 miles above sea level and has no definite top. Instead it merges gradually into the region called the exosphere, which in turn blends into interplanetary space. Traces of atmosphere extend for literally thousands of miles higher. Therefore, "top of the atmosphere" is a theoretical concept rather than a reality, with no true boundary between atmosphere and outer space.

How air temperature changes with altitude can be seen in Figure 3-6. Beginning at sea level, temperature first decreases steadily with increasing altitude through the troposphere. It then remains constant through the tropopause and for some distance into the stratosphere. At an altitude of about 12 miles (20 kilometers), air temperature begins increasing with increasing altitude, reaching a maximum at 30 miles (48 kilometers), at the bottom of the mesosphere. Then the temperature decreases with increasing altitude all through the mesosphere, reaching a minimum at the top of that layer at an altitude of 50 miles (80 kilometers). Temperature remains constant for several miles into the thermosphere and then begins to increase until, at an altitude of 125 miles (200 kilometers), it is higher than the maximum temperature in the troposphere. In the exosphere, the normal concept of temperature no longer applies.

Each warm zone in this temperature gradient has a specific source of heat. At ground level the heat source is the visible portion of sunlight, the energy of which is absorbed by the lowest layer of the troposphere as well as by Earth's surface. The heat absorbed by the ground is conducted upward into the troposphere. The warm zone at the stratopause is near the top of the ozone layer, where ozone is absorbing the ultraviolet portion of sunlight and thereby warming the atmosphere. In the thermosphere, various atoms and molecules also absorb ultraviolet rays and are thus split and heated. The cold areas that separate these warm zones are cold because they lack sources of heat.

Although there are many interesting physical relationships in the stratosphere, mesosphere, thermosphere, and exosphere, our attention in this book is directed almost entirely to the troposphere because storms and essentially all the other phenomena we call "weather" occur here. Only occasionally do we consider atmospheric conditions above the troposphere.

PRESSURE

Atmospheric pressure can be thought of, for simplicity's sake, as the weight of the overlying air. (In Chapter 5, we explore the concept of pressure in much greater detail.) The taller the column of air above an object, the greater the air pressure exerted on that object, as Figure 3-8 shows. Because air is highly compressible, the lower layers of the atmosphere are compressed by the air above, and this compression increases both the pressure exerted by the lower layers and the density of these layers. Air in the upper layers is subjected to less compression and therefore exerts a lower pressure and has a lower density.

Air pressure is normally highest at sea level (and below) and decreases rapidly with increasing altitude. The change of pressure with altitude is not constant, however. As a generalization, pressure decreases upward at a decreasing rate, as Figure 3-9 shows. Table 3-2 expresses the pressure at various altitudes as a percentage of sea-level pressure. From it we see that at 10 miles (16 kilometers) up, for instance, atmospheric pressure is only 10 percent of its sea-level value; this is just another way of saying that most of the mass of the atmosphere is found relatively close to the ground.

One-half of all the gas molecules making up the atmosphere lie below 3.5 miles (5.6 kilometers), and 90 percent of them are concentrated in the first 10 miles (16 kilometers) above sea level (Figure 3-2). Pressure becomes so slight in the upper layers that, above about 50 miles (80 kilometers), there is not enough to register

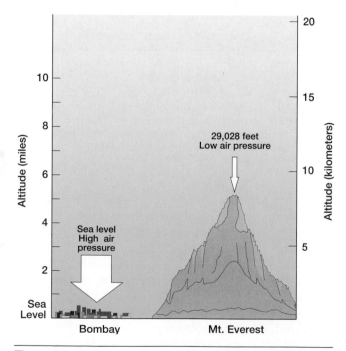

Figure 3-8 Atmospheric pressure normally is highest at sea level and diminishes rapidly with increasing altitude.

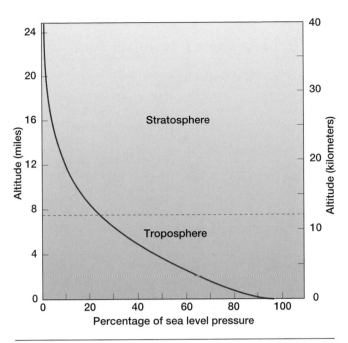

Figure 3-9 Air pressure decreases with increasing altitude but not at a constant rate.

on an ordinary barometer, the instrument used to measure air pressure. Above this level, atmospheric molecules are so scarce that air pressure is less than that in the most perfect laboratory vacuum at sea level.

COMPOSITION

The principal gases of the atmosphere have a remarkably uniform vertical distribution throughout the lowest 50 miles (80 kilometers) or so of the atmosphere. This zone of homogenous composition is referred to as the **homosphere** (Figure 3-10). The sparser atmosphere

T A B L E 3 - 2 Atmospheric Pressure at Various Altitudes Expressed as a Percentage of Pressure at Sea Level		
Altitude		*Percentage of*
Miles	*Kilometers*	*sea-level pressure*
0	0	100
3.5	5.5	50
10	16	10
20	32	1
30	48	0.1
50	80	0.001
60	96	0.00001

above this zone does not display such uniformity; rather, the gases tend to be layered in accordance with their molecular masses—nitrogen below, with oxygen, helium, and hydrogen successively above. This higher zone is called the **heterosphere**.

Water vapor also varies in its vertical distribution. Most is found near Earth's surface, with a general diminishment with increasing altitude. Above 10 miles (16 kilometers) above sea level, the temperature is so low that any moisture formerly present in the air has already frozen into ice. At these altitudes, therefore, there is rarely enough moisture to provide the raw material to make even a wisp of a cloud. If you have done any flying, you may recall the remarkable sight of a cloudless sky overhead once the plane breaks through the top of a solid cloud layer below.

Two other vertical compositional patterns are worthy of mention here:

1. The ozone layer, which, as stated above, lies between 9 and 30 miles (15 and 48 kilometers) up, is sometimes called the **ozonosphere**. Despite its name, the ozone layer is not composed primarily of ozone. It gets its name because that is where the concentration of ozone relative to other gases is at its maximum. Even in the section of the ozonosphere where the ozone attains its greatest concen-

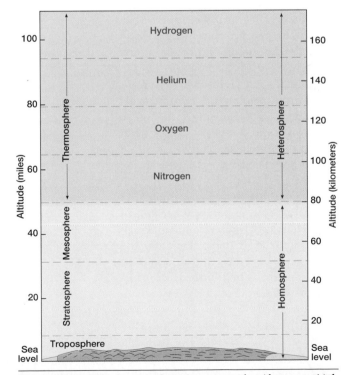

Figure 3-10 The homosphere is a zone of uniform vertical distribution of gases. In the heterosphere, however, the distribution lacks this uniformity.

PEOPLE AND THE ENVIRONMENT

DEPLETION OF THE OZONE LAYER

As mentioned in the text, ozone is a form of oxygen that has three atoms (O_3) rather than the more common two (O_2). It is created in the upper atmosphere by the action of solar radiation on oxygen molecules. Sunlight splits apart O_2 molecules, and then some of the free oxygen atoms combine with unsplit O_2 molecules to form O_3 (Figure 3-1-A).

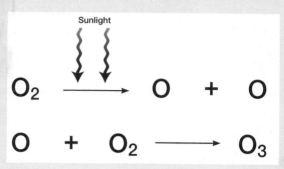

Figure 3-1-A Sunlight splits oxygen molecules (O_2) into free oxygen atoms (O), some of which combine with other O_2 molecules to form ozone (O_3).

Although ozone constitutes less than 0.002 percent of the volume of the atmosphere, its role is critical to life on Earth. About 90 percent of all atmospheric ozone is found in the ozonosphere (see Figure 3-12 in the main text). This ozone is considered "good" because it absorbs certain wavelengths of solar radiation. (The other 10 percent of ozone occurs in the troposphere and is "bad," but that is another story, one we put off until Chapter 4.) Thus stratospheric ozone molecules shield Earth by absorbing most of the potentially dangerous ultraviolet radiation found in sunlight. Ultraviolet radiation is biologically destructive in many ways. It causes skin cancer and cataracts, suppresses the human immune system, diminishes the yield of many crops, disrupts the aquatic food chain by killing microorganisms on the ocean surface, and doubtless causes other negative effects still undiscovered.

The ozone layer is a fragile shield, however. It is relatively thin to begin with and has been getting thinner in recent years, apparently because of the release of certain synthetic chemicals into the air. In the laboratory and anyplace else at Earth's surface, these chemicals, called **chlorofluorocarbons (CFCs)**, are odorless, nonflammable, noncorrosive, and nontoxic. For this reason, scientists originally believed CFCs could not possibly have any effect on the environment. They are widely used in refrigeration and air conditioning (the cooling liquid Freon is a CFC), in foam and plastic manufacturing, and in aerosol sprays.

Although extremely stable and inert in the lower atmosphere, CFCs are broken down by ultraviolet radiation once they reach the ozonosphere. Under certain circumstances, the chlorine released from the CFC breaks ozone molecules down to oxygen molecules (Figure 3-1-B). As many as 100,000 ozone molecules can be removed from the atmosphere for every chlorine molecule released. Stratospheric ozone depletion has been correlated with increased levels of ultraviolet radiation reaching ground level in Antarctica, Australia, mountainous regions of Europe, central Canada, and New Zealand.

Not only is the ozone layer thinning, in some places it has temporarily disappeared entirely. A "hole" in the layer has developed over Antarctica every year since 1979, and that hole has persisted for a longer and longer time every year (Figure 3-1-C). In 1988 an ozone hole was found over the Arctic for the first time, and it too has lasted longer and longer each year since then. Figure 3-1-D summarizes the causes and consequences of ozone depletion.

In response to these alarming discoveries, several countries (including the United States and Canada) banned the use of CFCs in aerosol sprays in 1978. A major international treaty (the Montreal Protocol on Substances That Deplete the Ozone Layer) was promulgated in 1987 to set timetables for phasing out the production of the major ozone-depleting substances. About 100 countries, including all major producers of ozone-depleting substances, have ratified the proposal.

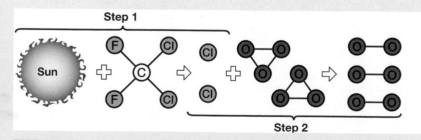

Figure 3-1-B The chemical breakdown of ozone. A typical CFC molecule has one carbon atom (C) bonded to two chlorine atoms (Cl) and two fluorine atoms (F). In step 1, sunlight breaks the bonds in CFCs, releasing chlorine atoms. In step 2, chlorine atoms engage in a complex chemical reaction with ozone and the result is the breakdown of ozone to oxygen. Chlorine atoms are unchanged by the reaction and can repeat the process. Thus a single chlorine atom can destroy tens of thousands of ozone molecules.

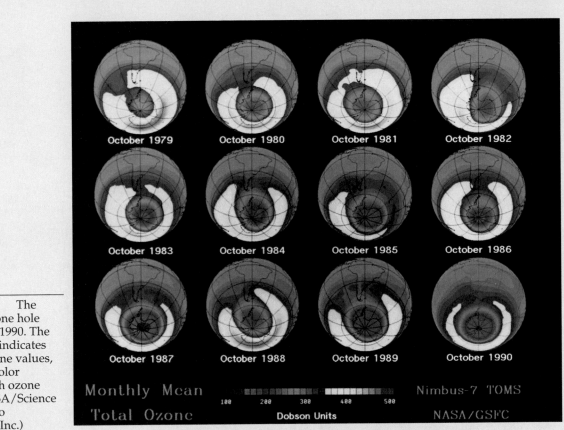

Figure 3-1-C The Antarctic ozone hole from 1979 to 1990. The purple color indicates very low ozone values, and yellow color indicates high ozone values. (NASA/Science Source/Photo Researchers, Inc.)

The treaty and its amendments stipulate that, by 1999, the production of all ozone-depleting chemicals will be decreased by 50 percent from the 1987 level. The Du Pont Company, the world's largest producer of CFCs (25 percent of total world output), voluntarily decided in 1988 to phase out all production of these substances within the next few years.

Since the late 1980s, global atmospheric concentrations of the major ozone depleters have continued to increase, though at a diminishing rate. Even with a worldwide ban on CFC use, it is estimated that the reservoir of CFCs in the atmosphere will persist for 50 to 100 years, which means that CFCs will continue to cause stratospheric ozone depletion long after their production and use have ceased.

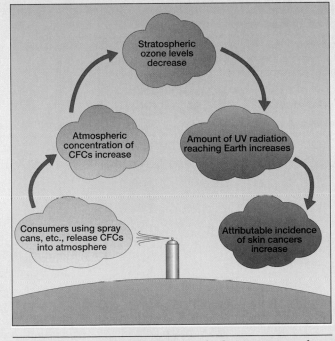

Figure 3-1-D Stratospheric ozone depletion: cause and effect.

tration, at about 15 miles (25 kilometers) above sea level, this gas accounts only for about 12 parts per million of the atmosphere.

2. The **ionosphere** is a deep layer of electrically charged molecules and atoms (which are called "ions") in the middle and upper mesosphere and the lower thermosphere, between about 40 and 250 miles (60 and 400 kilometers). The ionosphere is significant because it aids long-distance communication by reflecting radio waves back to Earth. It is also known for its auroral displays, such as the northern lights (Figure 3-11).

These two layers are shown in Figure 3-12, and Figure 3-13 shows the relationship between all the various "sphere" names.

HUMAN-INDUCED ATMOSPHERIC CHANGE

Some characteristics of the atmosphere are in a state of perpetual change and apparently have been throughout its existence. As with most other natural phenomena, the rate of change has been very slow, although the effects have been profound. One has only to look at the occurrence of past ice ages to recognize that the habitability of various parts of Earth has varied remarkably through time as temperature and precipitation conditions have changed.

As world population has exploded and technology has become more sophisticated, the human race has increasingly had an unintended and uncontrolled impact on the atmosphere, an impact seen all over the globe. The human impact, in simplest terms, consists of the introduction of "impurities" into the atmosphere at a pace previously unknown. As summarized in the report of the World Conference on the Changing Atmosphere: Implications for Global Security, held in 1988 in Toronto:

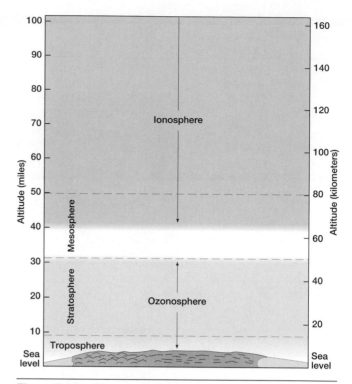

Figure 3-12   The ozonosphere contains significant concentrations of ozone. The ionosphere is a deep layer of ions, which are electrically charged molecules and atoms.

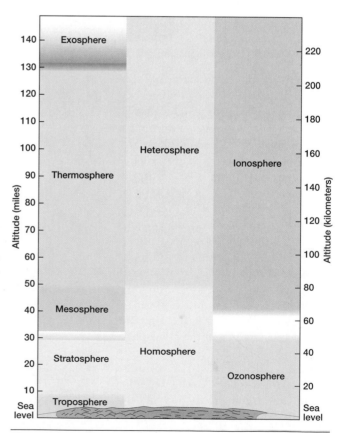

Figure 3-13   The vertical distribution pattern of the various atmospheric "spheres."

Figure 3-11   The Aurora Borealis as photographed over spruce trees in central Alaska, near Fairbanks. (Jack Finch/Science Photo Library/Photo Researchers, Inc.)

The earth's atmosphere is being changed at an unprecedented rate by pollutants resulting from human activities, inefficient and wasteful fossil fuel use and the effects of rapid population growth. . . . The best predictions (of global consequences) indicate potentially severe economic and social dislocation for present and future generations, which will worsen international tensions and increase risk of conflicts between and within nations.

Scattered throughout this book are boxed essays, all titled "People and the Environment," that focus on specific aspects of the accelerating impact that human activities have had on the atmosphere. In the first of these boxes, the spotlight is on the depletion of the ozone layer.

WEATHER AND CLIMATE

Our vast and invisible atmospheric envelope is energized by solar radiation, stimulated by earthly motions, and affected by contact with Earth's surface. The atmosphere reacts by producing an infinite variety of conditions and phenomena known collectively as **weather**. The term weather refers to short-run atmospheric conditions that exist for a given time in a specific area. It is the sum of temperature, humidity, cloudiness, precipitation, pressure, winds, storms, and other atmospheric variables for a short period of time. Thus we speak of the weather of the moment or the week or the season, or perhaps even of the year or the decade.

Weather is in an almost constant state of change, sometimes in seemingly erratic fashion. Yet in the long view, it is possible to generalize the variations into a composite pattern, which is termed **climate**. Climate is the aggregate of day-to-day weather conditions over a long period of time. It encompasses not only the average characteristics but also the variations and extremes. To describe the climate of an area requires weather information over an extended period, normally several decades at least.

Weather and climate, then, are related but not synonymous terms. The distinction between them is the difference between immediate specifics and protracted generalities. As the country philosopher said, "Climate is what you expect; weather is what you get." Or, stated more sarcastically, "It is the climate that attracts people, and the weather that makes them leave."

Weather and climate have direct and obvious influences on agriculture, transportation, and human life in general. Moreover, climate is a significant factor in the development of all major aspects of the physical landscape—soils, vegetation, animal life, hydrography, and topography.

Because climate and weather are generated in the atmosphere, our ultimate goal in studying the atmosphere is to understand the distribution and characteristics of climatic types over Earth. To achieve this understanding, we must consider in detail many of the processes that take place in the atmosphere. Our concern is primarily with long-run atmospheric conditions (climate), but we must also have appreciation for the dynamics involved in the momentary state of the atmosphere (weather).

THE ELEMENTS OF WEATHER AND CLIMATE

The atmosphere is a complex medium, and its mechanisms and processes are sometimes very complicated. Its nature, however, is generally expressed in terms of only a few variables, which are measurable. The data thus recorded provide the raw materials for understanding both temporary (weather) and long-term (climate) atmospheric conditions.

These variables can be thought of as the **elements** of weather and climate. The most important are (a) temperature, (b) moisture content, (c) pressure, and (d) wind (Table 3-3). These are the basic ingredients of weather and climate, the ones you hear about on the nightly weather report. Measuring how they vary in time and space makes it possible to decipher at least partly the complexities of weather dynamics and climatic patterns.

THE CONTROLS OF WEATHER AND CLIMATE

Variations in the climatic elements are frequent, if not continuous, over Earth. Such variations are caused by, or at least strongly influenced by, certain semipermanent attributes of our planet, which are often referred to as **controls**. The principal controls are briefly described below and are explained in more detail in subsequent chapters. Although they are discussed individually here, it should be emphasized that there often is much overlap and interaction among them, with widely varying effects.

TABLE 3-3
Major Elements and Controls of Weather and Climate

Elements	Controls
Temperature	Latitude
Pressure	Distribution of land and water
Wind	General circulation of the atmosphere
Moisture content	General circulation of the oceans
	Elevation
	Topographic barriers
	Storms

Latitude   We noted in Chapter 1 that the continuously changing positional relationship between the sun and Earth brings continuously changing amounts of sunlight, and therefore of radiant energy, to different parts of Earth's surface. Thus the basic distribution of heat over Earth is first and foremost a function of latitude, as indicated in Figure 3-14. In terms of elements and controls, we say that the control *latitude* influences the element *temperature*.

Distribution of Land and Water   Probably the most fundamental distinction concerning the geography of climate is that between continental climates and maritime (oceanic) climates. Oceans heat and cool more slowly and to a lesser degree than do landmasses, which means that maritime areas experience milder temperatures than do continental areas in both summer and winter. For example, Seattle, Washington, and Fargo, North Dakota, are at approximately the same latitude (47° N), with Seattle on the western coast of the United States and Fargo deep in the interior. Seattle has an average January temperature of 41°F (5°C), while the January average in Fargo is 7°F (-13°C). In the opposite season, Seattle has a July average temperature of 66°F (19°C), whereas in Fargo the July average is 71°F (22°C).

Also, oceans are a much more prolific source of atmospheric moisture; thus maritime climates are normally more humid than continental climates. The uneven distribution of continents and oceans over the world, then, is a prominent control of the elements moisture content and temperature.

General Circulation of the Atmosphere   The atmosphere is in constant motion, with flows that range from transitory local breezes to vast regional wind regimes. At the planetary scale, a semipermanent pattern of major wind and pressure systems dominates the troposphere and greatly influences most elements of weather and climate. As a simple example, most sur-

face winds in the tropics come from the east, whereas the middle latitudes are characterized by flows that are mostly from the west, as Figure 3-15 shows.

General Circulation of the Oceans   Somewhat analogous to atmospheric movements are the motions of the oceans (Figure 3-16). Like the atmosphere, the oceans have many minor motions but also a broad general pattern of currents. These currents assist in heat transfer by moving warm water poleward and cool water equatorward. Although the influence of currents on climate is much less than that of atmospheric circulation, the former is not inconsequential. For example, warm currents are found off the eastern coasts of continents, and cool currents occur off western coasts, a distinction that has a profound effect on coastal climates.

Elevation   We have already noted that three of the four weather elements—temperature, pressure, and moisture content—generally decrease upward in the troposphere and are therefore under the influence of the control altitude. This simple relationship between the three elements and the control has significant ramifications for many climatic characteristics, particularly in mountainous regions (Figure 3-17).

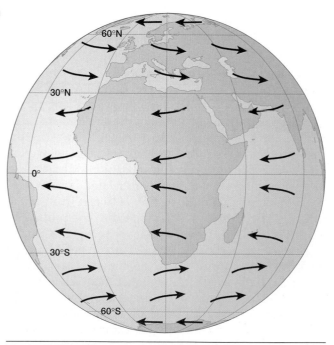

Figure 3-15   The general circulation of the atmosphere is an important climatic control. This diagram is a highly simplified version of the actual circulation, which is discussed in Chapter 5.

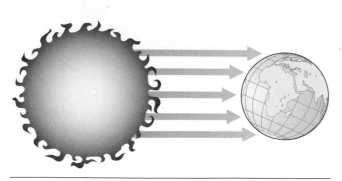

Figure 3-14   Solar energy coming to Earth. See Figure 1-24 and the accompanying text for a review of this topic.

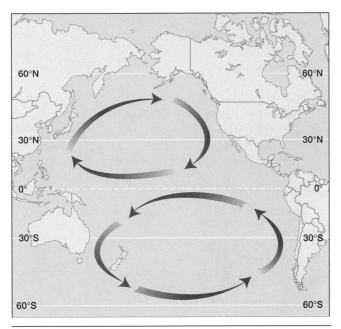

Figure 3-16 The general circulation of the oceans involves the movement of large amounts of warm and cool water. These oceanic currents have a significant climatic effect on neighboring landmasses.

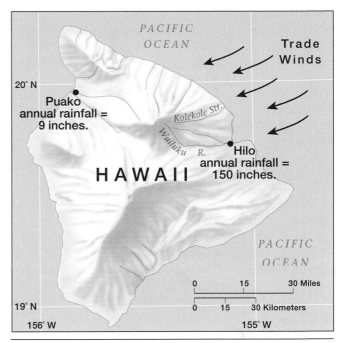

Figure 3-18 A topographic barrier as a control of climate. The difference in average annual rainfall in these two locations on the island of Hawaii is caused by the mountain range separating them. Moisture-laden trade winds coming in from the northeast drop their moisture when forced to rise by the eastern face of the mountains. The result is a very wet eastern side of the island and a very dry western side.

Figure 3-17 A hiker cooking dinner in the Ruwenzori Mountains of Zaire in central Africa. The water boils at a low temperature at these altitudes, so the soup is never really hot. (Ned Gillette/The Stock Market)

Topographic Barriers Mountains and large hills sometimes have prominent effects on one or more elements of climate by blocking or diverting wind flow (Figure 3-18). The side of a mountain range facing the wind (the "windward" side), for example, is likely to have a climate vastly different from that of the sheltered ("leeward") side.

Storms Various kinds of storms occur over the world; some have very widespread distribution, whereas others are localized (Figure 3-19). Although they often result from interactions among other climate controls, all storms create specialized weather circumstances and so are themselves considered to be a control. Indeed, some storms are prominent and frequent enough to affect not only weather but climate as well.

Figure 3-19 A prominent storm system over the British Isles is counterpointed by localized storms over North Africa, Sicily, Italy, and Greece. (European Space Agency / Science Photo Library / Photo Researchers, Inc.)

CHAPTER SUMMARY

The distinctive atmosphere of Earth is a shallow, encircling envelope of gases that makes life possible on our planet. Dominated by nitrogen and oxygen, which together constitute nearly 99 percent of the atmospheric volume, the atmosphere also contains small amounts of water vapor, carbon dioxide, ozone, and argon, traces of a few other gases, and solid and liquid particulate matter.

The vertical structure of the atmosphere is complex and variable. Temperature alternately decreases and increases with altitude through the four thermal layers: troposphere, stratosphere, mesosphere, thermosphere. Pressure decreases upward at a decreasing rate. The principal gases are distributed uniformly in the lower atmosphere (therefore called the homosphere) but are layered in accordance with their molecular masses at upper levels (in the heterosphere). Most of the ozone in the atmosphere is concentrated in a band called the ozonosphere, and electrically charged atoms and molecules are found in the ionosphere.

Weather refers to short-run atmospheric conditions; climate is the aggregate of day-to-day weather conditions over a long period of time. To describe the climate of an area requires weather information over an extended period. The four principal elements of weather and climate are temperature, moisture content, pressure, and wind. The main controls affecting these elements are latitude, global distribution of oceans and landmasses, circulation of air in the atmosphere and water in the oceans, altitude, topography, and storm systems.

KEY TERMS

air	ionosphere	stratosphere
chlorofluorocarbons (CFCs)	mesopause	thermal layer
climate	mesosphere	thermosphere
controls (of weather/climate)	ozone	tropopause
elements (of weather/climate)	ozone layer	troposphere
exosphere	ozonosphere	weather
heterosphere	particulate	
homosphere	stratopause	

REVIEW QUESTIONS

1. What benefits to human life are provided by Earth's atmosphere?
2. Describe the composition of the atmosphere.
3. Why is the question "How deep is the atmosphere?" difficult to answer?
4. Describe both the vertical distribution of water vapor in the atmosphere and its horizontal distribution near Earth's surface.
5. Explain how the amount of carbon dioxide present in the atmosphere affects climate.
6. What is ozone, and why is it important to life on Earth?
7. What is happening to the ozone layer?
8. In our study of physical geography, why do we concentrate primarily on the troposphere rather than on other zones of the atmosphere?
9. Describe the vertical variation of temperature in the atmosphere.
10. Describe how atmospheric pressure changes with increasing altitude.
11. What is the difference between weather and climate?
12. Name the four main elements of weather/climate and discuss the principal controls affecting them.

SOME USEFUL REFERENCES

BARRY, ROGER G., AND RICHARD J. CHORLEY, *Atmosphere, Weather and Climate*, 6th ed. New York: Routledge, 1992.

ELSOM, DEREK, *Atmospheric Pollution*. Oxford, England: Basil Blackwell, 1987.

GEDZELMAN, S. D., *The Science and Wonders of the Atmosphere*. New York: John Wiley and Sons, 1980.

GRAEDEL, T. E., AND P. J. CRUTZEN, "The Changing Atmosphere," *Scientific American*, vol. 259, September 1989, pp. 58–68.

GRIBBIN, JOHN, *The Hole in the Sky: Man's Threat to the Ozone Layer*. New York: Bantam New Age Books, 1989.

LYDOLPH, PAUL E., *Weather and Climate*. Totowa, NJ: Bowman & Allanheld, 1985.

MCELROY, M. B., AND J. B. SALAWITCH, "Changing Composition of the Global Stratosphere," *Science*, vol. 243, Feb. 10, 1989, pp. 763—770.

STOLE, RICHARD S., "The Antarctic Ozone Hole," *Scientific American*, vol. 258, January 1988, pp. 30-36.

4

INSOLATION AND TEMPERATURE

THE TEMPERATURE OF THE AIR AT ANY TIME AND AT any place in the atmosphere is the result of the interaction between a variety of complex factors. In this chapter, our attention is focused on the energetics of the atmosphere—the important processes involved in bringing insolation (incoming solar radiation) to the atmosphere, in determining the extent of heating (and cooling) that takes place, and in transferring heat from one place to another. Understanding these processes will help us understand the distribution of temperature over Earth.

(a)

THE IMPACT OF TEMPERATURE ON THE LANDSCAPE

All organisms have certain temperature tolerances, and most are harmed by wide fluctuations in temperature. Thus when the weather becomes particularly hot or cold, mobile organisms are likely to search for shelter and their presence in the landscape is diminished. Human beings also seek haven from temperature extremes, although they have other options (such as specialized clothing) that allow them to brave the elements. In a broader view, immoderate temperatures have a more profound effect on the landscape because both animals and plants often evolve in response to hot or cold climates. For this reason, the inventory of flora and fauna in any area of temperature extremes is likely to be determined by the capability of the various species to withstand the long-term temperature conditions. An example of such an adaptation is shown in Figure 4-1(a).

Most inorganic components of the landscape also are affected by long-run temperature conditions. For example, temperature is a basic factor in soil development, and repeated fluctuations of temperature are a prominent cause of the breakdown of exposed bedrock. The human-built landscape also is influenced by temperature considerations, as demonstrated by architectural styles and building materials. An example of such influence is shown in Figure 4-1(b).

(b)

Figure 4-1   (a) A yellow-headed collared lizard in the desert of western Colorado. It strikes an erect pose to keep its belly off the hot rock. (Rod Plank/Photo Researchers, Inc.) (b) The town museum in Igloolik in Canada's Northwest Territories is built in a style that is adjusted to the cold climate. (Robert Sememiuk/The Stock Market)

SOLAR ENERGY

The sun is the only important source of energy for Earth and its atmosphere. Millions of other stars radiate energy, but they are too far away to affect Earth. Energy is also released from inside Earth, primarily as radioactive minerals decay, but only in insignificant quantities. Tidal energy is also of minor importance. Thus the sun supplies essentially all the energy that

supports life on Earth and energizes most of the atmospheric processes.

The sun is a star of average size and average temperature, but its proximity to Earth gives it a far greater influence on our planet than that exerted by all other celestial bodies combined. The sun is a prodigious generator of energy. In a single second it produces more energy than the amount used by humankind since civilization began. The sun functions as an enormous thermonuclear reactor, producing energy by fusion, a process that burns only a very small portion of the sun's mass but provides an immense and continuous flow of radiant energy that is dispersed in all directions.

The radiant energy from the sun is in the form of electromagnetic waves. These are waves that can

A scorching scene in the Sahara Desert of southern Morocco. *(Jose Fuste Raga/The Stock Market)*

transport energy without requiring a medium (the presence of matter) to pass through. They traverse the great voids of space in unchanging form. The waves travel outward from the sun in straight lines at the speed of light—186,000 miles (300,000 kilometers) per second.

Electromagnetic waves are classified on the basis of wavelength, which can be thought of as the distance from the crest of one wave to the crest of the next (Figure 4-2). Wavelengths vary enormously, as the range in Figure 4-2 shows. For our purposes, the most important distinction is between shortwave and long-wave radiation, the dividing line between the two being a wavelength of about 4 micrometers (four millionths of a meter).

Only a tiny fraction of the sun's radiant output is intercepted by Earth. The waves travel through space without loss of energy, but since they are diverging from a spherical body their intensity continuously diminishes with increased distance from the sun (Figure 4-3). As a result of this intensity drop and the distance separating Earth from the sun, less than one two-billionth of total solar output reaches the outer limit of Earth's atmosphere, having traveled 93,000,000 miles (150,000,000 kilometers) in just over 8 minutes. Although it consists of only a minuscule portion of total solar output, in absolute terms the amount of solar energy Earth receives is enormous: the amount received in 1 second is approximately equivalent to all the electric energy generated on Earth in a week.

INSOLATION

As we learned in the discussion of Figure 2-22, electromagnetic waves of various lengths make up what is called the electromagnetic spectrum. For the physical geographer, only three areas of the spectrum are of importance (Figure 4-4):

1. Wavelengths that measure from about 0.01 to 0.4 micrometers are the **ultraviolet waves**, which are too short to be seen by the human eye. The sun is a prominent natural source of ultraviolet rays, and solar insolation reaching the top of our atmosphere contains a considerable amount. However, much of it is absorbed by the atmosphere, and the shortest ultraviolet rays do not reach Earth's surface, where they could cause considerable damage to most living organisms.
2. *Visible light* is concentrated entirely in the narrow band between about 0.4 and 0.7 micrometers. Only about 3 percent of all electromagnetic waves are in the visible light spectrum.
3. Between about 0.7 and 1000 micrometers are **infrared waves**, which are too long to be seen by

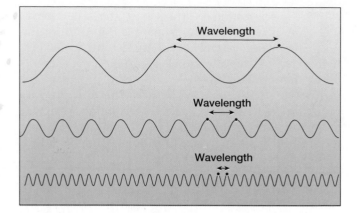

Figure 4-2   Electromagnetic waves can be of almost any length. The distance from one crest to the next is called the wavelength.

the human eye. They are generally emitted by hot objects and are sometimes referred to as "heat rays." They have a variety of uses that depend in part on their ability to pass through materials that block ordinary light rays but still produce heat inside the blocking material, as with infrared heat lamps. Earth radiation is entirely infrared, but only a small fraction of solar radiation is.

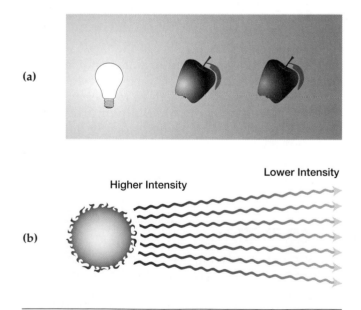

Figure 4-3   (a) The closer apple intercepts more rays from the lightbulb and therefore is better illuminated than the farther apple. **(b)** Electromagnetic waves from the sun spread out as they move outward from the sun. Thus their intensity decreases as they get farther and farther from the sun.

Wavelength

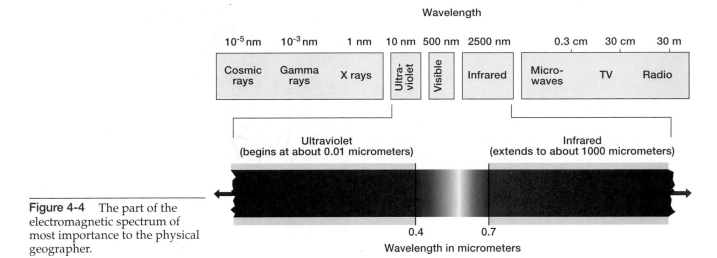

Figure 4-4   The part of the electromagnetic spectrum of most importance to the physical geographer.

Solar radiation is mainly visible light along with some shorter infrared and longer ultraviolet wavelengths (Figure 4-5). Terrestrial radiation is entirely in the infrared spectrum. A wavelength of about 4 micrometers is considered the boundary on the spectrum separating long waves from short ones; thus all terrestrial radiation is long-wave radiation, and almost all solar radiation is shortwave radiation. The longest terrestrial-radiation waves are approximately 20 times longer than the longest solar-radiation waves.

As a basic generalization, hot bodies radiate mostly shortwaves and cool bodies radiate mostly long waves. The sun is the ultimate hot body of our solar system, and so most of its radiation is in the shortwave part of the electromagnetic spectrum. For this reason, solar radiation is often referred to as **shortwave radiation**.

The total insolation received at the top of the atmosphere is believed to be constant when averaged over a year, although it may vary slightly with fluctuations in the sun's temperature. This constant amount of incoming energy—referred to as the **solar constant**—is slightly less than 2 calories per square centimeter per minute. (A calorie is the amount of heat required to raise the temperature of 1 gram of water by 1°C.) The solar constant is more properly given as slightly less than 2 langleys per minute because the unit of radiation intensity is the **langley**, which is equal to 1 calorie per square centimeter.

The entrance of insolation into the atmosphere is just the beginning of a complex series of events in the atmosphere and at Earth's surface. Some of the insolation is reflected off the atmosphere and bounces back into space, and the part that is not reflected is transformed in one way or another. Some passes through the atmosphere to Earth's surface; some does not. The mixed reception of solar energy waves and the energy cascade that results are discussed after a brief digression to define our terms.

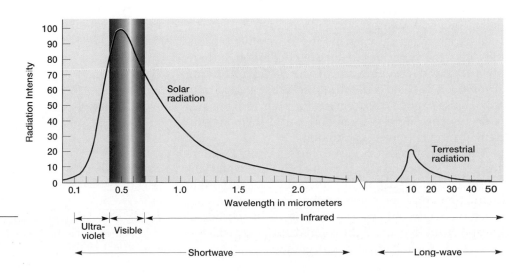

Figure 4-5   Comparison of solar and terrestrial radiation intensity.

BASIC PROCESSES IN HEATING AND COOLING THE ATMOSPHERE

Before looking at the events that occur as energy travels from the sun to Earth, let us look at the physical processes involved in the movement of heat energy. There are three ways in which heat energy can move from one place to another—by radiation, by conduction, and by convection. Once that energy gets to a destination, it can be absorbed, reflected, scattered, or transmitted.

RADIATION

Radiation is the process by which heat energy is emitted from a body. It involves the flow of radiant energy out of the body and through the air (Figure 4-6). All bodies radiate, but hotter bodies are more potent radiators than cooler ones. The hotter the object, the more intense its radiation and the shorter the wavelength of that radiation (Figure 4-7).

Temperature, however, is not the only control of radiation effectiveness. Objects at the same temperature may vary considerably in their radiating capability because the nature of the surface of the objects is an important determining factor. A body that emits the maximum amount of radiation possible—at every wavelength—is called a **blackbody**.

Both the sun and Earth function essentially as blackbodies, that is, as perfect radiators. They radiate with almost 100 percent efficiency for their respective temperatures. Because it is exceedingly hotter than Earth, however, the sun emits 2 billion times more energy than Earth.

ABSORPTION

Heat energy striking an object can be absorbed by the object like water into a sponge; this process is called **absorption.** When insolation strikes an object and is absorbed, the temperature of the object increases, as exemplified by the uncomfortably warm skier in Figure 4-8.

Different materials vary in their absorptive capabilities, with the variations depending in part on the temperature and wavelength of the radiation being absorbed. The basic generalization is that a good radiator is also a good absorber and a poor radiator is a poor absorber. Both the sun and Earth, then, are efficient absorbers as well as radiators.

Mineral materials (rock, soil) are generally excellent absorbers; snow and ice are poor absorbers; water surfaces vary in their absorbing efficiency. One important distinction concerns color. Dark-colored surfaces are much more efficient absorbers in the visible portion of the spectrum than are light-colored surfaces (which is why the dark-clothed skier is sweating).

Figure 4-6   In the heat-transfer process called radiation, heat travels through the air, here from the fire to the camper's feet and legs.

REFLECTION

For our purposes, **reflection** is the ability of an object to repel waves without altering either the object or the waves. Thus in some cases insolation striking a surface in the atmosphere or on Earth is bounced away, unchanged, in the general direction from which it came, much like a mirror reflection, where nothing is changed (Figure 4-8).

In this context, reflection is the opposite of absorption. If the wave is reflected, it cannot be absorbed. Hence, an object that is a good absorber is a poor reflector, and vice versa. A simple example of this principle is the existence of unmelted snow on a warm, sunny day. If it is to melt, the snow must absorb heat energy from the sun. Although the air temperature may be well above freezing, the snow does not melt

Figure 4-7   Hot bodies radiate with greater intensity than cold bodies. Thus solar radiation is much more potent than terrestrial radiation.

rapidly because its white surface reflects away a large share of the solar energy that strikes it.

SCATTERING

Particulate matter and gas molecules in the air sometimes deflect light waves and redirect them in a process known as **scattering** (Figure 4-8). This deflection involves a change in the direction of the light wave but no change in wavelength. Some of the waves are backscattered into space and thus are lost to Earth, but most of them continue through the atmosphere in altered directions.

The amount of scattering that takes place depends on the wavelength of the wave as well as on the size, shape, and composition of the molecule or particulate. Shorter waves are more readily scattered than longer ones. This means that the violets and blues in the visible part of the spectrum are more likely to be redirected than are the oranges and reds.

TRANSMISSION

Transmission is the process whereby a wave passes completely through a medium, as when light waves are transmitted through a pane of clear, colorless glass (Fig-

Figure 4-8 The fate of solar radiation that comes near or to Earth's surface. Some waves are scattered into space and therefore lost to Earth, and others are scattered but continue through the atmosphere in altered directions. Upon striking a surface, some waves are reflected. Others are absorbed, raising the temperature of the absorbing object. Transmission is illustrated by the shortwave insolation (sunlight) being transmitted through the car windows and into the car interior. Because glass allows transmission of only shortwave radiation, the longer waves reradiated by the heated upholstery are not able to escape through the windows.

ure 4-8). There is obviously considerable variability among mediums in their capacity to transmit rays. Earth materials, for example, are very poor transmitters of insolation; sunlight is absorbed at the surface of rock or soil and does not penetrate at all. Water, on the other hand, transmits sunlight well: even in very murky water, light penetrates some distance below the surface; and in clear water, sunlight may illuminate to considerable depths.

In some cases, transmission depends on the wavelength of the rays. For example, glass has high transmissivity for shortwave radiation but not for long waves. Thus heat builds up in a closed automobile left parked in the sun because shortwave insolation is transmitted through the window glass but the long waves that are reradiated from the interior of the car cannot escape in similar fashion (Figure 4-8).

The atmosphere transmits a considerable amount of shortwave solar radiation, but it is not nearly as effective a transmitter of long-wave terrestrial radiation (Figure 4-9). This characteristic is enhanced when the air contains clouds, water vapor, or dust. In simplest terms, solar energy readily penetrates to Earth's surface, but reradiated terrestrial energy is mostly "trapped" in the lower troposphere and much of it is reflected back toward the ground. This entrapment keeps Earth's surface and lower troposphere at a higher average temperature than would be the case if there were no atmosphere.

The circumstances just described are referred to as the **greenhouse effect** because it was long thought that greenhouses maintained heat in the same manner—the glass roof transmitting shortwave solar energy in but inhibiting the passage of long-wave radiation out. Recently it has been shown that this is not the full story; for example, greenhouses having windows made of

rock salt, which permits equal transmission of long waves and shortwaves, experienced a heat buildup approximately as great as that of ordinary glass greenhouses. Further investigation showed that glass greenhouses maintain high temperatures largely because the warm air in the building is trapped and does not dissipate through mixing with the cooler air outside. Thus the term *greenhouse effect* is based on a misconception, and the trapping of heat in the lower troposphere because of differential transmissivity for shortwaves and long waves should probably be called something else; *atmospheric effect* has been suggested, but greenhouse effect continues to be the customary term. Whatever the name, the principle is exceedingly important in understanding the heating of the atmosphere and is considered more fully later in this chapter.

Conduction

The movement of heat energy from one molecule to another without changes in their relative positions is called **conduction** (Figure 4-10). This process enables heat to be transferred from one part of a stationary body to another or from one object to a second object when the two are in contact.

Conduction comes about through molecular collision, as the blown-up view in Figure 4-10 illustrates. A hot molecule becomes increasingly agitated as heat is

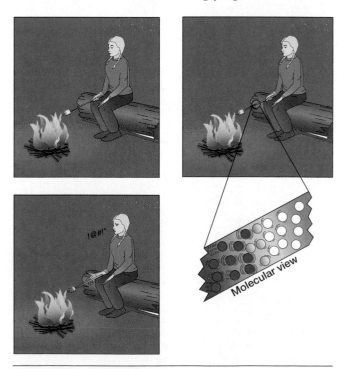

Figure 4-10   Energy is conducted from one place to another by molecular agitation. The tip of the marshmallow fork is in the flames and so becomes hot. This heat is then conducted the length of the metal fork and surprises the camper.

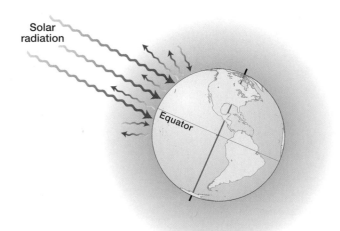

Figure 4-9   The atmosphere easily transmits shortwave radiation from the sun but is a poor transmitter of the longer waves radiated from Earth's surface. This selective transmission causes what we call the greenhouse effect.

added to it and collides against a cooler, calmer molecule, transferring energy to it. In this manner, the heat is passed from one place to another. The principle is that when two molecules of unequal temperature are in contact with one another, heat passes from the warmer to the cooler until they attain the same temperature.

The ability of different substances to conduct heat is quite variable. For example, most metals are excellent conductors, as can be demonstrated by pouring hot coffee into a metal cup and then touching your lips to the edge of the cup. The heat of the coffee is quickly conducted throughout the metal and burns the lips of the incautious drinker. On the other hand, hot coffee poured into a ceramic cup only very slowly heats the cup because such earthy material is a poor conductor.

Earth's land surface warms up rapidly during the day because it is a good heat absorber, and some of that warmth is transferred away from the surface by conduction. A small part is conducted deeper underground, but not much because earth materials are not good conductors. Most of this absorbed heat is transferred to the lowest portion of the atmosphere by conduction from the ground surface. Air, however, is a poor conductor, and so only the air layer touching the ground is heated very much. Physical movement of the air is required to spread the heat around.

Moist air is a slightly more efficient conductor than dry air. If you are outdoors on a winter day, you will stay warmer if there is little moisture in the air.

Convection

In **convection**, heat is transferred from one point to another by a moving liquid or gas. This method of heat transfer involves movement of the heated molecules from one place to another. Do not confuse this movement from one place to another with the back-and-forth vibratory movement of conduction. In convection the molecules physically move away from the heat source. In conduction, they do not.

If you have ever warmed your hands by holding them directly above a campfire, as in Figure 4-11, you have taken advantage of heat convection. The air molecules immediately above the flame are heated by the flame and then begin to rise until they hit the surface of your outstretched hands.

(Although the principal action in convection is vertical, there is some horizontal motion. When the convecting liquid or gas moves horizontally, we call the process **advection**.)

The heated air immediately above the fire rises because the heating has caused it to expand and therefore become less dense. Because it is now less dense than the nearby air that is not right above the fire, the heated air

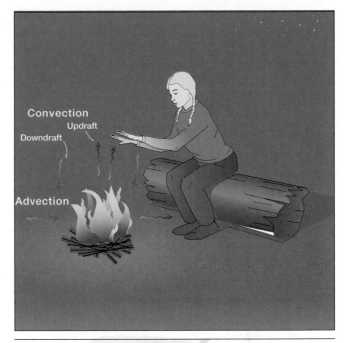

Figure 4-11 A convective updraft (red arrows) from the campfire warms the camper's hands. Cooler air (blue arrows) flows into the area just above the fire to replace the heated air that has risen. Advectional drafts are also shown.

rises. Then the surrounding air moves in to fill the empty space. Cooler air from above descends to replace that which has moved in, and a cellular circulation is established—up, out, down, and in, as Figure 4-11 shows.

A similar convective pattern frequently develops in the atmosphere. As far as our study of insolation is concerned, the important point to remember about convection is that it causes warm air to rise. Unequal heating (for a variety of reasons) may cause a parcel of surface air to become warmer than the surrounding air. The heated air expands and moves upward, in the direction of lowest pressure. The cooler surrounding air then moves in toward the heat source, and air from above sinks down to replace that which has moved in, thus establishing a convective system. The prominent elements of the system are an updraft of warm air and a downdraft of cool air. Convection is common in each hemisphere during its summer and throughout the year in the tropics.

Adiabatic Cooling and Warming

Whenever air ascends or descends, its temperature changes. This invariable result of vertical movement is due to the variation in pressure. When air rises, it expands because there is less air above it and so less pressure is exerted on it [Figure 4-12(a,b)]. When air descends, it is compressed because there is more air above it and so more pressure is exerted on it [Figure 4-12 (c,d)].

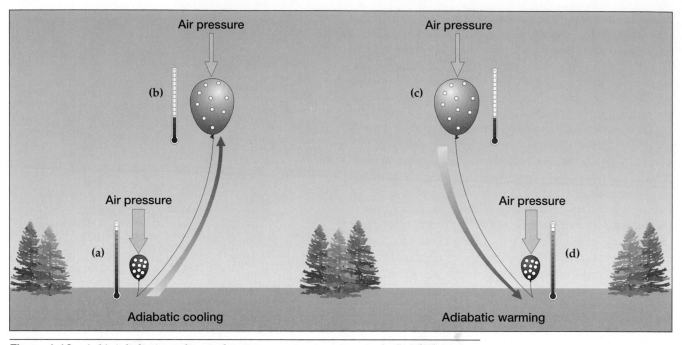

Figure 4-12   **(a,b)** Adiabatic cooling is due to expansion in rising air. **(c,d)** Adiabatic warming is due to compression in descending air. No heat transfer is involved in either process.

The expansion that occurs in rising air is a cooling process even though no heat is taken away. Spreading the molecules over a greater volume of space requires energy, and this energy comes from the molecules. The loss of energy slows them down and decreases their frequency of collision. The result is a drop in temperature. This is **adiabatic cooling**—cooling by expansion in rising air.

Conversely, when air descends, it must become warmer. The descent causes *compression* as the air comes under increasing pressure. The molecules draw closer together and collide more frequently. The result is a rise in temperature even though no heat is added from external sources. This is **adiabatic warming**—warming by compression in descending air.

LATENT HEAT

We shall see in Chapter 6 that the physical state of moisture in the atmosphere frequently changes—ice changes to liquid water, liquid water changes to water vapor, and so forth. The changes involve either the storage or the release of energy, depending on the process. The two most common state changes are **evaporation**, in which liquid water is converted to gaseous water vapor, and **condensation**, in which gaseous water vapor condenses to liquid water. In evaporation, energy is stored as **latent heat** (latent is from the Latin, "lying hidden"); in condensation, the latent heat is released.

THE HEATING OF THE ATMOSPHERE

We now turn to the specifics of atmospheric heating. What happens to solar radiation when it enters Earth's atmosphere? How is it received and distributed? What are the dynamics of converting electromagnetic waves to atmospheric heat?

Most of the insolation that enters the atmosphere does not heat it directly. About one-third of the total amount is reflected (or scattered) back into space, as Figure 4-13 shows; this radiation that is bounced back into space is called Earth's **albedo.** (Albedo is the technical term for the reflectivity of an object. The higher the albedo value, the more radiation the object reflects.) About 45 percent of the insolation passes on through the atmosphere to Earth's surface, leaving only 22 percent to heat the atmosphere directly, 3 percent heating the ozone layer, and 19 percent heating the rest of the atmosphere.

Let us now examine the details of that 45 percent absorbed by Earth's surface. As Figure 4-14 shows, about 12 percent of it is conducted back into the atmosphere, where it is dispersed by convection and contributes significantly to the heating of the atmosphere. Another 38 percent is reradiated as longwave terrestrial radiation. About 5 percent of this terrestrial radiation is transmitted through the atmosphere and lost to space. The rest is absorbed by the atmosphere, particularly by water vapor, dust,

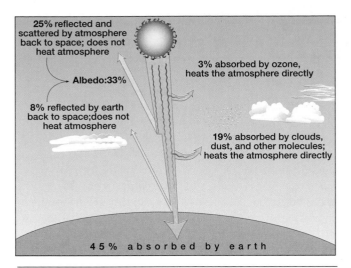

Figure 4-13   Earth's solar radiation budget.

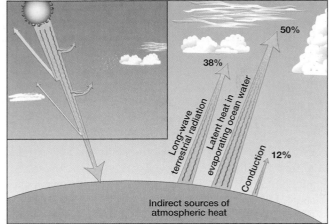

Figure 4-14   The 45 percent of solar radiation that reaches Earth's surface is redirected to the atmosphere by evaporation, radiation, and conduction.

carbon dioxide, and clouds. This absorbed radiation, in turn, is mostly reradiated back to Earth. This sequence of reradiation between Earth and atmosphere continues indefinitely, as the "greenhouse effect."

So far we have accounted for 12 + 38 = 50 percent of the insolation absorbed by Earth. The other 50 percent becomes latent heat locked up in water evaporated from oceans, lakes, and other bodies of water. More than three-fourths of all sunshine falls on a water surface when it reaches Earth. Much of this energy is utilized in evaporating water, and so the energy is passed into the atmosphere as latent heat stored in the resulting water vapor. This energy is subsequently released when condensation takes place, and it is a major source of heat energy for the atmosphere.

For the most part, then, the atmosphere is heated by Earth, although the sun is the original source of the energy. Thus there is an intimate link between troposphere temperatures and Earth surface conditions. The air temperature at any given time represents the balance between insolation and terrestrial radiation.

In the long run, there is an apparently perfect balance in this complex energy budget: As much energy leaves the atmosphere for space as enters it from the sun, and as much energy reaches Earth's surface from the atmosphere as leaves it for the atmosphere.

This complicated sequence of atmospheric heating has many ramifications. One of the most striking is that the atmosphere is heated mostly from below rather than from above. The result is a troposphere in which cold air overlies warm air. This unstable situation (explored further in Chapter 5) creates an

environment of almost constant convective activity and vertical mixing. If the atmosphere were heated directly from the sun, producing warm air at the top and cold air near Earth's surface, the situation would be stable, essentially without vertical air movements. The result would be a troposphere that was largely motionless, apart from the effects of the Earth's rotation.

SPATIAL AND SEASONAL VARIATIONS IN HEATING

The radiation budget we have been discussing is broadly generalized, and the indicated percentages represent an average for the whole world. There are, however, many latitudinal and vertical imbalances in this budget, and these are among the most fundamental causes of weather and climate variations. In essence, we can trace a causal continuum wherein radiation differences lead to temperature differences that lead to air-density differences that lead to pressure differences that lead to wind differences that often lead to moisture differences.

It has already been noted that world weather and climate differences are fundamentally caused by the unequal heating of Earth and its atmosphere. This unequal heating is the result of latitudinal and seasonal variations in how much energy is received by Earth.

LATITUDINAL DIFFERENCES

There are only a few basic reasons for the unequal heating of different latitudinal zones.

PEOPLE AND THE ENVIRONMENT

GLOBAL WARMING AND THE GREENHOUSE EFFECT

We have noted that *greenhouse effect* is a term of questionable appropriateness that refers to an important atmospheric process. In essence, the term alludes to the fact that the atmosphere inhibits the escape of long-wave terrestrial radiation, thereby increasing the air temperature, and that atmospheric impurities enhance this process. In the last decade this concept has received considerable attention from the media and the general public. There is some evidence that the atmosphere is warming, and there are strong indications that an intensified greenhouse effect is at least partly responsible.

It is apparent that there has been an increase in average global temperature of about 0.5°C during this century, with the 1980s being the warmest decade on record—although the record goes back for only about a hundred years, which is very short range for discerning

climatic trends. Despite the fact that there is still no clear evidence that a long-term heating trend is taking place, even the short-term results could be significant enough to warrant concern.

Humans did not create the greenhouse effect; it has been part of the basis of life on Earth since the atmosphere first formed. Without it, our planet would be a frozen mass, more than 30 C° colder on average than it is now. The greenhouse effect is natural, but we seem to have turned up the heat. It is well known that climate undergoes frequent natural fluctuations, becoming warmer or colder regardless of human activities. Such events as the episodic warming of seawater in the tropical Pacific is likely to be at least partly responsible for atmospheric warming.

There is, however, an increasing body of evidence that indicates that anthropogenic (human-induced) fac-

tors are responsible for recent temperature increases. The "culprits" are popularly referred to as "greenhouse gases"—carbon dioxide, water vapor, methane, nitrous oxide, ozone, and chlorofluorocarbons. All of these gases have a low capacity for transmitting long-wave radiation, and as their concentrations in the atmosphere increase, more terrestrial radiation is retained in the lower atmosphere, thereby raising the temperature.

Human activities are clearly responsible for the increasing release of most of these gases into the air. Chlorofluorocarbons (CFCs) are synthetic chemicals that were very popular for a variety of uses until recently (see the focus box on p. 66). Nitrous oxide comes from chemical fertilizers and automobile emissions. Methane is produced by grazing livestock and rice paddies and is a by-product of the combustion of wood, natural gas, coal, and oil. All these gases have

---

Angle of Incidence   The angle at which rays from the sun strike Earth's surface is called the **angle of incidence**. This angle is measured from a line drawn tangent to the surface, as Figure 4-15 shows. By this definition, a ray striking Earth's surface directly, when the sun is directly overhead, has an angle of incidence of 90°, a ray striking the surface at a slant has an angle of incidence smaller than 90°, and for a ray striking Earth at either pole the angle of incidence is zero.

Because Earth's surface is curved and because the positional relationship between Earth and the sun is always changing, the angle of incidence is also always changing. As mentioned in Chapter 1, this changing angle is the primary determinant of the intensity of solar radiation received at any spot on Earth. If a ray strikes Earth's surface directly, the energy is concentrated in a small area; if the ray strikes Earth not directly but obliquely, the energy is spread out over a larger portion of the surface. The more nearly perpendicular the ray (in other words, the

closer to 90° the incidence angle), the smaller the surface area heated by a given amount of insolation and the more effective the heating. The yellow areas on the globe's surface indicate the comparative spread of rays. It is clear that, considering the year as a whole, the insolation received by high-latitude regions is much less intense than that received by tropical areas (Figure 4-16).

Day Length   The duration of sunlight is another important factor in explaining latitudinal inequalities in heating. Longer days allow more insolation to be received and thus more heat to be absorbed. In tropical regions, this factor is relatively unimportant because the number of hours between sunrise and sunset does not vary significantly from one month to another; at the equator, of course, daylight and darkness are essentially equal in length (12 hours each) every day of the year. In middle and high latitudes, however, there are pronounced seasonal variations in day

been released at an accelerating rate in recent years.

Carbon dioxide (CO_2), however, appears to be the principal offender; some studies indicate that it accounts for about half the recent global warming. In the last quarter century, concentrations of carbon dioxide in the atmosphere have increased by about 25 percent. Carbon dioxide is a principal by-product of the combustion of anything containing carbon, particularly coal and petroleum. The world consumption of fossil fuels has been increasing at a rate of 2.0–2.5 percent per year recently, which is only about half the rate that prevailed before the energy crisis of the mid-1970s but is more than enough to continue accelerating the problem. Indeed, it is estimated that eliminating the warming trend would require a decrease of at least 50 percent in burning fossil fuels. Moreover, the forests of the world, particularly the tropical rainforests (see the focus box on p. 312), are rapidly being depleted. Trees are major absorbers of carbon dioxide,

and with fewer trees, more carbon dioxide floats into the atmosphere.

The long-term climatic result of the buildup of greenhouse gases is still unknown, and the predicted scenarios are wildly variable. It is likely that temperature and precipitation patterns will change. Presumably, in one scenario, heat and drought would become more prevalent in much of the midlatitudes, and milder temperatures would prevail in higher latitudes. Some arid lands might receive more rainfall. Ice caps would surely melt, and global sea levels would rise. Current living patterns over much of the world would be affected.

What can be done to ameliorate the situation? The bottom line is to reduce emissions, particularly from smokestacks and internal combustion engines. Coal and petroleum are major offenders; natural gas produces fewer emissions; solar, wind, and nuclear energy sources are "clean," insofar as carbon dioxide is concerned.

The United States is the world's leading producer of carbon dioxide; Russia ranks second. These two na-

tions combined yield about 45 percent of the world total. Serious efforts to reduce emissions are now being made by industrialized nations. Most developed countries (the United States, Canada, the nations of western Europe, Japan, Australia) are now burning less oil and coal than they did a decade ago. Developing countries, on the other hand, are increasing emission output as they attempt to build their own industrial infrastructures. Addressing the obvious economic needs of developing nations while attempting to curb carbon dioxide emissions presents a painful dilemma. These nations argue that industrialized nations created the problem and now want the developing countries to forego the benefits of industrialization in order to protect the atmosphere.

Global warming is a very complex issue, the parameters of which are still unclear. Its effects on the earthly environment and the human habitat, however, are likely to be profound.

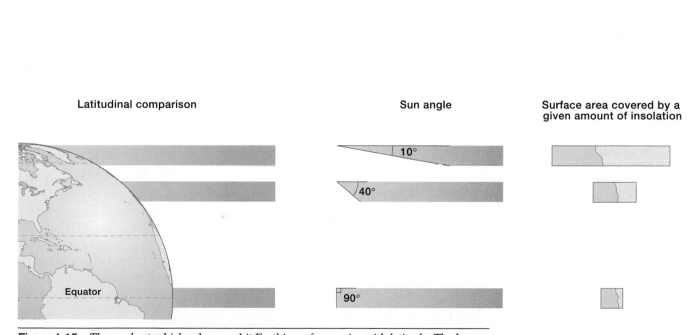

Figure 4-15 The angle at which solar rays hit Earth's surface varies with latitude. The larger the angle, the more concentrated the energy and therefore the more effective the heating.

Figure 4-16   The noon-time sun is very low in the sky during winter in the high latitudes. This is a November scene at the village of Mesters Vig in Greenland, at about 70 degrees north latitude. (Simon Fraser/Science Photo Library/Photo Researchers, Inc.)

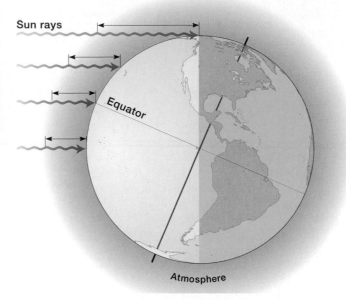

Figure 4-17   Atmospheric obstruction of sunlight. Low-angle rays must pass through more atmosphere than high-angle rays; thus the former are subject to more depletion through reflection, scattering, and absorption.

length. The conspicuous buildup of heat in summer in these regions is largely a consequence of the long hours of daylight, and the winter cold is a manifestation of limited insolation being received because of the short days.

Atmospheric Obstruction   We have already noted that clouds, particulate matter, and gas molecules in the atmosphere either absorb, reflect, or scatter insolation. The result is a reduction in the intensity of the energy received at Earth's surface. On the average, sunlight received at Earth's surface is only about half as strong as it is outside Earth's atmosphere.

This weakening effect varies from time to time and from place to place, depending on two factors: the amount of atmosphere the radiation has to pass through, as shown in Figure 4-17, and the transparency of that atmosphere (Figure 4-18). The distance a ray of sunlight travels through the atmosphere (commonly referred to as "path length") is determined by the angle of incidence. A large-angle ray (in other words, a nearly perpendicular ray) traverses a shorter course through the atmosphere than a small-angle one. A tangent ray (one having an incidence angle of zero) must pass through nearly 20 times as much atmosphere as a direct ray (one striking Earth at a 90° angle).

The effect of atmospheric obstruction on the distribution of solar energy at Earth's surface is to reinforce the pattern established by the varying angle of incidence. Solar radiation is more depleted of energy in the high latitudes than in the low latitudes; thus there are smaller losses of energy in the tropical atmosphere than in the polar atmosphere.

Latitudinal Radiation Balance   As the direct rays of the sun shift northward and southward across the equator during the course of the year, the belt of maximum solar energy swings back and forth through the tropics. Thus in the low latitudes, to about 28° N and 33° S, there is an energy surplus, with more incoming than outgoing radiation. In the latitudes north and south of these two parallels, there is an energy deficit, with more radiant loss than gain. The surplus of energy in low latitudes is directly related to the consistently large angle of incidence, and the energy deficit in high latitudes is associated with small angles.

Figure 4-19 shows the distribution of radiation around the world, and Figure 4-20 shows the same for the United States and parts of Canada, and Mexico. (The maps show the average receipt of langleys per day.) The variations are largely latitudinal, as is to be expected. The principal interruptions to the simple latitudinal pattern are based on the presence or absence of frequent cloud cover, where insolation is reflected, diffused, and scattered. In Figure 4-20, for example, it can be seen that radiation is greatest in the southwestern United States, where clouds are consistently sparse, and is least in the northwestern and northeastern corners of the country, where cloud cover is frequent.

Despite the variable pattern shown on the maps, there is a balance between incoming and outgoing radiation for the Earth/atmosphere complex as a whole; in other words, the net radiation balance for Earth is zero. The mechanisms for exchanging heat between

Figure 4-18 Sunrise on the island of Nauru in the central Pacific. Clouds and haze deplete the amount of insolation that reaches Earth's surface. (TLM photo)

the surplus and deficit regions involve the general circulation patterns of the atmosphere and oceans, which are discussed later in the book.

LAND AND WATER CONTRASTS

As mentioned above, the atmosphere is heated mainly by heat reradiated from Earth rather than by heat from the sun; thus the heating of Earth's surface is a primary control of the heating of the air above it. In order to comprehend variations in air temperatures, it is useful to understand how different kinds of surfaces react to solar energy. There is considerable variation in the absorbing and reflecting capabilities of the almost limitless kinds of surfaces found on Earth—soil, water, grass, trees, cement, rooftops, and so forth. Their varying receptivity to insolation in turn causes differences in the temperature of the overlying air.

By far the most significant contrasts are those between land and water surfaces. The generalization is that land heats and cools faster and to a greater degree than water (Figure 4-21 on page 90).

Heating   A land surface heats up more rapidly and reaches a higher temperature than a comparable water surface subject to the same insolation. In essence, a thin layer of land is heated to relatively high temperatures, whereas a thick layer of water is heated more slowly to moderate temperatures. There are several significant reasons for this difference:

1. Water has a higher **specific heat** than land. Specific heat is the amount of energy required to raise the temperature of 1 gram of a substance by 1°C. The specific heat of water is about five times as great as that of land, which means that water can absorb much more solar energy without its temperature increasing.
2. As mentioned above, sun rays penetrate water more deeply than they do land; that is, water is a better transmitter than land. Thus in water the heat is diffused over a much greater volume of matter, and maximum temperatures remain considerably lower than they do on land, where the heat is concentrated and maximum temperatures can be much higher.
3. Water is highly mobile, and so turbulent mixing and ocean currents disperse the heat both broadly and deeply. Land, of course, is immobile, and so heat is dispersed only by conduction (and land is a very poor conductor).
4. The unlimited availability of moisture on a water surface means that evaporation is much more prevalent than on a land surface. The latent heat needed for this evaporation is drawn from the water and its immediate surroundings, causing a drop in temperature. Thus the cooling effect of evaporation slows down any heat buildup on a water surface.

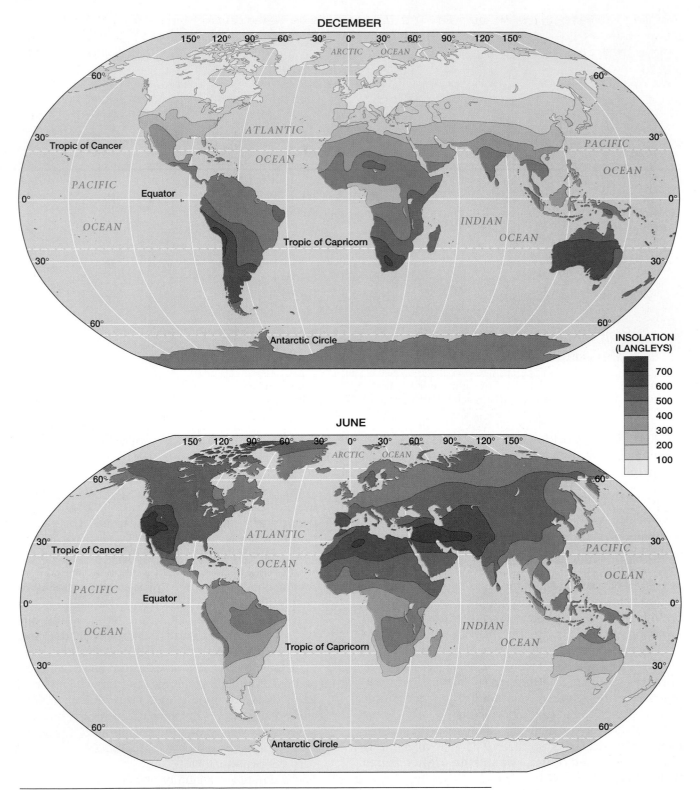

Figure 4-19 World distribution of average insolation in December and June. The pattern is determined mainly by latitude (note the low December levels in northern areas) and amount of cloudiness (note the high June levels in such cloud-free desert areas as the southwestern United States and southwestern Asia).

Cooling   When both are overlain by air at the same temperature, a land surface cools more rapidly and to a lower temperature than a water surface. During winter the shallow heated layer of land radiates its heat away quickly. Water loses its heat more gradually because the heat has been stored deeply and is brought only slowly to the surface for radiation. As the surface water cools, it sinks and is replaced by warmer upwellings from below. The entire water body must be cooled before the surface temperatures decrease significantly.

Implications   The significance of these contrasts between land and water heating and cooling rates is that both the hottest and coldest areas of Earth are found in the interiors of continents, distant from the influence of oceans. In the study of the atmosphere, probably no single geographic relationship is more important than the distinction between continental and maritime climates. A continental climate experiences greater seasonal extremes of temperature—hotter in summer, colder in winter—than a maritime climate.

These differences are shown in Figure 4-22, which portrays average monthly temperatures for San Diego and Dallas. These two cities are at approximately the same latitude and experience almost identical lengths of day and angles of incidence. Although their annual average temperatures are almost the same, the monthly averages vary significantly. Dallas, in the interior of the continent, experiences notably warmer summers and cooler winters than San Diego, which enjoys the moderating influence of an adjacent ocean.

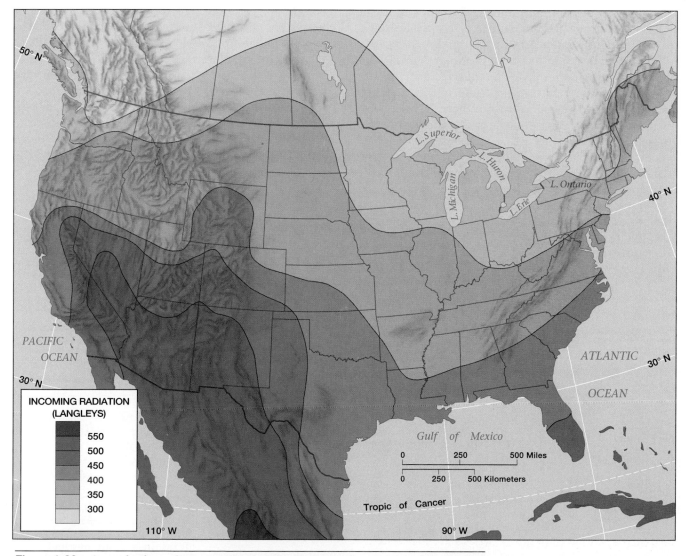

Figure 4-20   Annual solar radiation received in the 48 conterminous states of the United States and adjacent parts of Canada and Mexico. The sunny southwestern areas receive the most radiation; the cloudy northeastern and northwestern areas receive the least.

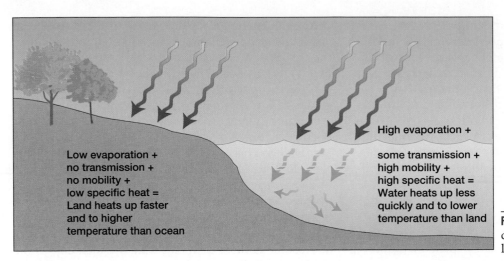

Low evaporation +
no transmission +
no mobility +
low specific heat =
Land heats up faster
and to higher
temperature than ocean

High evaporation +

some transmission +
high mobility +
high specific heat =
Water heats up less
quickly and to lower
temperature than land

Figure 4-21 Some contrasting characteristics of the heating of land and water.

The oceans, in a sense, act as great reservoirs of heat. In summer they absorb heat and store it. In winter they give off heat and warm up the air. Thus they function as a sort of global thermostatically controlled heat source, moderating temperature extremes.

The ameliorating influence of the oceans can also be demonstrated, on a totally different scale, by comparing latitudinal temperature variations in the Northern Hemisphere with those in the Southern Hemisphere. The former is often thought of as a "land hemisphere," because 39 percent of its area is land surface; the latter is a "water hemisphere," with only 19 percent of its area as land. Table 4-1 shows the average annual temperature range (difference in average temperature of the coldest and warmest months) for comparable parallels in each hemisphere. It is obvious that the land hemisphere has greater extremes.

TEMPERATURE: A MEASURE OF HEAT

Thus far we have referred to energy and heat interchangeably and have mentioned *temperature* several times without definition. It is important to clarify these terms and the concepts they represent.

Energy is the capacity to do work and can take various forms. **Heat** is one form of energy; it is associated with how fast the atoms in any solid, liquid, or gas are vibrating. The more heat energy any atom contains, the faster it vibrates. **Temperature** is an expression of the degree of hotness or coldness of a substance.[1]

---

[1]In this book, the adjectives *hot*, *cold*, *warm*, and *cool* are used frequently, and like other descriptive words that appear (*high, low, wet, dry, fast, slow*, to name a few), they are often used in a comparative sense; something is *hot* or *cold* in comparison with something else.

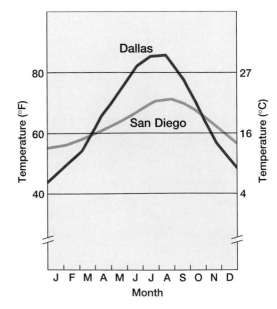

Dallas
Latitude: 32° 51' N
Annual average temperature: 65°F (18°C)

San Diego
Latitude: 32° 44' N
Annual average temperature: 63°F (17°C)

Figure 4-22 Annual temperature curves for San Diego, California, and Dallas, Texas. In both summer and winter, San Diego, situated on the coast, experiences milder temperatures than inland Dallas.

TABLE 4-1
Average Annual Temperature Range by Latitude, in degrees Celsius

Latitude	Northern Hemisphere	Southern Hemisphere
0	0	0
15	3	4
30	13	7
45	23	6
60	30	11
75	32	26
90	40	31

Source: From Frederick K. Lutgens and Edward J. Tarbuck, *The Atmosphere: An Introduction to Meteorology*, 4th ed. (Englewood Cliffs, NJ: Prentice Hall, 1989), p. 71. Used by permission of Prentice Hall.

Most people have a ready sensitivity to temperature. The nerves in our skin readily respond to variations in heat and cold and make us conscious of temperature changes. The sensitivity of our skin is not a reliable temperature indicator, however. Our nerves may be "misled" by other factors, such as the presence of moisture or air movement, factors that are quite separate from temperature but affect our perception of temperature. For example, on a cool day when a strong wind is blowing, we may perceive that the weather is colder than the temperature would indicate, simply because our body heat is dissipated so rapidly by the air movement. The term **sensible temperature** is applied to this phenomenon. It is the "sensation" of temperature that we feel in response to the total condition of the air around us; it may or may not be representative of the actual air temperature.

MECHANISMS OF HEAT TRANSFER

If there were not mechanisms for moving heat poleward in both hemispheres, the tropics would become progressively warmer until the amount of heat energy absorbed equaled the amount radiated from Earth's surface and the high latitudes would become progressively colder. Such temperature trends do not occur because there is a persistent shifting of warmth toward the high latitudes and the consequent cooling of the low latitudes. This shifting is accomplished by circulation patterns in the atmosphere and in the oceans. The broad-scale, or planetary, circulation of these two mediums moderates the buildup of heat in equatorial regions and the loss of heat in polar regions, thereby making both of those latitudinal zones more habitable than they would otherwise be. Both the atmosphere and the oceans act as enormous thermal engines, with their latitudinal imbalance of heat driving the currents of air and water, which, in turn, transfer heat and somewhat modify the imbalance.

ATMOSPHERIC CIRCULATION

Of the two mechanisms of global heat transfer, by far the more important is the general circulation of the atmosphere. Air moves in an almost infinite number of ways, but there is a broad planetary circulation pattern that serves as a general framework for moving warm air poleward and cool air equatorward. Some 75 to 80 percent of all horizontal heat transfer is accomplished by atmospheric circulation.

Our discussion of atmospheric circulation is withheld until Chapter 5, following consideration of some fundamentals concerning pressure and wind.

OCEANIC CIRCULATION

There is a close relationship between the general circulation patterns of the atmosphere and oceans. Various kinds of oceanic water movements are categorized as **currents,** and it is air blowing over the surface of the water that is the principal force driving the major ocean currents. In the other direction, the heat energy stored in the oceans has important effects on atmospheric circulation.

For our purposes in understanding heat transfer by the oceans, we are concerned primarily with the broad-scale surface currents that make up the general circulation of the oceans (Figure 4-23 on page 94). These major currents respond to changes in wind direction, but they are so broad and ponderous that the response time normally amounts to many months. In essence, ocean currents reflect average wind conditions over a period of several years, with the result that the major components of oceanic circulation are closely related to major components of atmospheric circulation.

The Basic Pattern   All the oceans of the world are interconnected. Because of the location of landmasses and the pattern of atmospheric circulation, however, it is convenient to visualize five relatively separate ocean basins—North Pacific, South Pacific, North Atlantic, South Atlantic, and South Indian. Within each of these basins, there is a similar pattern of surface current flow, based on a general similarity of prevailing wind patterns.

Despite variations based on the size and shape of the various ocean basins and on the season of the year, a single simple pattern of surface currents is character-

FOCUS

MEASURING TEMPERATURE

Instruments

The principle at work in a thermometer is that most substances expand when heated and contract when cooled. If this change in volume can be calibrated, it can provide a precise measurement of temperature changes.

Most thermometers are of the *liquid-in-glass* variety, in which a liquid is sealed in a glass tube that has an enlarged bulb at the bottom to store the liquid. The liquid, normally mercury or ethyl alcohol, expands much more than the glass when heated and contracts much more than the glass when cooled. Thus it rises or falls in the tube in response to temperature changes. The length of the column of liquid in the tube indicates the temperature of the surrounding air. The thermometer is calibrated at a few fixed temperatures (such as the freezing and boiling points of water), and then values between these extremes are interpolated.

A minimum thermometer records the lowest temperature in a given period (Figure 4-2-A). In this instrument, a small glass wire shaped like a tiny dumbbell (called an index) is placed in the column of alcohol (which is used instead of mercury because it is transparent and allows the index to be seen). The thermometer is kept horizontal. When the temperature decreases, the alcohol contracts and the tension in its concave surface is strong enough to drag the index down by the tension of its concave surface. When the temperature increases, the alcohol expands and rises in the tube. It does not pull the index with it because the concave surface does not generate enough tension in that direction. Thus, the index is left behind as clear evidence of the lowest temperature reached. After this value is read by the observer, the thermometer is reset by inverting it so that the index drops to the end of the column again.

A maximum thermometer has a narrow *constriction* in the glass tube just above the bulb. It also is kept horizontal. As the temperature rises, the mercury expands and is forced through the constriction one tiny drop at a time. When the temperature subsequently falls, however, nothing forces the mercury back through the constriction, and so the column cannot get shorter. The height of the column indicates the maximum temperature attained. The thermometer is reset by shaking or whirling it to force the mercury back into the bulb.

Temperature Scales

In the United States, three temperature scales are in concurrent use (Figure 4-2-B). Each permits a precise measurement, but the existence of three scales creates an unfortunate degree of confusion.

The Fahrenheit Scale   The scale most widely understood by the general public in the United States is the Fahrenheit scale (named after the eighteenth-century German physicist who devised it, Gabriel Daniel Fahrenheit). Public weather reports from the National Weather Service and the news media usually state temperatures in degrees Fahrenheit. The reference points on this scale are the sea-level freezing and boiling points of pure water, which are 32° and 212°, respectively. The United States is one of only a few countries that still use the Fahrenheit scale.

The Celsius Scale   In most other countries, the Celsius scale (named for the eighteenth-century Swedish astronomer who devised it, Anders

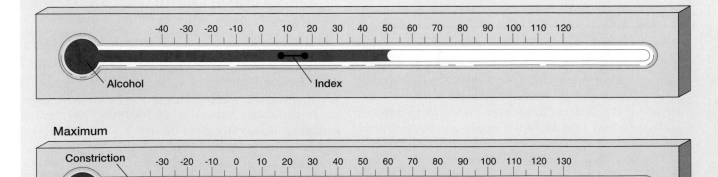

Figure 4-2-A   Minimum and maximum thermometers.

Celsius) is used either exclusively or predominantly. It is an accepted component of the International System of measurement (SI) because it is a decimal scale with 100 units (degrees) between the freezing and boiling points of water. The Celsius scale has long been used for scientific work in the United States and is now slowly being established to supersede the Fahrenheit scale in all usages in this country.

To convert from degrees Celsius to degrees Fahrenheit, the following formula is used:

$$\text{degrees Fahrenheit} = 9/5 \times (\text{degrees Celsius} + 32°)$$

To convert from degrees Fahrenheit to degrees Celsius, the formula is

$$\text{degrees Celsius} = 5/9 \times (\text{degrees Fahrenheit} - 32°)$$

The Kelvin Scale　For many scientific purposes, the Kelvin scale (named for the nineteenth-century British physicist William Thomson, Lord Kelvin) has long been used because it measures what are called absolute temperatures, which means that the scale begins at absolute zero, the theoretical point at which there is no temperature. The scale maintains a 100° range between the boiling and freezing points of water but has no negative values. Because this scale is not normally used by climatologists and meteorologists, we ignore it in this book except to compare it to the Fahrenheit and Celsius scales. On the Celsius scale, absolute zero is at –273°, and so the conversion is simple:

$$\text{degrees Celsius} = \text{degrees Kelvin} - 273°$$

$$\text{degrees Kelvin} = \text{degrees Celsius} + 273°$$

Temperature Data

At thousands of locations (called *stations*) throughout the world, temperature data are recorded on a regular basis. These data provide important raw material for contemporary weather reports and forecasts as well as for long-run climatic analyses.

Proper placing of the thermometer is necessary to obtain meaningful temperature readings. All official temperatures are taken in the shade so that what is measured is air temperature and not solar radiation. Good ventilation is also needed because thermometers sheltered from freely moving air do not indicate the true air temperature. Therefore, most official thermometers are mounted in an instrument shelter, consisting of a white

Figure 4-2-C　A weather instrument shelter. It is placed a few feet above the ground, is painted white so that it reflects direct sunlight, and has slatted sides to permit air circulation. (Dr. E. R. Degginger)

box with louvered sides that shields the instruments from sunshine and precipitation while permitting air circulation (Figure 4-2-C). Shelters are placed as far from buildings as practical, usually over grass, and mounted about 3 feet (1 meter) above the ground.

At some stations, the temperature is recorded every hour by a human observer, but more often a *thermograph* (a continuously recording thermometer) is in operation. At most stations only the highest and lowest temperatures are recorded for each 24-hour period, generally by means of maximum–minimum thermometers. The *daily mean temperature* is then calculated by averaging the 24 hourly readings or is estimated from the average of the maximum and minimum. From these data can also be calculated the *daily temperature range*, which is the number of degrees difference between the highest and lowest recorded values of the 24-hour period.

The *monthly mean temperature* for each station is determined by averaging the daily means for each day in the calendar month. The *annual mean temperature* represents the average of the 12 monthly means. The difference between the means of the warmest and coldest months is the *annual temperature range*.

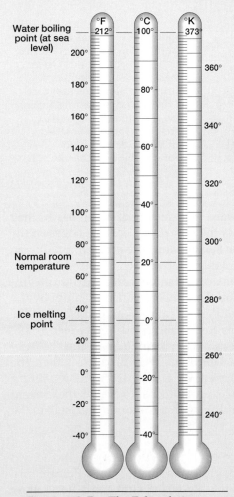

Figure 4-2-B　The Fahrenheit, Celsius, and Kelvin temperature scales.

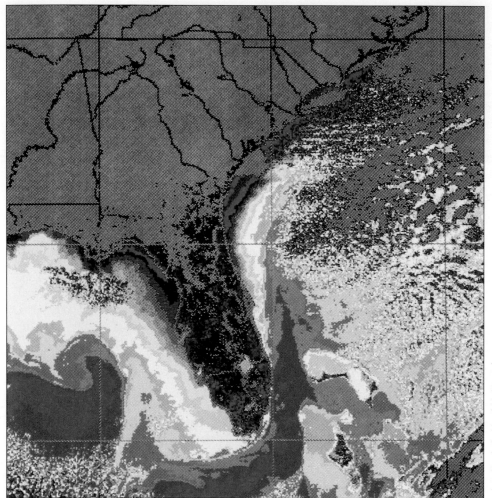

Figure 4-23 Warm water issuing from the Gulf of Mexico into the North Atlantic Ocean. The warm Gulf Stream flows northward along the east coast of the United States. The image was gathered by the NOAA 11 satellite in January of 1994. (NOAA/Science Photo Library/Photo Researchers, Inc.)

istic of all the basins (Figure 4-24). It consists of a series of enormous elliptical loops elongated east-west and centered approximately at 30° of latitude (except in the Indian Ocean, where it is centered on the equator). These loops, called **gyres**, flow clockwise in the Northern Hemisphere and counterclockwise in the Southern Hemisphere.

On the equatorward side of each subtropical (this word means "bordering on the tropics") gyre is the Equatorial Current, which moves steadily toward the west. The two equatorial currents have an average position 5° to 10° north or south of the equator, and are separated by the Equatorial Countercurrent, which is an east-moving flow approximately along the equator in each ocean. The Equatorial currents feed the Equatorial Contercurrent near its western margin in each basin. Water from the Equatorial Countercurrent in turn drifts poleward to feed the Equatorial Current near the eastern end of its path.

Near the western margin of each ocean basin, the general current curves poleward. As these currents ap-

proach the poleward margins of the ocean basins they curve east, and as they reach the eastern edges of the basins they curve back toward the equator, producing an incompletely closed loop in each basin.

The movement of these currents, although impelled partially by the wind, is caused mainly by the deflective force of Earth's rotation, which is called the *Coriolis effect*. This force (discussed in Chapter 5) dictates that the ocean currents are deflected to the right in the Northern Hemisphere and to the left in the Southern Hemisphere. A glance at the basic pattern (Figures 4-25, 4-26, and 4-27) shows that the current movement around the gyres responds precisely to the Coriolis effect.

Northern and Southern Variations In the two Northern Hemisphere basins—North Pacific and North Atlantic—the bordering continents lie so close together at the poleward basin margin that the bulk of the current flow is prevented from entering the Arctic Ocean. This effect is more pronounced in the Pacific

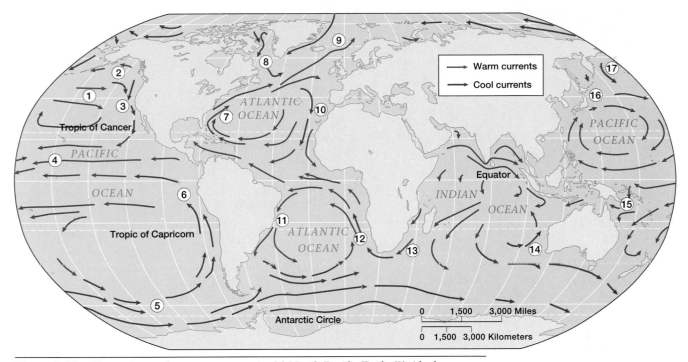

Figure 4-24 The major surface ocean currents: (1) North Pacific Drift; (2) Alaska Current; (3) California Current; (4) Equatorial Current; (5) West Wind Drift; (6) Humboldt Current; (7) Gulf Stream; (8) Labrador Current; (9) North Atlantic Drift; (10) Canaries Current; (11) Brazil Current; (12) Benguela Current; (13) Agulhas Current; (14) West Australian Current; (15) East Australian Current; (16) Kuroshio (Japan Current); (17) Oyashio (Kamchatka Current).

than in the Atlantic. The North Pacific has only very limited flow northward between Asia and North America, whereas in the North Atlantic a larger proportion of the flow escapes northward between Greenland and Europe.

In the Southern Hemisphere, the continents are far apart. Thus the poleward parts of the gyres in the South Pacific, South Atlantic, and South Indian oceans are connected as one continuous flow in the uninterrupted belt of ocean that extends around the world in the vicinity of latitude 60° S (Figure 4-24). This circumpolar flow is called the West Wind Drift.

Current Temperatures   Of utmost importance to our understanding of latitudinal heat transfer are the temperatures of the various currents. Each major current can be characterized as warm or cool relative to the surrounding water at that latitude. The generalized temperature characteristics are as follows:

1. Low-latitude currents (Equatorial Current, Equatorial Countercurrent) have warm water.
2. Poleward-moving currents on the western sides of ocean basins carry warm water toward higher latitudes.

3. Northern components of the Northern Hemisphere gyres carry warm water toward the north and east.
4. Southern components of the Southern Hemisphere gyres (generally combined into the West Wind Drift) are strongly influenced by Antarctic waters and are essentially cool.
5. Equatorward-moving currents on the eastern sides of ocean basins carry cool water toward the equator.

In summary, the general circulation of the oceans is a poleward flow of warm tropical water along the western edge of each ocean basin and an equatorward movement of cool high-latitude water along the eastern margin of each basin.

Rounding Out the Pattern   Two other relatively minor aspects of oceanic circulation are influential in heat transfer:

1. The northwestern portions of Northern Hemisphere ocean basins receive an influx of cool water from the Arctic Ocean (Figure 4-28 on page 98). A prominent cool current from the vicinity of Greenland comes southward along the Canadian coast, and a smaller

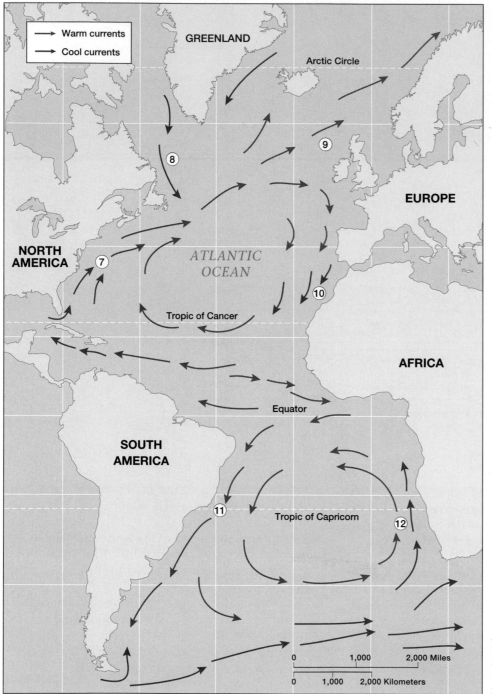

Warm currents
Cool currents

GREENLAND

Arctic Circle

⑧
⑨

EUROPE

NORTH
AMERICA

⑦

*ATLANTIC
OCEAN*

Tropic of Cancer

⑩

AFRICA

Equator

SOUTH
AMERICA

⑪
Tropic of Capricorn

⑫

0   1,000   2,000 Miles

0   1,000   2,000 Kilometers

Figure 4-25   The major surface
currents of the Atlantic Ocean
(see Figure 4-24 for current
names).

flow of cold water issues from the Bering Sea southward along the coast of Siberia to Japan.

2. Wherever an equatorward-flowing cool current
pulls away from a subtropical western coast, a
pronounced and persistent upwelling of cold
water occurs. This is most striking off South
America but is also notable off North America,
northwestern Africa, and southwestern Africa. It
is much less developed off the coast of western
Australia.

VERTICAL TEMPERATURE PATTERNS

As we study the geography of climate—or the geography of anything else, for that matter—most of our attention is directed to the horizontal dimension; in
other words, we are concerned with the spatial distribution of phenomena over the surface of Earth. From
time to time, however, we must focus on vertical patterns or processes because of their influence on surface
features. This statement is particularly pertinent with
regard to climate.

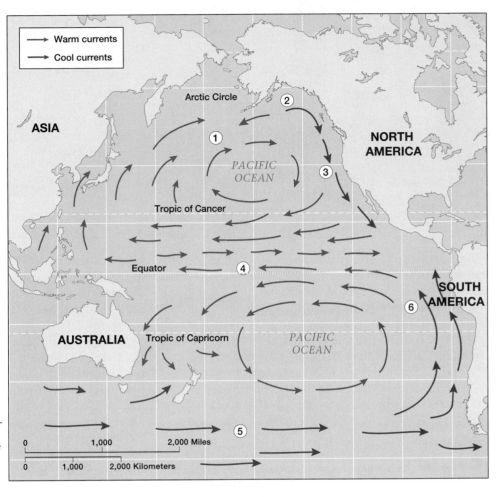

Figure 4-26 The major surface currents of the Pacific Ocean (see Figure 4-24 for current names).

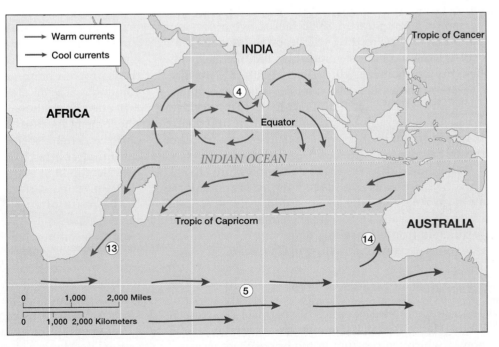

Figure 4-27 The major surface currents of the Indian Ocean (see Figure 4-24 for current names).

Figure 4-28   Cold-water movement in ocean basins of the Northern Hemisphere.

Temperature in the troposphere is relatively predictable. As we learned in Chapter 3, throughout the troposphere, under normal conditions, there is a general decrease in temperature with increasing altitude. However, there are many exceptions to this general statement. Indeed, the rate of vertical temperature decline can vary according to season, time of day, amount of cloud cover, and a host of other factors. In some cases, there is even an opposite trend, with the temperature increasing upward for a limited distance.

LAPSE RATE

The rate at which temperature decreases with height is variable, particularly in the lowest few hundred feet of the troposphere, but the normal expectable rate is about 3.6 °F per 1000 feet (6.5 °C per kilometer). This is the average **lapse rate,** or normal vertical temperature gradient. The lapse rate tells us that, if a thermometer measures a temperature 1000 feet above a previous measurement, the reading will be, on the average, 3.6 °F cooler; if the second measurement is 1000 feet below the first, the thermometer will register about 3.6 °F warmer.

Determining the lapse rate of a column of air involves measuring air temperature at various elevations. Then a graph of temperature as a function of height is drawn in order to get a temperature profile of the air column.

When measuring a lapse-rate temperature change, only the thermometer is moved; the air is at rest. If the air is moving vertically, expansion or contraction will cause an adiabatic temperature change, which is totally different from a lapse-rate temperature change. This concept is explored more fully in Chapter 6.

TEMPERATURE INVERSIONS

The most prominent exception to a normal lapse-rate condition [Figure 4-29(a)] is a **temperature inversion**, a situation in which temperature in the troposphere increases, rather than decreases, with increasing altitude. Inversions are relatively common in the troposphere but are usually of brief duration and restricted depth. They can occur near Earth's surface, as in Figure 4-29(b), or at higher levels, as in Figure 4-29(c).

Inversions influence weather and climate. As we shall see in Chapter 6, an inversion inhibits vertical air movements and greatly diminishes the possibility of precipitation. Inversions also contribute significantly to increased air pollution because they create stagnant air conditions that greatly limit the natural upward dispersal of urban/industrial pollutants.

Surface Inversions   The most readily recognizable inversions are those found at ground level (Figure 4-30). These are usually classified as **radiational inversions** because they result from rapid radiational cooling. They occur typically on a long, cold winter night when a land surface (an efficient radiator) rapidly emits long-wave radiation into a clear, calm sky. The ground is soon colder than the air above it and so now cools the air by conduction. In a relatively short time, the lowest few hundred feet of the troposphere become colder than the air above and a temperature inversion is in effect. Radiational inversions are primarily winter phenomena because there is only a short daylight period for incoming solar heating and a long night for radiational cooling. They are therefore much more prevalent in high latitudes than elsewhere.

An inverted surface temperature gradient can also be the result of an **advectional inversion**, in which there is a horizontal inflow of cold air into an area. This condition commonly is produced by cool maritime air blowing into a coastal locale [Figure 4-31(a)]. Advectional inversions are usually short-lived (typically overnight) and shallow. They may occur at any time of year, depending on the location of the relatively cold surface and on wind movement.

Another type of surface inversion results when cooler air slides down a slope into a valley, thereby displac-

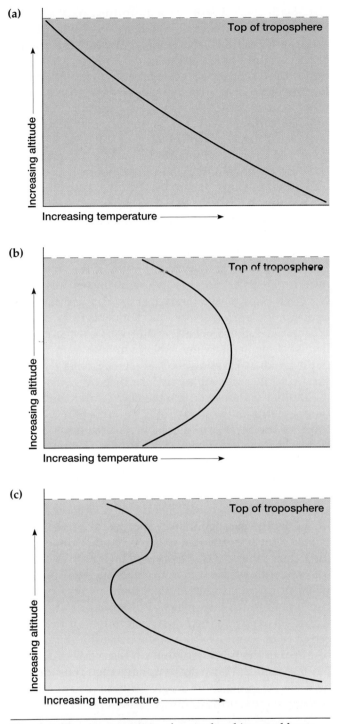

(a)

Increasing altitude

Increasing temperature →

Top of troposphere

(b)

Increasing altitude

Increasing temperature →

Top of troposphere

(c)

Increasing altitude

Increasing temperature →

Top of troposphere

Figure 4-29 A comparison of normal and inverted lapse rates. **(a)** Tropospheric temperature normally decreases with increasing altitude. **(b)** In a surface inversion, temperature increases with increasing altitude from ground level to some distance above the ground. **(c)** In an upper-air inversion, temperature first decreases with increasing altitude as in a normal lapse rate but then at some altitude well below the tropopause begins to increase with increasing altitude.

Figure 4-30 A typical January day in Mexico City, which has superseded Los Angeles as the smog capital of the world. (Wesley Bocxe/Photo Researchers, Inc.)

ing slightly warmer air [Figure 4-31(b)]. This fairly common occurrence during winter in some midlatitude regions is called a **cold-air-drainage inversion**.

Upper-Air Inversions Temperature inversions well above the ground surface nearly always are the result of air sinking from above. These **subsidence inversions** are usually associated with high-pressure conditions, which are particularly characteristic of subtropical latitudes throughout the year and of Northern Hemisphere continents in winter. A subsidence inversion can be fairly deep (sometimes several thousand feet), and its base is usually a few thousand feet above the ground, as low-level turbulence prevents it from sinking lower.

GLOBAL TEMPERATURE PATTERNS

The goal of this and the two succeeding chapters is to examine the world distribution pattern of climate. With the preceding pages as background, we now turn our attention to the worldwide distribution of temperature, the first of the four climate elements.

Maps of global temperature patterns display seasonal extremes rather than annual averages. January and July are the months of lowest and highest temperatures for most places on Earth, and so maps portraying the average temperatures of these two months provide a simple but meaningful expression of thermal conditions in winter and summer (Figures 4-32 and 4-33 on page 102). Temperature distribution is shown by means of **isotherms**, lines joining points of equal temperature. Temperature maps are based on monthly averages, which are based on daily averages; the maps do not show the maximum daytime heating or the

(a)

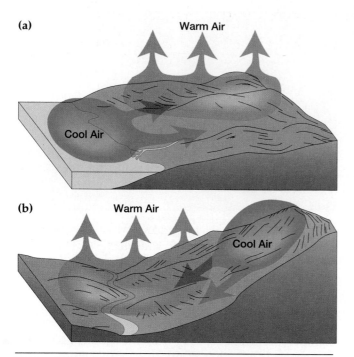

(b)

Figure 4-31   **(a)** Formation of an advectional inversion. **(b)** Formation of a cold-air-drainage inversion.

maximum nighttime cooling. Although the maps are on a very small scale, they permit a broad understanding of temperature patterns for the world.

PROMINENT CONTROLS OF TEMPERATURE

Gross patterns of temperature are controlled largely by four factors—altitude, latitude, land/water contrasts, and ocean currents.

Altitude   Because temperature responds sharply to altitudinal changes, it would be misleading to plot actual temperatures on a temperature map, as high-altitude stations would almost always be colder than low-altitude stations. The complexity introduced by hills and mountains would make the map more complicated and difficult to comprehend. Consequently, the data for most maps displaying world temperature patterns are modified by reducing the temperature to what it would be if the station were at sea level. This is done most simply by using the average lapse rate, a method that produces artificial temperature values but eliminates the complication of terrain differences. Maps plotted in this way are useful in showing world patterns, but they are not satisfactory for indicating actual temperatures for locations that are not close to sea level.

Latitude   Clearly the most conspicuous feature of any world temperature map is the general east-west trend of the isotherms, roughly following the paral-

lels of latitude. If Earth had a uniform surface and did not rotate, the isotherms probably would coincide exactly with parallels, showing a progressive decrease of temperature poleward from the equator. However, Earth does rotate, and it has ocean waters that circulate and land that varies in elevation. Consequently, there is no precise temperature correlation with latitude. Nevertheless, the fundamental cause of temperature variation the world over is insolation, which is governed primarily by latitude, and the general temperature patterns reflect latitudinal control.

Land/Water Contrasts   The different heating and cooling characteristics of land and water are also reflected conspicuously on a temperature map. Summer temperatures are higher over the continents than over the oceans, as shown by the poleward curvature of the isotherms over continents in the respective hemispheres (July in the Northern Hemisphere; January in the Southern). Winter temperatures are lower over the continents than over the oceans; the isotherms bend equatorward over continents in this season (January in the Northern Hemisphere; July in the Southern). Thus at both seasons, isotherms make greater north-south shifts over land than over water.

Another manifestation of the land/water contrast is the regularity of the isothermal pattern in the midlatitudes of the Southern Hemisphere, in contrast to the situation in the Northern Hemisphere. There is very little land in these Southern Hemisphere latitudes, and so contrasting surface characteristics are absent.

Ocean Currents   Some of the most obvious bends in the isotherms occur in near-coastal areas of the oceans, where prominent warm or cool currents reinforce the isothermal curves caused by land/water contrasts. Cool currents deflect isotherms equatorward, whereas warm currents deflect them poleward. Cool currents produce the greatest isothermal bends in the warm season: Note the January situation off the western coast of South America and the southwestern coast of Africa or the July conditions off the western coast of North America. Warm currents have their most prominent effects in the cool season: Witness the isothermal pattern in the North Atlantic Ocean in January.

SEASONAL PATTERNS

Apart from the general east-west trend of the isotherms, probably the most conspicuous feature of Figures 4-32 and 4-33 is the latitudinal shift of the isotherms from one map to the other. The isotherms follow the changing balance of insolation during the course of the year, moving northward from January to July and returning southward from July to January. Note, for example, the 50°F isotherm in southernmost

PEOPLE AND THE ENVIRONMENT

URBAN HEAT ISLANDS

In many parts of the world the population is largely urban, and in virtually every country there has been a long-term continuing increase in both the number of cities and their areal expansion. The proliferation and spread of cities have produced the most profound landscape modifications created by humans. Changes in topography, vegetation, hydrography, and soils are obvious and conspicuous. It is less apparent that the atmosphere is also affected by urban development; many aspects of weather and climate are significantly different in cities than in adjacent rural areas. Some of these effects have been recognized for the better part of two centuries—which is to say, since the beginnings of the industrial revolution accelerated the trend toward urbanization.

Compared with rural areas, cities have lower relative humidity, lower wind speeds, and higher thunderstorm frequency, cloudiness, fog, haze, particulate matter in the air, and temperature.

One of the most important and interesting aspects of human-induced climatic modification in cities is the concept of an **urban heat island**, which means simply that city temperatures are higher than in the surrounding rural lands. The heat island exists both in summer and in winter and is characteristic of both daytime and nighttime. It varies from a fraction of a degree to as much as 15°F (9°C). The heat differential is usually greatest in the evening because the harder surface materials in cities continue to radiate heat long after dark. Central business districts, with their massive buildings and sparse greenery, show the greatest temperature differential, but suburban areas clearly demonstrate the heat island effect. Often there is an abrupt drop in the temperature profile where suburb gives way to rural landscape (Figure 4-3-A).

The reasons for the development of a heat island are various; they are both passive and active:

1. Urban building materials—asphalt, concrete, bricks, and so forth—absorb and store more heat than do the "softer" surfaces of rural areas. The temperatures of the former rise more rapidly and reach a higher level.

Conversely, at night there is more heat to release by terrestrial radiation.

2. Much heat is trapped in or only slowly filtered out of the city because the vertical surfaces of buildings absorb, reflect, and reradiate among themselves as well as to ground level.

3. Urban building materials hold no water, and runoff after a rain rapidly dissipates. Moreover, there is a sparser vegetative cover, which means less transpiration. These conditions combine to ensure that evaporative cooling is at a minimum in cities.

4. The greenhouse effect is stronger over cities because there is a thicker blanket of particulate matter to absorb long-wave radiation from below and reradiate it downward.

5. A very important active factor in the creation of a heat island is the burning of fossil fuels in buildings and vehicles. This effect is most notable in winter, when the quantity of heat produced from combustion may exceed the amount of insolation reaching the ground.

Figure 4-3-A Temperature in and around an urban heat island. The air over the urban section is warmer than the air over the suburbs, and there is abrupt cooling at the suburban-rural boundary. (After William M. Marsh and Jeff Dozier, *Landscape: An Introduction to Physical Geography.* Reading, MA: Addison-Wesley, 1981, p. 63. By permission of the publisher.)

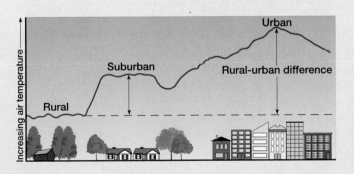

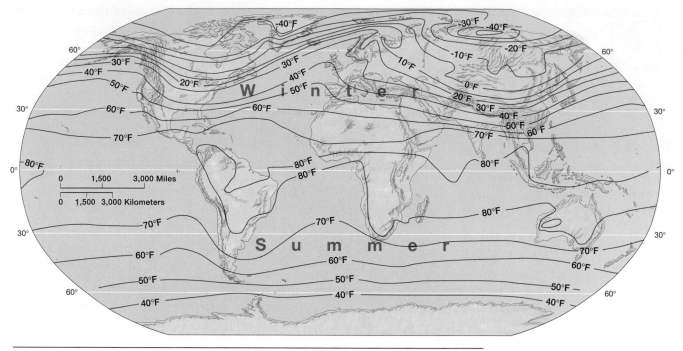

Figure 4-32   Average January sea-level temperatures in degrees Fahrenheit.

South America: in January (midsummer) it is positioned at the southern tip of the continent, whereas in July (midwinter) it is shifted considerably to the north. This isothermal shift is much more pronounced in high latitudes than in low and also much more pronounced over the continents than over the oceans. Thus tropical areas, particularly tropical oceans, show relatively small displacement of the isotherms from January to July, whereas over middle- and high-latitude landmasses an isotherm may migrate northward or southward more than 2,500 miles (4000 kilometers)—some 14° of latitude—as demonstrated in Figure 4-34.

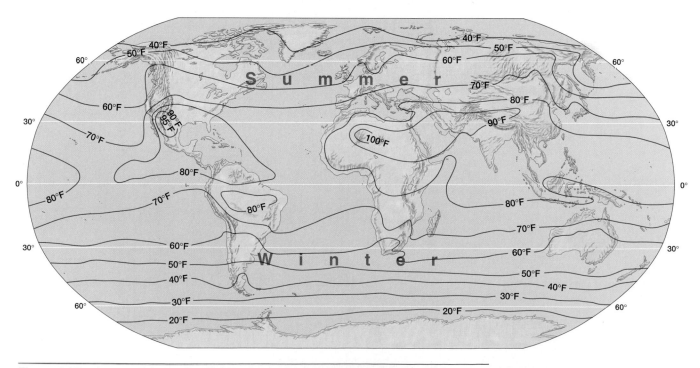

Figure 4-33   Average July sea-level temperatures in degrees Fahrenheit.

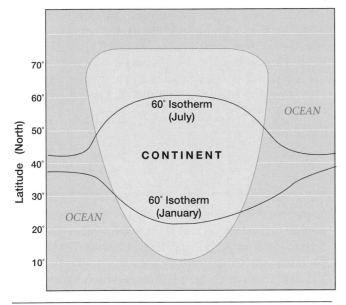

Figure 4-34 Idealized seasonal migration of the 60°F isotherm over a hypothetical Northern Hemisphere continent. The latitudinal shift is greatest over the interior of the continent and least over the adjacent oceans. For example, over the western ocean, the isotherm moves only from latitude 38°N in January to 42°N in July, but over the continent the change is from 22°N in January all the way to 60°N in July.

Isotherms are also more tightly packed in winter. This close line spacing indicates that the temperature gradient (rate of temperature change with horizontal distance) is steeper in winter than in summer, which,

in turn, reflects the greater contrast in radiation balance in winter. The temperature gradient is also steeper over continents than over oceans.

The coldest places on Earth are over landmasses in the higher latitudes. During July, the polar region of Antarctica is the dominant area of coldness. In January, the coldest temperatures occur many hundreds of miles south of the North Pole, in subarctic portions of Siberia, Canada, and Greenland. The principle of greater cooling of land than water is clearly demonstrated.

The highest temperatures also are found over the continents. The locations of the warmest areas in summer, however, are not equatorial. Rather they are in subtropical latitudes, where descending air maintains clear skies most of the time, allowing for almost uninterrupted insolation. Frequent cloudiness precludes such a condition in the equatorial zone. Thus the highest July temperatures occur in northern Africa and in the southwestern portions of Asia and North America, whereas the principal areas of January heat are in subtropical parts of Australia, southern Africa, and South America.

Average annual temperatures are highest in equatorial regions, however, because these regions experience so little winter cooling. Subtropical locations cool substantially on winter nights, and so their annual average temperatures are lower.

The ice-covered portions of the Earth—Antarctica and Greenland—remain quite cold throughout the year.

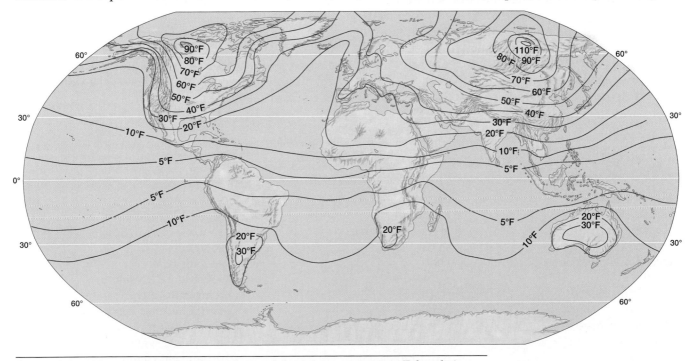

Figure 4-35 The world pattern of average annual temperature range in Fahrenheit degrees. The largest ranges occur in the interior of high-latitude landmasses.

ANNUAL TEMPERATURE RANGE

Another map useful in understanding the global pattern of air temperature is one that portrays the average annual range of temperatures (Figure 4-35). Annual temperature range is the difference between the average temperatures of the warmest and coldest months (normally July and January). The data are portrayed on the map by isolines that resemble isotherms.

Enormous seasonal variations in temperature occur in the interiors of high-latitude continents, and continental areas in general experience much greater ranges than do equivalent oceanic latitudes. At the other extreme, the average temperature fluctuates only slightly from season to season in the tropics, particularly over tropical oceans.

CHAPTER SUMMARY

The sun is the only important source of energy for Earth and its atmosphere. Radiant energy flows as electromagenetic waves from sun to Earth, mostly as shortwaves. In contrast, outgoing terrestrial radiation is mostly long waves.

The atmosphere is heated by a complex suite of processes set in motion by the arrival of insolation. Radiation, absorption, reflection, scattering, transmission, conduction, and convection all play important roles in the receipt and transfer of heat in Earth's atmosphere. Any vertical movement of air produces adiabatic cooling or heating, and any change in state of atmospheric moisture involves the storage or release of latent heat.

Most insolation received by the atmosphere does not contribute directly to its heating. Only about 22 percent of this solar energy is absorbed directly by the atmosphere, another 33 percent is reflected back out to space, and the remaining 45 percent is absorbed by the ground and then reradiated or conducted into the atmosphere or else bound up as latent heat in evaporated ocean water.

Air over landmasses heats and cools faster and to a greater degree than that over oceanic areas, which means that continental climates experience greater seasonal temperature variations than maritime climates.

Heat is a form of energy that is produced by molecular collision. Temperature is an expression of the degree of hotness or coldness. It is measured by a simple instrument called at thermometer, although there are several different scales of measurement.

Because the sun is high in the sky in low latitudes, there is a significant surplus of energy in the tropics and a corresponding deficit in polar regions. However, the general circulation patterns of both the atmosphere and the oceans function to ameliorate this global temperature imbalance by transferring heat latitudinally.

In the troposphere, there is usually a general decrease in temperature with increasing altitude, the average lapse rate being 3.6°F per 1000 feet (6.5°C per kilometer). The most notable exception to normal lapse-rate conditions involves temperature inversions of one sort or another.

The worldwide temperature pattern is broadly controlled by latitude, altitude, land/water contrasts, and surface ocean currents. The coldest places are over landmasses in high latitudes, whereas the warmest locales are over subtropical continents.

KEY TERMS

absorption
adiabatic cooling
adiabatic warming
advection
advectional inversion
albedo
angle of incidence
blackbody
cold-air-drainage inversion
condensation
conduction
convection
current

energy
evaporation
greenhouse effect
gyre
heat
infrared wave
isotherm
langley
lapse rate
latent heat
radiation
radiational inversion
reflection

scattering
sensible temperature
shortwave radiation
solar constant
specific heat
subsidence inversion
temperature
temperature inversion
transmission
ultraviolet wave
urban heat island

REVIEW QUESTIONS

1. Why is the sun so dominant a source of energy for Earth?
2. Why is so little solar energy received by Earth?
3. Distinguish between long-wave and shortwave radiation.
4. Explain the concept of the solar constant.
5. Explain the concept of a blackbody.
6. Explain the difference between conduction and transmission; between absorption and reflection.
7. What is the importance of the so-called greenhouse effect to the heating of Earth's surface?
8. "The atmosphere is mostly heated directly from the Earth rather than the sun." Comment on the validity of this statement.
9. What is the albedo?
10. Why do high latitudes receive less insolation than low latitudes?
11. Explain why land heats up faster than water.
12. Why do continental climates experience greater temperature extremes than maritime climates?
13. Describe the pattern of flow of major surface ocean currents in a typical ocean.
14. Is the temperature lapse rate of the troposphere constant? Explain.
15. What is the difference between a radiational inversion and an advective inversion?
16. Why are the hottest parts of Earth not in equatorial regions?
17. Explain the concept of urban heat island.

SOME USEFUL REFERENCES

ABRAHAMSON, DEAN E., (ed.), *The Challenge of Global Warming.* Washington, DC: Island Press, 1989.

ERT OLIN, B. R. DOOS, JILL JAGER, AND RICHARD A. WARRICK, *The Greenhouse Effect, Climatic Change, and Ecosystems.* New York: John Wiley & Sons Inc., 1986.

BALLING, ROBERT C., JR., *The Heated Debate: Greenhouse Predictions Versus Climatic Reality.* San Francisco: Pacific Research Institute, 1992.

EDGERTON, LYNNE T., *Global Warming and World Sea Levels.* Washington, DC: Island Press, 1991.

FALK, JIM, AND ANDREW BROWNLOW, *The Greenhouse Challenge: What's To Be Done?* Ringwood, Australia: Penguin Books Australia, 1988.

LUTGENS, F. K., AND E. J. TARBUCK, *The Atmosphere: An Introduction to Meteorology,* 4th ed. Englewood Cliffs, NJ: Prentice Hall, 1989.

OKE, T. R., *Boundary Layer Climates,* 2nd ed. New York: Methuen, 1987.

SCHNEIDER, STEPHEN H., *Global Warming: Are We Entering the Greenhouse Century?* San Francisco: Sierra Club Books, 1989.

JOEL B. SMITH AND DENNIS A. TIRPAK (EDS.), *The Potential Effects of Global Climate Change on the United States.* Washington, DC: Hemisphere Publishing, 1990.

WUEBBLES, DONALD L., AND JAE EDMUNDS, *Primer on Greenhouse Gases.* New York: Lewis Publishers, 1991.

YOUNG, L. B., *Sowing the Wind: Reflections on the Earth's Atmosphere.* Englewood Cliffs, NJ: Prentice Hall, 1990.

5

ATMOSPHERIC PRESSURE AND WIND

To THE LAYPERSON, ATMOSPHERIC PRESSURE IS THE most difficult climate element to comprehend. The other three—temperature, wind, and moisture—are more readily understood because our bodies are much more sensitive to them. We can "feel" heat, air movement, and humidity, and we are quick to recognize variations in these elements. Pressure, on the other hand, is a phenomenon of which we are usually unaware; its variations are considerably less noticeable to our senses. Pressure usually impinges on our sensitivity only when we experience rapid vertical movement, as in an elevator or an airplane. This quick rising creates an unpleasant sensation because of the difference in pressure inside and outside our ears, which is sometimes relieved only by "popping" them.

Despite its inconspicuousness, pressure is an important feature of the atmosphere. It is tied closely to the other weather elements, acting on them and responding to them. Pressure has an intimate relationship with wind: spatial variations in pressure are responsible for air movements. Hence, pressure and wind are often discussed together, as is done in this chapter.

THE IMPACT OF PRESSURE AND WIND ON THE LANDSCAPE

The influence of atmospheric pressure on the landscape is significant but indirect; this influence is manifested mostly by wind and temperature as these two elements respond to pressure changes. The impact of wind is more direct and explicit than the impact of temperature because wind has the energy to transport solid particles in the air and thus has a visible component to its activity. Vegetation may bend in the wind, and loose material of whatever kind may be shifted from one place to another. The results are nearly always short run and temporary, though, and usually have no lasting effect on the landscape except at a time of severe storm. Nevertheless, pressure and wind are major elements of weather and climate, and their interaction with other atmospheric components and processes cannot be overestimated.

THE NATURE OF ATMOSPHERIC PRESSURE

Gas molecules, unlike those of a solid or a liquid, are not strongly bound to one another. Instead, they are in continuous motion, colliding frequently with one an-

other and with any surfaces to which they are exposed. Consider a container in which a gas is confined, as in Figure 5-1. The molecules of the gas zoom around inside the container and collide again and again with the walls. The **pressure** of the gas is defined as the force the gas exerts on some specified area of the container walls.

The atmosphere is made up of gases, of course, and so **atmospheric pressure** is the force exerted by the gas molecules on some area of Earth's surface or on any other body, including yours. At sea level, the pressure exerted by the atmosphere is about 14.7 pounds per square inch (slightly more than 1 kilogram per square centimeter) (Figure 5-2). This value drops with increasing altitude because the farther away you get from Earth and its gravitational pull, the fewer gas molecules are present in the atmosphere.

The atmosphere exerts pressure on every solid or liquid surface it touches. The pressure is omnidirectional, which means it is exerted equally in all directions—up, down, sideways, and obliquely. In other words, atmospheric pressure is not simply a "weight" from above. This means that every square inch of any exposed surface—animal, vegetable, or mineral—at sea level is subjected to that much pressure. We are not sensitive to this ever-present burden of pressure because the fluids and cells of our bodies contain air at the same pressure; in other words, there is an exact balance between outward pressure and inward pressure.

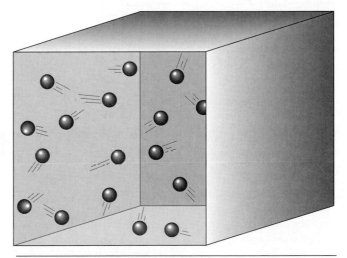

Figure 5-1 Gas molecules are always in motion. In this closed container they bounce around, colliding with one another and with the walls of the container. These collisions give rise to what is called the pressure exerted by the gas.

A windy sail in Charleston Bay, South Carolina. *(Gabe Palmer/The Stock Market)*

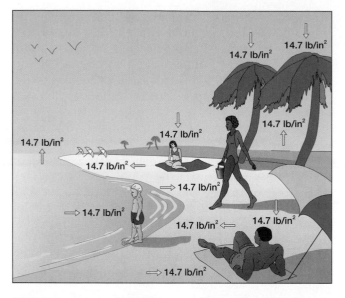

(a)

(b)

Figure 5-2 **(a)** Average atmospheric pressure at sea level is 14.7 pounds per square inch (slightly more than 1 kilogram per square centimeter). **(b)** Normal air pressure in a room will collapse a metal can if all the air within it is evacuated. (Copyright Leonard Lessen/Peter Arnold, Inc.)

PRESSURE, DENSITY, AND TEMPERATURE

Atmospheric pressure is closely related to the density and temperature of the atmosphere. Variations in any one of the three can cause variations in the other two. In this subsection, we examine first how the pressure varies with density and then how it varies with temperature.

Density and Pressure **Density** is the amount of matter in a unit volume. For example, if you have a 10-pound chunk of some material that is in the shape of a cube having an edge length of 1 foot, the density of that material is 10 pounds per cubic foot (10 lb/ft^3). The density of solid material is the same on Earth or the moon or in space, that of liquids varies very slightly from one place to another, and that of gases varies a lot with location. The reason gas density changes so easily is that a gas expands as far as the environmental pressure will allow. If you have 100 pounds of gas in a container that has a volume of 1 cubic foot, the gas density is 100 lb/ft^3. If you then transfer all the gas to a container having a volume of 5 cubic feet, the gas expands to fill the larger volume. There are the same number of gas molecules that were in the smaller container, but now they are spread out over a volume five times as large. Therefore the gas density in the larger container is only 20 lb/ft^3 (100 pounds divided by 5 cubic feet).

The density of a gas is proportional to the pressure on it. The reverse of this statement is also true: the pressure a gas exerts is proportional to its density. The denser the gas, the greater the pressure it exerts.

The atmosphere is held to Earth by the force of gravity, which prevents the gaseous molecules from escaping into space. This gravitational force is directly proportional to distance: the closer two bodies are to each other, the stronger the gravitational attraction between them. Thus the gravitational force between Earth and a low-altitude molecule is stronger than the force between Earth and a high-altitude molecule. As a result, at lower altitudes the gas molecules of the atmosphere are packed more densely together because of the stronger gravitational pull of Earth (Figure 5-3). Because the density is higher, there are more molecular collisions and therefore higher pressure at lower altitudes. At higher altitude, the air is less dense and there is a corresponding decrease in pressure. At any level in the atmosphere, then, the pressure is directly proportional to the air density at that altitude.

Temperature and Pressure If air is heated, as we noted in the preceding chapter, the molecules become more agitated and their speed increases. This increase in speed produces a greater force to their collisions and results in higher pressure. Therefore if other conditions remain the same (in particular, if volume is held constant), an increase in the temperature of a gas produces an increase in pressure and a decrease in temperature produces a decrease in pressure. Knowing this, you might conclude that the air pressure will be high on warm days and low on cold days. Such is usually not the case, however; warm weather generally is associated with low atmospheric

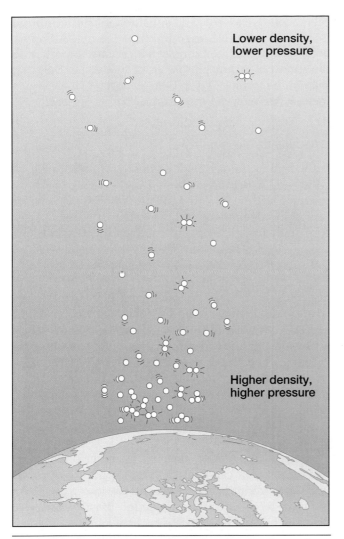

Figure 5-3 In the upper atmosphere, gas molecules are far apart and collide with each other infrequently, a condition that produces relatively low pressure. In the lower atmosphere, the molecules are closer together and there are many more collisions, a condition that produces high pressure.

Lower density, lower pressure

Higher density, higher pressure

pressure, and cool weather with high atmospheric pressure.

Complications The pressure of a gas is proportional to its density and temperature. This relationship can be explained by several equations collectively referred to as the gas law. The cause-and-effect association, however, is complex. For example, notice that we made the qualifying statement "if other conditions remain the same" in describing how temperature and pressure are related. When air is heated while no control is kept on its volume, it expands, which decreases its density. Thus the increase in temperature may be ac-

companied by a decrease in pressure caused by the decrease in density.

The point of all this is that atmospheric pressure is affected by both air density and air temperature and the relationship among the three variables is intricate. It is important for us to be alert to these linkages, but it is difficult to predict how a change in one will influence the others in a specific instance.

MAPPING PRESSURE WITH ISOBARS

Weather stations normally record atmospheric pressure, either continuously or periodically, in units called **millibars** (the instrument used to measure pressure is called a barometer), where 1 bar = 1000 millibars = 14.7 pounds per square inch, and so 1 millibar = 0.0147 pound per square inch.

Once pressure in millibars is plotted on a weather map, it is then possible to draw isolines of equal pressure called **isobars**, as shown in Figure 5-4. The pattern of the isobars reveals the horizontal distribution of pressure in the region under consideration. Prominent on such maps are roughly circular or oval areas characterized as being either "high pressure" or "low pressure." These highs and lows represent relative conditions—pressure that is higher or lower than that of the surrounding areas. It is important to keep the relative nature of pressure measurement in mind. For example, a pressure reading of 1005 millibars could be either "high" or "low" depending on the pressure of the adjoining areas.

While highs and lows represent the pressure extremes in any region, less extreme pressure areas can be recognized on weather maps by the arrangement of the isobars. Thus a "ridge" of high pressure may separate two isobars of low pressure, and a "trough" of low pressure may intervene between two isobars of high pressure, as demonstrated in Figure 5-5.

On most maps of air pressure, actual pressure readings are adjusted to represent pressures at a common elevation, usually sea level. This is done because pressure decreases rapidly with increasing altitude and consequently significant variations in pressure readings are likely at different weather stations simply because of differences in elevation.

As with other types of isolines, the relative closeness of isobars indicates the horizontal rate of pressure change, or **pressure gradient**. The gradient can be thought of as representing the "steepness" of the pressure slope, a characteristic that has a direct influence on the speed of the wind, the topic we turn to next.

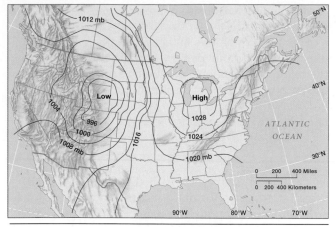

Figure 5-4   Isobars are lines connecting points of equal atmospheric pressure. When they have been sketched on a weather map, it is easy to determine the location of high-pressure and low-pressure centers. This simplified weather map shows pressure in millibars. Because 1000 millibars = 1 bar = 14.7 pounds per square inch, the 1015-millibar isobar, for instance, shows where the pressure is 1.015 bar = (14.7)(1.015) = 14.9 pounds per square inch.

THE NATURE OF WIND

The atmosphere is virtually always in motion. Air is free to move in any direction, its specific movements being shaped by a variety of factors. Some airflow is lackadaisical and brief, some is strong and persistent. Atmospheric motions often involve both horizontal and vertical displacement. Instead of being called wind, however, small-scale vertical motions are normally referred to as **updrafts** and **downdrafts**; large-scale verti-

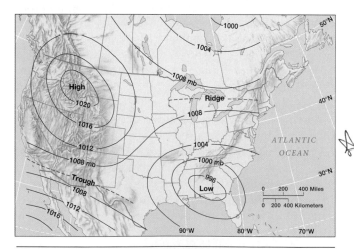

Figure 5-5   Isobars show the positions of pressure ridges and troughs as well as high-pressure and low-pressure centers. The high-pressure center on this map reaches a pressure of more than 1020 millibars, whereas the pressure of the ridge exceeds 1008 millibars. The low-pressure center has a pressure below 996 millibars, and the trough pressure is something less than 1008 millibars.

cal motions are **ascents** and **subsidences**; the term wind is applied only to horizontal movements. Although both vertical and horizontal motions are important in the atmosphere, much more air is involved in the latter than in the former. **Wind**, then, is horizontal air movement; it has been characterized as "air in a hurry."

DIRECTION OF MOVEMENT

Insolation is the ultimate cause of wind because all winds originate from the same basic sequence of events: unequal heating of different parts of Earth's surface brings about temperature gradients that generate pressure gradients, and these pressure gradients set air into motion. Winds represent nature's attempt to even out the uneven distribution of air pressure over Earth.

The generalization is that air flows from areas of high pressure to areas of low pressure. If Earth did not rotate and if there were no such thing as friction, that is precisely what would happen—a direct movement of air from a high-pressure region to a low-pressure region. However, rotation and friction both exist, and so this general statement is usually not completely accurate. The direction of wind movement is determined principally by the interaction of three factors: pressure gradient, the Coriolis effect, and friction.

Pressure Gradient   If there is higher pressure on one side of a parcel of air than on the other, the parcel will move from the higher toward the lower, as shown in Figure 5-6. If you visualize a high-pressure area as a pressure hill and a low-pressure area as a pressure valley, it is not difficult to imagine air flowing down this "pressure slope" in the same manner that water flows down a hill.

The pressure-gradient force acts at right angles to the isobars in the direction of the lower pressure. If there were no other factors to consider, that is the way the air would move. Such a flow rarely occurs in the atmosphere, however, and so there are clearly other important influences.

The Coriolis Effect   Because Earth rotates, any object moving freely near Earth's surface appears to deflect to the right in the Northern Hemisphere and to the left in the Southern Hemisphere. This **Coriolis effect** has an important influence on the direction of wind flow. If it were the only factor affecting wind direction, the wind would be shifted 90° from its pressure-gradient path and would flow parallel to the isobars, as Figure 5-7 shows.

There is an eternal battle, then, between the pressure-gradient force and the Coriolis effect: the Coriolis effect keeps the wind from flowing down a pressure gradient, and the pressure gradient pre-

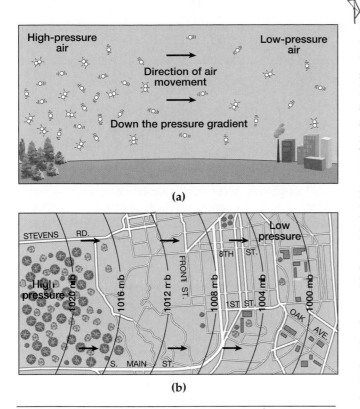

(a)

(b)

Figure 5-6 **(a)** A parcel of air bordered by high-pressure air on one side and low-pressure air on the other moves toward the low-pressure area. We say this movement is "down the pressure gradient." **(b)** If pressure gradient were the only force involved, air would flow down the pressure gradient, and that means the flow would be perpendicular to the isobars.

vents the Coriolis effect from turning the wind back up the pressure slope. Where these two factors are in balance, the wind moves parallel to the isobars and is called a **geostrophic wind**. In actuality, most winds are almost geostrophic in that they flow nearly parallel to the isobars. Only near the ground is there another significant factor—friction—to further complicate the situation.

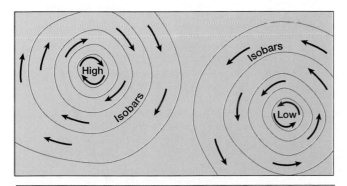

Figure 5-7 If the Coriolis effect were the only factor involved in air movement, air would flow parallel to isobars.

Friction In the lower portion of the troposphere, a third force influences wind direction—the force of friction. The frictional drag of Earth's surface acts both to slow down wind movement and to modify its direction. Instead of blowing perpendicular to the isobars (in response to the pressure gradient) or parallel to them (in response to the Coriolis effect), the wind takes an intermediate course between the two and crosses the isobars at some angle that is larger than 0° but less than 90° (Figure 5-8). As a general rule, the frictional influence is greatest near Earth's surface and diminishes progressively upward (Figure 5-9 on page 114). Thus the angle of wind flow across the isobars is greatest (closest to 90°) at low altitudes and becomes smaller at increasing elevations. The effect of friction (the friction layer) extends to only about 5000 feet (1500 meters) above the ground. Higher than that, most winds follow a geostrophic course.

CYCLONES AND ANTICYCLONES

Distinct and predictable wind-flow patterns develop around all high-pressure and low-pressure centers, patterns determined by pressure gradient, Coriolis effect, and friction. Eight circulation patterns are possible—four associated with high-pressure cells and four with low-pressure centers, as shown in Figure 5-10 on page 114.

High-Pressure Circulation Patterns A high-pressure center is known as an **anticyclone**, and the flow of air associated with it is described as being anticyclonic. The four patterns of anticyclonic circulation, shown in Figure 5-10 on page 114, are as follows:

1. In the upper air of the Northern Hemisphere, the winds move parallel to the isobars and clockwise.

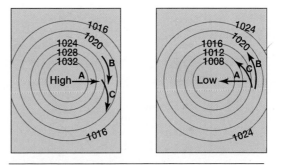

Figure 5-8 The direction of wind flow is determined by a combination of three factors. The pressure gradient (A) dictates that movement is perpendicular to the isobars. The Coriolis effect (B) causes movement parallel to the isobars. Friction (C) dictates that an intermediate course is followed, so that the direction of movement is across the isobars at some angle between 0° and 90°.

FOCUS

THE CORIOLIS EFFECT

Everyone is familiar with the unremitting force of gravity. Its powerful pull toward the center of Earth influences all vertical motion that takes place near Earth's surface. Much less well known, however, because of its inconspicuous nature, is a pervasive horizontal influence on earthly motions. All things that move over the surface of Earth or in Earth's atmosphere appear to drift sideways as a result of Earth's rotation. George Hadley described this apparent deflection in the 1730s, but it was not explained quantitatively until Gaspard G. Coriolis (1792–1843), a French civil engineer

and mathematician, did so a century later. The phenomenon is called the Coriolis effect in his honor (Figure 5-1-A).

The nature of the apparent deflection due to the Coriolis effect can be demonstrated by imagining a rocket fired at New York City from the North Pole. During the few minutes that the rocket is in the air, earthly rotation will have moved the target some miles to the east because the planet rotates from west to east. If the Coriolis effect was not included in the ballistic computation, the rocket would land some distance to the west of New York City (Figure 5-1-B).

To a person standing at the launch point and looking south, the uncorrected flight path appears to deflect to the right.

This rightward deflection (in the Northern Hemisphere) applies no matter in which direction the rocket moves. If a rocket were aimed at New York City from a location on the same parallel of latitude—say, northern California—there would also be a drift to the right (as viewed by a person at the launch sight and looking east). During the time the rocket is in motion, both the California firing point and the New York target rotate eastward. Because Earth rotates, the landing point of an

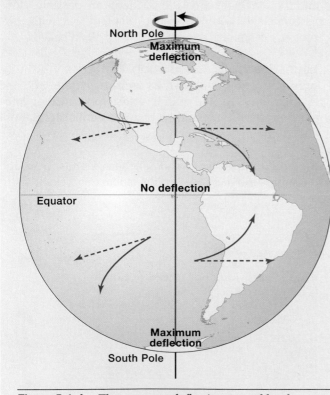

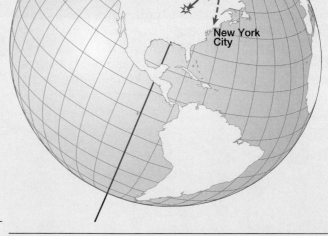

Figure 5-1-A   The apparent deflection caused by the Coriolis effect is to the right in the Northern Hemisphere and to the left in the Southern Hemisphere. The dashed lines represent the planned route, and the solid lines represent actual movement.

Figure 5-1-B   A rocket fired from the North Pole and aimed at New York City lands to the west of the target if the Coriolis effect is not considered when the flight path is computed.

uncorrected flight path is southwest of New York City (Figure 5-1-C).

The Coriolis principle applies to any freely moving object—ball, bullet, airplane, automobile, even a person walking. For these and other short-range movements, however, the deflection is so minor and so counterbalanced by other factors (such as friction, air resistance, and initial impetus) as to be insignificant. Long-range movements, on the other hand, can be significantly influenced by the Coriolis effect. The accurate firing of artillery shells or launching of rockets and spacecraft requires careful compensation for the Coriolis effect if the projectiles are to reach their targets.

The following are four basic points to remember:

1. Regardless of the initial direction of motion, any freely moving object appears to deflect to the right in the Northern Hemisphere and to the left in the Southern Hemisphere.
2. The apparent deflection is strongest at the poles and decreases progressively toward the equator, where there is zero deflection.
3. The Coriolis effect is proportional to the speed of the object, and so a fast-moving object seems to be deflected more than a slower one.
4. The Coriolis effect influences direction of movement only; it has no influence on speed.

The major importance of the Coriolis effect in our study of climate involves its influence on winds and ocean currents:

1. All winds are affected by the Coriolis principle.
2. Ocean currents are also deflected by the Coriolis effect.

The latter is an important component of the dynamics of the general circulation of the oceans, with Northern Hemisphere currents trending to the right and Southern Hemisphere currents to the left. The Coriolis drift is also a causative factor in the upwelling of cold water that takes place in subtropical latitudes where cool currents veer away from continental coastlines. The surface water that moves away from the shore is replaced by cold water rising from below.

One phenomenon that the Coriolis effect does not appear to influence is the circulation pattern of water that drains out of a washbowl. There is an old wives' tale that Northern Hemisphere washbowls drain clockwise and Southern Hemisphere washbowls counterclockwise. The time involved is so short and the speed of the water so slow, however, that Coriolis control cannot be postulated for these movements. The characteristics of the plumbing system, the shape of the washbowl, and pure chance are more likely to determine the flow patterns. A reader can test this hypothesis, of course, by filling and emptying several washbowls and recording the results. The author spent five months in Australia in 1992/93 and tested 100 washbowls. The results: 34 that emptied clockwise, 39 that emptied counterclockwise, and 27 that simply gushed down without a swirl in either pattern.

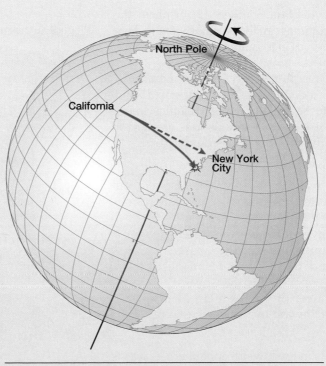

Figure 5-1-C   A rocket fired at New York City and launched from a point at the same latitude in California appears to curve to the right because both firing and target points rotate to the east while the rocket is in the air.

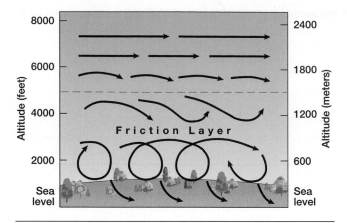

Figure 5-9 A vertical cross section of the atmosphere to show wind movement. Near Earth's surface, friction causes wind flow to be turbulent and irregular. At higher altitudes, where there is less friction, the lines of wind flow are much straighter.

2. In the friction layer of the Northern Hemisphere, there is a divergent clockwise flow, with the air spiraling out away from the center of the anticyclone.
3. In the upper air of the Southern Hemisphere, there is a counterclockwise movement parallel to the isobars.
4. In the friction layer of the Southern Hemisphere, the pattern is a mirror image of the Northern hemisphere case. The air diverges in a counterclockwise pattern.

Low Pressure Circulation Patterns Low-pressure centers are called **cyclones**, and the associated wind movement is said to be cyclonic. As with anticyclones, Northern Hemisphere cyclonic circulations are mirror images of their Southern Hemisphere counterparts:

5. In the upper air of the Northern Hemisphere, air movement parallels the isobars in a counterclockwise direction.
6. In the friction layer of the Northern Hemisphere, a converging counterclockwise flow exists.
7. In the upper air of the Southern Hemisphere, there is clockwise flow paralleling the isobars.
8. In the friction layer of the Southern Hemisphere, the winds move inward in a clockwise spiral.

Cyclonic patterns in the lower troposphere—parts 6 and 8 in Figure 5-10—may at first glance appear to be misleading because the arrows seem to defy the Coriolis effect. In the Northern Hemisphere, the arrows bend to the left, whereas we know that the Coriolis deflection is to the right. The point to remember is that the arrows portray the general flow pattern, but the Coriolis deflection pertains to individual parcels of air. Each individual parcel is deflected to the right of a di-

Northern Hemisphere upper-air pattern

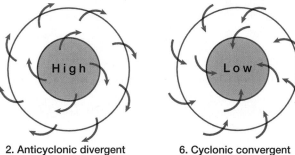

1. Anticyclonic geostrophic clockwise flow

5. Cyclonic geostrophic counterclockwise flow

Northern Hemisphere friction-layer pattern

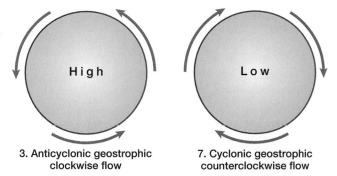

2. Anticyclonic divergent clockwise flow

6. Cyclonic convergent counterclockwise flow

Southern Hemisphere upper-air pattern

3. Anticyclonic geostrophic clockwise flow

7. Cyclonic geostrophic counterclockwise flow

Southern Hemisphere friction-layer pattern

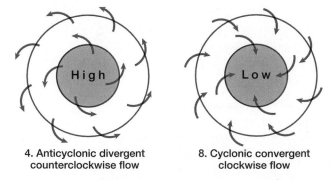

4. Anticyclonic divergent counterclockwise flow

8. Cyclonic convergent clockwise flow

Figure 5-10 The eight basic patterns of air circulation around pressure cells.

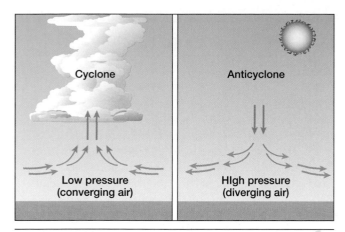

Figure 5-11 In an anticyclone (high-pressure cell), air subsides (descends) and diverges. In a cyclone (low-pressure cell), air converges and rises.

rect path across the isobars toward the center of the cyclone, but the resultant flow is an in-spiral to the left. Similar conditions prevail in the Southern Hemisphere, where a leftward Coriolis deflection has a rightward appearance.

Thus, large-scale winds do not blow from high pressure to low. Indeed, whether the wind is geostrophic or not, there is always high pressure on one side and low pressure on the other.

A prominent vertical component of air movement also is associated with pressure centers. As Figure 5-11 shows, air descends in anticyclones and rises in cyclones. Such motions are particularly notable in the lower troposphere. The anticyclonic pattern can be visualized as upper air sinking down into the center of the high and then diverging near the ground surface. Opposite conditions prevail in a low-pressure center, with the air converging horizontally into the cyclone and then rising.

WIND SPEED

Thus far, we have been considering the direction of wind movement and paying little attention to speed. Although some complications are introduced by inertia and other factors, it is accurate to say that the speed of wind flow is determined primarily by the pressure gradient. If the gradient is steep, the air moves swiftly; if the gradient is gentle, movement is slow. This relationship can be portrayed in the simple diagram of Figure 5-12. The closeness of the isobars indicates the steepness of the pressure gradient.

Over most of the world most of the time, surface winds are relatively gentle. As Figure 5-13 shows, for instance, annual average wind speed in North America is generally between 6 and 12 knots. (A knot is a unit of speed, equivalent to 1 nautical mile per hour. A nautical mile is a bit longer that a statute mile, the former being equal to 6076 feet or 1852 meters and the latter

being equal to 5280 feet or 1609 meters.) Cape Dennison in Antarctica holds the dubious distinction of being the windiest place on Earth, with an annual average wind speed of 38 knots.

The most persistent winds are usually in coastal areas or high mountains.

VERTICAL VARIATIONS IN PRESSURE AND WIND

Although the main topic of this chapter is the horizontal distribution of pressure and wind, it is worthwhile to note major features of the vertical pattern as well.

Atmospheric pressure, with only minor localized exceptions, decreases rapidly with height (Table 5-1 on page 118). The pressure change is most rapid at lower altitudes, the rate of decrease diminishing significantly above about 10,000 feet (3 kilometers). Prominent surface pressure centers (anticyclones and cyclones) often lean with height, which is to say that they are not absolutely vertical in orientation.

Wind speed is quite variable from one altitude to another and from time to time, usually increasing with height. Winds tend to move faster above the friction layer. As we shall see in subsequent sections, the very strongest tropospheric winds usually are found at intermediate levels in what are called jet streams or in violent storms near Earth's surface.

Winds in the upper portion of the troposphere have a simple generalized arrangement (Figure 5-14 on page 118). Over tropical regions, between the equator and 20° to 25° of latitude, winds aloft are generally from the east. Everywhere else in the upper troposphere, winds are generally westerly, except over polar areas where easterly winds again occur.

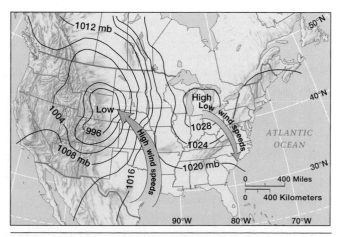

Figure 5-12 Wind speed is determined by pressure gradient, which is indicated by the spacing of isobars. Where isobars are close together, the pressure gradient is steep and wind speed is high; where isobars are far apart, the pressure gradient is gentle and wind speed is low.

PEOPLE AND THE ENVIRONMENT

WIND ENERGY

As the production of electricity becomes increasingly costly and fossil fuels become increasingly scarce, more attention is being paid to the harnessing of energy from sunlight, tides, and wind. These are particularly attractive resources because they are free, renewable, clean, and virtually unlimited. The problem is to devise the technology to harness them economically.

The first windmills apparently were used in the Middle East in the tenth century A.D. and in Europe in the twelfth century. A windmill is a simple form of turbine, a wheel turned by the force of the wind striking blades that are set at an angle on a horizontal shaft.

Until quite recently, most windmills were relatively small structures used to pump water and drive electric generators. Within the last decade, however, much attention has been paid to the development of large-scale electricity generation by the concentration of thousands of giant wind turbines in windy locations and connected to an electrical grid. These experimental *wind farms* have been constructed in various parts of the world. In some cases the wind turbines are fixed on 200-foot (60-meter) towers and swirl through a circle that has a diameter of as much as 300 feet (90 meters). Most turbines, however, are only about one-third this size. They usually have only two or three blades, although some are designed like giant eggbeaters without individual blades.

If the wind turbine is to produce energy, there must be steady wind, not too strong or too weak. A slight breeze will not turn the blades, and a powerful gust might severely damage

the machine. In a steady wind of 15 to 20 knots, large wind turbines can convert about one-third of the available wind energy to electricity. Their efficiency increases with greater wind speed (up to a point) and with larger blade size. The available energy is proportional to the cube of the wind speed. Thus a 20-knot wind can provide eight times more energy than a wind of 10 knots.

Wind farms are creatures of the 1980s; none existed prior to about 1982. The U.S. Congress enacted research and development legislation in 1980, and the U.S. Department of Energy has infused a modest amount of developmental money (about $40 million per year) into the industry. Various states, particularly California, have provided some financing and tax incentives. As a result, some major corporations, a few utility companies, and many small new firms have developed and installed turbines.

More than 90 percent of the active wind turbines in the United States are found in California, due in part to generous tax credits and to the availability of undeveloped land in windy

Figure 5-2-A A wind farm near Palm Springs, California. (Jim Corwin/Photo Researchers, Inc.)

locales. There were fewer than 200 turbines in the state in 1982, but a decade later there were nearly 20,000. The world's largest concentration, with some 6000 windmills, is at Altamont Pass, about 50 miles (80 kilometers) east of San Francisco; nearly 5000 each are found at San Gorgonio Pass, 100 miles (160 kilometers) east of Los Angeles, and Tehachapi Pass, 50 miles (80 kilometers) southeast of Bakersfield (Figure 5-2-A). In addition, there are half a dozen small wind farms elsewhere in the state. The combined output of these machines meets about 1 percent of the state's total electricity demand.

Generating electricity from wind turbines is an appealing activity because it does not deplete any resources or create pollution and, unlike solar energy, is not restricted to daytime use. Moreover, it usually does not interfere with other forms of land use; for example, most California wind farms are in pastoral areas, and cattle have no problem grazing among the windmills. The machines are very noisy, however, and so cannot be located near residential areas.

The long-term future of this new and rapidly developing industry is obscure. There are still technological and financial questions to be answered, however. By the early 1990s federal financial support had diminished considerably and California state tax advantages had been phased out. Nevertheless, the resource is there for the taking, its natural advantages undiminished. It seems only logical that further wind energy development will become important in the future.

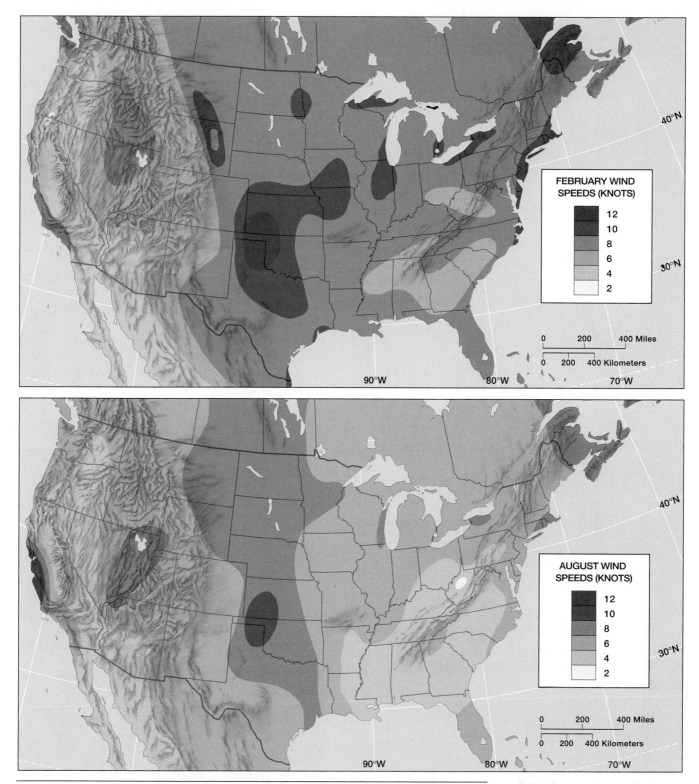

Figure 5-13 Average North American wind speeds in February and August. Wind speed tends to be higher in winter than in summer because pressure changes are more abrupt and storms are more frequent in winter than in summer. The Great Plains tend to have the highest speed winds in all seasons.

TABLE 5-1
How Atmospheric Pressure Varies with Altitude

Altitude		Pressure (millibars)
Miles	Kilometers	
43	70	0.06
37	60	0.23
31	50	0.78
25	40	2.87
22	35	5.75
19	30	12.0
16	25	25.5
12	20	55.3
11	18	75.6
10	16	104
8.7	14	142
7.4	12	194
6.2	10	265
5.6	9.0	308
5.0	8.0	356
4.3	7.0	411
3.7	6.0	472
3.1	5.0	540
2.5	4.0	617
2.2	3.5	658
1.9	3.0	701
1.6	2.5	747
1.2	2.0	795
0.9	1.5	846
0.6	1.0	899
0.3	0.5	955
0	0	1013

THE GENERAL CIRCULATION OF THE ATMOSPHERE

Earth's atmosphere is an extraordinarily dynamic medium. It is constantly in motion, responding to the various forces described previously as well as to a variety of more localized conditions. Some atmospheric motions are broadscale and sweeping; others are minute and momentary. Most important to an understanding of geography is the general pattern of circulation, which involves major semipermanent conditions of both wind and pressure. This circulation is the principal mechanism for both longitudinal and latitudinal heat transfer and is exceeded only by the global pattern of insolation as a determinant of world climates.

If Earth were a nonrotating sphere of uniform surface, we could expect a very simple circulation pattern (Figure 5-15). Insolational heating in the equatorial region would produce a girdle of low pressure around the world, and radiational cooling at the poles would develop a cap of high pressure in those areas. Surface winds in the Northern Hemisphere would flow directly down the pressure gradient from north to south, whereas those in the Southern Hemisphere would follow a similar gradient from south to north. Air would

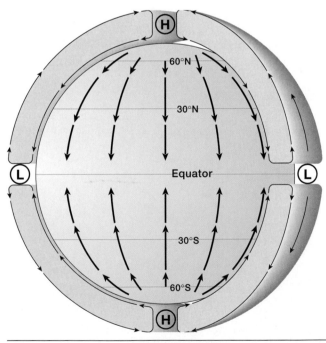

Figure 5-15 Wind circulation patterns would be simple if Earth's surface was uniform (no distinction between continents and oceans) and if the planet did not rotate. High pressure at the poles and low pressure at the equator would produce northerly surface winds in the Northern Hemisphere and southerly surface winds in the Southern Hemisphere.

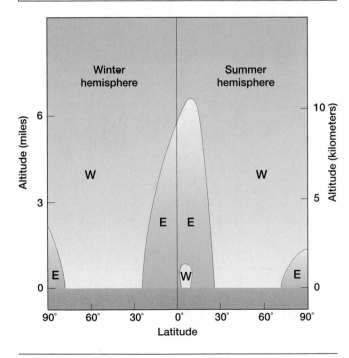

Figure 5-14　A generalized cross section through the troposphere, showing the horizontal component of air movement. Legend: E = easterly winds; W = westerly winds.

rise at the equator and flow toward the poles (south to north in the Northern Hemisphere; north to south in the Southern Hemisphere), where it would subside into the polar highs.

Earth does rotate, however, and in addition has an extremely varied surface. Consequently, the broad-scale circulation pattern of the atmosphere is exceedingly complex, although it can be generalized, as is done in Figures 5-16 and 5-17. The basic pattern has seven surface components, which are replicated north and south of the equator. From pole to equator, the names are as follows:

Polar high
Polar easterlies
Subpolar low
Westerlies
Subtropical high
Trade winds
Intertropical convergence zone

Tropospheric circulation is essentially a closed system, with neither a beginning nor an end, and so we can begin describing it almost anywhere. It seems logical, however, to begin in the subtropical latitudes of the five major ocean basins, as these areas serve as the "source" of the major surface winds of the planet.

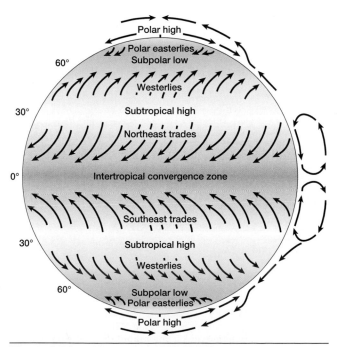

Figure 5-16 The seven surface components of the general circulation of the atmosphere, disregarding the effect of landmasses. (The vertical component is considerably exaggerated.)

SUBTROPICAL HIGHS

Each ocean basin has a large semipermanent high-pressure cell centered at about 30° of latitude called a **subtropical high (STH)** (Figure 5-18). These gigantic anticyclones, with an average diameter of perhaps 2000 miles (3200 kilometers), are usually elongated east-west and tend to be centered in the eastern portions of a basin. Their latitudinal positions vary from time to time, shifting a few degrees poleward in summer and a few degrees equatorward in winter.

The STHs are so persistent that each has been given a proper name (the Azores STH in the North Atlantic, for example). From a global standpoint, the STHs represent intensified cells of high pressure (and subsiding air) in two general ridges of high pressure that extend around the world in these latitudes, one in each hemisphere. The high-pressure ridges are significantly broken up over the continents, especially in summer when high land temperatures produce lower air pressure, but the STHs normally persist over the ocean basins throughout the year because temperatures and pressures there remain essentially constant. Associated with these high-pressure cells is a general subsidence of air from higher altitudes in the form of a broadscale, gentle downdraft.

Within an STH, the weather is nearly always clear, warm, and calm. We shall see in the next chapter that subsiding air is totally inimical to the development of clouds or the production of rain. Instead, these areas are characterized by warm, tropical sunshine and an absence of wind. Thus it comes as no surprise that these anticyclonic, subsiding-air regions coincide with most of the world's major deserts. These regions are sometimes called the **horse latitudes**, presumably because sixteenth- and seventeenth-century sailing ships were sometimes becalmed there and their cargos of horses were thrown overboard to conserve drinking water.

The air circulation pattern around an STH is typically anticyclonic: divergent clockwise in the Northern Hemisphere and counterclockwise in the Southern Hemisphere. A permanent feature of the STHs is a subsidence temperature inversion that covers wide areas in the subtropics. In essence the STHs can be thought of as gigantic wind-wheels whirling in the lower troposphere, fed with air sinking down from above and spinning off winds horizontally in all directions (Figure 5-19 on page 122). The winds are not dispersed uniformly around an STH, however; instead, they are concentrated on the northern and southern sides. Wind flow is less pronounced on the eastern and western sides of an STH because of the high-pressure ridge in those latitudes.

Although the global flow of air is essentially a closed circulation from a viewpoint at Earth's surface,

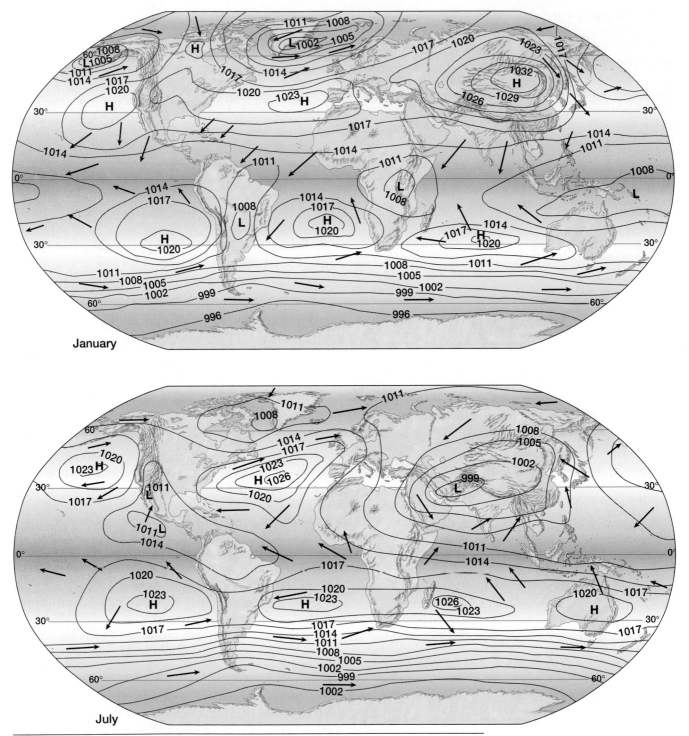

Figure 5-17 Average atmospheric pressure and wind direction in January and July. Pressure is reduced to sea-level values and shown in millibars. Arrows indicate generalized surface wind movements.

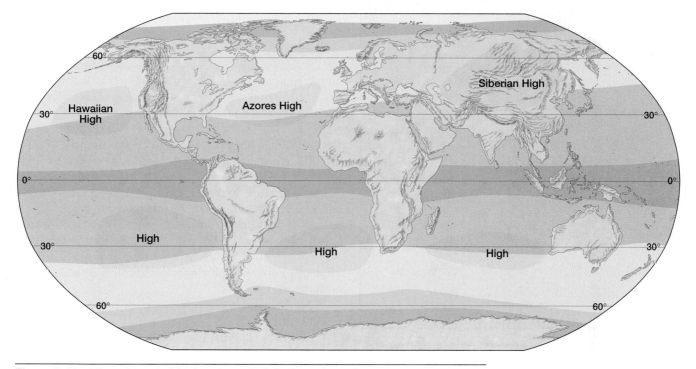

Figure 5-18   The subtropical highs, also called the horse latitudes.

the STHs can be thought of as the source of two of the world's three major surface wind systems: the trade winds and the westerlies (Figure 5-19).

TRADE WINDS

Issuing from the equatorward sides of the STHs and diverging toward the west and toward the equator is the major wind system of the tropics—the **trade winds**. These winds cover most of Earth between about latitude 25° N and latitude 25° S (Figure 5-20). They are particularly prominent over oceans but tend to be significantly interrupted and modified over landmasses. Because of the vastness of Earth in tropical latitudes and because most of this expanse is oceanic, the trade winds dominate more of the globe than any other wind system.

The trade winds are predominantly easterly; that is, they generally flow toward the west (winds are named for the direction *from* which they blow; an easterly wind blows from east to west, a westerly wind blows from the west, and so forth). There is also a latitudinal component to their movement, however. In the Northern Hemisphere, they usually come from the northeast (and are sometimes called the northeast trades); south of the equator, they are from the southeast (the southeast trades). There are exceptions to this general pattern, especially over the Indian Ocean where westerly winds sometimes prevail, but for the most part there is easterly flow above the tropical oceans.

Indeed, the trade winds are by far the most "reliable" of all winds. They are extremely consistent in both direction and speed, as Figure 5-21 shows. They blow most of the time in the same direction at the same speed, day and night, summer and winter (Figure 5-22 on page 124). This steadiness is reflected in their name: "trade winds" really means "winds of commerce." Mariners of the sixteenth century early recognized that the quickest and most reliable route for their sailing vessels from Europe to America lay in the belt of northeasterly winds of the southern part of the North Atlantic Ocean. Similarly, the trade winds were used by Spanish galleons in the Pacific Ocean, and the name became generally applied to these tropical easterly winds.

The trades originate as warming, drying winds capable of holding an enormous amount of moisture. As they blow across the tropical oceans, they evaporate vast quantities of moisture and therefore have a tremendous potential for storminess and precipitation. They do not release the moisture, however, unless forced to do so by being uplifted by a topographic barrier or some sort of pressure disturbance (Figure 5-23 on page 124). Low-lying islands in the trade wind zone often are desert islands because the moisture-laden winds pass over them without dropping any rain. If there is even a slight topographic irregularity, however, the air that is forced to rise may release abundant precipitation. Some of the wettest places in the world are windward slopes in the trade winds, such as in Hawaii.

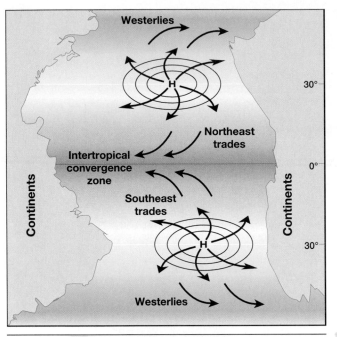

Figure 5-19   In a sense, subtropical highs are the source of surface trade winds and westerlies.

INTERTROPICAL CONVERGENCE ZONE

The northeast trades and the southeast trades come together in the general vicinity of the equator, although the latitudinal position shifts northward and southward following the sun. This shift is greater over land than over sea because the land heats more. The zone where air from the two hemispheres meets, illustrated in Figure 5-24, is usually called the **intertropical convergence (ITC) zone**, but it is also referred to by such names as equatorial front (a **front** is an area where unlike air masses come together), intertropical front, and **doldrums** (this last name is attributed to the fact that sailing ships were often becalmed in these latitudes).

The ITC is a zone of convergence and weak horizontal airflow characterized by feeble and erratic winds. It is a globe-girdling zone of low pressure, associated with instability and rising air. It is not a region of continuously ascending air, however. Almost all the rising air of the tropics ascends in the updrafts that occur in thunderstorms in the ITC zone, and these updrafts pump an enormous amount of sensible heat and latent heat of condensation into the upper troposphere, where much of it spreads poleward.

The ITC zone often appears as a well-defined, relatively narrow cloud band over the oceans (Figure 5-25). Over continents, however, it is likely to be more diffused and indistinct, although thunderstorm activity is common.

THE WESTERLIES

The fourth component of the general atmospheric circulation is represented by the arrows that issue from the poleward sides of the STHs in Figure 5-19. This is the great wind system of the midlatitudes, commonly called the **westerlies**. These winds flow basically from

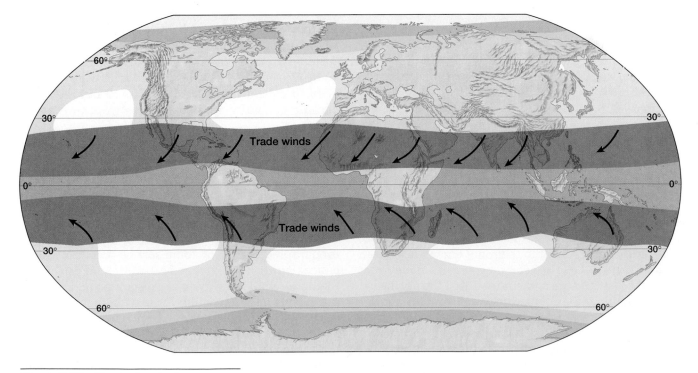

Figure 5-20   The trade winds.

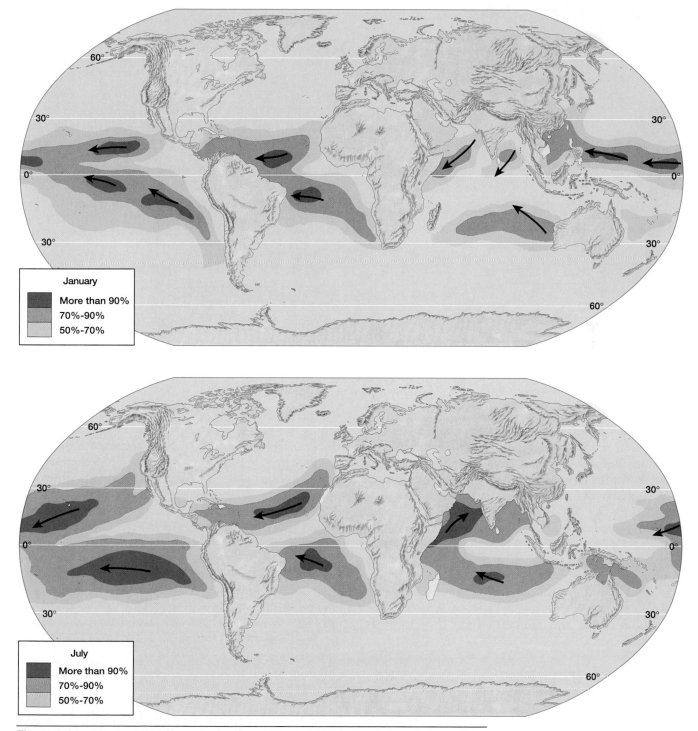

Figure 5-21   The trade winds are by far the most consistent of the major wind systems. These maps show their frequency of consistency for the midseason months of January and July. In the orange areas, for instance, the wind blows in the indicated direction more than 90 percent of the time.

Figure 5-22 Tropical coastal areas often experience the ceaseless movement of the trades. This breezy scene is at Cairns on the northeastern coast of Queensland, Australia. (TLM photo)

west to east around the world in the latitudinal zone between about 30° and 60° both north and south of the equator (Figure 5-26). Because the globe is smaller at these latitudes than in the tropics, the westerlies are less extensive than the trades; nevertheless they cover much of Earth.

Near the surface, the westerlies are much less constant and persistent than the trades, which is to say that in the westerlies surface winds often do not flow from the west but may come from any point of the compass. Near the surface there are interruptions and modifications of the westerly flow, which can be likened to eddies and countercurrents in a river (Fig-

Figure 5-23 Trade winds usually are heavily laden with moisture, but they do not produce clouds and rain unless forced to rise. Thus they may blow across a low-lying island with little or no visible effect. An island of greater elevation, however, causes the air to rise up the side of the mountain, and the result is usually a heavy rain.

ure 5-27). These interruptions are caused by surface friction, by topographic barriers, and especially by migratory pressure systems, which produce airflow that is not westerly. The winds aloft, however, which are geostrophic, blow very prominently from the west. Moreover, there are two remarkable cores of high-speed winds, one called the **polar front jet stream** and the other called the **subtropical jet stream,** at high altitudes in the westerlies (Figure 5-28). The belt of the westerlies can therefore be thought of as a meandering river of air moving generally from west to east around the world in the midlatitudes, with the jet streams as its fast-moving core.

The polar front jet stream, which usually occupies a position 30,000 to 40,000 feet (9 to 12 kilometers) high, is not in the center of the westerlies; it is displaced poleward, as Figure 5-28(c) shows (hence the "polar front" part of its name because of its location near the polar front). This jet stream is a feature of that part of the upper troposphere located over the area of greatest horizontal temperature gradient— that is, cold just poleward and warm just equatorward.

A jet stream is not a single wind; rather, it is a location of strong winds. Jet stream speed is variable. Sixty knots is generally considered as the minimum speed required for recognition as a jet stream, but speeds as much as five times that number have been recorded.

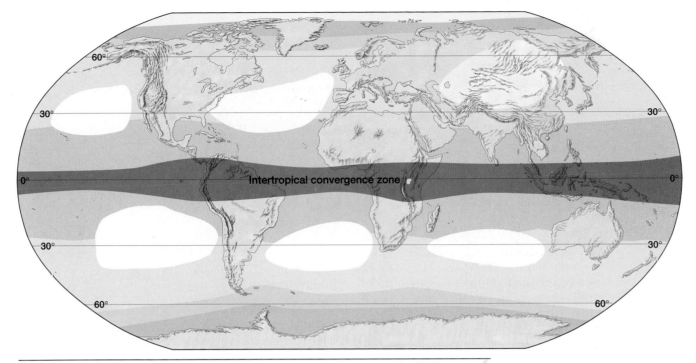

Figure 5-24 The intertropical convergence (ITC) zone, also known as the doldrums.

The polar front jet stream shifts its latitudinal position with some frequency, and this change has considerable influence on the path of the westerlies. Although the basic direction of movement is west to east, frequently sweeping undulations develop in westerlies flow and produce a meandering path that wanders widely to north and south (Figure 5-29 on page 128). These curves are very large and are generally referred to as *long waves* or **Rossby waves** (after the Chicago meteorologist C. G. Rossby, who first explained their nature). At any given time there are usually from three to six Rossby waves in the westerlies of each hemisphere. These waves can be thought of as separating cold polar air from warmer tropical air. When the polar front jet stream path is more directly west-east, there is a zonal nature to the weather, with

Figure 5-25 A well-defined band of cumulus clouds in the intertropical convergence zone of the Pacific Ocean. (Arvind Garg/Photo Researchers, Inc.)

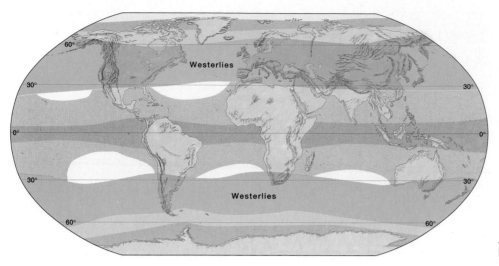

Figure 5-26 The westerlies.

cold air poleward of warm air. However, when the jet stream begins to oscillate and the Rossby waves develop significant amplitude (which means a prominent north-south component of movement, or "meridional" flow), cold air is brought equatorward and warm air moves poleward, bringing frequent and severe weather changes to the midlatitudes.

The subtropical jet stream is usually located at high altitudes—just below the tropopause, as Figure 5-28(b) shows—over the poleward margin of the subsiding air of the STH zone. It has less influence on surface weather patterns because there is less temperature contrast in the associated air streams. Sometimes, however, the polar front jet and the subtropical jet merge as shown in Figure 5-30 to produce a broad belt of high-speed winds in the upper troposphere, a condition that can intensify the weather conditions associated with either zonal or meridional flow of the Rossby waves.

All things considered, no other portion of Earth experiences such short-run variability of weather as the midlatitudes. These variations are caused not by the westerlies themselves but by the Rossby waves and by the migratory pressure systems and storms associated with westerly flow.

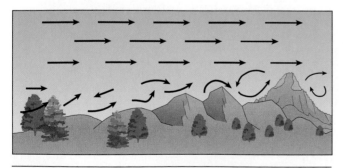

Figure 5-27 Near Earth's surface, topographic barriers and friction cause the direction of the westerlies to be erratic. In the upper air, however, these winds blow more consistently from the west.

POLAR HIGHS

Situated over both polar regions are high-pressure cells called **polar highs** (Figure 5-31). The Antarctic high, which forms over an extensive, high-elevation, very cold continent, is strong, persistent, and almost a permanent feature above the Antarctic continent. The Arctic high is much less pronounced and more transitory, particularly in winter. It tends to form over northern continental areas rather than over the Arctic Ocean. Air movement associated with these cells is typically anticyclonic. Air from above sinks down into the high and diverges horizontally near the surface, clockwise in the Northern Hemisphere and counterclockwise in the Southern Hemisphere.

POLAR EASTERLIES

The third broad-scale global wind system occupies most of the area between the polar highs and about 60° of latitude (Figure 5-32). The winds move generally from east to west and are called the **polar easterlies**. They are typically cold and dry but quite variable.

SUBPOLAR LOWS

The final surface component of the general pattern of atmospheric circulation is a zone of low pressure at about 50° to 60° of latitude in both Northern and Southern hemispheres (Figure 5-33). It is commonly called the **subpolar low** and often contains the **polar front**. It is a meeting ground and zone of conflict between the cold winds of the polar easterlies and the relatively warmer westerlies. The subpolar low of the Southern Hemisphere is virtually continuous over the uniform ocean surface of the cold seas surrounding Antarctica. In the Northern Hemisphere, however, the low-pressure zone is discontinuous, being interrupted by the continents. It is much more prominent in winter than in summer and is best developed

(a)

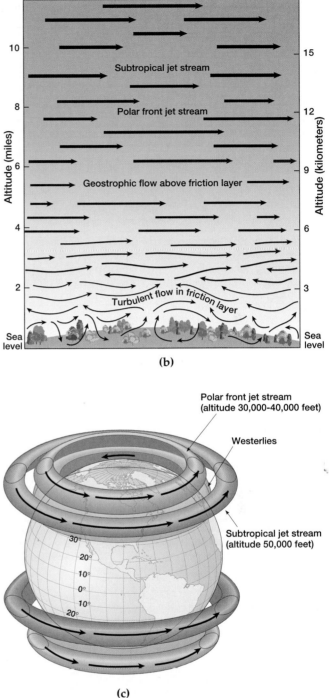

(b)

(c)

Figure 5-28 **(a)** A jet stream sometimes generates a distinctive cloud pattern that is conspicuous evidence of its presence, as in this photograph taken over the Nile Valley and Red Sea. Equatorward of the axis of the jet there is a tendency for air to rise, a condition that can produce thin clouds. Poleward of the axis, the air is clear. (NASA/ Headquarters) **(b)** A vertical cross section of the atmosphere cut parallel to the jet streams, showing the altitudes of the polar front jet stream and the subtropical jet stream. **(c)** Neither jet stream is centered in the band of the westerlies. The polar front jet stream is closer to the poleward boundary, and the subtropical jet stream is closer to the equatorward boundary of this wind system. Remember that the two streams are not at the same altitude, a detail that is obscured in this two-dimensional drawing. If you were in a spaceship looking down at Earth, the blue band would be closer to you than the green band.

over the northernmost reaches of the Pacific and Atlantic oceans.

The subpolar low area is characterized by rising air, widespread cloudiness, precipitation, and generally unsettled or stormy weather conditions. Many of the migratory storms that travel with the westerlies have their origin in the conflict zone of the polar front (Figure 5-34 on page 130).

MODIFICATIONS OF THE GENERAL CIRCULATION

There are many variations to the pattern discussed on the preceding pages, and all features of the general circulation may appear in altered form, much different from the idealized description. Indeed, components

sometimes disappear from sizable parts of the atmosphere where they are expected to exist. Even the troposphere sometimes "disappears". For example, during a high-latitude winter with very cold surface temperatures, the atmospheric temperature may steadily increase with height into the stratosphere; in such cases the troposphere cannot be identified. Nevertheless, the generalized pattern of global wind and pressure systems comprises the seven components

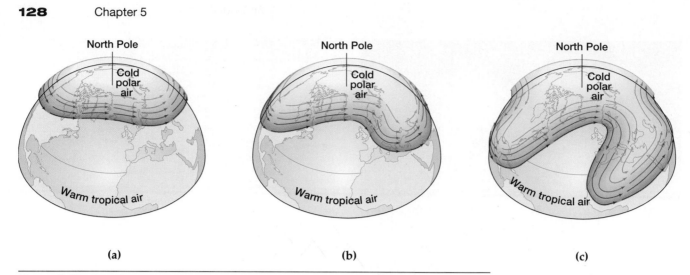

(a)   (b)   (c)

Figure 5-29 Rossby waves in the general flow (particularly the upper-air flow) of the westerlies. **(a)** When there are few waves and their amplitude (north-south component of movement) is small, cold air usually remains poleward of warm air. **(b)** This distribution pattern begins to change as the Rossby waves grow. **(c)** When the waves have great amplitude, cold air pushes equatorward and warm air moves poleward.

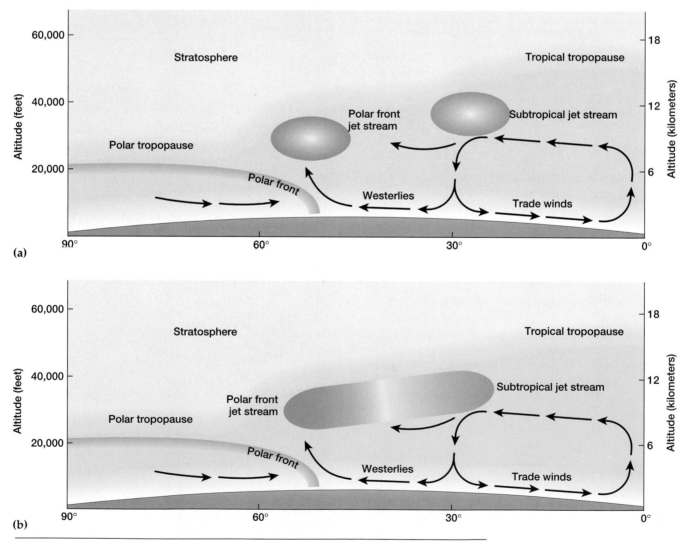

Figure 5-30 **(a)** A vertical cross section of the atmosphere cut perpendicular to the jet streams, showing the usual relative positions of the two jet streams. **(b)** The two jet streams sometimes merge, and the result is intensified weather conditions.

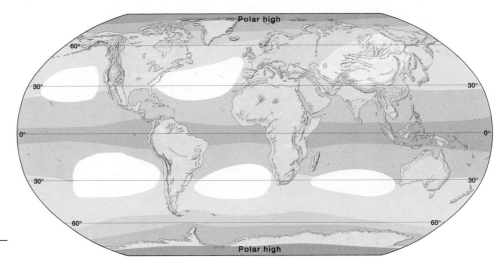

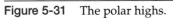

Figure 5-31 The polar highs.

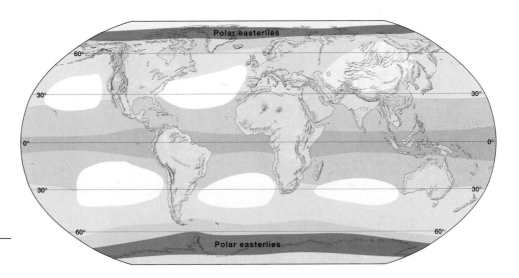

Figure 5-32 The polar easterlies.

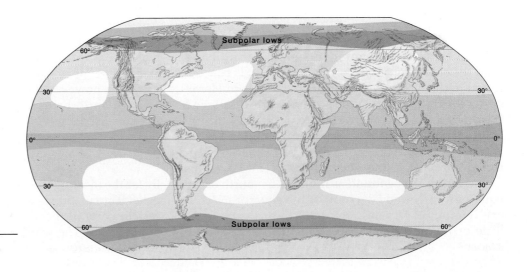

Figure 5-33 The subpolar lows.

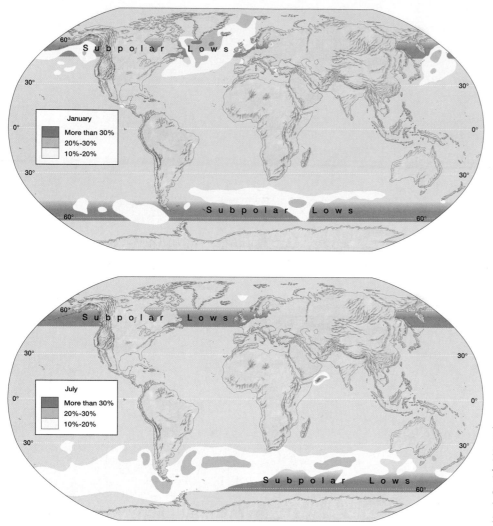

Figure 5-34 Frequency of gale-force winds over the oceans in January and July. It is clear from this map that the strongest oceanic winds are associated with activities along the polar front of the subpolar lows.

described above. In order to understand how real-world weather and climate differ from this general picture, it is necessary to discuss three important modifications of the generalized scheme.

SEASONAL VARIATIONS IN LOCATION

The seven surface components shift latitudinally with the changing seasons. When sunlight is concentrated in the Northern Hemisphere (Northern Hemisphere summer), all components are displaced northward; during the opposite season (Southern Hemisphere summer), everything is shifted southward. The displacement is greatest in the low latitudes and least in the polar regions. The ITC zone, for example, can be found as much as 25° north of the equator in July and 20° south of the equator in January (Figure 5-35), while the polar highs experience little or no latitudinal displacement from season to season.

The principal result of these annual north-south shifts is to extend summerlike conditions farther poleward during the summer and winterlike conditions farther equatorward during the winter. Weather is affected only minimally in equatorial and Arctic/Antarctic regions, but the effects can be quite significant in the midlatitudes and their fringes. For example, regions of mediterranean climate, which are centered at about 35° of latitude, have warm rainless summers; in winter, however, the belt of the westerlies is shifted equatorward, bringing changeable and frequently stormy weather to these regions.

MONSOONS

By far the most significant disturbance of the pattern of general circulation is the development of monsoons in certain parts of the world, particularly southern and eastern Asia. The word **monsoon** is derived from the Arabic (*mawsim*, season) and has come to mean a sea-

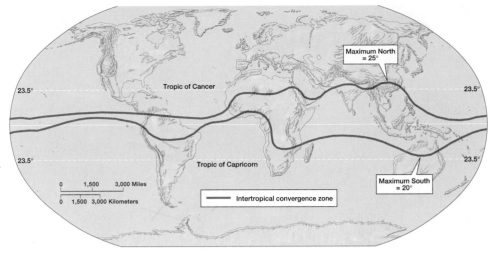

Figure 5-35 Average positions of the intertropical convergence zone at its seasonal extremes. The greatest variation in location is associated with monsoonal activity in Asia and Australia.

sonal reversal of winds, a general sea-to-land movement (called **onshore flow**) in summer and a general land-to-sea movement (called **offshore flow**) in winter. Associated with the monsoon wind pattern is a distinctive seasonal precipitation regime—heavy summer rains derived from the moist maritime air of the onshore flow and a pronounced winter dry season when continental air moving seaward dominates the circulation.

It would be convenient to explain monsoonal circulation on the basis of the unequal heating of continents and oceans. A strong thermal (in other words, heat-produced) low pressure generated over a continental landmass in summer would attract oceanic air onshore; similarly, a prominent thermal anticyclone in winter over a continent would produce an offshore circulation. It is clear that these thermally induced pressure differences contribute to monsoonal development, but they are not the whole story.

Monsoon winds essentially represent unusually large latitudinal migrations of normal trade winds and westerly flow. The explanation for such extensive migrations, however, is not clear, and we are left with the realization that the origin of monsoons is still not understood, although there is increasing evidence that it is associated with upper-air phenomena, particularly jet stream behavior.

It is difficult to overestimate the importance of monsoonal circulation to humankind. More than half of the world's population inhabits the regions in which climates are largely controlled by monsoons. Moreover, these are largely "underdeveloped" regions in which the majority of the populace depend upon agriculture for their livelihood. Their lives are intricately bound up with the reality of monsoonal rains, which are essential for both food production and cash crops (Figure

5-36). The failure, or even late arrival, of monsoonal moisture inevitably causes widespread starvation and economic disaster.

The characteristics of monsoons, on the other hand, are well known, and it is possible to describe the monsoonal patterns with some precision (Figure 5-37). There are two major monsoonal systems (one in South Asia; the other in East Asia); two minor systems (in Australia and West Africa), and several other regions where monsoonal tendencies develop (especially in Central America and the southeastern United States).

The most notable environmental event each year in South Asia is the annual "burst" of the summer monsoon, illustrated in Figure 5-38(a). In this first of the two major monsoonal systems, prominent onshore winds spiral in from the Indian Ocean, bringing life-giving rains to the parched subcontinent. In winter, South Asia is dominated by outblowing dry air diverging generally from the northeast. This flow is not very different from normal northeast trades except for its low moisture content.

Turning to the second of the two major monsoonal systems, we see that winter is the more prominent season in the East Asian monsoonal system, which primarily affects China, Korea, and Japan and is illustrated in Figure 5-38(b). A strong outflow of dry continental air, largely from the northwest, is associated with anticyclonic circulation around a massive thermal high-pressure cell over western China. The onshore flow of maritime air in summer is not as notable as that in South Asia, but it does bring southerly and southeasterly winds, as well as considerable moisture, to the region.

In one of the two minor systems, the northern quarter of the Australian continent experiences a distinct monsoonal circulation, with onshore flow from the north during the height of the Australian summer (De-

Figure 5-36   Flooding caused by summer monsoon rains in Dacca, Bangladesh. (Chip Hires/Gamma-Liaison, Inc.)

cember through March) and dry, southerly, offshore flow during most of the rest of the year. This system is illustrated in Figure 5-39(a).

The south-facing coast of West Africa is dominated by the second minor monsoonal circulation within about 400 miles (650 kilometers) of the coast. This system is shown in Figure 5-39(b). Moist oceanic air flows onshore from the south and southwest during summer, and dry, northerly, continental flow prevails in the opposite season.

VERTICAL COMPONENTS OF THE GENERAL CIRCULATION

Although the troposphere is a zone of considerable turbulence and vertical mixing, most of the broadscale air movements of the general circulation are horizontal. Important vertical components of the pattern are associated with the semipermanent belts and centers of high and low pressure, with air sinking into the highs and rising out of the lows.

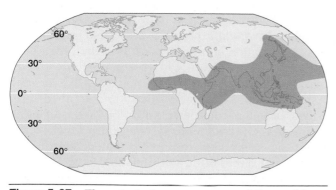

Figure 5-37   The principal monsoon areas of the world.

Apparently only the tropical regions have a complete vertical circulation cell. Such cells have been postulated for the middle and high latitudes, but observations indicate that the midlatitude and high-latitude cells either do not exist or are weakly and sporadically developed. The low-latitude cells—one north and one south of the equator—can be thought of as gigantic convection systems. The trade winds move toward the west and toward the equator, converging at the ITC zone. There much of the warm air rises to great heights, as Figure 5-40 shows, mostly in thunderstorm updrafts. In the upper troposphere (at about 50,000 feet = 15 kilometers), a poleward flow begins with strong **antitrade winds** moving toward the northeast in the Northern Hemisphere and toward the southeast in the Southern Hemisphere. This flow eventually becomes more westerly and encompasses the subtropical jet stream. At about 30° of latitude, much of the air subsides into the STHs, especially over the eastern parts of oceans and the general ridge of high pressure that circles the globe there. These two prominent tropical circulations are called **Hadley cells**, after George Hadley (1685—1768), an English meteorologist who first conceived the idea of enormous convective circulation cells in 1735.

LOCALIZED WIND SYSTEMS

The immediately preceding sections deal only with the broadscale wind systems that make up the global circulation and influence the climatic pattern of the world. There are many kinds of lesser winds, however, that are of considerable significance to weather and climate at a more localized scale. Such winds are the re-

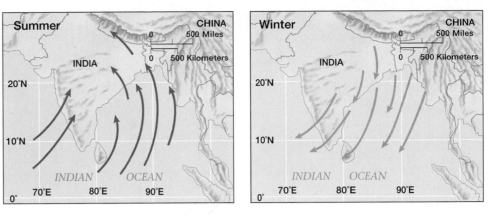

(a)

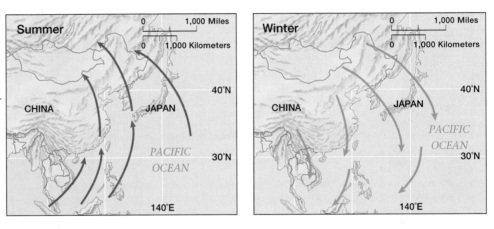

Figure 5-38 The two major monsoonal systems. **(a)** The South Asian monsoon is characterized by a strong onshore flow in summer and a somewhat less pronounced offshore flow in winter. **(b)** In East Asia, the outblowing winter monsoon is stronger than the inblowing summer monsoon.

(b)

sult of local pressure gradients that develop in response to topographic configurations in the immediate area, sometimes in conjunction with broadscale circulation conditions.

SEA AND LAND BREEZES

A common local wind system along tropical coastlines and to a lesser extent during the summer in midlatitude coastal areas is the cycle of **sea breezes** during the day and **land breezes** at night (Figure 5-41). (As is usual with winds, the name tells the direction from which the wind comes. A sea breeze blows from sea to land, and a land breeze blows from land to sea.) This is essentially a convectional circulation caused by the differential heating of land and water surfaces. The land warms up rapidly during the day, heating the air above by conduction and reradiation. This heating causes the air to expand and rise, creating low pressure that attracts surface breezes from over the adjacent wa-

ter body. Because the onshore flow is relatively cool and moist, it holds down daytime temperatures in the coastal zone and provides moisture for afternoon showers. Sea breezes are sometimes strong, but they rarely are influential for more than 10 to 20 miles (16 to 32 kilometers) inland.

The reverse flow at night is normally considerably weaker than the daytime wind. The land and the air above it cool more quickly than the adjacent water body, producing relatively higher pressure over land. Thus air flows offshore in a land breeze.

VALLEY AND MOUNTAIN BREEZES

Another notable daily cycle of airflow is characteristic of many hill and mountain areas. During the day, conduction and reradiation from the land surface cause air near the mountain slopes to heat up more than air over the valley floor (Figure 5-42). The heated air rises, creating a low-pressure area, and then cooler air from the

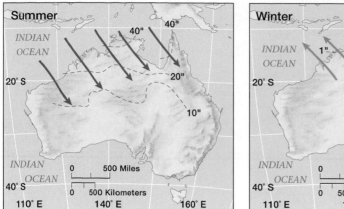

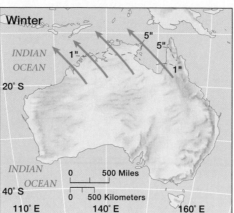

(a)

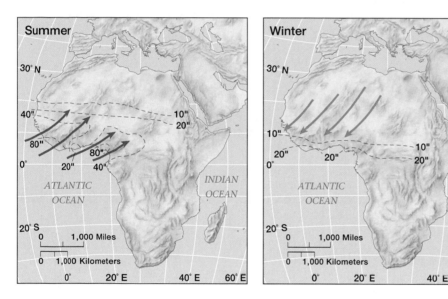

(b)

Figure 5-39 The two minor monsoonal systems. **(a)** In Australia, northwesterly summer winds bring the wet season to northern Australia; dry southeasterly flow dominates in winter. **(b)** In West Africa, summer winds are from the southwest and winter winds are from the northeast.

valley floor flows upslope from the high-pressure area to the low-pressure area. This upslope flow is called a **valley breeze**. The rising air often causes clouds to form around the peaks, and afternoon showers are common in the high country as a result. After dark, the pattern is reversed. The mountain slopes lose heat rapidly through radiation, which chills the adjacent air, causing it to slip downslope as a **mountain breeze**.

Valley breezes are particularly prominent in summer, when solar heating is most intense. Mountain breezes are often weakly developed in summer and are likely to be more prominent in winter. Indeed, a frequent winter phenomenon in areas of even gentle slope is **air drainage**, which is simply the nighttime sliding of cold air downslope to collect in the lowest spots; this is a modified form of mountain breeze.

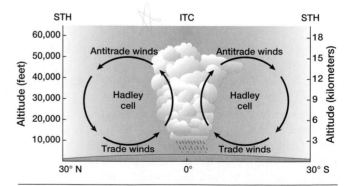

Figure 5-40 Distinct cells of vertical circulation occur in tropical latitudes; they are called Hadley cells. The equatorial air rises to some 40,000 or 50,000 feet (12 to 15 kilometers) before spreading poleward.

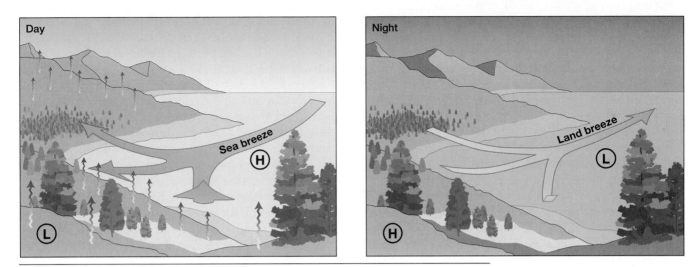

Figure 5-41 In a typical sea/land breeze cycle, diurnal heating over the land produces relatively low pressure there, and this low-pressure center attracts an onshore flow of air from the sea. Later, nocturnal cooling over the land causes high air pressure there, a condition that creates an offshore flow of air.

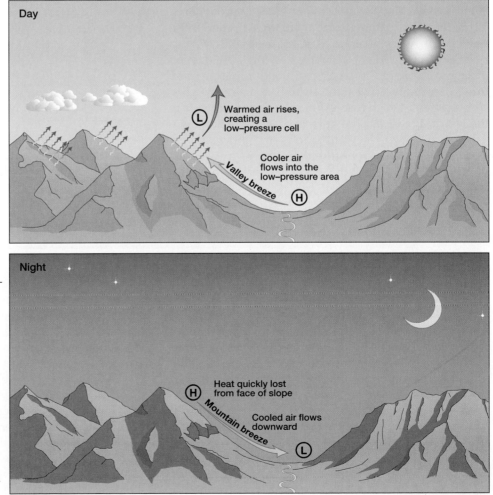

Figure 5-42 Daytime heating of the mountain slopes causes the air right above the slopes to warm and rise, creating an area where the air pressure is lower than over the valley. Cooler valley air then flows along the pressure gradient (in other words, up the mountain slope) in what is called a valley breeze. At night, the slopes radiate their heat away and as a result the air just above them cools. This cooler, denser air flows down into the valley in what is called a mountain breeze.

KATABATIC WINDS

Related to simple air drainage is the more general and more powerful spilling of air downslope in the form of **katabatic winds** (from the Greek *katabatik*, descending). These winds originate in cold upland areas and cascade toward lower elevations under the influence of gravity; they are sometimes referred to as *gravity-flow winds*. The air in them is dense and cold, and although warmed adiabatically as it descends, it is usually colder than the air it displaces in its downslope flow.

Katabatic winds are particularly common in Greenland and Antarctica, especially where they come whipping off the edge of the high, cold ice sheets. Sometimes a katabatic wind will become channeled through a narrow valley where it may develop high speed and considerable destructive power. An infamous example of this phenomenon is the **mistral**, which sometimes surges down France's Rhône Valley from the Alps to the Mediterranean Sea. Similar winds are called **bora** in the Adriatic region and **taku** in southeastern Alaska.

FOEHN/CHINOOK WINDS

Another downslope wind is called a **foehn** (pronounced as in "fern" but with a silent "r") in the Alps and a **chinook** in the Rocky Mountains. It originates only when a steep pressure gradient develops with high pressure on the windward side of a mountain and a low-pressure trough on the leeward side. Air moves down the pressure gradient, which means from the windward side to the leeward side, as shown in Figure 5-43. The downflowing air on the leeward side is dry and relatively warm: it has lost its moisture through precipitation on the windward side, and it is warm relative to the air on

the windward side because it contains all the latent heat of condensation given up by the condensing of the snow or rain that fell at the peak. As the wind blows down the leeward slope, it is further warmed adiabatically, and so it arrives at the base of the range as a warming, drying wind. It can produce a remarkable rise of temperature leeward of the mountains in just a few minutes. It is known along the Rocky Mountains front as a "snow-eater" because it not only melts the snow rapidly but also quickly dries the resulting mud.

Similar winds in California are known as **Santa Anas**. They are noted for high speed, high temperature, and extreme dryness. Their presence provides ideal conditions for wildfires. Virtually every year they make headlines by fanning large brush fires that destroy hundreds of homes in late summer and fall and occasionally in spring.

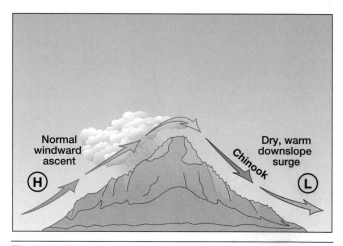

Figure 5-43   A chinook is a rapid downslope movement of relatively warm air. It is caused by a pressure gradient on the two faces of a mountain.

CHAPTER SUMMARY

Atmospheric pressure, air density, and air temperature are interdependent: a variation in any one of the three can cause a change in the other two.

Wind, which is defined as horizontal air movement, is a response to a pressure difference. The direction of wind movement is determined by the pressure gradient, the Coriolis effect, and friction. There are eight basic pressure-center circulation patterns, four around low-pressure centers (cyclones) and four around high-pressure centers (anticyclones). There is also a vertical component: rising air associated with cyclones and descending air in anticyclones.

With only minor exceptions, atmospheric pressure decreases rapidly with height. The pressure gradient is steepest at lower elevations. Wind velocity is variable from place to place and time to time, but it generally increases with in-

creasing altitude. Overall, surface wind flow is much more complex than that in the upper troposphere.

The general circulation pattern of the atmosphere has seven surface components, which occur in both the Northern and Southern hemispheres. Moving from pole to equator, these components are polar highs, polar easterlies, subpolar lows, westerlies, subtropical highs, trade winds, and the intertropical convergence zone.

All of these components are displaced northward in the Northern Hemisphere summer and southward in the Northern Hemisphere winter. This displacement is greatest in the low latitudes and least in the polar regions. The most notable surface-wind modification of the general circulation pattern is afforded by the monsoonal circulations that develop over southern and eastern Asia. Clear-cut cells of convective circulation occur in the tropics, with air sinking into the subtropical highs and rising out of the intertropical convergence zone.

Localized wind systems—sea and land breezes, valley and mountain breezes, katabatic winds, and foehn/chinook winds—are sometimes much more prominent than general circulation flows.

KEY TERMS

air drainage	horse latitudes	pressure
anticyclone	intertropical convergence (ITC)	pressure gradient
antitrade wind	zone	Rossby wave
ascent	isobar	Santa Ana
atmospheric pressure	katabatic wind	sea breeze
bora	land breeze	subpolar low
chinook	millibar	subsidence
Coriolis effect	mistral	subtropical high (STH)
cyclone	monsoon	subtropical jet stream
density	mountain breeze	taku
doldrums	offshore flow	trade wind
downdraft	onshore flow	updraft
foehn	polar easterlies	valley breeze
front	polar front	westerlies
geostrophic wind	polar front jet stream	wind
Hadley cell	polar high	

REVIEW QUESTIONS

1. Explain how atmospheric pressure is related to air density and air temperature.
2. What is the relationship of millibars to isobars?
3. Why do winds not simply flow down a pressure gradient?
4. What is the relationship of the Coriolis effect to geostrophic winds?
5. Describe the four patterns of anticyclonic flow that occur in the atmosphere.
6. Contrast the vertical movements of air in high-pressure and low-pressure cells.
7. What does the spacing of isobars tell about wind speed?
8. Does atmospheric pressure always decrease with increasing height? Explain.
9. Why are upper-air winds usually stronger than surface winds?
10. Discuss the location and persistence of subtropical high-pressure cells.
11. Why do trade winds cover such a large part of the globe?
12. Differentiate between trade winds and antitrade winds.
13. What are the principal differences between trade winds and westerlies?
14. Describe the jet streams of the westerlies.
15. What are the principal factors that interfere with the normal pattern of general atmospheric circulation?
16. What are the principal characteristics of monsoonal circulations?
17. In what ways are sea breezes and valley breezes similar?

SOME USEFUL REFERENCES

DJURIC, DUSAN, *Weather Analysis*. Englewood Cliffs, NJ: Prentice Hall, Inc., 1994.

HARE, F.K., *The Restless Atmosphere*, rev. ed. New York: Harper & Row Publishers, 1961.

HARMAN, JAY R., *Synoptic Climatology of the Westerlies: Process and Patterns*. Washington, DC: Association of American Geographers, 1991.

HASTENRATH, STEFAN, *Climate and Circulation of the Tropics*. Dordrecht, Netherlands: D. Reidel Publishing Company, 1985.

HIDY, G.M., *The Winds*. New York: Van Nostrand Reinhold Company, 1967.

RAMAGE, C.S., *Monsoon Meteorology*. New York: Academic Press, Inc., 1971.

REITER, E.R., *Jet Streams*. New York: Doubleday & Co., Inc., 1967.

6

ATMOSPHERIC MOISTURE

THE FOURTH ELEMENT OF WEATHER AND CLIMATE IS moisture. This might seem a familiar feature because everyone knows what water is. In actuality, however, most atmospheric moisture occurs not as liquid water but rather as water vapor, which is much less conspicuous and much less familiar.

One of the most distinctive attributes of water is that it occurs in the atmosphere in three physical states, as Figure 6-1 shows: solid (snow, hail, sleet, ice), liquid (rain, water droplets), and gas (water vapor). Of the three states, the gas state is the most important insofar as the dynamics of the atmosphere are concerned.

THE IMPACT OF ATMOSPHERIC MOISTURE ON THE LANDSCAPE

When the atmosphere contains enough moisture, water vapor may condense to form haze, fog, cloud, rain, sleet, hail, or snow, producing a skyscape that is both visible and tangible.

Precipitation produces dramatic short-run changes in the landscape whenever rain puddles form, streams and rivers flood, or snow and ice blanket the ground. The long-term effect of atmospheric moisture is even more fundamental. Water vapor stores energy that can galvanize the atmosphere into action, as, for example, in the way rainfall and snowmelt in soil and rock are an integral part of weathering and erosion. In addition, the presence or absence of precipitation is critical to the survival of almost all forms of terrestrial vegetation.

WATER VAPOR AND THE HYDROLOGIC CYCLE

Water vapor is a colorless, odorless, tasteless, invisible gas that mixes freely with the other gases of the atmosphere. It is a minor constituent of the atmosphere, with the amount present being quite variable from place to place and from time to time. It is virtually absent in some places but constitutes as much as 4 percent of the total atmospheric volume in others. Essentially, water vapor is restricted to the lower troposphere. More than half of all water vapor is found within 1 mile (1.6 kilometers) of Earth's surface, and only a tiny fraction exists above about 4 miles (6.4 kilometers).

The erratic distribution of water vapor in the atmosphere reflects the ease with which moisture can change from one state to another at the pressures and temper-

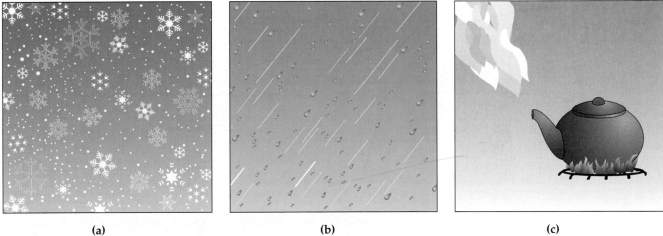

(a)   (b)   (c)

Figure 6-1 Depending on temperature, water can exist in any one of three physical states. **(a)** When the temperature is below freezing, water exists in the solid state we know as ice. **(b)** When the temperature is above freezing but below the boiling point, water exists in the familiar liquid state. **(c)** When the temperature is above the boiling point, water exists in the gas state and is called water vapor. The cloud you see coming out of the spout is steam, which is technically a liquid. Water vapor is invisible, as indicated by the empty region between the tip of the spout and the beginning of the steam cloud. The heat of the stove causes the liquid water in the kettle to boil and become water vapor. This vapor comes out of the spout but immediately changes back to liquid (the steam cloud) because it is no longer near the heat source and so has cooled.

A double rainbow following a thunderstorm in central Connecticut. *(Randy O'Rourke/ The Stock Market)*

atures found in the lower troposphere. Moisture can leave Earth's surface as a gas and return as a liquid or a solid. Indeed, there is a continuous interchange of moisture between Earth and the atmosphere (Figure 6-2). This unending circulation of our planet's water supply is referred to as the **hydrologic cycle**, and its essential feature is that liquid water (primarily from the oceans) evaporates into the air, condenses to the liquid (or solid) state, and returns to Earth as some form of precipitation. The movement of moisture through the cycle is intricately related to many atmospheric phenomena and is an important determinant of climate because of its role in rainfall distribution and temperature modification.

EVAPORATION

The conversion of moisture from liquid to gas—in other words, from water to water vapor—is called **evaporation.** This process involves molecular escape: molecules of water escape from the liquid surface into the surrounding air (Figure 6-3). The molecules of liquid water are continuously in motion and frequently collide with one another. Evaporation can take place at any temperature, but higher temperatures cause molecules to move faster and collide more forcefully. The

impact of such collisions near the water surface may provide sufficient energy to allow the molecules to break free from the water and enter the air.

The energy absorbed by the escaping molecules is stored as latent heat of vaporization and is released as latent heat of condensation when the vapor changes back to a liquid. Because heat leaves the water as latent heat in the evaporating molecules, the remaining water is cooled, and this process is called *evaporative cooling*. The effect of evaporative cooling is experienced when a swimmer leaves a swimming pool on a dry, warm day. The dripping wet body immediately loses moisture through evaporation to the surrounding air, and the skin feels the consequent drop in temperature.

The amount and rate of evaporation from a water surface depend on three factors: the temperature (of both air and water), the amount of water vapor already in the air, and whether the air is still or moving.

TEMPERATURE

The water molecules in warm water are more agitated than those in cool water; thus there is more evaporation from the former.

Warm air also promotes evaporation. Just as high water temperature produces more agitation in the molecules of liquid water, so high air temperature pro-

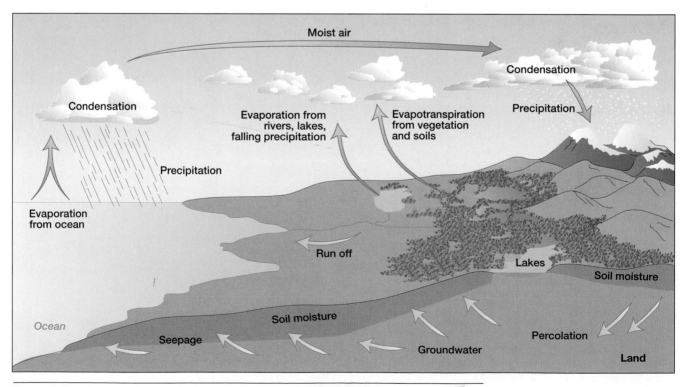

Figure 6-2   The hydrologic cycle is a continuous interchange of moisture between the atmosphere and Earth.

Water molecules cannot keep vaporizing and entering the air without limit, however. As we learned in Chapter 5, each gas in the atmosphere exerts a pressure, and the sum of all the pressures exerted by the individual gases is what we call atmospheric pressure. The pressure exerted by water vapor in the air is called **vapor pressure.** At any given air temperature, there is a maximum vapor pressure that water vapor molecules can exert. When there are enough water vapor molecules in air to exert the maximum vapor pressure at any given temperature, we say that the air is **saturated** with water vapor. When this maximum vapor pressure is exceeded, some water vapor molecules must leave the air and become liquid. More and more vapor molecules condense to the liquid state until the vapor pressure exerted by the remaining vapor molecules is right at the maximum value again.

The higher the air temperature, the higher the maximum vapor pressure. In other words, the warmer the air, the more water vapor it can hold before becoming saturated (Figure 6-4).

STILL VS. MOVING AIR

If the air overlying a water surface is almost saturated with water vapor, very little further evaporation can take place. Under these conditions, the rate at which water molecules go from the liquid state to the gas

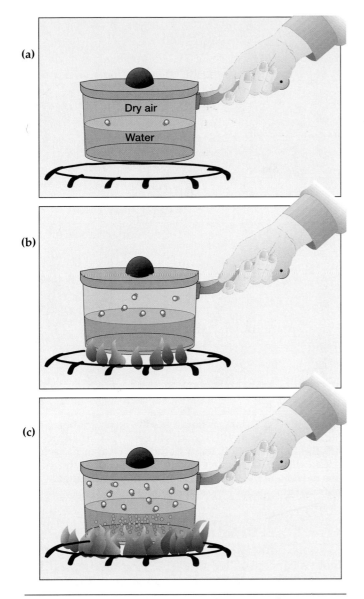

Figure 6-3 Evaporation involves the escape of water molecules from a liquid surface into the air. It can take place at any temperature, but high temperatures increase the energy of the molecules in the liquid and therefore accelerate the rate of evaporation. **(a)** With the burner off, only a very few molecules evaporate from the surface. **(b)** As the temperature rises, all the molecules become more agitated and more and more of them can break free from the water surface and enter the air. **(c)** Continued heating increases molecular activity so much that the air may become saturated with water vapor.

duces more agitation in the molecules of all the gases making up the air. The more energetic molecules in warm air bounce around more than do the molecules in cool air. As these bouncing molecules in warm air above a body of water hit the liquid surface, they give some of their energy to the liquid molecules. The energized liquid molecules then evaporate into the warm air.

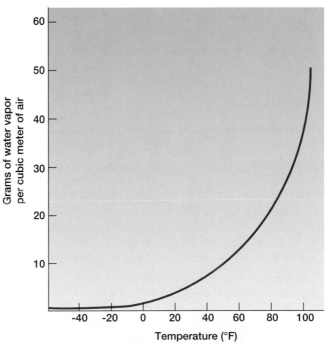

Figure 6-4 The maximum amount of water vapor a given volume of air can hold increases as the temperature increases.

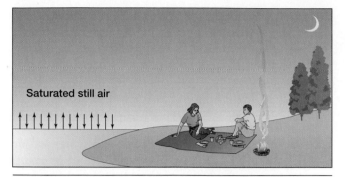

Saturated still air

Figure 6-5   When the air is calm and saturated, the number of water molecules evaporating from the water surface is equal to the number of water-vapor molecules condensing out. The net result in this case is zero evaporation.

state is about the same as the rate at which they go from the gas state to the liquid state (Figure 6-5). If the air remains calm and the temperature doesn't change, there is no net evaporation. If the air is in motion, however, through windiness and/or turbulence, the water vapor (and all other) molecules in it are dispersed more widely (Figure 6-6). This dispersing of vapor molecules originally in the air at the air–water interface means that that air is now farther from being saturated than it was when it was still. Because the air is now farther from being saturated, the rate of evaporation increases.

To summarize, the rate of evaporation from a water surface is determined by the temperature of the water, the temperature of the air, and the degree of windiness. Higher temperatures and greater windiness cause more evaporation.

Evapotranspiration

Although most of the water that evaporates into the air comes from bodies of water, a relatively small amount comes from the land. This evaporation from land has two sources: (1) soil and other inanimate surfaces, and (2) plants. The amount of moisture that evaporates from soil is relatively minor, and thus most of the land-derived moisture present in the air comes from plants. The process whereby plants give up moisture through their leaves is called **transpiration,** and so the combined process of water vapor entering the air from land sources is called **evapotranspiration.** Thus the water vapor in the atmosphere was put there through evaporation from bodies of water and evapotranspiration from land surfaces.

Whether a given land location is wet or dry depends on the rates of evapotranspiration and precip-

(a)

(b)

Figure 6-6   **(a)** As long as the air is still, the net evaporation rate from the lake surface remains at zero. **(b)** Once a breeze puts the air into motion, many of the water-vapor molecules in it are dispersed (to parcels of air that are not saturated). This movement means the air over the lake is no longer saturated and more evaporation can take place.

itation. To analyze these rates, we need to know about a concept called **potential evapotranspiration.** This is the amount of evapotranspiration that would occur if the ground at the location in question were sopping wet all the time. To determine a value for the potential evapotranspiration at any location, data on temperature, vegetation, and soil characteristics at that location are added to the actual evapotranspiration value in a formula that results in an estimate of the maximum evapotranspiration that could result under local environmental conditions if the moisture were available.

In locations where the precipitation rate exceeds the potential evapotranspiration rate, a water surplus accumulates in the ground. In many parts of the world, however, there is no groundwater surplus, except locally and/or temporarily, because the potential evapotranspiration rate is higher than the precipitation rate. Where potential evapotranspiration exceeds actual precipitation, there is no water available for storage in soil and in plants; dry soil and brown vegetation are the result.

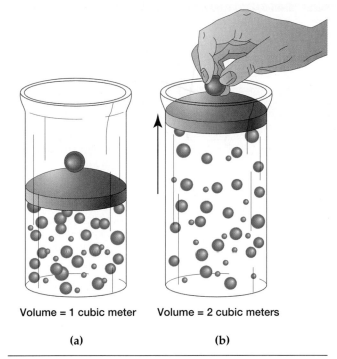

Volume = 1 cubic meter   Volume = 2 cubic meters

(a)   **(b)**

Figure 6-7   Absolute humidity changes as air volume changes. **(a)** With the piston in this position, the container has a volume of 1 cubic meter. If the amount of water vapor present in this volume is 10 grams, the absolute humidity is 10 grams per cubic meter. **(b)** When the volume is increased to 2 cubic meters, the 10 grams of water vapor is distributed over twice the volume, resulting in an absolute humidity of only 5 grams per cubic meter.

MEASURES OF HUMIDITY

The amount of water vapor in the air is referred to as **humidity**. It can be measured and expressed in a number of ways, each useful for certain purposes.

ABSOLUTE HUMIDITY

A direct measure of the water vapor content of air is **absolute humidity,** which is the amount of water vapor in a given volume of air. Absolute humidity is normally expressed in grams of vapor per cubic meter of air (1 gram is approximately 0.035 ounces, and 1 cubic meter is about 35 cubic feet) and so changes as the volume of air being considered changes. For instance, suppose we have 1 cubic meter of air in a container that has a movable piston as its top, as in Figure 6-7(a). If there are 10 grams of water vapor in the container, the absolute humidity is 10 grams per cubic meter. If we then pull the piston up until the volume is 2 cubic meters, as in Figure 6-7(b), we now have 10 grams of water vapor in 2 cubic meters, which means the absolute humidity is

10 grams/2 cubic meters = 5 grams per cubic meter

Thus if the volume of air changes, the value of the absolute humidity also changes, even though there is no change in the amount of water vapor present.

Absolute humidity has a limiting value that depends on temperature. The colder the air, the less water vapor it can hold, as we saw in Figure 6-4. Thus warm air is capable of a much higher absolute humidity than cold air.

SPECIFIC HUMIDITY

The mass of water vapor in a given mass of air is called the **specific humidity** and is usually expressed in grams of vapor per kilogram of air. Specific humidity has the advantage of changing only as the quantity of water vapor varies; it is not affected by variations in air volume the way absolute humidity is. Specific humidity is particularly useful in studying the characteristics and movements of air masses (Chapter 7).

Both absolute humidity and specific humidity are direct indications of the potential for precipitation in a parcel of air. They are measures of the quantity of water that could be extracted by precipitation.

RELATIVE HUMIDITY

The most familiar of humidity measures is **relative humidity**, which is the amount of water vapor in the air at a given temperature compared with the amount that could be there if the air were saturated at that temperature. Relative humidity is not a direct measure of amount of water vapor the way absolute and specific humidity are. Rather, it is a ratio that is expressed as a percentage (by multiplying the ratio by 100). In essence, it is the percentage of saturation.

To see how relative humidity is calculated, suppose a certain volume of air at 70°F can contain 80 grams of water vapor when saturated. If it is a dry 70° day, that volume might contain only 40 grams of water vapor, giving it a relative humidity of

(40 grams/80 grams) × 100 = 50%

One way of looking at this number is to say that this air is 50 percent of the way to being saturated.

Relative humidity is determined by the balance between water vapor content, on the one hand, and the air's capacity for holding water vapor, on the other. In turn, the capacity for holding water vapor is determined primarily by temperature. If water vapor is added to the air by evaporation, the relative humidity increases because the top number of the ratio increases; if water vapor is removed from the air by condensation or dispersal, the relative humidity decreases because the top number of the ratio decreases. If the air's capacity for holding water decreases (normally

FOCUS

THE HEAT INDEX

Sensible temperatures may be significantly influenced by humidity, which can make the weather seem either colder or warmer than it actually is. Humidity can make us feel either warmer than it is or colder than it is, but it is more likely to impinge on our lives in hot weather. Quite simply, high humidity makes hot weather seem hotter.

The National Weather Service has developed a **heat index** that combines temperature and relative humidity to produce an "apparent temperature" that quantifies how hot the air feels to one's skin. A sample heat index chart is given in Figure 6-1-A.

To this index has been added a general heat stress index to indicate heat-related dangers at various apparent temperatures, as shown in the accompanying tabulation.

General Heat Stress Index		
Danger category	*Heat index*	*Heat syndrome*
IV. Extreme danger	Above 130 °F (54 °C)	Heat/sunstroke highly likely with continued exposure
III. Danger	105–130 °F (40–54 °C)	Sunstroke, heat cramps, or heat exhaustion likely; heatstroke possible with prolonged exposure and/or physical activity
II. Extreme caution	90–105 °F (32–40 °C)	Sunstroke, heat cramps, and heat exhaustion possible with prolonged exposure and/or physical activity
I. Caution	80–90 °F (27–32 °C)	Fatigue possible with prolonged exposure and/or physical activity

Temperature (F°)	10	20	30	40	50	60	70	80	90	
104	98°	104°	110°	120°	132°	*	*	*	*	
102	97°	101°	108°	117°	125°	*	*	*	*	
100	95°	99°	105°	110°	120°	132°	*	*	*	
98	93°	97°	101°	106°	110°	125°	*	*	*	IV
96	91°	95°	98°	104°	105°	120°	128°	*	*	
94	89°	93°	95°	100°	105°	111°	120°	*	*	
92	87°	90°	92°	96°	100°	106°	115°	122°	*	
90	85°	88°	90°	92°	96°	100°	106°	114°	122°	
88	82°	86°	87°	89°	93°	95°	100°	106°	115°	III
86	80°	84°	85°	87°	90°	92°	96°	100°	109°	
84	78°	81°	83°	85°	86°	89°	91°	95°	99°	
82	77°	79°	80°	81°	84°	86°	89°	91°	95°	II
80	75°	77°	78°	79°	81°	83°	85°	86°	89°	
78	72°	75°	77°	78°	79°	80°	81°	83°	85°	I
76	70°	72°	75°	76°	77°	77°	77°	78°	79°	
74	68°	70°	73°	74°	75°	75°	75°	76°	77°	

Relative Humidity (%)

* Beyond the capacity of the Earth's atmosphere to hold water vapor

Figure 6-1-A How hot it feels (heat index).

because the temperature drops), the relative humidity increases because the bottom number of the ratio decreases. If the air's capacity for holding water increases (usually because the temperature increases), the relative humidity decreases because the bottom number of the ratio increases.

See Appendix IV for a table used in determining relative humidity.

The relationship between relative humidity and air temperature is portrayed in Figure 6-8, which demonstrates the fluctuations in these two variables during a typical day. It is important to note that there is no variation in the amount of water vapor in the air all day long. In the early morning, the temperature is low and the relative humidity is high because the air's capacity for holding moisture is low. As the air warms up during the day, the relative humidity declines because the warmer air has a higher capacity for moisture than the cooler air. With the approach of evening, air temperature decreases, the air's moisture-holding capacity diminishes, and relative humidity rises.

RELATED HUMIDITY CONCEPTS

Two other concepts related to relative humidity are useful in a study of physical geography; they are discussed in the next two subsections.

Dew Point As we have seen, when air is cooled, the maximum amount of water vapor it can hold decreases. If the amount of water vapor in a sample of air is kept constant, therefore, cooling can mean that formerly unsaturated air becomes saturated. According to Figure 6-4, for instance, 1 cubic meter of air at 80 °F (26 °C) is saturated when it contains 25 grams of water

vapor. Suppose we have 1 cubic meter of air at 80 °F but containing only 10 grams of water vapor. As we cool this volume of air while keeping the amount of water vapor constant, the air remains unsaturated until its temperature is 45 °F (9 °C). At that temperature, according to Figure 6-4, the air is saturated. When this saturation point is reached, condensation begins.

The temperature at which saturation is reached is called the **dew point**. The dew point varies, of course, with the moisture content of the air. If we had 20 grams of water vapor in the example we just worked through, the dew point would be at 70 °F (21 °C) rather than at 45°.

Sensible Temperature The term **sensible temperature** refers to the temperature sensed by a person's body. It involves not only the actual air temperature but also other atmospheric conditions, particularly relative humidity and wind, that influence our perception of heat and cold.

On a warm, humid day, the air seems hotter than the thermometer indicates and the sensible temperature is said to be high. This is because the air is near saturation, and so perspiration on the human skin does not evaporate readily. Thus there is little evaporative cooling and the air seems warmer than it actually is. On a warm, dry day, evaporative cooling is effective and thus the air seems cooler than it actually is; in this case, we say that the sensible temperature is low.

On a cold, humid day, the coldness seems more piercing because body heat is conducted away more rapidly in damp air; the sensible temperature is again described as low. On a cold, dry day, body heat is not conducted away as fast. The temperature seems warmer than it actually is, and we say that the sensible temperature is high.

The amount of wind movement also affects sensible temperature, primarily by its influence on evaporation and the convecting away of body heat.

CONDENSATION

Condensation is the opposite of evaporation. It is the process whereby water vapor is converted to liquid water. In other words, it is a change in state from gas to liquid. In order for condensation to take place, the air must be saturated. In theory, this saturated state can come about through the addition of water vapor to the air, but in practice it is usually the result of the air being cooled to a temperature below the dew point.

Saturation alone is not enough to cause condensation, however. The characteristic of water called surface tension makes it virtually impossible to grow droplets of

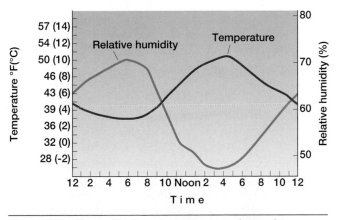

Figure 6-8 Typically there is an inverse relationship between temperature and relative humidity on any given day. As the temperature rises, the relative humidity decreases. Thus relative humidity tends to be lowest in midafternoon and highest just before dawn.

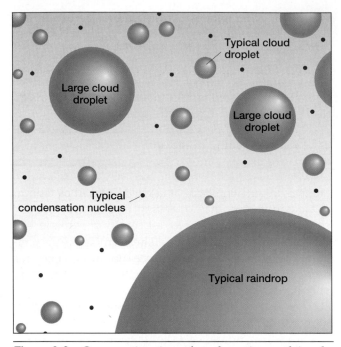

Figure 6-9 Comparative sizes of condensation nuclei and precipitation particles. The former can be particles of dust, smoke, salt, pollen, bacteria, or any other submicroscopic matter found in the air.

pure water. Because surface tension inhibits an increase in surface area, it makes it very difficult for molecules to enter a droplet. (On the other hand, molecules can easily leave a small droplet by evaporation, thereby decreasing its area.) Thus it is necessary to have a surface on which condensation can take place. If no such surface is available, no condensation occurs. In such a situation, the air becomes supersaturated if cooling continues.

Normally, plenty of surfaces are available for condensation. At ground level, availability of a surface is obviously no problem. In the air above the ground, there is also usually an abundance of "surfaces," as represented by tiny particles of dust, smoke, salt, pollen, and other compounds. (The most common particles around which condensation takes place are bacteria that are either blown off plants or thrown into the air by ocean waves.) Most of these various particles are submicroscopic and therefore invisible to the naked eye (Figure 6-9). They are most concentrated over cities, seacoasts, and volcanoes, which are the source of much particulate matter, but are present in lesser amounts throughout the troposphere. They are referred to as **hygroscopic particles** or **condensation nuclei**, and they serve as collection centers for water molecules.

As soon as the air temperature cools to the dew point, water vapor molecules begin to condense around condensation nuclei. The droplets grow rapidly as more and more water vapor molecules stick to them, and as they become larger, they bump into one another and coalesce (which means the colliding

droplets stick together). Continued growth can make them large enough to be visible, forming haze or cloud particles. The diminutive size of these particles can be appreciated by realizing that a single raindrop may contain a million or more condensation nuclei plus all their associated moisture.

Clouds often are composed of water droplets even when their temperature is below freezing. Although water in large quantity freezes at 32°F (0°C), if it is dispersed as fine droplets, it can remain in liquid form at temperatures as cold as -40°F (-40°C). Water that persists in liquid form at temperatures below freezing is said to be supercooled. Supercooled droplets are important to condensation because they promote the growth of ice particles in cold clouds by freezing around them or by evaporating into vapor from which water molecules are readily added to the ice crystals.

ADIABATIC PROCESSES

One of the most significant facts in physical geography is that the only way in which large masses of air can be cooled to the dew point is by expansion as the air masses rise. Thus the only prominent mechanism for the development of clouds and the production of rain is adiabatic cooling. As we noted in Chapter 4, when air rises, its pressure decreases, and so it expands and cools adiabatically.

As a parcel of unsaturated air rises, it cools at the relatively steady rate of 5.5°F per 1000 feet (10°C per kilometer). This is known as the **dry adiabatic lapse rate.** (The term is a misnomer: the air is not necessarily dry; it simply is not saturated.) If the air mass rises high enough, it cools to the dew point, condensation begins, and clouds form. The altitude at which this occurs is known as the **lifting condensation level (LCL).** Under normal circumstances the LCL is clearly visible as the base of the clouds that form (Figure 6-10).

As soon as condensation begins, latent heat is released (this heat was stored originally as the latent heat of vaporization). If the air continues to rise, cooling continues but release of the latent heat slackens the rate of cooling. This diminished rate of cooling is called the **saturated adiabatic lapse rate** (Figure 6-11) and depends on temperature and pressure but averages about 3.3°F per 1000 feet (5°C per kilometer).

Adiabatic warming occurs when air descends. This increase in temperature increases the capacity of the air for holding moisture and thus causes saturated air to become unsaturated. Therefore, any descending air warms at the dry adiabatic lapse rate.

In any consideration of adiabatic temperature changes, remember that we are dealing with air that is rising or descending. The adiabatic lapse rates are not to be confused with the average lapse rate described in

Figure 6-10 The flat bottoms of these clouds represent the lifting condensation level. (Alan L. Detrick/Photo Researchers, Inc.)

Chapter 4, which pertains to still air (Figure 6-12). To understand the difference between the two lapse rates, consider a rising balloon that has one thermometer on the inside and one taped to the outside, as in Figure 6-13. The interior thermometer records temperature changes in the air inside the balloon, which is the rising air; therefore this thermometer is recording the adiabatic lapse rate. The exterior thermometer records temperature changes in the still air through which the

balloon is passing; this thermometer is recording the normal lapse rate.

CLOUDS

Clouds are collections of minute droplets of water or tiny crystals of ice. They are the visible expression of condensation and provide perceptible evidence of other things happening in the atmosphere. They provide at a glance some understanding of the present weather and often are harbingers of things to come. At any given time, about 50 percent of Earth is covered by clouds, the basic importance of which is that they are the source of precipitation. Not all clouds precipitate, but all precipitation comes from clouds.

Clouds are also important because of their influence on radiant energy. They receive both insolation from above and terrestrial radiation from below, and then either absorb, reflect, scatter, or reradiate this energy. The function of clouds in the global energy budget is important.

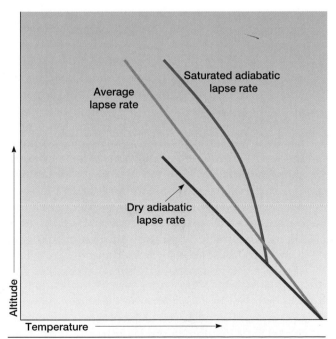

Figure 6-11 Relationships of hypothetical vertical cooling rates. Unsaturated rising air cools at the dry adiabatic lapse rate. Saturated rising air cools at the saturated adiabatic lapse rate. Nonrising air experiences vertical cooling at various rates, generalized as the average lapse rate.

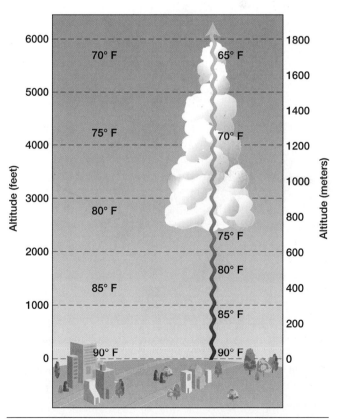

Figure 6-12 Comparison of lapse rates. The column of temperatures above the buildings on the left represents the vertical temperature gradient in a nonrising column of air, which is generalized as the average lapse rate (3.6 °F per 1000 feet = 5.1 °C per kilometer). The red arrow on the right represents rising air that is cooling at the dry adiabatic lapse rate (5.5 °F per 1000 feet = 10 °C per kilometer). When condensation begins, the air continues to rise but the rate of cooling is reduced to the saturated adiabatic lapse rate (3.3 °F per 1000 feet = 5 °C per kilometer).

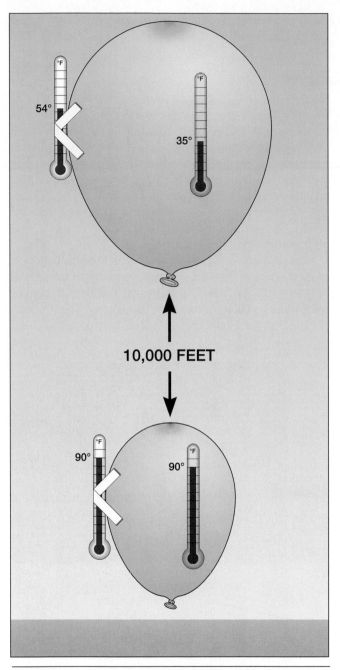

Figure 6-13 This rising balloon contains a parcel of rising air that is being cooled adiabatically as its volume increases with increasing altitude. The thermometer inside the balloon shows the temperature changes associated with this adiabatic cooling. The thermometer outside the balloon shows the vertical temperature differences in the nonrising air. Thus the reading on the interior thermometer changes from 90 °F at the bottom to 35 °F at the top, whereas the reading on the exterior one changes from 90 °F to 54 °F.

Classifying Clouds   Although clouds occur in an almost infinite variety of shapes and sizes, certain general forms recur commonly. Moreover, the various cloud forms are normally found only at certain generalized altitudes, and it is on the basis of these two factors—form and altitude—that clouds are classified (Table 6-1).

The international classification scheme for clouds recognizes three forms (Figure 6-14). **Cirriform clouds** (Latin *cirrus*, a lock of hair) are thin and wispy and composed of ice crystals rather than water droplets. **Stratiform clouds** (Latin *stratus*, spread out) appear as grayish sheets that cover most or all of the sky, rarely being broken up into individual cloud units. **Cumuliform clouds** (Latin *cumulus*, mass or pile) are massive and rounded, usually with a flat base and limited horizontal extent but often billowing upward to great heights.

These three cloud forms are subclassified into ten types based on shape (Figure 6-15). The types overlap, and cloud development frequently is in a state of change, so that one type may evolve into another. Three of the ten types are purely of one form, and these are called **cirrus clouds, stratus clouds,** and **cumulus clouds.** The other seven types are combinations of these three. Cirrocumulus clouds, for example, have the wispiness of cirrus clouds and the puffiness of cumulus clouds.

As the final detail of the international classification scheme, the ten cloud types are divided into four families on the basis of altitude:

1. *High clouds* are generally found above 20,000 feet (6 kilometers). Because of the small amount of water vapor and low temperature at such altitudes, these clouds are thin, white, and composed of ice crystals. Included in this family are **cirrus, cirrocumulus,** and **cirrostratus.** These high clouds often are harbingers of an approaching weather system or storm.

2. *Middle clouds* normally occur between about 6500 and 20,000 feet (2 and 6 kilometers). They may be either stratiform or cumiliform and are composed of liquid water. Included types are **altocumulus** and **altostratus.** The puffy altocumulus clouds usually indicate settled weather conditions, whereas the lengthy altostratus often are associated with changing weather.

3. *Low clouds* usually are below 6500 feet (2 kilometers). They sometimes occur as individual clouds but more often appear as a general overcast. Low cloud types include **stratus, stratocumulus,** and **nimbostratus.** These low clouds often are widespread and are associated with somber skies and drizzly rain.

4. A fourth family, *clouds of vertical development*, grow upward from low bases to heights of as much as 60,000 feet (15 kilometers). Their horizontal spread is usually very restricted. They indicate very active vertical movements in the air. The relevant types are **cumulus,** which usually indicate fair weather, and **cumulonimbus,** which are storm clouds.

TABLE 6-1
The International Classification Scheme For Clouds

Family	Type	Form	Characteristics
High	Cirrus	Cirriform	
	Cirrocumulus	Cirriform	Thin, white, icy
	Cirrostratus	Cirriform	
Middle	Altocumulus	Cumuliform	Layered or puffy;
	Altostratus	Stratiform	made of liquid water
Low	Stratus	Stratiform	
	Stratocumulus	Stratiform	General overcast
	Nimbostratus	Stratiform	
Vertical	Cumulus	Cumuliform	
	Cumulonimbus	Cumuliform	Tall, narrow, puffy

(a)

(b)

(c)

Figure 6-14 **(a)** The classic cirrus clouds referred to as "mare's tails" because of their shape. (Joyce Photographics/ Photo Researchers, Inc.) **(b)** A low stratus cloud overcast near Needles, Utah. (Claudia Parks/The Stock Market) **(c)** Puffs of fair-weather cumulus clouds over eastern Colorado. (Henry Lansford/Photo Researhcers, Inc.)

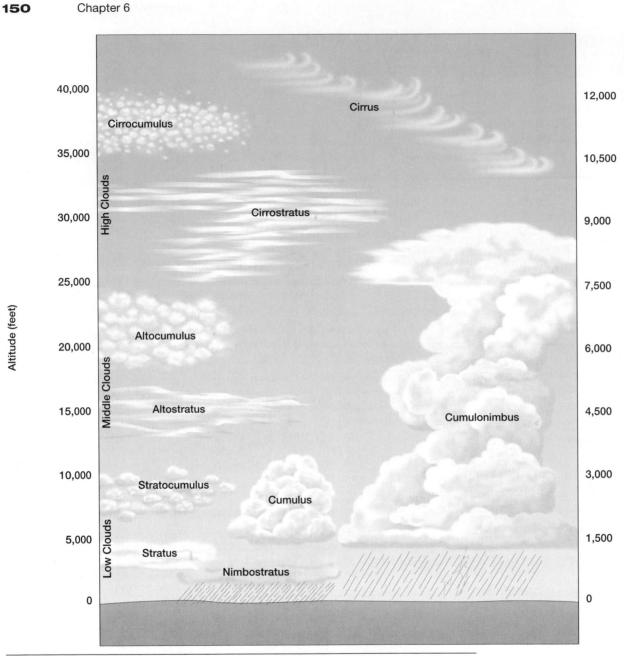

Figure 6-15   Typical shapes and altitudes of the ten principal cloud types.

Precipitation comes only from clouds that have "nimb" in their name, specifically nimbostratus or cumulonimbus. Normally these types develop from other types; that is, cumulonimbus clouds develop from cumulus clouds, and nimbostratus clouds develop from stratus clouds.

Fog

From a global standpoint, fogs represent a minor form of condensation. Their importance to humans is disproportionately high, however, because they can hin-

der visibility enough to make surface transportation hazardous or even impossible (Figure 6-16). **Fog** is simply a cloud on the ground. There is no physical difference between a cloud and fog, but there are important differences in how each forms. Most clouds develop as a result of adiabatic cooling in rising air, but only rarely is uplift involved in fog formation. Instead, most fogs are formed either when air at Earth's surface cools to below its dew point or when enough water vapor is added to the air to saturate it.

There are four generally recognized types of fog (Figure 6-17):

Figure 6-16   An advection fog enveloping the harbor at Ketchikan, Alaska. (TLM photo)

1. A **radiation fog** results when the ground loses heat through radiation, usually at night. The heat radiated out of the ground rises through the lowest layer of air and into higher areas. The air closest to the ground cools as heat flows conductively from it to the relatively cool ground, and fog condenses in the cooled air at the dew point.

2. An **advection fog** develops when warm, moist air moves horizontally over a cold surface, such as snow-covered ground or a cold ocean current. Air moving from sea to land is the most common source of advection fogs.

3. An **upslope fog**, or **orographic fog** (from the Greek *oro*, mountain), is created by adiabatic

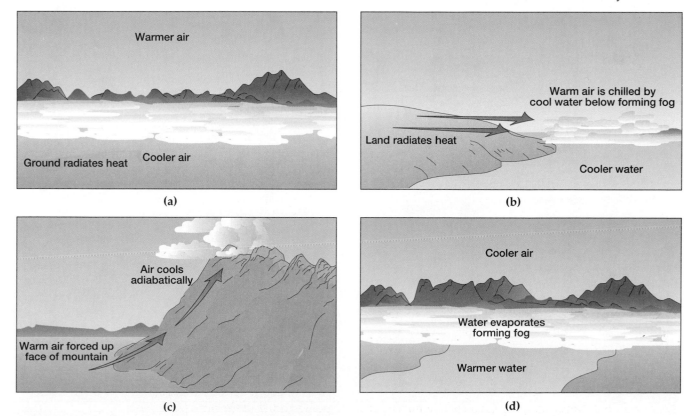

Figure 6-17   The four principal types of fog: **(a)** radiation; **(b)** advection; **(c)** upslope (orographic); **(d)** evaporation.

cooling when humid air climbs a topographic slope.

4. An **evaporation fog** results when water vapor is added to cold air that is already near saturation.

As can be seen in Figure 6-18, areas of heavy fog in North America are mostly coastal. The western-mountain and Appalachian fogs are mostly radiation fogs. Areas of minimal fog are in the Southwest, Mexico, and the Great Plains in both the United States and Canada, where available atmospheric moisture is limited and winds are strong.

DEW

Dew also originates from terrestrial radiation. Nighttime radiation cools objects (grass, pavement, automobiles, or whatever) at Earth's surface, and the adjacent air is in turn cooled by conduction. If the air is cooled enough to reach saturation, tiny beads of water collect on the cold surface of the object. If the temperature is below freezing, ice crystals (*white frost*) rather than water droplets are formed (Figure 6-19).

THE BUOYANCY OF AIR

Because most condensation and precipitation are the result of rising air, conditions that promote or hinder upward movements in the troposphere are obviously of great importance to weather and climate. Depending on how buoyant it is, air rises more freely and extensively under some circumstances than under others. Therefore the concept of air buoyancy is one of the most significant in physical geography.

STABILITY

The tendency of any object to rise in a fluid is called the **buoyancy** of that object. For the present discussion, you should picture a given parcel of air (one having imaginary boundaries) as being the object and the surrounding air as being the fluid (Figure 6-20). As with other gases (and liquids, too), air tends to seek its own level. This means that a parcel of air moves vertically until it reaches a level at which the surrounding air is of equal density. Said another way, if a parcel of air is warmer, and thus less dense, than the surrounding air, it tends to rise. If a parcel is cooler, and therefore more

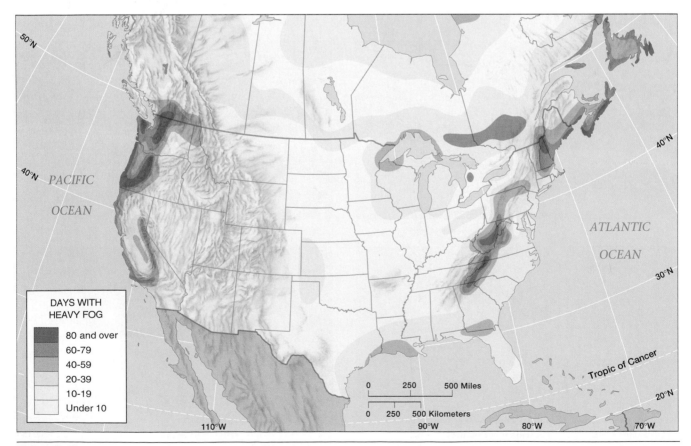

Figure 6-18   Distribution of fog in North America.

(a)   (b)

Figure 6-19   **(a)** Dewdrops on a wild daisy. (Lawrence Pringle/ Photo Researchers, Inc.)
(b) White frost emphasizes the delicate lacework of a spiderweb in Yellowstone Park's
Hayden Valley. (TLM photo)

dense, than the surrounding air, it tends either to sink or at least to resist uplift. Thus we say that warm air is more buoyant than cool air.

If a parcel of air resists vertical movement, it is said to be **stable** (Figure 6-21). If stable air is forced to rise, perhaps by coming up against a mountain slope, it does so only as long as the force is applied. Once the

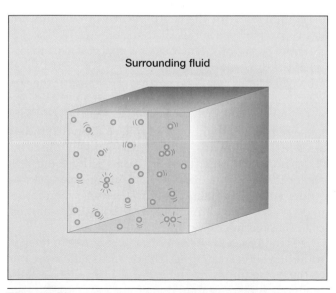

Figure 6-20   A volume of air enclosed by imaginary boundaries. The air enclosed by the boundaries is the buoyant "object," and all the surrounding air is the "fluid."

force is removed, the air sinks back to its former position. In other words, stable air is nonbuoyant. When unstable air comes up against the same mountain, it continues to rise once it has passed the peak.

In the atmosphere, high stability is promoted when cold air is beneath warm air, a condition most frequently observed during a temperature inversion. With colder, denser air below warmer, lighter air, upward movement is unlikely. A cold winter night is a typical, highly stable situation, although high stability can also occur in the daytime. Because it does not rise and therefore stays at an essentially constant pressure and volume, highly stable air obviously provides little opportunity for adiabatic cooling unless there is some sort of forced uplift. Highly stable air is normally not associated with cloud formation and precipitation.

Air is said to be **unstable** if it either rises without any external force other than the buoyant force or continues to rise after such an external force has ceased to function. In other words, unstable air is buoyant. When a mass of air is heated enough so that it is warmer than the surrounding air, it becomes unstable. This is a typical condition on a warm summer afternoon. The unstable air rises until it reaches an altitude where the surrounding air has similar temperature and density, which is referred to as the *equilibrium level* (Figure 6-22). While ascending, it will be cooled adiabatically. In this situation, clouds are likely to form.

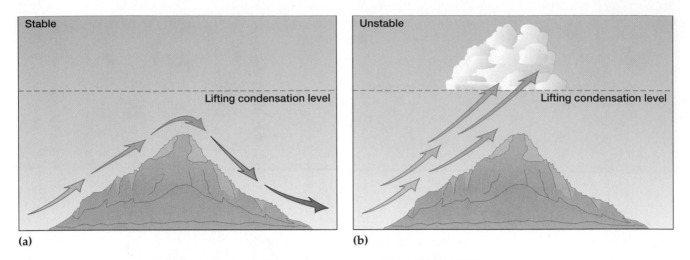

Stable

Unstable

Lifting condensation level

Lifting condensation level

(a)

(b)

Figure 6-21 **(a)** As stable air blows over a mountain, it rises only as long as it is forced to do so by the mountain slope. On the leeward side, it moves downslope. **(b)** When unstable air is forced up a mountain slope, it is likely to continue rising of its own accord until it reaches surrounding air of similar temperature and density; if it rises to the condensation level, clouds form.

There is an intermediate condition, called **conditional instability**, between absolute stability and absolute instability. An air parcel is conditionally unstable when its adiabatic lapse rate is somewhere between the dry and saturated adiabatic rates. Left alone, such a parcel acts like stable air. When forced to rise, however, it does so only until condensation begins, at which altitude the release of latent heat may provide enough buoyancy to make the parcel unstable. It then behaves the way unstable air does, rising until it reaches an altitude where the surrounding air has density and temperature similar to its own.

Figure 6-22 On a summer day, the air just above the blacktop surface gets warmer than the air above the grass. This unstable warmer air rises until it reaches an altitude where the surrounding air has similar temperature and density.

DETERMINING AIR STABILITY

The various degrees of air stability, from highly stable to unstable, are related to lapse rates in Figure 6-23. An accurate determination of the stability of any mass of air depends on temperature measurements, but a rough indication often can be obtained simply by observing the state of the sky (primarily the cloud forms).

Determination of Stability via Temperature and Lapse Rate

The temperature of the rising air can be compared with the temperature of surrounding nonrising air by a series of thermometer readings at different elevations. The rising air cools (at least initially) at the dry adiabatic lapse rate of 5.5 °F per 1000 feet (10 °C per kilometer). The normal lapse rate of the surrounding (nonrising) air depends on many things and may be either faster or slower than the dry adiabatic lapse rate. If the normal lapse rate of the surrounding air is less than the dry adiabatic lapse rate of the rising air, as in parts (a) and (b) of Figure 6-23, the rising air is cooler than the surrounding air and therefore stable. Under such conditions, the air rises only when forced to do so. Once the lifting force is removed, the air ceases to rise and may sink.

Instability results if the normal lapse rate of the surrounding air is greater than the dry adiabatic lapse rate of the rising air, as in Figure 6-23(d). In this case, the rising air is warmer than the air around it and so rises until it reaches an elevation where the surrounding air is of similar temperature and density.

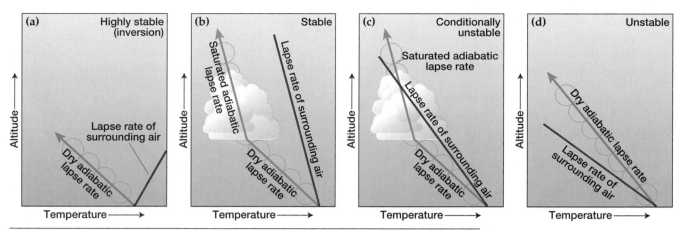

Figure 6-23 In these drawings the "lapse rate of surrounding air" shows the temperature of the atmosphere at different elevations, whereas the "dry adiabatic lapse rate" and the "saturated adiabatic lapse rate" show the temperature change of a rising parcel of air. At any given elevation, if the temperature of the parcel is warmer than that of the surrounding air, the parcel will be unstable; if the temperature of the rising parcel of air is the same as or cooler than the surrounding air, the parcel will be stable. **(a)** A surrounding-air lapse rate line sloping to the right represents a temperature inversion, which means that the parcel of air is highly stable. **(b)** If the lapse rate of the surrounding air is less than the dry adiabatic lapse rate of the parcel of air, conditions are stable, and the parcel rises only if forced to do so. **(c)** If the surrounding-air lapse rate is between the dry and the saturated adiabatic rates of the parcel, the parcel is conditionally unstable; it rises only if forced until condensation begins and after that rises of its own accord. **(d)** If the lapse rate of the surrounding air is greater than the dry adiabatic lapse rate of the parcel of air, conditions are unstable and the parcel rises because of its buoyancy.

The situation becomes more complicated if the rising air is cooled to its dew point. Reaching this point causes condensation to begin, and the rising air then cools at a slower rate (the saturated adiabatic lapse rate) because latent heat of condensation is released. This situation increases the tendency toward instability and reinforces the rising trend.

Visual Determination of Stability The cloud pattern is often indicative of air stability. Unstable air is associated with distinct updrafts, which are likely to produce vertical clouds (Figure 6-24). Thus the presence of cumulus clouds suggests instability, and a towering cumulonimbus cloud is an indicator of pronounced instability. Horizontally developed clouds,

Figure 6-24 Multiple cumulus clouds in the foreground with a cumulonimbus developing an anvil top in the background. This is a localized thunderstorm near Wiluna, Western Australia. (TLM photo)

most notably stratiform, are characteristic of stable air forced to rise, and a cloudless sky is a strong indicator of stable air that is immobile.

Regardless of stability conditions, no clouds form unless the air is cooled to the dew point. Thus the mere absence of clouds is not certain evidence of stability; it is only an indication. By the same token, unstable air can produce a cloudless sky if the dew point is not reached.

The general features of stable and unstable air are summarized in Table 6-2.

PRECIPITATION

All precipitation originates in clouds, but most clouds do not yield precipitation. Exhaustive experiments have demonstrated that condensation alone is insufficient to produce raindrops. The tiny water droplets that make up clouds cannot fall to the ground as rain because their size makes them very buoyant and the normal turbulence of the atmosphere keeps them aloft. Even in still air, their fall would be so slow that it would take many days for them to reach the ground from even a low cloud. Besides that, most droplets would evaporate in the drier air below the cloud before they made a good start downward.

Despite these difficulties, rain and other forms of precipitation are commonplace in the troposphere. What is it, then, that produces precipitation in its various forms?

THE PROCESSES

An average-sized raindrop contains several million times as much water as the average-sized water droplet found in any cloud. Consequently, great multitudes of droplets must join together in order to form a drop large enough to overcome both turbulence and evaporation and thus be able to fall to Earth under the influence of gravity.

It is still not well understood why most clouds do not produce precipitation, but two mechanisms are believed to be principally responsible for producing precipitation particles: (1) ice-crystal formation, and (2) collision and coalescence of water droplets.

Ice-Crystal Formation   Many clouds or portions of clouds extend high enough to have temperatures well below the freezing point of liquid water. In this situation, ice crystals and supercooled water droplets often coexist in the cloud. These two types of particles are in direct competition for the water vapor that is not yet condensed. There is lower vapor pressure around the ice crystals, which means that if there is just enough water vapor in the air to keep a liquid droplet in equilibrium (here the term *in equilibrium* means that, for any one droplet, the evaporation rate is equal to the condensation rate), there is more than enough moisture to keep the ice in equilibrium. Thus the ice crystals attract most of the vapor, and the water droplets, in turn, evaporate to replenish the diminishing supply of vapor, as shown in Figure 6-25. Therefore the ice crystals grow at the expense of the water droplets until the crystals are large enough to fall. As they descend through the lower, warmer portions of the cloud, they pick up more moisture and become still larger. They may then either precipitate from the cloud as snowflakes or melt and precipitate as raindrops.

Precipitation by ice-crystal formation was first proposed by the Swedish meteorologist Tor Bergeron more than half a century ago. It is now known as the **Bergeron process** and is believed to account for the majority of precipitation outside of tropical regions.

Collision/Coalescence   In many cases, particularly in the tropics, cloud temperatures are too high for ice crystals to form. In such clouds, rain is produced by the collision and coalescing (merging) of water droplets. As mentioned earlier, condensation alone cannot yield rain because it produces lots of small

TABLE 6-2 Characteristics of Stable and Unstable Air	
Stable Air	*Unstable Air*
Nonbuoyant, remains immobile unless some outside agent exerts a force on it	Buoyant, rises even when no outside force is acting on it
Causes either no clouds or stratiform clouds	Causes either no clouds (if dew point not reached) or cumuliform clouds
Leads either to no precipitation or to light drizzles	Leads either to no precipitation or to showery precipitation

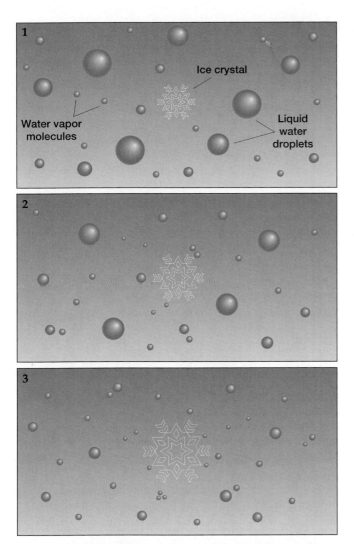

Figure 6-25   Precipitation by means of ice-crystal formation in clouds (the Bergeron process). Ice crystals grow by attracting water vapor to themselves, causing the liquid water droplets that make up the cloud to evaporate in order to replenish the water vapor supply. The process of growing ice crystals and shrinking cloud droplets may continue until the ice crystals are large and heavy enough to fall. (Particle sizes are greatly exaggerated in this and Figure 6-26.)

droplets but no large drops. Thus, droplets, the first things formed during condensation, must coalesce into drops large enough to fall. In a sense, the role of condensation (vapor molecules change to liquid molecules) is to enable droplets to grow large enough to coalesce (liquid droplets merge together to form larger and larger droplets, which eventually are large enough to be called drops).

Not all collisions result in coalescence, however. Apparently coalescence is assured only if atmospheric electricity is favorable, that is, only if a positively charged droplet collides with a negatively charged one.

Different-sized droplets fall at different speeds. Larger ones fall faster, overtaking and often coalescing with smaller ones, which are swept along in the descent (Figure 6-26). This sequence of events favors the continued growth of the larger particles. The larger they grow, the faster they grow.

Collision/coalescence is the process most responsible for precipitation in the tropics, and it also produces much precipitation in the middle latitudes.

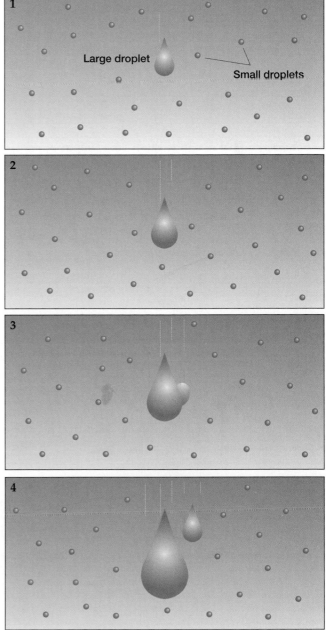

Figure 6-26   Raindrops forming by collision and coalescence. Large droplets fall more rapidly than small ones, coalescing with some and sweeping others along in their descending path. As droplets become larger during descent, they sometimes break apart.

PEOPLE AND THE ENVIRONMENT

ACID RAIN

One of the most vexing and perplexing environmental problems of the latter part of the twentieth century has been the rapidly increasing intensity, magnitude, and extent of **acid rain**. This term refers to the deposition of either wet or dry acidic materials from the atmosphere on Earth's surface. Although most conspicuously associated with rainfall, the pollutants may fall to Earth with snow, sleet, hail, or fog, or in the dry form of gases or particulate matter.

Sulfuric and nitric acids are the principal culprits recognized thus far. Although there is no universal agreement on the exact origin and processes involved, evidence indicates that the principal human-induced sources are sulfur dioxide emissions from smokestacks and nitrogen oxide exhaust from motor vehicles (Figure 6-2-A). These and other emissions of sulfur and nitrogen compounds are expelled into the air, where they may be wafted hundreds or even thousands of miles by winds. During this time they may mix with atmospheric moisture to form the sulfuric and nitric acids that sooner or later are precipitated.

Acidity is measured on a pH scale based on the relative concentration of hydrogen ions (Figure 6-2-B). The scale ranges from 0 to 14, where the lower end represents extreme acidity (battery acid has a pH of 1) and the upper end extreme alkalinity. (Alkalinity is the opposite of acidity; a substance that is very acidic can also be characterized as being of very low alkalinity, and a highly alkaline substance has a very low acidity. The alkaline chemical lye, for instance, has a pH of 13). The pH scale is a logarithmic scale,

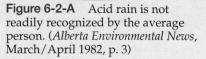

Figure 6-2-A Acid rain is not readily recognized by the average person. (*Alberta Environmental News*, March/April 1982, p. 3)

which means that a difference of one whole number on the scale reflects a tenfold change in absolute values.

Rainfall in clean, dust-free air has a pH of about 5.6, slightly acidic. Increasingly, however, precipitation with a pH of less than 4.5 (the level below which most fish perish) is being recorded, and an acid fog with a record low of 1.7 was measured in California in 1982.

Many parts of Earth's surface have naturally alkaline soil or bedrock that neutralizes acid precipitation. Soils de-

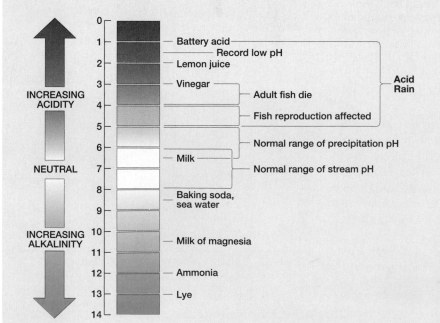

Figure 6-2-B The pH scale.

veloped from limestone, for example, contain calcium carbonate, which can neutralize acid. Granitic soils, on the other hand, have no neutralizing component (Figure 6-2-C).

Although the long-term effects of acid precipitation on human health and agricultural production have not yet been determined, it is known to be a major hazard to the environment. The most conspicuous damage is being done to aquatic ecosystems. Several hundred lakes in the eastern United States and Canada have become biological deserts in the last quarter century, primarily due to acid rain. The precise effects of acid rain on forests are not clearly understood, but evidence keeps mounting that it is the major culprit in forest diebacks currently taking place on every continent except Antarctica. Even buildings and monuments are being destroyed; acid deposition has caused more erosion on Athens' marble Parthenon in the last 30 years than took place in the previous 30 centuries.

One of the great complexities of the situation is that much of the pollution is deposited at great distances from its source. Downwind locations receive unwanted acid deposition from upwind origins. Thus Scandinavians and Germans complain about British pollution; Canadians blame U.S. sources; New Englanders accuse the Midwest. It is understandably difficult to persuade people in Ohio to finance expensive cleanup costs that will benefit forests in Maine.

One of the thorniest issues in North American international relations is the Canadian dissatisfaction with the approach of the U.S. government toward mitigation of acid rain. In general, acid rain is viewed by Canadians as their gravest environmental concern, and it is believed that about half the acid rain that falls on Canada comes from U.S.

sources, particularly older coal-burning power plants in the Ohio and Tennessee river valleys. Since the Clean Air Act was passed by the U.S. Congress in 1970, coal-fired power plants have invested more than $60 billion in mitigation efforts, resulting in a 25 percent decrease in sulfur emissions despite an almost doubling of the amount of coal being used. Still, the problem has magnified, and heroic efforts and staggering costs are obviously required to further sanitize emissions. In Canada, the federal and provincial governments have bitten the bullet by legislating a requirement for 50 percent reduction in acid rain emissions by the year 1994. In 1990 the U.S. Congress finally promulgated comparable legislation with amend-

ments to the Clean Air Act. This paved the way for a bilateral agreement between the United States and Canada, signed in 1991, which codified the principle that countries are responsible for the effects of their air pollution on one another. Specifics of the agreement include a permanent cap placed on sulfur dioxide emissions in both countries, a reduction schedule for emissions of nitrogen oxide, and stricter standards for new motor vehicles.

Acid rain is now clearly in the public consciousness. Governments and citizen groups are becoming increasingly mobilized. Although the costs of reducing the acid rain problem will be enormous, it is clear that the costs of not doing so will be exponentially greater.

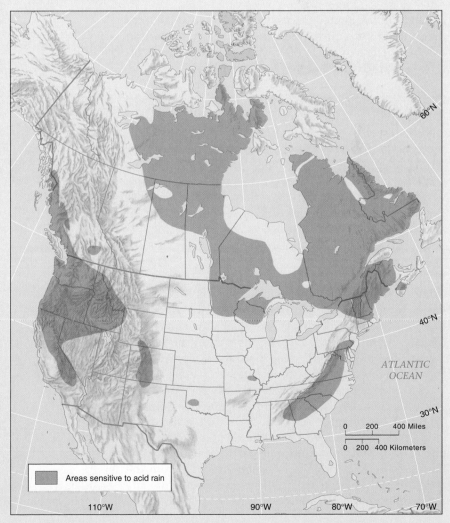

Figure 6-2-C Areas in the United States and Canada particularly sensitive to acid rain because of a scarcity of natural buffers.

FORMS OF PRECIPITATION

Several forms of precipitation can result from the processes just described, depending on air temperature and turbulence.

Rain   By far the most common and widespread form of precipitation is **rain**, which consists of drops of liquid water. Most rain is the result of condensation and precipitation in ascending air that has a temperature above freezing, but some results from the thawing of ice crystals as they descend through warmer air.

Meteorologists often make a distinction among "rain," which goes on for a relatively long time, "showers," which are relatively brief and involve large drops, and "drizzle," which consists of very small drops and usually lasts for some time.

Snow   The general name given to solid precipitation in the form of ice crystals, small pellets, or flakes is **snow**. It is formed when water vapor is converted directly to ice without an intermediate water stage. However, the vapor may have evaporated from supercooled cloud droplets.

Sleet   In the United States, **sleet** refers to small raindrops that freeze during descent and reach the ground as small pellets of ice. In other countries, the term is often applied to a mixture of rain and snow.

Glaze   **Glaze** is rain that turns to ice the instant it collides with a solid object. Raindrops fall through a shallow layer of subfreezing air near the ground. Although the drops do not freeze in the air (in other words, they do not turn to sleet), they become supercooled while in this cold layer and are instantly converted to an icy surface when they alight. The result can be a thick coating of ice that makes both pedestrian and vehicular travel hazardous as well as breaking tree limbs and transmission lines.

Hail   The precipitation form with the most complex origin is **hail**, which consists of either small pellets or larger lumps of ice. Hailstones are usually composed of roughly concentric layers of clear and cloudy ice (Figure 6-27). The cloudy portions contain numerous tiny air bubbles among small crystals of ice, whereas the clear parts are made up of large ice crystals.

Hail is produced in cumulonimbus clouds as a result of vertical air currents (updrafts and downdrafts) (Figure 6-28). In order for hail to form from a cloud, the cloud must have a lower part that is at a temperature above 32°F (0°C) and an upper part that is below this temperature. Updrafts carry water droplets from the above-freezing layer or small ice particles from the lowest part of the below-freezing layer upward, where they grow by collecting moisture from supercooled cloud droplets. When the particles become too large to be supported in the air, they fall, gathering more moisture on the way down. If they encounter a sufficiently strong updraft, they may be carried skyward again, only to fall another time. This sequence may be repeated several times, as indicated by the spiral path in Figure 6-28.

A hailstone normally continues to grow whether it is rising or falling in the below-freezing layer, providing it passes through portions of the cloud that contain supercooled droplets. If there is considerable supercooled moisture available, the hailstone, which is ice, becomes surrounded by a wet layer that freezes relatively slowly, producing large ice crystals and forcing the air out of the water. The result is the clear-ice rings in Figure 6-27. If there is a more limited supply of supercooled droplets, the water may freeze almost instantly around the hailstone. This fast freezing produces small crystals with tiny air bubbles trapped among them, forming the opaque rings of ice shown in Figure 6-27.

Hailstones may be formed in a single descent and still consist of concentric shells of varying degrees of opaqueness. In this case, the layering apparently is due to variations in the rate of accumulation of supercooled droplets, with the accumulation rate based on differences in the amount of supercooled water in various parts of the descent path.

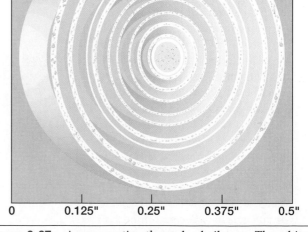

| 0 | 0.125" | 0.25" | 0.375" | 0.5" |

Figure 6-27 A cross section through a hailstone. The white rings are clear, perfectly transparent ice made up of large crystals. The cloudiness in the opaque rings is the result of trapped air bubbles and small ice crystals.

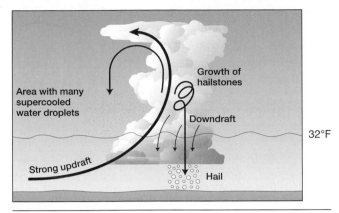

Figure 6-28 Hail is produced in cumulonimbus clouds that are partly at a temperature above the freezing point of water and partly at a temperature below the freezing point of water. The curved and spiral arrows indicate paths a hailstone takes as it is forming. (After John Oliver, *Climatology: Selected Applications*. New York: V. H. Winston and Sons, 1981, p. 140.)

The eventual size of a hailstone depends on the amount of supercooled water in the cloud, the strength of the updrafts, and the total length (up, down, and sideways) of the path taken by the stone through the cloud. The largest known hailstone, which fell in Kansas in 1970, weighed 1.67 pounds (766 grams).

ATMOSPHERIC LIFTING AND PRECIPITATION

The role of rising air and adiabatic cooling has been stressed in this chapter. Only through these events can any significant amount of precipitation originate. It remains for us to consider the causes of rising air. There are four principal types of atmospheric lifting. One type is spontaneous, and the other three require the presence of some external force. More often than not, however, the various types operate in conjunction.

Convective Lifting   Because of unequal heating of different surface areas, a parcel of air near the ground may be warmed by conduction more than the air around it, as we saw, for instance, in Figure 6-22 on page 154. The density of the heated air drops as the air expands, and so the parcel rises toward a lower-density layer, in a typical convective situation, as shown in Figure 6-29(a). This spontaneous uplift is particularly notable if unstable air is involved, which is often the case on a warm summer day. The pressure of the unstable air drops as the air rises, and so it cools adiabatically to the dew point. Condensation begins and a cumulus cloud forms. With the proper humidity, temperature, and stability conditions, the cloud is likely to grow into a towering cumulonimbus thunderhead,

with a downpour of showery raindrops and/or hailstones, accompanied sometimes by lightning and thunder.

An individual convective cell is likely to cover only a small horizontal area, although sometimes multiple cells are formed very close to each other, close enough to form a much larger cell. **Convective precipitation** typically is showery, with large raindrops falling fast and furiously but for only a short duration. It is particularly associated with the warm parts of the world and warm seasons.

In addition to the spontaneous uplift just described, various kinds of forced uplift can trigger formation of a convective cell if the air tends toward instability. Thus convective uplift often accompanies other kinds of uplift.

Orographic Lifting   Topographic barriers that block the path of horizontal air movements are likely to cause large masses of air to travel upslope, as Figure 6-29(b) shows. This kind of forced ascent can produce **orographic precipitation** if the ascending air is cooled to the dew point. As we learned above, if significant instability has been triggered by the upslope motion, the air keeps rising when it reaches the top of the slope and the precipitation continues. More often, however, the air descends the leeward side of the barrier. As soon as it begins to move downslope, adiabatic cooling is replaced by adiabatic warming and condensation/precipitation ceases. Thus the windward slope of the barrier is the wet side, the leeward slope is the dry side, and the term **rain shadow** is applied to both the leeward slope and the area beyond as far as the drying influence extends.

An extreme example of the windward/leeward contrast in orographic precipitation is found on the island of Kauai in Hawaii (Figure 6-30), a small island consisting mostly of one prominent mountain, Mt. Waialeale, the peak of which is 5243 feet (1598 meters) above sea level. Persistent northeastern trade winds bring prodigious amounts of rain to the windward slope, where a weather station records an average of 476 inches (12,090 millimeters) of rain annually. Only 15 miles (24 kilometers) away, however, on the leeward side of the mountain, the average precipitation is only 20 inches (518 millimeters).

Orographic precipitation can occur at any latitude, any season, any time of day. The only requisite conditions are a topographic barrier and moist air to move over it. Orographic precipitation is likely to be prolonged because there is a relatively steady upslope flow of air.

Frontal Lifting   When unlike air masses meet, they do not mix. Rather, a zone of discontinuity called a *front* is established between them, and the warmer air rises over the cooler air, as shown in Figure 6-29(c). As the warmer air is forced to rise, it may be cooled to the dew point with resulting clouds and precipitation. Precipitation that results from this process is referred to as **frontal precipitation.**

We shall discuss frontal precipitation in greater detail in the next chapter. It tends to be widespread and protracted, but frequently it is also associated with convective showers.

Frontal activity is most characteristic of the midlatitudes, and so frontal precipitation is particularly notable in those regions, which are meeting grounds of cold polar air and warm tropical air. It is less significant in the high latitudes and rare in the tropics because those regions contain air masses that tend to be like one another.

Convergent Lifting   Less common than the other three types, but nevertheless significant in some situations, is convergent lifting and the accompanying **convergent precipitation**, illustrated in Figure 6-29(d). Whenever air converges, the result is a general uplift because of the crowding. This forced uplift enhances instability and is likely to produce showery precipitation. It is frequently associated with cyclonic storm systems and is particularly characteristic of the low lat-

itudes. It is common, for example, in the intertropical convergence (ITC) zone (discussed in Chapter 5) and is notable in such tropical disturbances as hurricanes and easterly waves.

GLOBAL DISTRIBUTION OF PRECIPITATION

The most important geographic aspect of atmospheric moisture is the spatial distribution of precipitation. The broadscale zonal pattern is based on latitude, but many other factors are involved and the overall pattern is complex. This concluding section of the chapter focuses on a series of maps that illustrate worldwide and United States precipitation distribution. A major cartographic device used on these maps is the **isohyet,** a line joining points of equal quantities of precipitation.

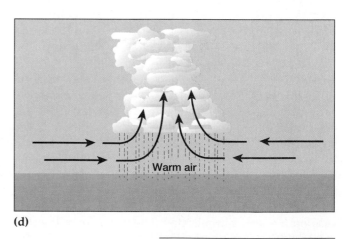

(c)

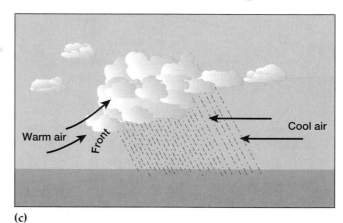

(d)

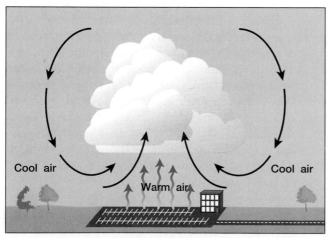

(a)

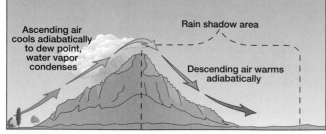

(b)

Figure 6-29   The four basic types of atmospheric lifting and precipitation: **(a)** convective; **(b)** orographic; **(c)** frontal; **(d)** convergent.

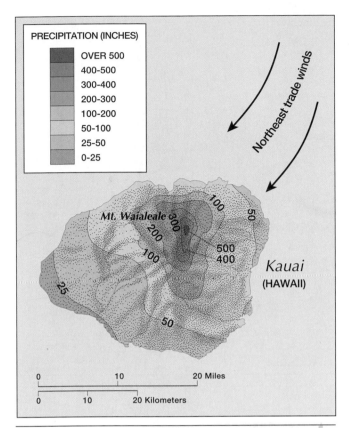

Figure 6-30 Mount Waiaieale on the Hawaiian island of Kauai receives an enormous amount of precipitation on its windward (northeastern) slopes from the persistent trade winds that blow upslope. On the leeward side of the island, the downslope windflow produces very little precipitation. The isolines are isohyets, and the values given are annual rainfall in inches.

AVERAGE ANNUAL PRECIPITATION

The amount of precipitation on any part of Earth's surface is determined by the nature of the air mass involved and the degree to which that air is uplifted. The humidity, temperature, and stability of the air mass are mostly dependent on where the air originated (over land or water, in high or low latitudes) and on the trajectory it has followed. Whether or not uplifting takes place and the amount of that uplifting are determined largely by zonal pressure patterns, topographic barriers, and storms and other atmospheric disturbances.

The most conspicuous feature of the worldwide annual precipitation pattern is that the tropical latitudes contain most of the wettest areas (Figure 6-31). The warm trade winds are capable of carrying enormous amounts of moisture, and where they are forced to rise very heavy rainfall is usually produced. Equatorial regions particularly reflect these conditions, as warm, moist, unstable air is uplifted in the ITC zone, where warm ocean water easily vaporizes. Considerable precipitation also results where trade winds are forced to

rise by topographic obstacles. As the trades are easterly winds, it is the eastern coasts of tropical landmasses—for example, the east coast of Central America, northeastern South America, and Madagascar—where this orographic effect is most pronounced. Where the normal trade-wind pattern is modified by monsoons, the onshore trade-wind flow may reverse direction. Thus the wet areas on the western coast of southeastern Asia, India, and what is called the Guinea Coast of West Africa are caused by the onshore flow of southwesterly winds that are nothing more than trade winds diverted from a "normal" pattern by the South Asian and West African monsoons.

The only other regions of high annual precipitation shown on the world map are narrow zones along the western coasts of North and South America between 40° and 60° of latitude. These areas reflect a combination of frequent onshore westerly airflow, considerable storminess, and mountain barriers running perpendicular to the direction of the prevailing westerly winds. The presence of these north-south mountain ranges near the coast restricts the precipitation to a relatively small area and creates a pronounced rain shadow effect to the east of the ranges.

The principal regions of sparse annual precipitation on the world map are found in three types of locations:

1. Dry lands are most prominent on the western sides of continents in subtropical latitudes (centered at 25° or 30°). High-pressure conditions dominate at these latitudes, particularly on the western sides of continents, which are closer to the normal positions of the subtropical high-pressure cells. High pressure means sinking air, which is not conducive to condensation and precipitation. These dry zones are most extensive in North Africa and Australia, primarily because of the blocking effect of landmasses or highlands to the east. (The presence of such landmasses prevents moisture from coming in from that direction.)

2. Dry regions in the midlatitudes are most extensive in central and southwestern Asia, but they also occur in western North America and southeastern South America. In each case, the dryness is due to lack of access for moist air masses. In the Asian situation, this lack of access is essentially a function of distance from any ocean where onshore airflow might occur. In North and South America, there are rain shadow situations in regions of predominantly westerly airflow.

3. In the high latitudes there is not much precipitation anywhere. Water surfaces are scarce and cold, and so little opportunity exists for moisture to evaporate into the air. As a result, polar air masses have low absolute humidities and precipitation is slight.

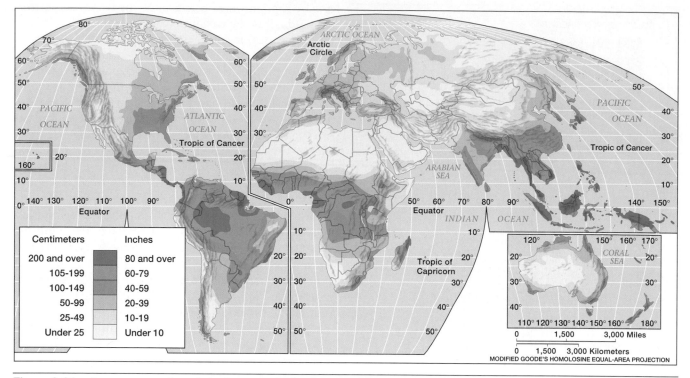

Figure 6-31   Average annual precipitation over the land areas of the world.

These regions are referred to accurately as "cold deserts."

One further generalization on precipitation distribution is the contrast between continental margins and interiors. Because coastal regions are much closer to moisture sources, they usually receive more precipitation than interior regions.

Seasonal Precipitation Patterns

A geographic understanding of climate requires knowledge of seasonal as well as annual precipitation patterns. Over most of the globe, the amount of precipitation received in summer is considerably different from the amount received in winter. This variation is most pronounced over continental interiors, where strong summer heating at the surface induces greater instability and the potential for greater convective activity. Thus in interior areas most of the year's precipitation occurs during summer months and winter is generally a time of anticyclonic conditions with diverging airflow.

Coastal areas often have a more balanced seasonal precipitation regime, which is again a reflection of their nearness to moisture sources.

Maps of average January and July precipitation provide a reliable indication of winter and summer rain/snow conditions (Figure 6-32). Prominent generalizations that can be derived from the maps include the following:

1. The seasonal shifting of major pressure and wind systems, a shifting that follows the sun (northward in July and southward in January), is mirrored in the displacement of wet and dry zones. This is seen most clearly in tropical regions, where the heavy rainfall belt of the ITC zone clearly migrates north and south in different seasons.
2. Summer is the time of maximum precipitation over most of the world. Northern Hemisphere regions experience heaviest rainfall in July, and Southern Hemisphere locations receive most precipitation in January. The only important exceptions to this generalization occur in relatively narrow zones along western coasts between about 35° and 60° of latitude, as illustrated for the United States in Figure 6-33. The same abnormal trend occurs in South America, New Zealand, and southernmost Australia.
3. The most conspicuous variation in seasonal precipitation is found, predictably, in monsoon regions (principally southern and eastern Asia, northern Australia, and West Africa), where summer tends to be very wet and winter is generally dry.

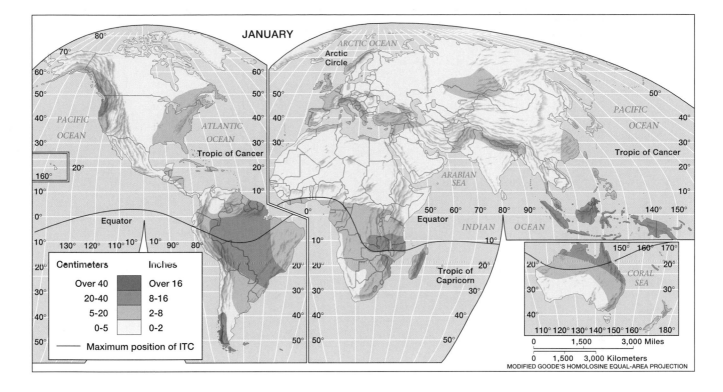

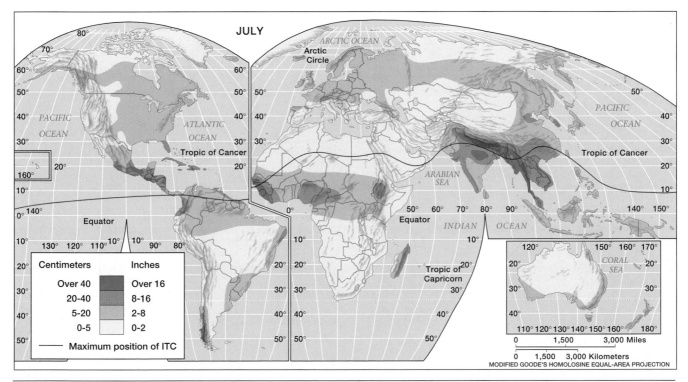

Figure 6-32 Average January and July precipitation over the land areas of the world. The red line marks the location of the intertropical convergence (ITC) zone.

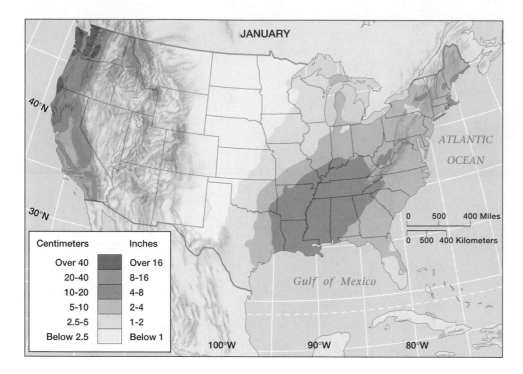

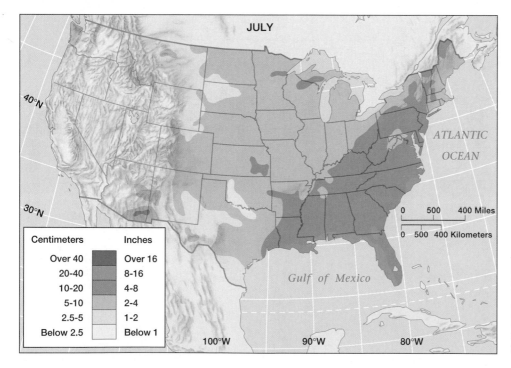

Figure 6-33 Average January and July precipitation in the United States. Winter precipitation is heaviest in the Pacific Northwest; summer rainfall is greatest in the Southeast.

PRECIPITATION VARIABILITY

The maps considered thus far all portray average conditions. The data on which they are based were gathered over decades, and thus the maps represent abstraction rather than reality. In any given year or any given season, the amount of precipitation may or may not be similar to the long-term average. This variation from average is another important facet of climate and is often critical for the organisms (including human beings) inhabiting a region.

Figure 6-34 reveals that regions of normally heavy precipitation experience the least variability and normally dry regions experience the most.

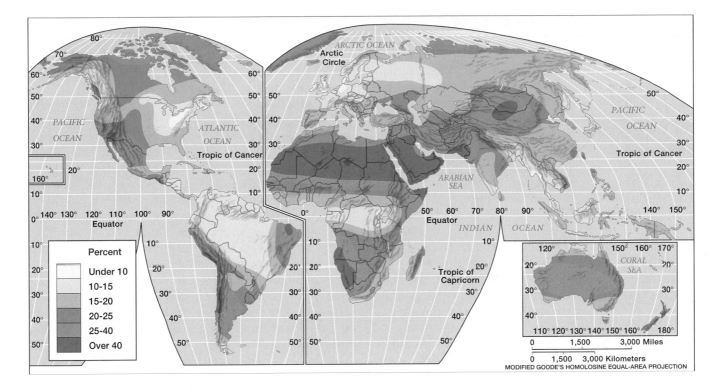

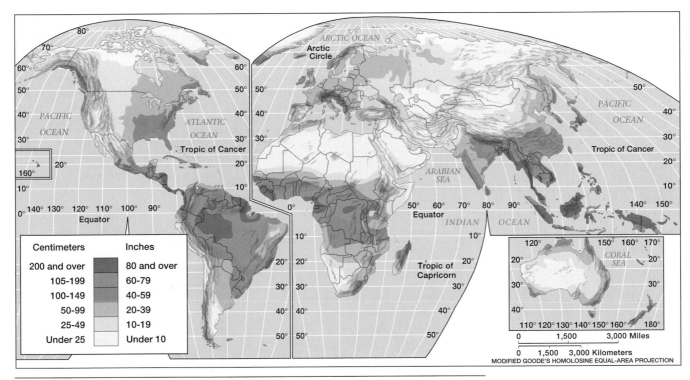

Figure 6-34 Precipitation variability. **(a)** Percentage departure from average precipitation. Dry regions (such as northern Africa, the Arabian peninsula, southwestern Africa, central Asia, and much of Australia) experience greater variability than humid areas (such as the eastern United States, northern South America, central Africa, and western Europe). **(b)** Worldwide annual average precipitation. When we compare maps (a) and (b), we see that regions of normally heavy precipitation experience the least variability and normally dry regions experience the most.

CHAPTER SUMMARY

Atmospheric moisture occurs in three physical states—solid, liquid, and gaseous—and can pass freely from one to another. The hydrologic cycle is a continuous interchange between the atmosphere and Earth, with liquid water evaporating into the gaseous air and subsequently returning to Earth as some form of falling moisture.

Evaporation is the process whereby liquid moisture (water) changes to gaseous moisture (water vapor); it is accomplished by molecular escape, and energy is stored as latent heat in the vapor molecules.

The amount of water vapor in the air is referred to as humidity, which can be measured and expressed in several ways. Absolute humidity is a direct measure of the amount of water vapor in a given volume of air; specific humidity is the mass of water vapor in a given mass of air; relative humidity is the amount of water vapor in the air compared with the amount that could be there if the air were saturated. The dew point is the temperature at which the air becomes saturated.

Condensation is the opposite of evaporation. Water vapor condenses to water, with a release of energy (latent heat) in the process. Most condensation takes place as a result of cooling by expansion in rising air. This adiabatic cooling is responsible for the formation of most clouds and, ultimately, of most precipitation. Clouds consist of visible collections of tiny water droplets or ice crystals; they are classified into three forms, four families, and ten types. Fog and dew are other notable forms of condensation.

How much vapor condenses at any given time and place is strongly influenced by the buoyancy (stability) of the air, which is an expression of horizontal density and temperature relationships.

Precipitation in all its forms (rain, snow, sleet, glaze, hail) results from a complex set of processes involving condensation nuclei, supercooled water droplets, and minute ice crystals. These processes are set in motion when large masses of air are forced to rise by convective lifting, orographic lifting, frontal uplift, or convergent lifting.

KEY TERMS

absolute humidity	dew point	rain
acid rain	dry adiabatic lapse rate	rain shadow
advection fog	evaporation	relative humidity
altocumulus cloud	evaporation fog	saturated adiabatic lapse rate
altostratus cloud	evapotranspiration	saturated air
Bergeron process	fog	sensible temperature
buoyancy	frontal precipitation	sleet
cirriform cloud	glaze	snow
cirrocumulus cloud	hail	specific humidity
cirrostratus cloud	heat index	stable air
cirrus cloud	humidity	stratiform cloud
condensation	hydrologic cycle	stratocumulus cloud
condensation nuclei	hygroscopic particle	stratus cloud
conditional instability	isohyet	transpiration
convective precipitation	lifting condensation level (LCL)	unstable air
convergent precipitation	nimbostratus cloud	upslope fog
cumuliform cloud	orographic fog	vapor pressure
cumulonimbus cloud	orographic precipitation	water vapor
cumulus cloud	potential evapotranspiration	
dew	radiation fog	

REVIEW QUESTIONS

1. How does evaporation involve the storage of heat?
2. What factors determine how much evaporation takes place at any particular place and time?
3. What is meant by evapotranspiration?
4. Distinguish among absolute humidity, specific humidity, and relative humidity.
5. Explain how dew point is a temperature concept.
6. Explain sensible temperature.
7. How is it possible for air to become supercooled?
8. What is the difference between adiabatic lapse rate and average lapse rate?
9. Why is it not possible for descending air to warm at the saturated adiabatic lapse rate?
10. What are the circumstances in which clouds form?
11. Identify the four families of clouds.
12. Describe the four principal types of fogs.
13. How is atmospheric stability related to adiabatic temperature changes?
14. How is hail related to atmospheric instability?
15. Is it possible to have a rain shadow in association with frontal precipitation? Explain.
16. What are the causes of three types of dry regions on a global basis?
17. Contrast the January and July rainfall distribution pattern for central Africa.

SOME USEFUL REFERENCES

BATTAN, L. J., *Fundamentals of Meteorology*, 2nd ed. Englewood Cliffs, NJ: Prentice-Hall, 1984.

HIDORE, JOHN J., AND JOHN E. OLIVER, *Climatology—An Atmospheric Science*. New York: Macmillan Publishing Company, 1993.

LUDLAM, F. H., *Clouds and Storms*. University Park, PA: Pennsylvania State University Press, 1980.

MORAN, JOSEPH M., AND MICHAEL D. MORGAN, *Meteorology—The Atmosphere and the Science of Weather*, 4th ed. New York: Macmillan Publishing Company, 1993.

NEIBURGER, M., J. G. EDINGER, AND W. D. BONNER, *Understanding Our Atmospheric Environment*. San Francisco: W. H. Freeman, 1973.

PARK, CHRIS, *Acid Rain*. New York: Methuen, 1988.

SCHAEFER, V. J., AND J. A. DAY, *A Field Guide to the Atmosphere*. Boston: Houghton Mifflin, 1981.

7

TRANSIENT ATMOSPHERIC FLOWS AND DISTURBANCES

IN CHAPTER 5 WE EXPLORED THE GENERAL CIRCULATION of the atmosphere, mainly the broadscale wind and pressure systems of the troposphere. Now we sharpen our focus a bit.

Over most of Earth, particularly in the midlatitudes, day-to-day weather conditions are accompanied by phenomena that are more limited than general-circulation phenomena in both magnitude and permanence. These more limited phenomena involve the flow of air masses as well as a variety of atmospheric disturbances usually referred to as storms. Both flows and disturbances are secondary features of the general circulation of the atmosphere. They move in and with the general circulation as migratory entities that persist for a relatively short time before dissipating.

Air masses and storms, then, are transient and temporary, but in some parts of the world they are so frequent and so dominating that their interactions are major determinants of weather and, to a lesser extent, of climate.

THE IMPACT OF STORMS ON THE LANDSCAPE

Storms usually are very dramatic phenomena. Under ordinary (that is, nonstormy) conditions, the atmosphere tends to be quietly inconspicuous, but storms bring excitement. The combination of expansive clouds, swirling winds, and abundant precipitation—often accompanied by thunder and lightning—that characterizes many storms makes us acutely aware of the awesome power latent in the gaseous environment we inhabit.

Even the mildest atmospheric disturbance is likely to influence our immediate activities, as when, for example, we seek shelter from wind or rain. The landscape may be significantly transformed in the short run, particularly if the storm is a violent one. Flooded streets, windblown trees, and darkened skies are prominent examples of the changes that occur.

The long-run impact of storms on the landscape often is equally notable. Damage varies with the intensity of the disturbance, but such things as uprooted trees, accelerated erosion, flooded valleys, destroyed buildings, and decimated crops are likely to result. Most storms also have a positive long-term effect on the landscape, however, as they promote diversity in the vegetative cover, increase the size of lakes and ponds, and stimulate plant growth via the moisture they add to the ground.

AIR MASSES

Although the troposphere is a continuous body of mixed gases that surrounds the planet, it is by no means a uniform blanket of air. Instead, it is composed of many large, variable parcels of air that are distinct from one another. As we saw in Chapter 6, such parcels are referred to as **air masses**.

CHARACTERISTICS

To be distinguishable, an air mass must meet three requirements:

1. It must be large. A typical air mass is more than 1000 miles (1600 kilometers) across and several miles deep (Figure 7-1).
2. It must have uniform properties in the horizontal dimension. This means that at any given altitude in the mass, its physical characteristics—primarily temperature, humidity, and stability—are relatively homogeneous. As an example, this homogeneity is shown for temperature in Figure 7-1.
3. It must be a recognizable entity and travel as one. Thus it must be distinct from the surrounding air, and when it moves, it must retain its original characteristics and not be torn apart by differences in airflow.

ORIGIN

An air mass develops its characteristics by remaining over a uniform land or sea surface long enough to acquire the temperature/humidity/stability characteristics of the surface. The stagnation needs to last for only a few days if the underlying surface has prominent temperature and moisture characteristics. For the air to stagnate, it needs to be stable (high pressure). Highly unstable air does not form a true air mass because it is constantly moving upward, away from the surface whose characteristics it is to acquire. Most air masses, then, are associated with anticyclonic conditions.

The formation of air masses is normally associated with what are called **source regions**. The idea is that certain parts of Earth's surface are particularly well suited to generate air masses. Such regions must be extensive, physically uniform, and associated with air that is stationary, or anticyclonic. Ideal source regions are ocean surfaces and extensive flat land areas that have a uniform covering of snow, forest, or desert.

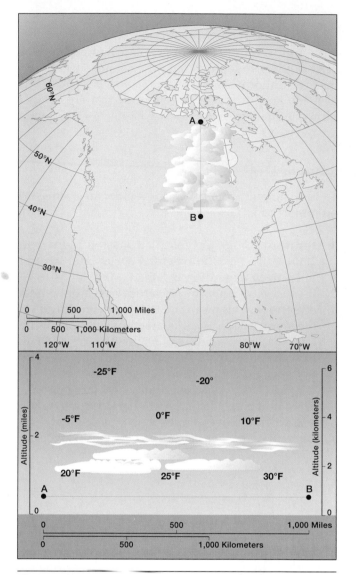

Figure 7-1 A schematic representation of a large air mass developed over central Canada. The cross section along line *AB* shows temperatures for various parts of the air mass. The general uniformity of temperatures in the horizontal dimension is clear.

Figure 7-2 portrays the principal recognized source regions for air masses that affect North America. Warm air masses can form in any season over the waters of the southern North Atlantic, the Gulf of Mexico/ Caribbean Sea, and the southern North Pacific and in summer over the deserts of the southwestern United States and northwestern Mexico. Cold air masses develop over the northern portions of the Atlantic and Pacific oceans and over the snow-covered lands of north-central Canada.

It may well be that the concept of source regions is of more theoretical value than actual value. A broader view, one subscribed to by many atmospheric scientists, holds that air masses can originate almost anywhere in the low or high latitudes but rarely in the midlatitudes (Figure 7-3).

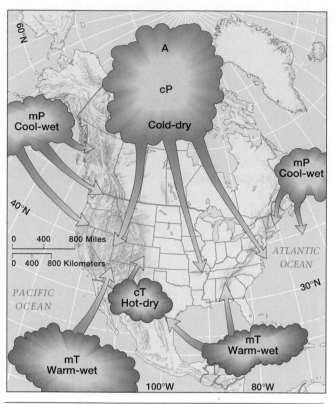

Figure 7-2 Major air masses that affect North America and their generalized paths. (For an explanation of the air-mass codes A, cP, mP, cT, and mT, see Table 7-1.)

CLASSIFICATION

Air masses are classified on the basis of source region. The latitude of the source region correlates directly with the temperature of the air mass, and the nature of the surface strongly influences the humidity content of the air mass. Thus a low-latitude air mass is warm or hot; a high-latitude one is cool or cold. If the air mass develops over a continental surface, it is likely to be dry; if it originates over an ocean, it must be moist.

A one- or two-letter code generally is used to identify air masses. Although some authorities recognize other categories, the basic classification is sixfold, as shown in Table 7-1.

MOVEMENT AND MODIFICATION

Some air masses remain in their source region for long periods, even indefinitely. In such cases, the weather associated with the air mass persists with little variation. Our interest, however, is in masses that leave their source region and move into other regions, particularly into the midlatitudes.

When an air mass departs from its source region, its structure begins to change, owing in part to thermal modification (heating or cooling from below), in part to dynamic modification (uplift, subsidence, conver-

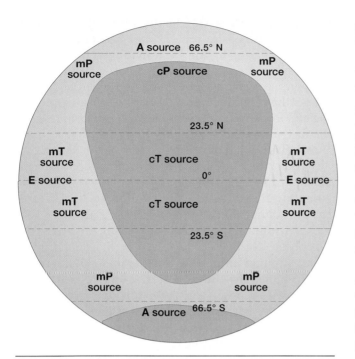

Figure 7-3   Schematic world map of air-mass source regions. Continents are shown in green and oceans in blue. The tropics and subtropics are important source regions, as are the high latitudes. Air masses do not originate in the middle latitudes except under unusual circumstances.

gence, turbulence), and perhaps also in part to addition or subtraction of moisture.

Once it leaves its source area, an air mass modifies the weather of the regions into which it moves: it takes source-region characteristics into other regions. A classic example of this modification is displayed in Figure 7-4, which diagrams a situation that may occur one or more times every winter. A midwinter outburst of continental polar (cP) air from northern Canada sweeps down across the central part of North America. With a source region temperature of -50°F (-46°C) around Great Bear Lake, the air mass has warmed to -30°F (-34°C) by the time it reaches Winnipeg, Manitoba, and it continues to warm as it moves southward, as shown in Figure 7-4.

Throughout its southward course the air mass becomes warmer, but it also brings some of the coldest weather that each of these places will receive all winter. Thus the air mass is modified, but it also modifies the weather in all regions it passes through. (In summer, cP air is much less well developed and prominent, but occasionally it provides cooling relief to portions of eastern and central North America.)

Temperature, of course, is only one of the characteristics modified by a moving air mass. There are also modifications in humidity and stability.

TABLE 7-1
Simplified Classification of Air Masses

Type	Code	Source regions	Source-region properties
Arctic/Antarctic	A	Antarctica, Arctic Ocean and fringes, and Greenland	Very cold, very dry, very stable
Continental polar	cP	High-latitude plains of Eurasia and North America	Cold, dry, very stable
Maritime polar	mP	Oceans in vicinity of 50°-60° latitude	Cool, moist, relatively unstable
Continental tropical	cT	Low-latitude deserts	Hot, very dry, unstable
Maritime tropical	mT	Tropical and subtropical oceans	Warm, moist, of variable stability
Equatorial	E	Oceans near the equator	Warm, very moist, unstable

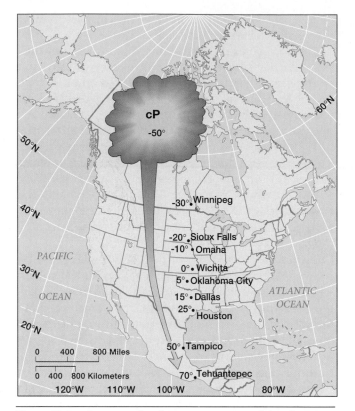

Figure 7-4 An example of temperatures resulting from a strong midwinter outburst of cP air from Canada. All temperatures are in degrees Fahrenheit.

NORTH AMERICAN AIR MASSES

The North American continent is a prominent area of air mass interaction. The lack of mountains trending east to west permits polar air to sweep southward and tropical air to flow northward unhindered by terrain, particularly over the eastern two-thirds of the continent (Figure 7-5). Major north-south trending mountain ranges in the western part of the continent, however, impede the movement of the Pacific air masses, causing significant modification of their characteristics.

Continental polar air masses develop in central and northern Canada, and Arctic (A) air masses originate farther north. These two masses are similar to each other except that the latter is even colder and drier than the former. They are dominant features in winter with their cold, dry, stable nature.

Maritime polar (mP) air that affects North America normally originates as cold, dry, cP air and then moves off the land. Once over water, this air either stagnates or else moves slowly over the North Pacific or North Atlantic, acquiring its mP characteristics only in its low-altitude portion. Consequently, the lower layers are cool, moist, and relatively unstable, whereas conditions at higher altitudes in the air mass may be cold, dry, and stable. Pacific mP air normally brings wide-

spread cloudiness and heavy precipitation to the mountainous coastal regions, but it is often severely modified in crossing the western ranges. By the time it reaches the interior of the continent, it has moderate temperatures and clear skies. In summer, the ocean is colder than the land, and so Pacific mP air produces fog and low stratus clouds along the coast but takes no distinctive weather conditions to the interior.

Air masses that develop over the North Atlantic are also cool, moist, and unstable. Except for occasional incursions into the mid-Atlantic coastal region, however, Atlantic mP air does not affect North America because the prevailing circulation is westerly.

Maritime tropical (mT) air from the Atlantic/Caribbean/Gulf of Mexico is warm, moist, and unstable. It strongly influences weather and climate east of the Rockies in the United States, southern Canada, and much of Mexico, serving as the principal precipitation source in this broad region. It is more prevalent and extensive in summer than in winter, bringing periods of uncomfortable humid heat.

Pacific mT air originates over water in areas of anticyclonic subsidence, and so it is cooler, drier, and more stable than Atlantic mT air. The influence of the former is felt only in the southwestern United States and northwestern Mexico, where in winter this air pro-

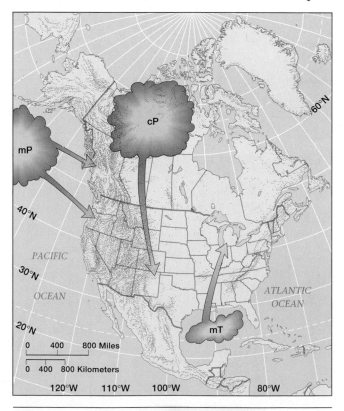

Figure 7-5 In North America, air masses can move freely north and south because there are no topographic barriers, but east-west movement is impeded by major mountain ranges aligned perpendicular to such flow.

duces some coastal fog and occasional moderate orographic rainfall if forced to ascend mountain slopes. It is also the source of some summer rains in the southwestern interior.

Continental tropical (cT) air is relatively unimportant in North America because its source region is not extensive and consists of varied terrain. There is no winter air mass genesis in this northern Mexico/southwestern United States source region, but in summer, hot, very dry, unstable cT air develops. It surges into the southern Great Plains area on occasion, bringing heat waves and drought conditions.

Equatorial (E) air affects North America only in association with hurricanes. It is similar to mT air except that E air provides an even more copious source of rain than does mT air because of high humidity and much instability.

FRONTS

When unlike air masses meet, they do not mix readily; instead, a boundary zone called a **front** develops between them (Figure 7-6). A front should not be thought of as a simple two-dimensional boundary surface or as an extensive zone of gradual transition. Rather, it consists of a relatively narrow but nevertheless three-dimensional zone of discontinuity within which the properties of the air change rapidly. A typical front is several miles or even tens of miles wide.

The frontal concept was developed by Norwegian meteorologists during World War I, and the term *front* was coined because these scientists considered the clash between unlike air masses to be analogous to a confrontation between opposing armies along a battle front. As the more "aggressive" air mass advances at the expense of the other, there is some mixing of the two within the frontal zone, but for the most part the air masses retain their separate identities as one is displaced by the other. Thus the basic significance of a front is that it functions as a barrier between two air masses, preventing their mingling except in the narrow transition zone.

Because the most conspicuous difference between air masses is usually temperature, a front usually separates warm air from cool air. At any given altitude or at any given latitude, there is warm air on one side of the front and cool air on the other, with a fairly steep temperature gradient through the front (Figure 7-7). Air masses may also have different densities, humidity levels, and stability, of course. Frequently all these factors have a steep gradient through the front.

An important attribute of fronts is that they lean, as Figure 7-6 shows, and this leaning allows air masses to be uplifted and adiabatic cooling or warming to take place. Indeed, fronts lean so much that they are much closer to horizontal than vertical. The normal slope of a front varies between about 1:50 and 1:300, with an average of about 1:150 (Figure 7-8). This last ratio, for instance, means that 150 miles (240 kilometers) away from the surface position of the front, the front is at a

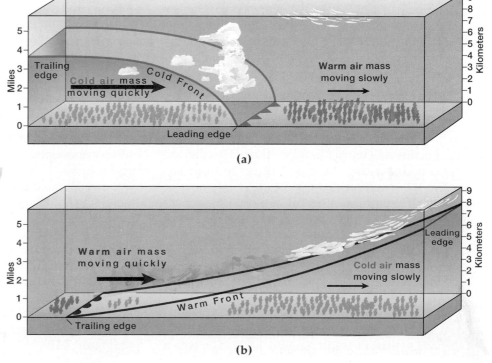

Figure 7-6 When unlike air masses come together, a front is formed. A front always slopes so that warmer air overlies cooler air. This means that a cold front leans "backward," with the air of its leading edge closer to the ground than the air of its trailing edge, and a warm front leans "forward," with the air of its leading edge farther above the ground than the air of its trailing edge. **(a)** When a cold air mass is actively underriding a warm air mass, a cold front is formed. **(b)** When a warm air mass is actively overriding a cold air mass, a warm front results.

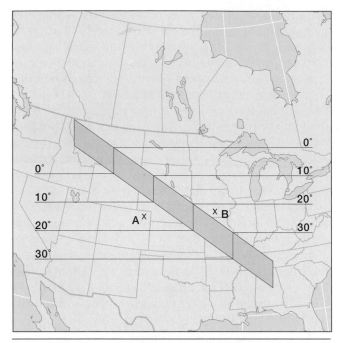

Figure 7-7 A vertical cross section through a cold front illustrates the steep temperature gradient typically found along a front. At **B**, the temperature is 22 °F, while at **A**, at the same latitude and only a short distance from **B**, the temperature is only 15 °F.

height of only 1 mile (1.6 kilometers) above the ground. Because of this very low angle of slope (less than 1°), the steepness shown in most vertical diagrams of fronts is greatly exaggerated. A front always slopes so that warmer air overlies cooler air, which means that a cold front leans "backward" so that its lower altitude part [the "leading edge" in Figure 7-6(a)] precedes its higher altitude ("trailing edge") part, whereas a warm front leans "forward" so that the higher altitude part of the front [the "leading edge" in Figure 7-6(b)] is ahead of the lower altitude part ("trailing edge").

In some cases, a front remains stationary for a few hours or even a few days. More commonly, however, it is in more or less constant motion, shifting the position of the boundary between the air masses but maintaining its function as a barrier between them. Usually one air mass is displacing the other; thus the front advances in the direction dictated by the movement of the more active air mass. Regardless of which air mass is advancing, it is always the warmer that rises over the cooler. The warmer, lighter air is inevitably forced aloft, and the cooler, denser mass functions as a wedge upon which the lifting occurs.

WARM FRONTS

A front that brings warm air is called a **warm front**. Its slope is gentle, averaging about 1:200. The warm

air ascends over retreating cool air, with the temperature of the warm air decreasing adiabatically as the air rises. The usual result is clouds and precipitation, as Figure 7-9 shows. Because the frontal uplift is very gradual, clouds form slowly and there is not much turbulence. High-flying cirrus clouds may signal the approaching front many hours before it arrives. As the front comes closer to where the first cirrus clouds formed, the clouds become lower, thicker, and more extensive, typically developing into altocumulus or altostratus. Precipitation usually occurs broadly; it is likely to be protracted and gentle, without much convective activity. If the rising air is inherently unstable, however, precipitation can be showery and even violent. As illustrated in Figure 7-9, most precipitation falls ahead of the ground-level position of the moving front.

The ground-level position of a warm front is portrayed on a weather map either by a red line (if color is used) or (more typically) by a solid black line along which black semicircles are located at regular intervals, with the semicircles extending in the direction toward which the front is moving (Figure 7-10).

COLD FRONTS

A front that brings cold air is called a **cold front**. Because there is friction between the ground surface and the lowest portion of the front, the advance of the lower portion is slowed relative to the upper portion. As a result, a cold front tends to become steeper as it moves forward and usually develops a protruding "nose" a few hundred yards above the ground (Figure 7-11). The average cold front is twice as steep as the average warm front, as Figure 7-6 indicates. Moreover, cold fronts normally move faster than warm fronts, this difference represented by the black arrows of different lengths in Figure 7-6. This combination of steeper slope and faster advance leads to rapid lifting of the warm air ahead of the cold front. The rapid lifting makes the warm air very unstable, and the result is blustery and violent weather along the cold front (Figure 7-12). Vertically developed clouds are common, with considerable turbulence and showery precipitation. Both clouds and precipitation tend to be concentrated along and immediately behind the ground-level position of the front. Precipitation is usually of higher intensity but shorter duration than that associated with a warm front.

On a weather map, the ground-level position of a cold front is shown either by a blue line or by a solid black line studded at intervals with solid triangles that extend in the direction toward which the front is moving (Figure 7-10).

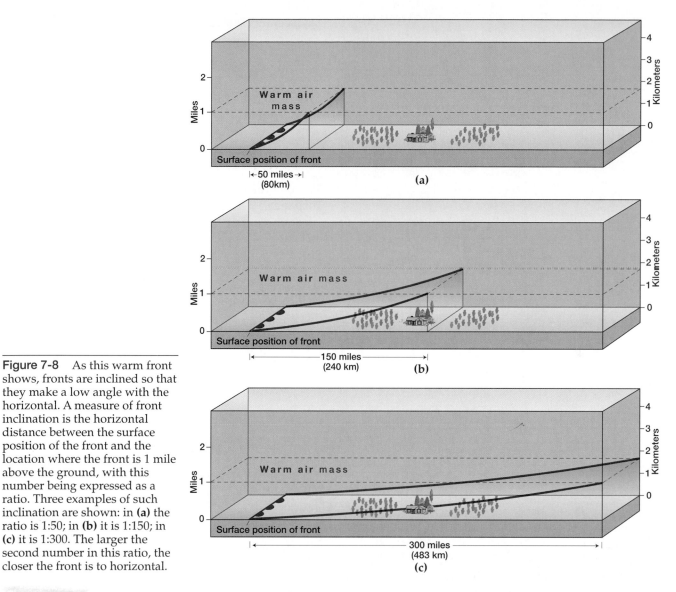

Figure 7-8 As this warm front shows, fronts are inclined so that they make a low angle with the horizontal. A measure of front inclination is the horizontal distance between the surface position of the front and the location where the front is 1 mile above the ground, with this number being expressed as a ratio. Three examples of such inclination are shown: in **(a)** the ratio is 1:50; in **(b)** it is 1:150; in **(c)** it is 1:300. The larger the second number in this ratio, the closer the front is to horizontal.

STATIONARY FRONTS

When neither air mass displaces the other, their common boundary is called a **stationary front**. Weather is not readily predictable along such a front, but often gently rising warm air produces the limited precipitation associated with a warm front. As Figure 7-10 shows, stationary fronts are portrayed on a weather map by a combination of warm and cold front symbols.

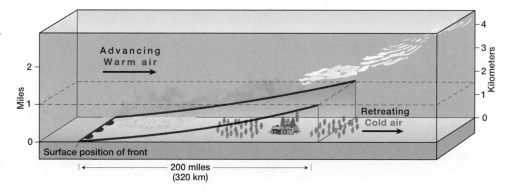

Figure 7-9 Along a warm front, warm air rises above cooler air. Often widespread cloudiness and precipitation develop along and in advance of the ground-level position of the front. Higher and less dense clouds often are dozens or hundreds of miles ahead of the ground-level position of the front.

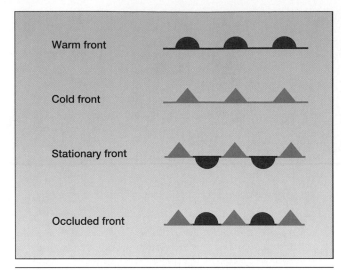

Figure 7-10 Weather map symbols for fronts.

OCCLUDED FRONTS

A fourth type of front, called an **occluded front**, is formed when a cold front overtakes a warm front. The development of occluded fronts is discussed later in this chapter.

ATMOSPHERIC DISTURBANCES

We now turn our attention to the various kinds of disturbances that occur within the general circulation. Most of these disturbances involve unsettled and even violent atmospheric conditions and are referred to as storms. Some, however, produce calm, clear, quiet weather that is quite the opposite of stormy. The following are common characteristics of the two types of disturbances (stormy and calm):

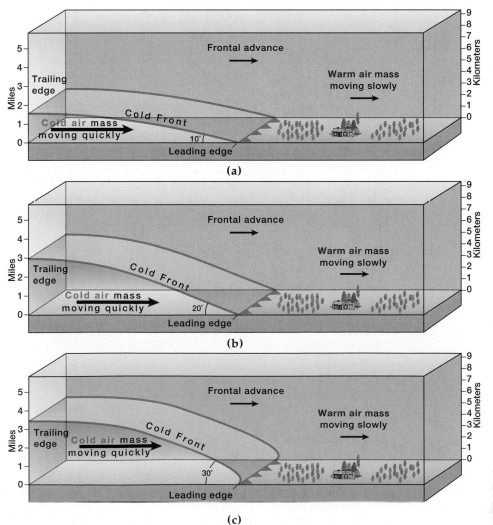

Figure 7-11 As a cold front advances, its slope tends to become steeper because of friction between air and ground.

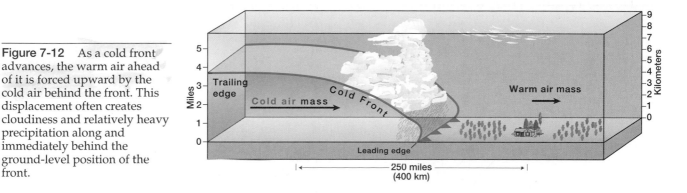

Figure 7-12 As a cold front advances, the warm air ahead of it is forced upward by the cold air behind the front. This displacement often creates cloudiness and relatively heavy precipitation along and immediately behind the ground-level position of the front.

1. They are smaller than the components of the general circulation, although they are extremely variable in size.
2. They are migratory and transient.
3. They have a relatively brief duration, persisting for only a few minutes, a few hours, or a few days.
4. They produce characteristic and relatively predictable weather conditions.

MAJOR MIDLATITUDE DISTURBANCES

The middle latitudes are the principal battleground of tropospheric phenomena: where polar and tropical air masses meet and come into conflict, where most fronts occur, and where weather is most dynamic and changeable from season to season and from day to day. Many kinds of atmospheric disturbances are associated with the midlatitudes, but two of these—extratropical cyclones and extratropical anticyclones—are much more important than the others because of their size and prevalence.

Extratropical Cyclones Probably most significant of all atmospheric disturbances are **extratropical cyclones** (the adjective meaning that they occur outside the tropics). Throughout the midlatitudes they dominate weather maps, are basically responsible for most day-to-day weather changes, and bring precipitation to much of the populated portions of the planet. Consisting of large, migratory low-pressure systems, they are usually called depressions in Europe and sometimes referred to as either lows or wave cyclones in the United States.

Extratropical cyclones are associated primarily with air mass convergence and conflict in regions between about 35° and 70° of latitude. Thus they are found almost entirely within the zone of westerly winds. Their general path of movement is toward the east, which explains why weather forecasting in the midlatitudes is essentially a west-facing vocation.

Because each extratropical cyclone differs from all others in greater or lesser detail, any description must be a general one only. The discussions that follow, then, pertain to "typical," or idealized, conditions. Moreover, these conditions are presented as Northern Hemisphere phenomena. For the Southern Hemisphere, the patterns of isobars, fronts, and wind flow should be visualized as mirror images of the Northern Hemisphere patterns.

Characteristics A typical mature extratropical cyclone has a diameter of 1000 miles (1600 kilometers) or so. It is essentially a vast cell of low-pressure air, with ground-level pressure in the center typically between 990 and 1000 millibars. The system [shown by closed isobars on a weather map, as in Figure 7-13(a)] usually tends toward an oval shape, with the long axis trending northeast-southwest. Often a clear-cut pressure trough extends southwesterly from the center.

Extratropical cyclones have a normal circulation pattern, converging counterclockwise in the Northern Hemisphere. This wind-flow pattern attracts cool air from the north and warm air from the south. The convergence of these unlike air masses characteristically creates two fronts: a cold front that extends southwesterly from the center of the cyclone and runs along the pressure trough described in the preceding paragraph, and a warm front extending eastward from the center and running along another pressure trough. These two fronts divide the cyclone into a cool sector north and west of the center and a warm sector to the south and east. At ground level, the cool sector is the larger of the two, but aloft the warm sector is more extensive. This size relationship exists because both fronts lean over the cool air. Thus the cold front slopes upward toward the northwest, and the warm front slopes upward toward the northeast, as Figure 7-13(b) shows.

Because warm air rises along both fronts, the typical result is two zones of cloudiness and precipitation that overlap around the center of the storm and extend outward in the general direction of the fronts. Along and immediately behind the ground-level position of the cold front (the steeper of the two fronts) a band of cumuliform clouds usually yields showery

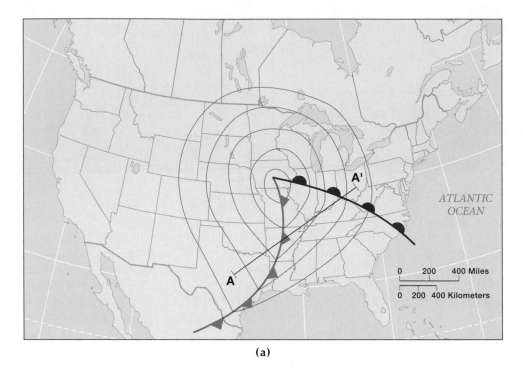

(a)

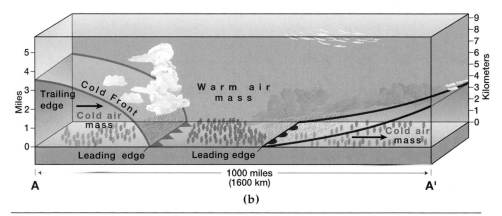

(b)

Figure 7-13 A map **(a)** and a cross section **(b)** of a typical mature extratropical cyclone. In the Northern Hemisphere, there is usually a cold front trailing to the southwest and a warm front extending toward the east. Arrows in **(b)** indicate the direction of frontal movement.

precipitation. The air rising more gently along the more gradual slope of the warm front produces a more extensive expanse of horizontally developed clouds, perhaps with widespread, protracted, low-intensity precipitation. In both cases, most of the precipitation originates in the warm air rising above the fronts and falls down through the front to reach the ground in the cool sector.

This precipitation pattern does not mean that all the cool sector has unsettled weather and that the warm sector experiences clear conditions throughout. Al-though most frontal precipitation falls within the cool sector, the general area to the north, northwest, and west of the center of the cyclone is frequently cloudless as soon as the cold front has moved on. Thus much of the cool sector is typified by clear, cold, stable air. In contrast, the air of the warm sector is often moist and tending toward instability, and so thermal convection and surface-wind convergence may produce sporadic thunderstorms. Also, sometimes one or more squall lines of intense thunderstorms develop in the warm sector in advance of the cold front.

Movement Extratropical cyclones are essentially transient features, on the move throughout their existence. There are four kinds of movement involved (Figure 7-14):

1. The whole system moves as a major disturbance in the westerlies, traversing the middle latitudes generally from west to east. The rate of movement averages 20 to 30 miles (32 to 48 kilometers) per hour, which means that the storm can cross the United States in some 3 to 4 days. It moves faster in winter than in summer.
2. The system has a cyclonic wind circulation, with air generally converging counterclockwise from all sides in the Northern Hemisphere.
3. The cold front normally advances faster than the storm is moving. Thus it swings counterclockwise around its pivot in the center, increasingly moving into and displacing the warm sector.
4. The warm front usually advances more slowly than the storm. This difference in speed causes the front to lag behind and has the effect of its seeming to swing clockwise in the direction of the advancing cold front. (This is only an apparent motion, however. In reality, the warm front is moving west to east just like every other part of the system.)

Life Cycle A typical extratropical cyclone develops from origin to maturity in 3 to 6 days, and from maturity to dissipation in about the same length of time. Most begin as waves (hence the alternative name *wave cyclone*) along the polar front, which as we learned in Chapter 5, is the name given to the contact zone between unlike air masses in the component of the general circulation called the subpolar low. The opposing air flows normally have a relatively smooth linear motion on either side of the polar front [Figure 7-15(a)]. On occasion, however, the smooth frontal surface may be distorted into a wave shape [Figure 7-15(b)]. Various ground factors—such as topographic irregularities, temperature contrasts between sea and land, or the influence of ocean currents—apparently can initiate a wave along the front, but it is believed that the most common cause of cyclogenesis (the birth of cyclones) is upper-air conditions in the vicinity of the polar-front jet stream.

There appears to be a close relationship between upper-level airflow and ground-level disturbances. When the upper airflow is *zonal*—by which we mean relatively straight from west to east—ground-level cyclonic activity is unlikely. When winds aloft behave as shown in Figure 7-16, meandering north to south (this motion is called *meridinal airflow*), large waves of alternating pressure troughs and ridges are formed and cy-

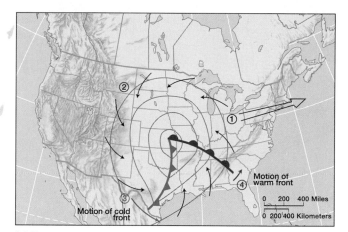

(a) Day 1

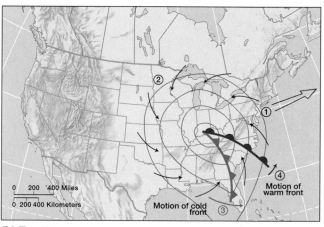

(b) Day 2

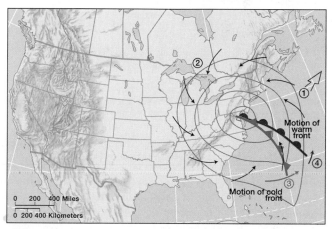

(c) Day 3

Figure 7-14 **(a)** Four varieties of motion occur in a typical extratropical cyclone: (1) The entire system moves west to east in the general flow of the westerlies; (2) there is cyclonic counterclockwise air flow; (3) the cold front advances; (4) the warm front advances. **(b&c)** Days 2 and 3: Because the cold front is traveling faster than the warm front, the angle between the two decreases over time.

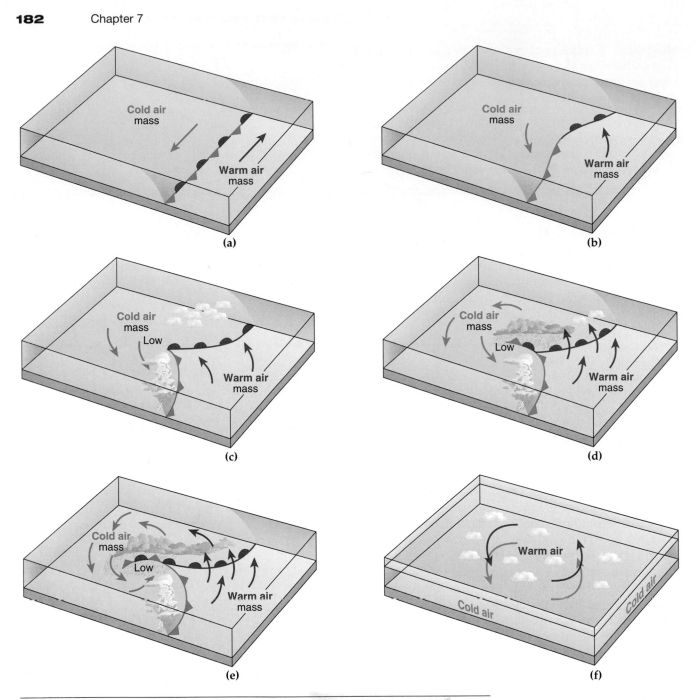

Figure 7-15 Schematic representation of the life cycle of an extratropical cyclone. **(a)** Front develops. **(b)** Wave appears along front. **(c)** Cyclonic circulation is well developed. **(d)** Occlusion begins. **(e)** Occluded front is fully developed. **(f)** Cyclone dissipates. (After Edward J. Tarbuck and Frederick K. Lutgens, *Earth Science*, 4th ed. Columbus, OH: Charles E. Merrill, 1985, p. 363.)

clonic activity at ground level is intensified. As seen in Figure 7-16, most extratropical cyclones are centered below the polar front jet stream axis and downstream from an upper-level pressure trough.

A cyclone is unlikely to develop at ground level unless there is divergence above it. In other words, the convergence of air near the ground must be supported by divergence aloft. Such divergence can be related to

changes in either speed or direction of the wind flow, but it nearly always involves broad north-to-south meanders in the Rossby waves and the jet stream (Figure 7-17).

Cyclogenesis also occurs on the leeward side of mountains. A low-pressure area drifting with the westerlies becomes weaker when it crosses a mountain range. As it ascends the range, the column of air com-

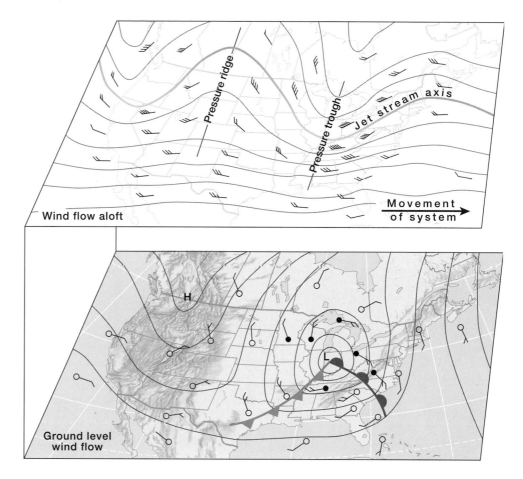

Figure 7-16 A typical winter situation in which the upper-level airflow is meridional (meanders north and south), creating standing waves aloft and cyclonic flow at ground level. (After Frederick K.Lutgens and Edward J. Tarbuck, *The Atmosphere: An Introduction to Meteorology,* 4th ed. Englewood Cliffs, NJ: Prentice Hall, 1989, p. 248.)

presses and spreads, slowing down its counterclockwise spin. When descending the leeward side, the air column stretches vertically and contracts horizontally. This shape change causes it to spin faster and may initiate cyclonic development, even if it was not a full-fledged cyclone before. This chain of events happens with some frequency in winter on the eastern flanks of the Rocky Mountains, particularly in Colorado, and with lesser frequency on the eastern side of the Appalachian Mountains, in North Carolina and Virginia. Cyclones formed in this way typically move toward the east and northeast, and often bring heavy rain or snowstorms to the northeastern United States and southeastern Canada.

Ultimately, the storm dissipates because the cold front overtakes the warm front. As the two fronts come closer and closer together [Figure 7-15(c)], the warm sector at the ground is increasingly displaced, forcing more and more warm air aloft. When the cold front catches up with the warm front, warm air is no longer in contact with Earth's surface and an occluded front is formed [Figure 7-15(d,e)]. This occlusion process usually results in a short period of intensified precipitation and wind until eventually all the warm sector is forced aloft and the ground-level low-

pressure center is surrounded on all sides by cool air. This sequence of events weakens the pressure gradient and shuts off the storm's energy so that it dies out [Figure 7-15(f)].

Occurrence and Distribution At any given time from 6 to 15 extratropical cyclones exist in the Northern Hemisphere midlatitudes, and an equal number in the Southern Hemisphere. They occur at scattered but irregular intervals throughout the zone of the westerlies. One such system is shown in Figure 7-18.

In each hemisphere, these migratory disturbances are more numerous, better developed, and faster moving in winter than in summer. Also, they follow much more equatorward tracks in winter. In the Southern Hemisphere, the Antarctic continent provides a prominent year-round source of cold air, and so vigorous cyclones are almost as numerous in summer as in winter. The summer storms are farther poleward than their winter cousins, however, and are mostly over the Southern Ocean. Thus they have little effect on land areas.

The route of a cyclone is likely to be undulating and erratic, although, as stated above, it moves generally from west to east.

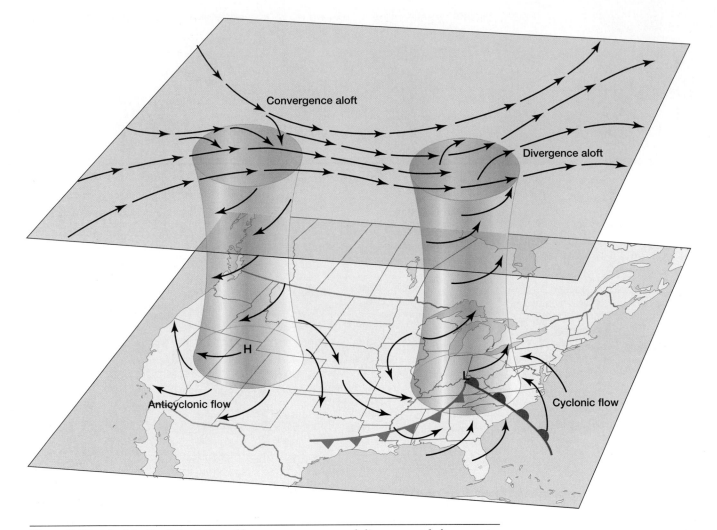

Figure 7-17 An idealized depiction of how convergence and divergence aloft support anticyclonic and cyclonic circulation at ground level. (After Frederick K. Lutgens and Edward J. Tarbuck, *The Atmosphere: An Introduction to Meteorology*, 4th ed. Englewood Cliffs, NJ: Prentice Hall, 1989, p. 229.)

Extratropical Anticyclones Another major disturbance in the general flow of the westerlies is the **extratropical anticyclone**, frequently referred to simply as a *high* (H). This is an extensive, migratory high-pressure cell of the midlatitudes (Figure 7-19). Typically it is larger than an extratropical cyclone and moves generally west to east with the westerlies.

As with any other high-pressure center, an extratropical anticyclone has air converging into it from above, subsiding, and diverging at the surface, clockwise in the Northern Hemisphere and counterclockwise in the Southern Hemisphere. No air-mass conflict or convergence is involved, and so anticyclones contain no fronts (the fronts shown in Figure 7-19 are technically outside the high-pressure system). The weather is clear and dry with little or no opportunity for cloud

formation. Wind movement is very limited near the center of an anticyclone but increases progressively outward. Particularly along the eastern margin (the leading edge) of the system, there may be strong winds. In winter, anticyclones are characterized by very low temperatures.

Anticyclones move toward the east either at the same rate as or a little slower than extratropical cyclones. Unlike cyclones, however, anticyclones are occasionally prone to stagnate and remain over the same region for several days. This stalling brings clear, stable, dry weather to the affected region, which enhances the likelihood that air pollutants will become concentrated. Such stagnation may block the eastward movement of cyclonic storms, causing protracted precipitation in some other region while the anticyclonic region remains dry.

Figure 7-18 A conspicuous extratropical cyclonic system is focused over the British Isles in this late summer satellite false-color image. (University of Dundee/Science Photo Library/Photo Researchers, Inc.)

Relationships of Cyclones and Anticyclones

Extratropical cyclones and anticyclones alternate with one another in irregular sequence around the world in the midlatitudes (Figure 7-20). Each can occur inde-

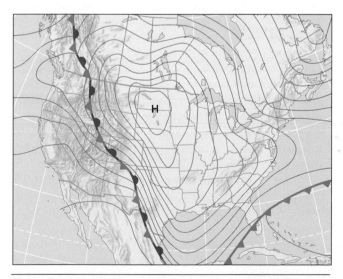

Figure 7-19 A typical well-developed extratropical anticyclone centered over the Dakotas. Both fronts shown here are considered to be outside the high-pressure system.

pendently of the other, but there is often a functional relationship between them. This relationship can be seen when an anticyclone closely follows a cyclone, as diagrammed in Figure 7-21. The winds diverging from the eastern margin of the high fit into the flow of air converging into the western side of the low. It is easy to visualize the anticyclone as a polar air mass having the cold front of the cyclone as its leading edge.

MAJOR TROPICAL DISTURBANCES: HURRICANES

The low latitudes are characterized by monotony—the same weather day after day, week after week, month after month. Almost the only breaks are provided by transient atmospheric disturbances, of which by far the most significant are **tropical cyclones**. These storms are known by different names in different places: **hurricanes** in North and Central America, **typhoons** in the western North Pacific, **baguios** in the Philippines, and simply **cyclones** in the Indian Ocean and Australia. Whatever the name, these are intense, revolving, rain-drenched, migratory, destructive storms that occur erratically in certain regions of the tropics and subtropics (Figure 7-22).

Characteristics Hurricanes (as tropical storms are usually referred to in this chapter) consist of prominent low-pressure centers that are essentially circular and have a steep pressure gradient outward from the center (Figure 7-23). As a result, strong winds spiral inward. Winds must reach a speed of 64 knots (74 miles or 119 kilometers per hour) in order for the storm to be officially a hurricane, but winds in a well-developed hurricane often double that speed and occasionally triple it.

Hurricanes are considerably smaller than extratropical cyclones, typically having a diameter of between 100 and 600 miles (160 and 1000 kilometers). A remarkable feature is the nonstormy **eye** in the center of the storm (Figure 7-24 on page 190). The winds do not converge to a central point but rather reach their highest speed at the **eye wall**, which is the edge of the eye. The eye has a diameter of from 10 to 25 miles (16 to 40 kilometers) and is a singular area of calmness in the maelstrom that whirls around it.

Updrafts are common throughout the hurricane (except in the eye), becoming most prominent around the eye wall. Near the top of the storm, outward flow dominates except in the eye, where a downdraft inhibits cloud formation.

The weather pattern within a hurricane is relatively symmetrical around the eye. Bands of dense cumulus and cumulonimbus clouds spiral in from the edge of the storm to the eye wall, producing heavy rains that

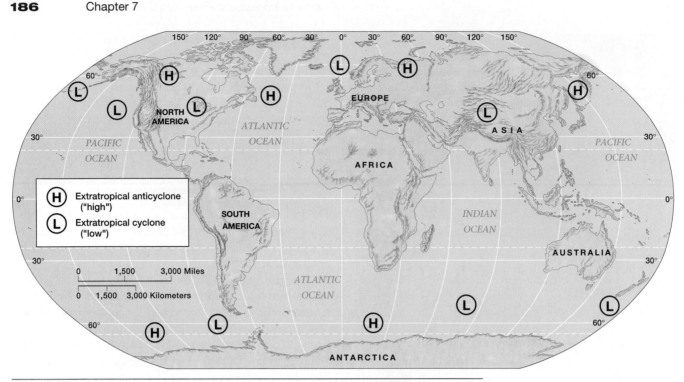

Figure 7-20 At any given time, the midlatitudes are dotted with extratropical cyclones and anticyclones. This map depicts a hypothetical situation in January.

generally increase in intensity inward. The clouds of the eye wall tower to heights that may exceed 10 miles (16 kilometers). Within the eye, there is no rain and no low clouds, and scattered high clouds may part to let in intermittent sunlight.

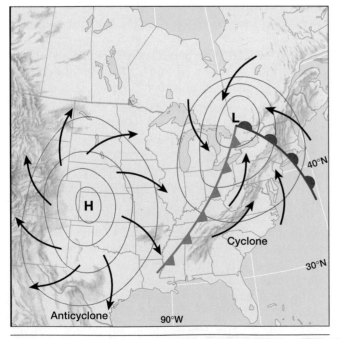

Figure 7-21 Extratropical cyclones and anticyclones often occur in juxtaposition in the middle latitudes.

Origin Hurricanes form only over warm oceans in the tropics at least a few degrees north or south of the equator (Figure 7-25 on page 190). Because the Coriolis effect is so minimal near the equator, no hurricane has ever been observed to form within 3° of it, no hurricane has ever been known to cross it, and the appearance of hurricanes closer than some 8° or 10° of the equator is very rare. More than 80 percent originate in or just on the poleward side of the intertropical convergence zone.

Hurricanes require an enormous supply of energy, which is provided by the latent heat released when water vapor condenses. An average mature hurricane produces in a day approximately as much energy as is generated by all electric utility plants in the United States in a year.

The exact mechanism of formation is not clear, but hurricanes always grow from some preexisting disturbance in the tropical troposphere. Easterly waves, discussed below, provide low-level convergence and lifting that catalyze the development of many hurricanes. Even so, fewer than 10 percent of all easterly waves grow into hurricanes. Hurricanes can evolve only where there are no significant wind changes with height, which implies that temperatures at low altitudes are reasonably uniform over a wide area.

Incipient perturbations in trade-wind flow are called *tropical disturbances* by the U.S. National Weather Service. About 100 of these are identified each year over the tropical North Atlantic, but only a few develop into hurricanes. Three degrees of tropical disturbances are recognized, on the basis of wind speed:

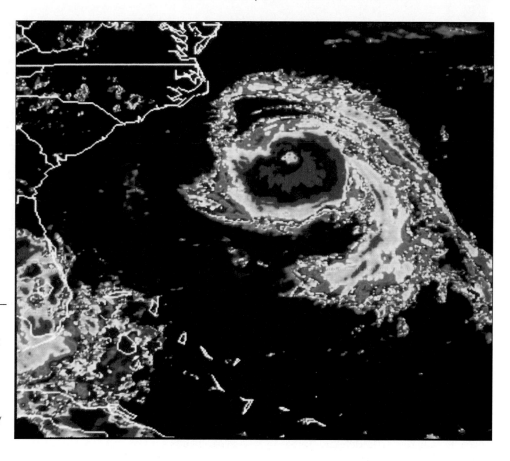

Figure 7-22 A color-coded satellite image of Hurricane Emily off the southeastern coast of the United States in August 1993. A band of thick clouds is shown (red and black) swirling around the eye (yellowish-orange) of the storm. (NOAA/Science Photo Library/Photo Researchers, Inc.)

1. A tropical depression has winds of 33 knots (38 miles or 61 kilometers per hour) or less.
2. A tropical storm has winds between 34 and 63 knots (39 and 73 miles or 63 and 117 kilometers per hour).
3. A hurricane has winds of 64 knots (74 miles or 119 kilometers per hour) or more.

Movement Hurricanes occur in a half-dozen low-latitude regions. They are most common in the North Pacific basin, originating largely in two areas: east of the Philippines, and west of southern Mexico and Central America. The third most notable region of hurricane development is in the west central portion of the North Atlantic basin, extending into the Caribbean Sea and Gulf of Mexico. These ferocious storms are also found in the western portion of the South Pacific and all across the South Indian Ocean, as well as in the North Indian Ocean both east and west of the Indian peninsula. They are totally absent from the South Atlantic and from the southeastern part of the Pacific, apparently because the water is too cold. The strongest and largest hurricanes typically are those of the China Sea; only a few storms in other parts of the world have attained the size and intensity of the large East Asian typhoons.

Once formed, hurricanes follow irregular tracks within the general flow of the trade winds. A specific path is very difficult to predict in advance, but the general pattern of movement is highly predictable. Roughly one-third of all hurricanes travel east to west without much latitudinal change. The rest, however, begin on an east-west path and then curve prominently poleward, where they either dissipate over the adjacent continent or become enmeshed in the general flow of the midlatitude westerlies (Figure 7-26 on page 191).

In one region—the southwestern Pacific Ocean north and northeast of New Zealand—there is a

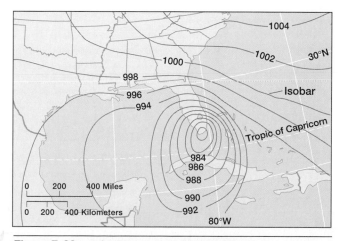

Figure 7-23 A hurricane as it might appear on a simplified weather map. The isobars show a very steep gradient around the low-pressure center.

FOCUS

EL NIÑO

One of the basic tenets of physical geography is the interrelatedness of the various elements of the environment. In a pedagogic situation, however, such as this textbook, we often isolate individual components in order to explicate more clearly their characteristics and functions. In doing so, we run the risk of masking complex environmental interactions.

Nowhere is environmental interrelatedness shown better than at the air–ocean interface—that vast horizontal surface that marks the meeting of atmosphere and hydrosphere over 70 percent of Earth. These two great fluid systems interact in intricate and often perplexing fashion so that cause and effect become almost meaningless concepts.

A prize example of this complexity is *El Niño,* an anomalous oceanographic/weather phenomenon of the eastern equatorial Pacific, particularly along the coast of Ecuador and Peru. In an *El Niño* event, abnormally warm waters appear at the surface of the ocean, displacing the cold, nutrient-rich upwellings that usually prevail and disrupting productive offshore fisheries. This event occurs once every few years at about Christmastime; hence the name *El Niño* (Spanish: "the Christ child").

Most of the time, the South Pacific trade winds blow toward the west, from the persistent subtropical high in the southeastern Pacific toward an equally persistent low-pressure system over Indonesia and northern Australia. This airflow drags warm surface water westward, raising sea levels in the Indonesian region by as much as 2 feet (60 centimeters) and turning the tropical western Pacific into an immense storehouse of energy. Meanwhile, the tropical South American coast experiences land breezes and an offshore upwelling of cold, nutrient-rich ocean water.

In an *El Niño* year, the southeastern trades mysteriously slacken or reverse direction. This reversal triggers changes in ocean circulation as the warm water that has accumulated in the tropical western Pacific begins a 2½-month journey back toward South America, lowering sea levels in the west and raising them in the east. Eventually the normally cool waters and upwelling off the coast of Ecuador and Peru are overridden by this warm surface flow, in an *El Niño* event.

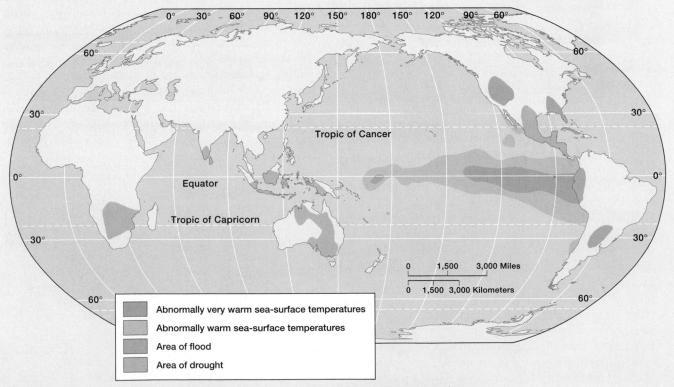

Figure 7-1-A   Some major weather events associated with *El Niño* during its 1982/83 outbreak. Abnormally warm ocean water in the eastern equatorial Pacific is the most readily recognizable characteristic. The widespread occurrence of unusual droughts and floods was peculiar to the 1982/83 episode.

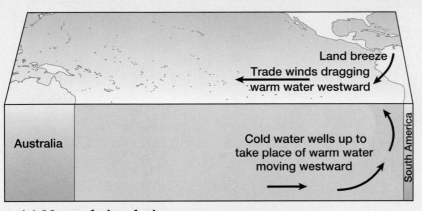

(a) Normal circulation

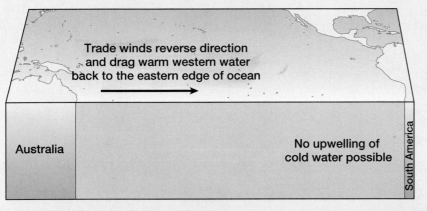

(b) Circulation during El Niño

Figure 7-1-B **(a)** Normal current conditions in the South Pacific (sometimes referred to as *La Niña*). **(b)** These conditions either slow or reverse during an *El Niño* event.

At sporadic intervals, the advent of *El Niño* coincides with a large-scale fluctuation in sea-level atmospheric pressure. Known as the Southern Oscillation, this fluctuation is a gigantic seesaw pattern of alternating atmospheric pressures in the eastern and western tropical Pacific caused by differences in water temperature. When the sea-level air pressure is abnormally high over the eastern tropical Pacific, it is abnormally low over the western tropical Pacific, and vice versa. When *El Niño* and the Southern Oscillation coincide, unusual atmospheric and oceanic conditions are more frequent and more intense than when either event occurs alone.

El Niño has been documented since 1726, but the connection with the Southern Oscillation was not recognized until the 1960s. And it was not until 1982/83 that the connection was suddenly brought to world attention by a series of weather events called "the most disastrous in recorded history." During a period of several months, there were crippling droughts in Australia, India, Indonesia, the Philippines, Mexico, Central America, and southern Africa; devastating floods in the western and southeastern United States, Cuba, and northwestern South America; destructive tropical cyclones in parts of the Pacific (Tahiti, Hawaii) where they are normally extremely rare; and a vast sweep of

warm (temperatures as much as 14°F above normal) water over an 8000-mile (12,800-kilometer) stretch of the equatorial Pacific, which caused massive dieoffs of fish, seabirds, and coral (Figure 7-1-A). Directly attributable to these events were more than 1500 human deaths, damage estimated at nearly $9 billion, and vast ecological havoc.

Although the connection was not always detectable, all these happenings were ascribed to the *El Niño*/Southern Oscillation combination, now often identified by the acronym ENSO (Figure 7-1-B).

But what causes *El Niño*? And what causes some *El Niño*-related disturbances to be more severe than others? The theories are many, but the triggering mechanisms are still unclear. In 1985 a 10-year, multinational research program, involving three satellites and a fleet of research vessels, was launched to probe tropical weather phenomena, with emphasis on ENSO. By 1994 sufficient data had been gathered by two international collaborations—the Tropical Ocean Global Atmosphere Program (TOGA) and the Coupled Atmosphere-Ocean Research Experiment (COARE)—to develop models that can predict the onset of an ENSO event several months in advance. This is only the first step, however. Each *El Niño* has different characteristics, and its effect on nontropical weather patterns is quite variable.

A clearer understanding of cause and effect, of countercause and countereffect, will undoubtedly be forthcoming eventually. Meanwhile, the essential coupling between atmospheric and oceanic circulations has been reaffirmed, and the ordered cycle of environmental interactions in a single Earth system has been revalidated.

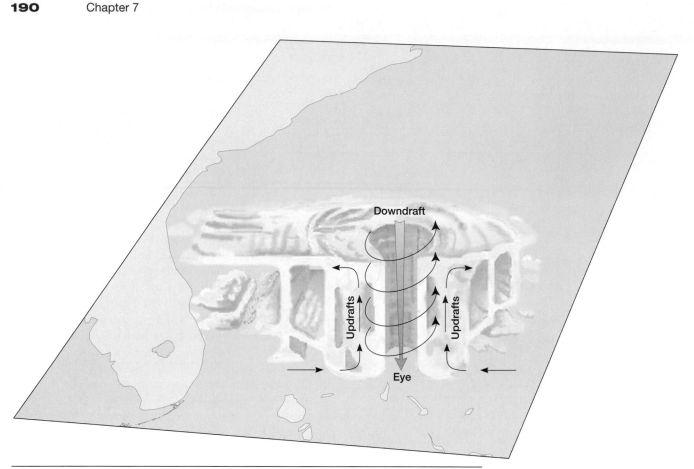

Figure 7-24   An idealized cross section through a well-developed hurricane. Air spirals into the storm horizontally and rises rapidly to produce towering cumulus and cumulonimbus clouds that yield torrential rainfall. In the center of the storm is the eye, where air movement is downward.

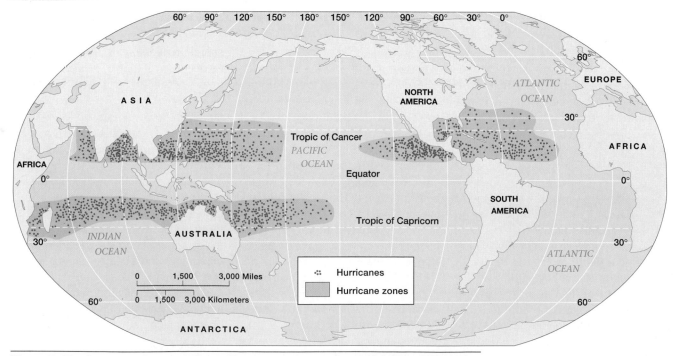

Figure 7-25   Location of origin points of hurricanes during the period 1952–1971. (After W. M. Gray, "Tropical Cyclone Genesis," *Atmospheric Science Paper 234*. Department of Atmospheric Science, Colorado State University, Fort Collins, 1975.)

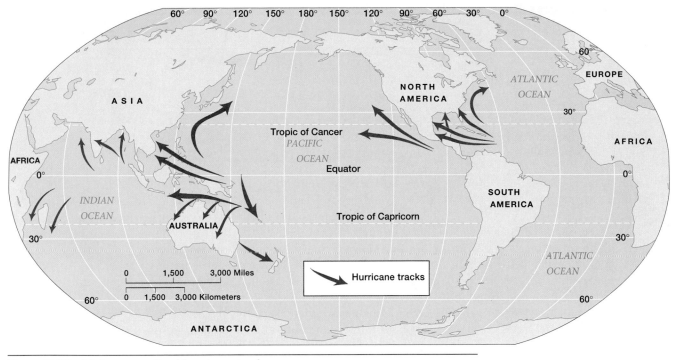

Figure 7-26   Major generalized hurricane tracks.

marked variation from this general flow pattern. Hurricanes in this part of the Pacific usually move, erratically, from northwest to southeast. When they strike an island, therefore, such as Fiji or Tonga, they approach from the west, a situation not replicated anywhere else in the world at such a low latitude. These seemingly aberrant tracks apparently result from the simple fact that the tropospheric westerlies extend quite far equatorward in the Southwest Pacific and hurricanes there are basically steered by the general circulation pattern.

Whatever their trajectory, hurricanes do not last long. The average hurricane exists for only about a week, with 4 weeks as the maximum duration (Figure 7-27). The longer lived hurricanes are those that remain over tropical oceans. As soon as a hurricane leaves the ocean and moves over land, it begins to die, for its energy source (warm, moist air) is cut off. If it stays over the ocean but moves into the midlatitudes, it dies as it penetrates the cooler environment. It is not unusual for a tropical hurricane that moves into the midlatitudes to diminish in intensity but grow in areal size until it develops into an extratropical cyclone that travels with the westerlies.

In most regions there is a marked seasonality to hurricanes (Figure 7-28 on page 194). They are largely restricted to late summer and fall, presumably because this is the time that ocean temperatures are highest and the intertropical convergence zone is shifted farthest poleward.

Damage and Destruction   Hurricanes are best known for their destructive capabilities. Some of the destruction comes from high winds and torrential rain,

but the overwhelming cause of damage and loss of life is high seas. The low pressure in the center of the storm allows the ocean surface to bulge as much as 3 feet (1 meter). To this is added a storm surge of wind-driven water as much as 25 feet (7.5 meters) above normal tide level when the hurricane pounds into a shoreline. Thus a low-lying coastal area can be severely inundated, and most hurricane-related deaths are drownings.

The greatest hurricane disaster in United States history occurred in 1900 when Galveston Island, Texas, was overwhelmed by a 20-foot (6-meter) storm surge that killed 6000 people, nearly one-sixth of Galveston's population. In other regions, hurricane devastation has been much greater. The flat deltas of the Ganges and Brahmaputra rivers in

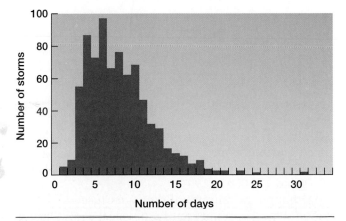

Figure 7-27   Duration of North Atlantic/Caribbean hurricanes. Most persist for from 3 to 12 days.

PEOPLE AND THE ENVIRONMENT

HURRICANES ALLEN AND ANDREW—CONTRASTS IN DEVASTATION

During the twentieth century, only three hurricanes in the North Atlantic/Caribbean/Gulf of Mexico region have rated 5 on the Saffir–Simpson hurricane scale. The first storm of the 1980 hurricane season, *Allen*, was one of them.

Hurricane Allen is generally considered the mightiest Caribbean storm ever recorded, with surface winds of 185 miles (297 kilometers) per hour recorded on one island. A hurricane-hunter aircraft that penetrated the storm measured gusts of 215 miles (345 kilometers) per hour as well as the second-lowest central pressure ever measured in a hurricane. By any standard, Allen had the

potential to leave enormous devastation in its wake.

It made its appearance on August 3 at latitude 12° N and about halfway between Africa and Trinidad (Figure 7-2-A). Wind speeds began in excess of 74 miles (119 kilometers) per hour and continued to gain force as the storm bore down on the heavily populated island of Barbados. It swung north of Barbados and then threaded its way west between St. Lucia and St. Vincent, causing considerable damage on all three islands but hitting none of them directly. In its path across the Caribbean, it veered west and missed Haiti, then north and mostly avoided Jamaica, then west

again and missed western Cuba. This put it on a direct line to crash into the crowded tourist resorts of Mexico's Yucatan Peninsula, but again it dodged northward, avoiding Yucatan. At about the 90th meridian it swerved farther north, this new trajectory sending it toward Galveston and Houston on the Texas coast.

By Saturday morning, August 9, Allen's massive cloud formations almost filled the entire Gulf, from Louisiana to Yucatan and over to the eastern coast of the rest of Mexico. Then suddenly it veered away from the track toward Houston and aimed at Corpus Christi. Finally, in the predawn hours of Sunday, Allen left

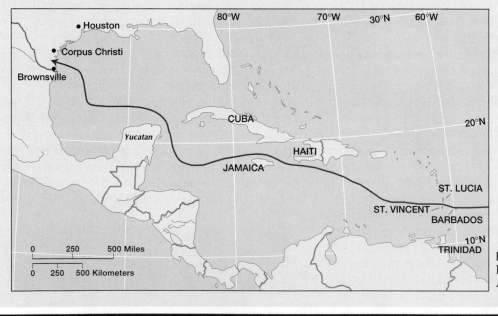

Figure 7-2-A The path of Hurricane Allen, early August 1980.

Bangladesh have been subjected to enormous losses of human life from Indian Ocean cyclones: 300,000 deaths in 1737; another 300,000 in 1876; 500,000 in 1970; and 175,000 in 1991.

Although the amount of damage caused by a hurricane depends in part on the physical configuration of the landscape and the population size and density

of the affected area, storm strength is the most important factor. In the United States a scale has been established to rank the relative intensity of hurricanes, ranging from 1 to 5, with 5 being the most severe (Table 7-2).

Destruction and tragedy are not the only legacies of hurricanes, however. Such regions as northwestern

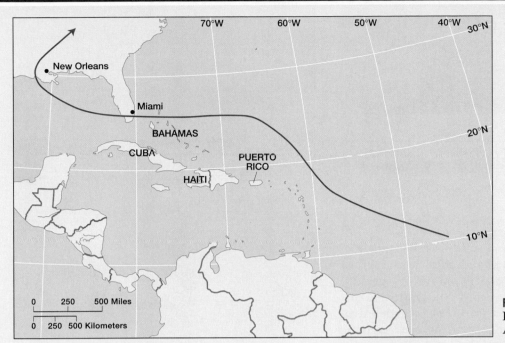

Figure 7-2-B   The path of Hurricane Andrew, late August 1992.

the Gulf and crossed the shoreline on the least populated portion of the Texas coast—between Corpus Christi and Brownsville, the perfect place for minimizing damage to humankind!

Once over land, the hurricane died rapidly. It caused flooding in southern Texas and brought rain as far north as Kansas, but by Tuesday, August 12, it was no longer on the weather map.

Allen was widely heralded as the "storm of the century." It had the potential for taking thousands of lives and causing billions of dollars worth of property damage. In actuality, it caused about 275 deaths in the Caribbean islands and 4 on the Texas coast. Not an inconsequential result, of course, but significantly less devastating than what might well have been.

In marked contrast is the devastation caused by Hurricane *Andrew* in August 1992. Andrew followed a more straightforward trajectory out of the southern North Atlantic at about latitude 25°N, moving directly west across the Bahamas (Figure 7-2-B). It reached hurricane strength on August 22 and crashed into the Florida coast just south of Miami in the early hours of August 24. The eye of the storm passed directly over the National Hurricane Center at Coral Gables (partially disabling the center) with sustained winds of 138 miles (220 kilometers) per hour and gusts of more than 160 miles (256 kilometers) per hour. Andrew crossed the Florida peninsula, cutting a swath of destruction as it went, then entered the Gulf of Mexico, and on August 26

slammed into the central coast of Louisiana.

At the time, Andrew was the most devastating natural disaster ever to strike the United States (this distinction has now passed to the Northridge, California, earthquake of 1994) and is still the most destructive hurricane in U.S. history. "Only" 62 lives were lost (43 in Florida, 15 in Louisiana, and 4 in the Bahamas) because most people heeded the warnings to evacuate. Property damage exceeded $30 billion in Florida and another $2 billion in Louisiana. To put that in perspective, the damage total for Florida alone was more than twice as great as the combined damage from 1989 Hurricane Hugo in the Carolinas and the 1989 Loma Prieta earthquake in northern California.

Mexico, northern Australia, and southeastern Asia rely on tropical storms for much of their water supply. Hurricane-induced rainfall is often a critical source of moisture for agriculture. Although crops within the immediate path of the storm may be devastated by winds and flooding, a much more extensive area may be nurtured by the life-giving rains.

MINOR ATMOSPHERIC DISTURBANCES

Several kinds of lesser atmospheric disturbances are common in various parts of the world. Some are locally of great significance, and some are destructive. All occur at a much more localized scale than do tropical and extratropical cyclones and anticyclones.

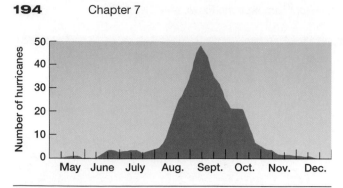

Figure 7-28 Seasonality of North Atlantic/Caribbean hurricanes. Fall is hurricane season, with a prominent maximum in September.

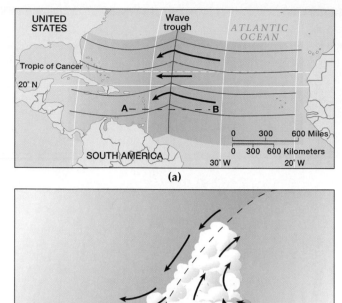

(a)

(b)

Figure 7-29 Diagrammatic map view **(a)** and cross section **(b)** of an easterly wave. The arrows indicate general direction of airflow.

Easterly Waves An **easterly wave** is a long but weak, migratory, low-pressure system that may occur almost anywhere between about 5° and 30° of latitude (Figure 7-29). Easterly waves are usually several hundred miles long and nearly always oriented north-south. They drift slowly westward in the flow of the trade winds, bringing characteristic weather with them. Ahead of the wave is fair weather with divergent airflow. Behind the wave, convergent conditions prevail, with moist air being uplifted to yield convective thunderstorms and sometimes widespread cloudiness.

There is little or no temperature change with the passage of easterly waves, which sometimes intensify into hurricanes.

Thunderstorms A **thunderstorm,** defined as a violent convective storm accompanied by thunder and lightning, is usually localized and short lived. It is always associated with vertical air motion, considerable humidity, and instability, a combination that produces a towering cumulonimbus cloud and (nearly always) showery precipitation.

Thunderstorms sometimes occur as individual clouds, produced by nothing more complicated than thermal convection; such developments are commonplace in the tropics and during summer in much of the midlatitudes. Thunderstorms also are frequently found in conjunction with other kinds of storms, however, or are associated with other mechanisms that can trigger unstable uplift. Thus thunderstorms often accompany hurricanes, tornadoes, fronts (especially cold fronts) in extratropical cyclones, and orographic lifting that forces unstable buoyancy.

The uplift, by whatever mechanism, of warm, moist air must release enough latent heat of condensation to

TABLE 7-2			
Saffir-Simpson Hurricane Scale			
Category	Central pressure (millibars)	Winds (kilometers per hour)	Storm surge (meters)	Damage
1	> 979	119—153	1.2—2.5	Minimal
2	965—979	154—177	1.6—2.4	Moderate
3	945—964	178—209	2.5—3.6	Extensive
4	920—944	210—250	3.7—5.4	Extreme
5	<920	> 250	> 5.4	Catastrophic

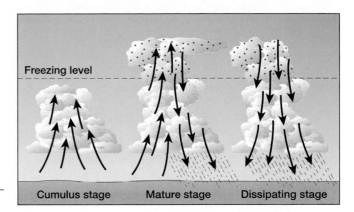

Figure 7-30   Sequential development of a thunderstorm cell.

Freezing level

Cumulus stage   Mature stage   Dissipating stage

sustain the continued rise of the air. In the early stage of thunderstorm formation (Figure 7-30), called the *cumulus stage*, updrafts prevail and the cloud grows. Above the freezing level, supercooled water droplets and ice crystals coalesce: when they become too large to be supported by the updrafts, they fall. These falling particles drag air with them, initiating a downdraft. When the downdraft with its accompanying precipitation leaves the bottom of the cloud, the thunderstorm enters the *mature stage*, in which updrafts and downdrafts coexist as the cloud continues to enlarge. The mature stage is the most active time, with heavy rain often accompanied by hail, blustery winds, lightning, thunder, and the growth of an anvil top composed of ice crystals on the massive cumulonimbus cloud. Eventually, downdrafts dominate and the *dissipating stage* is reached, with light rain ending and turbulence ceasing.

Thunderstorms are most common where there are high temperatures, high humidity, and high instability, a combination typical of the intertropical convergence zone. There is a general decrease in thunderstorm frequency away from the equator, and they are virtually unknown poleward of 60° of latitude (Figure 7-31).

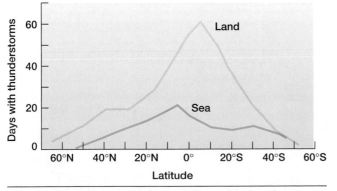

Figure 7-31   Average number of days with thunderstorms, as generalized by latitude. Most thunderstorms are in the tropics. Land areas experience many more thunderstorms than ocean areas because land warms up much more in summer.

There is much greater frequency of thunderstorms over land than water because summer temperatures are higher over land and most thunderstorms occur in the summer.

Tornadoes   Although very small and localized, the **tornado** is the most destructive of all atmospheric disturbances. It is the most intense vortex in nature: a deep low-pressure cell surrounded by a violently whirling cylinder of wind (Figure 7-32 on page 199). These are tiny storms, generally less than a quarter of a mile (400 meters) in diameter, but they have the most extreme pressure gradients known—as much as a 100-millibar difference from the center of the tornado to the air immediately outside the funnel. This extreme pressure difference produces winds of extraordinary speed. Exactly how high these speeds are is unknown because any tornado that has come close enough to an anemometer (an instrument for measuring wind speed) to be recorded has blown the instrument to bits. Maximum wind speed estimates range from 200 to 500 miles (320 to 800 kilometers) per hour. Air sucked into the vortex also rises at an inordinately fast rate. Such a storm clearly has the capability of transporting a little girl to the land of Oz!

Tornadoes occur at random. They usually originate a few hundred feet above the ground, the rotating vortex becoming visible when upswept water vapor condenses into a funnel-shaped cloud. The tornado generally advances at 15 to 30 miles (25 to 50 kilometers) per hour along an irregular track that generally extends from southwest to northeast in the United States. Sometimes the funnel sweeps along the ground, devastating everything in its path, but its trajectory is usually twisting and dodging and includes frequent intervals in which the funnel lifts completely off the ground and then touches down again nearby. A tornado lasts from a few minutes to an hour or so, with maximum longevity recorded at about 8 hours.

The dark, twisting funnel of a tornado contains not only cloud but also sucked-in dust and debris. Damage is caused largely by the strong winds and swirling up-

FOCUS

HOW NEW TECHNOLOGIES HELP TO FORECAST SEVERE OKLAHOMA STORMS

STEPHEN STADLER, OKLAHOMA STATE UNIVERSITY

The volatile combination of mT air from the Gulf of Mexico, cP air from Canada, cT air from the desert, and the cold, dry southwesterly flow of the polar-front jet stream makes the Oklahoma sky the world's most prolific breeder of tornadoes. Three new technologies in place in Oklahoma allow the improved identification of the atmospheric conditions that spawn severe thunderstorms and tornadoes: NEXRAD, wind profilers, and the Oklahoma Mesonetwork.

Virtually all tornadoes are generated by severe thunderstorms. The basic requirement is **vertical wind shear**—a significant change in wind direction from the bottom to the top of the storm. On a tornado day, low-level winds are southerly and jet-stream winds are southwesterly, and

this difference in direction causes turbulence on the boundary between the two systems. The conventional updrafts that become thunderstorms reach several miles up into the atmosphere, and the wind-shear turbulence is translated into a rotation that strengthens as it becomes taller. The rotation typically has a diameter of about 6 miles (10 kilometers) and is called a **mesocyclone.** More than half of all mesocyclones formed result in a tornado.

NEXRAD (for next-generation radar) can dissect the internal workings of a storm and thereby determine its severity. The operating principle on which the system works is the **Doppler effect**, familiar to anyone who has ever been stopped at a railroad track as a train crosses a

road. As the train approaches, the pitch of its whistle seems to get higher and higher. After the train passes, the whistle pitch seems to get lower and lower as the train gets farther and farther away. In reality, the whistle always has the same pitch; the apparent change depends on whether the sound source is approaching or receding from the listener.

Electromagnetic waves (light waves, radio waves, microwaves, and so forth) are also subject to the Doppler effect. NEXRAD transmits microwaves through the atmosphere toward a target storm. Raindrops, ice crystals, and hail reflect some of the microwaves back to the NEXRAD site, and how much of the transmitted signal is reflected gives an estimate of the intensity of the storm.

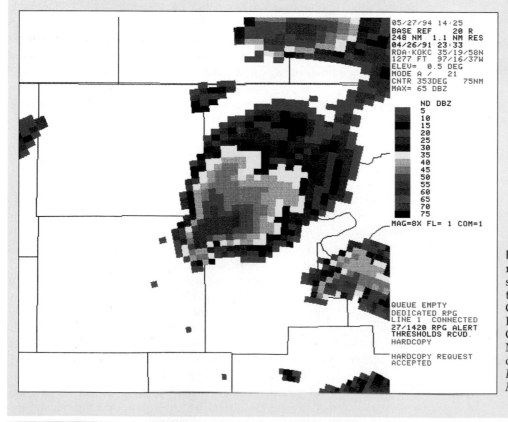

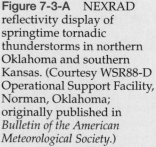

Figure 7-3-A NEXRAD reflectivity display of springtime tornadic thunderstorms in northern Oklahoma and southern Kansas. (Courtesy WSR88-D Operational Support Facility, Norman, Oklahoma; originally published in *Bulletin of the American Meteorological Society*.)

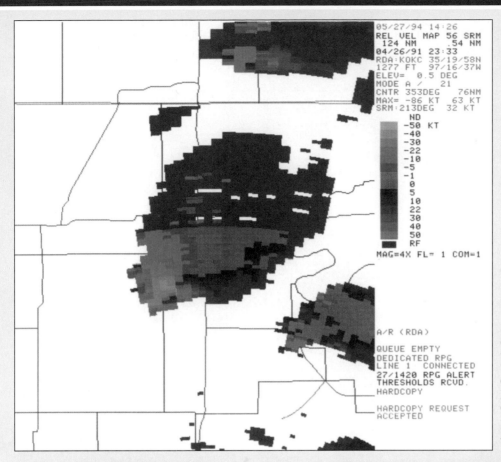

```
05/27/94 14:26
REL VEL MAP 56 SRM
   124 NM       .54 NM
04/26/91 23:33
RDA:KOKC 35/19/58N
1277 FT   97/16/37W
ELEV=  0.5 DEG
MODE A /     21
CNTR 353DEG    76NM
MAX= -86 KT   63 KT
SRM:213DEG   32 KT
         ND
       -50 KT
       -40
       -30
       -22
       -10
        -5
        -1
         0
         5
        10
        22
        30
        40
        50
        RF
MAG=4X FL= 1 COM=1
```

```
A/R (RDA)

QUEUE EMPTY
DEDICATED RPG
LINE 1   CONNECTED
27/1420 RPG ALERT
THRESHOLDS RCVD.
HARDCOPY

HARDCOPY REQUEST
ACCEPTED
```

Figure 7-3-B NEXRAD display of radial velocity in the storms shown in Figure 7-3-A. (Courtesy WSR88-D Operational Support Facility, Norman, Oklahoma; originally published in *Bulletin of the American Meteorological Society*.)

Figure 7-3-A shows a NEXRAD reflectivity display of severe thundersorms in northern Oklahoma and southern Kansas, the type of information also obtainable from any conventional radar system. However, tornado funnels are usually only a few hundred yards across, and it is rare that radar can detect them using reflectivity alone. NEXRAD enjoys an advantage over conventional radar in that NEXRAD can detect motion toward and away from the radar site to within 2 miles per hour. Figure 7-3-B is a display of the same storms as shown by Figure 7-3-A but using NEXRAD's motion-detection capability. In Figure 7-3-B, notice the southern portion of the large storm at the center. Light green, indicating wind movement in excess of 50 knots away from the radar site, and light orange, indicating wind movement in excess of 50 knots toward the site, exist side by side. This is a strong counterclockwise

rotation that is the hallmark of a mesocyclone. A radar operator would issue a tornado warning based on this display.

A second technology operating in Oklahoma is the *vertical wind profiler*. To estimate the likelihood of storms hours in advance, it is necessary to have some idea of wind conditions aloft. For the last 50 years, the United States has employed a network of balloon-launch sites spaced sparsely over the landscape. (Oklahoma, for instance, has one site.) Twice a day, instrument balloons are launched and data are radioed back from the troposphere and stratosphere. The data are excellent for monitoring movement of Rossby waves and air masses, but features relevant to thunderstorm formation can "slip through the cracks" of the balloon network.

The wind profiler network solves this problem. A wind profiler is a Doppler radar unit that resembles a chain-link fence turned on its side. It

sends radar energy upward to sense wind speed and direction in the first few miles of the atmosphere. Not only are profiler sites more closely spaced than balloon sites, but the profilers can provide a wind profile every few minutes instead of only once every 12 hours. This increased frequency is helpful in severe thunderstorm situations, where atmospheric conditions can change drastically from one minute to the next.

The *Oklahoma Mesonetwork* is a system of 111 solar-powered automated weather stations that radio readings each 15 minutes. Before the Mesonetwork was in place, there were only a dozen hourly reporting stations in Oklahoma with considerable distance between them. The Mesonetwork is the most extensive state weather network in the United States and presents forecasters with new possibilities in short-term forecasting.

(Continued)

HOW NEW TECHNOLOGIES HELP TO FORECAST SEVERE OKLAHOMA STORMS

Continued

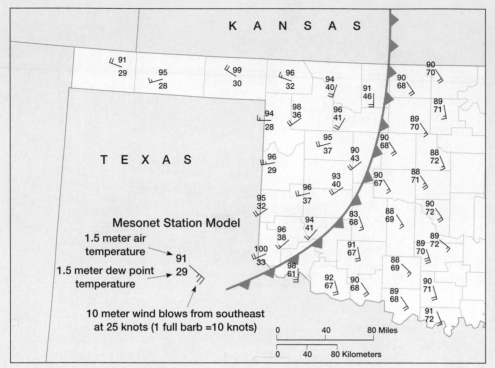

Figure 7-3-C Oklahoma Mesonetwork depiction of a dryline that could generate severe storms. (Courtesy Oklahoma Mesonetwork Project.)

Tornado-generating thunderstorms frequently occur along the **dryline,** the boundary between mT and cT air and a zone in which surface air streams are forced to converge and lift. The dryline can generate some of the largest tornadic thunderstorms on Earth. Therefore, its exact position is critical in forecasting where storms will begin. The Mesonetwork, with its average spacing of 18 miles (30 kilometers) between stations, gives forecasters a precise view of dryline location. Figure 7-3-C shows a dryline in western Oklahoma. Note the relatively high dewpoint temperatures east of the line in the mT air and the higher air temperatures and lower dewpoint temperatures in the cT air west of the line. The wind data show air streams are converging and, by implication, rising along the dryline. In this situation, severe thunderstorms are most likely just to the east of the dryline.

draft, but structures often explode as a result of the abrupt pressure contrast when the center of the storm (very low pressure) passes over a closed building (relatively high pressure inside the building).

As with most other storms, the exact mechanism of tornado formation is not well understood. They usually develop in the warm, moist, unstable air associated with an extratropical cyclone. A tornado is most often spawned along a squall line that precedes a rapidly advancing cold front or along the cold front. Tornadoes commonly are associated with thunderstorms, usually developing directly out of the base of an intense cumulonimbus thunderhead cloud. Not all thunderstorms spawn tornadoes, but with the appropriate conditions (unseasonably warm and humid air near the ground, cold air at middle altitudes, and strong upper-altitude jet stream winds) severe thunderstorms and tornadoes are likely to occur.

Spring and early summer are favorable for tornado development because of the considerable air mass contrast present in the midlatitudes at that time; a tornado can form in any month, however (Figure 7-33). Most occur in midafternoon, at the time of maximum heating.

Tornadoes do occur in the middle latitudes and subtropics, but more than 90 percent are reported from the United States (Figure 7-34). Such concentra-

Figure 7-32   A tornado funnel over central Kansas. (Howard Bluestein/Photo Researchers, Inc.)

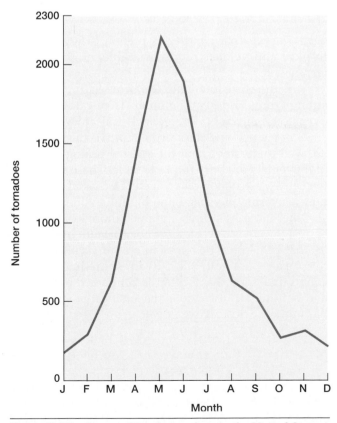

Figure 7-33   Seasonality of tornadoes in the United States.

tion in a single area presumably reflects optimum environmental conditions, with the relatively flat terrain of the central and southeastern United States providing an unhindered zone of interaction between prolific source regions for Canadian cP and Gulf mT air masses. Between 800 and 1200 tornadoes are recorded annually in the United States, but the actual total may be considerably higher than that because many small tornadoes that occur briefly in uninhabited areas are not reported.

True tornadoes apparently are restricted to land areas. Similar-appearing funnels over the ocean ("waterspouts") have a lesser pressure gradient, gentler winds, and reduced destructive capability.

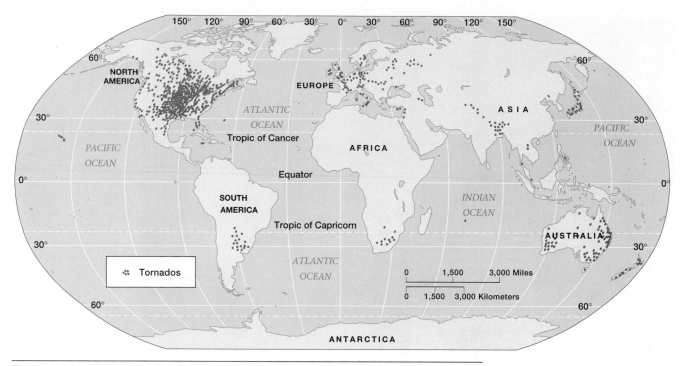

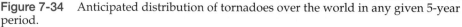

Figure 7-34 Anticipated distribution of tornadoes over the world in any given 5-year period.

CHAPTER SUMMARY

Large parcels of air in the troposphere that have relatively uniform horizontal physical characteristics are referred to as air masses. When air masses move away from their source regions, they cause significant weather changes as they go.

When unlike air masses meet, a front—a zone of unsettled, sometimes stormy, weather—is established between them. A warm front is the leading edge of an advancing warm-air mass; a cold front is the leading edge of an advancing cold-air mass that is displacing warm air; a stationary front is the boundary between two passive air masses; an occluded front is formed when a cold front overtakes a warm front.

Air masses and fronts are prominent components of major migratory pressure systems called extratropical cyclones and extratropical anticyclones. These systems dominate midlatitude circulation, particularly in winter. In extratropical cyclones, which are battlegrounds of tropical and polar air, dynamically and dramatically changing weather conditions prevail, whereas extratropical anticyclones represent stable, nonstormy interludes.

Hurricanes are low-latitude storms characterized by strong winds and heavy rainfall. They can form only over ocean surfaces. The general trajectory of hurricane movement and the season of hurricane formation are well understood, but the specific timing and path of movement are very unpredictable. These storms often cause destruction and tragedy in coastal areas, but they also bring life-sustaining rainfall to inland areas.

Other kinds of storms are much more localized. An easterly wave is a weak migratory system in the tropics that sometimes develops into a hurricane. A thunderstorm is a small, short-lived, sometimes violent convective storm that is often accompanied by thunder and lightning. A tornado is the smallest of storms but has immense destructive power.

KEY TERMS

air mass	extratropical cyclone	stationary front
baguio	eye	thunderstorm
cold front	eye wall	tornado
cyclone	front	tropical cyclone
Doppler effect	hurricane	typhoon
dryline	mesocyclone	vertical wind shear
easterly wave	occluded front	warm front
extratropical anticyclone	source region	

REVIEW QUESTIONS

1. What are the distinguishing characteristics of an air mass?
2. What regions of Earth are least likely to produce air masses? Why?
3. Why are maritime polar air masses from the Atlantic Ocean less important to the United States than mP air masses from the Pacific Ocean?
4. Explain the differences between a warm front and a cold front.
5. Distinguish between a stationary front and an occluded front.
6. Explain the various motions associated with an extratropical cyclone.
7. Explain what happens to all four weather elements when a cold front passes.
8. What is the relationship between upper-level airflow and the formation of surface disturbances in the midlatitudes?
9. What parts of the world are most affected by extratropical cyclones?
10. Why are there no fronts in an extratropical anticyclone?
11. Describe the typical weather conditions associated with an extratropical anticyclone.
12. Why are there no fronts in a tropical cyclone?
13. Describe the weather pattern within a hurricane.
14. Do hurricanes have any beneficial effects? Explain.
15. Why are thunderstorms more common over land than over water?
16. Contrast easterly waves and tornadoes.
17. Which type of storm is most common over the world? Why?

SOME USEFUL REFERENCES

CARLSON, T. N., *Mid-Latitude Weather Systems*. New York: Routledge, Chapman & Hall, 1991.

DIAZ, HENRY F., AND VERA MARKGRAF (eds.), *El Niño: Historical and Paleoclimatic Aspects of the Southern Oscillation*. New York: Cambridge University Press, 1993.

EAGLEMAN, J. R., *Meteorology: The Atmosphere in Action*. New York: Van Nostrand, 1980.

———, *Severe and Unusual Weather*, 2nd ed. Lenexa, KS.: Trimedia, 1990.

KESSLER, EDWIN, ed., *The Thunderstorm in Human Affairs*. Norman: University of Oklahoma Press, 1988.

LOCKHART, GARY, *The Weather Companion*. New York: John Wiley & Sons, 1988.

MUSK, LESLIE F., *Weather Systems*. Cambridge, England: Cambridge University Press, 1988.

PIELKE, ROGER A., *The Hurricane*. London: Routledge, 1990.

SIMPSON, R.H., AND H. RIEHL, *The Hurricane and Its Impact*. Baton Rouge: Louisiana State University Press, 1981.

8

CLIMATIC ZONES
AND TYPES

THE BASIC GOAL OF THE GEOGRAPHIC STUDY OF CLImate, as with the geographic study of anything else, is to understand its spatial characteristics—its distribution over Earth. Such understanding is exceedingly difficult to achieve for climate, however, because so many variables are involved. It is relatively easy to describe and even to map the distribution of such uncomplicated phenomena as giraffes, voting patterns, and a host of other concrete entities or simple relationships. Climate, however, is a result of a number of elements that are, for the most part, continuously and independently variable. Temperature, for instance, is one of the simplest climatic elements, and yet it fluctuates in time and place so frequently that, for example, in the *National Atlas of the United States* (U.S. Department of the Interior, Geological Survey, Washington, DC, 1970), 56 different maps are used to portray various aspects of temperature.

Although temperatures can be measured precisely, some form of abstraction is necessary for tabulation and analysis. Most widely used is the arithmetic mean (or average). For some purposes, however, the most probable temperature or some expression of variation (for example, standard deviation) is more useful, and for still other purposes extremes are more important than averages. The point at issue is that climate involves almost continuous variation, not only of temperature but also of a host of other factors. Thus a satisfactory synthesis of all these data involves a great deal of generalization and subjectivity.

CLIMATIC CLASSIFICATION

To cope with the great diversity of information encompassed by the concept of climate, the most meaningful climatic characteristics must be selected and some systematic way must be found to classify them. Classifiers generally have chosen temperature and precipitation as the most significant and understandable features of climate, as well as the most available, and that is the method we follow in this chapter.

THE PURPOSE OF CLASSIFYING CLIMATES

In this book we have a clear-cut and straightforward requirement: we need a classification scheme that is useful as a learning tool. Specifically, we need a device to simplify, organize, and generalize a vast array of data into a comprehensible system that helps us understand the distribution of climates over Earth.

Suppose a geography student in Georgia is asked to describe the climate of southeastern China. The stu-

dent is likely to be bewildered by such an assignment and feel incapable of a satisfactory performance without considerable research. A world map that displays a reputable climatic classification, however, can show the student that southeastern China has the same climate as the southeastern United States. Thus the student's familiarity with the home climate in Georgia provides a basis for understanding the general characteristics of the climate of southeastern China.

MANY CLASSIFICATION SCHEMES

The earliest known climatic classification scheme originated with the ancient Greeks in the first or second century B.C. Although the "known world" was very small at that time, as indicated in Figure 8-1, some Greek scholars were aware of the shape and approximate size of Earth. They knew that at the southern limit of their world, along the Nile River and the southern coast of the Mediterranean, the climate was much hotter and drier than on the islands and northern coast of that sea. At the other end of the world known to the Greeks, along the Danube River and the northern coast of the Black Sea, things were much colder, especially in winter. So the Greeks spoke of three climatic zones: the Temperate Zone of the midlatitudes, in which they lived (Athens is at 38° N latitude), the Torrid Zone of the tropics to the south, and the Frigid Zone to the north. Because they knew that Earth is a sphere, they suggested that the Southern Hemisphere has similar Temperate and Frigid zones, making five in all.

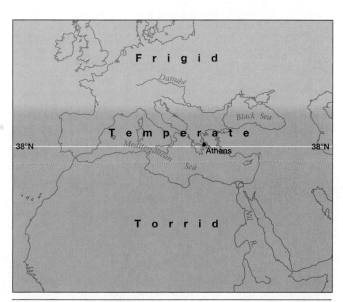

Figure 8-1 The climate classification scheme of the ancient Greeks.

Snow-covered trees near Sun Valley, Idaho. *(David Stoecklein; The Stock Market)*

For many centuries, this classification scheme was handed down from scholar to scholar. Gradually these five climatic zones were confused with, and eventually their climates ascribed to, the five astronomical zones of the Earth, bounded by the Tropics of Cancer and Capricorn and the Arctic and Antarctic circles (Figure 8-2). This revision put the equatorial rainy zone in with the hot arid region in the Torrid Zone, extended the Temperate Zone to include much of what the Greeks had called Frigid, and moved the Frigid Zone poleward to the polar circles. This simplistic but unrealistic classification scheme persisted for more than a thousand years and was finally discarded only in the twentieth century.

Today we recognize five basic climate zones in the world: equatorial warm-wet, tropical hot-dry, subtropical warm temperate, midlatitude cool temperate, and high-latitude cold (Figure 8-3). The equatorial and tropical zones are differentiated by rainfall amount and frequency. The two temperate zones differ primarily in whether summer or winter is the dominant season (in Atlanta, for instance, houses are designed to be cool in summer, whereas in Minneapolis keeping warm in winter is more important; thus summer is the dominant season in the former and winter is the dominant season in the latter). The cold zone has hardly any summer—not enough to grow important crops. Many climatologists have refined this five-zone scheme by subdividing the zones into various types and subtypes, with most schemes relying primarily on natural vegetation as the principal climate indicator.

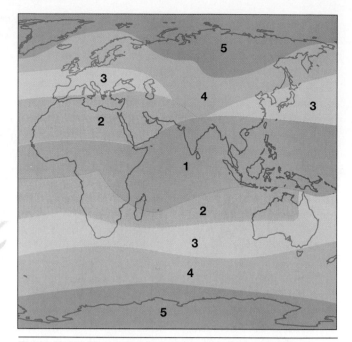

Figure 8-3 The major climatic zones of the Old World, as recognized today: (1) equatorial warm wet; (2) tropical hot dry; (3) subtropical warm temperate; (4) midlatitude cool temperate; (5) high-latitude cold.

THE KÖPPEN SYSTEM

Classification schemes having the greatest pedagogic value share three important attributes:

1. They are relatively simple to comprehend and to use.
2. They show some sort of orderly pattern over the world.
3. They give some indication of zone genesis.

The **Köppen system** meets these criteria reasonably well and is by far the most widely used climatic classification system. Wladimir Köppen (1846–1940) was a Russian-born German climatologist who was also an amateur botanist. The first version of his scheme appeared in 1918, and he continued to modify and refine it for the rest of his life, the last version being published in 1936. The Köppen system uses a numerical basis of classification (either average temperatures or average amounts of precipitation), and zone boundaries are determined by vegetation patterns. Thus the boundaries in the Köppen system represent climatic expressions of floristic limits.

The system uses as a database only the mean annual and monthly values of temperature and precipitation, combined and compared in a variety of ways. Consequently, the necessary statistics are commonly tabulated and easily acquired. Data for any location (called a "station") on Earth can be used to determine the precise classification of that place, and the areal

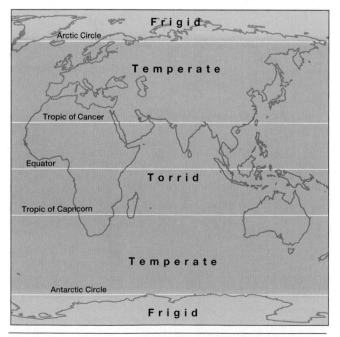

Figure 8-2 Eventually the Greek classification scheme was expanded and correlated with Earth's five astronomical zones to depict a world comprising five climate zones.

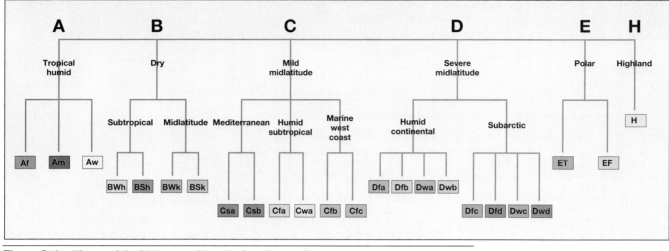

Figure 8-4   The modified Köppen climatic classification.

extent of any recognized climatic type can be determined and mapped. This means that the classification system is functional at both the specific and the general level.

Köppen defined four of his five major climatic zones by temperature characteristics, the fifth (the B zone) on the basis of moisture. He then subdivided each zone into types according to various temperature and precipitation relationships. The system's distinctive feature is a symbolic nomenclature to designate the various climatic types. This nomenclature consists of a combination of letters, with each letter having a precisely defined meaning.

The Köppen system has been widely used, although often modified, and its terminology has been even more widely adopted. There are some deficiencies in the system, however, as in any other climatic classification scheme. Köppen himself was unsatisfied with his last version and did not consider it a finished product. Thus many geographers and climatologists have used the Köppen system as a springboard to devise systems of their own or to modify the "final" Köppen classification.

THE MODIFIED KÖPPEN SYSTEM

The system of climatic classification used in this book is **modified Köppen**. It encompasses the basic design of the Koppen system but with a variety of minor modifications. Some of these modifications follow the lead of Glen Trewartha, late geographer/climatologist at the University of Wisconsin. No attempt is made to distinguish between pure Köppen and modified Köppen in this text; our goal is to comprehend the general pattern of world climate, not to learn a specific system or to nitpick about boundaries.

In the modified system, we have Köppen's five major climatic zones (A through E) plus a sixth zone (H) called highland climates, as shown in Table 8-1. Table 8-2 gives the derivation of the 20 code letters used in the system as well as the climate variable defined by each letter. Figure 8-4 is a schematic of the system, and Figure 8-5 on pages 208–209 shows the global distribution of all these climates.

As Figure 8-4 shows, the first five of the six zones are broken down into various types and subtypes. A brief description of each is given in Table 8-3.

CLIMOGRAPHS

Probably the most useful tool in a general study of world climatic classification is a simple graphic representation of monthly temperature and precipitation for a specific weather station. Such a graph is called a **climograph,** and a typical one is shown in Figure 8-6 on page 208. The customary climograph has 12 columns, one for each month, with a temperature scale on the left side and a precipita-

TABLE 8-1	
The Six Zones of the Modified Köppen Scheme	
Zone	*Description*
A	Tropical humid
B	Dry; evaporation exceeds precipitation
C	Mild midlatitude
D	Severe midlatitude
E	Polar
H	Highland

TABLE 8-2
Code Letters of the Modified Köppen Scheme

Letter	Derivation	Definition
First letters:		
A	Alphabetical	Average temperature of each month above 64°F (18°C)
B	Alphabetical	Average annual precipitation below 30 in. (76 cm)
C	Alphabetical	Average temperature 64–27°F [18–(-3)°C] in coldest month, above 50°F (10°C) in warmest month
D	Alphabetical	4–8 months with average temperature above 50°F (10°C)
E	Alphabetical	No month with average temperature above 50°F (10°C)
H	Highland	Significant climatic changes in short horizontal distances due to altitudinal variations
Second Letters:		
f	German *feucht*, moist	In A climates: average rainfall of each month at least 2.5 in. (6 cm); in C and D climates, is neither s category nor w category
F	Frost	No month with average temperature above 32°F (0°C)
m	Monsoon	Only 1-3 months with average rainfall below 2.5 in. (6 cm)
s	Summer dry	Driest summer month has below 1/3 the average precipitation of wettest winter month
S	Steppe, semiarid	Average annual precipitation 15–30 in. (38–76 cm) in low latitudes, 10–25 in. (25–64 cm) in midlatitudes, with no pronounced seasonal concentration
T	Tundra	At least 1 month with average temperature 32–50°F (0–10°C)
w	Winter dry	In A climates, 3–6 months with average rainfall below 2.5 in. (6 cm); in C and D climates, driest winter month has less than 1/10 the average precipitation of wettest summer month
W	German *Wüste*, desert	Average annual precipitation below 15 in. (38 cm) in low latitudes, below 10 in. (25 cm) in midlatitudes
Third letters:		
a	Alphabetical	Average temperature of warmest month above 72°F (22°C)
b	Alphabetical	Average temperature of warmest month below 72°F (22°C); at least 4 months with average temperature above 50°F (10°C)
c	Alphabetical	Average temperature of warmest month below 72°F (22°C); fewer than 4 months with average temperature above 50°F (10°C)
d	Alphabetical	Average temperature of coldest month below –36°F (–38°C)
h	German *heiss*, hot	Average annual temperature above 64°F (18°C)
k	German *kalt*, cold	Average annual temperature below 64°F (18°C)

tion scale on the right. Average monthly temperatures are connected by a curved line in the upper portion of the diagram, and average monthly precipitation is represented by bars extending upward from the bottom.

The value of a climograph is twofold: (1) it displays precise details of important aspects of the climate of a specific place, and (2) it can be used to classify the climate of that place.

TABLE 8-3
Climate Types and Subtypes in the Modified Köppen System

Type or subtype	Letter code	Description
Tropical wet	Af	No dry season
Tropical monsoonal	Am	Monsoonal; short dry season with heavy rains in other months
Tropical savanna	Aw	Dry season in winter (low-sun season)
Subtropical desert	BWh	Low-latitude true desert
Subtropical steppe	BSh	Low-latitude dry
Midlatitude desert	BWk	Midlatitude true desert
Midlatitude steppe	BSk	Midlatitude dry
Mediterranean	Csa	Mild midlatitude with dry, hot summer
Mediterranean	Csb	Mild midlatitude with dry, warm summer
Humid subtropical	Cfa	Mild midlatitude with no dry season and hot summer
Humid subtropical	Cwa	Mild midlatitude with dry winter and hot summer
Marine west coast	Cfb	Mild midlatitude with no dry season and warm summer
Marine west coast	Cfc	Mild midlatitude with no dry season and cool summer
Humid continental	Dfa	Humid midlatitude with severe winter, no dry season, and hot summer
Humid continental	Dfb	Humid midlatitude with severe winter, no dry season, and warm summer
Humid continental	Dwa	Humid midlatitude with severe winter, dry winter, and hot summer
Humid continental	Dwb	Humid midlatitude with severe, dry winter and warm summer
Subarctic	Dfc	Humid midlatitude with severe winter, no dry season, and cool summer
Subarctic	Dfd	Humid midlatitude with severe, very cold winter and no dry season
Subarctic	Dwc	Humid midlatitude with severe, dry winter and cool summer
Subarctic	Dwd	Humid midlatitude with severe, dry, very cold winter
Tundra	ET	Polar tundra with no true summer
Ice cap	EF	Polar ice cap
Highland	H	Highland

WORLD DISTRIBUTION OF MAJOR CLIMATIC TYPES AND SUBTYPES

Most of the remainder of this chapter is devoted to a discussion of the climatic types and subtypes. Our attention is focused primarily on three questions:

1. Where are the various climatic types and subtypes?
2. What are the characteristics of each?
3. Why are these characteristics found in these locations?

TROPICAL HUMID CLIMATES (ZONE A)

The tropical humid climates occupy almost all the land area of Earth within some 15–20° of the equator, in both the Northern and Southern hemispheres (Figure 8-7). This globe-girdling belt of A climates is interrupted slightly here and there by mountains or small regions of aridity, but it dominates the equatorial regions and extends poleward to beyond the 25° parallel in some windward coastal lowlands.

The A climates are noted not so much for warmth as for lack of coldness. These are the only truly win-

A TROPICAL HUMID CLIMATES

	Af	Tropical wet climate
	Am	Tropical monsoonal climate
	Aw	Tropical savanna climate

B DRY CLIMATES

	BWh	Subtropical desert
	BWk	Midlatitude desert
	BSh	Subtropical steppe
	BSk	Midlatitude steppe

C MILD MIDLATITUDE CLIMATES

	Cfa	Humid subtropical, without dry season, hot summers
	Cwa Cwb	Humid subtropical, winter-dry
	Cfb Cfc	Marine west coast, without dry season, warm to cool summers
	Csa Csb	Mediterranean summer-dry

D SEVERE MIDLATITUDE CLIMATES

	Dfa Dwa	Humid continental, hot summers
	Dfb Dwb	Humid continental, warm summers
	Dfc Dwc	Subarctic, cool summers
	Dfd Dwd	Subarctic, very cold winter

E POLAR CLIMATES

	ET	Tundra
	EF	Ice cap

H HIGHLAND

	H	Cold climates due to elevation

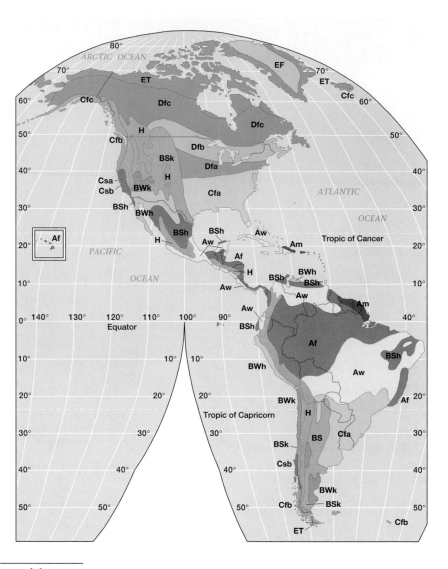

Figure 8-5 Climate zones, types, and subtypes of the modified Köppen system.

terless climates of the world. They are characterized by moderately high temperatures throughout the year, as is to be expected from their near-equatorial location. The sun is high in the sky every day of the year, and even the shortest days are not appreciably shorter than the longest ones. These are climates of perpetual warmth (although they do not experience the world's highest temperatures). The fundamental character of the A climates, then, is molded by their latitudinal location.

The second typifying characteristic of the tropical humid climates is of course the prevalence of moisture. Although not universally rainy, much of the A zone is among the wettest in the world. Warm, moist, unstable air masses frequent the oceans of these latitudes, and the intertropical convergence (ITC) zone is in the A zone for much of the year. Moreover, onshore winds and thermal convection are commonplace phenomena. Thus the A

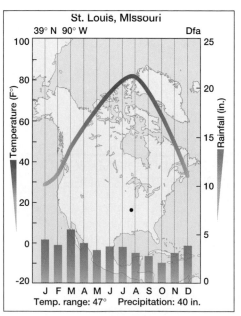

Figure 8-6 A typical climograph.

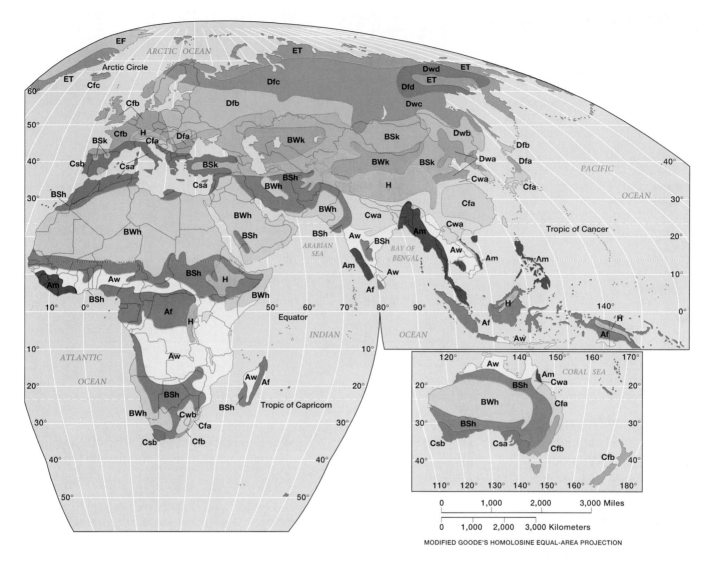

zone has not only abundant sources of moisture but also abundant mechanisms for uplift. High humidity and considerable rainfall are expectable results.

Table 8-4 summarizes A-zone characteristics.

The tropical humid climates are classified into three types on the basis of annual rainfall. The tropical wet type (Af in the modified Köppen system) has abundant rainfall (above 2.5 inches) every month of the year. The tropical monsoonal type (Am) has a distinct dry season in which the monthly rainfall average is below 2.5 inches and a very rainy wet season in which the monthly average is much higher. The tropical savanna type (Aw) is characterized by a longer dry season (3–6 months) and a prominent but not extraordinary wet season.

Tropical Wet Climate (Af)   Climates of the Af type (f = *feucht*, German for moist) characteristically occur in an east-west sprawl astride the equator, extending some 5–10° poleward on either side (Figure 8-7). In some eastern-coast situations, they may extend as much as 25° away from the equator. The largest areas of Af climate occur in the upper Amazon basin of South America, the northern Zaire (Congo) basin of Africa, and the islands of the East Indies.

The single most descriptive word that can be applied to the tropical wet climate is *monotonous* because

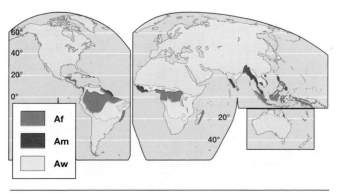

Figure 8-7   Distribution of A climates.

TABLE 8-4
Summary of A Climates: Tropical Humid

Type	Location	Temperature	Precipitation	What controls the climate
Tropical wet (Af)	5–10° either side of equator; farther poleward on eastern coasts	Warm all year; very small ATR; small DTR; high sensible temperature	60–100 in. (152–254 cm) annually; no dry season; many thunderstorms	Latitude; ITC zone; trade-wind convergence; onshore wind flow
Tropical monsoonal (Am)	Windward tropical coasts of Asia, Latin America, Guinea coast of Africa	Similar to Af with slightly larger ATR; hottest weather just before summer monsoon	100–200 in. (254–508 cm) annually; very heavy in summer; short winter dry season	Seasonal wind-direction reversal associated with ITC zone movement; jet stream fluctuation; continental pressure changes
Tropical savanna (Aw)	Fringing Af between 25° N and S	Warm to hot; moderate ATR and DTR	35–70 in. (90–180 cm) annually; distinct wet and dry seasons	Seasonal shifting of tropical wind and pressure belts, especially ITC zone

All A climates lie within 25° of the equator, with the vast majority within 20°. All are winterless, with no monthly average temperature below 64° (18°C). All are always very humid and have high rainfall totals.

ATR = annual temperature range; DTR = daily temperature range.

---

it is a seasonless climate, with endless repetition of the same weather day after day after day. Warmth prevails, with every month having an average temperature close to 80°F (27°C), as Figure 8-8 shows. The terms *summer* and *winter* are meaningless because the annual temperature range (the difference between the average temperatures of the coolest and warmest months) is minuscule, typically only 2 or 3°F (1 or 1.5°C) and only rarely over 8°F (4°C), by far the smallest annual temperature range of any climatic type. This seasonless condition in Af regions has given rise to the saying, "Night is the winter of the tropics."

The words *winter* and *summer* are used in discussing tropical climates, though, as in the "winter monsoon"

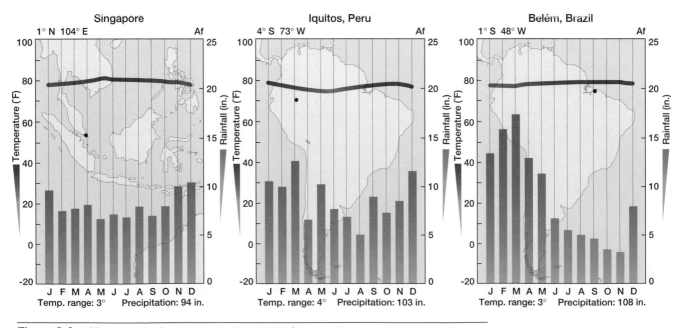

Figure 8-8 Climographs for representative tropical wet stations.

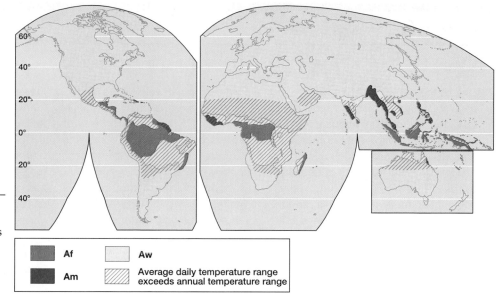

Figure 8-9 Land portions of the world where the average daily temperature range exceeds the average annual temperature range. This characteristic is typical of A climates but rare in other climates.

Af

Am

Aw

Average daily temperature range exceeds annual temperature range

and the "summer monsoon." In this context, these adjectives mean that time of the year when places that are farther from the equator are indeed cold. Alternately, we refer to the "high-sun season" (summer) and the "low-sun season" (winter), where *high* and *low* are strictly relative terms.

Daily temperature variations are somewhat greater than annual ones, although still not impressive. This is one of the few climates in which the average daily temperature range exceeds the average annual temperature range (Figure 8-9). On a typical afternoon, the temperature rises to the high 80s °F (low 30s °C), dropping to the middle or low 70s °F (low 20s °C) in the coolest period just before dawn. The temperature rarely extends much into the 90s °F (mid-30s °C), even on the hottest days, and equally unusual is much cooling at night.

Regardless of the thermometer reading, however, the weather feels warm in this climate because high humidity makes for high sensible temperatures, except perhaps where a sea breeze blows. Both absolute and relative humidity are notable, and rain can be expected just about every day—sometimes twice or three times a day (Figure 8-10). Rainfall is of the unstable, showery variety, usually coming from thunderstorms that yield heavy rain for a short time. A typical morning dawns bright and clear. Cumulus clouds build up in the forenoon and develop into cumulonimbus thunderheads, producing a furious convectional rainstorm in early afternoon. Then the clouds usually disperse, so that by late afternoon there is a partly cloudy sky and a glorious sunset. The clouds often recur at night to create a nocturnal thunderstorm, followed by dispersal once again. The next day dawns bright and clear, and the sequence repeats.

Each month receives several inches of rain, and the annual total normally is between 60 and 100 inches (150 and 250 centimeters), although in some locations it is considerably greater. Yearly rainfall in the Af climate is exceeded by that of only one other type of climate (tropical monsoonal).

Another characteristic of an Af climate is lack of wind. Except along coastlines, where sea breezes may be frequent or even persistent, the Af climate is poorly ventilated.

Why these climatic conditions occur where they do is relatively straightforward. The principal climatic control is latitude. A sun high in the sky throughout the year makes for relatively uniform insolation through the year, and so there is little opportunity for seasonal temperature variation. This extensive heating produces considerable thermal convection, which ac-

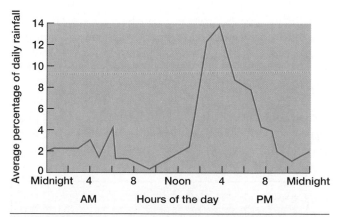

Figure 8-10 A typical daily pattern of rainfall in an Af climate. This example, from Malaysia, shows a heavy concentration in midafternoon with a small secondary peak at or shortly before dawn.

FOCUS

ENVIRONMENTAL RELATIONSHIPS

One of the most important themes in physical geography is the intertwining relationships of the various components of the environment. Time after time we note situations in which one aspect of the environment affects another—sometimes conspicuously, sometimes subtly. Although such effects often can be seen on a microscale, in this book your attention is directed mostly toward broadscale patterns.

In its own right, climate is a major component of the environment, but it takes on added importance because of its notable effects on other components. It is premature at this point to analyze these myriad relationships because thus far we have explored the fundamentals of only one "sphere"—the atmosphere. It is useful, however, to give at this point an example of some of the environmental associations that interest the physical geographer.

Let us consider a single climatic type, the tropical wet (Af), as an exemplar of these relationships. Specific locations where a tropical wet climate is found–equatorial zones and windward coasts in tropical latitudes–are characterized by fairly specific environmental conditions. It is tempting to state that the Af climate "creates" these conditions, but the cause-and-effect relationship is not that simple. It is proper, however, to note the *correlation* between climatic type and other environmental conditions. If climate is not the cause, it at least has a prominent influence on those conditions.

Flora

Because of the high temperatures and high humidity, regions having an Af climate normally are covered with natural vegetation that is unexcelled in luxuriance and variety. For the most part, this vegetation takes the form of a tropical rainforest, or **selva**, which is a broadleaf evergreen forest with numerous tree species (Figure 8-1-A). Many of the trees are very tall, and their intertwining tops form an essentially continuous canopy that prohibits sunlight from shining on the forest floor. Often shorter trees form a second and even a third partial canopy at lower elevations. Most of

Figure 8-1-A   The diverse and luxuriant growth of a tropical rainforest. This scene is from Caribbean National Forest in Puerto Rico. (Jeff Greenberg/dMRp/Photo Researchers, Inc.)

counts for a portion of the raininess. More important, trade-wind convergence in the ITC zone leads to widespread uplift of warm, humid, unstable air. Also, persistent onshore winds along trade-wind (that is, east-facing) coasts provide a consistent source of moisture and add another mechanism for precipitation—orographic ascent. Indeed, the maximum poleward extent of Af climate is found along such trade wind coasts, as in Central America and Madagascar.

Tropical Monsoonal Climate (Am)   The tropical monsoonal climate (Am type: m = monsoon) is most extensive on the windward (west-facing) coasts of southeastern Asia [primarily India, Myanmar (formerly Burma), and Thailand], but it also occurs in more restricted coastal regions of western Africa, northeastern South America, the Philippines, and some islands of the East Indies (Figure 8-7).

The distinctiveness of the Am climate is shown primarily in its rainfall pattern (Figure 8-11). During the high-sun season, an enormous amount of rain falls in association with the summer monsoon. It is not unusual to have more than 30 inches (75 centimeters) of rain in each of two or three months. The annual total for a typical Am station is between 100 and 200 inches (250

WORLD RECORD

Highest Average Annual Precipitation

460 inches (1168 centimeters) at Mount Waialeale, Hawaii, an Af climate

WORLD RECORD

Highest One-Day Rainfall Total

74 inches (188 centimeters) 15 March 1952 at Cilaos, La Réunion (an island in the Indian Ocean), an Af climate

the trees are smooth-barked and have no low limbs, although there is a profusion of vines and hanging plants that entangle the trunks and dangle from higher limbs. The dimly lit forest floor is relatively clear of growth because lack of sunlight inhibits survival of bushes and shrubs.

Where much sunlight reaches the ground, as along the edge of a clearing or banks of a stream, a maze of undergrowth can prosper. This sometimes impenetrable tangle of bushes, shrubs, vines, and small trees is called a *jungle*.

Fauna

Any Af region is the realm of flyers, crawlers, creepers, and climbers. Larger species, particularly hoofed animals, are not common. Birds and monkeys inhabit the forest canopy, often in great quantity and diversity. Snakes and lizards are common both on the forest floor and in the trees. Rodents are sometimes numerous at ground level, but the sparser population of larger mammals typically is secretive and nocturnal. Aquatic life,

particularly fish and amphibians, is usually abundant. Invertebrates, especially insects and arthropods, are characteristically superabundant.

Soil

The copious, warm, year-round rains provide an almost continuous infiltration of water downward, with the result that soils are usually deep but highly leached and infertile. Leaves, twigs, flowers, and branches frequently fall from the trees to the ground, where they are rapidly decomposed by the abundant earthworms, ants, bacteria, and microfauna of the soil. The accumulated litter is continuously incorporated into the soil, where some of the nutrients are taken up by plants and the remainder are carried away by the infiltrating water. Laterization (rapid weathering of mineral matter and speedy decomposition of organic matter) is the principal soil-forming process; it produces a thin layer of fertile topsoil that is rapidly used by plants, and a deep subsoil that is largely an infertile mixture of such insoluble constituents as iron, aluminum, and

magnesium compounds. These minerals typically impart a reddish color to the soil. River floodplains tend to develop soils of higher fertility because of flood-time deposition of silt.

Hydrography

The abundance of runoff water on the surface feeds well-established drainage systems. There is usually a dense network of streams, most of which carry both a great deal of water and a heavy load of sediment. Lakes are not common because there is enough erosion to drain them naturally. Where the land is very flat, swamps sometimes develop through inadequate drainage and the rapid growth of vegetation.

The environmental relations summarized above represent some of the more conspicuous features associated with a tropical wet climate. They may not apply to any specific location, but they serve as generalizations for such regions. Similar generalizations can be made, with varying validity, for each of the other climatic types and subtypes discussed in this chapter.

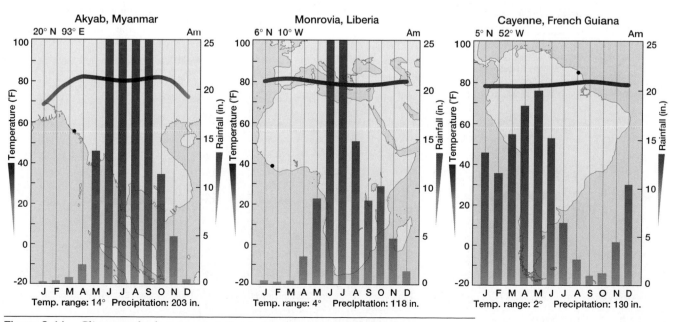

Figure 8-11 Climographs for representative tropical monsoonal stations.

and 500 centimeters). An extreme example is Cherrapunji (in the Khasi hills of Assam, in eastern India), which has an annual average of 425 inches (1065 centimeters); Cherrapunji has been inundated with 84 inches (210 centimeters) in three days, with 366 inches (930 centimeters) in one month, and with a memorable 1042 inches (2647 centimeters) in its record year.

During the low-sun season, Am climates are dominated by offshore winds. The monsoon during this season produces little precipitation; from 1 to 4 months may record less than 2.5 inches (6 centimeters) per month, and 1 or 2 months may be rainless.

A lesser distinction of the Am climate is its annual temperature curve. Although the annual temperature range may be only slightly greater than in a tropical wet climate, the highest Am temperatures normally occur in late spring, prior to the onset of the summer monsoon. The heavy cloud cover of the wet monsoon period shields out some of the insolation, resulting in slightly lower temperatures in summer than in spring.

Apart from these monsoonal modifications, the climatic characteristics of Af and Am locations are similar.

Tropical Savanna Climate   The most extensive of the A climates, the tropical savanna (Aw type: w = winter dry) generally lies both to the north and to the south of the Af and Am areas (Figure 8-7). As Figure 8-5 shows, it is widespread in Africa to the east of the tropical wet region, extending poleward to about latitudes 30° S and 10° N. It occurs broadly in South America and southern Asia and to a lesser extent in northern Australia, Central America, and the Caribbean islands.

WORLD RECORD

Greatest 1-Year Precipitation

1042 inches (2647 centimeters) in 1860/61 at Cherrapunji, India, an Am climate

The distinctive characteristic of the Aw climate is its clear-cut seasonal alternation of wet and dry periods (Figure 8-12). This characteristic is explained by the fact that Aw climates lie between unstable, converging air on their equatorial side and stable, anticyclonic air on their poleward side. As the global wind and pressure systems shift latitudinally with the sun over the course of a year, Aw regions experience extreme contrasts in weather. During the low-sun season (winter), all wind and air-pressure systems shift toward the opposite hemisphere, so that savanna regions are dominated by subtropical high-pressure conditions of subsiding and diverging air, which produce clear skies. In summer, the systems shift in the opposite direction, bringing the ITC zone and its wet tropical weather patterns into the Aw region (Figure 8-13). The poleward limits of Aw climate are approximately equivalent to the poleward maximum migration of the ITC zone.

Annual rainfall totals in the tropical savanna climate are generally less than in the other two A climates, as Table 8-4 shows; typical Aw annual averages are between 35 and 70 inches (90 and 180 centimeters). The high-sun season is conspicuously wet, with one to four months receiving at least 10 inches (25 centimeters) of

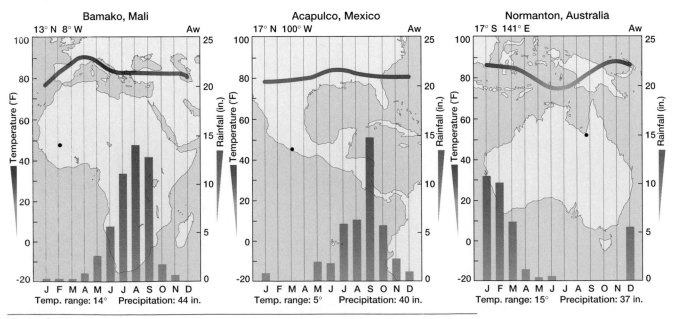

Figure 8-12   Climographs for representative tropical savanna stations.

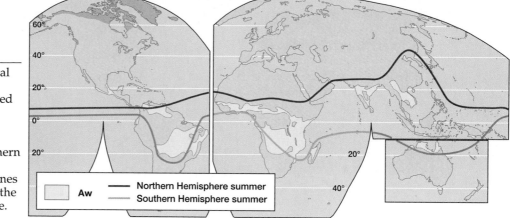

Figure 8-13 The intertropical convergence zone migrates widely during the year. The red line shows its northern boundary in the northern hemisphere summer, and the orange line indicates its southern boundary in the southern hemisphere summer. These lines approximately coincide with the poleward limits of Aw climate.

rain. The low-sun season, on the other hand, is distinctly a time of drought, sometimes with three or four rainless months. Although both tropical monsoonal and tropical savanna climates have dry winters, total annual rainfall is much greater in Am regions than in Aw regions.

Average annual temperatures in Aw regions are about the same as in Af regions but with a somewhat greater month-to-month variation in the former. Winter is a little cooler and summer a little hotter than in Af regions, giving an annual temperature range generally between 5 and 15°F. As in tropical monsoon climates, the hottest time of the year is likely to be in late "spring," just before the onset of the summer rains.

The tropical savanna climate can be thought of as having three seasons rather than two. The wet season is much like the wet season anywhere else in the tropics, with high sensible temperatures, muggy air, and frequent convective showers. The early part of the dry season is a period of clearing skies and slight cooling. The later part of the dry season is a time of fire: wildfires, fueled by the desiccated grass and shrubs, are common almost every year during this season.

DRY CLIMATES (ZONE B)

The dry climates cover about 30 percent of the land area of the world (Figure 8-14), more than any other climatic zone. Although at first glance their distribution pattern appears erratic and complex, it actually has a considerable degree of predictability.

The arid regions of the world (other than in the Arctic) generally are the result of lack of air uplift rather than lack of moisture in the air. Most desert areas do not lack precipitable moisture; rather, they lack mechanisms for the upward air motion necessary for cloud formation and precipitation. Vertical motion is suppressed by persistent stability, which is due mainly to the subsidence associated with subtropical high-pressure cells and secondarily to subsidence in the lee of mountain barriers. High-altitude temperature inversions often develop, further inhibiting the likelihood of updrafts and adiabatic cooling.

The largest expanses of dry areas are in subtropical latitudes, where thermodynamic subsidence is widespread, and in the western and central portions of continents because anticyclonic conditions are more

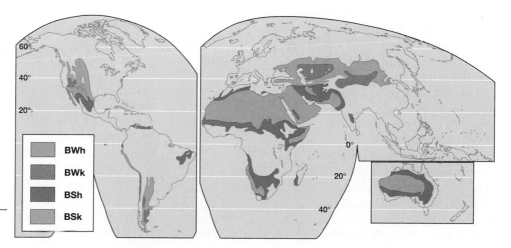

Figure 8-14 Distribution of B climates.

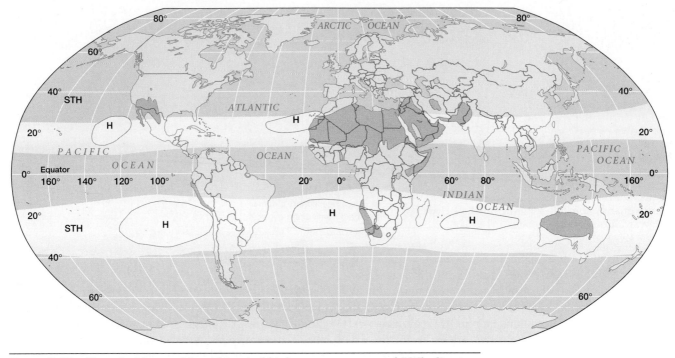

Figure 8-15   The coincidence of the subtropical high-pressure zones and BWh climates is striking.

prevalent there. (Desert conditions also occur over extensive ocean areas, and it is quite reasonable to refer to marine deserts.) In the midlatitudes, particularly in central Asia, the B climates are found in areas that are separated from sources of surface moisture either by great distances or by topographic barriers.

The concept of dry climate is a complex one because it involves the balance between precipitation and evapotranspiration and depends not only on rainfall but also on temperature. The basic generalization is that higher temperature engenders greater potential evapotranspiration, and so hot regions can receive more precipitation than cool ones but still be drier.

As Table 8-3 and Figure 8-5 show, the two main categories of B climates are desert and steppe. Deserts are arid, generally receiving less than 10 inches (25 centimeters) of precipitation annually, and steppes are semiarid, typically receiving 10 to 25 inches (25 to 63 centimeters) of precipitation annually. Normally the deserts of the world are large core areas of aridity surrounded by a transitional fringe of steppe that is slightly less dry.

The two B climates are further classified into four types: subtropical desert (BWh), subtropical steppe (BSh), midlatitude desert (BWk), and midlatitude steppe (BSk). Our discussion here focuses on the deserts because they represent the epitome of dry conditions—the arid extreme. Most of what is stated

about deserts applies to steppes but in modified intensity.

Table 8-5 gives an overview of all the B climates.

Subtropical Desert Climate (BWh)   In both the Northern and Southern hemispheres, subtropical deserts (BWh climate: W = *Wüste*, German for desert; h = *heiss*, German for hot) lie either in or very near the band of the subtropical high (Figure 8-15). Arid conditions reach to the western coasts of all continents in these latitudes. Subsidence is weaker on the western sides of the subtropical highs, with the result that, except in North Africa, which is sheltered from oceanic influence by the Arabian peninsula, the aridity does not extend to the eastern coast of any continent in these latitudes.

Climographs for three typical BWh locations are shown in Figure 8-16.

The enormous expanse of BWh climate in North Africa (the Sahara) and southwestern Asia (the Arabian Desert) represents more desert area than is found in the rest of the world combined. Such an extensive development is explained by the year-round presence of anticyclonic conditions and the remoteness from any upwind source of moisture. The adjacency of Asia makes Africa a continent without an eastern coast north of 10° N latitude, and so the BWh climate extends from coast to coast in Africa. This climate is also very expansive in Australia (50 percent of the conti-

TABLE 8-5
Summary of B Climates: Dry

Type	Location	Temperature	Precipitation	What controls the climate
Subtropical desert (BWh)	Centered at latitudes 25–30° on western sides of continents, extending into interiors; most extensive in northern Africa and southwestern Asia	Very hot summers, relatively mild winters; enormous DTR, moderate ATR	Rainfall scarce, unreliable, intense; little cloudiness	Subtropical anticyclonic subsidence; rain shadow of mountains; cold ocean currents
Subtropical steppe (BSh)	Fringing BWh except on west	Similar to BWh but more moderate	Semiarid	Similar to BWh
Midlatitude desert (BWk)	Central Asia; western interior of United States; Patagonia	Hot summers, cold winters; very large ATR, large DTR	Meager, erratic, mostly showery; some winter snow	Distant from sources of moisture; some rain shadow effects
Midlatitude steppe (BSk)	Peripheral to BWk; transitional to more humid climates	Similar to BWk but slightly more moderate	Semiarid; some winter snow	Similar to BWk

B climates are widespread in the subtropics, where they are associated with anticyclonic conditions. In the midlatitudes they occur in extreme continental locations that are remote from sources of moisture. Deserts are arid, steppes are semiarid, and evaporation exceeds precipitation in both.

nental area) because the mountains that parallel the eastern coast of that continent are just high enough to prevent Pacific winds from penetrating; most of Australia is in the rain shadow of those eastern highlands.

Subtropical deserts have a much more restricted longitudinal extent in southern Africa, South America, and North America, but they are elongated latitudinally along the coast because of the presence of cold

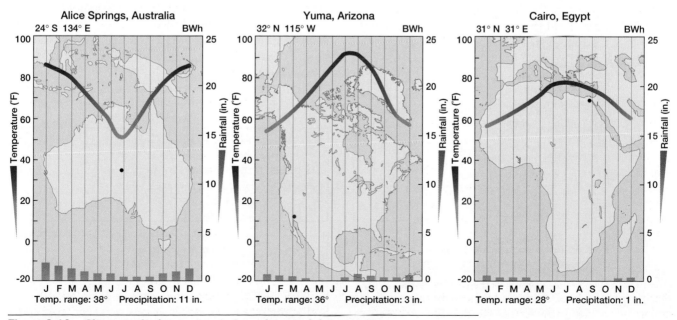

Figure 8-16 Climographs for representative subtropical desert stations.

offshore waters, which chill the overlying air and inhibit precipitation. The greatest elongation occurs along the western side of South America, where the Atacama Desert is not only the "longest" but also the driest of the dry lands. The Atacama is sandwiched in a double rain shadow position (Figure 8-17): moist winds from the east are kept out of this region by one of the world's great mountain ranges (the Andes), and Pacific air is thoroughly chilled and stabilized as it passes over the world's most prominent cold current (the Humboldt).

The distinctive climatic characteristic of deserts is lack of moisture, and three adjectives are particularly applicable to precipitation conditions in subtropical deserts: *scarce, unreliable, intense.*

1. *Scarce*—Subtropical deserts are the most nearly rainless regions on Earth. According to unofficial records, some have experienced several consecutive years without a single drop of moisture falling from the sky. Most BWh regions, however, are not totally without precipitation. Annual totals of between 2 and 8 inches (5 and 20 centimeters) are characteristic, and some places receive as much as 15 inches (38 centimeters).

2. *Unreliable*—An important climatic axiom is that the less the mean annual precipitation, the greater its variability. The very concept of an "average" yearly rainfall in a BWh location is misleading because of year-to-year fluctuations. Yuma, Arizona, for example, has a long-term average rainfall of 2.7 inches (6.5 centimeters), but within the last decade it has received as little as 0.5 inch (1.3 centimeters) and as much as 7 inches (18 centimeters) in a given year. In short, percentage deviation from the norm tends to be very high in BWh climates.

3. *Intense*—Most precipitation in these regions falls in vigorous convective showers that are localized and of short duration. Thus the rare rains may bring brief floods to regions that have been bereft of surface moisture for months.

Temperatures in BWh regions also have certain distinctive characteristics. The combination of low-latitude location and lack of cloudiness permits a great deal of insolation to reach the surface, and nocturnal terrestrial radiation is likewise appreciable. Summers are interminably long and blisteringly hot, with monthly averages in the middle to high 90s °F (high 30s °C). Midwinter months have average temperatures in the 60s °F (high teens °C), which gives moderate annual temperature ranges of 15–25 °F. Daily temperature ranges, on the other hand, are sometimes astounding. Summer days are so hot that the nights do not have time to cool off significantly, but during the transition seasons of spring and fall a 50 °F fluctuation between the heat of the afternoon and the cool of the following dawn is not unusual.

Subtropical deserts experience considerable windiness during daylight, but the air is usually calm at

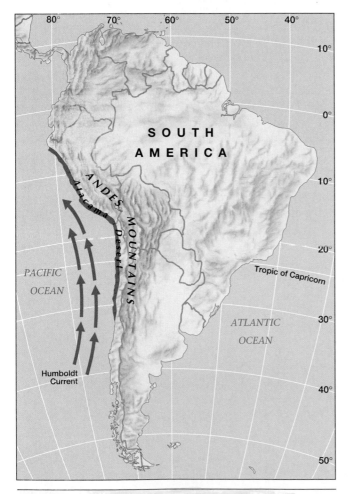

Figure 8-17   The Atacama Desert is in a double rain shadow: The Andes Mountains to the east block the movement of moist air from the Atlantic, and the cold Humboldt Current to the west inhibits the flow of moist Pacific air inland.

WORLD RECORD

Highest Temperature
136°F (58°C) on 13 September 1922 at El Azizia, Libya, a BWh climate

---

WORLD RECORD

Greatest Temperature Range in One Day
100°F (56°C) in 1927 at In-Salah, Algeria, a BWh climate

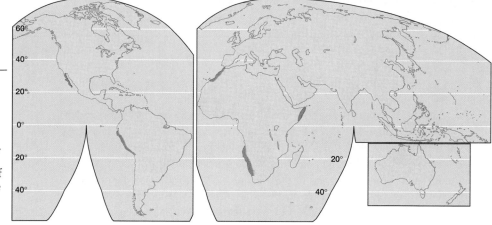

Figure 8-18 Cool, foggy west coast deserts are found along coasts paralleled by cool ocean currents and cold upwellings. Such deserts are mostly in subtropical west coast locations, with two exceptions: they are absent from the western coast of Australia, and they occur on the eastern coast of the "horn" of Africa (Somalia).

night. The daytime winds are apparently related to rapid daytime heating and strong convective activity, which accelerates surface currents. The persistent winds are largely unimpeded by soil and vegetation, with the result that a great deal of dust and sand is frequently carried along.

Specialized and unusual temperature conditions prevail along western coasts in subtropical deserts (Figure 8-18). The cold waters offshore (the result of currents and upwellings) chill any air that moves across them. This cooling produces high relative humidity as well as frequent fog and low stratus clouds. Precipitation almost never results from this advective cooling, however, and the influence normally extends only a few miles inland. The immediate coastal region, however, is characterized by such abnormal desert conditions as relatively low summer temperatures

(typical hot-month averages in the low 70s °F, or low 20s °C), continuously high relative humidity, and greatly reduced annual and daily temperature ranges, as indicated by a comparison of Figures 8-16 and 8-19.

Subtropical Steppe Climate (BSh)   The BSh climates (S = steppe; h = *heiss*, German for hot) characteristically surround the BWh climates (except on the western side), the former separating the latter from the more humid climates beyond. Temperature and precipitation conditions are not significantly unlike those just described for BWh regions except that the extremes are more muted in the steppes (Figure 8-20). Thus rainfall is somewhat greater and more reliable, and temperatures are slightly moderated. Moreover, the meager precipitation tends to have a seasonal concentration. On the equatorward side of the desert, rain occurs in the high-sun season; on the poleward side, it is concentrated in the low-sun season.

Midlatitude Desert Climate (BWk)   The BWk climates (W = *Wüste*, German for desert; k = *kalt*, German for cold) occur primarily in the deep interiors of continents, where they are either far removed geographically or blocked from oceanic influence (Figure 8-14). The largest expanse of midlatitude dry climates, in central Asia, is (1) distant from any ocean and (2) protected by massive mountains on the south from any contact with the Indian summer monsoon. In North America, high mountains closely parallel the western coast and moist maritime air from the Gulf of Mexico affects the eastern half of the continent; as a result the dry climates are displaced well to the west. The only other significant BWk region is in southern South America, where the desert reaches all the way to the eastern coast of Patagonia (southern Argentina). This anomalous situation has developed because the continent is so narrow in these latitudes that all of it lies in the rain shadow of the Andes to the west and the cool Falkland Current to the east.

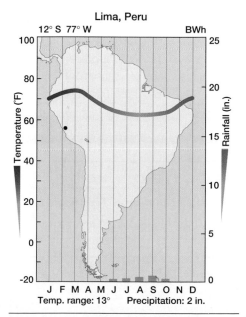

Lima, Peru
12° S 77° W   BWh

Temp. range: 13°   Precipitation: 2 in.

Figure 8-19 Climograph for a cool west coast desert station.

FOCUS

THE ARIZONA MONSOON

In the Phoenix office of the National Weather Service, there is an annual pool in which the date of the onset of the "Arizona monsoon" is predicted. Southern Arizona being one of the driest parts of the United States, this name is a humorous reference to the beginning of southern Arizona's "wet season," a meteorological reality but by no means a true monsoon (Figure 8-2-A).

Southern Arizona usually experiences a significant increase in moisture (both higher humidity and higher rainfall) in middle to late summer (Figure 8-2-B). This increase is caused by an incursion of air from the Gulf of Mexico, a flow that is related to the summertime northward shift of trade wind circulation and brings maritime tropical air deep into the continent. The moist influx generally continues for from 3 to 5 weeks before the

Douglas, Arizona
31° N 109° W

Figure 8-2-B Monthly rainfall chart for Douglas, Arizona. The concentration in middle and late summer is conspicuous.

easterly flow shifts back equatorward. During this interval, portions of the parched southwestern deserts receive unaccustomed thunderstorms, relatively high humidity, and even morning dew, as well as strong winds and slightly lowered temperatures.

Although the rainfall pattern has a distinctly monsoonal appearance, the other attributes of a monsoonal situation, such as seasonal reversal of windflow, are missing.

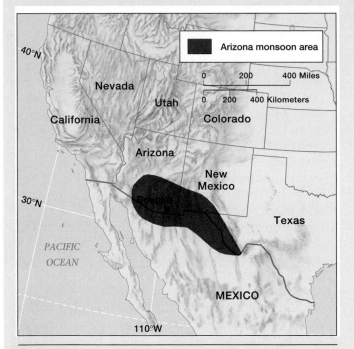

Figure 8-2-A The region principally affected by the "Arizona monsoon."

WORLD RECORD

Highest Annual Average Temperature

94°F (34°C) at Dallol, Ethiopia, a BWh climate

WORLD RECORD

Highest Average Daily Maximum Temperature

106°F (41°C) at Dallol, Ethiopia, a BWh climate

WORLD RECORD

Least Annual Average Precipitation

0.03 inch (0.8 millimeter) at Arica, Chile, a BWh climate

WORLD RECORD

Longest Dry Spell

14 years, 4 months, 1903–1918 at Arica, Chile, a BWh climate

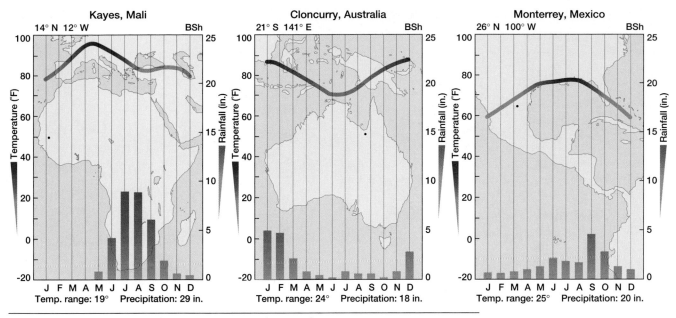

Figure 8-20　Climographs for representative subtropical steppe stations.

Climographs for three typical BWk locations are shown in Figure 8-21.

Precipitation in midlatitude (BWk) deserts is much like that of subtropical (BWh) deserts—meager and erratic. Differences lie in two aspects: seasonality and intensity. Most BWk regions receive the bulk of their precipitation in summer, when warming and instability are common. Winter is usually dominated by low temperatures and anticyclonic conditions. Although most BWk precipitation is of the unstable, showery variety, there are also some periods of general overcast and protracted drizzle.

The principal climatic differences between midlatitude and subtropical deserts are in temperature, especially winter temperature (Figure 8-22), with BWk regions having severely cold winters. The average cold-month temperature is normally below freezing, and some BWk stations have 6 months with below-freezing averages. This cold produces average annual temperatures that are much lower than in BWh regions and greatly increases the annual temperature range; variations exceeding 50°F between January and July average temperatures are not uncommon.

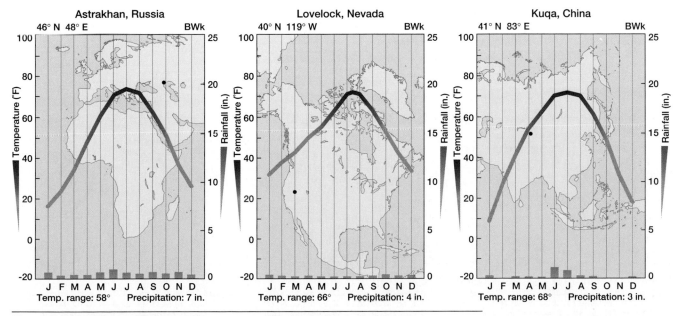

Figure 8-21　Climographs for representative midlatitude desert stations.

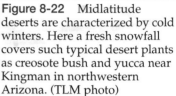

Figure 8-22 Midlatitude deserts are characterized by cold winters. Here a fresh snowfall covers such typical desert plants as creosote bush and yucca near Kingman in northwestern Arizona. (TLM photo)

Midlatitude Steppe Climate (BSk)   As in the subtropics, midlatitude steppes (BSk climate: S = steppe; k = *kalt*, German for cold) generally occupy transitional positions between deserts and humid climates. Typically midlatitude steppes have more precipitation than midlatitude deserts and lesser temperature extremes.

Figure 8-23 presents typical BSk climographs.

In western North America, the steppe climate is much more extensive in area than is the desert, so that the desert-core-and-steppe-fringe model is not followed exactly. Semiarid conditions are broadly preva-

lent; only in the interior southwest of the United States is the climate sufficiently arid to be classified as desert.

MILD MIDLATITUDE CLIMATES (ZONE C)

We noted previously that the middle latitudes, which extend approximately from 30° to 60° N and S, have the greatest weather variability over the short run of days or weeks. Seasonal contrasts are also marked in these latitudes, which lack both the constant heat of the tropics and the almost-continuous cold of the polar regions. The midlatitudes are a region of air mass contrast, with fre-

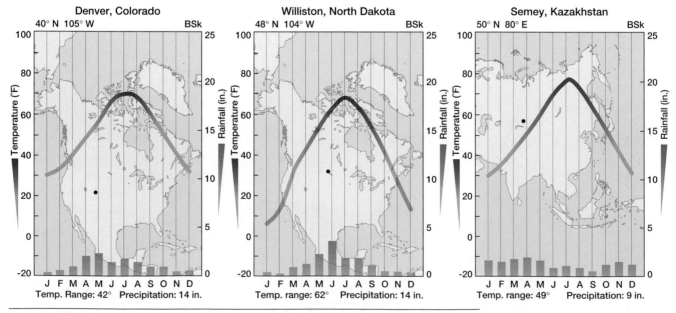

Figure 8-23 Climographs for representative midlatitude steppe stations.

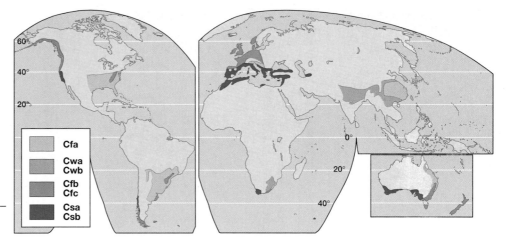

Figure 8-24 Distribution of C climates.

quent alternating incursions of tropical and polar air producing more convergence than anywhere else except at the equator. This air mass conflict creates a kaleidoscope of atmospheric disturbances and weather changes. The seasonal rhythm of temperature is usually more prominent than that of precipitation. Whereas in the tropics the seasons are characterized as wet and dry, in the midlatitudes they are clearly summer and winter.

The mild midlatitude climates occupy the equatorward margin of the middle latitudes, occasionally extending into the subtropics and being elongated poleward in some western coastal areas (Figure 8-24). They constitute a transition between warmer tropical climates and severe midlatitude climates.

Summers in the C climates are long and usually hot; winters are short and relatively mild. These zones, in contrast to the A-climate zones, experience occasional winter frosts and therefore do not have a year-round growing season. Precipitation is highly variable in the C climates, with regard to both total amount and seasonal distribution. Year-round moisture deficiency is not characteristic, but there are sometimes pronounced seasonal moisture deficiencies.

The C climates are subdivided into three types, primarily on the basis of precipitation seasonality, and secondarily on the basis of summer temperatures: mediterranean (Csa, Csb), humid subtropical (Cfa, Cwa), and marine west coast (Cfb, Cfc). Table 8-6 presents an overview.

TABLE 8-6
Summary of C Climates: Mild Midlatitude

Type	Location	Temperature	Precipitation	What controls the climate
Mediterranean (Csa, Csb)	Centered at 35° latitude on western sides of continents; limited east-west extent except in Mediterranean Sea area	Warm/hot summers; mild winters; year-round mildness in coastal areas	Moderate [15–25 in. (38–64 cm) annually], nearly all in winter; much sunshine, some coastal fog	STH subsidence and stability in summer; westerly winds and cyclonic storms in winter
Humid subtropical (Cfa, Cwa)	Centered at 30° latitude on eastern sides of continents; considerable east-west extent	Summers warm/hot, sultry; winters mild to cold	Abundant [40–60 in. (100–150 cm) annually], mostly rain; summer maxima but no true dry season	Westerly winds and storms in winter; moist onshore flow in summer; monsoons in Asia
Marine west coast (Cfb, Cfc)	Latitudes 40–60° on western sides of continents; limited inland extent except in Europe	Very mild winters for the latitude; generally mild summers; moderate ATR	Moderate to abundant, mostly in winter; many days with rain; much cloudiness	Westerly flow and oceanic influence year-round

C climates occupy the equatorward margin of the middle latitudes but may extend considerably poleward along western coasts. Summers are long and usually hot; winters are short and relatively mild; precipitation conditions are varied.

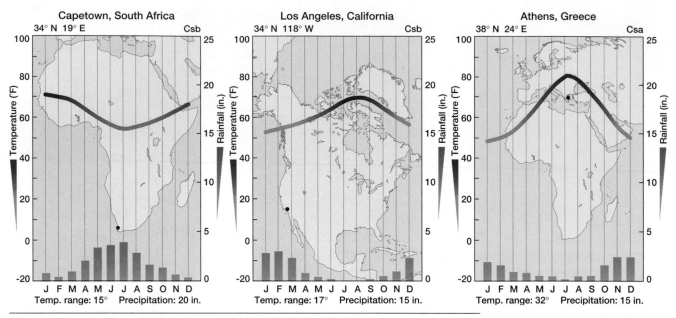

Figure 8-25 Climographs for representative mediterranean stations.

Mediterranean Climate (Csa, Csb)

The two Cs climates (s = summer dry; a = summer hot; b = summer warm) are sometimes referred to as dry subtropical, but the more widely used designation is mediterranean. (The proper terminology is without capitalization because it is a generic term for a type of climate; Mediterranean with the capital M refers to a specific region around the Mediterranean Sea.)

Cs climates are found on the western side of continents, centered at about latitudes 35° N and 35° S. Three representative locations are shown in Figure 8-25. With one exception, all mediterranean regions are small, being restricted mostly to coastal areas either by interior mountains or by limited landmasses. These small regions are in central and southern California, central Chile, the southern tip of Africa, and the two southwestern "corners" of Australia. The only extensive area of mediterranean climate is around the borderlands of the Mediterranean Sea.

Cs climates have three distinctive characteristics:

1. The modest annual precipitation falls in winter, summers being virtually rainless.
2. Winter temperatures are unusually mild for the midlatitudes, and summers vary from hot to warm.
3. Clear skies and abundant sunshine are typical, especially in summer.

Average annual precipitation is slight, ranging from about 15 inches (38 centimeters) on the equatorward margin to about 25 inches (64 centimeters) at the poleward margin. The midwinter rainfall is from 3 to 5 inches (8 to 13 centimeters) per month, and two or three midsummer months are dry. Only one other climatic type, marine west coast, has such a concentration of precipitation in winter (Figure 8-26).

Most mediterranean climate is classified as Csa, which means that summers are hot, with midsummer monthly averages between 75 and 85°F (24 and 29°C) and frequent maxima above 100°F (38°C). Average cold-month temperatures are about 50°F (10°C), with occasional minima below freezing.

Coastal mediterranean areas have much milder summers than inland mediterranean areas as a result of sea breezes; these coastal climates are classified as Csb (Figure 8-27). In the coastal areas, the average hot-month temperature is between 60 and 70°F (16 and 21°C). Csb winters are slightly milder than Csa winters, the former having cold-month averages of about 55°F (13°C). Csb regions also have higher humidity, frequent nocturnal fog, and occasional low stratus overcast.

The genesis of mediterranean climates is clear-cut. These regions are dominated in summer by dry, stable, subsiding air from the eastern portions of subtropical highs (STHs). In winter, the wind and pressure belts shift equatorward, and mediterranean regions come under the influence of the westerlies, with their migratory extratropical cyclones and associated fronts. Almost all precipitation comes from these cyclonic storms, except for occasional tropical influences in California.

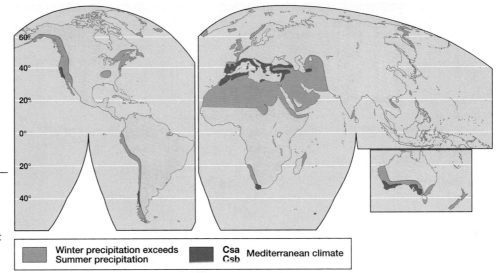

Figure 8-26 Lands that have precipitation maxima in winter rather than in summer are mostly associated with mediterranean and marine west coast climates and regions adjacent to them.

Winter precipitation exceeds Summer precipitation

Csa Csb   Mediterranean climate

Humid Subtropical Climate (Cfa, Cwa)

Whereas mediterranean climates are found on the western side of continents, the climate at the same latitude but on the eastern side is classified as humid subtropical (Cfa and Cwa types: f = *feucht*, German for moist; a = summer hot; w = winter dry). This climate covers a more extensive area, both latitudinally and longitudinally. In some places it reaches equatorward to almost 15° of latitude and extends poleward to about 40°. Its east-west extent is greatest in North America, Asia, and South America, as Figure 8-24 shows.

The humid subtropical climates differ from mediterranean climates in several important respects, as shown in Table 8-7. Summer temperatures in humid subtropi-

cal regions are generally warm to hot, with the highest monthly averages between 75 and 80°F (24 and 27°C). This is not dissimilar to the situation in Cs climates, but the humid subtropical regions are characterized by much higher humidity in summer and so sensible temperatures are higher. Cfa days tend to be hot and sultry, and often night brings little relief. Winter temperatures are mild on the average, but winter is punctuated by cold waves that bring severe weather for a few days at a time. Minimum temperatures can be 10–20°F lower in the humid subtropics than in mediterranean regions, which means that killing frosts occur much more frequently in the former, especially in North America and Asia. The importance of this fact can be shown by agricultural adjustments. For example, the northernmost

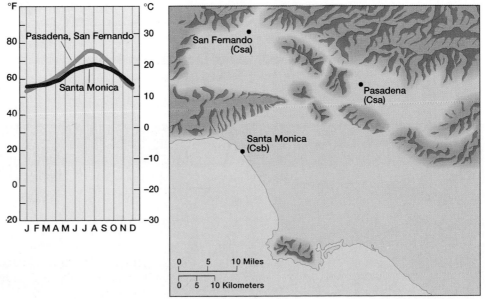

Figure 8-27 There are often significant temperature differences between coastal and inland mediterranean areas. This climograph shows the annual temperature curves for three stations in Southern California: Santa Monica (a coastal station; Csb climate) and Pasadena and San Fernando (inland stations having exactly the same average temperature for each month; Csa climate). The physiographic map shows the topographic and coastal relationships of the three stations.

FOCUS

CLIMATIC DISTRIBUTION IN AFRICA: A PRACTICALLY PERFECT PATTERN

Africa is a fascinating continent for geographic study for many reasons. It has unusual geologic features, a hydrographic pattern unlike that of any other continent, a remarkable fauna, a long and dramatic human history, an extraordinary diversity of cultures, extremely volatile politics, and various other unusual attributes that attract attention.

For certain aspects of physical geography, however, Africa is particularly interesting because of its normality. It is the only continent that is approximately bisected by the equator and so has the same latitudinal extent in both the Northern and Southern hemispheres. This equal distribution of land north and south of the equator produces some textbook examples of environmental patterns. Most notable in this regard is the distribution of climatic types, which can be displayed across the map of Africa as a model almost without blemish (Figure 8-3-A). For the geographer seeking predictable patterns, this display is a joy to behold.

In the center of the continent is a hot, wet core region (Af), and at the

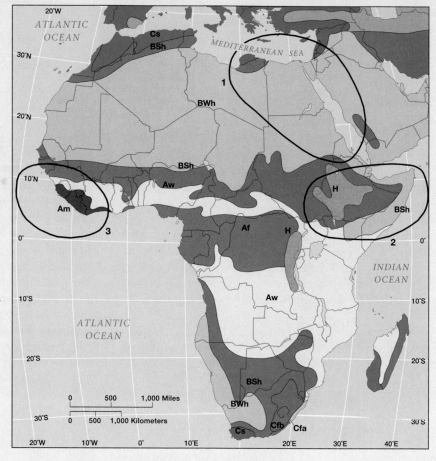

Figure 8-3-A The climatic regions of Africa. The pattern is mostly regular and "normal." The three black ovals mark the principal exceptions to that normality: (1) a greatly expanded area of B climate in the northeast; (2) an unusual and complex climatic pattern in East Africa; and (3) monsoonal development along the Guinea coast of West Africa.

TABLE 8-7
Principal Differences Between Humid Subtropical Climates and Mediterranean Climates

Characteristic	Humid subtropical	Mediterranean
Precipitation seasonality	Summer maximum	Winter maximum
Dry season	None or only short winter dry season	Distinct summer dry season
Total precipitation	Relatively abundant	Sparse
Winter temperatures	Generally mild; occasional cold spells	Generally mild
Annual temperature range	15–40°F (-10–4°C)	10–35°F (-12–2°C)

proper subtropical latitudes are two hot, dry core regions (BWh). Between these precipitation extremes are the two expectable transition climates (Aw and BSh), and north and south of the BWh deserts the transition (BSh) again occurs, leading into the moister C climates of the middle latitudes. It is a practically perfect pattern through the A climates and the subtropical B climates into the fringe of the C climates on the poleward extremities of the continent (Figure 8-3-B).

The inevitable anomalies, for the most part, can be explained fairly easily:

1. Most striking is the vast extent of dry climate in North Africa. The expectation is for desert to extend from the western coast for a considerable distance inland, but in this case it carries all the way to the eastern border of the continent. Desert covers the entire continent because the Asian landmass immediately to the east of Africa excludes maritime air and preserves the characteristic of continentality clear across North Africa.

2. The distribution of climatic types in the eastern equatorial portion, from about latitudes 10° N to 10° S,

is quite different from what might be expected. The expected Af climate is replaced by Aw, H, BSh, and BWh. This anomaly is due in part to the higher elevations of most of East Africa, which produces the lower temperatures typical of a highland zone, and in part to the airflow pattern along the coast north of about 5° S latitude. Influenced by the South Asian monsoon, winds blow parallel to this coast (northeasterly in winter and southwesterly in summer) throughout the year, which means that maritime air rarely is carried inland and dryness prevails.

3. The climate along the western Guinea Coast is Am rather than the expected Af or Aw. Although the full story of monsoon origin is not clear, there is a definite seasonal reversal of wind flow in this region, with maritime southwesterlies bringing a heavy summer rainy season and dry northeasterlies dominating for the remainder of the year.

Any explanation of the general distribution of climatic types over the world varies from the simple to the complex. If one is trying to comprehend the pattern, Africa is probably the best place to begin.

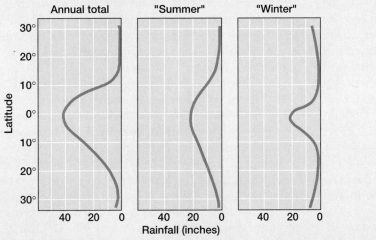

Figure 8-3-B Schematic diagram of rainfall variation along the central meridian (20° East longitude) of Africa. The left portion represents the annual total; the central portion portrays the high-sun season in both hemispheres; the right portion is representative of the low-sun season in both hemispheres.

limit of commercial citrus production in the eastern United States (Florida, a Cfa climate) is at about 29° N latitude, but in the western United States (California, a Cs climate) it is at 38° N latitude, 625 miles (1000 kilometers) farther north (Figure 8-28).

The mild-summer characteristics of Csb coasts have no counterpart in the humid subtropical regions because the offshore waters in the east are warm rather than cool. Mediterranean climates are all adjacent to cool currents, whereas warm currents wash the humid subtropical coasts.

Annual precipitation is generally abundant and in some places copious, with a general decrease from east

to west (Figure 8-29). Averages mostly are between 40 and 60 inches (100 and 150 centimeters) although some locations record as little as 30 inches (75 centimeters) and some receive up to 100 inches (250 centimeters). Summer is usually the time of precipitation maxima, associated with onshore flow of maritime air and frequent convection. Winter is a time of diminished precipitation, but it is not really a dry season except in China, where monsoonal conditions dominate. The rain and occasional snow of winter are the result of extratropical cyclones passing through the region. In the North American and Asian coastal areas, a late summer/autumn bulge in the precipitation curve is due to rainfall from tropical cyclones.

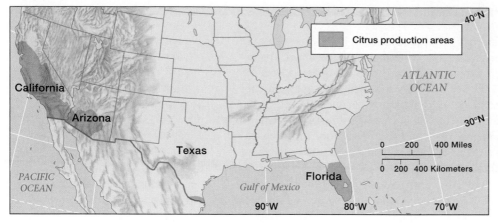

Figure 8-28 Distribution of citrus production in the conterminous United States. Winter cold spells prohibit commercial citrus growing north of central Florida in the eastern United States; the lack of severe freezes in much of California allows citrus to be grown much farther north. (After U.S. Census of Agriculture, 1992.)

Figure 8-30 shows three typical climographs for the humid subtropical climate.

Marine West Coast Climate (Cfb, Cfc)

As the name implies, marine west coast climates (Cfb and Cfc types: f = *feucht*, German for moist; b = summer warm; c =summer cool) are situated on the western side of continents, which is a windward location in the latitudes occupied by these climates, between about 40° and 65°. Only in the Southern Hemisphere, where landmasses are small in these latitudes (New Zealand and southernmost South America, for example), does this oceanic climate extend across to eastern coasts.

The most extensive area of marine west coast climate is in western and central Europe, where the maritime influence can be carried some distance inland without hindrance from topographic barriers. The North American region is much more restricted interiorward by the presence of formidable mountain ranges that run perpendicular to the direction of on-shore flow. These regions are influenced by the westerlies throughout the year, and the persistent onshore air movement ensures that the maritime influence permeates throughout. This air current creates an extraordinarily temperate climate considering the latitude: lack of extreme temperatures, consistently high humidity, much cloudiness, and a high proportion of days with some precipitation.

Three typical Cfb climographs are shown in Figure 8-31. The oceanic influence ameliorates temperatures most of the time; this moderation is particularly noticeable when the daily and seasonal maxima and minima are considered. Isotherms tend to parallel the coastline rather than following their "normal" east-west paths. Indeed, temperatures on a western coast in these latitudes decrease poleward less than half as fast as on an eastern coast. Average hot-month temperatures are generally between 60 and 70°F (16 and 21°C), with cold months averaging between 35 and 45°F (2 and 7°C). There are occasionally very hot days (Seattle

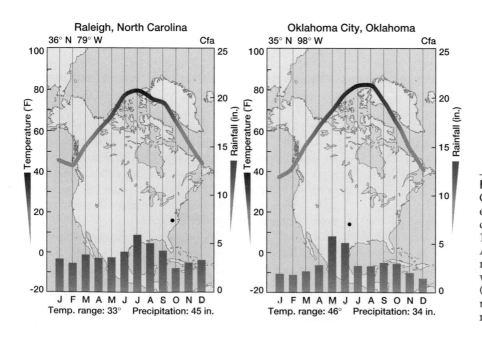

Figure 8-29 Annual precipitation in Cfa regions generally decreases from east to west, or interiorward into a continent. Raleigh, North Carolina, 150 miles (240 kilometers) from the Atlantic coast, receives 45 inches (1145 millimeters) of precipitation annually, whereas Oklahoma City, 1100 miles (1760 kilometers) from the coast, receives only 34 inches (865 millimeters).

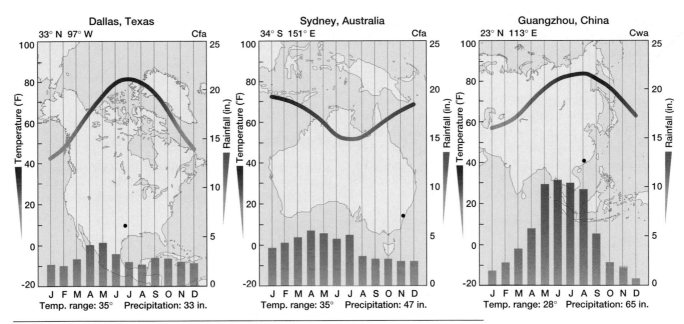

Figure 8-30 Climographs for representative humid subtropical stations.

has recorded a temperature of 98°F, or 37°C, and Paris has reached 100°F, or 38°C), but prolonged heat waves are unknown. Similarly, very cold days occur upon occasion, but low temperatures rarely persist. Frosts are relatively infrequent, except in interior European locations; London, for example, at 52° N latitude, experiences freezing temperatures on fewer than half the nights in January. There is also an abnormally long growing season for the latitude; around Seattle, for instance, the growing season is a month longer than that around Atlanta, a city lying 14° of latitude farther equatorward.

Marine west coast climates are among the wettest of the middle latitudes, although the total amount of Cfb/Cfc precipitation is not remarkable except in upland areas. Some localities receive as little as 20 inches (50 centimeters), but a range of between 30 and 50 inches (75 and 125 centimeters) is more typical. Much higher totals are recorded on exposed slopes, reaching 100–150 inches (250–375 centimeters) in some places. Snow is uncommon in the lowlands, but higher, west-facing slopes receive some of the heaviest snowfalls in the world.

Perhaps more important than total precipitation to an understanding of the character of the marine west

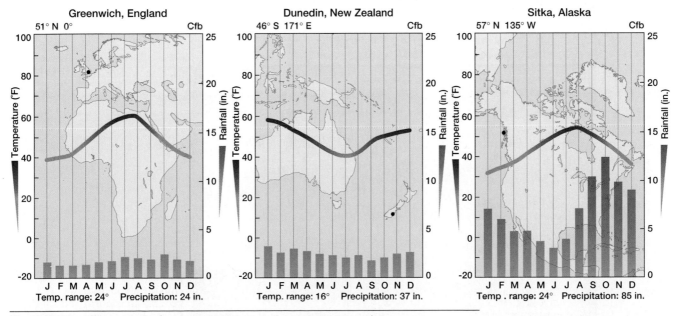

Figure 8-31 Climographs for representative marine west coast stations.

coast climate is precipitation frequency. Rainfall probability and reliability are high, but intensity is low. Drizzly frontal precipitation is characteristic. Humidity is high, with much cloudiness. Seattle, for example, receives only 43 percent of the total possible sunshine each year, in contrast to 70 percent in Los Angeles, and London has experienced as many as 72 consecutive days with rain. (Is it any wonder that the umbrella is that city's civic symbol?) Indeed, some places on the western coast of New Zealand's South Island have recorded 325 rainy days in a single year.

THE IDEALIZED PATTERN OF THE MILD CLIMATES

It should be clear by now that there is a predictable pattern in the location of climatic types, based primarily on latitude and on the general atmospheric and oceanic circulation. Figure 8-32 summarizes the idealized distribution of the mild (A, B, and C) climates.

SEVERE MIDLATITUDE CLIMATES (ZONE D)

The severe midlatitude climates occur only in the Northern Hemisphere (Figure 8-33) because the Southern Hemisphere has limited landmasses at the appro-

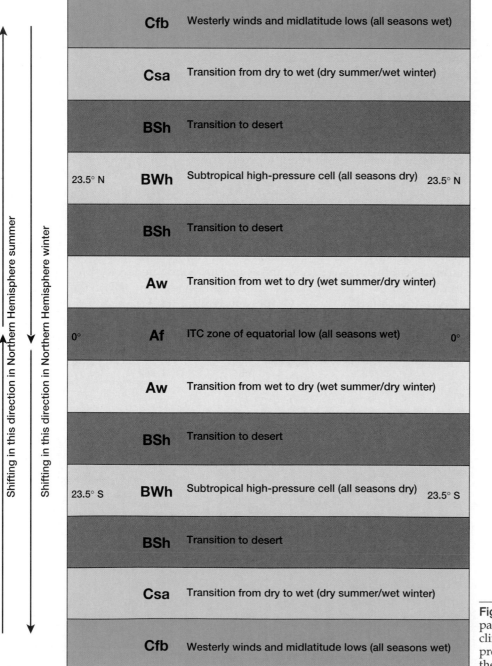

Figure 8-32 The idealized pattern of the mild (A, B, C) climates. Note that the progressions north and south of the equator are mirror images.

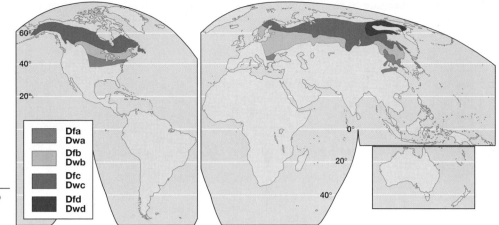

Figure 8-33   Distribution of D climates.

priate latitudes—between about 40° and 70°. This climatic zone extends broadly across North America (encompassing the northeastern United States and much of Canada and Alaska) and Eurasia (from eastern Europe through most of Russia to the Pacific Ocean).

"Continentality," by which is meant remoteness from oceans, is a keynote in the D climates. Landmasses are broader at these latitudes than anywhere else in the world. Even though these climates extend to the eastern coasts of the two continents, they experience little maritime influence because the general atmospheric circulation is westerly.

The most conspicuous result of continental dominance of D climates is broad annual temperature fluctuation. These climates have four clearly recognizable seasons: a long, cold winter, a relatively short summer that varies from warm to hot, and transition periods in spring and fall. Annual temperature ranges are very large, particularly at more northerly locations, where winters are most severe. Precipitation is moderate, although throughout the D climates it exceeds the potential evapotranspiration. Summer is the time of precipitation maximum, but winter is by no means completely dry and snow cover lasts for many weeks or months.

The severe midlatitude climates are subdivided into two types on the basis of temperature. The humid continental type (which is further classified into subtypes Dfa, Dfb, Dwa, Dwb) has long, warm summers. The subarctic type (having subtypes Dfc, Dfd, Dwc, and Dwd) is characterized by short summers and very cold winters.

Table 8-8 summarizes the D climates.

TABLE 8-8
Summary of D Climates: Severe Midlatitude

Type	Location	Temperature	Precipitation	What controls the climate
Humid continental (Dfa, Dfb, Dwa, Dwb)	Northern Hemisphere only; latitudes 35–55° on eastern sides of continents	Warm/hot summers; cold winters; much day-to-day variation; large ATR	Moderate to abundant [20–50 in. (50–125 cm) annually]; with summer maxima; diminishes interiorward and poleward	Westerly winds and storms, especially in winter; monsoons in Asia
Subarctic (Dfc, Dfd, Dwc, Dwd)	Northern Hemisphere only, latitudes 50–70° across North America and Eurasia	Long, dark, very cold winters; brief, mild summers; enormous ATR	Meager [5–20 in. (13–50 cm) annually], with summer maxima; light snow in winter but little melting	Pronounced continentality; westerlies and cyclonic storms alternating with prominent anti-cyclonic conditions

D climates occur only in the broad northern landmasses between about 50 and 70° of latitude. Winters are long, dark, and unremittingly cold; summers short and mild. Precipitation is meager, but evaporation is small.

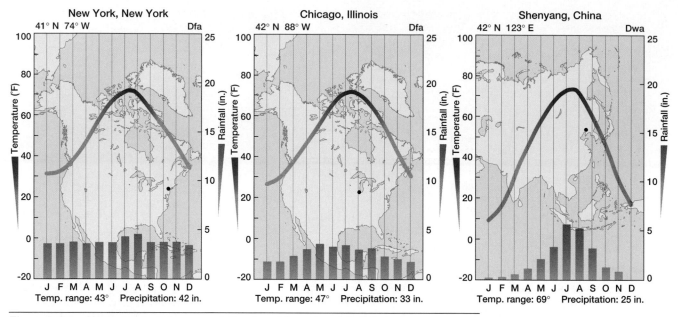

Figure 8-34   Climographs for representative humid continental stations.

Humid Continental Climate (Dfa, Dfb, Dwa, Dwb)

The latitudinal range of humid continental climate (Dfa, Dfb, Dwa, and Dwb types: f = *feucht*, German for moist; a = summer hot; b = summer warm; w = winter dry) in North America and Asia is between 35° and 55°. The European region spreads from 35° to 60° in central Europe and tapers easterly through Russia and into Siberia. Figure 8-34 shows some typical climographs.

This climatic type is dominated by the westerly wind belt throughout the year, which means that there are frequent weather changes associated with the passage of migratory pressure systems, especially in winter. Variability, then, both seasonal and daily, is a prominent characteristic.

Hot-month temperatures generally average in the mid-70s °F (mid-20s °C), and so summers are as warm as those of the humid subtropical climate to the south, although summer is shorter. The average cold-month temperature is usually between 10 and 25°F (12 and 4°C), with from 1 to 5 months averaging below freezing. Winter temperatures decrease rapidly northward in the humid continental climates, as Figure 8-35 shows, and the growing season diminishes from about 200 days on the southern margin to about 100 days on the northern edge.

Despite their name, precipitation is not copious in humid continental climates. Annual totals average between 20 and 40 inches (50 and 100 centimeters), with the highest values on the coast and a general decrease interiorward. There is also a decrease from south to north (Figure 8-36). Both these trends reflect increasing distance from warm moist air masses. Summer is distinctly

the wetter time of the year, but winter is not totally dry, and in coastal areas the seasonal variation is muted. Summer rain is mostly convective or monsoonal in origin. Winter precipitation is associated with extratropical cyclones, and much of it falls as snow. During a typical winter, snow covers the ground for only 2 or 3 weeks in the southern part of these regions, but for as long as 8 months in the northern portions (Figure 8-37).

Day-to-day variability and dramatic changes are prominent features of the weather pattern. These are regions of cold waves, heat waves, blizzards, thunderstorms, tornadoes, and other dynamic atmospheric phenomena.

Subarctic Climate (Dfc, Dfd, Dwc, Dwd)

The subarctic climate (Dfc, Dfd, Dwc, Dwd types: f = *feucht*, German for moist; c = summer cool; d = winter very cold; w = winter dry) occupies the higher midlatitudes, generally between 50° and 70°. As Figure 8-5 shows, this climate occurs as two vast, uninterrupted expanses across the broad northern landmasses: from western Alaska across Canada to Newfoundland, and across Eurasia from Scandinavia to easternmost Siberia. The name **boreal** (which means "northern" and comes from Boreas, mythological Greek god of the north wind) is sometimes applied to this climatic type in Canada; in Eurasia it is often called **taiga,** after the Russian name for the forest in the region where this climate occurs. Figure 8-38 shows some typical climographs.

The key word in the subarctic climate is winter, which is long, dark, and bitterly cold. In most places, ice begins to form on the lakes in September or October and doesn't

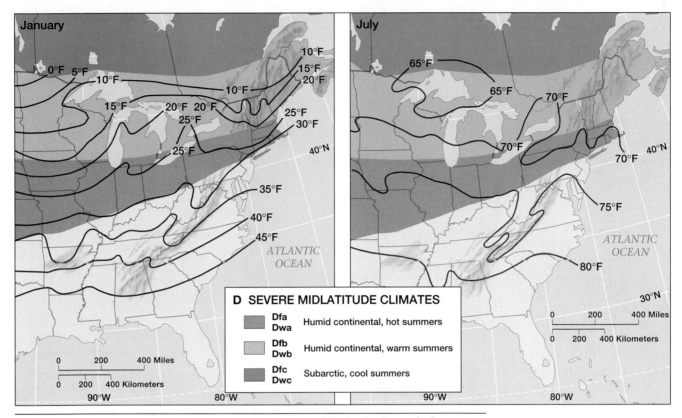

Figure 8-35   North-south temperature variation in the midlatitudes is much sharper in winter than in summer. These maps of the eastern United States show a very steep north-south January temperature gradient but only limited north-south temperature contrasts in July.

thaw until May or later. For 6 or 7 months the average temperature is below freezing, and the coldest months have averages below -36°F (-38°C). The world's coldest temperatures, apart from the Antarctic and Greenland ice caps, are found in the subarctic climate; the records are -90°F (-68°C) in Siberia and -82°F (-62°C) in Alaska.

Summer warms up remarkably despite its short duration. Although the intensity of the sunlight is low (because of the small angle of incidence), summer days are very long and nights are too short to permit much radiational cooling. Average hot-month temperatures are typically in the high 50s or low 60s°F (mid-teens or low 20s°C), but occasional frosts may occur in any month. Annual temperature ranges in this climate are the largest in the world. Variations from average hot-month to average cool-month temperatures frequently exceed 80°F (45°C) and in some places are more than 100°F (50°C) The

WORLD RECORD

Greatest Average Annual Temperature Range
112°F (62°C) at Yakutsk, Siberia, a Dfd climate

absolute annual temperature variation (fluctuation from the very coldest to the very hottest ever recorded) sometimes reaches unbelievable magnitude; the world record is 188°F (-90° to +98°) in Verkhoyansk, Siberia.

Spring and fall are brief transition seasons that slip by rapidly, usually in April/May and September/October. Summer is short, and winter is dominant.

Precipitation is usually meager in the subarctic climates. Annual totals range from only 5 inches (13 centimeters) to about 20 inches (50 centimeters), with the higher values occurring in coastal areas. The low temperatures allow for little moisture in the air, and anticyclonic conditions predominate. Despite these sparse totals, the evaporation rate is low and the soil is frozen for much of the year, so that moisture is adequate to support a forest. Summer is the wet season, and most precipitation comes from scattered convective showers. Winter experiences only light snowfalls (except near the coasts), which may accumulate to depths of 2 or 3 feet (60 to 90 centimeters). The snow that falls in October is likely to be still on the ground in May because little melts over the winter. Thus a continuous thin snow cover exists for many months despite the sparseness of actual snowfall.

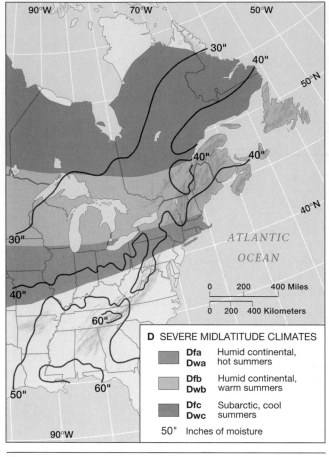

Figure 8-36 Annual precipitation in the Dfa portion of eastern North America. The isohyet values are in inches of moisture. Precipitation generally decreases inland and northward.

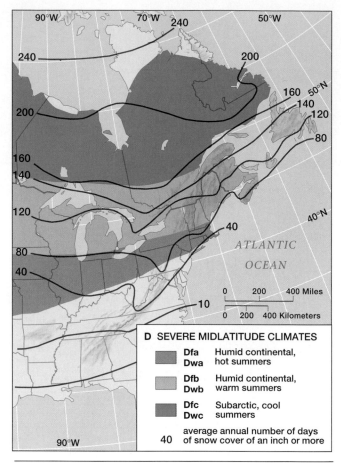

Figure 8-37 The average duration of snow cover in the Dfa portion of eastern North America. The numbers on the isolines represent the average annual number of days with a snow cover of 1 inch (2.5 centimeters) or more.

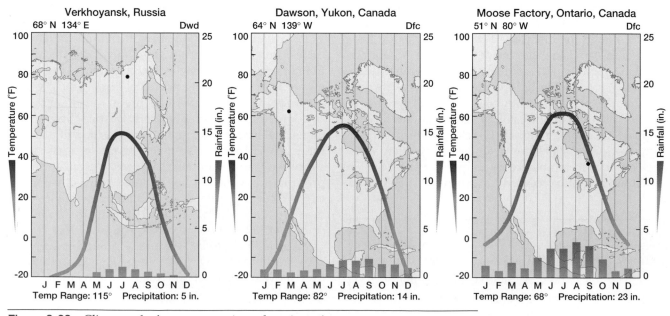

Figure 8-38 Climographs for representative subarctic stations.

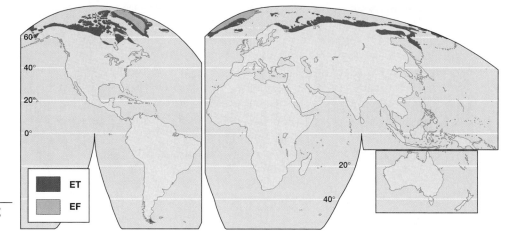

Figure 8-39   Distribution of E climates.

POLAR CLIMATE (ZONE E)

Being farthest from the equator, as Figure 8-39 shows, the polar climates are the most remote from the heat of the sun and receive inadequate insolation for any significant warming. By definition, no month has an average temperature of more than 50°F (10°C) in a polar climate. If the wet tropics represent conditions of monotonous heat, the polar climates are known for their enduring cold. They have the coldest summers and the lowest annual and absolute temperatures in the world. They are also extraordinarily dry, but evaporation is so minuscule that the group as a whole is classified as humid.

The two types of polar climates are distinguished by summer temperature. The tundra climate (ET) has at least one month with an average temperature exceeding the freezing point. The ice cap climate (EF) does not.

Table 8-9 summarizes the E climates.

Tundra Climate (ET)   The name **tundra** originally referred to the low, ground-hugging vegetation of high-latitude and high-altitude regions, but the term has been adopted to refer to the climate of the high-latitude regions as well. The generally accepted equatorward edge of the tundra climate (ET climate: T, of course, for tundra) is the 50°F (10°C) isotherm for the average temperature of the warmest month. This same isotherm corresponds approximately with the poleward limit of trees, so that the boundary between D and E climates (in other words, the equatorward boundary of the tundra climate) is the "treeline."

At the poleward margin, the ET climate is bounded by the isotherm of 32°F (0°C) for the warmest month, which approximately coincides with the extreme limit for growth of any plant cover. More than for any other climatic type, the delimitation of the tundra climate demonstrates Köppen's contention

		TABLE 8-9 **Summary of E Climates: Polar**		
Type	*Location*	*Temperature*	*Precipitation*	*What controls* *the climate*
Tundra (ET)	Fringes of Arctic Ocean; small coastal areas in Antarctica	Long, cold, dark winters; brief, cool summers; large ATR, small DTR	Very sparse [less than 10 in. (25 cm) annually], mostly snow	Latitude; distance from sources of heat and moisture; extreme seasonal contrasts in sunlight/darkness
Ice cap (EF)	Antarctica and Greenland	Long, dark, windy, bitterly cold winters; cold, windy summers; large ATR, small DTR	Very sparse [less than 5 in. (13 cm) annually], all snow	Latitude; distance from sources of heat and moisture; extreme seasonal contrasts in sunlight/darkness; polar anticyclones

E climates are found in Antarctica, Greenland, various Arctic islands, and the Arctic fringes of North America and Eurasia. They are lands of enduring cold; no month has an average temperature of more than 50°F (10°C), and most are much colder than that. Precipitation is scanty, but the evaporation rate is extremely low. Strong winds are common.

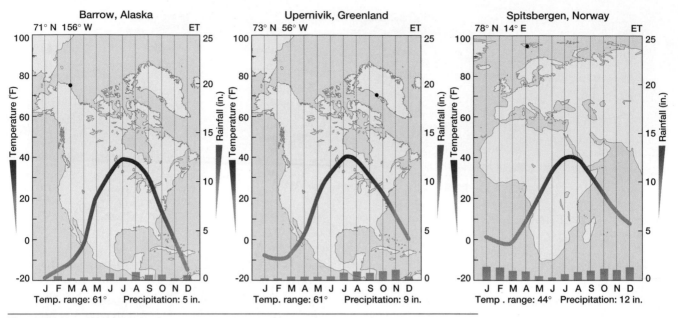

Figure 8-40 Climographs for representative tundra stations.

that climate is best delimited in terms of plant communities.

Figure 8-40 shows typical climographs for a tundra climate.

Long, cold, dark winters and brief, cool summers characterize the tundra. Only 1 to 4 months experience average temperatures above freezing, and the average hot-month temperature is in the 40s°F (between 4 and 10°C). Freezing temperatures can occur at any time, and frosts are likely every night except in midsummer. Although an ET winter is bitterly cold, it is not as severe as in the subarctic climate farther south because the ET climate is less continental. Coastal stations in the tundra often have cold-month average temperatures of only about 0°F (-18°C), whereas an inland ET location is more likely to average -25° or -30°F (-32° or -35°C). Annual temperature ranges are fairly large, commonly between 40 and 60°F. Daily temperature ranges are small because the sun is above the horizon for most of the time in summer and below the horizon for most of the time in winter; thus nocturnal cooling is limited in summer, and daytime warming is almost nonexistent in winter.

Moisture availability is very restricted in ET regions despite the proximity of an ocean. The air is simply too cold to hold much moisture, and so the absolute humidity is almost always very low. Moreover, anticyclonic conditions are common, with little air uplift to air condensation. Annual total precipitation is generally less than 10 inches (25 centimeters) but is somewhat greater in eastern Arctic Canada. Generally, more precipitation falls in the warm season than in winter, although the total amount in any month is small, and the month-to-month variation is minor. Winter snow is often dry and granular; it appears to be more than it actually is because there is no melting and because winds swirl it horizontally even when no snow is falling. Radiation fogs are fairly common throughout ET regions, and sea fogs are sometimes prevalent for days along the coast.

Ice Cap Climate (EF) The most severe of Earth's climates is restricted to Greenland (all but the coastal fringe) and most of Antarctica, the combined extent of these two regions amounting to more than 9 percent of the world's land area. The EF climate is one of perpetual frost (F = frost) where vegetation cannot grow, and the landscape consists of a permanent cover of ice and snow. Figure 8-41 shows typical climographs.

WORLD RECORD

Lowest Absolute Temperature

-130°F (-88°C) on 21 July , 1983, at Vostok, Antarctica, an EF climate

---

WORLD RECORD

Highest Average Annual Wind Speed

38 knots at Cape Dennison, Antarctica, an EF climate

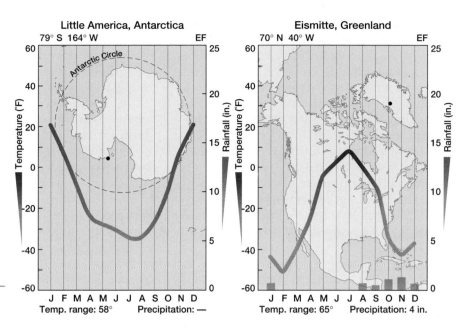

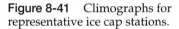

Figure 8-41 Climographs for representative ice cap stations.

The extraordinary severity of EF temperatures is emphasized by the fact that both Antarctica and Greenland are ice plateaus, so that relatively high altitude is added to high latitude as a thermal factor. All months have average temperatures below freezing, and in the most extreme locations the average temperature of the warmest month is below 0°F (–18°C). Cold-month temperatures average between –30 and –60°F (–34 and –51°C), and extremes well below –100°F (–73°C) have been recorded at interior Antarctic weather stations.

The air is chilled so intensely from the underlying ice that strong surface temperature inversions prevail most of the time. Heavy, cold air often flows downslope as a vigorous katabatic wind. A characteristic feature of the ice cap climate, particularly in Antarctica, is strong winds and blowing snow.

Precipitation is very limited. These regions are polar deserts; most places receive less than 5 inches (13 cen-

timeters) of moisture annually. The air is too dry and too stable, with too little likelihood of uplift, to permit much precipitation. Evaporation, of course, is minimal, and so moisture may be added to the ice.

HIGHLAND CLIMATE (ZONE H)

Highland climate is not defined in the same sense as all the others we have just studied. Climatic conditions in mountainous areas have almost infinite variations from place to place, and many of the differences extend over very limited horizontal distances. Köppen did not recognize highland climate as a separate zone, but most of the researchers who have modified his system have added such a category. Highland climates are delimited in this book to identify relatively high uplands (mountains and plateaus) having complex local climate variation in small areas (Figure 8-42).

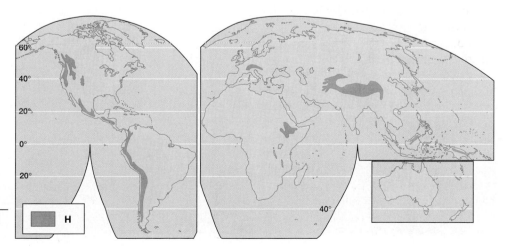

Figure 8-42 Distribution of H climate.

The climate of any highland location is usually closely related to that of the adjacent lowland, particularly with regard to seasonality of precipitation. Some aspects of highland climate, however, differ significantly from that of the surrounding lowlands.

With highland climates, latitude becomes less important as a climatic control than altitude and exposure. The critical climatic controls on a mountain slope are usually relative elevation and angle of exposure to sun and wind (Figure 8-43).

Altitude variations influence all four elements of weather and climate. A vertical temperature gradient of about 3.5°F (2.3°C) per 1000 feet (300 meters) generally prevails. Atmospheric pressure also decreases rapidly with increased elevation. Air movement is less predictable in highland areas, but it tends to be brisk and abrupt, with many local wind systems. Precipitation is characteristically heavier in highlands than in surrounding lowlands, so that the mountains usually stand out as moist islands on a rainfall map (Figure 8-44).

Altitude is more significant than latitude in determining climate in highland areas, so that a pattern of vertical zonation is usually present. The steep vertical gradients of climatic change are expressed as horizontal bands along the slopes. An increase of a few hundred feet in elevation may be equivalent to a journey of several hundred miles poleward insofar as temperature and related environmental characteristics are concerned. Vertical zonation is particularly prominent in tropical highlands (Figure 8-45).

Exposure—whether a slope, peak, or valley faces windward or leeward—has a profound influence on climate. Ascending air on a windward face brings a strong likelihood of heavy precipitation, whereas a leeward location is sheltered from moisture or has predominantly downslope wind movement with limited opportunity for precipitation. The angle of exposure to sunlight is also a significant factor in determining climate, especially outside the tropics. Slopes that face equatorward receive direct sunlight, which makes them warm and dry (through more rapid evapotranspiration); adjacent slopes facing poleward may be much cooler and moister simply because of a smaller angle of solar incidence and more shading. Similarly, west-facing slopes receive direct sunlight in the hot afternoon, but east-facing slopes are sunlit during the cooler morning hours.

Changeability is perhaps the single most conspicuous characteristic of highland climate. The thin, dry air permits rapid influx of insolation by day and rapid loss of radiant energy at night, and so daily temperature ranges are very large, with frequent and rapid oscillation between freeze and thaw. Daytime upslope winds and convection cause rapid cloud development and abrupt storminess. Travelers in highland areas are well advised to be prepared for sudden changes from hot to cold, from wet to dry, from clear to cloudy, from quiet to windy, and vice versa.

THE GLOBAL PATTERN IDEALIZED

From the basic characteristics and distribution of the various climatic types, it is possible to construct a model of the climate distribution on a hypothetical continent (Figure 8-46).

Figure 8-43 Mountain climates are quite variable because of changes in altitude and exposure. This is a summer scene near Rainy Pass in the North Cascade Montains of Washington. (TLM photo)

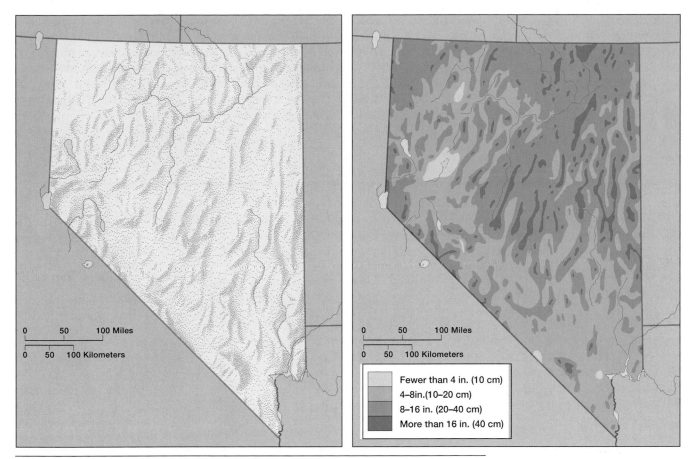

Figure 8-44   Comparison of annual precipitation and topography in Nevada. Where there are mountains, there is more precipitation; where there are basins, there is less precipitation. The relationship is simple, straightforward, and typical of highland environments.

Legend (for right-hand map):
- Fewer than 4 in. (10 cm)
- 4–8 in. (10–20 cm)
- 8–16 in. (20–40 cm)
- More than 16 in. (40 cm)

The model portrays the generalized distribution of five of the six zones and most of the types and subtypes; highland climate is not included because its location is determined solely by topography. Zones A, C, D, and E are defined by temperature, which means

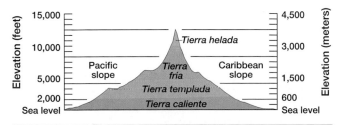

Figure 8-45   Vertical climate zonation is particularly noticeable in tropical mountainous areas. This diagram idealizes the situation at about 15°N latitude in Guatemala and southern Mexico. *Tierra caliente* (hot land) is a zone of high temperatures, dense vegetation, and tropical agriculture. *Tierra templada* (temperate land) is an intermediate zone of slopes and plateaus and temperatures most persons would find comfortable. *Tierra fría* (cold land) is characterized by warm days and cold nights, and its agriculture is limited to hardy crops. *Tierra helada* (frozen land) is a zone of cold weather throughout the year.

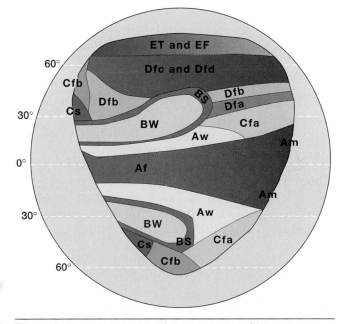

Figure 8-46   The presumed arrangement of Köppen climatic types on a hypothetical continent.

that their boundaries are strongly latitudinal because they are determined by insolation. The B zone is defined by moisture conditions, and its distribution cuts across those of the thermally defined zones.

Such a model is a predictive tool. One can state, with some degree of assurance, that at a particular latitude and a general location on a continent, a certain climate is likely to occur. Moreover, the locations of the climatic types and subtypes relative to one another can be more clearly understood when they are all shown together this way. The real world holds many refinements and modifications to this global pattern, of course, but the model shows the general alignments with considerable validity.

CHAPTER SUMMARY

The fundamental geographic aspect of climate is its global distribution.

Many climatic classification schemes have been devised, but the one most widely used for pedagogic purposes is the modified Köppen system, which is based on temperatures and precipitation amounts and patterns. The modified Köppen system recognizes six major climatic zones:

1. Tropical humid (A)
2. Dry (B)
3. Mild midlatitude (C)
4. Severe midlatitude (D)
5. Polar (E)
6. Highland (H)

The system is comprehensive, empirical, and logical, but perhaps its greatest asset as a learning tool is the relative predictability of the pattern it displays.

KEY TERMS

boreal	modified Köppen system	tundra
climograph	selva	
Köppen system	taiga	

REVIEW QUESTIONS

1. In our study of physical geography, why are we interested in climatic classification?
2. Are the five climatic zones of the ancient Greeks of any relevance today? Explain.
3. Explain the basic concept of the Köppen system.
4. What is shown on a climograph?
5. Explain the differences between the three types of tropical humid climates.
6. Distinguish between desert and steppe climates.
7. Why are B climates much more extensive in North Africa than in any other subtropical location?
8. Why are dry climates usually displaced toward the western sides of continents?
9. What are the specialized climatic characteristics of west coast subtropical deserts? What causes these anomalies?
10. Why are mediterranean climates less extensive than humid subtropical climates?
11. What are the distinctive characteristics of mediterranean climates?
12. In what climatic types is winter the season of precipitation maximum?
13. What causes the relatively mild temperatures of marine west coast climates?
14. What is meant by the phrase "continentality is a keynote in D climates"?
15. Why is precipitation so sparse in E climates?

SOME USEFUL REFERENCES

BARRY, R. G., *Mountain Weather and Climate*, 2nd ed. New York: Routledge, 1992.

GERRARD, A. J., *Mountain Environments: An Examination of the Physical Geography of Mountains*. Cambridge, MA: MIT Press, 1990.

LOCKWOOD, JOHN, *World Climatic Systems*. Baltimore: Edward Arnold, 1985.

LYDOLPH, PAUL E., *The Climate of the Earth*. Totowa, NJ: Rowman & Allenheld, 1985.

NIEUWOLT, S., *Tropical Climatology: An Introduction to the Climates of the Low Latitudes*. Chichester, England: John Wiley & Sons, 1982.

SECRETARIAT OF THE UNITED NATIONS CONFERENCE ON DESERTIFICATION ed., *Desertification: Its Causes and Consequences*. Elmsford, NY: Pergamon Press, 1977.

TREWARTHA, G. T. , *The Earth's Problem Climates*. Madison: University of Wisconsin Press, 1981.

9

THE HYDROSPHERE

THE HYDROSPHERE IS AT ONCE THE MOST PERVASIVE and the least well defined of the four "spheres" of Earth's physical environment. It includes the surface water in oceans, lakes, rivers, and swamps; all underground water; frozen water in the form of ice, snow, and high-cloud crystals; water vapor in the atmosphere; and the moisture temporarily stored in plants and animals.

The hydrosphere overlaps significantly with the other three spheres. Liquid water, ice, and even water vapor occur in the soil and rocks of the lithosphere. Water vapor and cloud particles composed of liquid water and ice are important constituents of the lower portion of the atmosphere, and water is a critical component of every living organism of the biosphere. It is through moisture, then, that the interrelationships of the four spheres are most conspicuous and pervasive.

THE NATURE OF WATER: COMMONPLACE BUT UNIQUE

Water is the most distinctive substance found on Earth. It set the stage for the evolution of life and is still an essential ingredient of all life today. It is the most abundant substance on the face of Earth, with surface water occupying more than 70 percent of the surface area of the planet.

Water is made of molecules containing one atom of oxygen and two atoms of hydrogen (H_2O), as Figure 9-1 shows. Pure water has no color, no taste, and no smell. It turns to a solid at 32°F (0°C) and to a vapor (at sea level) at 212°F (100°C). The density of liquid water is 1 gram per cubic centimeter, and it is an extremely good solvent.

Apparently the amount of moisture in existence is finite and remains constant through time. In other words, there is present on Earth today as much water as there ever was or ever will be. This water changes from one form to another and moves from one place to another, but it is neither created nor destroyed. Theoretically it is possible that some of the water of your morning shower was used by Jesus to wash his disciples' feet 2000 years ago or was drunk by a tyrannosaur 70 million years ago.

As we learned in Chapter 6, the water of Earth is found naturally in three states: as a liquid, as a solid, and as a gas (Figure 9-2). The great majority of the world's moisture, however, is in the form of liquid wa-

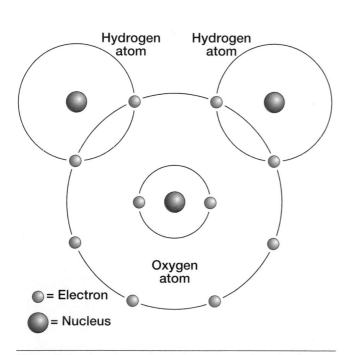

Figure 9-1   A water molecule is made up of two hydrogen atoms and one oxygen atom.

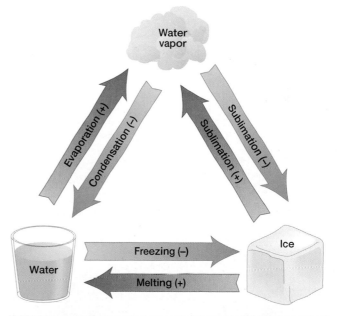

Figure 9-2   The three states in which water on Earth is found and the processes that cause it to change from one state to another. The plus and minus signs indicate gain or loss of heat in the process. In evaporation, for instance, water gains heat as it changes from liquid to vapor; in condensation, the water vapor loses heat as it becomes a liquid. This heat gained or lost is the latent heat we learned about in Chapter 4.

ter, which can be changed to the gaseous form (water vapor) by *evaporation* or to the solid form (ice) by *freezing*. Water vapor can be converted to liquid water by *condensation* or to ice by **sublimation** (which is defined as the process whereby a substance goes either from the gaseous state directly to the solid state or from the solid state directly to the gaseous state without ever passing through the liquid state). Ice can be converted to liquid water by *melting* or to water vapor by sublimation. In each of these processes there is either a gain or a loss of heat, as Figure 9-2 indicates.

Life is impossible without water; every living thing depends on it. Watery solutions in living organisms dissolve or disperse nutrients for nourishment. Chemical reactions that can take place only in a solution release energy from the nutrients. Most waste products are carried away in solutions. Indeed, the total mass of every living thing is more than half water, the proportion ranging from about 60 percent for some animals to more than 95 percent for some plants.

Water has many unusual properties. One of the most striking is its liquidity at the temperatures found at most places on Earth's surface. No other common substance is liquid at ordinary Earth temperatures. The liquidity of water greatly enhances its versatility as an active agent in the atmosphere, lithosphere, and biosphere.

Another environmentally important characteristic of water is its great heat capacity. When water is warmed, it can absorb an enormous amount of energy with only a small rise in temperature (Figure 9-3). Water's heat capacity is exceeded by that of no other common substance except ammonia. The practical result, as we saw in Chapter 4, is that bodies of water are very slow to warm up during the day or in summer and very slow to cool off during the night or in winter. Thus they have a moderating effect on surrounding temperatures by serving as reservoirs of warmth during winter and having a cooling influence in summer.

Most substances contract as they get colder no matter what the change in temperature. When water becomes colder, however, it contracts only until it cools to 39°F (4°C) and then expands as it cools from 39°F to its freezing point of 32°F (0°C). This expansion can break up rocks and is an important component of weathering, which is the name given to the disintegration of rock exposed to the weather. That water expands as it approaches freezing also makes ice less dense than water. As a result, ice floats on and near the surface of water. If it were denser than water, ice would sink to the bottom of lakes and oceans, where melting would be virtually impossible, and eventually many water bodies would become ice choked.

Water normally responds to the pull of gravity and moves downward, but it is also capable of moving upward under certain circumstances. This is because, as we learned in Chapter 6, water has extremely high surface tension and consequently water molecules tend to stick together (Figure 9-4). In addition, they also wet any surfaces with which they come in contact. Surface tension combined with wetting ability allows water to climb upward (Figure 9-5). This climbing capability is

Before heating:

After 10 calories of heat added to each substance:

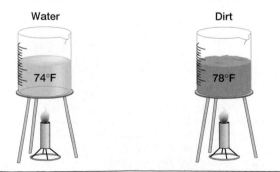

Figure 9-3 Water is capable of absorbing an enormous amount of heat energy with only a small temperature increase. Only ammonia has a greater heat capacity. Here equal volumes of water and dirt are being heated, with 10 calories of heat energy added to each. The water temperature rises by 4 °F and the dirt temperature by 8 °F.

Figure 9-4 A steady-handed person can hold a drop of water between two fingers because the attraction of one water molecule for another provides the water with strong surface tension.

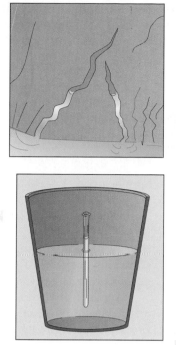

Figure 9-5 Capillarity. Water in narrow spaces can move upward against the pull of gravity because of a combination of surface tension (one water molecule sticks to another) and wetting ability (the water molecules stick to the walls of their container). Here water rises in a small capillary tube and in narrow cracks in a rock.

Enlarged area (upper left)

most notable in situations where water is confined in small pore spaces or narrow tubes. Confined in this way, water can sometimes climb upward for many inches or even feet, in an action called **capillarity**. Capillarity enables water to circulate upward through rock, soil, and the roots and stems of plants.

Of all water's attributes, however, perhaps the most significant is its ability to dissolve other substances. Water can dissolve almost any substance, and it is sometimes referred to as the "universal solvent." It functions in effect as a weak acid, dissolving some substances quickly and in large quantities, other substances slowly and in minute quantities. As a result, water in nature is nearly always impure, by which we mean that it contains various other chemicals in addition to its hydrogen and oxygen atoms. As water moves through the atmosphere, on the surface of Earth, and in soil, rocks, plants, and animals, it carries with it a remarkable diversity of dissolved minerals and nutrients as well as tiny solid particles in suspension.

THE HYDROLOGIC CYCLE

This unique substance—water—essential to life and finite in amount, is distributed very unevenly on, in, and above Earth. The great bulk of all moisture, more than 99 percent, is in storage—in oceans, lakes, and streams, locked up as glacial ice, or held in rocks below Earth's surface (Figure 9-6). The proportional amount of moisture in these various storage reservoirs is relatively constant over thousands of years. Only during an "ice age" is there a notable change in these components: during periods of glaciation, the volume of the oceans becomes smaller as the ice sheets grow and the level of atmospheric water vapor diminishes; then during deglaciation the ice melts, the volume of the oceans increases as the meltwater flows into them, and there is an increase in atmospheric water vapor.

The remaining small fraction—less than 1 percent—of Earth's total moisture is involved in an almost continuous sequence of movement and change, the effects of which are absolutely critical to life on this planet. This tiny por-

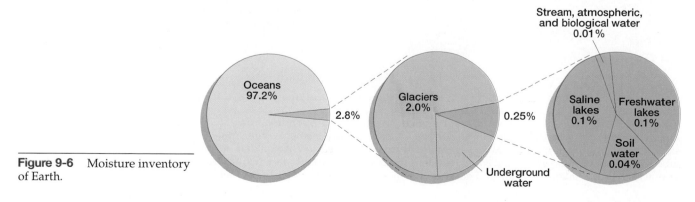

Figure 9-6 Moisture inventory of Earth.

tion of Earth's water supply moves from one storage area to another—from ocean to air, from air to ground, and so on—in what we call the **hydrologic cycle** (Chapter 6). This cycle can be viewed as a series of storage areas interconnected by various transfer processes, in which there is a ceaseless interchange of moisture in terms of both its geographic location and its physical state (Figure 9-7).

Liquid water on Earth's surface evaporates to become water vapor in the atmosphere. That vapor then condenses and precipitates, either as liquid water or as ice, back onto the surface. This precipitated water then runs off into storage areas and later evaporates into the atmosphere once again. As this is a closed, circular system, we can begin the discussion at any point. It is perhaps clearest to start with the movement of moisture from Earth's surface into the atmosphere.

SURFACE-TO-AIR WATER MOVEMENT

Most of the moisture that enters the atmosphere from Earth's surface does so through evaporation. (Transpiration is the source of the remainder.) The oceans, of course, are the principal source of water for evaporation. They occupy 71 percent of Earth's surface, have unlimited moisture available for ready evaporation, and are extensive in low latitudes, where considerable heat and wind movement facilitate evaporation. As a result, an estimated 84 percent of all evaporated mois-

ture is derived from ocean surfaces (Table 9-1). (The 16 percent that comes from land surfaces includes the twin processes of evaporation and transpiration.)

Water vapor from evaporation remains in the atmosphere a relatively short time—usually only a few hours or days. During that interval, however, it may move a considerable distance, either vertically through *convection* or horizontally through advection driven by wind currents.

AIR-TO-SURFACE WATER MOVEMENT

Sooner or later—usually sooner—water vapor in the atmosphere condenses to liquid water or sublimates to ice to form cloud particles. Under the proper circumstances (see Chapter 6), the clouds may drop precipitation in the form of rain, snow, sleet, or hail. As Table 9-1 shows, 77 percent of this precipitation falls into the oceans and 23 percent falls onto land.

Over several years, total worldwide precipitation is approximately equal to total worldwide evaporation/transpiration. Although precipitation and evaporation/transpiration balance in time, they do not balance in place. As Table 9-1 shows, evaporation exceeds precipitation over the oceans whereas the opposite is true over the continents. This imbalance is explained by the advection of moist maritime air onto land areas, so that there is less moisture available for precipitation over the ocean.

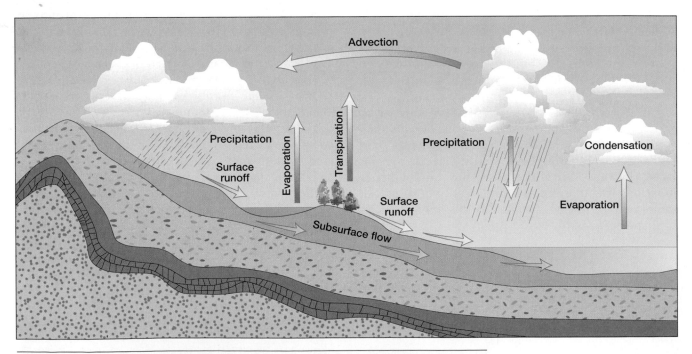

Figure 9-7 The hydrologic cycle. The two major components are evaporation from surface to air and precipitation from air to surface. Other important elements of the cycle include transpiration of moisture from vegetation to atmosphere, surface runoff and subsurface flow of water from land to sea, condensation of water vapor to form clouds from which precipitation may fall, and advection of moisture from one place to another.

TABLE 9-1 Comparisons of Moisture Balance of Continents and Oceans			
	Percentage of total world surface area	Percentage of total world precipitation received	Percentage of total world evaporation/ transpiration that occurs from the surface
Oceans	71	77	84
Continents	29	23	16

Except for coastal spray and storm waves, the only route by which moisture moves from sea to land is via the atmosphere.

MOVEMENT ON AND BENEATH EARTH'S SURFACE

Looking at Table 9-1, you may wonder why the oceans do not dry up and the continents become flooded, there being 7 percent more water leaving the ocean than precipitating back in to it and 7 percent more water falling on land than evaporating off it. The reason such drying up and flooding do not take place is **runoff,** that portion of Earth's circulating moisture that moves in the liquid state from land to sea.

The 77 percent of total global precipitation that falls on the ocean is simply incorporated immediately into the water already there; the 23 percent that falls on land goes through a more complicated series of events. Rain falling on a land surface either collects on that surface, runs off if the surface is a slope, or infiltrates the ground. Any water that pools on the surface eventually either evaporates or sinks into the ground, runoff water eventually ends up in the ocean, and infiltrated water is either stored temporarily as soil moisture or percolates farther down to become part of the underground water supply. Much of the soil moisture eventually evaporates or transpires back into the atmosphere, and much of the underground water eventually reappears at the surface via springs. Then, sooner or later, and in one way or another, most of the water that reaches the surface evaporates again, and the rest is incorporated into streams and rivers and becomes runoff flowing into the oceans.

As already mentioned, this runoff water from continents to oceans amounts to 7 percent of all moisture circulating in the global hydrologic cycle. It is this runoff that balances the excess of precipitation over evaporation taking place on the continents and that keeps the oceans from drying up and the land from flooding.

DURATION OF THE CYCLE

Although the hydrologic cycle is a closed system and is believed to have an unvarying total capacity, there is enormous variation in the cycling of individual molecules of water. A particular molecule of water may be stored in oceans and deep lakes, or as glacial ice, for thousands of years without moving through the cycle, and one trapped in rocks buried deep beneath Earth's surface may be excluded from the cycle for hundreds of thousands of years.

However, whatever water is moving through the cycle is in almost continuous motion. Runoff water can travel hundreds of miles to the sea in only a few days, and moisture evaporated into the atmosphere may remain there for only a few minutes or hours before it is precipitated back to Earth. Indeed, at any given moment the atmosphere contains only a few days' potential precipitation.

THE OCEANS

Despite the facts that (a) most of Earth's surface is oceanic and (b) the vast majority of all water is in the oceans (Figure 9-6), our knowledge of the seas has until recently been very limited. The ocean is a hostile environment for most air-breathing creatures, particularly humans. Thus only with great care can humans venture beneath the sea's surface, and only within the last three decades or so has sophisticated equipment been available to catalog and measure details of the maritime environment.

HOW MANY OCEANS?

From the broadest viewpoint, there is but one ocean. This "world ocean" has a surface area of 139 million square miles (360 million square kilometers) and contains 317 million cubic miles (1.32 billion cubic kilometers) of salt water. It spreads over almost three-fourths of Earth's surface, interrupted here and there by continents and islands. Although tens of thousands of bits of land protrude above the blue waters, the world ocean is so vast that half a dozen continent-sized portions of it are totally devoid of islands, without a single piece of land breaking the surface of the water. It is one or more of these large expanses of water we are usually referring to when we use the term "ocean."

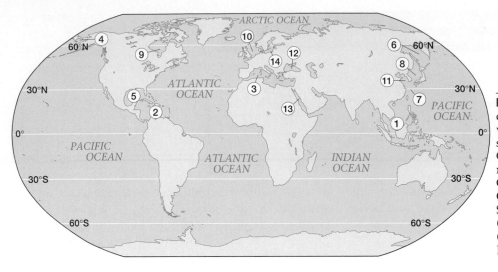

Figure 9-8   The four principal oceans and the major seas of the world. The numbers indicate seas: (1) South China Sea; (2) Caribbean Sea; (3) Mediterranean Sea; (4) Bering Sea; (5) Gulf of Mexico; (6) Sea of Okhotsk; (7) East China Sea; (8) Sea of Japan; (9) Hudson Bay; (10) North Sea; (11) Yellow Sea; (12) Black Sea; (13) Red Sea; (14) Baltic Sea.

In generally accepted usage, the world ocean is divided into four principal parts—the Pacific, Atlantic, Indian, and Arctic oceans (Figure 9-8). (Some people refer to the waters around Antarctica as the *Antarctic Ocean* or the *Great Southern Ocean*, but this distinction is not generally accepted.) The boundaries of the four oceans are not everywhere precise, and around some of their margins are partly landlocked smaller bodies of water called *seas, gulfs, bays,* and other related terms. Most of these smaller bodies can be considered as portions of one of the major oceans, although a few are so narrowly connected (Black Sea, Mediterranean Sea, Hudson Bay) to a named ocean as to deserve separate consideration, as Figure 9-8 and Table 9-2 show. This nomenclature is further clouded by the term *sea,* which is used sometimes synonymously with *ocean,* sometimes to denote a specific smaller body of water around the edge of an ocean, and occasionally to denote an inland body of water.

The *Pacific Ocean* [Figure 9-9(a)] is twice as large as any other body of water on Earth. Five of the seven continents are on its fringes. It occupies about one-third of the total area of Earth, more than all the world's land surfaces combined. It contains the greatest average depth of any ocean as well as the deepest known oceanic trenches. Although the Pacific extends almost to the Arctic Circle in the north and a few degrees beyond the Antarctic Circle in the south, it is largely a tropical ocean. Its greatest girth is in equatorial regions; almost one-half of the 24,000-mile (38,500-kilometer) length of the equator is in the Pacific. The character of this ocean often belies its tranquil name, for it houses some of the most disastrous of all storms and most of the world's major volcanoes are either in it or around its edge.

The *Atlantic Ocean* is slightly less than half the size of the Pacific [Figure 9-9(b)]. Its latitudinal extent is roughly the same as that of the Pacific, but its east-west spread is only about half as great. Its average depth is also only a little less than that of the Pacific.

The *Indian Ocean* [Figure 9-9(c)] is a little smaller than the Atlantic, and its average depth is slightly less than that of the Atlantic. Nine-tenths of its area is south of the equator.

The *Arctic Ocean* [Figure 9-9(d)] is much smaller and shallower than the other three. It is connected to the Pacific by a relatively narrow passageway between Alaska and Siberia, but it has a broad and indefinite connection with the Atlantic between North America and Europe.

CHARACTERISTICS OF OCEAN WATERS

Wherever they are found, the waters of the world ocean have many similar characteristics, but they also show significant differences from place to place. The differences are particularly notable in the surface layers, down to a depth of a few hundred feet (about 100 meters). Below this level, our limited knowledge indicates considerable uniformity in a cold, dark environment.

Chemical Composition   Seawater contains dissolved minerals, which make up about 3.5 percent of its total bulk. Almost all known minerals are found to some extent in seawater, but by far the most important are sodium and chlorine, which form sodium chloride—the common salt we know as table salt. (In the language of chemistry, "salts" are substances that contain various minerals. Table salt, for instance, contains the mineral sodium, and potassium salts contain the mineral potassium.) Table 9-3 lists the principal salts in seawater.

TABLE 9-2
Oceans and Major Seas of the World

	Approximate surface area		Percentage of water present on Earth[a]
	Square miles	Square kilometers	
Pacific Ocean	64,186,000	166,884,000	46
Atlantic Ocean	31,862,000	82,841,000	23
Indian Ocean	28,350,000	73,710,000	20
Arctic Ocean	5,427,000	14,110,000	4
South China Sea	1,150,000	2,990,000	–
Caribbean Sea	971,000	2,525,000	–
Mediterranean Sea	969,000	2,519,000	–
Bering Sea	875,000	2,275,000	–
Gulf of Mexico	600,000	1,560,000	–
Sea of Okhotsk	550,000	1,430,000	–
East China Sea	480,000	1,248,000	–
Sea of Japan	405,000	1,053,000	–
Hudson Bay	318,000	827,000	–
North Sea	222,000	577,000	–
Yellow Sea	220,000	572,000	–
Black Sea	190,000	494,000	–
Red Sea	175,000	455,000	–
Baltic Sea	160,000	416,000	–

[a]Values are omitted for the 14 seas because of their minimal size relative to the size of the world's oceans.

The **salinity** of seawater is a measure of the concentration of dissolved salts, which are mostly sodium chloride but also include salts containing magnesium, sulfur, calcium, and potassium. The average salinity of seawater is about 35 parts per thousand, or 3.5 percent of total weight.

The geographic distribution of surface salinity varies. At any given location on the ocean surface, the salinity depends on how much evaporation is going on and how much fresh water (primarily from rainfall and stream discharge) is being added. Where the evaporation rate is high, so is salinity; where the inflow of fresh water is high, salinity is low. Typically the lowest salinities are found where rainfall is heavy and near the mouths of major rivers. Salinity is highest in partly landlocked seas in dry, hot regions because here the evaporation rate is high and stream discharge is minimal. As a general pattern, salinity is low in equatorial regions because of heavy rainfall, cloudiness, and humidity, all of which inhibit evaporation, and also because of considerable river discharge. Salinity rises to a general maximum in the subtropics, where precipitation is low and evaporation extensive, and decreases to a general minimum in the polar regions, where evaporation is minimal and there is considerable inflow of fresh water from rivers and ice caps.

Temperature   As is to be expected, surface seawater temperatures generally decrease with increasing latitude. The temperature often exceeds 80°F (26°C) in equatorial locations and decreases to 28°F (-2°C), the average freezing point for seawater, in Arctic and Antarctic seas. (Dissolved salts lower the freezing point of the water from the 32°F (0°C) of pure water.) The western sides of oceans are nearly always warmer than the eastern margins because of the movement of major ocean currents. This pattern of warmer western parts is due to the contrasting effects of poleward-moving warm currents on the west and equatorward-moving cool currents on the east.

Density   Seawater density varies with temperature, degree of salinity, and depth. High temperature produces low density, and high salinity produces high density. Deep water has high density because of low temperature and because of the pressure of the overlying water.

Surface layers of seawater tend to contract and sink in cold regions, whereas in warmer areas deeper waters tend to rise to the surface. Surface currents also affect this situation, particularly by producing an upwelling of colder, denser water in some localities.

TABLE 9-3
Principal Components of 6 Gallons[a] of Seawater

5 gallons (20 quarts or 18.9 liters) water

20 cups salt

½ cup magnesium sulfate (Epsom salts)

1 tablespoon calcium chloride

2 teaspoons potassium chloride

1 teaspoon sodium bicarbonate (baking soda)

1 pinch sodium borate (borax)

[a] Some 22.7 liters.

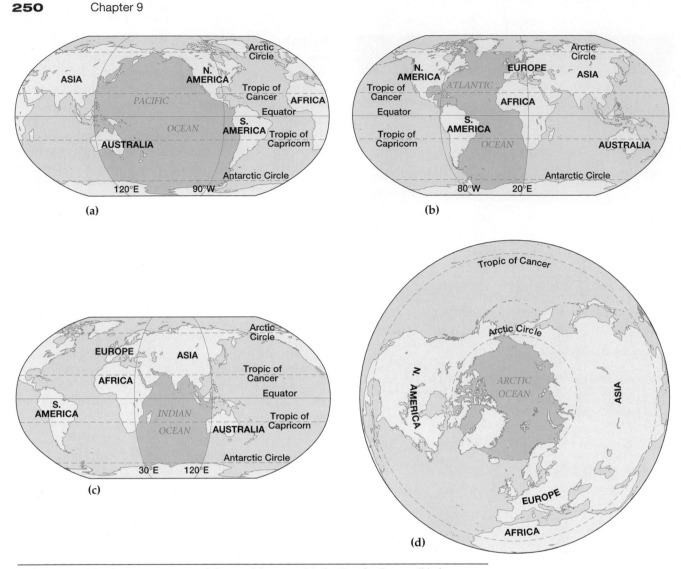

Figure 9-9 The four major parts of the world ocean: **(a)** the Pacific Ocean; **(b)** the Atlantic Ocean; **(c)** the Indian Ocean; **(d)** the Arctic Ocean.

MOVEMENT OF OCEAN WATERS

The liquidity of the ocean keeps it in continuous motion, and this motion can be grouped under three headings: waves, currents, and tides. The movement of almost anything over the surface—wind, a boat, a swimmer—can set surface water into motion. The ocean surface is almost always ridged with swells and waves, therefore, and below the surface there is the less conspicuous restlessness of currents. Disturbances in that part of Earth's crust that underlies ocean water can also trigger significant movements in the water, and the gravitational attraction of nearby heavenly bodies causes the greatest movements of all: the tides.

Waves mostly consist of a change in the shape of the ocean surface, with little displacement of water. Currents involve a considerable displacement of water, particularly horizontally, but also vertically and obliquely. By far the greatest vertical movements of ocean waters are due to tides.

Tides On the shores of the world ocean, almost everywhere, the sea level fluctuates regularly. For about 6 hours each day, the water rises, and then for about 6 hours it falls. These rhythmic oscillations have continued unabated, day and night, winter and summer, for eons. Tides are essentially bulges in the sea surface in some places that are compensated by sinks in the surface at other places. Thus tides are primarily vertical motions of the water. In shallow-water areas around the margins of the oceans, however, the vertical oscillations of the tides may produce significant horizontal water movements as well.

Currents As we learned in Chapter 4, the world ocean contains a variety of currents that shift vast quantities of water both horizontally and vertically. Some of these currents are set in motion by contrasts in temperature and salinity. Surface currents are caused primarily by wind flow. All currents are likely to be in-

FOCUS

TIDES

It is a law of physics that every body in the universe exerts an attractive force on every other body. Thus Earth exerts an attractive force on the moon, and the moon exerts an attractive force on Earth. The same is true for Earth and the sun. It is the moon's pulling on Earth and the sun's pulling on Earth that cause tides.

The strength of the force is inversely proportional to the distance between the two bodies, and so, the sun being 93 million miles from Earth and the moon 240,000 miles, the moon produces a greater percentage of Earth's tides than does the sun. The breakdown is that 44 percent of any daily tide is caused by the sun and 56 percent is caused by the moon. To keep things simple, let us first discuss lunar tides alone, ignoring solar tides for the moment.

There is a bulge in the world ocean on the side of Earth facing the moon because the water on that side, being closer to the moon than is the planet, is pulled toward the moon more than is the planet . On the opposite side of Earth, facing away from the moon, there is a similar bulge because the planet (which is closer to the moon than is the far-side water) is pulled toward the moon and thus away from

the far-side ocean. These two bulges produce simultaneous high tides on opposite sides of Earth. At the same time, there are compensating low tides halfway between the two bulges.

As Earth rotates eastward, the tidal progression appears to move westward. The tides rise and fall twice in the interval between two "rising" moons, an interval that is about 50 minutes longer than a 24-hour day. The combination of Earth's rotation and the moon's revolution around Earth means that Earth makes about 12° more than a full rotation between each rising of the moon. Thus two complete tidal cycles have a duration of about 24 hours and 50 minutes. This means that on all oceanic coast-

lines there are normally two high tides and two low tides every 25 hours.

The magnitude of tidal fluctuation is quite variable in time and place, but the sequence of the cycle is generally similar everywhere. From its lowest point, the water rises gradually for about 6 hours and 13 minutes, so that there is an actual movement of water toward the coast in what is called a *flood tide*. At the end of the flooding period, the maximum water level, *high tide*, is reached. Soon the water level begins to drop, and for the next 6 hours and 13 minutes there is a gradual movement of water away from the coast, this movement being

(Continued)

Figure 9-1-A   Juxtaposition of the sun, moon, and Earth accounts for variations in Earth's tidal range. The three basic positional relationships are illustrated here. (1) When the moon and sun are neither aligned nor at right angles to each other, we have normal levels of high tides on both sides of Earth. (2) When the sun, Earth, and moon are positioned along the same line, spring tides (the highest high tides) are produced. (3) When the line joining Earth and the moon forms a right angle with the line joining Earth and the sun, neap tides (the lowest high tides) result.

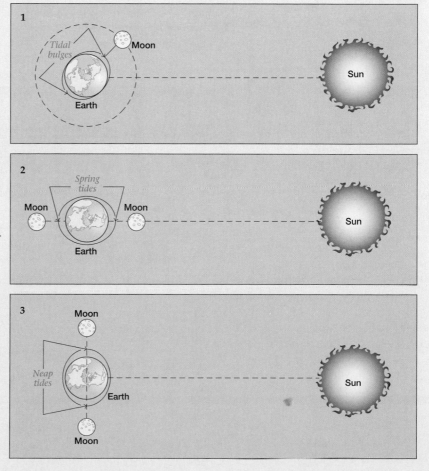

TIDES

(Continued)

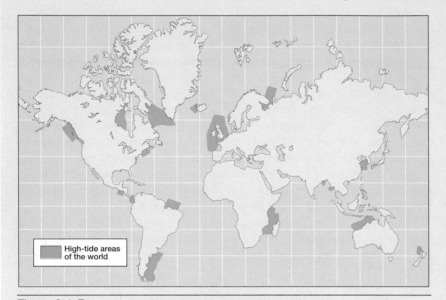

Figure 9-1-B Areas with tidal ranges exceeding 13 feet (4 meters). The pattern is not a predictable one, as it depends upon a variety of unrelated factors, particularly shoreline and sea-bottom configuration. (After J. L. Davies, *Geographical Variation in Coastal Development*, 2nd ed., New York: Longman, 1980, p. 179.)

tween Earth and the sun and when Earth is between the moon and sun. In either case, this is a time of higher than usual tides, called spring tides. (The name has nothing to do with the season; think of water "springing" up to a very high level.)

When the sun and moon are located at right angles to one another with respect to Earth, their individual gravitational pulls are diminished because they are now pulling at right angles to each other. This right-angle pulling results in a lower-than-normal tidal range called a neap tide. The sun-moon alignment that causes neap tides generally takes place twice a month at about the time of first-quarter and third-quarter moons.

Tidal range is also affected by the moon's nearness to Earth. The moon follows an elliptical orbit in its

called an *ebb tide*. When the minimum water level (*low tide*) is reached, the cycle begins again.

The vertical difference in elevation between high and low tide is called the **tidal range**. Changes in the relative positions of Earth, moon, and sun induce periodic variations in tidal ranges, as shown in Figure 9-1-A. The greatest range (in other words, the highest tide) occurs when the three bodies are positioned in a straight line, which usually occurs twice a month near the times of the full and new moons. When thus aligned, the joint gravitational pull of the sun and moon is along the same line, so that the combined pull is a maximum. This is true both when the moon is be-

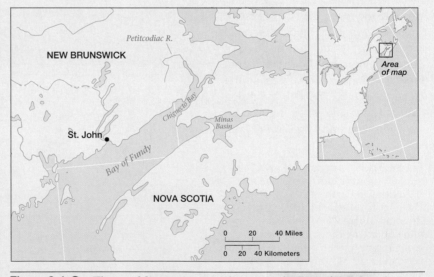
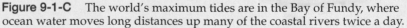

Figure 9-1-C The world's maximum tides are in the Bay of Fundy, where ocean water moves long distances up many of the coastal rivers twice a day.

Figure 9-1-D   A tidal bore moving up the Petitcodiac River in New Brunswick, Canada. (J. H. Robinson/Photo Researchers, Inc.)

revolution around Earth, the nearest point (called perigee) being about 31,200 miles (50,000 kilometers), or 12 percent, closer than the farthest point (*apogee*). During perigee, tidal ranges are greater than during apogee.

Tidal range fluctuates all over the world at the same times of the month. There are, however, enormous variations in range along different coastlines (Figure 9-1-B). Midocean islands experience tides of only 2 or 3 feet (0.6 or 0.9 meter), whereas continental seacoasts have greater tidal ranges, the amplitude being greatly influenced by the shape of the coastline and the configuration of the sea bottom beneath coastal waters. Along most coasts, there is a moderate tidal range of 5–10 feet (1.5–3 meters). Some partly landlocked seas, such as the Mediterranean, have almost negligible tides. Other places, such as the northwestern coast of Australia, experience enormous tides of 35 feet (10.5 meters) or so.

The greatest tidal range is found at the upper end of the Bay of Fundy in eastern Canada, where a 50-foot (15-meter) water-level fluctuation twice a day is not uncommon, and a wall of seawater (called a **tidal bore**) several inches to several feet in height rushes up the Petitcodiac River in New Brunswick for many miles (Figures 9-1-C and 9-1-D).

Tidal variation is exceedingly small in inland bodies of water. Even the largest lakes usually experience a tidal rise and fall of no more than 2 inches (5 centimeters). Effectively, then, tides are important only in the world ocean, and they are normally noticeable only around its shorelines.

fluenced by the size and shape of the particular ocean, the configuration and depth of the sea bottom, and the Coriolis effect.

Some currents involve subsidence of surface waters downward; other vertical flows bring an upwelling of deeper water to the surface. Geographically speaking, however, the most important currents are the major horizontal flows that make up the general circulation of the various oceans (see Figure 4-24).

Waves   To the casual observer, the most conspicuous motion of the ocean is provided by waves. Most of the sea surface is in a state of constant agitation, with wave crests and troughs bobbing up and down most of the time. Moreover, around the margin of the ocean, waves of one size or another lap, break, or pound on the shore in endless procession.

In point of fact, most of this movement is like running in place from the water's point of view, with little forward progress. Waves in the open ocean are mostly just shapes, and the movement of a wave across the sea surface is a movement of form rather than of substance or, to say the same thing another way, of energy rather than matter. Individual water particles make only small oscillating movements. Only when a wave "breaks" does any significant shifting of water take place. Waves are discussed in detail in Chapter 18.

PERMANENT ICE

Second only to the world ocean as a storage reservoir for moisture is the solid portion of the hydrosphere—the ice of the world, as we learn from a glance back at Figure 9-6. Although minuscule in comparison with the amount of water in the oceans, the moisture content of ice at any given time is more than twice as large as the combined total of all other types of storage (underground water, surface waters, soil moisture, atmospheric moisture, and biological water).

The ice portion of the hydrosphere is divided between ice on land and ice floating in the ocean, with the land portion being the larger. Ice on land is found as alpine glaciers, ice sheets, and ice caps, all of which are studied in Chapter 20. Approximately 10 percent of the land surface of Earth is covered by ice (Figure 9-10). It is estimated that there is enough water locked up in this ice to feed all the rivers of the world at their present rate of flow for nearly 900 years.

Oceanic ice has various names, depending on size, as Table 9-4 shows. Despite the fact that some oceanic ice freezes directly from seawater, all forms of oceanic ice are composed entirely of fresh water because the salts present in the seawater in its liquid state are never taken up into the ice crystals when that water

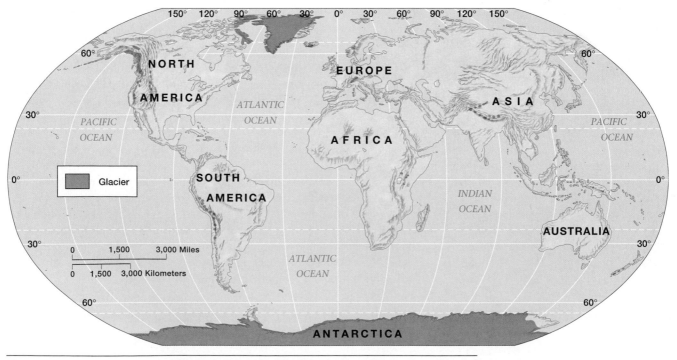

Figure 9-10   Glacial ice covers about 10 percent of Earth's surface. This ice is confined primarily to Antarctica and Greenland, with small amounts also found at high altitudes in the Canadian Rockies, Andes, Alps, and Himalayas.

TABLE 9-4
Oceanic Ice Forms

Name	Definition
Ice pack	An extensive and cohesive mass of floating ice
Ice shelf	A massive portion of an ice sheet that projects out over the sea
Ice floe	A large, flattish mass of ice that breaks off from larger ice bodies and floats independently
Iceberg	A chunk of floating ice that breaks off an ice shelf or glacier

being essentially doubled by increased freezing around their margins.

There are a few small ice shelves in the Arctic, mostly around Greenland, but several gigantic shelves are attached to the Antarctic ice sheet, most notably the Ross ice shelf of some 40,000 square miles (100,000 square kilometers). Some Antarctic ice floes are enormous; the largest ever observed was ten times as large as the state of Rhode Island.

A relatively small proportion of the world's ice occurs beneath the land surface as ground ice. This type of ice occurs only in areas where the temperature is continuously below the freezing point, and so it is restricted to high-latitude and high-elevation regions (Figure 9-13). Most permanent ground ice is **permafrost**, which is permanently frozen subsoil. It is widespread in northern Canada, Alaska, and Siberia and found in small patches in many high mountain areas. Some ground ice is aggregated as veins of frozen water, but most of it develops as ice crystals in the spaces between soil particles.

freezes. The largest ice pack covers most of the surface of the Arctic Ocean (Figure 9-11); on the other side of the globe, an ice pack fringes most of the Antarctic continent (Figure 9-12). Both of these packs become greatly enlarged during their respective winters, their areas

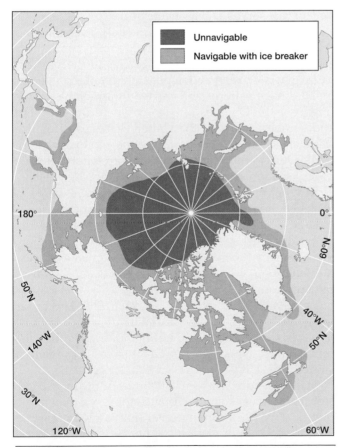

Figure 9-11 The largest ice pack on Earth covers most of the Arctic Ocean, making that body of water essentially unnavigable. Powerful icebreaker ships allow passage from the Atlantic to the Pacific via this northern route, the fabled "Northwest Passage" of early European explorers.

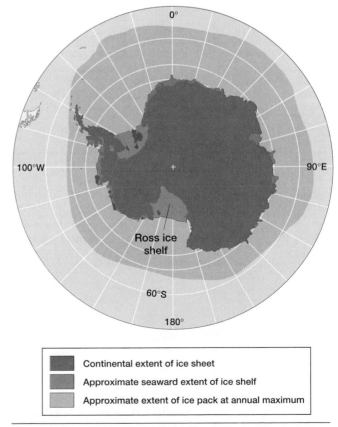

Figure 9-12 Maximum extent of ice in Antarctica today. The ice sheet covers land, and the ice shelf and ice pack are oceanic ice.

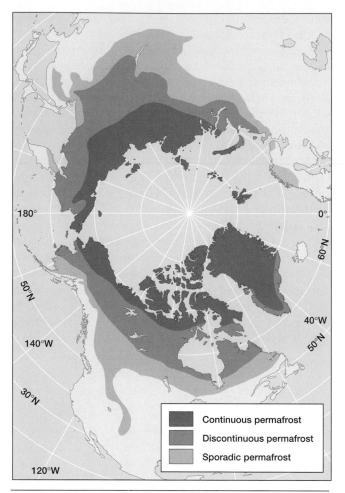

Figure 9-13   Extent of permafrost in the Northern Hemisphere. All the high-latitude land areas and some of the adjacent midlatitude land areas are underlain by permafrost.

SURFACE WATERS

Surface waters represent only about 0.25 percent of the world's total moisture supply (Figure 9-6), but from the human viewpoint they are of incalculable value. Lakes, swamps, and marshes abound in many parts of the world, and all but the driest parts of the continents are seamed by rivers and streams.

LAKES

Lakes have been called "wide places in rivers." In even simpler terms, a **lake** is a body of water surrounded by land. No minimum or maximum size is attached to this definition, although the word *pond* is often used to designate a very small lake. Well over 90 percent of the surface water of the continents is contained in lakes.

With a few exceptions, the origin of lakes is not related to stream activity. Most lakes are fed and drained by streams, but lake genesis is usually due to other fac-

tors. Two conditions are necessary for the formation and continued existence of a lake: (1) some sort of natural basin having a restricted outlet, and (2) sufficient inflow of water to keep the basin at least partly filled.

Most of the world's lakes contain fresh water, but some of the largest lakes are saline. Indeed, more than 40 percent of the lake water of the planet is salty, with the lake we call the Caspian Sea containing more than three-quarters of the total volume of all the world's nonoceanic saline water. (In contrast, Utah's famous Great Salt Lake contains less than 1/2500 the volume of the Caspian.) Any lake that has no natural drainage outlet, either as a surface stream or as a sustained subsurface flow, will become saline.

Most small salt lakes and some large ones are *ephemeral*, which means that they contain water only sporadically and are dry much of the time because they are in dry regions with insufficient inflow to maintain them on a permanent basis.

The water balance of most lakes is maintained by surface inflow, sometimes combined with springs and seeps below the lake surface. A few lakes are fed entirely by springs. Most freshwater lakes have only one stream that serves as a drainage outlet.

Lakes are distributed very unevenly over the land (Figure 9-14 and Table 9-5). They are most common in regions that were visited by a glacier in the past because glacial erosion and deposition deranged the normal drainage patterns and created innumerable basins (Figure 9-15). One has only to compare the northern and southern parts of the United States to recognize this fact (Figure 9-16). Most of Europe and the northern part of Asia demonstrate a similar correlation between past glaciation and present-day lakes.

Some parts of the world notable for lakes were not glaciated, however. For example, the remarkable series of large lakes in eastern and central Africa was created by Earth's crust splitting apart and by volcanic action, and the many thousands of small lakes in Florida were formed by sinkhole collapse when rainwater dissolved calcium from the limestone bedrock.

Lake Baykal (often spelled Baikal) in Siberia is by far the world's largest freshwater lake in terms of volume of water, containing considerably more water than the combined contents of all five Great Lakes in the central United States. It is also the world's deepest lake.

Most lakes are temporary features of the landscape. Not many have been in existence for more than a few thousand years, a time interval that is momentary in the grand scale of geologic time. Inflowing streams bring sediment to fill lakes up, outflowing streams cut channels progressively deeper to drain them, and as the lake becomes shallower a continuous increase in plant growth accelerates the infilling. Thus the destiny of most lakes is to disappear (Figure 9-17).

TABLE 9-5
The World's Largest Lakes Ranked by Surface Area

Rank	Name	Continent	Square miles	Square kilometers	Water
			Area		
1	Caspian Sea	Asia	143,250	372,450	Salt
2	Lake Superior	North America	31,700	82,420	Fresh
3	Lake Victoria	Africa	26,700	69,400	Fresh
4	Lake Huron	North America	23,000	59,800	Fresh
5	Lake Michigan	North America	22,300	58,000	Fresh
6	Aral Sea	Asia	14,025	36,300	Salt
7	Lake Tanganyika	Africa	12,650	33,000	Fresh
8	Lake Baykal	Asia	12,200	31,700	Fresh
9	Great Bear Lake	North America	12,100	31,500	Fresh
10	Lake Malawi	Africa	11,550	30,000	Fresh
11	Great Slave Lake	North America	11,300	29,400	Fresh

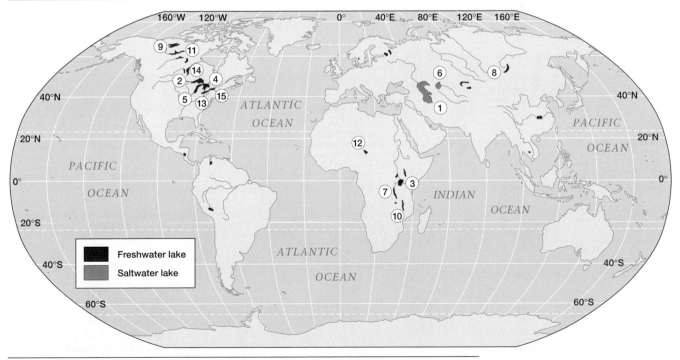

Figure 9-14 The world's largest lakes: (1) Caspian Sea; (2) Lake Superior; (3) Lake Victoria; (4) Lake Huron; (5) Lake Michigan; (6) Aral Sea; (7) Lake Tanganyika; (8) Lake Baykal; (9) Great Bear Lake; (10) Lake Malawi; (11) Great Slave Lake; (12) Lake Chad; (13) Lake Erie; (14) Lake Winnipeg; (15) Lake Ontario.

Figure 9-15 Glaciation is responsible for the formation of St. Mary Lake in Montana's Glacier National Park. (TLM photo)

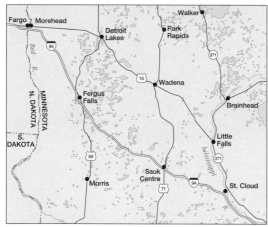

(a)

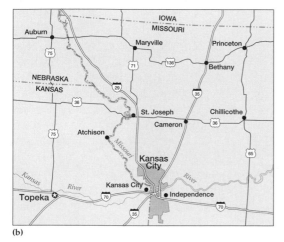

(b)

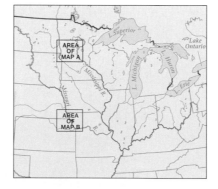

Figure 9-16 There is an abundance of natural lakes in the region north of the Ohio and Missouri rivers, a result of glacial action. South of these rivers, however, there was no glaciation and consequently natural lakes are almost unknown.

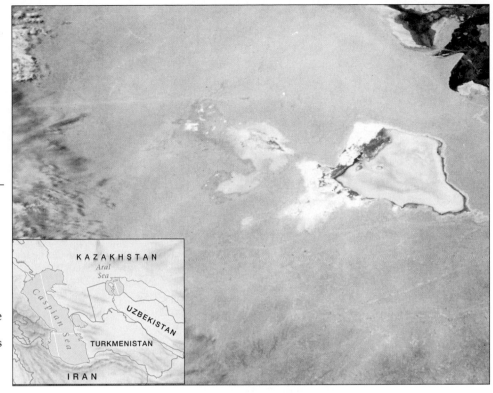

Figure 9-17 A quarter of a century ago, the Aral Sea was the world's fourth largest lake. However, as irrigation needs have forced farmers to drain more and more water from the rivers feeding the Aral, its volume has decreased by two-thirds in the past 25 years. As shown in this satellite image, the lake has now split into two bodies of water. If present trends continue, the Aral will cease to exist by the year 2010. (NASA Headquarters)

One of the most notable things people have done to alter the natural landscape is to produce artificial lakes, or *reservoirs*. Such lakes have been created largely by the construction of dams, ranging from small earth mounds heaped across a gully to immense concrete structures blocking the world's major rivers (Figure 9-18). Some reservoirs are as large as medium-sized natural lakes. The creation of artificial lakes has had immense ecological and economic consequences, not all of them beneficial.

SWAMPS AND MARSHES

Closely related to lakes but less numerous and containing a much smaller volume of water are swamps and marshes, flattish places that are submerged in water at least part of the time but are shallow enough to permit the growth of water-tolerant plants (Figure 9-19). The conceptual distinction between the terms is that a **swamp** has a plant growth that is dominantly trees whereas a **marsh** is vegetated primarily with grasses and rushes. Both are usually associated with coastal plains, broad river valleys, or recently glaciated areas. Sometimes they represent an intermediate stage in the infilling of a lake.

RIVERS AND STREAMS

Although containing only a small proportion of the world's water at any given time, rivers and streams are an extremely dynamic component of the hydrologic cycle. (Although the terms are basically interchangeable, in common usage a "stream" is smaller than a "river." Geographers, however, call any flowing water a stream, no matter what its size.) They provide the means by which the land surface drains and by which water, sediment, and dissolved chemicals are moved ever seaward. The occurrence of rivers and streams is closely, but not absolutely, related to precipitation patterns. Humid lands have many rivers and streams, most of which flow year round; dry lands have few, almost all of which are ephemeral (which means they dry up for part of the year).

Tables 9-6 and 9-7 list the world's longest and largest rivers, and Figure 9-20 shows the drainage basins feeding them. (A *drainage* basin is all the land area drained by a river and its tributaries.) A mere two dozen great rivers produce one-half of the total stream discharge of the world. The mighty Amazon yields nearly 20 percent of the world total, 4.5 times the discharge of the second-ranking river, the Zaire. Indeed, the discharge of the Amazon is three times as great as the total combined discharge of all rivers in the United States. The Mississippi is North America's largest river by far, with a drainage basin that encompasses about 40 percent of the total area of the 48 conterminous states and a flow that amounts to about one-third of the total discharge from all other rivers of the nation.

Figure 9-18 One of the largest reservoirs in the West is Lake Mead, behind Hoover Dam on the Nevada–Arizona border. (Lowell Georgia/Photo Researchers, Inc.)

Figure 9-19 Marshes are particularly numerous along the poorly drained South Atlantic and Gulf of Mexico coasts of the United States. This is Loxahatchee Marsh in Florida. (Mark E. Gibson/The Stock Market)

UNDERGROUND WATER

Beneath the land surface is another important component of the hydrosphere—underground water. As Figure 9-6 shows, the total amount of underground water is about 2.5 times that contained in lakes and streams. Moreover, underground water is much more widely distributed than surface water. Whereas lakes and rivers are found only in restricted locations, underground water is almost ubiquitous, occurring beneath

| | | | | TABLE 9-6 | | |
| | | | | The World's Longest Rivers | | |

| | | | | Length | |
Rank	Name	Continent	Empties into	Miles	Kilometers
1	Nile	Africa	Mediterranean Sea	4130	6600
2	Amazon	South America	Atlantic Ocean	3900	6200
3	Missouri-Mississippi	North America	Gulf of Mexico	3740	6000
4	Yangtze	Asia	East China Sea	3400	5450
5	Ob-Irtysh	Asia	Kara Sea	3400	5450
6	Hwang Ho	Asia	Yellow Sea	2900	4650
7	Amur	Asia	Tartar Strait	2800	4500
8	Zaire	Africa	Atlantic Ocean	2700	4300
9	Lena	Asia	Laptev Sea	2700	4300
10	Mackenzie	North America	Beaufort Sea	2635	4200
11	Mekong	Asia	South China Sea	2600	4150
12	Niger	Africa	Gulf of Guinea	2600	4150
13	Yenisy	Asia	Kara Sea	2500	4000
14	Paraná	South America	Atlantic Ocean	2450	3900
15	Volga	Europe	Caspian Sea	2300	3700

TABLE 9-7
The World's Largest Rivers Ranked by Average Discharge and Drainage Area

Name	Continent	Average discharge (cubic meters per second)	Rank by discharge volume	Drainage area (square kilometers)	Rank by drainage area
Amazon	South America	175,000	1	5,800,000	1
Zaire	Africa	40,000	2	4,000,000	2
Yangtze	Asia	22,000	3	1,900,000	9
Brahmaputra	Asia	20,000	4	935,000	20
Ganges	Asia	19,000	5	1,060,000	18
Yenisy	Asia	17,000	6	2,600,000	5
Mississippi	North America	17,000	7	3,200,000	3
Orinoco	South America	17,000	8	880,000	24
Lena	Asia	15,500	9	2,510,000	6
Paraná	South America	15,000	10	2,200,000	8
St. Lawrence	North America	14,000	11	1,300,000	14
Irrawaddy	Asia	13,600	12	430,000	35
Ob	Asia	12,500	13	2,500,000	7
Mekong	Asia	11,000	14	800,000	26
Tocantins	South America	10,000	15	910,000	23

the land surface throughout the world. Its quantity is sometimes limited, its quality is sometimes poor, and its occurrence is sometimes at great depth, but almost anywhere on Earth one can dig deep enough and find water.

More than half of the world's underground water is found within about half a mile (800 meters) of the surface. Below that depth, the amount of water generally decreases gradually and erratically. Although water has been found at depths below 6 miles (10 kilome-

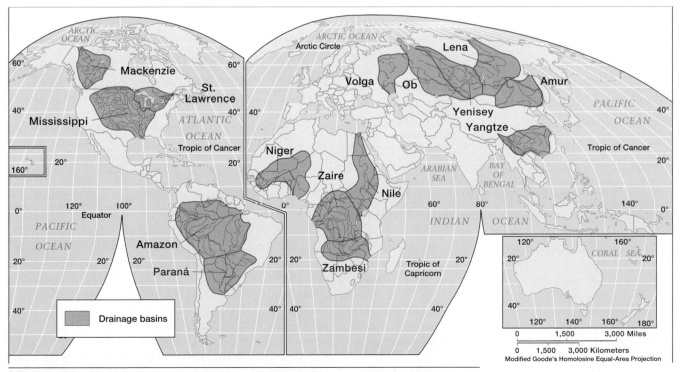

Figure 9-20 The world's largest drainage basins are scattered over the four largest continents, in all latitudes.

ters), it is almost immobilized because the pressure exerted by overlying rocks is so great and openings are so few and small.

Almost all underground water comes originally from above. Its source is precipitation that either percolates directly into the soil or else seeps downward eventually from lakes and streams.

Once the moisture gets underground, any one of several things can happen to it, depending largely on the nature of the soil and rocks it infiltrates. The quantity of water that can be held in subsurface material (rock or soil) depends on the **porosity** of the material, which is the percentage of the total volume of the material that consists of voids (pore spaces or cracks) that can fill with water. The more porous a material is, the greater the amount of open space it contains and the more water it can hold.

Porosity is not the only factor affecting underground water flow. If water is to move through rock or soil, the pores must be connected to one another and be large enough for the water to move through them. The ability to transmit underground water (as opposed to just hold it, as in the definition of porosity) is termed **permeability**, and this property of subsurface matter is determined by the size of pores and by their degree of interconnectedness. The water moves by twisting and turning through these small, interconnected openings. The smaller and less connected the pore spaces, the less permeable the material and the slower the water moves.

The rate at which water moves through rock depends on both porosity and permeability. For example, clay is usually of high porosity because it has a great many **interstices** (openings) among the minute flakes that make up the clay, but it generally has low permeability because the interstices are so tiny that the force of molecular attraction binds the water to the clay flakes and holds it in place. Thus, clay typically is very porous but relatively impermeable and consequently can trap large amounts of water and keep it from draining.

Underground water is stored in, and moves slowly through, moderately to highly permeable rocks called **aquifers** (from the Latin, *aqua*, water, and *ferre*, to bear). The rate of movement of the water varies with the situation. In some aquifers the flow rate is only a few inches a day; in others, it may be several hundred feet per day. A "rapid" rate of flow would be 40–50 feet (12–15 meters) per day.

Impermeable materials composed of components such as clay or very dense rock, which hinder or prevent water movement, are called **aquicludes** (Figure 9-21).

The general distribution of underground water can probably best be understood by visualizing a

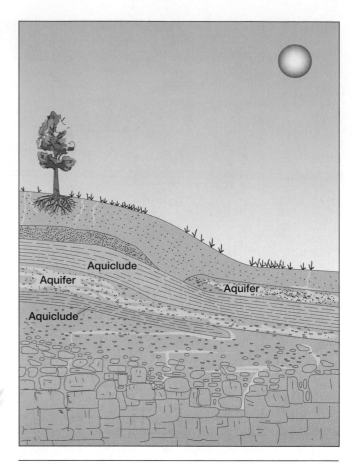

Figure 9-21 An aquifer is a rock structure that is permeable and/or porous enough to hold water, whereas an aquiclude has a structure that is too dense to allow water to penetrate it.

vertical subsurface cross section. Usually at least three and often four **hydrologic zones** are arranged one below another. From top to bottom, these layers are called the zone of aeration, the zone of saturation, the zone of confined water, and the waterless zone.

ZONE OF AERATION

The topmost band, the **zone of aeration,** is a mixture of solids, water, and air. Its depth can be quite variable, from a few centimeters to hundreds of meters. The interstices in this zone are filled partly with water and partly with air. The amount of water fluctuates considerably with time. After a rain the pore spaces may be saturated with water, but the water may drain away rapidly. Some of the water evaporates, but much is absorbed by plants, which later return it to the atmosphere by transpiration. Water that molecular attraction cannot hold seeps downward into the next zone.

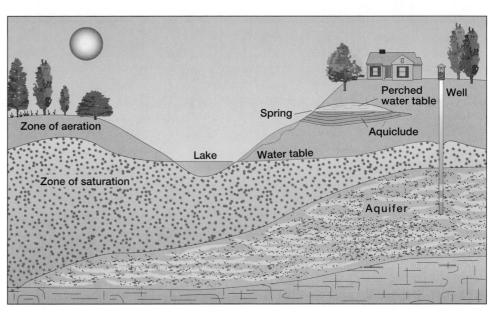

Figure 9-22   Relationship between the water table and associated features. Wherever the table touches Earth's surface, surface water is found, either as standing water (lakes and marshes) or as flowing water (streams and springs).

ZONE OF SATURATION

Immediately below the zone of aeration is the **zone of saturation**, in which all pore spaces in the soil and cracks in the rocks are fully saturated with water. The moisture in this zone is called **groundwater**; it seeps slowly through the ground following the pull of gravity and guided by rock structure. The top of the saturated zone is referred to as the **water table** (Figure 9-22). The orientation and slope of the water table usually conform roughly to the slope of the land surface above, nearly always approaching closer to the surface in valley bottoms and being more distant from it beneath a ridge or hill. Where the water table intersects Earth's surface, water flows out. A lake, swamp, marsh, or permanent stream is almost always an indication that the water table reaches the surface there. In humid regions the water table is higher than in arid regions, which means that the zone of saturation is nearer the surface in humid regions. Some desert areas have no saturated zone at all.

Sometimes a localized zone of saturation develops above an aquiclude, and this configuration forms a **perched water table**.

A well dug into the zone of saturation fills with water up to the level of the water table (Figure 9-23). When water is taken from the well faster than it can flow in from the saturated rock, the water table drops in the immediate vicinity of the well, in the approximate shape of an inverted cone. This striking feature is called a **cone of depression**. If many wells are withdrawing water faster than it is being replenished naturally, the water table may be significantly depressed over a large area.

Water percolates slowly through the saturated zone along tiny parallel paths. Gravity supplies much of the energy for groundwater percolation, leading it from areas where the water table is high toward areas where it is lower, that is, toward surface streams or lakes. Percolation flow channels are not always downward, however. Often the flow follows a curving path and then turns upward (against the force of gravity) to enter the stream or lake from below. This trajectory is possible because saturated-zone water at any given height is under greater pressure beneath a hill than beneath a stream valley. Thus the water moves toward points where the pressure is least.

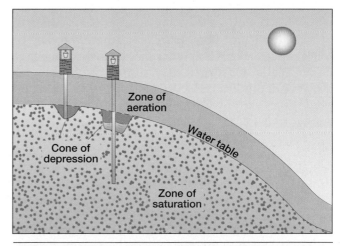

Figure 9-23   The well on the left reaches just to the water table; it has created a small cone of depression and will have an unreliable water supply if the cone becomes any deeper. The well on the right extends deep into the saturated zone; it has created a larger cone of depression but should provide a reliable water supply for some time to come.

PEOPLE AND THE ENVIRONMENT

M I N I N G   G R O U N D W A T E R   I N   T H E   G R E A T   P L A I N S

In most parts of the world where groundwater occurs, it has been accumulating for a long time. Rainfall and snowmelt seep and percolate downward into aquifers, where the water may be stored for decades or centuries or millennia. Only in recent years have most of these aquifers been discovered and tapped by humans. They represent valuable sources of water that can supplement surface water resources. Underground water has been particularly utilized by farmers to irrigate in areas that contain insufficient surface water.

The accumulation of underground water is tediously slow. Its use by humans, however, can be distressingly rapid. In many parts of the U.S. Southwest, for example, the recharge (replenishment) rate averages only 0.2 inch (0.5 centimeter) per year, but it is not uncommon for a farmer to pump 30 inches (75 centimeters) per year. Thus yearly pumpage is equivalent to 150 years' recharge. This rate of groundwater use can be likened to mining, be-

cause a finite resource is being removed with no hope of replenishment. Almost everywhere in the world that underground water is being utilized on a large scale, the water table is dropping steadily and often precipitously.

A classic example of groundwater mining is seen in the southern and central parts of the Great Plains, where the largest U.S. aquifer, the Ogallala, underlies 225,000 square miles (585,000 square kilometers) of eight states. The Ogallala formation consists of a series of limey and sandy layers that function as a gigantic underground reservoir ranging in thickness from a few inches in parts of Texas to more than 1000 feet (300 meters) under the Nebraska Sandhills (Figure 9-2-A). Water has been accumulating in this aquifer for some 30,000 years. At the midpoint of the twentieth century, it was estimated to contain 1.4 billion acre-feet (456 trillion gallons, or 1.7 quadrillion liters) of water, an amount roughly equivalent to the volume of one of the larger Great Lakes.

Farmers began to tap the Ogallala in the early 1930s. Before the end of that decade, the water table already was dropping. After World War II the development of high-capacity pumps, sophisticated sprinklers, and other technological innovations encouraged the rapid expansion of irrigation based on Ogallala water. Water use in the region has almost quintupled since 1950. The results of this accelerated usage have been spectacular. Above ground, there has been a rapid spread of high-yield farming into areas never before cultivated (especially in Nebraska) and a phenomenal increase in irrigated crops in all eight Ogallala states. Beneath the surface, however, the water table is sinking ever deeper. Farmers who once obtained water from 50-foot (15-meter) wells now must bore to 150 or 250 feet (45 or 75 meters), and as the price of energy skyrockets, the cost of pumping increases operating expenses enormously. Some 170,000 wells tap the Ogallala; many are al-

The lower limit of the zone of saturation is marked by the absence of pore spaces and therefore the absence of water. This boundary may be a single layer of impermeable rock, or it may simply be that the increasing depth has created so much pressure that no pore spaces exist in any rocks at that level.

ZONE OF CONFINED WATER

In many, but not most, parts of the world, a third hydrologic zone lies beneath the zone of saturation, separated from it by impermeable rock. This **zone of confined water** contains one or more aquifers into which water can infiltrate. Sometimes aquifers alternate with impermeable layers (aquicludes). Water cannot penetrate an aquifer in this deep zone by infiltration from above because of the impermeable barrier, and so any water the zone contains must have percolated along the aquifer from a more distant area where no aquiclude interfered. Characteristically, then,

an aquifer in the confined water zone is a sloping or dipping layer that reaches to, or almost to, the surface at some location, where it can absorb infiltrating water. The water works its way down the sloping aquifer from the catchment area, building up considerable pressure in its confined situation.

If a well is drilled from the surface down into the confined aquifer, which may be at considerable depth, the confining pressure forces water to rise in the well. The elevation to which the water rises is known as the **piezometric surface.** In some cases, the pressure is enough to allow the water to rise above the ground, as shown in Figure 9-24. This free flow of water is called an **artesian well**. If the confining pressure is sufficient to push the water only partway to the surface and it must be pumped the rest of the way, the well is **subartesian**.

Unlike the distribution of groundwater, which is closely related to precipitation, the distribution of confined water is quite erratic over the world. Confined water underlies many arid or semiarid regions

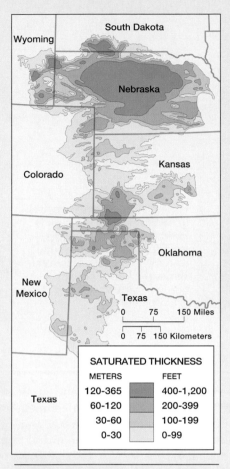

Figure 9-2-A The Ogallala aquifer. Darker areas indicate greater thickness of the water-bearing strata.

SATURATED THICKNESS

METERS	FEET
120-365	400-1,200
60-120	200-399
30-60	100-199
0-30	0-99

ready played out, and most of the rest must be deepened annually.

Anguish over this continuously deteriorating situation is widespread. Some farmers are shifting to crops that require less water. Others are adopting water- and energy-conserving measures that range from a simple decision to irrigate less frequently to the installation of sophisticated machinery that uses water in the most efficient fashion. Many farmers have faced or will soon face the prospect of abandoning irrigation entirely. During the next four decades, it is estimated that 5 million acres (2 million hectares) now irrigated will revert to dry-land production. Other farmers concentrate on high-value crops before it is too late, hoping to make a large profit and then get out of farming.

Water conservation is further complicated by the obvious fact that groundwater is no respecter of property boundaries. A farmer who is very conservative in his or her water use must face the reality that less-careful neighbors are pumping from the same aquifer and that their profligacy may seriously diminish the water available to everyone.

The situation varies from place to place. The Nebraska Sandhills have the most favorable conditions. The aquifer is deepest there, previous water use was minimal, and there is a relatively rapid recharge rate. Indeed, for the 13-county area that makes up the bulk of the Sandhills, withdrawal averages only about 10 percent of recharge, a remarkable situation. In contrast, the 13 counties of southwestern Kansas have a withdrawal rate 22 times the recharge rate; if present use patterns continue, southwestern Kansas will have no more Ogallala water by about the year 2019.

There is no way to escape the inevitable: Ogallala water is a finite resource, and sooner or later it will disappear no matter what conservation techniques are used. Doomsday is a decade away in some areas, perhaps half a century in others. Irrigation is still on an upward trend in areas where the aquifer is deepest and less used, as in much of Nebraska. For most of the region, however, Ogallala water is becoming decreasingly available and increasingly costly.

that are poor in surface water or groundwater, thus providing a critical resource for these dry lands (Figure 9-25).

WATERLESS ZONE

At some depth below the surface, there is no water because the overlying pressure increases the density of the rock and so there are no pores. This **waterless zone** generally begins several miles beneath the land surface.

Figure 9-24 An artesian system. Surface water penetrates the aquifer in the recharge area and infiltrates downward. It is confined to the aquifer by impermeable strata (*aquicludes*) above and below. If a well is dug through the upper aquiclude into the aquifer, the confining pressure forces the water to rise in the well. In an artesian well, the pressure forces the water to the surface; in a subartesian well, the water is forced only partway to the surface and must be pumped the rest of the way.

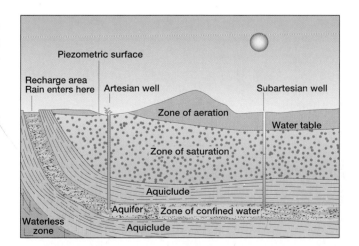

Figure 9-25   An artesian well in Australia's Great Artesian Basin, the largest and most productive source of confined water in the world. This scene shows the Pilliga Bore in northwestern New South Wales. (TLM photo)

CHAPTER SUMMARY

The hydrosphere encompasses all moisture in, on, and above Earth. Water is the most common substance of the habitable zone of Earth and is absolutely essential for all life forms. It has many unusual properties, one of the most notable being its ability to dissolve other substances.

Almost all of Earth's moisture is in semipermanent storage in the oceans and other bodies of water, as well as in glaciers, underground reservoirs, and living organisms. The rest (less than 1 percent of the total) circulates from sea to air to land to sea in the hydrologic cycle.

More than 97 percent of all moisture is contained in the world ocean, which generally is subdivided into four major parts—Pacific, Atlantic, Indian, and Arctic. Ocean water varies in chemical composition, temperature, and density from one location to another. Oceanic waters are in almost continuous movement due to tides, currents, and waves.

About 2 percent of the world's moisture is locked up in ice. Most of this is in land ice (glaciers), and a small part is in floating sea ice.

Surface waters contain only a tiny fraction of the world's total moisture supply, primarily in lakes, and to a lesser extent in swamps, marshes, rivers, and streams. Underground water is more widely distributed than surface water, but its availability and quality vary considerably from place to place.

KEY TERMS

aquiclude	marsh	swamp
aquifer	perched water table	tidal bore
artesian well	permafrost	tidal range
capillarity	permeability	waterless zone
cone of depression	piezometric surface	water table
groundwater	porosity	zone of aeration
hydrologic cycle	runoff	zone of confined water
hydrologic zone	salinity	zone of saturation
interstices	subartesian well	
lake	sublimation	

REVIEW QUESTIONS

1. Water has several unique characteristics. Describe three of them.
2. Can water ever move upward, against the pull of gravity? Explain.
3. Explain the role of evaporation in the hydrologic cycle.
4. What is the relationship between transpiration and evaporation?
5. How long does it take an average water molecule to go through the hydrologic cycle?
6. Where is most of the world's fresh water found?
7. How many oceans are there? Why is this a difficult question to answer?
8. Is the Pacific Ocean significantly different from other oceans? Explain.
9. Why does salinity vary in different parts of the world ocean?
10. Which are more important in the hydrologic cycle, lakes or streams? Explain.
11. Distinguish among a lake, a swamp, and a marsh.
12. Explain the concept of water table.
13. Explain the relationship between confined water and groundwater.
14. Distinguish between an artesian well and a subartesian well.

SOME USEFUL REFERENCES

CHORLEY, R. J., *Introduction to Geographical Hydrology*. London: Methuen, 1971.

COUPER, ALASTAIR D., (ed.), *The Times Atlas of the Oceans*. New York: Van Nostrand Reinhold, 1983.

CZAYA,EBERHARD, *Rivers of the World*. New York: Van Nostrand Reinhold, 1984.

FREEZE, R. ALLEN, AND JOHN A. CHERRY, *Groundwater*. Englewood Cliffs, NJ: Prentice Hall, 1979.

LEOPOLD, L. B., *Water: A Primer*. San Francisco: W. H. Freeman, 1974.

OLSEN, R. E., *A Geography of Water*. Dubuque, IA: Wm. C. Brown, 1970.

POSTEL, SANDRA, *Last Oasis: Facing Water Scarcity*. New York: W. W. Norton, 1992.

10

CYCLES AND PATTERNS IN THE BIOSPHERE

O F THE FOUR PRINCIPAL COMPONENTS OF OUR EARTHLY environment, the biosphere has the boundaries that are hardest to pin down. The atmosphere consists of the envelope of air that surrounds the planet, the lithosphere is the solid portion, and the hydrosphere encompasses the various forms of water. These three spheres are distinct from one another and easy to visualize. The biosphere, on the other hand, impinges spatially on the other three. It consists of the incredibly numerous and diverse array of organisms—plants and animals—that populate our planet. Most of these organisms exist at the interface between atmosphere and lithosphere, as Figure 10-1 shows, but some live largely or entirely within the hydrosphere or the lithosphere, and others move relatively freely from one sphere to another.

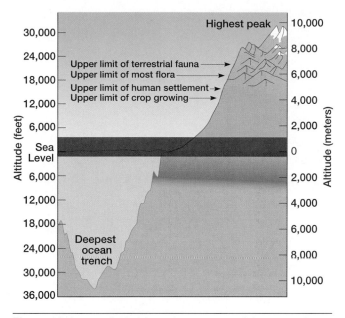

Figure 10-1 A schematic representation of the vertical extent of the biosphere. Most plants and animals live within the zone shaded in red. A few animal species live at the bottom of the ocean, however, and some airborne bacteria, wind-blown invertebrates, and stray birds appear in the air above the highest mountains.

THE IMPACT OF PLANTS AND ANIMALS ON THE LANDSCAPE

Originally, vegetation grew in profusion over most of the land surface of the planet. Today, native plants are still widespread in parts of the world sparsely populated by humans, as Figure 10-2(a) shows, but much of the vegetation in populated areas has been removed, and much that persists has been adulterated and modified by human introduction of crops, weeds, and ornamental plants [Figure 10-2(b)].

Native animal life is much less apparent in the landscape than plant life and often is more conspicuous by sound (especially birds and insects) than by sight. Still, we should keep in mind that wildlife usually is shy and reclusive, and its absence may be more apparent than real. Moreover, most species of animals are tiny and therefore less noticeable in our normal scale of vision.

Both plants and animals interact with other components of the natural landscape (such as soil, landforms, and water) and may be important influences in the development and evolution of these components.

THE GEOGRAPHIC APPROACH TO THE STUDY OF ORGANISMS

Even the simplest living organism is an extraordinarily complex entity. When a student sets out to learn about an organism—whether it be alga or anteater, tulip or turtle—she or he embarks on a complicated quest for knowledge. An organism differs in many ways from other aspects of the environment, but most significantly in that it is alive and in that its survival depends on an enormously intricate set of life processes.

To learn about even a single organism is a truly herculean task. It is much easier to understand a cloud than a chrysanthemum, simpler to comprehend a moraine than a moose, less difficult to perceive a tornado than a tarantula. If our goal as geographers were a complete understanding of the world's organisms, we would need at least the life span of a Methuselah. As beneficial as such knowledge would be, however, a student of geography can be a bit less ambitious. With organisms, as with every other feature of the world, the geographer must focus on certain aspects rather than on a complete comprehension of the whole.

The geographic viewpoint is the viewpoint of broad understanding, whether we are dealing with plants and animals or with anything else. This does not mean that we ignore the individual organism; rather it means that we seek generalizations and patterns and assess their overall significance. Here, as elsewhere, the geographer is interested in distributions and relationships.

Sunlight filtering through a coniferous forest in western Oregon. *(Sanford/Agliolo/The Stock Market)*

(a)

(b)

Figure 10-2 **(a)** Many parts of Earth are still covered with native vegetation, with little or no human impact in evidence. This forested scene is near Marlinton, West Virginia. (TLM photo) **(b)** A small natural pothole of wetland remains, but most of the natural vegetation in this scene has been displaced by crops. This is wheat country in the Canadian province of Alberta. (Bates Littlemales/Earth Scenes)

BIOGEOCHEMICAL CYCLES

The web of life comprises a great variety of organisms coexisting in a diversity of associations. Processes and interactions within the biosphere are exceedingly intricate. Organisms survive only through a bewildering complex of systemic flows of energy, water, and nutrients. These flows are different in different parts of the world, in different seasons of the year, and under various local circumstances.

It is generally believed that, for the last billion years or so, Earth's atmosphere and hydrosphere have been composed of approximately the same balance of chemical components we live with today. This constancy implies a planetwide steady-state condition in which the various chemical elements have been maintained by cyclic passage through the tissues of plants and animals—first absorbed by an organism and then returned to the air/water/soil through decomposition. These grand cycles, which sustain all life on our planet, have continued unperturbed for millennia, at rates and scales almost too vast to conceptualize. In recent years, however, the rapid growth of the human population and the accompanying ever-accelerating rate at which we consume Earth resources have had a deleterious effect on every one of these cycles. None of the damage is yet irreparable, but the threat that such disruption will produce irreversible harm to the biosphere is increasing.

The sun is the basic energy source on which all life ultimately depends. Solar energy can ignite life processes in the biosphere only through **photosynthesis**, the production of organic matter by chlorophyll-containing bacteria and plants.

Only about 0.1 percent of the solar energy that reaches Earth is fixed in photosynthesis. More than half of that total is used immediately in the plant's own respiration, and the remainder is temporarily stored. Eventually this remainder enters a food chain. An increasing proportion of this energy is being diverted to the direct support of one species—*Homo sapiens*.

If the biosphere is to function properly, its components must be recycled continually. In other words, after one organism uses a component, that component must be converted, at the expense of some solar energy, to a reusable form. For some components, this conversion can be accomplished in less than a decade; for others, it may require hundreds of millions of years.

THE FLOW OF ENERGY

Solar energy is of course fundamental to life on Earth. Although readily absorbed by some substances, solar energy is also readily reradiated. Thus it is difficult to store and easy to lose. Happily, most places on Earth receive a daily renewal of the supply.

The biosphere is a temporary recipient of a small fraction of this solar energy, which is fixed (made stable) by green plants via photosynthesis. In the presence of sunlight, a green plant takes carbon dioxide from the air and combines it with water to form the energy-rich carbohydrate compounds we know as sugars, and in the process the energy of the sunlight is locked up as chemical energy in the sugars:

$$CO_2 + H_2O \xrightarrow{\text{light}} \text{carbohydrates} + O_2$$

This chemical energy then flows through the biosphere as animals eat either the photosynthesizing plants or other animals that had eaten those plants earlier (Figure 10-3).

THE HYDROLOGIC CYCLE

The most abundant single substance in the biosphere, by far, is water. It is the medium of life processes and the source of their hydrogen. Most organisms contain considerably more water in their mass than anything else, as Table 10-1 shows. Every living thing depends on keeping its water supply near normal. For example, humans can survive without food for 2 months or more, but they can live without water for only about a week.

All living things require water to carry out their life processes. Watery solutions dissolve nutrients and carry them to all parts of an organism. Through chemical reactions that take place in a watery solution, the organism converts nutrients to energy or to materials it needs to grow or to repair itself. In addition, the organism needs water to carry away waste products.

There are two ways in which water is found in the biosphere: (1) in residence, with its hydrogen chemically bound into plant and animal tissues, and (2) in transit, as part of the transpiration/respiration stream. As we learned in Chapter 9, the movement of water from one sphere to another is called the hydrologic cycle (or sometimes the water cycle).

THE CARBON CYCLE

Carbon is one of the basic elements of life and a part of all living things. The biosphere contains a complex mixture of carbon compounds, more than half a million in total. These compounds are in a continu-

TABLE 10-1	
Water Content in Some Plants and Animals	
Organism	*Percentage water in body mass*
Human	65
Elephant	70
Earthworm	80
Ear of corn	70
Tomato	95

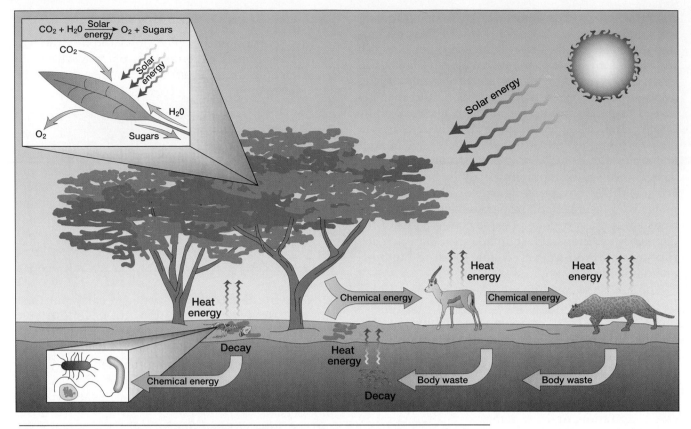

$$CO_2 + H_2O \xrightarrow{\text{Solar energy}} O_2 + \text{Sugars}$$

Figure 10-3 Energy flow in the biosphere. Plants use solar energy in photosynthesis, trapping that energy in the sugar molecules they manufacture. Grazing and browsing animals then acquire that energy when they eat the plants. Other animals eat the grazers/browsers and thereby acquire some of the energy originally in the plants. Body wastes from all the animals, the bodies of the animals once they die, and dead plant matter that has fallen to the ground all return energy to the soil. As all this waste and dead matter decays, it gives off energy in the form of heat. Thus the energy originally produced in nuclear reactions in the sun ends up as random heat energy in the universe.

ous state of creation, transformation, and decomposition.

The main **carbon cycle** is conversion from carbon dioxide to living matter and back to carbon dioxide (Figure 10-4). This conversion is initiated when carbon dioxide from the atmosphere is photosynthesized into carbohydrate compounds (assimilation), as shown in the photosynthesis equation we learned earlier. Some of the carbohydrates are then consumed directly by the plant, and this consumption generates more carbon dioxide to be released through the leaves or roots (plant respiration). The rest of the carbohydrates manufactured in photosynthesis become part of the plant tissue, which is either consumed by animals or eventually decomposed by microorganisms. Plant-eating animals convert some of the consumed carbohydrates back to carbon dioxide and exhale it into the air (animal respiration); the remainder is decomposed by microorganisms after the animal dies. The carbohydrates acted upon by microorganisms are ultimately oxidized

into carbon dioxide and returned to the atmosphere (soil respiration). This is a true cycle, or, more precisely, a complex of interlocking cycles; carbon moves constantly from the inorganic reservoir to the living system and back again. A similar cycle takes place in the ocean.

The carbon cycle operates relatively rapidly (the time measured in years or centuries), and only a small proportion (thought to be less than 1 percent) of the total quantity of carbon on or near Earth's surface is part of the cycle at any given moment. The overwhelming bulk of near-surface carbon has been concentrated over millions of years in geologic deposits—such as coal, petroleum, and carbonate rocks—composed of dead organic matter that accumulated mostly on sea bottoms and was subsequently buried. Carbon from this reservoir is normally incorporated into the cycle very gradually, mostly by normal rock weathering. In the last century and a half, however, humans have added consid-

Figure 10-4 The carbon cycle. Carbon from the carbon dioxide in the atmosphere is used by plants to make the carbon-containing sugars formed during photosynthesis. Through various paths, these compounds eventually are again converted to carbon dioxide and returned to the atmosphere.

erable carbon dioxide to the atmosphere by extracting and burning fossil fuels (coal, oil, gas) containing carbon fixed by photosynthesis many millions of years ago. This rapid acceleration of the rate at which carbon is freed and converted to carbon dioxide is likely to have far-reaching effects on the biosphere.

THE OXYGEN CYCLE

Oxygen is a building block in most organic molecules and consequently makes up a significant proportion of the atoms in living matter.

Although our atmosphere is now rich in oxygen, it was not always so. Earth's earliest atmosphere was oxygen poor; indeed, in the early days of life on this planet, about 3.4 billion years ago, oxygen was poisonous to living cells. Evolving life had to develop mechanisms to neutralize or, better still, exploit its poisonous presence. This exploitation was so successful that most life now cannot function without oxygen.

The oxygen now in the atmosphere is largely a by-product of vegetable life, as the equation for photosynthesis shows. Thus, once life could sustain itself in the presence of high amounts of oxygen in the air, primitive plants made possible the evolution of higher plants and animals by providing molecular oxygen for their metabolism.

The **oxygen cycle** (Figure 10-5) is extremely complicated and is summarized only briefly here. Oxygen occurs in many chemical forms and is released into the atmosphere in a variety of ways. Most of the oxygen in the atmosphere is molecular oxygen produced when plants decompose water molecules in photosynthesis. Some atmospheric oxygen is bound up in water molecules that came from evaporation or plant transpiration, and some is bound up in the carbon dioxide released during animal respiration. Much of this carbon dioxide and water is eventually recycled through the biosphere via photosynthesis.

Other sources of oxygen for the oxygen cycle include atmospheric ozone, oxygen involved in the ox-

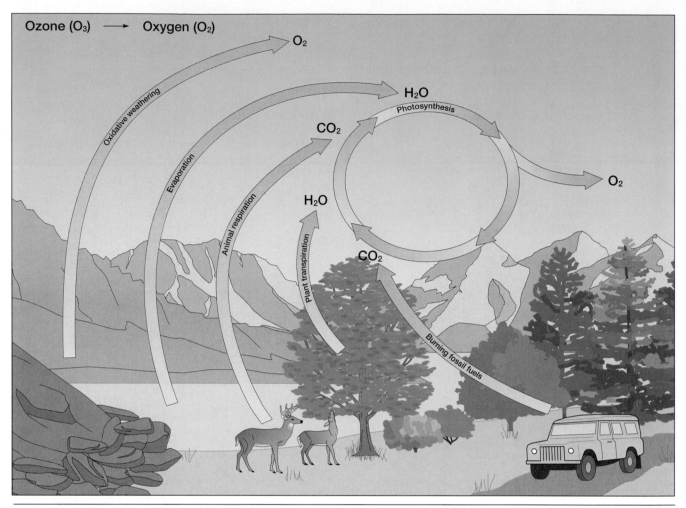

Figure 10-5   The oxygen cycle. Molecular oxygen is essential for almost all forms of life. It is made available to the air through a variety of processes and is recycled in a variety of ways.

idative weathering of rocks, oxygen stored in and sometimes released from carbonate rocks, and various other processes, including some (such as the burning of fossil fuels) that are human-induced.

THE NITROGEN CYCLE

Although nitrogen gas is an apparently inexhaustible component of the atmosphere (air is about 78 percent N_2), only certain species of soil bacteria and blue-green algae can use this essential nutrient in its gaseous form. For the vast majority of living organisms, atmospheric nitrogen is usable only after it has been converted to nitrogen compounds (nitrates) that can be used by plants (Figure 10-6). This conversion process is called **nitrogen fixation**, and the overall process is called the **nitrogen cycle**. Some nitrogen is fixed in the atmosphere by lightning and cosmic radiation, and some is fixed in the ocean by marine organisms, but the amount involved in these processes is minimal. It is

nitrogen fixation by soil microorganisms and associated plant roots that provides most of the usable nitrogen for Earth's biosphere.

Once atmospheric nitrogen has been fixed into an available form (nitrates), it is assimilated by green plants, some of which are eaten by animals. The animals then excrete nitrogenous wastes in their urine. These wastes, as well as the dead animal and plant material, are attacked by bacteria, and nitrite compounds are released as a further waste product. Other bacteria convert the nitrites to nitrates, making them available again to green plants. Still other bacteria convert some of the nitrates to nitrogen gas (N_2) in a process called **denitrification**, and the gas becomes part of the atmosphere. This atmospheric nitrogen is then carried by rain back to Earth, where it enters the soil/plant portion of the cycle once more.

Human activities have produced a major modification in the natural nitrogen cycle. The synthetic manufacture of nitrogenous fertilizers and widespread introduction of nitrogen-fixing crops (such as alfalfa, clover, and soybeans) have significantly changed the balance between fixation and denitrification. The short-term result has been an excessive accumulation of nitrogen compounds in many lakes and streams. This buildup of nitrogen depletes the oxygen supply of the water and upsets the natural balance; the long-term results are still not known.

OTHER MINERAL CYCLES

Although carbon, oxygen, and nitrogen—along with hydrogen—are the principal chemical components of the biosphere, many other minerals are critical nutrients for plants and animals. Most notable among these *trace minerals* are phosphorus, sulfur, and calcium, but more than a dozen others are occasionally significant.

Some nutrients are cycled along gaseous pathways, which primarily involves an interchange between biota and the atmosphere/ocean environment, as we just saw in the carbon, oxygen, and nitrogen cycles. Other nutrients follow sedimentary pathways, which involve interchange between biota and the Earth/ocean environment (Figure 10-7). Elements with sedimentary cycles include calcium, phosphorus, sulfur, copper, and zinc.

In a typical sedimentary cycle, the element is weathered from bedrock into the soil. Some of it is then washed downslope with surface runoff or percolated into the groundwater supply. Much of it reaches the ocean, where it may be deposited in the next round of sedimentary rock formation. Some, however, is ingested by aquatic organisms and later released into the cycle again through waste products and dead organisms.

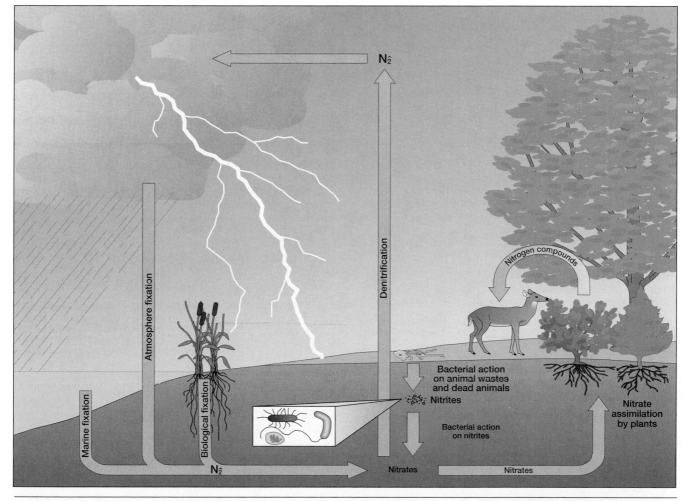

Figure 10-6   The nitrogen cycle. Atmospheric nitrogen is fixed into nitrates in various ways, and the nitrates are then assimilated by green plants, some of which are eaten by animals. Dead plant and animal materials, as well as animal wastes, contain various nitrogen compounds, and these compounds are acted on by bacteria so that nitrites are produced. The nitrites are then converted by other bacteria to nitrates, and thus the cycle continues. Still other bacteria denitrify some of the nitrates, releasing free nitrogen into the air again.

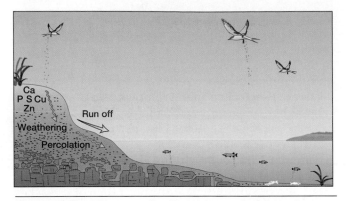

Figure 10-7   The general sedimentary cycle for nutrients. Elements are weathered from bedrock and then carried either into the soil by percolation or into the ocean by runoff. Some of the elements weathered in this way are quickly returned to the cycle by organisms; some take the much slower route via sedimentary rock formation.

The general conclusion is that the amounts of biotic nutrients available on Earth are finite. These nutrients move over and over through cycles that are extremely variable from place to place, and some of the cycles are either damaged or modified by human interference.

FOOD CHAINS

The unending flows of energy, water, and nutrients through the biosphere are channeled in significant part by direct passage from one organism to another in pathways referred to as **food chains**. A food chain is a simple concept, as Figure 10-8 shows: organism A is eaten by organism B, which thereby absorbs A's energy and nutrients; organism B is eaten by organism C, with similar results; organism C is eaten by D; and so on and on.

In nature, however, the matter of who eats whom may be extraordinarily complex, with a bewildering

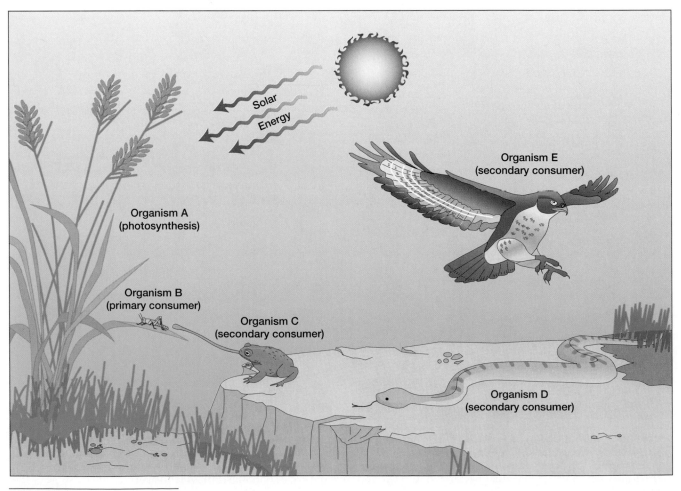

Figure 10-8   A food chain.

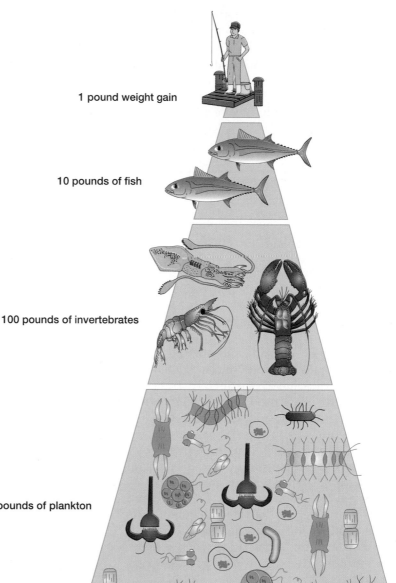

1 pound weight gain

10 pounds of fish

100 pounds of invertebrates

1000 pounds of plankton

Figure 10-9 A food pyramid. It takes half a ton (500 kilograms) of plankton (which are microscopic marine plants and animals) to provide a 1-pound (0.5-kilogram) weight gain for a human. (Adapted from Life Nature Library/ *Ecology,* drawing by Otto van Eersel. New York: Time-Life Books, 1963, p. 37.)

number of interlaced strands. Therefore "chain" probably is a misleading word in this context because it implies an orderly linkage of equivalent units. It is more accurate to think of each link as an energy transformer that ingests all the energy of the preceding link, uses some of that energy for its own sustenance, and then passes the balance on to the next link.

The fundamental unit in any food chain is the plants that trap solar energy through photosynthesis. The plants are then eaten by herbivorous (plant-eating; *herba* is Latin for plant; *vorare,* to devour) animals (also called herbivores), which are referred to as primary consumers. The herbivores then become food for other animals, carnivores (*carne,* is Latin for meat), which are referred to as secondary consumers or predators. There may be many levels of secondary consumers in a food chain, as Figure 10-8 shows.

A food chain can also be conceptualized as a **food pyramid** because the number of energy-trapping organisms is much, much larger than the number of primary consumers, the number of primary consumers is larger than the number of secondary consumers, and so on up the pyramid (Figure 10-9). There are usually several levels of carnivorous secondary consumers, each succeeding level consisting of fewer and usually larger animals. The final consumers at the top of the pyramid are usually the largest and most powerful predators in the area (Figure 10-10).

The consumers at the apex of the pyramid do not constitute the final link in the food chain, however. When they die, they are fed upon by tiny (mostly microscopic) organisms that function as decomposers, returning the nutrients to the soil to be recycled into yet another food pyramid.

THE SEARCH FOR A MEANINGFUL CLASSIFICATION SCHEME

When a geographer studies any set of phenomena, he or she attempts to group individual members of the set in some meaningful fashion. In some cases, the geographer borrows classification schemes from specialists in other disciplines; often, however, schemes devised for other purposes are not particularly useful for geographic studies, and the geographer must develop different ones.

The systematic study of plants and animals is primarily the domain of the biologist, and many biological classification schemes have been devised. By far the most significant and widely used is the binomial (two names) system originally developed by the Swedish botanist Carolus Linnaeus in the eighteenth century. This system focuses primarily on the morphology (structure and form) of organisms and groups them on the basis of structural similarity. The Linnaean scheme is generally useful for geographers, but it has certain shortcomings that preclude its total acceptance. Its principal disadvantage for geographic use is that it is based entirely on anatomic similarities, whereas geographers are more interested in distribution patterns and habitat preferences.

It would be nice to be able to say that geographers have come up with a more appropriate classification scheme, but such is not the case, nor is a universally accepted geographic classification of organisms ever likely to be developed. Too many subjective decisions would have to be made, making widespread agreement on any scheme very unlikely.

Seeking Pertinent Patterns

Among the life forms of our planet are perhaps 600,000 species of plants and more than twice that many species of animals. With such an overwhelming diversity of organisms, how can we study their distributions and relationships in any meaningful manner? A logical approach is to decide on some generalizing procedures and useful groupings and then consider the patterns that emerge.

The term **biota** refers to the total complex of plant and animal life. The basic subdivision of biota separates **flora,** or plants, from **fauna,** or animals. In this book we recognize a further fundamental distinction—between oceanic biota and terrestrial (living on land; *terrenus* is Latin for earth) biota.

The inhabitants of the oceans are generally divided into three groups—*plankton* (floating plants and animals), *nekton* (animals, such as fish and marine mammals, that swim freely), and *benthos* (animals and plants that live on or in the ocean bottom). Although these marine life forms are fascinating, and despite the fact that 70 percent of Earth's surface is oceanic, in this book we pay scant attention to oceanic biota, primarily because of constraints of time and space. The terrestrial biota, much more diverse than its oceanic counterpart, is the focus of our interest in the remainder of our study of the biosphere.

ECOSYSTEMS AND BIOMES

In our search for organizing principles that help us comprehend the biosphere, two concepts are of particular value—ecosystem and biome.

Ecosystem: A Concept for All Scales

The term **ecosystem** is a contraction of the phrase *ecological system*. An ecosystem includes all the organisms in a given area, but it is more than simply a community of plants and animals existing together. The ecosystem concept encompasses the totality of interactions among the organisms and between the organisms and the nonliving portion of the environment in the area under consideration. The nonliving portion of the environment includes soil, rocks, water, sunlight, and atmosphere, but it essentially can be considered as nutrients and energy.

An ecosystem, then, is fundamentally an association of plants and animals along with the surrounding nonliving environment and all the interactions in which the organisms take part. The concept is built around the flow of energy among the various components of the ecosystem, which is the essential determinant of how a biological community functions (Figure 10-11).

This functional, ecosystemic concept is very attractive as an organizing principle for the geographic study of the biosphere. It must be approached with caution, however, because of the various scales at which it can be applied. There is an almost infinite variety in the magnitude of ecosystems we might study. At one extreme of scale, for example, we can conceive

Figure 10-10 A lynx pouncing on a snowshoe hare in a Montana forest. (Alan Carey/Photo Researchers, Inc.)

of a global ecosystem that encompasses the entire biosphere; at the other end of the scale might be the ecosystem of a fallen log or of the underside of a rock or even of a drop of water.

If we are going to identify and understand broad distributional patterns in the biosphere, we must focus only on ecosystems that can be recognized at a useful scale.

BIOME: A SCALE FOR ALL BIOGEOGRAPHERS

Among terrestrial ecosystems, the type that provides the most appropriate scale for understanding world distribution patterns is called a **biome,** defined as any large, recognizable assemblage of plants and animals in functional interaction with its environment. A biome is usually identified and named on the basis of its dominant vegetation, which normally constitutes the bulk of the biomass (the total weight of all organisms—plant and animal) in the biome, as well as being the most obvious and conspicuous visible component of the landscape.

There is no universally recognized classification system of the world's biomes, but scholars commonly accept 11 major types (discussed in greater detail in Chapter 11):

Tropical rainforest
Tropical deciduous forest
Tropical scrub
Tropical savanna
Desert
Mediterranean woodland and shrub

Midlatitude grassland
Midlatitude deciduous forest
Boreal forest
Tundra
Mountains and ice caps

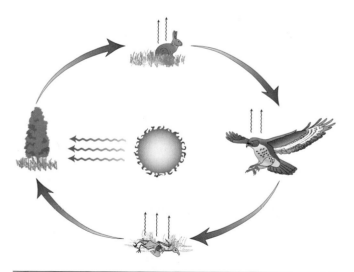

Figure 10-11 The flow of energy in a simple ecosystem. Energy from the sun is trapped by the grass during photosynthesis. The grass then is eaten by a rabbit, which is eaten by a hawk, which then dies. The energy originally contained in the photosynthetic products made by the grass passes through stages during which it is bound up in the body molecules of the rabbit and the hawk but ultimately becomes heat energy lost from the live animals and from the decaying dead matter.

A biome comprises much more than merely the plant association that gives it its name. A variety of other kinds of vegetation usually grows among, under, and occasionally over the dominant plants. Diverse animal species also occupy the area. Often, significant and even predictable relationships exist between the biota (particularly the flora) of a biome and the associated climate and soil types.

On any map showing the major biome types of the world, the regional boundaries are somewhat arbitrary. Biomes do not occupy sharply defined areas in nature, no matter how sharp the demarcations may appear on a map. Normally the communities merge more or less imperceptibly with one another through **ecotones**—transition zones of competition in which the typical species of one biome intermingle with those of another (Figure 10-12).

ENVIRONMENTAL RELATIONSHIPS

The survival of plants and animals depends on an intimate and sometimes precarious set of relationships with other elements of the environment. The details of these relationships vary with different species, but we can generalize about many of them. They can be discussed at various scales of generalization. However, you must remember that a generalization that is true at one scale may be quite invalid at another. For example, if we are considering global or continental patterns of biotic distribution, we are concerned primarily with gross generalizations that deal with average conditions, seasonal characteristics, latitudinal extent, zonal winds, and other broadscale factors. If our interest is instead a small area, such as an individual valley or a single hillside, we are more concerned with such localized environmental factors as degree of slope, direction of exposure, and permeability of topsoil.

Whatever the scale, there are nearly always exceptions to the generalizations, and the smaller the scale, the more numerous the exceptions. (Remember that a small-scale map shows a relatively large portion of Earth's surface, and a large-scale map shows a relatively small portion.) Thus in a region that is generally humid there are probably many localized sites that are extremely dry, such as cliffs or sand dunes. And in even a very dry desert there are likely to be several places that are always damp, such as an oasis or a spring.

Throughout the following discussion of environmental relationships, keep in mind that both *intraspecific* competition (among members of the same species) and *interspecific* competition (among members of different species) are at work. Both plants and animals compete with one another as they seek light, water, nutrients, and shelter in a dynamic environment.

THE INFLUENCE OF CLIMATE

At almost any scale, the most prominent environmental constraints on biota are exerted by various climatic factors.

Light   No plants can survive without light. (It is essentially for this reason that vegetation is absent from deeper ocean areas, where light does not penetrate.) We have already discussed the basic process—photosynthesis—whereby plants produce stored chemical energy; this process is activated by light.

Light can have a significant effect on plant shape, as Figure 10-13 shows. In places where the amount of light is restricted, such as in a dense forest, trees are likely to be very tall but have limited lateral growth. In areas that have less-dense vegetation, more light is available, and as a result trees are likely to be expansive in lateral spread but truncated vertically.

Another important light relationship involves how much light an organism receives during any 24-hour period. This relationship is called **photoperiodism**. Except in the immediate vicinity of the equator, the sea-

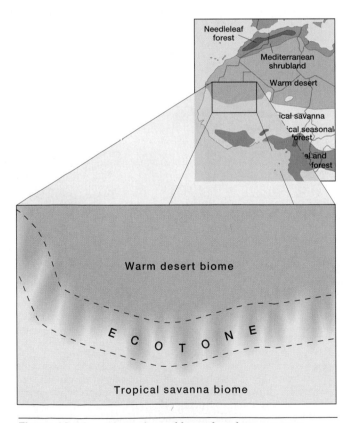

Figure 10-12   A hypothetical boundary between two biomes of the world. The irregular boundary between the two shows much interfingering, or interdigitation. This transition zone is called an ecotone.

Figure 10-13 In an open stand (left), there is abundant light all around the tree and it responds by broad lateral growth rather than vertical growth. Under crowded conditions (right), less light reaches the tree and it elongates upward rather than spreading laterally.

sonal variation in the photoperiod becomes greater with increasing latitude. Fluctuation in the photoperiod stimulates seasonal behavior—such as flowering, leaf fall, mating, and migration—in both plants and animals.

Moisture The broad distribution patterns of the biota are governed more significantly by the availability of moisture than by any other single environmental factor. A prominent trend throughout biotic evolution has been the adaptation of plants and animals to either excesses or deficiencies in moisture availability (Figure 10-14). The availability of water is not wholly dependent on climate—precipitation versus evaporation—but it is very largely determined by these atmospheric dynamics.

Temperature The temperature of the air and the soil is also important to biotic distribution patterns. Fewer species of both plants and animals can survive in cold regions than in areas of more moderate temperatures. Plants, in particular, have a limited tolerance for low temperatures because they are continuously exposed to the weather, and they experience tissue damage and other physical disruption when their cellular water freezes. Animals in some instances are able to avoid the bitterest cold by moving around to seek shelter. Even so, the cold-weather areas of high latitudes and high elevations have a limited variety of animals and plants.

Wind The influence of wind on biotic distributions is more limited than that of the other climatic factors. Where winds are persistent, however, they often serve as a constraint.

The principal negative effect of wind is that it causes excessive drying by increasing evaporation from exposed surfaces, thus causing a moisture deficiency (Figure 10-15). In cold regions wind escalates the rate at which animals lose body heat.

The sheer physical force of wind can also be influential: a strong wind can uproot trees, modify plant forms, and increase the heat intensity of wildfires. On the positive side, wind sometimes aids in the dispersal of biota by carrying pollen, seeds, lightweight organisms, and flying creatures.

Figure 10-14 Even the most stressful environments often contain distinctive and conspicuous plants. These large cacti—organ pipe on the left and saguaro on the right—prosper in the parched Sonoran desert along the Arizona–Mexico border. (TLM photo)

PEOPLE AND THE ENVIRONMENT

WILDFIRES IN YELLOWSTONE

During the summer of 1988, Yellowstone National Park experienced the most devastating wildfires within human memory. Beginning in late June, 13 major fires burned out of control within a burn perimeter that covered nearly half of the park's 2.2 million acres (880,000 hectares). On a single day (August 20), more area was torched than had burned in any previous decade since the park was created in 1872.

The park's fire-management policy, in place since 1976, decreed that human-caused fires were to be suppressed but natural fires were to be left alone unless they threatened human life, private property, or significant natural resources. The theory behind this let-burn policy is that the ecosystem contained in any large national park should be allowed to function naturally and remain as pristine as possible, both in appearance and in process. Natural processes—drought, flood, insect infestation, wildfire, and so forth—should not be interfered with. During the 11 years prior to 1988 that this policy had been in effect, 235 nature-caused fires had been allowed to burn, with the largest destroying only 7400 acres (3000 hectares).

However, the summer of 1988 was unique: rainless, hot, and windy. Regardless of the fire-management policy, Yellowstone was ripe for conflagration in 1988. Eight of the 13 major fires were caused by humans, and attempts to suppress these were made from the outset. The other five were not fought until late July, at which time an all-out effort at suppression was begun. At maximum deployment, nearly 10,000 firefighters were involved in suppression, at a cost of $120 million (ten times Yellowstone's annual budget).

This prodigious effort was largely ineffectual. The horrendous weather conditions produced firestorms that were totally uncontrollable. About 1 million acres (400,000 hectares) was affected by the fires, and it is thought that firefighters "saved" less than 2000 acres (800 hectares), although some buildings (most notably Old Faithful Inn) were spared because of the suppression efforts.

Indeed, suppression often caused more harm than benefit. A burned forest may return to its original condition in 100 years, but a bulldozer scar made in creating a fireline may persist for ten times that long. In any case, on September 11, a quarter of an inch of rain and snow fell and accomplished what the firefighters could not: the fires were quenched. (Figure 10-1-A).

The Yellowstone fires received massive media attention. The American public mourned the "devastation" of its most famous national park. The National Park Service was excoriated for its "shortsighted" fire management policy, and political repercussions were widespread.

Certainly, the conflagration produced negative results. Large expanses of forest and grassland were killed, and thousands of small animals un-

Figure 10-1-A Firefighter mopping up some of the last remnants of Yellowstone's fires in 1988. (TLM photo)

doubtedly died. However, the body count of large animals was fewer than 400, mostly elk (of which there are more than 30,000 in the park). This is not a significantly larger total than the number of road-killed animals in Yellowstone in an average summer. Tourism dropped about 10 percent in July, 30 percent in August, and 75 percent during the first 3 weeks of September. Most important locations were briefly evacuated, and the entire park was closed to visitors for only 2 days.

On balance, however, the positive results of the fires far outweigh the negative ones. Habitat diversity will be greatly enhanced. Much of Yellowstone was covered only with an old-growth lodgepole pine forest, but one result of the massive, uneven burning of 1988 will be a mosaic of vegetation that has been missing for a long time.

The fires are an overall boon to the ecosystem. Many of Yellowstone's coniferous trees reproduce by means of cones that require great heat before they can open and drop their seeds.

Fires are necessary to recycle bound-up nutrients; without them, the forest would slowly starve. Thus the decomposition of vast quantities of organic matter has given the ecosystem a gigantic nutrient fix that will greatly boost primary productivi-

ty. Since the fires, this nourishment has been particularly manifested by a vigorous growth of grasses, herbs, and shrubs (Figure 10-1-B). The vegetative growth will support large populations of arthropods and rodents, which will invigorate the entire food web.

For a few years, or even decades, tourists will see a considerable amount of burned forest. In the long run, however, the Yellowstone ecosystem will be much healthier for this ordeal by fire. And even the tourist businesses that lost money in 1988 have more than made it back as visitors flocked to Yellowstone as never before, to see the effects of the fires for themselves. Indeed, October 1988 was the best October in tourist industry history, and total visitation has set new records every year since 1988.

Debate continues over fire-management policy, although two inde-

pendent teams of scientists and land managers appointed to review past strategy and recommend changes agreed that some sort of let-burn policy is appropriate in a large wilderness park. By 1992 the new policy was in place: it is similar to the pre-1988 policy of nonsuppression for "prescribed" fires (those that meet certain stipulated criteria).

Many lessons are to be learned from Yellowstone's trial by fire. Perhaps the most important is the necessity of balanced analysis. The natural and significant role of fire in a primeval ecosystem is incontrovertible. Where human activities and human values are involved, however, objectivity fades and difficult decisions must be made. Contingency plans are necessary, snap judgments are to be avoided, and a long-range viewpoint is imperative.

Figure 10-1-B This photograph was taken from the same spot as Figure 10-1-A, some 2 years later. The vegetation of the forest floor has regrown thoroughly. It will take much longer for the trees to regenerate. (TLM photo)

Figure 10-15   Some plants are remarkably adaptable to environmental stress. In this timberline scene from north-central Colorado, there is no question about the prevailing wind direction. Persistent wind from the right has so desiccated this subalpine fir that branches are able to survive only by growing toward the left. This preferrred growth direction places the trunk of the tree between them and the wind. (TLM photo.)

Edaphic Influences

Soil characteristics, known as **edaphic factors** (from *edaphos*, Greek for ground or soil), also influence biotic distributions. These factors are direct and immediate in their effect on flora but usually indirect in their effect on fauna. Soil is a major component of the habitat of any vegetation, of course, and its characteristics significantly affect rooting capabilities and nutrient supply. Especially significant are soil texture, soil structure, humus content, chemical composition, and the relative abundance of soil organisms.

Topographic Influences

In global biome patterns, general topographic characteristics are the most important factor affecting distribution. For example, the assemblage of plants and animals in a plains region is very different from that in a mountainous region. At a more localized scale, the factors of slope and drainage are likely to be significant, primarily the steepness of the slope, its orientation with regard to sunlight, and the porosity of the soil on the slope.

Wildfire

Most environmental factors that affect the distribution of plants and animals are passive, and their influences are slow and gradual. Occasionally, however, abrupt and catastrophic events, such as floods, earthquakes, volcanic eruptions, landslides, insect infestations, and droughts, also play a significant role. By far the most important of these is wildfire (Figure 10-16). In almost all portions of the continents, except for the always-wet regions where fire simply cannot start and the always-dry regions where there is an insufficiency of combustible vegetation, uncontrolled natural fires have occurred with surprising frequency (Figure 10-17). Fires generally result in complete or partial devastation of the flora and the killing or driving away of all or most of the fauna. These results, of course, are only temporary; sooner or later, vegetation sprouts and animals return. At least in the short run, however, the composition of the biota is changed, and if the fires occur with sufficient frequency the change may be more than temporary.

Wildfire can be very helpful to the seeding or sprouting of certain plants and the maintenance of certain plant interactions. In some cases, grasslands are sustained by relatively frequent natural fires, which inhibit the encroachment of tree seedlings. Moreover, many plant species, particularly certain trees, such as the California redwood and the southern yellow pine, scatter their seeds only after the heat of a fire has caused the cones or other types of seedpods to open.

Predictable Correlations

There are recognizable and, in many cases, predictable patterns of biotic distributions, based on environmental relationships. Although these patterns exist at the species level, they are much more apparent at higher levels of organization—groups, communities, and particularly biomes.

In terms of broad distribution patterns, climate and vegetation have a particularly close correlation. To illustrate this principle at its most fundamental level, we might consider the distribution of forests over the world. Where, under natural conditions, do forests grow? There are many facets to the answer to this question, but by far the most important is the availability of moisture.

Figure 10-16   Wildfires are commonplace in many parts of the world. This ground fire in the "Top End" of the Northern Territory of Australia is almost an annual occurrence under natural conditions. (TLM photo)

A relative abundance of moisture during the growing season usually implies that trees flourish and that a forest is likely to be the dominant plant association. Trees depend so much on the availability of moisture primarily because, unlike other plants, they must have a mechanism for transporting mineral nutrients a relatively great distance from the place of acquisition (roots) to the place of need (leaves). Such transport can take place only in a dilute solution; therefore much water is needed by trees throughout the growing season.

Other plant forms can flourish in areas of relatively high precipitation, but they rarely become dominant because they are shaded out by trees. The broad generalization, then, is that trees in particular and forests in general are usually found wherever there is a relative abundance of precipitation. Low-growing plants dominate the landscape only in areas where trees cannot flourish, which means mostly in regions of relatively sparse precipitation.

The relationship between climate and faunal distributions is less pervasive and obvious, but it is still important. If we examine the global distribution of fur-bearing animals, for example, we find that climatic correlations do exist but are not so distinct as their floral counterparts. There are several hundred species of mammals whose skin is covered with a fine, soft, thick, hairy coat referred to as fur. These mammals range in size from tiny mice and moles to the largest bears. An examination of their ranges reveals three generalized habitat preferences:

1. Many fur-bearing species live in high-latitude and/or high-elevation locations, where winters are long and cold.
2. A number of fur-bearing species live in aquatic environments.
3. The remainder are widely scattered over the continents, occupying a considerable diversity of habitats.

From a quick look at the evidence, we can conclude tentatively that many fur-bearing species, including all those with the heaviest, thickest fur, live in regions where cold temperatures are common. The climatic correlation in this case is partial and indistinct.

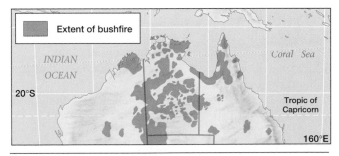

Figure 10-17   The widespread occurrence of bushfires is demonstrated by this map of the northern half of Australia, which shows the extent of areas burned in the 1974/75 fire season. (After R. H. Luke and A. G. McArthur, *Bushfires in Australia.* Canberra: Australia Government Publishing Service, 1978.)

CHAPTER SUMMARY

The biosphere consists of all plant and animal life forms on Earth. It overlaps a great deal with the other three environmental spheres.

All components of the web of life depend upon energy from the sun, water, and nutrients. These three ingredients are continuously being cycled through the biosphere as they are absorbed by organisms for sustenance and then returned to the atmosphere/hydrosphere/lithosphere by decomposition. The most prominent of these biogeochemical cycles involve water, carbon, oxygen, and nitrogen.

Floral/faunal relationships can be described by a food chain or a food pyramid. Plants, the first link in the chain and the bottom of the pyramid, convert solar energy to the chemical energy locked up in sugars produced during photosynthesis. Herbivores, animals that eat the plants, are called primary consumers in the chain/pyramid, and carnivores, animals that eat the herbivores, are called secondary consumers.

Because the combined number of plant and animal species approaches 2 million, the biosphere is a daunting challenge for analysis. Useful organizing concepts for geographic study include the ecosystem (the totality of interactions among organisms and their environment in a given area) and the biome (a major assemblage of biota in functional interaction with its environment).

Climate influences vegetation distributions more than any other environmental factor. Climate is also of considerable significance to faunal distribution, but the most important factor in faunal distribution is the vegetation pattern. Other environmental elements are also influential in biotic distributions, but they are much less important than climate.

KEY TERMS

biome	edaphic factor	nitrogen fixation
biota	fauna	oxygen cycle
carbon cycle	flora	photoperiodism
denitrification	food chain	photosynthesis
ecosystem	food pyramid	
ecotone	nitrogen cycle	

REVIEW QUESTIONS

1. What is the importance of photosynthesis to the flow of energy through the biosphere?
2. Describe the basic steps in the carbon cycle.
3. Most of the carbon at or near Earth's surface is not involved in any short-term cycling. Explain.
4. Explain the differences between nitrogen fixation and denitrification.
5. What is the relationship between a food chain and a food pyramid?
6. Explain the difference between ecosystem and ecotone, between biome and biomass, and between ecosystem and biome.
7. Explain how both photosynthesis and photoperiodism are dependent on sunlight.
8. What are the beneficial effects of wildfire?
9. Why do trees require so much more moisture to survive than grass?

SOME USEFUL REFERENCES

BRIGGS, JOHN C., *Centres of Origin in Biogeography*. Leeds, England: Biogeography Study Group, 1984.

HENGEVELD, R., *Dynamic Biogeography*. New York: Cambridge University Press, 1990.

JEFFREY, C., *Biological Nomenclature*. London: Edward Arnold, 1989.

JONES, G. E., *The Conservation of Ecosystems and Species*. New York: Croom Helm, 1987.

LIETH, HELMUT, *Patterns of Primary Production in the Biosphere*. New York: Hutchinson Ross, 1978.

MACARTHUR, R. H., *Geographical Ecology: Patterns in the Distribution of Species*. Princeton, NJ: Princeton University Press, 1984.

MIEKLE, H. W., *Patterns of Life: Biogeography in a Changing World*. Boston: Unwin Hyman, 1989.

MYERS, A. A., AND P. GILLER, (eds.), *Analytical Biogeography*. New York: Chapman & Hall, 1988.

SIMMONS, I. G., *Biogeographical Processes*. Winchester, MA: Allen & Unwin, 1983.

11

TERRESTRIAL FLORA AND FAUNA

THE MOST BASIC STUDIES OF ORGANISMS MADE BY GEO-graphers are usually concerned with distribution, and distribution therefore is our main focus in this chapter. Here we are concerned with such questions as "What is the range of a certain species or group of plants/animals?" "What are the reasons behind this distribution pattern?" and "What is the significance of the distribution?"

NATURAL DISTRIBUTIONS

At the most fundamental level, the natural distribution of any species or group of organisms is determined by four conditions: evolutionary development, migration/dispersal, reproductive success, and extinction.

EVOLUTIONARY DEVELOPMENT

The Darwinian theory of natural selection, sometimes referred to as "survival of the fittest," explains the origin of any species as a normal process of descent, with modification, from parent forms. The progeny best adapted to the struggle for existence survive, whereas those less well adapted perish. This long, slow, and essentially endless process accounts for the development of all organisms.

To understand the distribution of any species or genus, therefore, we begin with a consideration of where it evolved. In some cases, there was a very localized beginning for a genus; in other cases, similar evolutionary development at several scattered localities led to the same genus. Unfortunately, the evidence on which case applies in a given situation is not always clear. As an example of these two extremes, consider the contrast in apparent origin of two important groups of plants—acacias and eucalypts (Figures 11-1 and 11-2). Acacias are an extensive genus of shrubs and low-growing trees represented by numerous species found in low-latitude portions of every continent that extends into the tropics or subtropics. Eucalypts, on the other hand, are a genus of trees native only to Australia and a few adjacent islands. Acacias apparently evolved in a number of different localities, but all evidence points to eucalypts originating only on a single small continent.

MIGRATION/DISPERSAL

Throughout the millennia of Earth's history, organisms have always moved from one place to another. Animals possess active mechanisms for locomotion—legs, wings, fins, and so on—and their possibilities for migration are obvious. Plants are also mobile, however. Although most individual plants become rooted and therefore fixed in location for most of their life, there is much opportunity for passive migration, particularly in the seed stage. Wind, water, and animals are the principal natural mechanisms of seed dispersal.

The contemporary distribution pattern of many organisms is often the result of natural migration or dispersal from an original center(s) of development. Among thousands of examples that could be used to illustrate this process are the following:

1. The cattle egret (*Bubulcus ibis*) apparently originated in southern Asia but during the last few centuries has spread to other warm areas of the world, particularly Africa (Figure 11-3). In recent decades, a change in land use in South America has caused a dramatic expansion in the cattle egret's range. At least as early as the nineteenth century, some cattle egrets crossed the Atlantic from West Africa to Brazil, but they were unable to find suitable ecological conditions and thus did not become established. The twentieth-century introduction of extensive cattle raising in tropical South America apparently provided the missing ingredient, and egrets quickly adapted to the newly suitable habitat. Their descendants spread northward throughout the subtropics and are now common inhabitants of the Gulf coastal plain in the southeastern United States, are well established in California, and occur as far north as southeastern Canada. Also within this century, cattle egrets have dispersed at the other end of their "normal" range to enter northern Australia and spread across that continent.

2. The coconut palm (*Cocos nucifera*) is believed to have originated in southeastern Asia and adjacent Melanesian islands. It is now extraordinarily widespread along the coasts of tropical continents and islands all over the world. Most of this dispersal apparently has come about because coconuts, the large hard-shelled seeds of the plant, can float in the ocean for months or years without losing their fertility. Thus they wash up on beaches throughout the world and colonize successfully if environmental conditions are right (Figure 11-4). This natural dispersion was significantly augmented by human help in the Atlantic region, particularly by the deliberate transport of coconuts from the Indian and Pacific ocean areas to the West Indies.

Caribou swimming the Porcupine River in northeastern Alaska. (*Michael Male/Photo Researchers, Inc.*)

Figure 11-1 Hundreds of species of acacias grow in semiarid and subhumid portions of the tropics. This scene from central Kenya shows acacias in tree form, although lower shrub forms are more common. (TLM photo)

REPRODUCTIVE SUCCESS

A key factor in the continued survival of any biotic population is reproductive success. Poor reproductive success can come about for a number of reasons—heavy predation (a fox eats quail eggs from a nest), climatic change (heavy-furred animals perishing in a climate that once was cold but has warmed up for some reason or other), failure of food supply (a string of unusually cold winters keeps plants from setting seed), and on and on.

Changing environmental conditions are also likely to favor one group over another, as when warming waters on the fringes of the Arctic Ocean allowed cods to expand their range at the expense of several other types of fishes. Thus reproductive success is usually the limiting factor that allows one competing population to flourish while another languishes (Figure 11-5).

EXTINCTION

The range of a species can be diminished by the dying out of some or all of the population. The history of the biosphere is replete with examples of such range diminution, varying from minor adjustments in a

Figure 11-2 The original forests of Australia were composed almost entirely of species of eucalyptus. This scene is in eastern New South Wales, near Eden. (TLM photo)

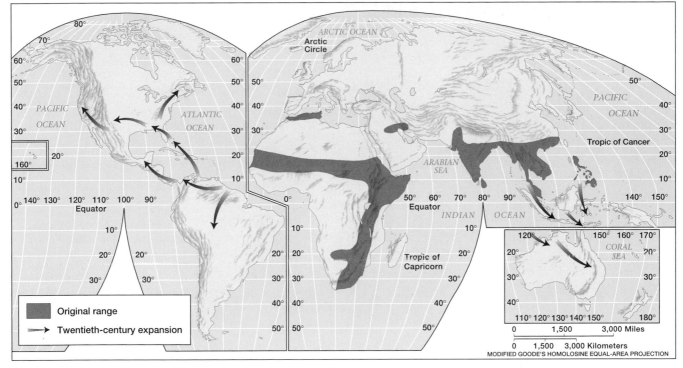

(a)

(b)

Figure 11-3 Natural dispersal of an organism. **(a)** During the twentieth century, cattle egrets expanded from their Asian/African range into new habitats in the Americas and Australia. **(b)** Cattle egrets have found a happy home among the cattle herds of Texas. (Alan D. Carey/Photo Researchers, Inc.)

small area to extinction over the entire planet. The point is that evolution is a continuing process; no species is likely to be a permanent inhabitant of Earth, and during the period of its ascendancy, there is apt to be a great deal of distributional variation within a species, part of which is caused by local extinctions.

One of the simplest and most localized examples of this process is **plant succession,** in which one type of

Figure 11-4 Coconuts have a worldwide distribution in tropical coastal areas, in part because the nut can float long distances and then take root if it finds a favorable environment. This sprouting example is from the island of Vanua Levu in Fiji. (TLM photo)

Figure 11-5 Once nearly exterminated, American bison now occur in large numbers in areas of suitable habitat. Under natural conditions they have a high level of reproductive success. This scene is in Elk Island National Park in western Canada. (TLM photo)

vegetation is replaced naturally by another. Plant succession is a normal occurrence in a host of situations; a very common one involves the infilling of a lake (Figure 11-6). As the lake gradually fills with sediments and organic debris, the aquatic plants at the bottom are slowly choked out, while the sedges, reeds, and mosses of the shallow edge waters become more numerous and extensive. Continued infilling further diminishes the aquatic habitat and allows for the increasing encroachment of low-growing land plants, such as grasses and shrubs. As the process continues, trees move in to colonize the site, replacing the grasses and shrubs and completing the transition from lake to marsh to meadow to forest.

A similar series of local animal extinctions would accompany the plant succession because of the significant habitat changes. Lake animals would be replaced by marsh animals, which in turn would be replaced by meadow and forest animals.

Plant succession is not to be confused with extinction. Extinction is permanent, but species succession is not. Although a particular plant species may not be growing at a given time in a given location, it may reappear quickly if environmental conditions change and if there is an available seed source. Extinction means a species is extinct over the entire world, eliminated forever from the landscape. Extinction has taken place many times in Earth's history. Indeed, it is estimated that *half a billion* species have become extinct during the several-billion-year life of our planet. Probably the most dramatic example is the disappearance of the dinosaurs. For many millions of years those gigantic reptiles were the dominant life forms of our planet, and yet in a relatively short period of geo-logic time they were all wiped out. The reasons behind their extermination are imperfectly understood, but the fact remains that there have been innumerable such natural extinctions of entire species in the history of the world.

TERRESTRIAL FLORA

The natural vegetation of the land surfaces of Earth is of interest to the geographer for three reasons:

1. Over much of the planet, the terrestrial (land-dwelling) flora is the most significant visual component of the landscape. Plants often grow in such profusion that they mask all other elements of the environment. Topography, soils, animal life, and even water surfaces are often obscured or obliterated by plants. Only in areas of rugged terrain, harsh climate, or significant human activities are plants not likely to dominate the landscape.

2. Vegetation is a sensitive indicator of other environmental attributes. Floristic characteristics typically reflect subtle variations in sunlight, temperature, precipitation, evaporation, drainage, slope, soil conditions, and other natural parameters. Moreover, the influence of vegetation on soil, animal life, and microclimatic characteristics is frequently pervasive.

3. Vegetation often has a prominent and tangible influence on human settlement and activities. In some cases, it is a barrier or hindrance to human endeavor; in other instances, it provides an important resource to be exploited or developed.

capable of sustaining life despite whatever may happen to the above-surface portion of the organism.

The survival capability of a species also depends in part on its reproductive mechanism. Plants that endure seasonal climatic fluctuations from year to year are called **perennials**, whereas those that perish during times of climatic stress (such as winter) but leave behind a reservoir of seeds to germinate during the next favorable period are called **annuals**.

Plant life varies remarkably in form, from microscopic algae to gigantic trees. Most plants, however, have common characteristic features—roots to gather nutrients and moisture and anchor the plant, stems and/or branches for support and for nutrient transportation from roots to leaves, leaves to absorb and convert solar energy for sustenance and to exchange gases and transpire water, and reproductive organs for regeneration.

ENVIRONMENTAL ADAPTATIONS

Despite the hardiness of most plants, there are definite tolerance limits that govern their survival, distribution, and dispersal. During hundreds of millions of years of development, plants have evolved a variety of protective mechanisms to shield against harsh environmental conditions and to enlarge their tolerance limits. Two prominent adaptations to environmental stress are xerophytic adaptations and hygrophytic ones.

Xerophytic Adaptations   The descriptive term for plants that are structurally adapted to withstand protracted dry conditions is **xerophytic** *(xero* is Greek for dry; *phyt-* comes from *phuto-,* Greek for plant). Xerophytic adaptations can be grouped into four general types:

1. Roots are modified in shape or size to enable them to seek widely for moisture. Sometimes taproots extend to extraordinary depths to reach subterranean moisture. Also, root modification may involve the growth of a large number of thin hairlike rootlets to penetrate tiny pore spaces in soil (Figure 11-7).
2. Stems are sometimes modified into fleshy, spongy structures that can store moisture. Plants with such fleshy stems are called **succulents**; most cacti are prominent examples.
3. Leaf modification takes many forms, all are designed to decrease transpiration. Sometimes a leaf surface is hard and waxy to inhibit water loss, or white and shiny to reflect insolation and thus reduce evaporation. Still more effective is for the plant to have either tiny leaves or no leaves at all. In many types of dry-land shrubs, leaves have been replaced by thorns, from which there is virtually no transpiration.

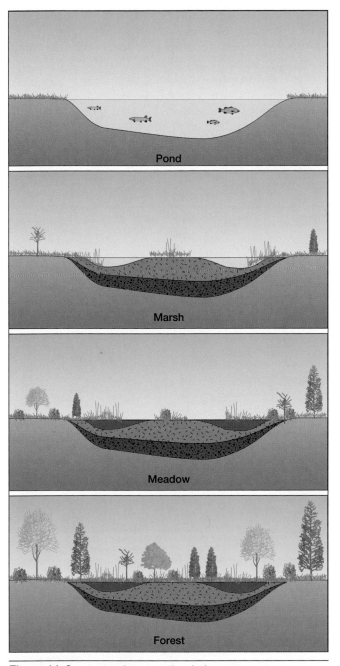

Figure 11-6   A simple example of plant succession: infilling of a small lake. Over time, successional colonization by different plant associations changes the area from pond to marsh to meadow to forest.

CHARACTERISTICS OF PLANTS

Despite the fragile appearance of many plants, most varieties are remarkably hardy. Although exposed to the elements and without any ability to seek shelter, plants are capable of surviving the harshest environmental circumstances. They survive, and often flourish, in the wettest, driest, hottest, coldest, and windiest places on Earth. Much of their survival potential is based on a subsurface root system that is

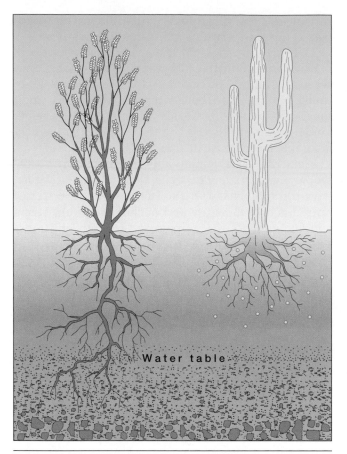

Figure 11-7 Desert plants have evolved various mechanisms for survival in an arid climate. Some plants produce a long taproot that penetrates deeply in search of the water table. More common are plants that have no deep roots but rather have myriad small roots and rootlets that seek any moisture available near the surface.

4. Perhaps the most remarkable floristic adaptation to aridity is not structural but involves the plant's reproductive cycle. Many xerophytic plants lie dormant for years without perishing. When rain eventually arrives, these plants promptly initiate and pass through an entire annual cycle of germination, flowering, fruiting, and seed dispersal in only a few days, then lapse into dormancy again if the drought resumes.

Hygrophytic Adaptations Some plants are particularly suited to a wet terrestrial environment. Distinction is sometimes made between **hydrophytes** (species living more or less permanently immersed in water, such as the water lilies in Figure 11-8(a) and **hygrophytes** (moisture-loving plants that generally require frequent soakings with water, as do many ferns, mosses, and rushes), but both groups are often identified by the latter term.

Hygrophytes are likely to have extensive root systems to anchor them in the soft ground, and hygrophytic trees often develop a widened, flaring trunk near the ground to provide better support [Figure 11-8(b)]. Many hygrophytic plants that grow in standing or moving water have weak, pliable stems that can withstand the ebb and flow of currents rather than standing erect against them; the buoyancy of the water, rather than the stem, provides support for the plant.

(a)

(b)

Figure 11-8 Some types of plants flourish in a totally aqueous environment. **(a)** These lily pads virtually cover the surface of a bay of Cold Lake, on the Alberta-Saskatchewan border in western Canada. (TLM photo) **(b)** Many hygrophytic trees, such as these cypress in Louisiana, have wide, flaring trunks near the ground to provide firmer footing in the wet environment. (Eastcott/Momatiuk/Photo Researchers, Inc.)

THE CRITICAL ROLE OF COMPETITION

As important as climatic, edaphic, and other environmental characteristics are to plant survival, a particular species will not necessarily occupy an area just because all these conditions are favorable. Plants are just as competitive as animals or used-car salespersons. Of the dozens of plant species that might be suitable for an area, only one or a few are likely to survive.

This is not to say that all plants are mutually competitive; indeed, thousands of ecologic niches can be occupied without impinging on one another. However, most plants draw their nutrients from the same soil and their energy from the same sun, and what one plant obtains cannot be used by another.

FLORISTIC TERMINOLOGY

To continue with our consideration of plants, we need a specialized vocabulary. Here are some of the terms used in the discussion.

First of all, plants can be divided into two categories: those that reproduce through spores and those that reproduce through seeds. The former include two major groups:

1. **Bryophytes** include the true mosses, peat mosses, and liverworts. Presumably they have never in geologic history been very important among plant communities, except in localized situations.
2. **Pteridophytes** are ferns, horsetails, and club mosses (which are not true mosses). During much of geologic history, great forests of tree ferns, giant horsetails, and tall club mosses dominated continental vegetation, but they are less important today.

Plants that reproduce by means of seeds are encompassed in two broad groups:

1. The more primitive of the two groups, the **gymnosperms** (naked seeds), carry their seeds in cones, and when the cones open, the seeds fall out. (For this reason, gymnosperms are sometimes called conifers.) Gymnosperms were more important in the geologic past; today the only large surviving gymnosperms are cone-bearing trees such as pines.
2. **Angiosperms** (vessel seeds) are the flowering plants. Their seeds are encased in some sort of protective body, such as a fruit, nut, or pod. Trees, shrubs, grasses, crops, weeds, and garden flowers are angiosperms. Along with a few

conifers, they have dominated the vegetation of the planet for the last 50 or 60 million years.

Several other terms are commonly used to describe vegetation, as summarized in Figure 11-9. Their definitions are not always precise, but their meanings generally are clear.

1. One fundamental distinction is made on the basis of stem or trunk composition. **Woody plants** have stems composed of hard fibrous material, whereas **herbaceous plants** have soft stems. Woody plants are mostly trees and shrubs; herbaceous plants are mostly grasses, forbs, and lichens.
2. With trees, whether or not a plant loses its leaves sometime during the year is an important distinguishing characteristic. An **evergreen** tree is one that sheds its leaves on a sporadic or successive basis but always appears to be fully leaved. A **deciduous** tree is one that experiences an annual period in which all leaves die and usually fall from the tree, due to either a cold season or a dry season.

Hardwood
Deciduous
Broadleaf
Angiosperm

Softwood
Coniferous
Needleleaf
Gymnosperm

Figure 11-9 Terminology used in describing plants. Although the terminology is somewhat confusing, there are conspicuous differences between hardwood (angiosperm) and softwood (gymnosperm) trees. The most obvious difference is in general appearance.

3. Trees are also often described in terms of leaf shapes. **Broadleaf** trees have leaves that are flat and expansive in shape, whereas **needleleaf** trees are adorned with thin slivers of tough, leathery, waxy needles rather than typical leaves. Almost all needleleaf trees are evergreen, and the great majority of all broadleaf trees are deciduous, except in the rainy tropics, where everything is evergreen.

4. **Hardwood** and **softwood** are two of the most unsatisfactory terms in biogeography, but they are widely used in everyday parlance, and so we must not ignore them. Hardwoods are angiosperm trees that are usually broad-leaved and deciduous. Their wood has a relatively complicated structure, but it is not always hard. Softwoods are gymnosperms; nearly all are needleleaf evergreens. Their wood has a simple cellular structure, but it is not always soft.

SPATIAL GROUPINGS OF PLANTS

A geographer trying to understand the floristic characteristics of the environment looks sometimes at individual plants but more commonly at their spatial groupings. Over most land surfaces, the natural vegetation occurs in considerable variety, but regardless of species diversity, plant associations usually can be classified on the basis of dominant members, dominant appearance, or both.

The Inevitability of Change

In considering the major vegetational associations, we should remember that the floristic pattern of Earth is impermanent. The plant cover that exists at any given time and place may be in a state of constant change or may be relatively stable for millennia before experiencing significant changes. Sooner or later, however, change is inevitable. Sometimes the change is slow and orderly, as when a lake is filled in, as described earlier, or when there is a long-term trend toward different climatic conditions over a broad area. On occasion, however, the change is abrupt and chaotic, as in the case of a wildfire.

At some time in Earth's history, all parts of the present landmasses were newly created and therefore unvegetated. Such new land is first occupied temporarily by some plant association that soon gives way to another and then another and another. Many complicated changes occur until some sort of floristic stability is attained. Each succeeding association alters the local environment, making possible the establishment of the next association. The general sequential trend is toward taller plants and greater stability in species composition. The longer plant succession continues, the more slowly change takes place because more advanced associations usually contain species that live a relatively long time.

Eventually a plant association of constant composition comes into being. In other words, a point is reached where change is no longer noticeable, and each succeeding generation of the association is much like its predecessor. This stable association is generally referred to as the **climax vegetation**, and the various stages leading up to it are called **seral associations.**

The implication of the term *climax vegetation* is that the dominant plants of a climax association have demonstrated that, of all possibilities for that particular situation, they can compete the most successfully. Thus they represent the optimal floristic cover for that environmental context. The climax vegetation presumably persists unchanged for an indefinite period until the next environmental disturbance. The climax vegetation is, then, an association in equilibrium with prevailing environmental conditions. When these conditions change, the climax stage is disturbed and another succession sequence is initiated.

Identifying and Mapping Associations

The geographer attempting to recognize spatial groupings of plants faces some significant difficulties. Plant associations that are similar in appearance and in environmental relationships can occur in widely separated localities and are likely to contain totally different species. At the other extreme, exceedingly different plant associations can often be detected within a very small area. As the geographer tries to identify patterns and recognize relationships, generalization invariably is needed. This generalization must accommodate gradations, ecotones, interdigitations, and other irregularities. When associations are portrayed on maps, therefore, their boundaries in nearly all cases represent approximations.

Another special problem facing the student who would learn about Earth's environment is that in many areas the natural vegetation has been completely removed or replaced through human interference. Forests have been cut, crops planted, pastures seeded, and urban areas paved. Over extensive areas of Earth's surface, therefore, climax vegetation is the exception rather than the rule. Most world maps that purport to show natural vegetation ignore human interference and are actually maps of theoretical natural vegetation, in which the mapmaker makes assumptions about what the natural vegetation would be if it had not been modified by human activity.

The Major Floristic Associations There are many ways to classify plant associations. For broad geographical purposes, emphasis is usually placed on the structure and appearance of the dominant plants. The major associations generally recognized (Figure 11-10) include the following:

1. **Forests** consist of trees growing so close together that their individual leaf canopies generally overlap. This means that the ground is largely in shade, a condition that usually precludes the development of much undergrowth. Forests require considerable annual precipitation and can survive in widely varying temperature zones. Except where moisture is inadequate or the growing season very short, forests are likely to become the climax vegetation association in any area.

2. **Woodlands** are tree-dominated plant associations in which the trees are spaced more widely apart than in forests and do not have interlacing canopies (Figure 11-11). Undergrowth may be either dense or sparse, but it is not inhibited by lack of sunlight. Woodland environments generally are drier than forest environments.

3. **Shrublands** are plant associations dominated by relatively short woody plants, generally called *shrubs* or *bushes*. Shrubs take a variety of forms, but most have several stems branching near the ground and a leafy foliage that begins very close to ground level. Trees and grasses may be interspersed with the shrubs but are less prominent in the landscape. Shrublands have a wide latitudinal range, but they are generally restricted to semiarid or arid locales.

4. **Grasslands** may contain scattered trees and shrubs, but the landscape is dominated by grasses and forbs (broadleaf herbaceous plants). Prominent types of grassland include **savanna,** low-latitude grassland characterized by tall grasses; **prairie,** midlatitude grassland characterized by tall grasses; and **steppe,** midlatitude grassland characterized by short grasses and bunchgrasses. Grasslands are associated with semiarid and subhumid climates.

5. **Deserts** are typified by widely scattered plants with much bare ground interspersed. *Desert* is actually a climatic term, and desert areas may have a great variety of vegetation, including grasses, succulent herbs, shrubs, and straggly trees (Figure 11-12). Some extensive desert areas comprise loose sand, bare rock, or extensive gravel, with virtually no plant growth.

6. **Tundra,** as we noted in Chapter 8, consists of a complex mix of very low plants, including grasses, forbs, dwarf shrubs, mosses, and lichens, but no trees. Tundra occurs only in the perennially cold climates of high latitudes or high altitudes.

7. **Wetlands** have a much more limited geographic extent than the associations described above. They are characterized by shallow standing water all or most of the year, with vegetation rising above the water level. The most widely distributed wetlands are *swamps* (with trees as the dominant plant forms) and *marshes* (with grasses and other herbaceous plants dominant).

Vertical Zonation In Chapter 8 we learned that mountainous areas often have a distinct pattern of **vertical zonation** in vegetation patterns (Figure 11-13). Significant elevational changes in short horizontal distances cause various plant associations to exist in relatively narrow zones on mountain slopes. This zonation is largely due to the effects of elevation on temperature and precipitation.

The essential implication is that elevation changes are the counterpart of latitude changes; in other words, to travel from sea level to the top of a tall tropical peak is roughly equivalent environmentally to a horizontal journey from the equator to the Arctic. This elevation/latitude relationship is shown most clearly by how the elevation of the upper treeline (the elevation above which trees are unable to survive) varies with latitude (Figure 11-14).

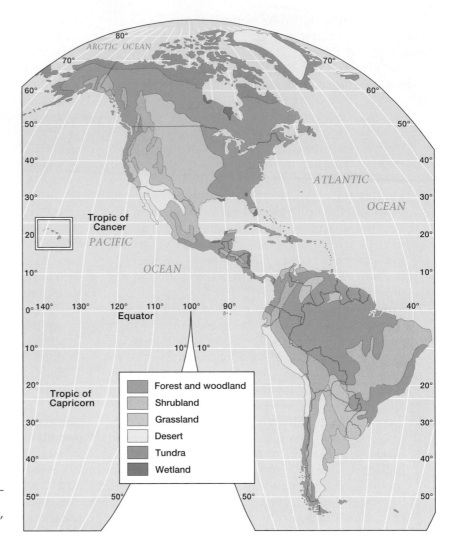

Figure 11-10 The major natural vegetation associations. The purple represents mountains, and the white areas are ice-covered.

Figure 11-11 A pinon/juniper woodland scene in central Arizona. (C. Prescott-Allen/Earth Scenes)

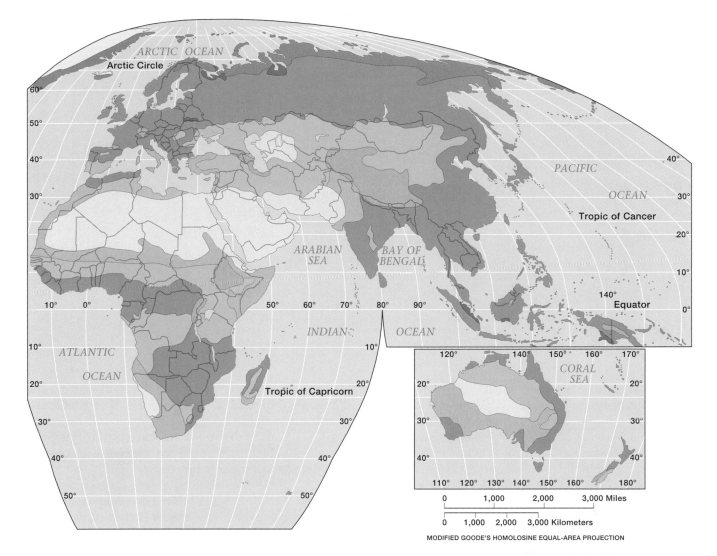

MODIFIED GOODE'S HOMOLOSINE EQUAL-AREA PROJECTION

Figure 11-12 Some deserts contain an abundance of plants, but they are usually spindly and always xerophytic. This scene is near Phoenix, Arizona. (TLM photo)

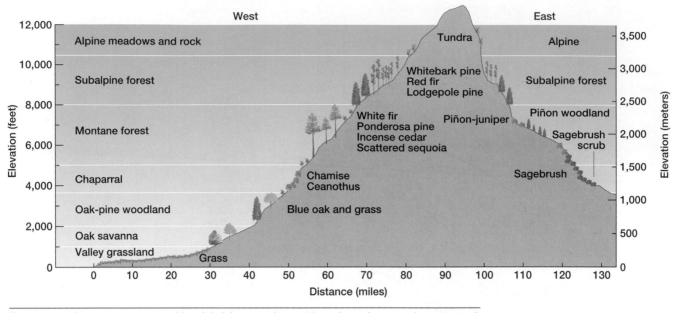

Figure 11-13   A west-east profile of California's Sierra Nevada, indicating the principal vegetation at different elevations on the western (wet) and eastern (dry) sides of the range.

An interesting detail of vertical zonation is that the elevation/latitude graph for the Southern Hemisphere is different from that of the Northern Hemisphere. For example, between latitudes 35° S and 40° S in Australia and New Zealand, the treeline is below 6000 feet (1800 meters); at comparable latitudes in North America, the treeline is nearly twice as high. The reason for this significant discrepancy is not understood.

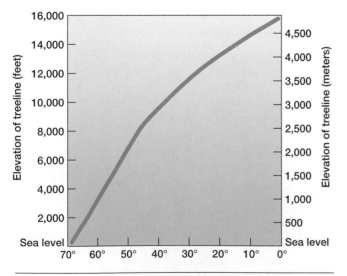

Figure 11-14   Treeline elevation varies with latitude. This graph for the Northern Hemisphere shows that trees cease to grow at an elevation of about 16,000 feet (5000 meters) in the equatorial Andes of South America but at only 10,000 feet (3000 meters) at 40° N latitude in Colorado. At 70° N latitude in northern Canada, the treeline is at sea level.

Treeline variation represents only one facet of the broader design of vertical zonation in vegetation patterns. All vegetation zones are displaced downward with increasing distance from the equator. This principle accounts for the significant vegetational complexity found in all mountainous areas.

Local Variations   Each major vegetation association extends over a large area of Earth's surface. Within a given association, however, there are also significant local variations caused by a variety of local environmental conditions, as illustrated by the following two examples.

1. *Exposure to sunlight.* The direction in which a sloped surface faces is often a critical determinant of vegetation composition, as illustrated by Figure 11-15. Exposure has many aspects, but one of the most pervasive is simply the angle at which sunlight strikes the slope. If the sun's rays arrive at a high angle, they are much more effective in heating the ground and thus in evaporating available moisture. Such a sun slope (called an **adret slope**) is hot and dry, and its vegetation not only is sparser and smaller than that on adjacent slopes having a different exposure to sunlight but is also likely to have a different species composition.

    The opposite condition is a **ubac slope**, which is oriented so that sunlight strikes it at a low angle and is thus much less effective in heating and evaporating. This cooler condition produces more luxuriant vegetation of a richer diversity.

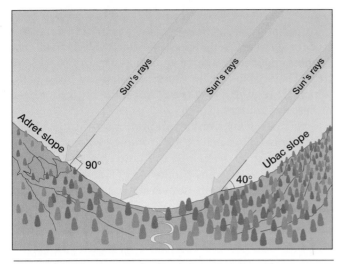

Figure 11-15 A typical adret/ubac situation. The noon sun rays strike the adret slope at approximately a 90° angle, a condition that results in maximum heating. The same rays strike the ubac slope at approximately a 40° angle, with the result that the heating is spread over a large area and is therefore less intense.

The difference between adret and ubac slopes decreases with increasing latitude, presumably because the ubac flora becomes impoverished under the cooler conditions that prevail as latitude increases, and now the relative warmth of the adret surface encourages plant diversity.

2. *Valley-bottom location.* In mountainous areas where a river runs through a valley, the vegetation associations growing on either side of the river have a composition that is significantly different from that found higher up on the slopes forming the valley. This floral gradient is sometimes restricted to immediate stream-side locations and sometimes extends more broadly over the valley floor.

The difference is primarily a reflection of the perennial availability of subsurface moisture near the stream and is manifested in a more diversified and more luxuriant flora. This vegetative contrast is particularly prominent in dry regions, where streams may be lined with trees even though no other trees are to be found in the landscape. Such anomalous stream-side growth is called **riparian vegetation** (Figure 11-16).

TERRESTRIAL FAUNA

Animals occur in much greater variety than plants over Earth. As objects of geographical study, however, they have been relegated to a place of lesser importance for at least two reasons:

1. Animals are much less prominent than plants in the landscape. Apart from extremely localized situations—such as waterfowl flocking on a lake or an insect plague attacking a crop—animals tend to be secretive and inconspicuous.

2. Environmental interrelationships are much less clearly evidenced by animals than by plants. This

Figure 11-16 Riparian vegetation is particularly prominent in dry lands. This stream in northern Queensland (Australia) would be inconspicuous in this photograph if it were not for the trees growing along it. (TLM photo)

is due in part to the inconspicuousness of animals, which renders them more difficult to study, and in part to the fact that animals are mobile and therefore more able to adjust to environmental variability.

This is not to say that fauna is inconsequential for students of geography. Under certain circumstances, wildlife is a prominent element of physical geography; and in some regions of the world, it is an important resource for human use and/or a significant hindrance to human activity. Moreover, it is increasingly clear that animals are sometimes more sensitive indicators than plants of the health of a particular ecosystem.

CHARACTERISTICS OF ANIMALS

The diversity of forms of animal life is not realized by most people. We commonly think of animals as being relatively large and conspicuous creatures that run across the land or scurry through the trees, seeking to avoid contact with humankind. In actuality, the term *animal* encompasses not only the larger, more complex forms but also hundreds of thousands of species of smaller and simpler organisms that may be inconspicuous or even invisible. The variety of animal life is so great that it is difficult to find many unifying characteristics. The contrast, for example, between an enormous elephant and a microscopic protozoan is so extreme as to make their kinship appear ludicrous. Animals really have only two universal characteristics, and even these are so highly modified in some cases as to be almost unrecognizable:

1. Animals are motile, which means that they are capable of self-generated movement.
2. Animals must eat plants and/or other animals for sustenance. They are incapable of manufacturing their food from air, water, and sunlight, which plants can do.

ENVIRONMENTAL ADAPTATIONS

As with plants, animals have evolved slowly and diversely through eons of time. Evolution has made it possible for animals to diverge remarkably in adjusting to different environments. Just about every existing environmental extreme has been met by some (or many) evolutionary adaptations that make it feasible for some (or many) animal species to survive and even flourish.

Physiological Adaptations   The majority of animal adaptations to environmental diversity have been physiological, which is to say that they are anatomical and/or metabolic changes. A classic example is the

size of fox ears. Ears are prime conduits for body heat in furred animals, as they provide a relatively bare surface for its loss. Arctic foxes [Figure 11-17(a)] have unusually small ears, which minimizes heat loss; desert foxes [Figure 11-17(b)] possess remarkably large ears, which are a great advantage during the blistering heat of desert summers.

The catalog of similar adaptations is almost endless: webbing between toes to make swimming easier, dense fur to keep out the cold, broad feet that won't sink in soft snow, increase in size and number of sweat glands to aid in evaporative cooling, and a host of others.

(b)

(b)

Figure 11-17   **(a)** An Arctic fox near Cape Churchill in the Canadian province of Manitoba. Its tiny ears are an adaptation to conserve body heat in a cold environment. (Copyright Dan Guravich/Shil Photo Researchers, Inc.) **(b)** Desert foxes have large ears, the better to radiate away body heat. This is a bat-eared fox in northern Kenya. (TLM photo)

Behavioral Adaptations   An important advantage that animals have over plants, in terms of adjustment to environmental stress, is that the former can move about and therefore can modify their behavior in order to minimize the stress. Animals can seek shelter from heat, cold, flood, or fire; they can travel far in search of relief from drought or famine; they can shift from daytime (diurnal) to nighttime (nocturnal) activities to minimize water loss during hot seasons. Such techniques as migration (periodic movement from one region to another), hibernation (spending winter in a dormant condition), and estivation (spending a dry/hot period in a torpid state) are behavioral adaptations employed regularly by many species of animals.

Reproductive Adaptations   Harsh environmental conditions are particularly destructive to the newly born. As partial compensation for this factor, many species have evolved specialized reproductive cycles or have developed modified techniques of baby care. During lengthy periods of bad weather, for example, some species delay mating or postpone nest building. If fertilization has already taken place, some animal reproductive cycles are capable of almost indefinite delay, resulting in a protracted egg or larval stage or even of total suspension of embryo development until the weather improves.

If the young have already been born, they sometimes remain longer than usual in nest, den, or pouch, and the adults may feed them for a longer time. When good weather finally returns, some species are capable of hastened estrus (the period of heightened sexual receptivity by the female), nest building, den preparation, and so on, and the progeny produced may be in greater-than-normal numbers.

COMPETITION AMONG ANIMALS

Competition among animals is even more intense than that among plants because the former involves not only indirect competition in the form of rivalry for space and resources but also the direct antagonism of predation. Animals with similar dietary habits compete for food and occasionally for territory. Animals in the same area also sometimes compete for water. And animals of the same species often compete for territory and for mates. Across this matrix of ecological rivalry is spread a prominent veneer of predator/prey relationships.

Many animals live together in social groups of varying sizes, generally referred to as *herds, flocks,* or *colonies.* This is a common, but by no means universal, behavioral characteristic among animals of the same species (Figure 11-18), and sometimes a social group encompasses several species in a communal relationship, as zebras, wildebeest, and impalas living together on an East African savanna. Within such groupings there may be a certain amount of cooperation among unrelated animals, but competition for both space and resources is likely to be prominent as well.

Competition among animals is a major part of the general struggle for existence that characterizes natural relationships in the biosphere. Individual animals are concerned either largely or entirely with their own survival (and sometimes with that of their mates), in response to normal primeval instincts. In some species, this concern is broadened to include their own young, although such maternal (and, much more rarely, paternal) instinct is by no means universal among animals. Still fewer species show

Figure 11-18   A pride of Kenya lions at rest. Lions are very gregarious mammals. (George Holton/Photo Researchers, Inc.)

FOCUS

ALIEN ANIMALS IN AUSTRALIA

Over the centuries, many foreign animal species have been introduced into Australia, either by accident or by design, and have become established in the wild. Because native animals are generally unaggressive and vulnerable, it was relatively easy for an introduced exotic to become established. In many instances the exotics have become the dominant animals, outcompeting or preying on the native wildlife.

The earliest exotic introduction was the dingo, a coyote-sized dog apparently brought to Australia perhaps five or ten thousand years ago by Aboriginal migrants from Southeast Asia, and a well-established member of the fauna for many centuries. The most notable recent introduction was that of the European rabbit, whose spread from an initial advent of 24 animals near Melbourne in 1859 to a half-continent plague within 50 years is the classic scare story of all mammalian importations.

The diffusion of the European fox over Australia was even more expansive; the rabbit never moved into the northern third of the continent, whereas the fox has spread to almost every corner of the land.

The proliferation of feral livestock in Australia is even more spectacular than the story of either the rabbit or the fox. All six of the common midlatitude barnyard ungulates (horses, donkeys, cattle, sheep, goats, pigs) have been bred as domesticated livestock in considerable numbers over most of the settled parts of the continent. In addition, two subtropical livestock varieties—dromedaries and water buffaloes—were brought to Australia in limited numbers. Seven of these species (all except sheep) have on occasion escaped from confinement or deliberately been turned loose and have established sizable free-ranging feral populations in various parts of the country (Figure 11-1-A).

At the present time Australia has more feral horses, donkeys, cattle, goats, water buffaloes, and camels than any other country in the world and ranks second (to the United States) in number of feral pigs. The total number of these animals in the country varies from year to year but is in the vicinity of three to four million. There are also uncountable numbers of feral dogs and cats.

The presence of such a vast population of exotic wildlife is almost universally deplored. Most objections, especially those of ranchers, are on

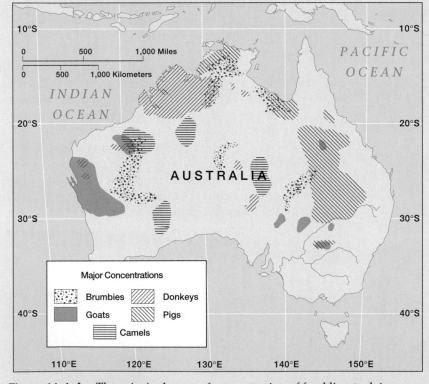

Figure 11-1-A   The principal areas of concentration of feral livestock in Australia. (Brumbies are feral horses.)

individual concern for the group, as represented in colonies of ants or prides of lions. For the most part, however, animal survival is a matter of every creature for itself, with no individual helping another and no individual deliberately destroying another apart from normal predatory activities.

COOPERATION AMONG ANIMALS

Symbiosis is the arrangement in which two dissimilar organisms live together. There are three principal forms:

1. **Mutualism** involves a mutually beneficial relationship between the two organisms, as exemplified by

Figure 11-1-B   The cane toad is a prominent pest in northeastern Australia. (Klaus Uhlenhut/Animals/Earth Science)

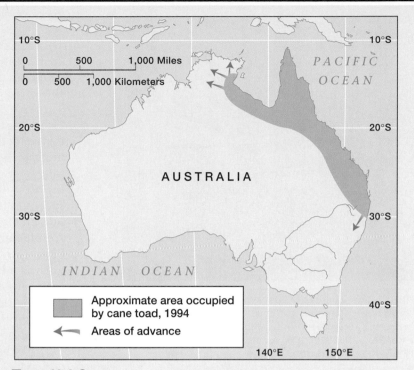

Figure 11-1-C   The spreading menace of the cane toad in Australia.

economic grounds. Dingoes prey on sheep; rabbits and feral ungulates inhibit ranching operations, primarily by consuming feed and water intended for livestock; foxes, feral dogs, and feral cats are devastating predators upon such native wildlife as ground-nesting birds, lizards, and vulnerable marsupials.

Consequently, enormous efforts have been, and continue to be, expended to control these exotics by such measures as poisoning, trapping, shooting, and building of barrier fences. The implacable opposition of the ranching industry to these animals is thoroughly understandable from an economic point of view. And even when considered from the loftier ethic of the integrity of the continental ecosystem, it would seem the better part of wisdom to beware of the persistence of such exotic animals in the wild. One of the broad, usually ignored, lessons of history is that humankind's tampering with the biota usually has unsatisfactory results.

One other exotic animal is causing increasing problems in Australia. In 1935 the South American cane toad

(Figure 11-1-B) was introduced to serve as a "natural" control for two invertebrate pests infesting the sugarcane districts of the northeastern coastal country. This large toad [the record specimen is 8.5 inches (22 centimeters) long and weighs nearly 3 pounds (1.35 kilograms)], has an insatiable appetite and prodigious reproductive capabilities—a female can produce 24,000 eggs per month.

For various reasons, the cane toad did not control the sugarcane pests for which it was intended. It has, however, become a major destroyer of virtually any animal its size or smaller. It has wiped out most of the native frogs, as well as a sizable share of the smaller reptiles, tiny marsupials, and terrestrial invertebrates, in its adopt-

ed habitat. Moreover, it is highly toxic to most carnivorous vertebrates, and so any predator that eats the toad is likely to die as a result.

Most disturbing, perhaps, is the inexorable expansion of this adaptable creature's range. It is now found throughout eastern and central Queensland, has spread widely into the Northern Territory, and is moving southward into New South Wales (Figure 11-1-C). Presumably it will not be able to survive in cool or arid climates, but it has had a devastating effect on the local ecosystems in its areas of proliferation and has the awesome potential of being the most destructive exotic animal to threaten Australia since the European rabbit.

the tickbirds that invariably accompany rhinoceroses (Figure 11-19). The birds aid the rhino by removing insects that infest the latter's skin, and the birds benefit by having a readily available supply of food.

2. **Commensalism** involves two dissimilar organisms living together with no injury to either, as repre-

sented by burrowing owls sharing the underground home of prairie dogs.

3. **Parasitism** involves one organism living on or in another, obtaining nourishment from the host, which is usually weakened and sometimes killed by the actions of the parasite.

Figure 11-19 There is often a symbiotic relationship between hoofed animals and insectivorous birds. In this scene from Zambia a red-billed oxpecker rests on the back of a warthog after searching for ticks. (Leonard Lee Rue II/Photo Researchers, Inc.)

KINDS OF ANIMALS

The vast majority of animals are so tiny and/or secretive as to be either invisible or extremely inconspicuous. Their size and habits, however, are not valid indicators of their significance for geographic study. Very minute and seemingly inconsequential organisms sometimes play exaggerated roles in the biosphere—as carriers of disease, or as hosts of parasites, sources of infection, or providers of scarce nutrients. For example, no geographic assessment of Africa, however cursory, can afford to ignore the presence and distribution of the small tsetse fly and the tiny protozoan called *Trypanosoma*, which together are responsible for the transmission of trypanosomiasis (sleeping sickness), a widespread and deadly disease for humans and livestock over much of the continent.

Zoological classification is much too detailed to be useful in most geographical studies. Therefore, presented below is a brief summation of the principal kinds of animals that might be recognized in a general study of physical geography.

Invertebrates Animals without backbones are called **invertebrates**. More than 90 percent of all animal species are encompassed within this broad grouping, as is made graphically clear by Figure 11-20. Invertebrates include worms, sponges, mollusks, various marine animals, and a vast host of creatures of microscopic or near-microscopic size. Very prominent among invertebrates are the *arthropods*, a group that in-

clude insects, spiders, centipedes, millipedes, and crustaceans (shellfish).

Vertebrates **Vertebrates** are animals that have a backbone that protects the main nerve (or spinal) cord. Geographers generally follow biologists in recognizing five principal groups of vertebrates:

1. Fishes are the only vertebrates that can breathe under water (a few species are also capable of breathing in air). Most fishes inhabit either fresh water or salt water only, but some species are capable of living in both environments, and several species spawn in freshwater streams but live most of their lives in the ocean.
2. Amphibians are semiaquatic animals. When first born, they are fully aquatic and breathe through gills; as adults, they are air-breathers by means of lungs and through their glandular skin. Most amphibians are either frogs or salamanders.
3. Most reptiles are totally land-based. Ninety-five percent of all reptile species are either snakes or lizards. The remainder are mostly turtles and crocodilians.
4. Birds are believed to have evolved from reptiles; indeed, they have so many reptilian characteristics that they have been called "feathered reptiles." There are more than 9000 species of birds, all of which reproduce by means of eggs. Birds are so adaptable that some species can live almost anywhere on Earth's surface. They are endothermic, which means that, regardless of the temperature of the air or water in which they live, they maintain a constant body temperature.
5. Mammals are distinguished from all other animals by several internal characteristics as well as by two prominent external features. The external features are that only mammals produce milk with which they feed their young, and only mammals possess true hair. (Some mammals have very little hair, but no creatures other than mammals have any hair.) Mammals are also notable for being endothermic. Thus the body temperature of mammals and birds stays about the same under any climatic conditions, which enables them to live in almost all parts of the world.

The great majority of all mammals are *placentals*, which means that their young grow and develop in the mother's body, nourished by an organ known as the *placenta*, which forms a vital connecting link with the mother's bloodstream (Figure 11-21). A small group of mammals (about 135 species) are *marsupials*, whose females have pouches in which the young, which are born in a very undeveloped condition, live for several weeks

INVERTEBRATES

Arthropods

All other Invertebrates

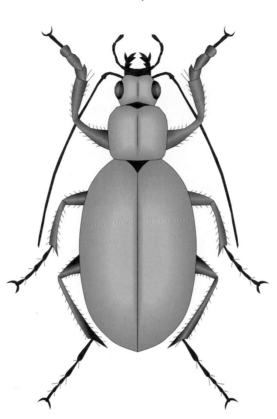

VERTEBRATES

Fishes   Birds   Reptiles   Mammals   Amphibians

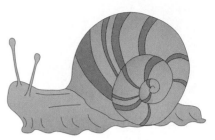

Figure 11-20   Relative abundance of different types of animals. The size of each animal indicates the comparative number of species in that group.

Figure 11-21   Most large land animals are placental mammals, as exemplified by this bull moose in Yellowstone Park. (TLM photo)

or months after birth (Figure 11-22). The most primitive of all mammals are the *monotremes*, of which only two types exist (Figure 11-23); they are egg-laying mammals.

ZOOGEOGRAPHIC REGIONS

The distribution of animals over the world is much more complex and irregular than that of plants, primarily because animals are mobile and therefore capable of more rapid dispersal. As with plants, however, the broad distributions of animals are reflective of the general distribution of energy and of food diversity. Thus the richest faunal assemblages are found in the permissive environment of the humid tropics, and the dry lands and cold lands have the sparsest representations of both species and individuals.

Figure 11-22 A red kangaroo joey (baby) peers out from its mother's pouch. (Tom McHugh/Photo Researchers, Inc.)

When considering the global patterns of animal geography, most attention is usually paid to the distribution of terrestrial vertebrates, with other animals being given only casual notice. The classical definition of world zoogeographic regions is credited to the nineteenth-century British naturalist A. R. Wallace, whose scheme is based on the work of P. L. Sclater. As shown in Figure 11-24, nine zoogeographic regions are generally recognized, but you should understand that this or any other system of faunal regions represents average conditions and cannot portray some common pattern in which different groups of animals fit precisely. It is simply a composite of many diverse distributions of contemporary fauna.

The Ethiopian Region is primarily tropical or subtropical and has a rich and diverse fauna. It is separated from other regions by an oceanic barrier on three sides and a broad desert on the fourth. Despite its isolation, however, the Ethiopian Region has many faunal affinities with the Oriental and Palearctic regions. Its vertebrate fauna is the most diverse of all the zoogeographic regions and includes the greatest number of mammalian families, many of which have no living relatives outside Africa.

The Oriental Region is separated from the rest of Asia by mountains. Its faunal assemblage is generally similar to that of the Ethiopian Region, with somewhat less diversity. The Oriental Region has some endemic groups (*endemic* means these groups are found nowhere else) and a few species that are found only in the Oriental, Palearctic, and Australian regions. Many brilliantly colored birds live in the Oriental Region,

(a)

(b)

Figure 11-23 There are only two kinds of monotremes, or egg-laying mammals in existence—the echidna and the duckbill platypus. **(a)** The echidna is found only in Australia and New Guinea. (TLM photo) **(b)** The platypus is totally aquatic in its lifestyle. (Tom McHugh, Taronga Park Zoo, Sydney, Australia/Photo Researchers, Inc.)

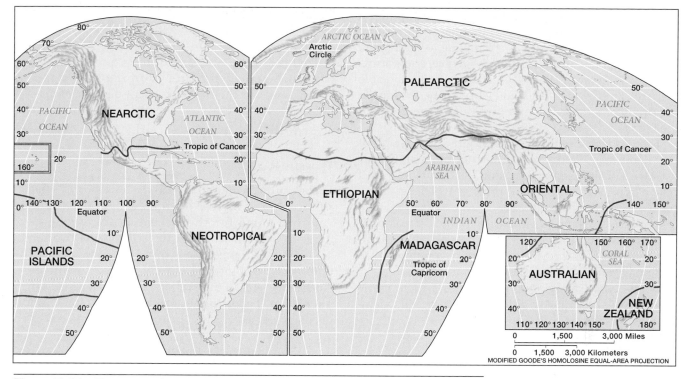

Figure 11-24   Zoogeographic regions of the world.

and reptiles are numerous, with a particularly large number of venomous snakes.

The Palearctic Region includes the rest of Asia, all of Europe, and most of North Africa. Its fauna as a whole is much poorer than that of the two regions previously discussed, which is presumably a function of its location in higher latitudes with a more rigorous climate. This region has many affinities with all three bordering regions, particularly the Nearctic. Indeed, the Palearctic has only two minor mammal families (both rodents) that are endemic, and almost all its birds belong to families that have a very wide distribution.

The Nearctic Region consists of the nontropical portions of North America. Its faunal assemblage (apart from reptiles, which are well represented) is relatively poor and is largely a transitional mixture of Palearctic and Neotropical groups. It has few important groups of its own except for freshwater fishes. The considerable similarities between Palearctic and Nearctic fauna have persuaded some zoologists to group them into a single superregion, the Holarctic, which had a land connection in the recent geologic past (the Bering land bridge), across which considerable faunal dispersal took place (Figure 11-25).

The Neotropical Region encompasses all of South America and the tropical portion of North America. Its fauna is rich and distinctive, which reflects both a variety of habitats and a considerable degree of isolation from other regions. Neotropic faunal evolution often

followed a path different from that in other regions. It contains a larger number of endemic mammal families than any other region. Moreover, its bird fauna is exceedingly diverse and conspicuous.

The Australian Region is restricted to the continent of Australia and some adjacent islands, particularly New Guinea. Its fauna is by far the most distinctive of any major region, primarily due to its lengthy isolation from other principal landmasses. Its vertebrate fauna is noted

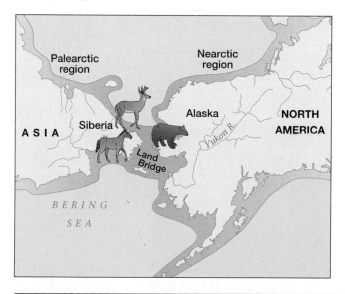

Figure 11-25   The Bering Land Bridge facilitated the interchange of animals between the Palearctic and Nearctic regions. Most of the dispersal was from the former to the latter.

for its paucity, but the lack of variety is made up for by uniqueness. There are only nine families of terrestrial mammals, but eight of them are unique to the region. The bird fauna is varied, and both pigeons and parrots reach their greatest diversity here. There is a notable scarcity of freshwater fishes and amphibians. Within the region there are many significant differences between the fauna of Australia and that of New Guinea.

The Madagascar Region, restricted to the island of that name, has a fauna very different from that of nearby Africa. The Madagascan fauna is dominated by a relic assemblage of unusual forms in which primitive primates (lemurs) are notable.

The New Zealand Region has a unique fauna dominated by birds, with a remarkable proportion of flightless types. It has almost no terrestrial vertebrates (no mammals; only a few reptiles and amphibians).

The Pacific Islands Region includes a great many far-flung islands, mostly quite small. Its faunal assemblage is very limited.

THE MAJOR BIOMES

The major biomes of the world (Figure 11-26) are described below. As we learned in Chapter 10, most biomes are named for their dominant vegetation association, but the biome concept also encompasses fauna as well as interrelationships with soil, climate, and topography.

TROPICAL RAINFOREST

You are probably familiar with the term *tropical rainforest* because it is used so often these days in the media. Another name for this biome is **selva** (the Portuguese/Spanish word for forest), and the two terms are used interchangeably in this discussion. The distribution of this biome is closely related to climate—consistent rainfall and relatively high temperatures. Thus there is an obvious correlation with the location of Af and some Am climatic regions (Figure 11-27).

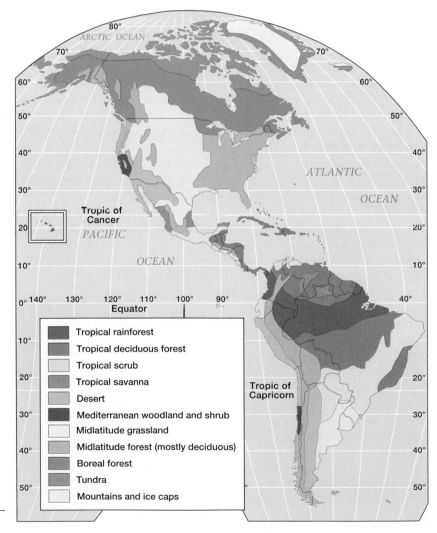

Figure 11-26   Major biomes of the world.

Figure 11-27 World distribution of tropical rainforest. The colors show the three A climates from Figure 8-7; the black diagonal lines indicate the rainforest, which occurs in tropical wet (Af) and tropical monsoonal (Am) climates.

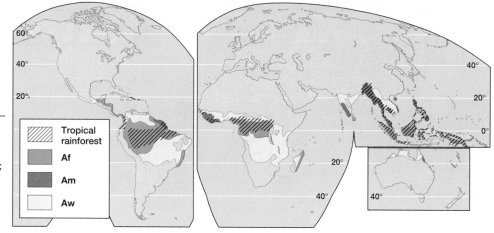

The rainforest is probably the most complex of all terrestrial ecosystems. It contains a bewildering variety of trees growing in close conjunction. Mostly they are tall, high-crowned, broadleaf species that never experience a seasonal leaf fall because the con-

cept of seasons is unknown in this environment of continuous warmth and moistness. The selva has a layered structure (Figure 11-28); the second layer down from the top usually forms a complete canopy of interlaced branches that provides continuous

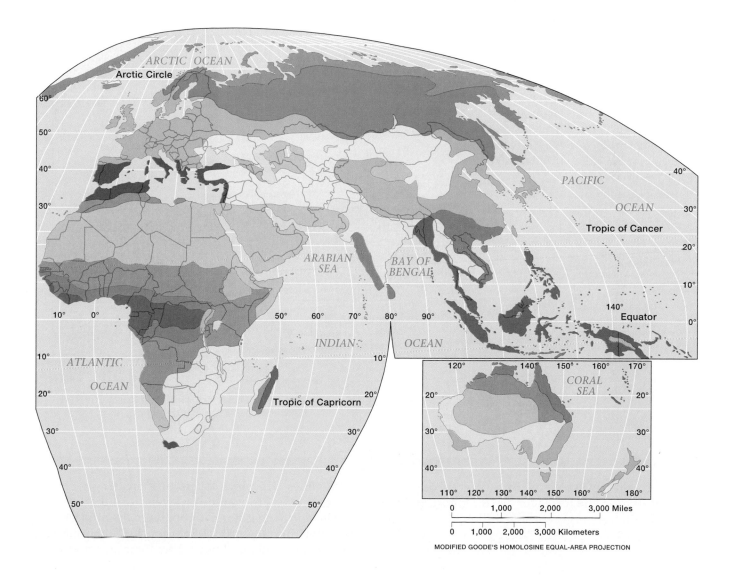

MODIFIED GOODE'S HOMOLOSINE EQUAL-AREA PROJECTION

PEOPLE AND THE ENVIRONMENT

RAINFOREST REMOVAL

Tropical rainforests, the extremely diverse biome found in 50 countries on five continents, are the climax vegetation over nearly 3 billion acres (1.2 billion hectares), about 8.3 percent of Earth's land surface. Biologists believe that rainforests are the home of perhaps half the world's species, about five-sixths of which have not yet been described and named.

Throughout most of history, rainforests were considered remote, inaccessible, unpleasant places, and as a consequence they were little affected by human activities. In the present century, however, rainforests have been exploited and devastated at an accelerating pace, and in the last decade or so, tropical deforestation has become one of Earth's most serious environmental problems. The rate of deforestation is spectacular—some 51 acres (21 hectares) per minute; 74,000 acres (30,000 hectares) per day; 27 million acres (11 million hectares) per year. More than half of the original African rainforest is now gone, about 45 percent of Asia's rainforest no longer exists, and the proportion in Latin America is approaching 40 percent.

The current situation varies in the five major rainforest regions:

1. The rate of deforestation is highest in southern and southeastern Asia, primarily associated with commercial timber exploitation, especially for teak and mahogany.

2. The current rate of deforestation is relatively low in central Africa.

3. Timber harvesting and agricultural expansion are responsible for a continuing high rate of forest clearing in West Africa. Nigeria has lost about 90 percent of its forests; Ghana, 80 percent.

4. Deforestation of the Amazon region as a percentage of the total area of rainforest has been moderate (about 5 percent of the total has been cleared), but it continues at an accelerating pace.

5. Very rapid deforestation persists in Central America, mostly due to expanded cattle ranching (Figure 11-2-A).

As the forest goes, so goes its habitability for both indigenous peoples and native animal life. In the mid-1980s it was calculated that tropical deforestation was responsible for the extermination of one species per day; by the mid-1990s it is estimated that the rate was two species per hour. Moreover, loss of the forests contributes to accelerated soil erosion, drought, flooding, water quality degradation, declining agricultural productivity, and greater poverty for rural inhabitants. In addition, atmospheric carbon dioxide continues to be increased because there are fewer trees to absorb it and because burning trees as a way of clearing forest releases more carbon to the air. Other

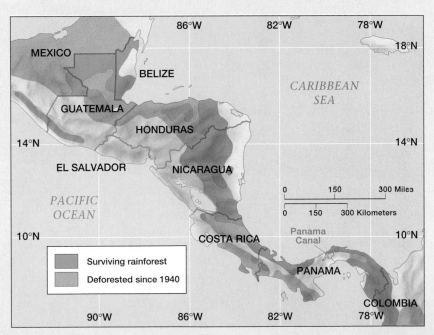

Figure 11-2-A Shrinkage in the Central American rainforest during the last half-century. This region has one of the fastest rates of deforestation in the world.

shade to the forest floor. Bursting through the canopy to form the top layer are the forest giants—tall trees that often grow to great heights above the general level. Beneath the canopy is an erratic third layer of lower trees able to survive in the shade.

Sometimes still more layers of increasingly shade-tolerant trees grow at lower levels.

Undergrowth is normally sparse in the tropical rainforest because the lack of light precludes the survival of most green plants. Only where there are gaps in the

Figure 11-2-B When rainforests are cleared for agriculture, the result often is puny crops and accelerated soil erosion. This scene is from central Thailand. (Holt Studios International [Nigel Cattlin]/Photo Researchers, Inc.)

broad-scale climatic alterations have been postulated, but these are still speculative.

The irony of tropical deforestation is that the anticipated economic benefits are usually illusory. Much of the forest clearing, especially in Latin America, is in response to the social pressure of overcrowding and poverty in societies where most of the people are landless. The governments throw open "new lands" for settlement in the rainforest. The settlers clear the land for crop or livestock. The result almost always is an initial nutrient pulse of high soil productivity, followed in only 2 or 3 years by a pronounced fertility decline as the nutrients are quickly leached and cropped out of the soil, weed species rapidly invade, and erosion becomes rampant (Figure 11-2-B). Sustainable agriculture generally can be expected only with continuous heavy fertilization, a costly procedure.

The forests, of course, are renewable. If left alone, they can regenerate, providing there are seed trees in the vicinity and the soil has not been stripped of all its nutrients. The loss of biotic diversity, however, is a much more serious problem because extinction is irreversible. Valuable potential resources—pharmaceutical products, new food crops, natural insecticides, industrial materials—may disappear before they are even discovered. Wild plants and animals that could be combined with domesticated cousins to impart resistance to disease, insects, parasites, and other environmental stresses may also be lost.

Much concern has been expressed about tropical deforestation, and some concrete steps have been taken. The development of agroforestry (planting crops with trees, rather than cutting down the trees and replacing them with crops) is being fostered in many areas. In Brazil,

which has by far the largest expanse of rainforest, some 46,000 square miles (119,000 square kilometers) of reserves have been set aside, and Brazilian law requires that any development in the Amazon region leave half the land in its natural state. In 1985 a comprehensive world plan, sponsored by the World Bank, the World Resources Institute, and the United Nations Development Programme, was introduced. It proposed concrete, country-by-country strategies to combat tropical deforestation. It is an $8 billion, five-year project, dealing with everything from fuel-wood scarcity to training foresters. Only small parts of the program have been started; its price tag makes full implementation highly unlikely.

Meanwhile, the sounds of ax, chain saw, and bulldozer continue to be heard throughout the tropical forest lands.

canopy, as alongside a river, does light reach the ground, resulting in the dense undergrowth associated with a jungle. Epiphytes like orchids and bromeliads hang from or perch on tree trunks and branches. Vines and lianas often dangle from the arching limbs.

The interior of the rainforest, then, is a region of heavy shade, high humidity, windless air, continuous warmth, and an aroma of mold and decomposition. As plant litter accumulates on the forest floor, it is acted upon very rapidly by plant and animal decomposers.

Figure 11-28 Mist rises from the forest canopy after heavy rains in the upper reaches of the Amazon basin in Ecuador. (Dr. Morley Read / Science Photo Library)

The upper layers of the forest are areas of high productivity, and there is a much greater concentration of nutrients in the vegetation than in the soil. Indeed, most selva soil is surprisingly infertile.

Rainforest fauna is largely **arboreal** (tree-dwelling) because the principal food sources are in the canopy rather than on the ground. Large animals are generally scarce on the forest floor, although there are vast numbers of invertebrates. The animal life of this biome is characterized by creepers, crawlers, climbers, and flyers—monkeys, arboreal rodents, birds, tree snakes and lizards, and multitudes of invertebrates.

TROPICAL DECIDUOUS FOREST

The distribution of the tropical deciduous forest biome is shown in Figure 11-29. The locational correlation of this biome with specific climatic types is irregular and fragmented, indicating complex environmental relationships.

There is structural similarity between the selva and the tropical deciduous forest, but several important differences are usually obvious (Figure 11-30). In the tropical deciduous forest, the canopy is less dense, the trees are somewhat shorter, and there are fewer layers, all these details being a response to either less total precipitation or less periodic precipitation. As a result of a pronounced dry period that lasts for several weeks or months, many of the trees shed their leaves at the same time, allowing light to penetrate to the forest floor. This light produces an understory of lesser plants that often grows in such density as to produce classic jungle conditions. The diversity of tree species is not as great in this biome as in the selva, but there is a greater variety of shrubs and other lesser plants.

The faunal assemblage of the tropical deciduous forest is generally similar to that of the rainforest. Although there are more ground-level vertebrates than in the selva, arboreal species are particularly conspicuous in both biomes.

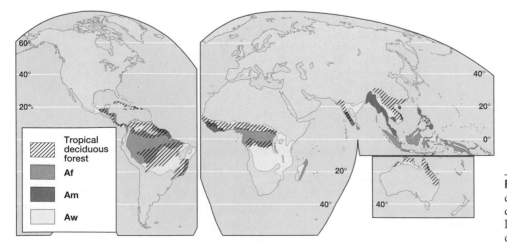

Figure 11-29 World distribution of tropical deciduous forest (black diagonal lines) compared with distribution of A climates.

Figure 11-30 A tropical deciduous forest scene at about 7000 feet (2100 meters) elevation in central Mexico. (Tom McHugh/Photo Researchers, Inc.)

TROPICAL SCRUB

The tropical scrub biome is widespread in drier portions of the A climatic realm, covering extensive areas in the tropics and subtropics (Figure 11-31). It is dominated by low-growing, scraggly trees and tall bushes, usually with an extensive understory of grasses (Figure 11-32). The trees range from 10 to 30 feet (3 to 9 meters) in height. Their density is quite variable, with the trees sometimes growing in close proximity to one another but often spaced much more openly. Species diversity is much less than in the selva and tropical deciduous forest biomes; frequently just a few species comprise the bulk of the taller growth over vast areas. In the more tropical and wetter portions of the tropical scrub biome, most of the trees and shrubs are evergreen; elsewhere most species are deciduous. In some areas a high proportion of the shrubs are thorny.

The fauna of tropical scrub regions is notably different from that of the two biomes previously discussed. There is a moderately rich assemblage of ground-dwelling mammals and reptiles, and of birds and insects.

TROPICAL GRASSLAND

As Figure 11-33 shows, there is an incomplete correlation between the distribution of the tropical grassland biome (also called the savanna biome) and that of the Aw (tropical savanna) climate. The correlation tends to be most noticeable where seasonal rainfall contrasts are greatest, a condition particularly associated with the broad-scale annual shifting of the intertropical convergence (ITC) zone.

Savanna lands are dominated by tall grasses (Figure 11-34). Sometimes the grasses form a complete ground cover, but sometimes there is bare ground among dispersed tufts of grass in what is called a bunchgrass pattern. The name *savanna* without any modifier usually refers to areas that are virtually without shrubs or trees, but this type of savanna is not the most common. In most cases, a wide scattering of both types dots the grass-covered terrain, and this mixture of plant forms is often referred to as parkland or park savanna.

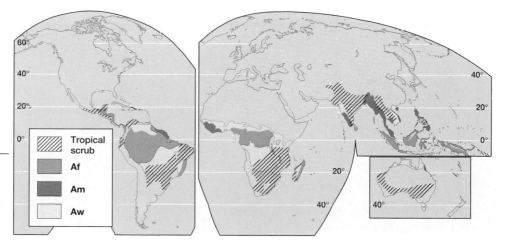

Figure 11-31 World distribution of tropical scrub (diagonal lines) compared with distribution of A climates. Most tropical scrub is found in Aw climates.

Figure 11-32 A thorn scrub scene in northern Namibia. (TLM photo)

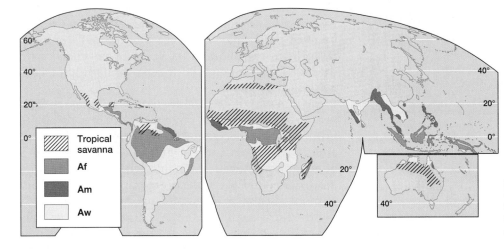

Figure 11-33 World distribution of tropical grassland (diagonal lines) compared with distribution of A climates.

Figure 11-34 A variety of ungulates occur in large herds on the Serengeti Plain of Tanzania. (Philip Kahl, Jr./Photo Researchers, Inc.)

In much of the savanna, the vegetation is degenerate because it has been created by human interference with natural processes. A considerable area of tropical deciduous forest and tropical scrub, and perhaps some tropical rainforest, has been converted to savanna over thousands of years through fires set by humans and through the grazing and browsing of domestic animals.

The savanna biome has a very pronounced seasonal rhythm. During the wet season, the grass grows tall, green, and luxuriant. At the onset of the dry season, the grass begins to wither, and before long the above-ground portion is dead and brown. At this time, too, many of the trees and shrubs shed their leaves. The third "season" is the time of wildfires. The accumulation of dry grass provides abundant fuel, and most parts of the savannas experience natural burning every year or so. The recurrent grass fires are stimulating for the ecosystem, as they burn away the unpalatable portion of the grass without causing significant damage to shrubs and trees. When the rains of the next wet season arrive, the grasses spring into growth with renewed vigor.

Savanna fauna varies from continent to continent. The African savannas are the premier "big game" lands of the world, with an unmatched richness of large animals, particularly ungulates (hoofed animals) and carnivores (meat eaters), but also including a remarkable diversity of other fauna. The Latin American savannas, on the other hand, have only a sparse population of large wildlife, with Asian and Australian areas intermediate between these two extremes.

DESERT

In previous chapters we noted a general decrease in precipitation as one moves away from the equator in the low latitudes. This progression is matched by a gradation from the selva biome of the equator to the desert biome of the subtropics. The desert biome also occurs extensively in midlatitude locations in Asia, North America, and South America (Figure 11-35).

Desert vegetation is surprisingly variable (Figure 11-36). It consists largely of drought-resisting plants with structural modifications that allow them to conserve moisture and drought-evading plants capable of hasty reproduction during brief rainy times. The plant cover is usually sparse, with considerable bare ground dotted by a scattering of individual plants. Typically the plants are shrubs, which occur in considerable variety, each with its own mechanisms to combat the stress of limited moisture. Succulents are common in the drier parts of most desert areas, and many desert plants have either tiny leaves or no leaves at all, as a moisture-conserving strategy. Grasses and other herbaceous plants are widespread but sparse in desert areas. Despite the dryness, trees can be found sporadically in the desert, especially in Australia.

Animal life is exceedingly inconspicuous in most desert areas, leading to the erroneous idea that animals are nonexistent. In actuality, most deserts have a moderately diverse faunal assemblage, although the variety of large mammals is limited. A large proportion of desert animals avoid the principal periods of desiccating heat (daylight in general and the hot season in particular) by resting in burrows or crevices during the day and prowling at night.

Generally speaking, life in the desert biome is characterized by an appearance of stillness. In favorable times (at night, and particularly after rains) and in favored places (around water holes and oases), however, there is a great increase in biotic activities, and sometimes the total biomass is of remarkable proportions.

MEDITERRANEAN WOODLAND AND SHRUB

As Figure 11-37 shows, the mediterranean woodland and shrub biome is found in six widely scattered and relatively small areas of the midlatitudes, all of which

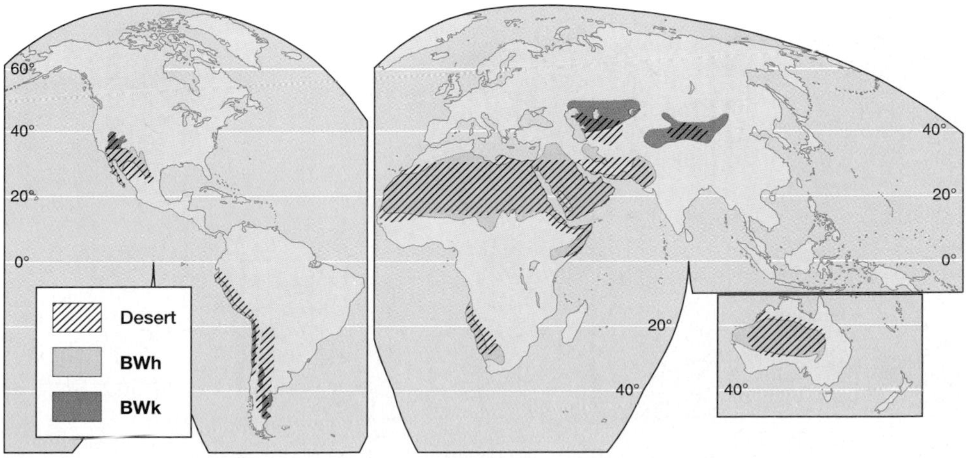

Figure 11-35 World distribution of desert (diagonal lines) compared with distribution of B climates.

Figure 11-36 Natural vegetation typically is sparse and spindly in deserts. Here a scattered growth of creosote bushes punctuates the otherwise bare ground of the landscape near Palm Springs in southern California. (Tom McHugh/Photo Researchers, Inc.)

experience the pronounced dry summer/wet winter precipitation typical of mediterranean climates. In this biome, the dominant vegetation associations are physically very similar to each other but taxonomically quite varied. The biome is dominated mostly by a dense growth of woody shrubs, known as a *chaparral* in North America but having other names in other areas. A second significant plant association of mediterranean regions is an open grassy woodland, in which the ground is almost completely grass-covered but has a considerable scattering of trees as well.

The plant species vary from region to region. Oaks of various kinds are by far the most significant genus in the Northern Hemisphere mediterranean lands, sometimes occurring as prominent medium-sized trees but also appearing as a more stunted, shrubby growth. In all areas, the trees and shrubs are primarily broadleaf evergreens. Their leaves are mostly small and have a leathery texture or waxy coating, which inhibits water loss during the long dry season. Moreover, most plants have deep roots.

Summer is a virtually rainless season in mediterranean climates, and so summer fires are relatively common. Many of the plants are adapted to rapid recovery after a wildfire has swept over the area. Indeed, as noted in Chapter 10, some species have seeds that are released for germination only after the heat of a fire has caused their seedpods to open. Part of the seasonal rhythm of this biome is that winter floods sometimes follow summer fires, as slopes left unprotected by the burning away of grass and lower shrubs are susceptible to abrupt erosive runoff if the winter rains arrive before the vegetation has a chance to resprout (Figure 11-38).

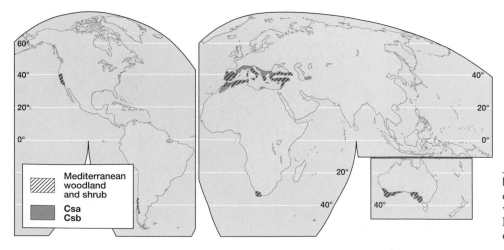

Figure 11-37 World distribution of mediterranean woodland and shrub (diagonal lines) compared with distribution of Cs climates.

(a)

(b)

AUSTRALIA

(c)

(d)

Figure 11-38 Wildfires are a natural and recurrent event in many biomes, but the seasonal rhythm in mediterranean lands is particularly notable. These four photographs record the annual pattern in the Mount Lofty Ranges of South Australia. **(a)** Winter is moist, mild, and green. **(b)** Early summer is hot, dry, and brown. **(c)** Then comes the fire season. **(d)** Late summer is the black time. (TLM photos)

The fauna of this biome is not particularly distinctive. Seed-eating, burrowing rodents are common, as are some bird and reptile groups. There is a general overlap of animals between this biome and adjacent ones.

MIDLATITUDE GRASSLAND

Vast grasslands occur widely in the midlatitudes of North America and Eurasia (Figure 11-39). In the Northern Hemisphere the locational coincidence between this biome and the steppe climatic type is very pronounced. The smaller Southern Hemisphere areas (mostly the *pampa* of Argentina and the *veldt* of South Africa) have less distinct climatic correlations.

The vegetation typical of a grassland biome is a general response either to a lack of precipitation suf-ficient to support larger plant forms or to the frequency of fires (both natural and human-induced) that prevent the growth of tree or shrub seedlings. In the wetter areas of a grassland biome, the grasses grow tall and the term *prairie* is often applied. In drier regions, the grasses are shorter; such growth is often referred to as *steppe* (Figure 11-40). Sometimes a continuous ground cover is missing, and the grasses grow in discrete tufts as bunchgrass or tussock grass.

Most of the grass species are perennials, lying dormant during the winter and sprouting anew the following summer. Trees are mostly restricted to riparian locations, whereas shrubs and bushes occur sporadically on rocky sites. Grass fires are fairly common in summer, which helps to explain the relative scarcity of shrubs. The woody plants cannot tolerate fires and generally can survive only on dry slopes where there is little grass cover to fuel a fire.

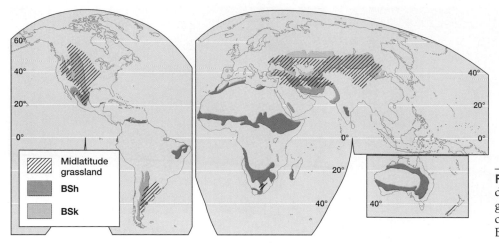

Figure 11-39 World distribution of midlatitude grassland (diagonal lines) compared with distribution of BS climates.

Grasslands provide extensive pastures for grazing animals, and before encroachment by humans drastically changed population sizes, the grassland fauna comprised large numbers of relatively few species. The larger herbivores often were migratory prior to human settlement. Many of the smaller animals spend all or part of their lives underground, where they find some protection from heat, cold, and fire.

MIDLATITUDE DECIDUOUS FOREST

Extensive areas on all Northern Hemisphere continents, as well as more limited tracts in the Southern Hemisphere, were originally covered with a forest of largely broadleaf deciduous trees (Figure 11-41). Except in hilly country, a large proportion of this forest has been cleared for agriculture and other types of human use, so that very little of the original natural vegetation remains.

The forest is characterized by a fairly dense growth of tall broadleaf trees with interwoven branches that provide a complete canopy in summer (Figure 11-42). Some smaller trees and shrubs exist at lower levels, but for the most part, the forest floor is relatively barren of undergrowth. In winter the appearance of the forest changes dramatically owing to the seasonal fall of leaves.

Trees species vary considerably from region to region, although most are broadleaf and deciduous. The principal exception is in eastern Australia, where the forest is composed almost entirely of varieties of eucalyptus, which are broadleaf evergreens. Northern Hemisphere regions have a northward gradational mixture with needleleaf evergreen species. An unusual situation in the southeastern United States finds extensive stands of pines (needleleaf evergreens) rather than deciduous species occupying most of the well-drained sites above the valley bottoms. In the Pacific Northwest

Figure 11-40 Pronghorn antelope on a short-grass plain in the midlatitude grassland of central Wyoming. (TLM photo)

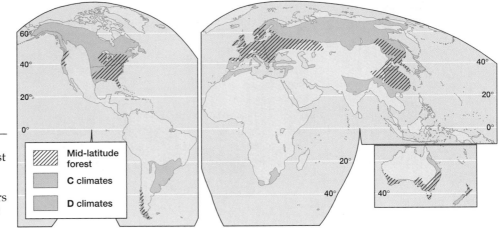

Figure 11-41 World distribution of midlatitude forest (diagonal lines) before human activities made significant changes in this biome. The colors indicate distribution of C and D climates.

//// Mid-latitude forest

☐ C climates

☐ D climates

of the United States, the forest association is primarily evergreen coniferous rather than broadleaf deciduous.

This biome generally has the richest assemblage of fauna to be found in the midlatitudes, although it does not have the diversity to match that of most tropical biomes. It has (or had) a considerable variety of birds and mammals, and in some areas, reptiles and amphibians are well represented. Summer brings a diverse and active population of insects and other arthropods. All animal life is less numerous (partly due to migrations and hibernation) and less conspicuous in winter.

BOREAL FOREST

One of the most extensive biomes is the **boreal forest**, sometimes called *taiga* after the Russian word for the northern fringe of the boreal forest in that country (paralleling the way these synonyms are used to de-scribe climates, as mentioned in Chapter 8). The boreal forest occupies a vast expanse of northern North America and Eurasia (Figure 11-43). There is very close correlation between the location of the boreal forest biome and the subarctic climatic type, with a similar correlation between the locations of the tundra climate and the tundra biome.

This great northern forest contains perhaps the simplest assemblage of plants of any biome (Figure 11-44). Most of the trees are conifers, nearly all needleleaf evergreens, with the important exception of the tamarack or larch, which drops its needles in winter. The variety of species is limited to mostly pines, firs, and spruces extending broadly in homogeneous stands. In some places, the coniferous cover is interrupted by areas of deciduous trees. These deciduous stands are also of limited variety (mostly birch, poplar, and aspen), and often represent a seral situation following a forest fire.

Figure 11-42 Much of the northeastern United States was originally covered with a deciduous forest. This forest persists today mostly in hilly or mountainous areas. This view is in the Shenandoah Mountains of Virginia. (TLM photo)

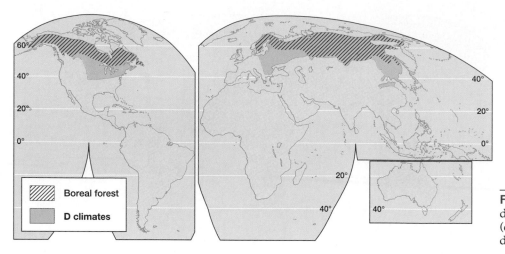

Figure 11-43 World distribution of boreal forest (diagonal lines) compared with distribution of D climates.

The trees grow taller and more densely near the southern margins of this biome, where the summer growing season is longer and warmer. Near the northern margins, the trees are spindly, short, and more openly spaced. Undergrowth is normally not dense beneath the forest canopy, but a layer of deciduous shrubs sometimes grows in profusion. The ground is usually covered with a complete growth of mosses and lichens, with some grasses in the south and a considerable accumulation of decaying needles over all.

Poor drainage is typical in summer, due partially to permanently frozen subsoil, which prevents downward percolation of water, and partially to the derangement of normal surface drainage by the action of glaciers during the recent Ice Age (Pleistocene). Thus bogs and swamps are numerous, and the ground generally is spongy in summer. During the long winters, of course, all is frozen.

The immensity of the boreal forest gives an impression of biotic productivity, but such is not the case. Harsh climate, floristic homogeneity, and slow plant growth produce only a limited food supply for animals. Faunal species diversity is limited, although the number of individuals of some species is astounding. With relatively few animal species in such a vast biome, populations sometimes fluctuate enormously within the space of only a year or so. Mammals are represented prominently by fur-bearers and by a few species of ungulates. Birds are numerous and fairly diverse in summer, but nearly all migrate to milder latitudes in winter. Insects are totally absent in winter but superabundant during the brief summer.

Figure 11-44 The boreal forest contains trees that generally are short, close-growing, and of uniform species composition. This spruce forest surrounds Shady Lake near Prince Albert in the Canadian province of Saskatchewan. (TLM photo)

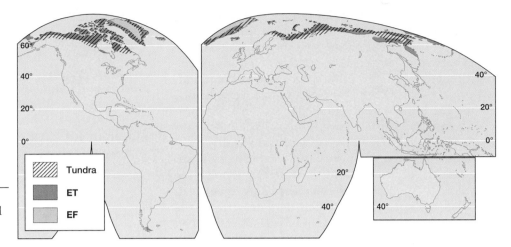

Figure 11-45 World distribution of tundra (diagonal lines) compared with distribution of E climates.

TUNDRA

The tundra is essentially a cold desert in which moisture is scarce and summers so short and cool that trees are unable to survive. This biome is distributed across the top of the Northern Hemisphere (Figure 11-45). The plant cover consists of a considerable mixture of species, many of them in dwarf forms (Figure 11-46). Included are grasses, mosses, lichens, flowering herbs, and a scattering of low shrubs. These plants often occur in a dense, ground-hugging arrangement, although some places have a more sporadic cover with considerable bare ground interspersed. The plants complete their annual cycles hastily during the brief summer, when the ground is often moist and waterlogged because of inadequate surface drainage and particularly inadequate subsurface drainage.

Animal life is dominated by birds and insects during the summer. Extraordinary numbers of birds flock to the tundra for summer nesting, migrating southward as winter approaches. Mosquitoes, flies, and other insects proliferate astoundingly during the short warm season, laying eggs that can survive the bitter winter. Other forms of animal life are scarcer—a few species of mammals and freshwater fishes but almost no reptiles or amphibians.

HUMAN MODIFICATION OF NATURAL DISTRIBUTION PATTERNS

Thus far in our discussion of the biosphere, our attention has been focused on "natural" conditions, that is, events and processes that take place in nature with-

Figure 11-46 Tundra vegetation consists of a mixture of ground-hugging species, as seen here in central Alaska, with the Alaska Range in the background. (Charlie Ott/Photo Researchers, Inc.)

out the aid or interference of human activities. Such natural processes have been going on for millennia, and their effects on floral and faunal distribution patterns normally have been very slow and gradual. The pristine environment, uninfluenced by humankind, experiences its share of abrupt and dramatic events, to be sure, but environmental changes generally proceed at a very leisurely pace. When *Homo sapiens* appear, however, the tempo changes dramatically.

People are capable of exerting extraordinary influences on the distribution of plants and animals. Not only is the magnitude of the changes likely to be great, but also the speed with which they are effected is sometimes exceedingly rapid. In broadest perspective, humankind exerts three types of direct influences on biotic distributions—physical removal of organisms, habitat modification, and artificial translocation of organisms.

PHYSICAL REMOVAL OF ORGANISMS

One of humankind's most successful skills is in the elimination of other living things. As human population increases and spreads over the globe, there is often a wholesale removal of native plants and animals to make way for the severely modified landscape that is thought necessary for civilization. The natural plant and animal inhabitants are cut down, plowed up, paved over, burned, poisoned, shot, trapped, and otherwise eradicated in actions that have far-reaching effects on overall distribution patterns.

HABITAT MODIFICATION

Habitat modification is another activity in which humankind excels. The soil environment is changed by farming, grazing, engineering, and construction practices; the atmospheric environment is degraded by the introduction of impurities of various kinds; the waters of the planet are impounded, diverted, and polluted. All these deeds influence the native plants and animals in the affected areas (Figure 11-47).

ARTIFICIAL TRANSLOCATION OF ORGANISMS

People are capable of elaborate rearrangement of the natural complement of plants and animals in almost every part of the world. This is shown most clearly with domesticated species—crops, livestock, pets. There is now, for example, more corn than native grasses in Iowa, more cattle than native gazelles in the Sudan, more canaries than native thrushes in Detroit.

In our study of physical geography, however, we are not concerned with domesticated conditions; our interest is in the natural state. Even so, humans have accounted for many introductions of wild plants and animals into "new" habitats; such organisms are called **exotics** in their new homelands.

In some cases, the introduction of exotic species was deliberate. A few examples among a great many include the taking of prickly-pear cactus from Arizona to

Figure 11-47 An overgrazed range in the Colorado high country illustrates human-induced habitat modification caused by overstocking of livestock in the area to the right of the fence. The locale is near Central City. (TLM photo)

Australia, crested wheat grass from Russia to Kansas, European boar from Germany to Tennessee, pronghorn antelope from Oregon to Hawaii, and red fox from England to New Zealand. Frequently, however, the introduction of an exotic species was an accidental result of human carelessness. The European flea, for example, has become one of the most widespread creatures on Earth because it has been an unseen accompaniment to human migrations all over the world.

Similarly, the English sparrow and European brown rat have been inadvertently introduced to all inhabited continents by traveling as stowaways on ships.

One other type of human-induced translocation of animals involves the deliberate release or accidental escape of livestock to become established as a "wild" (properly termed *feral*) population. This has happened in many parts of the world, most notably in North America and Australia (Figure 11-48).

Figure 11-48 Burros are prominent feral animals in much of our southwestern desert area. This group is in the Panamint Mountains of southern California. (TLM photo)

CHAPTER SUMMARY

The natural distribution pattern of any biotic group can usually be explained by the group's evolutionary development, migration/dispersal history, reproductive success, and trend toward extinction.

Most terrestrial vegetation is hardy, having evolved various protective mechanisms, mostly physiological and reproductive, as a shield against such environmental hardships as inadequate moisture (xerophytic adaptations), excessive moisture (hygrophytic adaptation), and temperature extremes. Competition, both intraspecific and interspecific, is critical to plant survival.

Plants are categorized into four groups: bryophytes and pteridophytes are relatively simple spore-bearing plants, whereas gymnosperms and angiosperms are more complex seed-bearing plants.

The principal major floristic associations include forests, woodlands, shrublands, grasslands, deserts, tundra, and wetlands.

Terrestrial fauna has also evolved various adaptations—physiological, behavioral, and reproductive—to cope with diverse environmental limitations. Competition among animals involves both indirect rivalry for territory and resources and the direct antagonism of predation. Among some animals, however, there are mutually beneficial relationships called symbiotic relationships. For general geographic purposes, useful faunal groups are invertebrates, fishes, amphibians, reptiles, birds, and mammals. Worldwide, nine zoogeographical regions are generally recognized.

The major terrestrial biomes are tropical rainforest, tropical deciduous forest, tropical scrub, tropical grassland (savanna), desert, mediterranean woodland and shrub, midlatitude grassland, midlatitude deciduous forest, boreal forest, and tundra.

Natural distribution patterns of biota often are severely altered by human activities through physical removal, habitat modification, and artificial translocation.

KEY TERMS

adret slope	gymnosperm	selva
angiosperm	hardwood	seral association
annual	herbaceous plant	shrubland
arboreal	hydrophyte	softwood
boreal forest	hygrophyte	steppe
broadleaf	invertebrate	succulent
bryophyte	mutualism	symbiosis
climax vegetation	needleleaf	tundra
commensalism	parasitism	ubac slope
deciduous	perennial	vertebrate
desert	plant succession	vertical zonation
evergreen	prairie	wetland
exotic	pteridophyte	woodland
forest	riparian vegetation	woody plant
grassland	savanna	xerophyte

REVIEW QUESTIONS

1. Explain the concept of plant succession.
2. What are some basic characteristics that distinguish plants from animals?
3. Describe some typical xerophytic adaptations of plants.
4. Name a tree that is an example of each of the following: gymnosperm; angiosperm; coniferous; deciduous; evergreen; hardwood; softwood.
5. Explain the concept of climax vegetation.
6. What is the difference between forest and woodland?
7. Distinguish among savanna, prairie, and steppe.
8. Why do geographers study plants more than they study animals?
9. Explain the statement, "Altitude compensates for latitude."
10. What is riparian vegetation?
11. Distinguish among the three ways (physiological, behavioral, reproductive) animals adapt to the environment.
12. Distinguish between mutualism and parasitism.
13. List and briefly describe the principal groups of vertebrates.
14. Select one zoogeographic region and describe its principal characteristics.

SOME USEFUL REFERENCES

COLE, M. M. , *The Savannas: Biogeography and Geobotany.* London: Academic Press, 1986.

COLLINS, MARK (ed.), *The Last Rain Forests: A World Conservation Atlas.* Don Mills, Ontario: Oxford University Press, 1991.

COLLINSON, A. S. , *Introduction to World Vegetation*, 2nd ed. London: Unwin Hyman, 1988.

FURLEY, P. A., AND W. W. NEWEY, *Geography of the Biosphere.* London: Butterworths, 1983.

JARVIS, P. J., *Plant and Animal Introductions.* New York: Blackwell, 1989.

LIDICKER, WILLIAM Z., AND ROY L. CALDWELL, *Dispersal and Migrations.* New York: Hutchinson Ross, 1982.

MORAIN, STANLEY A., *Systematic and Regional Biogeography.* New York: Van Nostrand Reinhold, 1984.

MYERS, A. A., AND P. S. GILLER (eds.), *Analytical Biogeography: An Integrated Approach to the Study of Animal and Plant Distributions.* New York: Chapman & Hall, 1988.

PARK, CHRIS C., *Tropical Rainforests.* New York: Routledge, 1992.

PEARS, N., *Basic Biogeography.* London: Longman, 1985.

STONEHOUSE, B., *Polar Ecology.* New York: Blackie & Son, 1988.

TIVY, JOY, *Biogeography: A Study of Plants in the Ecosphere*, 3rd ed. London: Longman Group, 1993.

WILCOCK, C. C., *Plant–Animal Interactions.* New York: Blackie & Son, 1989.

12

SOILS

THE FINAL MAJOR COMPONENT IN OUR STUDY OF Earth's environment is the lithosphere. This fourth sphere is just as complex as the atmosphere, biosphere, or hydrosphere but contrasts with these other realms in its enormity and particularly in its seeming stability.

It is easy to observe change in the three other spheres—clouds forming, flowers blooming, rivers flowing—but the dynamics of the lithosphere, with a few spectacular exceptions, operate with such incredible slowness that Earth's crust often appears changeless. Most laypersons consider the phrase "the everlasting hills" a literal expression aptly describing the permanence of Earth's topography, whereas in reality the phrase is merely hyperbole that fails to recognize the remarkable alterations that take place over time.

A few lithospheric processes, such as earthquakes and volcanic eruptions, occur abruptly, but the vast majority operate so slowly as to be unrecognizable by the casual observer. Thus the student of topographic change must search for clues about the ponderous processes that shape the land, clues from the past that help us interpret the landscape of the present and predict that of the future. Our goal in the remaining chapters of this book is to understand the contemporary character of Earth's surface and to explain the processes that have caused it to be as it is.

SOIL AND REGOLITH

Although "lithosphere" encompasses all of the planet, from surface to core, the part that holds our attention here is soil, the topmost layer. Soil is the essential medium in which all terrestrial life is nurtured. Almost all land plants sprout from this precious medium that is spread so thinly across the continental surfaces, with an average worldwide depth of only about 6 inches (15 centimeters).

Despite the implication of the well-known simile "as common as dirt," soil is extremely complex. It is an infinitely varying mixture of weathered mineral particles, decaying organic matter, living organisms, gases, and liquid solutions.

Preeminently, however, soil is a zone of plant growth. Although the concept almost defies definition, **soil** can be conceptualized as a relatively thin surface layer of mineral matter that normally contains a considerable amount of organic material and is capable of supporting living plants. It occupies that part of the outer skin of Earth that extends from the surface down to the maximum depth to which living organisms penetrate, which means basically the area occupied by plant roots. Soil is characterized by its ability to produce and store plant nutrients, an ability made possible by the interactions of such diverse factors as water, air, sunlight, rocks, plants, and animals.

Although thinly distributed over the land surface, soil functions as a fundamental interface where atmosphere, lithosphere, hydrosphere, and biosphere meet. The bulk of most soils is inorganic material, and so soil is usually classified as part of the lithosphere, but its relationship to the other three spheres is both intimate and complex.

Soil development begins with the physical and chemical disintegration of rock exposed to the atmosphere and to the action of water percolating down from the surface. This disintegration is called weathering. As we shall learn in Chapter 15, the basic result of weathering is the weakening and breakdown of solid rock, the fragmentation of coherent rock masses, the making of little rocks from big ones. The principal product is a layer of loose inorganic material called **regolith** ("blanket rock") because it lies like a blanket over the unfragmented rock below (Figure 12-1). Normally the regolith has a crude gradation of particle sizes, with the largest and least fragmented pieces at the bottom, immediately adjacent to the bedrock.

Above the regolith, usually but not invariably, is soil. (Geographers usually consider soil as being separate from regolith, but some earth scientists regard soil as part of the regolith.) Soil is composed largely of finely fragmented mineral particles, the ultimate product of weathering. It normally also contains an abundance of living plant roots, dead and rotting plant parts, microscopic plants and animals both living and dead, and a variable amount of air and water. Soil is not the end product of a process, but rather a stage in a never-ending continuum of physical/chemical/biotic activities (Figure 12-2).

SOIL AS A COMPONENT OF THE LANDSCAPE

The surface of the lithosphere usually, but not always, is represented by soil. Although extremely pervasive, soil is an inconspicuous component of the landscape because it is beneath our feet and its presence normally is masked either by vegetation or by human constructed features.

We recognize soil in the landscape mostly by its color, which often can be seen through the filigree of plant life that covers it. A second main aspect of soil—its depth—become obvious only where some of its vertical dimension is exposed by gully erosion, road cuts, or some other type of excavation.

Deep black soils in the floodplain of the Skagit River in the state of Washington. *(Jim Foster/The Stock Market)*

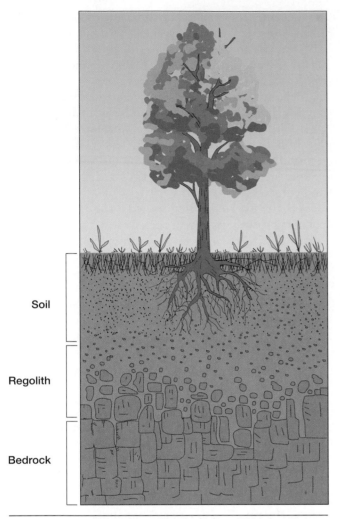

Figure 12-1 Vertical cross section from surface to bedrock, showing the relationship between soil and regolith. The distinction between the two is that soil contains organic matter as well as disintegrated mineral matter from weathered rocks and can therefore support plant life. Regolith is weathered rock material only; it contains no organic matter.

SOIL-FORMING FACTORS

Five variables are the principal soil-forming factors: geology, climate, topography, biology, and time. Geology, topography, and time are passive factors, and climate and biology are active factors.

THE GEOLOGIC FACTOR

The source of the rock fragments that make up soil is **parent material**, which may be either bedrock or loose sediments transported from elsewhere by water, wind, or ice. Whatever the parent material, it is sooner or later disintegrated and decomposed at and near Earth's surface, providing the raw material for soil formation.

The nature of the parent material often influences the characteristics of the soil that develops from it, and, particularly in the early stages of soil formation, this factor sometimes dominates all others. The chemical composition of parent material obviously is reflected in the resulting soil, and parent-material physical characteristics also may be influential in soil development, particularly in terms of texture and structure. Bedrock that weathers into large particles (as does sandstone, for example) normally produces a coarse-textured soil, one easily penetrated by air and water to some depth. Bedrock that weathers into minute particles (shale, for example) yields soil containing very few pore spaces for air and water penetration.

Young soils are likely to be very reflective of the rocks or sediments from which they were derived. With the passage of time, however, other soil-forming factors become increasingly important, and the significance of the parent material diminishes. Eventually the influence of the parent material may be completely obliterated, so that sometimes it is impossible to ascertain the nature of the rock from which the soil evolved.

THE CLIMATIC FACTOR

Temperature and moisture are the climatic variables of greatest significance to soil formation. As a basic generalization, both the chemical and biological processes in soil are usually accelerated by high temperatures and abundant moisture and are slowed by low temperatures and lack of moisture. One predictable result is that soils tend to be deepest in warm, humid regions and shallowest in cold, dry regions.

It is difficult to overemphasize the role of moisture moving through the soil. The flow is mostly downward because of the pull of gravity, but it is sometimes lateral in response to drainage opportunities and sometimes, in special circumstances, even upward. Wherever and however water moves, it always carries dissolved chemicals in solution and usually also carries tiny particles of matter in suspension. Thus moving water is ever engaged in rearranging the chemical and physical components of the soil, as well as contributing to the variety and availability of plant nutrients.

In terms of general soil characteristics, climate is likely to be the most influential factor in the long run. This generalization has many exceptions, however, and when soils are considered on a local scale, climate is likely to be less prominent as a determinant.

THE TOPOGRAPHIC FACTOR

Slope and drainage are the two main features of topography that influence soil characteristics. Wherever soil develops, its vertical extent undergoes continuous, if

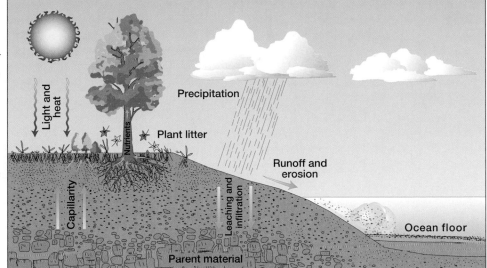

Figure 12-2   Soil develops through a complex interaction of physical, chemical, and biological processes. Parent-material bedrock weathers to regolith, and then plant litter combines with the regolith to form soil. Some of that soil washes to the ocean floor, where, over the eons of geologic time, it is transformed to sedimentary rock. Someday that ocean floor will be uplifted above sea level and the exposed sedimentary rock will again be weathered into soil.

usually very slow, change. This change comes about through a lowering of both the bottom and the top of the soil layer (Figure 12-3). The bottom slowly gets deeper as weathering penetrates deeper into the regolith and parent material and as plant roots extend to greater depths. At the same time, the soil surface is being lowered by sporadic removal of its uppermost layer through normal erosion, which is the removal of individual soil particles by running water, wind, and gravity.

Where the land is flat, soil tends to develop at the bottom more rapidly than it is eroded away at the top. This does not mean that the downward development is speedy; rather it means that surface erosion is extraordinarily slow. Thus, the deepest soils are usually on flat land. Where slopes are relatively steep, surface erosion is more rapid than soil deepening, with the result that such soils are nearly always thin and immaturely developed (Figure 12-4).

Because most soils are well drained, moisture relationships are relatively unremarkable factors in soil de-

velopment. Some soils have inefficient natural drainage, however, a condition that imparts significantly different characteristics. For example, a waterlogged soil tends to contain a high proportion of organic matter, and the biological and chemical processes that require free oxygen are impeded (because air is the source of the needed oxygen and a waterlogged soil contains essentially no air). Most ill-drained soils are in valley bottoms or in some other flat locale because soil drainage is usually related to slope.

In some cases, such subsurface factors as permeability and the presence or absence of impermeable layers are more influential than slope.

THE BIOLOGICAL FACTOR

From a volume standpoint, soil is about half mineral matter and about half air and water, with only a small fraction of organic matter. However, the organic fraction, consisting of both living and dead plants and animals, is of utmost importance. The biological factor in

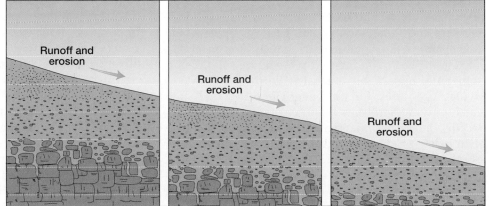

Figure 12-3   Over time, the extent of soil undergoes slow, continuous change. The bottom of the soil layer is lowered by the break-up of parent material as weathering processes extend deeper into the regolith and bedrock. At the same time, the top of the soil layer can be lowered through erosional processes.

Figure 12-4 Slope is a critical determinant of soil depth. On flat land, soil normally develops more deeply with the passage of time because there is very little erosion washing away the topmost soil. On a slope, the rate of erosion is equal to or greater than the rate at which soil is formed at the bottom of the soil layer, with the result that the soil remains shallow.

particular gives life to the soil and makes it more than just dirt. Every soil contains a quantity (sometimes an enormous quantity) of living organisms, and every soil incorporates some (sometimes a vast amount of) dead and decaying organic matter.

Vegetation of various kinds growing in soil performs certain vital functions. Plant roots, for instance, work their way down and around, providing passageways for drainage and aeration, as well as being the vital link between soil nutrients and the growing plants. Many kinds of animals contribute to soil development as well. Even such large surface-dwelling creatures as elephants and bison affect soil formation by compaction with their hooves, rolling in the dirt, grazing the vegetation, and dropping excreta. Many smaller animals spend most or all of their lives in the soil layer, tunneling here and there, moving soil particles upward and downward, and providing passageways for water and air (Figure 12-5). Mixing and plowing by soil fauna is sometimes remarkably extensive. Earthworms, in particular, are noted for the beneficial effect that their earth-moving activities have on plant growth. Ants, worms, and all other land animals fertilize the soil with their waste products and contribute their carcasses for eventual decomposition and incorporation into the soil.

Another important component of the biological factor is the microorganisms, both plant and animal, that occur in uncountable billions. An estimated three-

Figure 12-5 Like other burrowing animals, prairie dogs contribute to soil development by bringing subsoil to the surface and providing passageways for air and moisture to get underground. (Tom McHugh/Photo Researchers, Inc.)

quarters of a soil's metabolic activity is generated by microorganisms. These microbes help release nutrients from dead organisms for use by live ones by decomposing organic matter into **humus**, a dark adhesive of minute particles, and by converting nutrients to forms usable by plants. Algae, fungi, protozoans, actinomycetes, and other minuscule organisms all play a role in soil development, but bacteria probably make the greatest contribution overall. This is because certain types of bacteria are responsible for the decomposition and decay of dead plant and animal material and the consequent release of nutrients into the soil.

THE CHRONOLOGICAL FACTOR

For soil to develop on a newly exposed land surface requires time, with the length of time needed varying according to the nature of the exposed parent material and the characteristics of the environment. Soil-forming processes are generally very slow, and tens of centuries may be required for an inch of soil to form on a newly exposed surface. A warm, moist environment is inducive to soil development. Normally of much greater importance, however, are the attributes of the parent material. For example, soil develops from sediments relatively quickly and from bedrock relatively slowly.

Most soil develops with geologic slowness—so slowly that changes are almost imperceptible within a human life span. It is possible, however, for a soil to be degraded, either through the physical removal associated with accelerated erosion or through depletion of nutrients, in only a few years (Figure 12-6). In the grand scale of geologic time, then, soil can be formed and re-formed, but in the dimension of human time, it is a nonrenewable resource.

Figure 12-6 Accelerated erosion cutting a deep gully in the Coast Ranges of central California. (Richard R. Hansen/ Photo Researchers, Inc.)

SOIL COMPONENTS

Soil is made up of a variety of natural components existing together in myriad combinations. All these components can be classified, however, into just a few main groups: inorganics, organics, air, and water.

INORGANIC MATERIALS

As mentioned above, the bulk of most soils is mineral matter, mostly in the form of small but macroscopic particles. Inorganic material also occurs as microscopic clay particles and as dissolved minerals in solution.

About half the volume of an average soil is small, granular mineral matter called sand and silt. These particles may consist of a great variety of minerals, depending on the nature of the parent material from which they were derived, and are simply fragments of the wasting rock. Most common are bits of quartz, which are composed of silica (SiO_2) and appear in the soil as very resistant grains of sand. Other prominent minerals making up sand and silt are some of the feldspars and micas.

The smallest particles in the soil are clay, which is usually a combination of silica and of oxides of aluminum and iron found only in the soil and not in the parent material. Clay has properties significantly different from those of larger (sand and silt) fragments. Most clay particles are colloidal in size, which means they are larger than molecules but too small to be seen with the naked eye. They are usually flat platelets, as Figure 12-7 illustrates, and therefore have a relatively large surface area. For this reason, clay has an important influence on chemical activity in the soil because many chemical reactions occur at the surfaces of soil particles. The platelets group together in loose, sheet-like assemblages, and water moves easily between these sheets. Substances dissolved in the water are at-

tracted to and held by the sheets (Figure 12-7). Since the sheets are negatively charged, they attract positively charged ions (**cations**). Many essential plant nutrients occur in soil solutions as cations, with the result that clay is an important reservoir for plant nutrients, just as it is for soil water.

ORGANIC MATTER

Although organic matter generally constitutes less than 5 percent of total soil volume, it has an enormous influence on soil characteristics and plays a fundamental role in the biochemical processes that make soil an effective medium of plant growth. Some of the organic matter is living organisms, some is dead but undecomposed plant parts and animal carcasses, some is totally decomposed and so has become humus, and some is in an intermediate stage of decomposition.

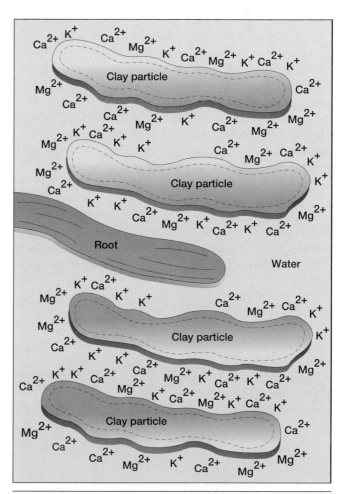

Figure 12-7 Clay particles offer a large surface area on which substances dissolved in soil water can cling. The particles are negatively charged and therefore attract cations (positively charged ions) from the water. These cations held by the clay are then absorbed by plant roots and become nutrients for the plant.

Apart from plant roots, evidence of the variety and bounty of organisms living in the soil may be inconspicuous or invisible, but most soils are seething with life. A single acre (0.4 hectare) may contain a million earthworms, and the total number of organisms in an ounce (28 grams) of soil is likely to exceed 100 trillion. Microorganisms far exceed more complex life forms, both in total numbers and in cumulative mass. They are active in rearranging and aerating the soil and in yielding waste products that are links in the chain of nutrient cycling. Some make major contributions to the decay and decomposition of dead organic matter, and others make nitrogen available for plant use.

Leaves, twigs, stalks, and other dead plant parts accumulate at the soil surface, where they are referred to collectively as **litter**. The eventual fate of most litter is decomposition, in which the solid parts are broken down into chemical elements, which are then either absorbed into the soil or washed away. In cold, dry areas, litter may remain undecomposed for a very long time; where the climate is warm and moist, however, decomposition may take place almost as rapidly as litter accumulates.

After most of the residues have been decomposed, a brown or black, gelatinous, chemically stable organic matter remains; this is referred to as humus. This "black gold" is of utmost importance to agriculture because it loosens the structure and lessens the density of the soil, thereby facilitating root development. Moreover, humus, like clay, is a catalyst for chemical reactions and a reservoir for plant nutrients and soil water.

Soil Air

Nearly half the volume of an average soil is made up of pore spaces (Figure 12-8). These spaces provide a labyrinth of interconnecting passageways, called *inter-* *stices*, among the soil particles. This labyrinth lets air and water penetrate into the soil. On the average, the pore spaces are about half filled with air and half with water, but at any given time and place, the amounts of air and water are quite variable, the quantity of one varying inversely with that of the other (Figure 12-9).

The characteristics of air in the soil are significantly different from those of atmospheric air. Soil air is found in openings generally lined with a film of water, and since this air exists in such close contact with water and is not exposed to moving air currents, it is saturated with moisture. Soil air is also very rich in carbon dioxide and poor in oxygen because plant roots and soil organisms remove oxygen from, and respire carbon dioxide into, the pore spaces. The carbon dioxide then slowly escapes into the atmosphere.

Soil Water

Water comes into the soil largely by percolation of rainfall and snowmelt, but some is also added from below when groundwater is pulled up above the water table by capillary action (Figure 12-10). Once water has penetrated the soil, it envelops in a film of water each solid particle that it contacts, and it either wholly or partially fills the pore spaces. Water can be lost from the soil by percolation down into the groundwater, by upward capillary movement to the surface followed by evaporation, or by plant use (transpiration).

Water performs a number of important functions in the soil. It is an effective solvent, dissolving essential soil nutrients and making them available to plant roots. These dissolved nutrients are carried downward in solution, to be partly redeposited at lower levels. This process, called **leaching**, tends to deplete the topsoil of soluble nutrients. Water is also required for many of the chemical reactions of clay and for the actions of the microorganisms that produce humus. In addition, it can

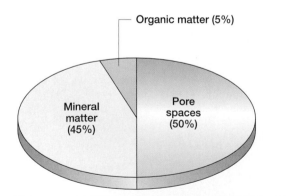

Figure 12-8 The best soil for plant growth is about half solid material (by volume) and about half pore spaces. Most of the solids are mineral matter, with only a small amount being organic. On average, about half of the pore spaces in an ideal soil are filled with air and the other half are filled with water.

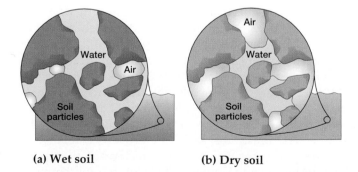

(a) Wet soil   **(b) Dry soil**

Figure 12-9 The relative amounts of water and air in soil pores vary from place to place and from time to time. **(a)** The interstices of wet soil contain much water and little air. **(b)** In dry soil there is much air and little water.

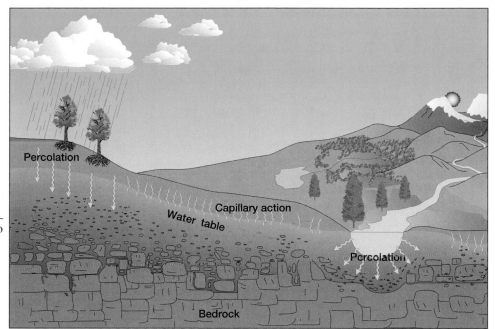

Figure 12-10 Water is added to the soil layer by the percolation of rainwater and snowmelt from above. Additional moisture enters the soil from below as groundwater is pulled up above the water table by capillary action.

have considerable influence on the physical characteristics of soil by moving particles around.

As water percolates into the soil, it picks up fine particles of mineral matter from the upper layers and carries them downward in a process called **eluviation** (Figure 12-11). These particles are eventually deposited at a lower level, and this deposition process is known as **illuviation**.

The moisture added to the soil by percolation of rainfall or snowmelt is diminished largely through evapotranspiration. The dynamic relationship between these two processes is referred to as the **soil–water balance**. It is influenced by a variety of factors, including soil and vegetation characteristics, but is primarily determined by temperature and humidity.

How much water is available to plants is much more important to an ecosystem than is the amount of precipitation. Much water derived from rainfall or snowmelt becomes unavailable to plants because of evaporation, runoff, deep infiltration, or other processes. At any given time and place, there is likely to be either a surplus or a deficit of water in the soil. Such a condition is only temporary, and it varies in response to changing weather conditions, particularly those related to seasonal changes. Generally speaking, warm weather causes increased evapotranspiration, which diminishes the soil–water supply, and cool weather slows evapotranspiration, allowing more moisture to be retained in the soil.

In a hypothetical Northern Hemisphere midlatitude location, January is a time of surplus water in the soil because low temperatures inhibit evaporation and there is little or no transpiration from plants. The soil is likely to be at or near **field capacity** at this time, which

Figure 12-11 In the process of eluviation, fine particles in upper soil layers are picked up by percolating water and carried deeper into the soil. In the process of illuviation, these particles are deposited in a lower soil layer.

FOCUS

FORMS OF SOIL MOISTURE

Four forms of soil moisture are generally recognized (Figure12-1-A).

Gravitational Water (Free Water)

Gravitational water is temporary in that it results from prolonged infiltration from above (usually due to prolonged precipitation) and is pulled downward by gravity, through the interstices toward the groundwater zone. Thus this water stays in the soil only for a short time and is not very effective in supplying plants because it drains away rapidly once the external supply ceases.

Gravitational water accomplishes significant functions during its passage through the soil, however. It is the principal agent of eluviation and illuviation and therefore is a translo-

Figure 12-1-A The four forms of soil moisture.

means that most of the pore spaces are filled with water. With the arrival of spring, temperatures rise and plant growth accelerates, so that both evaporation and transpiration increase. The soil-water balance tips from a water surplus to a water deficit. This deficit builds to a peak in middle or late summer, as temperatures reach their greatest heights and plants need maximum water. Heavy use and diminished precipitation may combine to deplete all the moisture available to plants, and the **wilting point** is reached. Thus the amount of soil moisture available for plant use is essentially the difference between field capacity and wilting point.

In late summer and fall, as air temperature decreases and plant growth slackens, evapotranspiration diminishes rapidly. At this time, the soil–water balance shifts once again to a water surplus, which continues through the winter. Then the cycle begins again.

Figure 12-12 illustrates the annual sequence just described. Such variation in the soil–water balance through time is called a **soil–water budget**.

SOIL PROPERTIES

As one looks at, feels, smells, tastes, and otherwise examines soils, various physical and chemical characteristics appear useful in describing, differentiating, and

Figure 12-1-B   Relative availability of soil moisture. Combined water is least available; capillary and gravitational water are most available. For any given soil, the amount of soil moisture available for plant use is the difference between field capacity and wilting point. (After Donald Steila, *The Geography of Soils: Formation, Distribution, and Management.* Englewood Cliffs, NJ: Prentice-Hall, 1976, p. 45. Used by permission.)

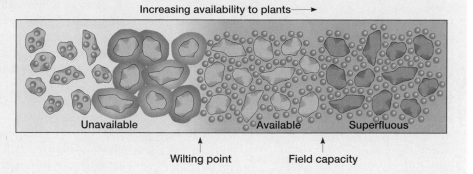

cation agent that makes the topsoil coarser and more open-textured and the subsoil denser and more compact.

Capillary Water (Water of Cohesion)

Capillary water, which remains after gravitational water has drained away, consists of moisture held at the surface of soil particles by surface tension, which is the attraction of water molecules for each other (the same property that causes water to form rounded droplets rather than dispersing in a thin film). Capillary water is by far the principal source of moisture for plants. In this form of soil moisture, the surface-tension forces are stronger than the

downward pull of gravity, and so this water is free to move about equally well in all directions in response to capillary tension. It tends to move from wetter areas toward drier ones, which accounts for the upward movement of capillary water when no gravitational water is percolating downward.

Hygroscopic Water (Water of Adhesion)

Hygroscopic water consists of a microscopically thin film of moisture bound rigidly to all soil particles by adhesion, which is the attraction of water molelcules to solid surfaces. Hygroscopic water adheres so tightly to the particles that it is normally unavailable to plants.

Combined Water

Combined water is least available of all. It is held in chemical combination with various soil minerals and is freed only if the chemical is altered.

How They Work Together

For plants, capillary water is the most important and gravitational water is largely superfluous (Figure 12-1-B). After gravitational water has drained away, the remaining volume of water represents the field capacity of the soil. If drought conditions prevail and the capillary water is all used up by plants or evaporated, the plants are no longer able to extract moisture from the soil, and the **wilting point** is reached.

---

classifying them. Some soil properties are easily recognized, but most can be ascertained only by precise measurement.

COLOR

The most conspicuous property of a soil is usually its color, but color is by no means the most definitive property. Soil color can provide clues about the nature and capabilities of the soil, but the clues are sometimes misleading. Soil scientists recognize 175 gradations of color. The standard colors are generally shades of black, brown, red, yellow, gray, and white. Soil color occasionally reflects the color of the unstained mineral

grains, but in most cases, color is imparted by stains on the surface of the particles; stains caused by either metallic oxides or organic matter.

Black or dark brown usually indicates a considerable humus content; the blacker the soil, the more humus it contains. Color gives a strong hint about fertility, therefore, because humus is an important catalyst in releasing nutrients to plants. Dark color is not invariably a sign of fertility, however, because it may be due to other factors, such as poor drainage or high carbonate content.

Reddish and yellowish colors generally indicate iron oxide stains on the outside of soil particles. These colors are most common in tropical and subtropical

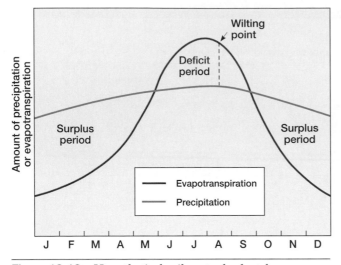

Figure 12-12 Hypothetical soil–water budget for a Northern Hemisphere midlatitude location. From January through May, there is more precipitation than evapotranspiration and consequently the soil contains a surplus of water, more than sufficient for any plant needs. From mid-May to mid-September, the evapotranspiration curve rises above the precipitation curve, indicating more water leaves the soil than enters it. About the first of August, so much water has been removed from the soil that plants begin to wilt. After mid-September, the evapotranspiration curve again dips below the precipitation curve and there is again a surplus of water in the soil.

regions, where many minerals are leached away by water moving under the pull of gravity, leaving insoluble iron compounds behind. In such situations, a red color bespeaks good drainage, and a yellowish hue suggests imperfect drainage. Red soils are also common in desert and semidesert environments, where the color is carried over intact from reddish parent

materials rather than representing a surface stain (Figure 12-13).

Gray or white soils may develop in varying environments. In humid areas, a light color implies so much leaching that even the iron has been removed, but in dry climates, it indicates an accumulation of salts. It may also indicate simply a lack of organic matter.

TEXTURE

All soils are composed of myriad particles of various sizes, as Figure 12-14 shows, although smaller particles usually predominate. Rolling a sample of soil about between the fingers can provide a feel for the principal particle sizes. Table 12-1 shows the standard classification scheme for particle sizes (in this scheme, the size groups are called **separates**). The gravel, sand, and silt separates are fragments of the weathered parent material and are mostly the grains of minerals found commonly in rocks, especially quartz, feldspars, and micas. These coarser particles are the inert materials of the soil mass, its skeletal framework. As noted above, only the clay particles take part in the intricate chemical activities that occur in the soil.

Because no soil is made up of particles of uniform size, the texture of any soil is determined by the relative amounts of the various separates present. The texture triangle (Figure 12-15) shows the standard classification scheme for soil texture; this scheme is based on the percentage of each separate by weight. Near the center of the triangle is **loam**, the name given to a texture in which none of the three principal separates dominates the other two. This fairly even-textured mix is generally the most productive for plants.

(a)

(b)

Figure 12-13 A red soil means different things in different places. **(a)** In this Mississipi forest, the soluble minerals have been leached away, leaving the insoluble iron to impart its reddish color to the soil. **(b)** In this Australian desert, the reddish color reflects the iron content of the underlying bedrock. (TLM photos)

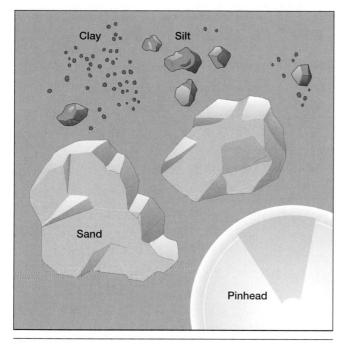

Figure 12-14 Comparative sizes of sand, silt, and clay. For scale, a portion of the head of a pin is shown in the lower right. (After Robert A. Muller and Theodore M. Oberlander, *Physical Geography Today: A Portrait of a Planet*, 2nd ed. New York: Random House, 1978. Used by permission.)

STRUCTURE

The individual particles of most soils tend to aggregate into clumps called **peds**, and it is these clumps that determine soil structure. The size, shape, and stability of peds have a marked influence on how easily water, air, and organisms (including plant roots) move through the soil, and consequently on soil fertility. Peds are classified on the basis of shape as spheroidal, platy, blocky, or prismatic, with these four shapes giving rise to seven generally recognized soil types (Figure 12-16). Aeration and drainage are usually facilitated by peds

of intermediate size; both massive and fine structures tend to inhibit these processes.

Some soils, particularly those composed largely of sand, do not develop a true structure, which is to say that the individual grains do not aggregate into peds. Silt and clay particles readily aggregate in most instances. Other things being equal, aggregation is usually greatest in moist soils and least in dry ones.

Structure is an important determinant of a soil's porosity and permeability. As we learned in Chapter 9, porosity refers to the amount of pore space between soil particles or between peds (Figure 12-17). We can define it as

$$porosity = volume\ of\ voids/total\ volume$$

Porosity is usually expressed as a percentage or a decimal fraction. It is a measure of a soil's capacity to hold water and air.

The relationship between porosity and permeability is not simple; that is, the most porous materials are not necessarily the most permeable. Clay, for example, is the most porous separate, but it is the least permeable because it soaks up water rather than allowing it to pass through.

SOIL CHEMISTRY

The effectiveness of soil as a growth medium for plants is based largely upon the presence and availability of nutrients, which are determined by an intricate series of chemical reactions. Soil chemistry is an extraordinarily complex subject that revolves primarily around microscopic particles and electrically charged atoms or groups of atoms called ions.

TABLE 12-1	
Standard U.S. Classification of Soil Particle Sizes	
Separate	*Diameter*
Gravel	Greater than 0.08 in. (2 mm)
Very coarse sand	0.04–0.08 in. (1–2 mm)
Coarse sand	0.02–0.04 in. (0.5–1 mm)
Medium sand	0.01–0.02 in. (0.25–0.5 mm)
Fine sand	0.004–0.01 in. (0.1–0.25 mm)
Very fine sand	0.002–0.004 in. (0.05–0.1 mm)
Coarse silt	0.0008–0.002 in. (0.02–0.05 mm)
Medium silt	0.00024–0.0008 in. (0.006–0.02 mm)
Fine silt	0.00008–0.00024 in. (0.002–0.006 mm)
Clay	Less than 0.00008 in. (0.002 mm)

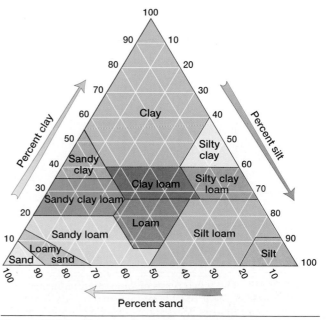

Figure 12-15 The standard soil-texture triangle.

COLLOIDS

Particles smaller than about 0.1 micrometer in diameter are called **colloids**. Inorganic colloids consist of clay in thin, crystalline, platelike forms created by the chemical alteration of larger particles; organic colloids represent decomposed organic matter in the form of humus; and both types are the chemically active soil particles. When mixed with water, colloids remain suspended indefinitely as a homogeneous, murky solution. Some have remarkable storage capacities, and consequently colloids are major determinants of the water-holding capacity of a soil. They function as a virtual sponge, soaking up water, whereas the soil particles that are too large to be classified as colloids can maintain only a surface film of water.

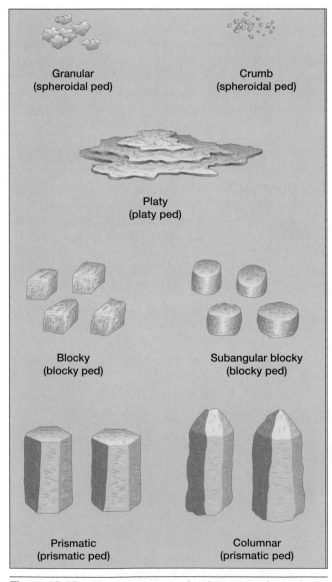

Figure 12-16   The seven types of soil structure formed from spheroidal, platy, blocky, and prismatic peds. (After *Soils Laboratory Exercise Source Book.* American Society of Agronomy, 1964.)

Both inorganic and organic colloids attract and hold great quantities of ions.

CATION EXCHANGE

As noted above, cations are positively charged ions. The minerals that form them are called **bases**, and they include such elements as calcium, potassium, and magnesium, which are all essential for soil fertility and plant growth.

Colloids carry mostly negative electrical charges on their surfaces, and these charges attract swarms of nutrient cations that would be leached from the soil if their ions were not retained by the colloids.

The combination of colloid and attached cations is called the **colloidal complex,** and it is a delicate mechanism. If it holds the nutrients strongly, they will not be leached away; yet if the bond is too strong, they cannot be absorbed by plants. Thus a fertile soil is likely to be one in which the cation–colloid attraction is intricately balanced.

Adding to the complexity of the situation is the fact that some types of cations are bound more tightly than others. Cations that tend to bond strongly in the colloid complex may replace those that bond less strongly. For example, basic ions are fairly easily replaced by metal ions or hydrogen ions, and this process is called cation exchange. The capability of a soil to attract and exchange cations is known as its **cation exchange capacity (CEC)**. As a generalization, the higher the CEC, the more fertile the soil. Soils with a high clay content have a higher CEC than more coarsely grained soils because the former have more colloids. Humus is a particularly rich source of high-CEC activity because humus colloids have a much higher CEC than do inorganic clay minerals. The most fertile soils, then, tend to be those with a notable clay and humus content.

Figure 12-17   The contrast between a porous soil and a nonporous one.

PEOPLE AND THE ENVIRONMENT

THE LAND–CAPABILITY CLASSIFICATION SYSTEM

Soil is important to humans in many ways, but by far its dominant significance is in its role as a growth medium for crops, pasture grasses, and commercial timber. The use—and abuse—of soil by farmers, pastoralists, and foresters thus becomes a matter of concern for us all.

One of the most useful soil-classification systems for the person on the land, called the land-capability classification system, was devised several decades ago by the Soil Conservation Service, which is the federal agency primarily concerned with soil use and erosion in the United States. This system is an interpretive grouping of soil types made primarily for agricultural purposes. Arable soils are grouped according to their potentialities and limitations for sustained production of the common cultivated crops. Nonarable soils are grouped according to their potentialities and limitations for the production of permanent vegetation (essentially grass and trees) and according to their risks of soil damage if mismanaged.

The scheme places all soils in one of eight capability classes. Limitations in use and risks of soil damage are least in class I soils and greatest in class VIII soils. Under good management, soils in the first four classes can sustain production of the common cultivated field crops. Soils in classes V through VII are primarily suited for native plants (grasses and trees), and soils in class VIII are generally unsuited for commercial production of anything.

In classifying a soil, the pedologist considers depth, texture, chemical characteristics, drainage, and degree of erosion, as well as the slope of the land surface. No specific assessment of fertility or productivity is attempted.

The system is easy to use and func-tional. A Soil Conservation Service pedologist makes an on-the-ground survey of a farm (or other management unit), recording soil characteristics on aerial photos of the property. She or he then groups the various soil types in classes and draws the boundaries of the classes on the air photos, as shown in Figure 12-2-A. The landowner is then given recommendations about the most productive and effective use of each parcel of land. It is up to the landowner's discretion whether or not to adopt the recommendations.

Land-Capability Classes

Class I. Soils with few limitations that can be cultivated with ordinary farming methods.

Class II. Soils with some limitations on plant choice or requiring moderate conservation practices.

Class III. Soils with severe limitations on plant choice or requiring special conservation practices or both.

Class IV. Soils with severe limitations on plant choice or requiring very careful management or both.

Class V. Soils with little or no erosion hazard that have other limitations that limit them to pasture, woodland, or wildlife uses and make them unsuitable for crop growing.

Class VI. Soils with severe limitations that make them generally unsuitable for cultivation.

Class VII. Soils with very severe limitations that make them unsuitable for cultivation and restrict their use to grazing, woodland, or wildlife with very careful management.

Class VIII. Soils that are not suitable for cultivation, grazing, or forestry and usable only for wildlife, recreation, or watershed protection.

Figure 12-2-A An example of the land-capability classification system as applied to a farm landscape. (U.S. Department of Agriculture)

ACIDITY/ALKALINITY

We have just learned about a class of chemicals called bases. Another class of chemicals are **acids.** Both acids and bases dissolve in water. Solutions that contain dissolved acids are described as being *acidic.* Those that contain dissolved bases are called *basic solutions,* and another word for *basic* in this context is *alkaline.* Thus any solution of chemicals can be characterized on the basis of its **acidity** or **alkalinity.**

Nearly all nutrients are provided to plants in solution. An overly alkaline soil solution is inefficient in dissolving minerals and releasing their nutrients. On the other hand, if the solution is highly acidic, the nutrients are likely to be dissolved and leached away too rapidly for plant roots to absorb them. The optimum situation, then, is for the soil solution to be neutral, neither too alkaline nor too acidic. The acidity/alkalinity of a soil is determined to a large extent by its CEC.

The chemist's symbol for the measure of the acidity/alkalinity of a solution is pH, which is based on the relative concentration of hydrogen ions (H^+) in the solution. The scale ranges from zero to 14: the lower end represents acidic conditions; higher numbers indicate alkaline conditions (Figure 12-18). Neutral conditions are represented by a value of 7, and it is soil having a pH of about 7 that is most suitable for the great majority of plants and microorganisms.

SOIL PROFILES

The development of any soil is expressed in two dimensions—depth and time. There is no straight-line relationship between depth and age, however; some soils deepen and develop much more rapidly than others.

There are four processes that deepen and age soils: *addition, loss, translocation, and transformation* (Figure 12-19). The five soil-forming factors discussed earlier—geologic, climatic, topographic, biological, chronological—influence the rate of these four processes, the result being the development of various soil horizons and the soil profile.

The vertical variation of soil properties is not random but rather an ordered layering with depth. Soil tends to have more or less distinctly recognizable layers, called **horizons,** each with different characteristics. The horizons are positioned approximately parallel with the land surface, one above the other, normally, but not always, separated by a transition zone rather than a sharp line. A vertical cross section (as might be seen in a road cut or the side of a trench dug in a field) from the Earth's surface down through the soil layers and into the parent material is referred to as a **soil profile.** The almost infinite variety of soils in the world usually are grouped and classified on the basis of differences exhibited in their profiles.

Figure 12-20 presents an idealized sketch of a well-developed soil profile, in which six horizons are differentiated:

- The *O horizon* is the surface layer, and in it organic matter, both fresh and decaying, makes up most of the volume. This horizon results essentially from litter derived from dead plants and animals. It is common in forests and generally absent in grasslands.
- The *A horizon,* colloquially referred to as *topsoil,* is a mineral horizon that also contains considerable organic matter. Usually dark, it is formed either at the surface or immediately below an O horizon. Seeds germinate mostly in the A horizon.
- The *E horizon* normally is lighter in color than either the overlying A or the underlying B horizon. It is essentially an eluvial layer from which clay, iron, and aluminum have been removed, leaving a concentration of abrasion-resistant sand or silt particles.

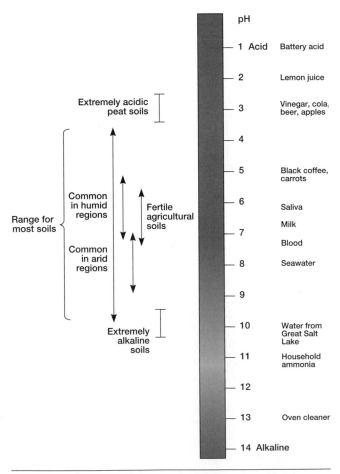

Figure 12-18 The standard pH scale.

- The **B horizon**, usually called *subsoil,* is a mineral horizon of illuviation where most of the materials removed from above have been deposited. A collecting zone for clay, iron, and aluminum, this horizon is usually of heavier texture, greater density, and relatively greater clay content than the A horizon.
- The **C horizon** is unconsolidated parent material (regolith) beyond the reach of plant roots and most soil-forming processes except weathering. It is lacking in organic matter.
- The **R horizon** is bedrock, with little evidence of weathering.

True soil, which is called **solum**, only extends down through the B horizon.

As we learned earlier in the chapter, time is a critical passive factor in profile development, but the vital active factor is surface water. If there is no surface water, from rainfall or snowmelt or some other source, to infiltrate the soil, there can be no profile development. Descending water carries material from the surface downward, from topsoil into subsoil, by eluviation and leaching. This transported material is mostly deposited a few inches or a few feet below the surface. In the usual pattern, topsoil (A) becomes a somewhat depleted horizon through eluviation and leaching, and subsoil (B) develops as a layer of accumulation due to illuviation.

A profile that contains all horizons is typical of a humid area on well-drained but gentle slopes in an environment that has been undisturbed for a long time. In many parts of the world, however, such idealized conditions do not pertain, and the soil profile may have one horizon particularly well-developed, one missing altogether, a fossil horizon formed under a different past climate, an accumulation of a hardpan (a very dense and impermeable layer), surface layers removed through accelerated erosion, or some other variation. Moreover, many soils are too young to have evolved a normal profile. A soil containing only an A horizon atop partially altered parent material (C horizon) is said to be *immature.* The formation of an illuvial B horizon is normally an indication of a *mature* soil.

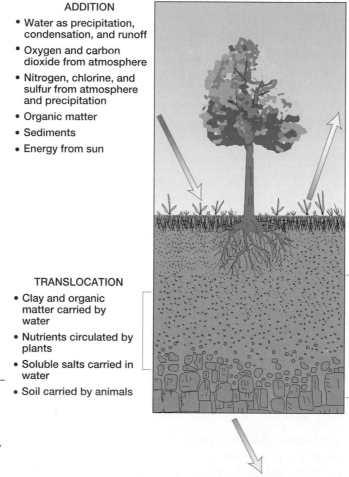

ADDITION
- Water as precipitation, condensation, and runoff
- Oxygen and carbon dioxide from atmosphere
- Nitrogen, chlorine, and sulfur from atmosphere and precipitation
- Organic matter
- Sediments
- Energy from sun

LOSS
- Water by evapotranspiration
- Nitrogen by denitrification
- Carbon as carbon dioxide from oxidation of organic matter
- Soil by erosion
- Energy by radiation

TRANSLOCATION
- Clay and organic matter carried by water
- Nutrients circulated by plants
- Soluble salts carried in water
- Soil carried by animals

TRANSFORMATION
- Organic matter converted to humus
- Particles made smaller by weathering
- Structure and concretion formation
- Minerals transformed by weathering
- Clay and organic matter reactions

LOSS
- Water and materials in solution or suspension

Figure 12-19 The four soil-forming processes: addition, loss, translocation, transformation. Geologic, climatic, topographic, biological, and chronological soil-forming factors influence the rate at which these four processes occur and therefore the rate at which soil is formed.

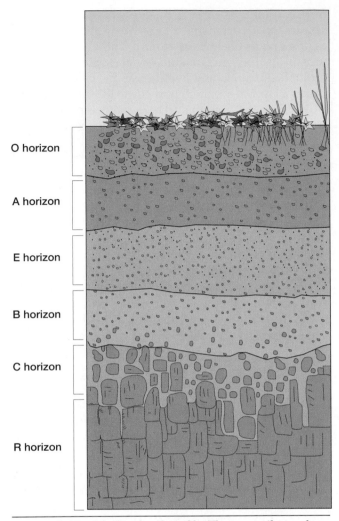

O horizon

A horizon

E horizon

B horizon

C horizon

R horizon

Figure 12-20 Idealized soil profile. The true soil, or solum, consists of the O, A, E, and B horizons.

PEDOGENIC REGIMES

Soil-forming factors and processes interact in almost limitless variations to produce soils of all descriptions. Fundamental to an understanding of soil classification and distribution is the realization that only five major **pedogenic** (that is, soil-forming) **regimes** exist: laterization, podzolization, gleization, calcification, and salinization. These regimes can be thought of as environmental settings in which certain physical/chemical/biological processes prevail.

LATERIZATION

Laterization is named for the brick-red color of the soil it produces (*later*: Latin, "brick"). The processes associated with this regime are typical of the warm, moist regions of the world, and a significant annual moisture surplus is a requisite condition. The soil formed by lat-

erization is most prominent, then, in the tropics and subtropics, in regions dominated by forest, shrub, and savanna vegetation.

A laterization regime is characterized by rapid weathering of parent material, dissolution of nearly all minerals, and speedy decomposition of organic matter. Probably the most distinctive feature of laterization is the leaching away of silica, the most common constituent of most rock and soil and a constituent that is usually highly resistant to being dissolved. That silica is indeed removed during laterization indicates the extreme effectiveness of chemical weathering and leaching under this regime. Most other minerals are also leached out rapidly, leaving behind primarily iron and aluminum oxides and barren grains of quartz sand. This residue normally imparts to the resulting soil the reddish color that gives this regime its name (Figure 12-21). The A horizon is highly eluviated and leached, whereas the B horizon has a considerable concentration of illuviated materials.

Because plant litter is rapidly decomposed in places where laterization is the predominant regime, little humus is incorporated into the soil. Even so, plant nutrients are not totally removed by leaching because the natural vegetation, particularly in a forest, quickly absorbs many of the nutrients in solution. If the vegetation is relatively undisturbed by human activities, this regime has the most rapid of nutrient cycles, and the soil is not totally impoverished by the speed of mineral decomposition and leaching. Where the forest is cleared for agriculture or some other human purpose, however, most base nutrients are likely to be lost from the cycle because the tree roots that would bring them up are gone. The soil then rapidly becomes impoverished, and hard crusts of iron and aluminum compounds are likely to form.

The general term applied to soils produced by laterization is **latosols**. These soils sometimes develop to depths of tens of feet because of the strong weathering activities and the fact that laterization continues year-round in these benign climates. Most latosols have little to offer as agricultural soils, for the reasons noted above, but laterization often produces such concentrations of iron and aluminum oxides that mining them can be profitable.

PODZOLIZATION

Podzolization is another regime named after the color of the soil it produces; in this case, gray (*podzol*: Russian for like ashes). It also occurs in regions having a positive moisture balance and involves considerable leaching, but beyond those two characteristics, it bears little similarity to laterization. Podzolization occurs

Figure 12-21 In the wet tropics, laterization is the dominant soil-forming regime, and most soils are reddish as a result of the prominence of iron and aluminum compounds. Lateritic soils are exposed in this road cut near Savusavu on the Fijian island of Vanua Levu. (TLM photo)

primarily in areas where the vegetation has limited nutrient requirements and where the plant litter is acidic. These conditions are most prominent in mid- and high-latitude locales having a coniferous forest cover. Thus podzolization is largely a Northern Hemisphere phenomenon because there is not much land in the higher midlatitudes south of the equator. The typical location for podzolization is under a boreal forest in subarctic climates, which is found only in the Northern Hemisphere.

In these cool regions, chemical weathering is slow but the abundance of acids, plus adequate precipitation, makes leaching very effective. Mechanical weathering from frost action is relatively rapid during the unfrozen part of the year. Moreover, much of the land was bulldozed by Pleistocene glaciers, leaving an abundance of broken rock debris at the surface. Bedrock here consists mostly of ancient crystalline rocks rich in quartz and aluminum silicates and poor in the alkaline mineral cations important in plant nutrition. The boreal forests, dominated by conifers, require little in the way of soil nutrients, and their litter returns few nutrient minerals when it decays. The litter is largely needles and twigs, which accumulate on the surface of the soil and decompose slowly. Microorganisms do not thrive in this environment, and so humus production is retarded. Moisture is relatively abundant in summer, so that leaching of whatever nutrient cations are present in the topsoil, along with iron oxides, aluminum oxides, and colloidal clays, is relatively complete.

Podzolization, then, produces soils that are shallow and acidic and have a fairly distinctive profile. There is usually an O horizon. The upper part of the A horizon is eluviated to a silty or sandy texture and is so leached as to appear bleached. It usually is the ashy, light gray color that gives this regime its name, a color imparted by its high silica content. The illuviated B horizon is a receptacle for the iron/aluminum oxides and clay minerals leached from above and has a sharply contrasting darker color (sometimes with an orange or yellow tinge). Soil fertility is generally low, and a crumbly structure makes the soil very susceptible to accelerated erosion if the vegetation cover is disturbed, whether by human activities or by such natural agencies as wildfire. Soils produced by podzolization often are referred to collectively as **podzols**.

GLEIZATION

Gleization is a regime restricted to waterlogged areas, normally in a cool climate. (The name comes from *glej*, Polish for "muddy ground.") Although occasionally widespread, it is generally much more limited in occurrence than laterization and podzolization. The poor drainage that produces a waterlogged environment can be associated with flat land, but it can also result from a topographic depression, a high water table, or various other conditions. In North America, gleization is particularly prominent in areas around the Great Lakes, where recent glacial deposition has interrupted preglacial drainage patterns.

The general term for soils produced by gleization is **gley soils**. They characteristically have a dark, highly organic A horizon, where decomposition proceeds slowly because bacteria are inhibited by the lack of oxygen in a waterlogged situation. This slow decay yields organic acids that cannot oxidize iron to produce reddish colors, and the pH is invariably low. The B horizon in a true gley soil is poorly developed, but where waterlogging is more limited, a distinctive B horizon may show various colors.

Gley soils are usually too acidic and oxygen-poor to be productive for anything but water-tolerant vegetation. If drained artificially, however, and fertilized with lime to counteract the acid, their fertility can be greatly enhanced.

CALCIFICATION

In semiarid and arid climates, where precipitation is less than potential evapotranspiration, leaching is either absent or transitory. Natural vegetation in such areas consists of grasses or shrubs. **Calcification** (so called because many calcium salts are produced in this regime) is the dominant pedogenic process in these regions, as typified by the drier prairies of North America, the steppes of Eurasia, and the savannas and steppes of the subtropics.

Both eluviation and leaching are restricted by the absence of percolating water, and so materials that would be carried downward in other regimes become concentrated in the soil where calcification is at work. Moreover, there is considerable upward movement of water by capillary action in dry periods. Calcium carbonate ($CaCO_3$) is the most important chemical compound active in a calcification regime. It is carried downward by limited leaching after a rain and often is concentrated in the B horizon to form a dense layer of hardpan, then brought upward by capillary water and by grass roots, and finally returned to the soil when the grass dies. Little clay is formed because of the limited amount of chemical weathering. Organic colloidal material, however, is often present in considerable quantity.

Where calcification takes place under undisturbed grassland, the resulting soils are likely to have remarkable agricultural productivity. Humus from decaying grass yields abundant organic colloidal material, imparting a dark color to the soil and contributing to a structure that can retain both nutrients and soil moisture. Grass roots tend to bring calcium up from the B horizon sufficiently to inhibit or delay the formation of calcic hardpans. Where shrubs are the dominant vegetation, roots are fewer but deeper, so that nutrients are brought up from deeper layers, and little accumulates at the surface with less humus being incorporated into the soil.

Where a calcification regime is operative in true deserts, the soils tend to be shallower and sandier, calcic hardpans may form near the surface, little organic matter accumulates either on or in the soil, and the soils are not very different from the parent material.

SALINIZATION

In arid and semiarid regions, it is fairly common to find areas with inadequate drainage, particularly in enclosed valleys and basins. Moisture is drawn upward and into the atmosphere by intense evaporation. The evaporating water leaves behind various salts in or on the surface of the soil, sometimes in such quantity as to impart a brilliant white surface color to the land, and the pedogenic regime is called **salinization**. These salts, which are mostly chlorides and sulfates of calcium and sodium, are toxic to most plants and soil organisms, and the resulting soil is able to support very little life apart from a few salt-tolerant grasses and shrubs.

Soils developed in a salinization (*salin*: Latin for salt) regime sometimes can be made productive through careful water management. Irrigation water can provide a blood to the land, but artificial drainage is equally necessary or else salt accumulation will be intensified. Indeed, human-induced salinization has ruined good agricultural land in various parts of the world many times in the past.

CLIMATE AND PEDOGENIC REGIMES

The regimes are distinguished primarily on the basis of climate as reflected in temperature and moisture availability (Figure 12-22) and secondarily on the basis of vegetation cover. In regions where there is normally a surplus of moisture—which is to say annual precipitation exceeds annual evapotranspiration—water movement in the soil is predominantly downward and leaching is a prominent process. In such areas where

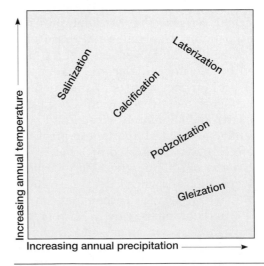

Figure 12-22   Temperature/moisture relationships for the five principal pedogenic regimes. Where temperature is high but precipitation low, salinization is the main method of soil formation. Where temperature is low but precipitation high, gleization predominates. Laterization occurs where both temperature and precipitation are high. Calcification and podzolization take place in less extreme environments.

temperatures are relatively high throughout the year, laterization is the dominant regime; where winters are long and cold, podzolization predominates; and where the soil is saturated most of the time due to poor drainage, gleization is notable. In regions having a moisture deficit, the principal soil moisture movement is upward (through capillarity) and leaching is limited. Calcification and salinization are the principal pedogenic regimes under these conditions.

SOIL CLASSIFICATION

One of the most significant products of scholarly studies is classification systems. If phenomena can be meaningfully classified, it becomes easier to remember them and to understand the relationships among them. Our consideration thus far has included various classifications. In no other subdiscipline of physical geography, however, is the matter of classification more complicated than with soil.

EARLIER CLASSIFICATION SCHEMES

Unlike the case with most other natural phenomena, soil classification has changed significantly over the last century, largely in response to increasing knowledge. Early global soil classification systems were mostly based on relationships between a soil and the parent material, which was then considered the primary factor in determining soil characteristics. New approaches introduced in the late 1800s, particularly by Russian soil scientists, focused on other aspects of soil genesis, especially the interrelated roles of climate and vegetation. Various worldwide classification schemes based on these pedogenetic aspects have been devised, most of them simply being modifications of previous systems.

In 1938, the U.S. Department of Agriculture gave its official blessing to a classification system that became widely accepted in the United States and in much of the rest of the world, again based largely on soil/climate/vegetation relationships. This system was generally popular with geographers because it was a genetic classification, which means that it was based on soil-forming conditions and processes. Thus the relationship between soil and other aspects of the environment was an integral part of the system, and the distribution of major soil categories could be fruitfully compared with the distribution of other major environmental complexes, particularly climate and vegetation.

THE SOIL TAXONOMY

During the 1940s, however, soil scientists became increasingly disenchanted with the 1938 classification, and a new system was developed slowly and laboriously by the Soil Survey Staff of the U.S. Department of Agriculture during the 1950s and 1960s. Originally referred to as the *Seventh Approximation* because it was the seventh version of the system, it is now known simply as the **Soil Taxonomy**. This system has now largely superseded the 1938 system among soil scientists in the United States and is increasingly being accepted (sometimes with modifications) in other countries. Other major classifications are also in use outside the United States, particularly in Canada, the United Kingdom, Russia, France, and Australia. Moreover, United Nations agencies have their own classification schemes, developed in the 1960s and 1970s. The acceptance of the Soil Taxonomy has been slow in part because it is such a radical departure from previous classification systems and in part because the terminology is confusing, at least initially. Even so, it is likely to become increasingly acceptable internationally and equally likely to continue to undergo revisions.

The basic characteristic that sets the Soil Taxonomy apart from previous systems is that it is *generic*, which means it is organized on the basis of observable soil characteristics. The focus is on the existing properties of a soil rather than on the environment, genesis, or

TABLE 12-2 Name Derivations of Soil Orders	
Order	*Derivation*
Alfisols	"al" for aluminum, "f" for iron (chemical symbol Fe), two prominent elements in these soils
Andisols	*and*esite, rock formed from type of magma in Andes Mountains volcanoes; soils high in volcanic ash
Aridisols	Latin *aridus*, dry; dry soils
Entisols	last three letters in "rec*ent*"; these are recently formed soils
Histosols	Greek *histos*, living tissue; these soils contain only organic matter
Inceptisols	Latin *inceptum*, beginning; young soils at the beginning of their "life"
Mollisols	Latin *mollis*, soft; soft soils
Oxisols	soils with large amounts of oxygen-containing compounds
Spodosols	Greek *spodos*, wood ash; ashy soils
Ultisols	Latin *ultimus*, last; soils that have had the last of their nutrient bases leached out
Vertisols	Latin *verto*, turn; soils in which material from O and A horizons falls through surface cracks and ends up below deeper horizons; the usual horizon order is inverted

properties it would possess under virgin conditions. The logic of such a generic system is theoretically impeccable: soil has certain properties that can be observed, measured, and at least partly quantified.

Like other logical generic systems, the Soil Taxonomy is a hierarchical system, which means that it has several levels of generalization, with each higher level encompassing several members of the level immediately below it. There are only a few similarities among all the members of the highest level category, but the number of similarities increases with each step downward in the hierarchy, so that in the lowest-level category, all members have mostly the same properties. (See Appendix VI for details.)

At the highest level (the smallest scale of generalization) of the Soil Taxonomy is soil *order*, of which only eleven are recognized worldwide (Table 12-2). The soil orders, and many of the lower-level categories as well, are distinguished from one another largely on the basis of certain diagnostic properties, which are often expressed in combination to form *diagnostic horizons*. The two basic types of diagnostic horizon are the **epipedon** (based on the Greek word *epi*, meaning over or upon), which is essentially the A horizon or the combined O/A horizon, and the **subsurface horizon**, which is roughly equivalent to the B horizon. (Note that all A and B horizons are not necessarily diagnostic, and so the terms and concepts are not synonymous.)

Soil orders are subdivided into *suborders*, of which about 50 are recognized in the United States. The third level consists of great groups, which number about 230 in the United States. Successively lower levels in the classification are subgroups, families, and series. About 172,000 soil series have been identified in the United States to date, and the list will undoubtedly be expanded in the future. For the purpose of comprehending general world distribution patterns, however, we need to concern ourselves only with orders and suborders.

The relationship between Soil Taxonomy categories and those of previous systems is vague because of the different criteria of classification. At the lowest level—the series—there is considerable agreement between the Soil Taxonomy and the 1938 system, but at

other levels the two diverge because of their different theoretical foundations. Many geographers have been slow to accept the Soil Taxonomy because they like the conspicuous correlations between the high-level categories of the 1938 system and global climate and vegetation patterns. In other words, the genetic bias of the 1938 system seemed well suited for broad-scale geographic uses, whereas the generic approach of the Soil Taxonomy yields categories that have less clear relationships with other environmental components. This is not to say that there are no relationships between Soil Taxonomy categories and pedogenetic factors, but rather that the relationships are not as apparent nor as easy to demonstrate in terms of global or continental patterns of distribution.

TABLE 12-3
Approximate Equivalents Between the Soil Taxonomy and the 1938 USDA Soil-Classification System

Soil Taxonomy order	1938 system component	
	Mostly included	*Significantly included*
Entisols	Lithosols, regosols	Alluvial soils, low-humic gley soils
Vertisols	Grumusols	
Andisols		
Inceptisols	Ando soils, brown forest soils, half-bog soils, *sols bruns acides*, tundra soils	Alluvial soils, brown podzolic soils, latosols, low-humic gley soils
Aridisols	Desert soils, red desert soils, sierozems, solonchak	Brown soils, calcisols, reddish-brown soils, solonetz soils
Mollisols	Brunizems, chernozems, chestnut soils, reddish prairie soils, rendzina soils	Alluvial soils, brown soils, calcisols, humic gley soils, reddish-chestnut soils, solonetz soils
Spodosols	Brown podzolic soils, ground water podzols, podzol soils	
Alfisols	Gray-brown podzolic soils, gray wooded soils	Degraded chernozems, noncalcic brown soils, reddish-brown soils
Planosols		Reddish-chestnut soils, solonetz soils
Ultisols	Groundwater laterite soils, humic latosols, reddish-brown lateritic soils, red-yellow podzolic soils	Latosols
Oxisols	Laterite soils	Latosols
Histosols	Bog soils	

Nevertheless, the Soil Taxonomy has superseded all previous classifications, at least in the United States, and geographers must accommodate to it and learn to appreciate its obvious merits. One method of easing the transition is to point out equivalent categories in the two systems. Thus geographers explaining world soil distribution patterns often list "great soil groups" (from the 1938 classification) that are approximately equivalent to orders or suborders of the Soil Taxonomy. Unfortunately, the search for environmental relationships sometimes blurs the searcher's judgment, with the result that equivalence may be perceived when it does not actually occur. Thus the reader is warned that the equivalents between categories in the two systems listed in Table 12-3 are only approximate.

THE MAPPING QUESTION

Maps are a basic tool of geographic study, and one of the fundamental problems that confronts any geographic inquiry is how the phenomena under study should be mapped. There are precise techniques for mapping features that are located at specific points or for mapping phenomena that are spread uniformly over an area. However, if the object of study is a generalized abstraction, its depiction on a map is necessarily imprecise, and choice of the mapping technique becomes more subjective. Because the higher levels of the Soil Taxonomy do not represent phenomena that actually exist but rather are generalized abstractions of average or typical conditions over broad areas, the selection of an appropriate mapping technique can significantly influence our understanding of the situation.

Most soil maps use the same timeworn technique of areal expression. If one soil type (at whatever level of generalization is being studied) is more common in an area than any other, that area is classified by the prevailing type and colored or shaded appropriately. Such a map is effective in indicating the principal type of soil in each region and is useful in portraying the general distribution of the major soil types.

Maps of this type are compiled through generalization of data, and the smaller the scale, the greater the generalization needed. Thus more intricate patterns can be shown on the larger scale map of the United States (Figure 12-23) than on the smaller scale world map (Figure 12-24). In either case, the map is only as good as its generalizations are meticulous.

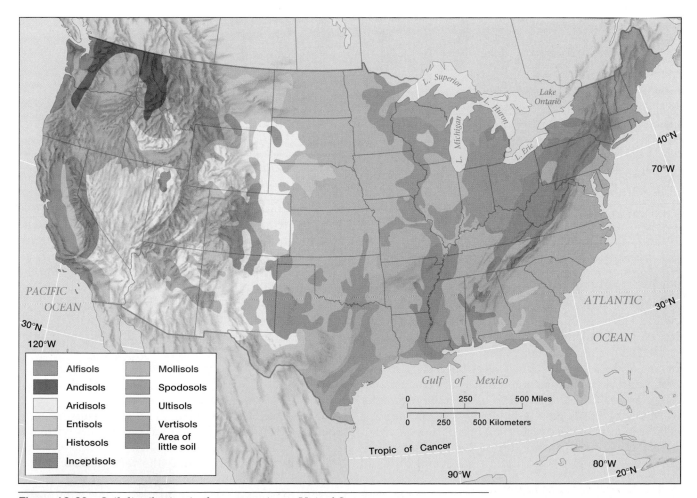

Figure 12-23 Soil distribution in the conterminous United States.

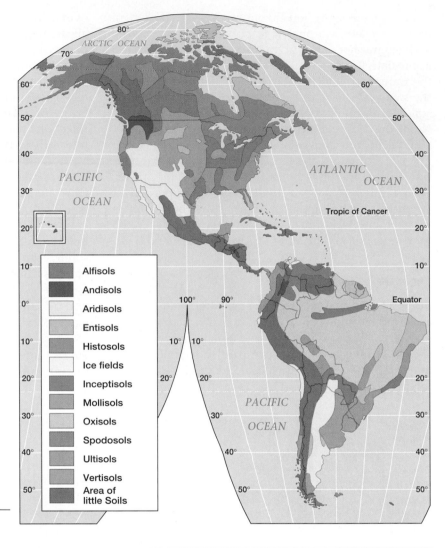

Figure 12-24 Soils of the world.

Legend:
- Alfisols
- Andisols
- Aridisols
- Entisols
- Histosols
- Ice fields
- Inceptisols
- Mollisols
- Oxisols
- Spodosols
- Ultisols
- Vertisols
- Area of little Soils

Soil maps according to the Soil Taxonomy often appear visually confusing and seem to lack understandable patterns, particularly when compared with similar maps based on earlier classification schemes. This confusion arises, at least in part, because the new system has a different logical framework, a framework that perhaps calls for a different mapping technique. An alternative method of cartographic expression has been proposed by geographer Philip Gersmehl in an attempt to portray the distribution patterns more clearly. His technique is to plot the proportional occurrence of particular soil orders in certain-size areas (he chose 1000 square kilometers as the mapping unit) by means of dots. This approach, as applied to 11 soil orders widespread in the United States, is demonstrated in the maps of the United States used in the remainder of this chapter. Its great advantage is in providing a clearer picture of the distribution of a particular order; its principal disadvantage lies in the difficulty of absorbing a unified perspective from several different maps.

There is no simple solution to the mapping of complex data on a small scale.

GLOBAL DISTRIBUTION OF MAJOR SOILS

Of the 11 orders of soils recognized in the Soil Taxonomy, 10 are arranged in a hierarchy in which each succeeding order represents a generally increased degree of weathering, particularly as expressed by mineral alteration and profile development (Figure 12-25). The eleventh order, Histosols, is essentially an organic soil that lies outside the concept of this hierarchy. Let us look at each order briefly now, beginning with the least weathered and working our way up from the bottom of Figure 12-25.

ENTISOLS

The least well developed of all soils, **Entisols** have experienced little mineral alteration and are virtually without pedogenic horizons. Their undeveloped state is usually a function of time (the very name of the order connotes recency); most Entisols are surface deposits that have not been in place long enough for pedogenetic processes to have had much effect. Some,

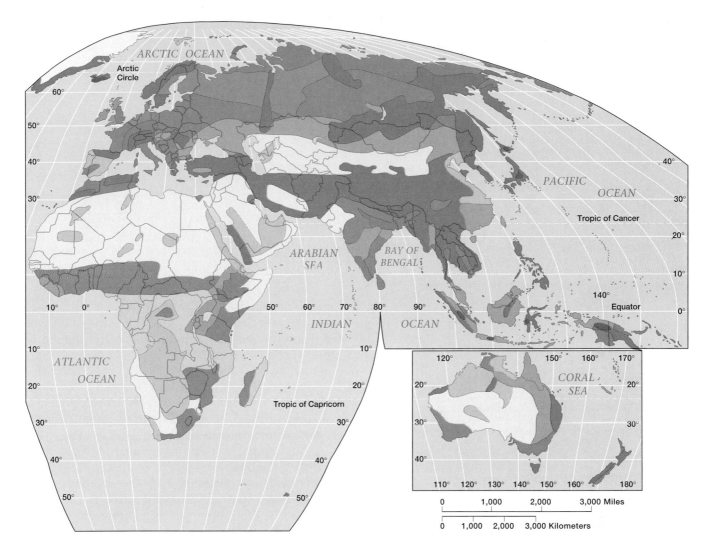

however, are very old, and in these soils the lack of horizon development is due to a mineral content that does not alter readily, to a very cold climate, or to some other factor totally unrelated to time.

The distribution of Entisols is therefore very widespread and cannot be specifically correlated with particular moisture or temperature conditions or with certain types of vegetation or parent materials (Figure 12-26). In the United States, Entisols are most prominent in the dry lands of the West but are found in most other parts of the country as well (Figure 12-27). They are commonly thin and/or sandy and have limited productivity, although those developed on recent alluvial deposits tend to be quite fertile (Figure 12-28).

VERTISOLS

Vertisols contain a large quantity of clay that becomes a dominant factor in the soil's development. The clay of Vertisols is described as "swelling" or "cracking" clay. This clay-type soil has an exceptional capacity for absorbing water: when moistened, it swells and expands; as it dries, deep, wide cracks form, sometimes an inch

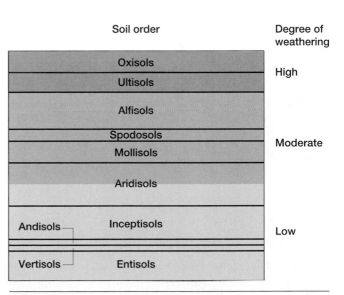

Figure 12-25 Soil order ranked by degree of weathering. The height of each band is proportional to the approximate worldwide areal extent of that order.

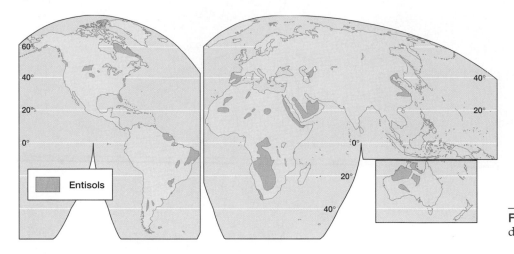

Figure 12-26 World distribution of Entisols.

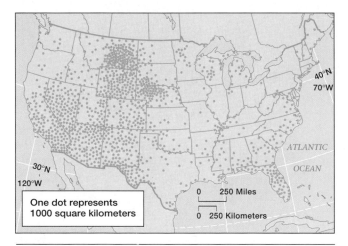

Figure 12-27 Distribution of Entisols in the conterminous United States. (After Philip J. Gersmehl, "Soil Taxonomy and Mapping," *Annals of the Association of American Geographers*, Vol. 67, September 1977, p. 423. By permission of the Association of American Geographers.)

(2.5 centimeters) wide and as much as a yard (1 meter) deep. Some surface material falls into the cracks, and more is washed in when it rains. When the soil is wetted again, more swelling takes place and the cracks close. This alternation of wetting and drying, expansion and contraction, produces a churning effect that mixes the soil constituents (the name *Vertisol* connotes an inverted condition), inhibits the development of horizons, and may even cause minor irregularities in the land surface (Figure 12-29).

An alternating wet and dry climate is needed for Vertisol formation because the sequence of swelling and contraction is necessary. Thus the wet/dry climate of tropical and subtropical savannas is ideal, but there must also be the proper parent material to yield the clay minerals. Consequently, Vertisols are widespread in distribution but are very limited in extent (Figure 12-30). The principal occurrences are in eastern Australia, India, and a small part of East Africa. They are uncommon in the United States, although prominent in some parts of Texas and California (Figure 12-31).

Figure 12-28 Profile of an Entisol in Thailand. As is characteristic of Entisols, differentiating horizons are difficult to discern. (Dr. Hari Eswaran, U. S. Department of Agriculture)

Figure 12-29 A dark Vertisol profile from Zambia. Many cracks are typically found in Vertisols. (U. S. Department of Agriculture)

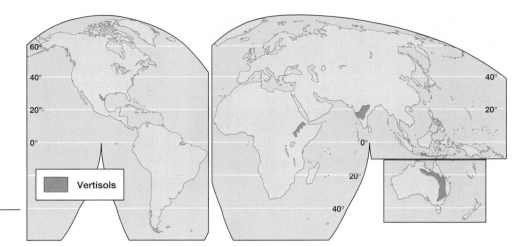

Figure 12-30 World distribution of Vertisols.

The fertility of Vertisols is relatively high, as they tend to be rich in nutrient bases. They are difficult to till, however, because of their sticky plasticity, and so they are often left uncultivated.

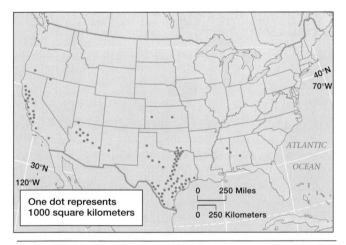

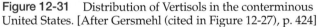

Figure 12-31 Distribution of Vertisols in the conterminous United States. [After Gersmehl (cited in Figure 12-27), p. 424]

ANDISOLS

Having developed from volcanic ash, **Andisols** have been deposited in relatively recent geological time. They are not highly weathered, therefore, and there has been little downward translocation of their colloids. There is minimum profile development, and the upper layers are dark. Their inherent fertility is relatively high.

Andisols are found primarily in volcanic regions of Japan, Indonesia, and South America, as well as in the very productive wheat lands of Washington, Oregon, and Idaho (Figure 12-32).

INCEPTISOLS

Another immature order of soils is the **Inceptisols**. Their distinctive characteristics are relatively faint, not yet prominent enough to produce diagnostic horizons (Figure 12-33). If the Entisols can be called youthful, the Inceptisols might be classified adolescent. They are primarily eluvial soils and lack illuvial layers.

Like Entisols, Inceptisols are widespread over the world in various environments (Figure 12-34). Also like Entisols, they include a variety of fairly dissimilar

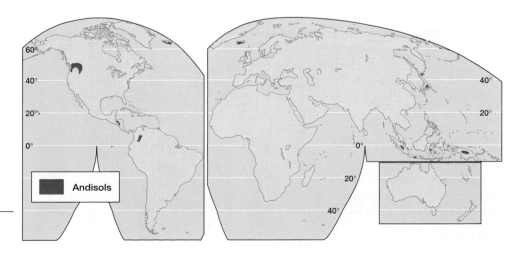

Figure 12-32 World distribution of Andisols.

Figure 12-33 Profile of a New Zealand Inceptisol with a distinctive B horizon of white pebbly material. (U. S. Department of Agriculture)

soils whose common characteristic is lack of maturity. They are most common in tundra and mountain areas but are also notable in older valley floodplains. Their world distribution pattern is very irregular. This is also true in the United States, where they are most typical of the Appalachian Mountains, the Pacific Northwest, and the lower Mississippi Valley (Figure 12-35).

ARIDISOLS

Nearly one-fifth of the land surface of Earth is covered with **Aridisols**, the most extensive spread of any soil order (Figure 12-36). They are preeminently soils of the dry lands, occupying environments that do not have enough water to remove soluble minerals from the soil. Thus their distribution pattern is largely correlated with that of desert and semidesert climate, as the distribution map for the United States shows (Figure 12-37).

Aridisols are typified by a thin profile that is sandy and lacking in organic matter, characteristics clearly associated with a dry climate and a scarcity of penetrating moisture (Figure 12-38). The epipedon is almost invariably light in color. There are various kinds of diagnostic subsurface horizons, nearly all distinctly alkaline. Most Aridisols are unproductive, particularly because of lack of moisture; if irrigated, however, some display remarkable fertility. The threat of salt accumulation is ever present, however.

MOLLISOLS

The distinctive characteristic of **Mollisols** is the presence of a mollic epipedon, which is a mineral surface horizon that is dark and thick, contains abundant humus and basic cations, and retains a soft character (rather than becoming hard and crusty) when it dries out (Figure 12-39). Mollisols can be thought of as transition soils that evolve in regions not dominated by ei-

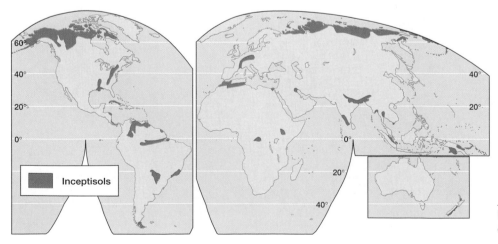

Figure 12-34 World distribution of Inceptisols.

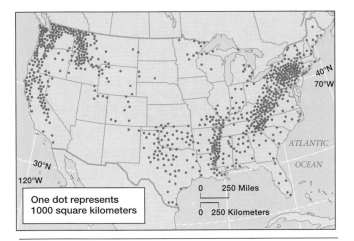

Figure 12-35 Distribution of Inceptisols in the conterminous United States. [After Gersmehl (cited in Figure 12-27), p. 423]

ther humid or arid conditions. They are typical of the midlatitude grasslands and are thus most common in central Eurasia, the North American Great Plains, and the pampas of Argentina (Figures 12-40 and 12-41).

The grassland environment generally maintains a rich clay/humus content in a Mollisol soil. The dense, fibrous mass of grass roots permeates uniformly through the epipedon and to a lesser extent into the subsurface layers. There is almost continuous decay of plant parts to produce a nutrient-rich humus for the living grass.

Mollisols on the whole are probably the most productive soil order. They are generally derived from loose parent material rather than from bedrock and tend to have favorable structure and texture for cultivation. Because they are not overly leached, nutrients are generally retained within reach of plant roots. Moreover, Mollisols provide a favored habitat for earthworms, which contribute to softening and mixing the soil.

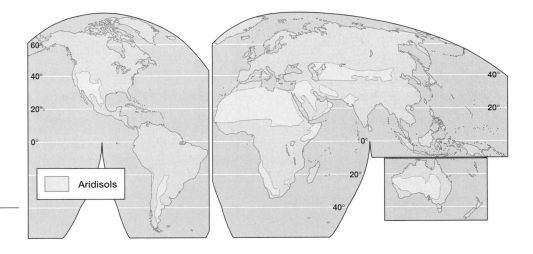

Figure 12-36 World distribution of Aridisols.

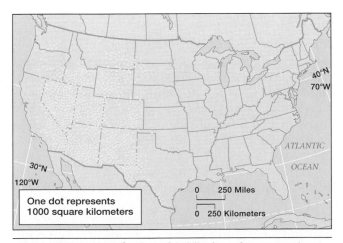

Figure 12-37 Distribution of Aridisols in the conterminous United States. [After Gersmehl (cited in Figure 12-27), p. 424]

Figure 12-38 The typical sandy profile of an Aridisol, in this case from New Mexico. (U. S. Department of Agriculture)

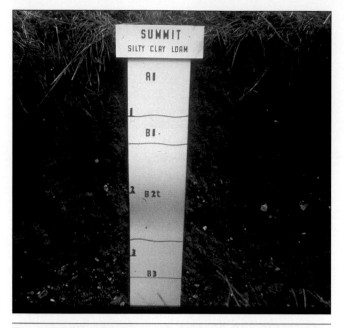

Figure 12-39 A South Dakota Mollisol with a typical mollic epipedon, a surface horizon that is dark and replete with humus. (U. S. Department of Agriculture)

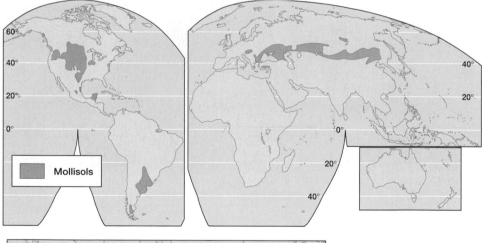

Figure 12-40 World distribution of Mollisols.

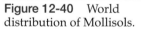

Figure 12-41 Distribution of Mollisols in the conterminous United States. [After Gersmehl (cited in Figure 12-27), p. 425]

SPODOSOLS

The key diagnostic feature of a **Spodosol** is a spodic subsurface horizon, an illuvial dark or reddish layer where organic matter, iron, and aluminum accumulate. The upper layers are light-colored and heavily leached (Figure 12-42). At the top of the profile is usually an O horizon of organic litter. Such a soil is a typical result of podzolization.

Spodosols are notoriously infertile. They have been leached of useful nutrients and are acidic throughout. They do not retain moisture well and are lacking in humus and often in clay.

Spodosols are most widespread in areas of coniferous forest where there is a subarctic climate (Figure 12-43). Alfisols, Histosols, and Inceptisols also occupy these regions, however, and Spodosols are sometimes found in other environments, such as poorly drained portions of Florida (Figure 12-44).

ALFISOLS

The most wide-ranging of the mature soils, **Alfisols** occur extensively in low and middle latitudes, as Figure 12-45 shows. They are found in a variety of temperature and moisture conditions and under diverse vegetation associations. By and large, they tend to be associated with transitional environments and are less characteristic of regions that are particularly hot or cold or wet or dry. Their global distribution is extremely varied. They are also widespread in the United States, with particular concentrations in the Midwest (Figure 12-46).

Alfisols are distinguished by a subsurface clay horizon and a medium to generous supply of plant nutrients and water. The epipedon is ochric (light colored), as Figure 12-47 shows, but beyond that, it has no char-

Figure 12-42 The profile of a Spodosol often shows a light-colored A horizon overlying a reddish B horizon. This example is from Quebec. (U. S. Department of Agriculture)

acteristics that are particularly diagnostic and can be considered an ordinary eluviated horizon. The relatively moderate conditions under which Alfisols develop tend to produce balanced soils that are reasonably fertile. Alfisols rank second only to Mollisols in agricultural productivity.

ULTISOLS

Ultisols are roughly similar to Alfisols except that the former are more thoroughly weathered and more completely leached of nutrient bases. They have experienced greater mineral alteration than any other soil in the midlatitudes, although they also occur in the low latitudes. Many pedologists believe that the ultimate fate of Alfisols is to degenerate into Ultisols.

Typically, Ultisols are reddish as a result of the significant proportion of iron and aluminum in the A horizon. Usually they have a fairly distinct layer of subsurface clay accumulation. The principal properties of Ultisols have been imparted by a great deal of weathering and leaching (Figure 12-48). Indeed, the connotation of the name (derived from the Latin *ultimos*) is that these soils represent the ultimate stage of weathering in the conterminous United States. The result is a fairly deep soil that is acidic, lacks humus, and has a relatively low fertility due to the lack of bases.

Ultisols have a fairly simple world distribution pattern (Figure 12-49). They are mostly confined to humid subtropical climates and to some relatively youthful tropical land surfaces. In the United States, they are restricted largely to the southeastern quarter of the country and to a narrow strip along the northern Pacific coast (Figure 12-50).

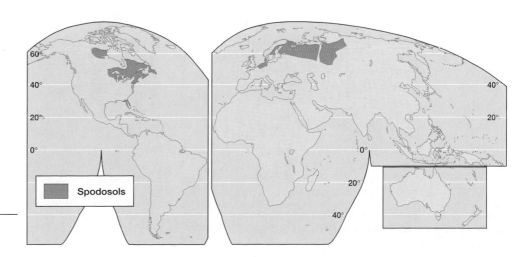

Figure 12-43 World distribution of Spodosols.

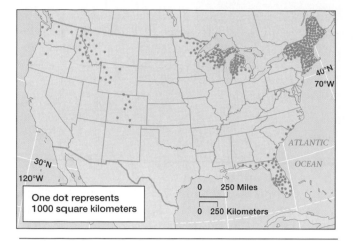

Figure 12-44 Distribution of Spodosols in the conterminous United States. [After Gersmehl (cited in Figure 12-27), p. 425]

Oxisols

The most thoroughly weathered and leached of all soils are the **Oxisols**, which invariably display a high degree of mineral alteration and profile development. They occur mostly on ancient landscapes in the humid tropics, particularly in Brazil and equatorial Africa, and to a lesser extent in Southeast Asia (Figure 12-51). The distribution pattern is often spotty, with Oxisols mixed with less-developed Entisols, Vertisols, and Ultisols. Oxisols are totally absent from the United States, except for Hawaii, where they are common.

Oxisols are essentially the products of laterization (and in fact were called *Latosols* in the older classification systems). They have evolved in warm, moist climates, although some are now found in drier regions, an indication of climatic change since the soils developed. The diagnostic horizon for Oxisols is a subsurface dominated

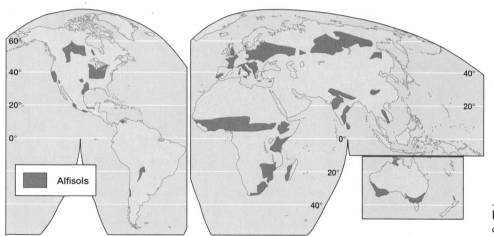

Figure 12-45 World distribution of Alfisols.

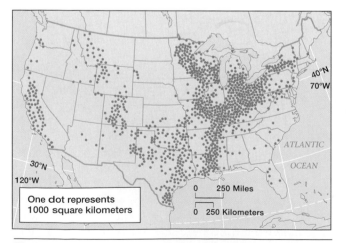

Figure 12-46 Distribution of Alfisols in the conterminous United States. [After Gersmehl (cited in Figure 12-27), p. 426]

Figure 12-47 An Alfisol from southern Texas. The reddish surface horizon in this profile is characteristic of Alfisols in drier localities. (U.S. Department of Agriculture)

Figure 12-48 A tropical Ultisol from Thailand. It is reddish throughout its profile, indicative of much leaching and weathering. (U.S. Department of Agriculture)

by oxides of iron and aluminum and with a minimal supply of nutrient bases (this is called an *oxic horizon*). These are deep soils but not inherently fertile (Figure 12-52). The natural vegetation is efficient in cycling the limited nutrient supply, but if the flora is cleared (to attempt agriculture, for example), the nutrients are rapidly leached out, and the soil becomes impoverished.

HISTOSOLS

Least important among the soil orders are the **Histosols**, which occupy only a small fraction of Earth's land surface, a much smaller area than any other order. These are organic, rather than mineral, soils and invariably are saturated with water all or most of the time. They may occur in any waterlogged environment but are most characteristic in mid- and high-latitude regions that experienced Pleistocene glaciation. In the

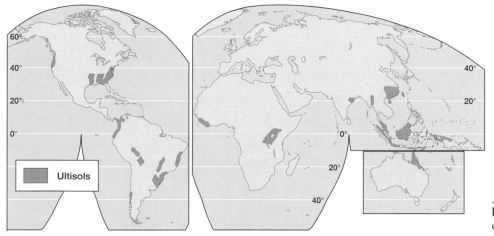

Figure 12-49 World distribution of Ultisols.

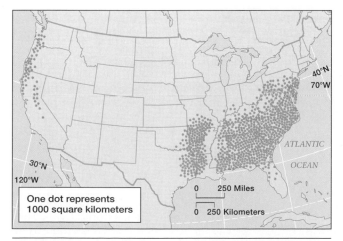

Figure 12-50 Distribution of Ultisols in the conterminous United States. [After Gersmehl (cited in Figure 12-27), p. 426]

United States, they are most common around the Great Lakes, but they also occur in southern Florida and Louisiana (Figure 12-53). Nowhere, however, is their occurrence extensive.

Some Histosols are composed largely of undecayed or only partly decayed plant material, whereas others consist of a thoroughly decomposed mass of muck (Figure 12-54). The lack of oxygen in the waterlogged soil slows down the rate of bacterial action, and the soil becomes deeper mostly by growing upward, that is, by more organic material being added from above.

Histosols are usually black, acidic, and fertile only for water-tolerant plants. If drained, they can be very productive agriculturally for a short while. Before long, however, they are likely to dry out, shrink, and oxidize, a series of steps that leads to compaction, susceptibility to wind erosion, and danger of fire.

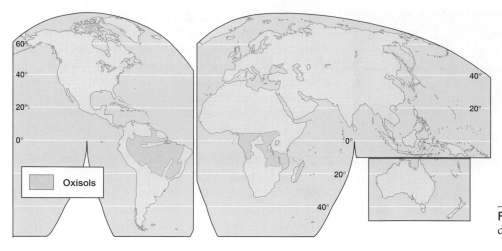

Figure 12-51 World distribution of Oxisols.

DISTRIBUTION OF SOILS IN THE UNITED STATES

The distributions of the various soils in the United States is quite different from that of the world as a whole. This difference is due to many factors, the most important being that the United States is essentially a midlatitude country, and it lacks significant expanses of area in the low and high latitudes. Table 12-4 compares the relative areas occupied by the eleven soil orders, nationally and globally, as well as the relative areas of suborders significant in the United States. The statistics are generalized estimates prepared by the Soil Conservation Service of the U.S. Department of Agriculture and should not be considered definitive.

Mollisols are much more common in the United States than in the world as a whole; they are the most prevalent soil order throughout the Great Plains and in much of the West. Also significantly more abundant in the United States than in other parts of the world are Inceptisols and Ultisols. Almost totally lacking from this country are Oxisols, and Aridisols and Entisols are proportionally less extensive.

Figure 12-52 Oxisols are impoverished tropical soils. The reddish color in this profile from Hawaii indicates considerable leaching. (U. S. Department of Agriculture)

TABLE 12-4
Approximate Proportional Extent of Soil Orders and Subotrders

Order	Suborder	United States	World
		Percentage of land area occupied	
Alfisols		13.4	14.7
	Aqualfs	1.0	
	Boralfs	3.0	
	Udalfs	5.9	
	Xeralfs	0.9	
Andisols			
Aridisols		11.5	19.2
	Argids	8.6	
	Orthids	2.9	
Entisols		5.7	12.5
	Aquents	0.2	
	Eluvents	0.3	
	Orthents	5.2	
Histosols		0.5	0.8
	Fibrists	0.2	
	Hemists	0.2	
	Saprists	0.1	
Inceptisols		18.2	15.8
	Andepts	1.9	
	Aquepts	11.3	
	Ochrepts	4.3	
	Umbrepts	0.7	
Mollisols		24.6	9.0
	Aquolls	1.3	
	Borolla	4.9	
	Udolls	4.7	
	Ustolls	8.9	
	Xerolls	4.8	
Oxisols			9.2
Spodosols		5.1	5.4
	Aquods	0.7	
	Orthods	4.4	
Ultisols		12.9	5.6
	Aqults	1.1	
	Humults	0.8	
	Udults	10.0	
	Xerults	1.0	
Vertisols		1.0	1.8
	Uderts	0.4	
	Usterts	0.6	

Source: After Nyle C. Brady, *The Nature and Properties of Soils*, 10th ed. New York: Macmillan, 1990.

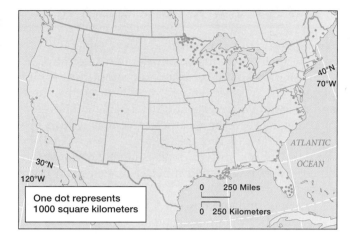

Figure 12-53 Distribution of Histosols in the conterminous United States. [After Gersmehl (cited in Figure 12-27), p. 427]

Figure 12-54 The dark color of this Histosol profile is typical. Histosols are organic soils. This example is from Brazil. (U.S. Department of Agriculture)

CHAPTER SUMMARY

Soil, the relatively thin surface layer of mineral matter that normally contains a considerable amount of organic material and is capable of supporting living plants, is an interface where atmosphere, hydrosphere, and biosphere interact with lithosphere. Soil development begins with the weathering of rock, which produces a layer of loose inorganic material called regolith. The soil proper consists largely of unconsolidated, finely fragmented mineral particles that occur at the top of the regolith.

The five fundamental soil-forming factors are geologic, climatic, topographic, biological, and chronological. These factors encompass the variables—chemical composition, environmental conditions, and time—that interact to produce a particular soil.

Soil is composed of four basic constituents. *Mineral matter* makes up the bulk of most soils. *Organic matter*, in the form of both living and dead organisms, plays a major role in the biochemical processes that promote plant growth. About half the total volume of most soils is made up of pore spaces that are filled with a mixture of *air* and *water*.

The most conspicuous property of most soils is color, but color generally is not a defining property. More important in terms of soil characteristics and capabilities are texture and structure. Texture refers to the size of individual soil particles, and structure concerns the aggregation of individual particles into larger masses or clumps.

The effectiveness of soil as a growth medium is based largely on the availability of nutrients, which in turn depends mostly on chemical characteristics and processes. The soil's colloidal complex is one of its most vital attributes, and the attraction and exchange of cations determine the inherent fertility of the soil.

Five processes, called pedogenic regimes—laterization, podzolization, gleization, calcification, and salinization—produce an infinite variety of soil types.

The most broadly accepted soil classification scheme in the United States is called the Soil Taxonomy. It is logical, generic, and comprehensive but has a daunting terminology.

KEY TERMS

acid
acidity
Alfisol
alkalinity
Andisol
Aridisol
base
calcification
cation
cation exchange capacity (CEC)
colloid
colloidal complex
eluviation
Entisol
epipedon
field capacity
gleization

gley soils
Histosol
horizon
humus
illuviation
Inceptisol
laterization
latosol
leaching
litter
loam
Mollisol
Oxisol
parent material
ped
pedogenic regimes
podzol

podzolization
regolith
salinization
separates
soil
soil profile
Soil Taxonomy
soil-water balance
soil-water budget
solum
Spodosol
subsurface horizon
Ultisol
Vertisol
wilting point

REVIEW QUESTIONS

1. What is the relationship between weathering and regolith?
2. Briefly describe the five principal soil-forming factors.
3. Explain the importance of parent material to the nature of the overlying soil.
4. Why does soil tend to be deepest on flat land?
5. Why are earthworms generally considered beneficial to humans?
6. Explain the importance of clay as a constituent of soil.
7. Describe the four forms of soil moisture.
8. Distinguish between field capacity and wilting point.
9. What can you learn about a soil from its color?
10. Distinguish between soil texture and soil structure.
11. Explain the difference between porosity and permeability.
12. What is a soil profile?
13. How does the Soil Taxonomy differ from previous soil classification schemes?
14. Why is it so difficult to portray soil distribution with reasonable accuracy on a small-scale map?
15. Describe the distribution and characteristics of one soil order.

SOME USEFUL REFERENCES

BRADY, NYLE C., *The Nature and Properties of Soils*, 10th ed. New York: Macmillan, 1990.

BUOL, S.W., F. D. HOLE, AND R. J. MCCRACKEN, *Soil Genesis and Classification*, 3rd ed. Ames: Iowa State University Press, 1989.

FANNING, DELVIN S., AND MARY C. B. FANNING, *Soil Morphology, Genesis, and Classification*. New York: John Wiley & Sons, 1989.

FINKL, CHARLES W., JR. (ed.), *Soil Classification*. New York: Hutchinson Ross, 1982.

FOTH, HENRY D., *Fundamentals of Soil Science*, 8th ed. New York: John Wiley & Sons, 1990.

ROSS, SHEILA, *Soil Processes: A Systematic Approach*. London: Routledge, 1989.

SOIL SURVEY STAFF, *Soil Taxonomy*. Soil Conservation Service, U.S. Department of Agriculture, Agriculture Handbook No. 436. Washington, DC: U.S. Government Printing Office, 1975.

STEILA, DONALD, AND THOMAS E. POND, *The Geography of Soils*, 2nd ed. Totowa, NJ: Rowman & Littlefield, 1989.

13

INTRODUCTION TO LANDFORM STUDY

THOUGHTFUL BEGINNING STUDENTS OF PHYSICAL geography may be daunted by the magnitude of their object of study because Earth is enormous from the human viewpoint. Its diameter of 8000 miles (12,800 kilometers) and circumference of 25,000 miles (40,000 kilometers) are distances well beyond our normal scale of living and thinking. To comprehend the nature of such a massive body is a colossal task.

Our endeavor is greatly simplified, however, because we can largely ignore the interior and concentrate our attention on the surface. Two factors make this focus possible: (1) the characteristics of Earth's interior are very imperfectly understood by scientists, and (2) the focus of geographic inquiry is primarily in humankind's zone of habitation.

THE UNKNOWN INTERIOR

Our knowledge of the interior of Earth is scanty and based entirely on indirect evidence. No human activity has explored more than a minute fraction of the vastness beneath the surface. No person has penetrated as much as one-thousandth of the radial distance from the surface to the center of Earth; the deepest existing mine shaft extends a mere 2.4 miles (3.8 kilometers). Nor have probes extended much deeper; the deepest known drill holes from which sample cores have been brought up have penetrated only a modest 7 miles (11 kilometers) into Earth. Earth scientists, in the colorful imagery of John McPhee, "are like dermatologists: they study, for the most part, the outermost two per cent of the earth. They crawl around like fleas on the world's tough hide, exploring every wrinkle and crease, and try to figure out what makes the animal move" (*Assembling California*, p. 36).

Even so, a considerable body of inferential knowledge concerning Earth's interior has been amassed by geophysical means, primarily by monitoring shock waves transmitted through Earth from earthquakes or from human-made explosions. Such seismic waves change their speed and direction whenever they cross a boundary from one type of material to another. Analysis of these changes, augmented by related data on Earth's magnetism and gravitational attraction, has enabled earth scientists to develop various models of Earth's internal structure.

Figure 13-1 is one rendition of this inferred structure. It has been deduced that Earth has a heavy inner core surrounded by three concentric layers of various composition and density. Starting at the surface and moving inward, these four regions are called the crust, the mantle, the outer core, and the inner core:

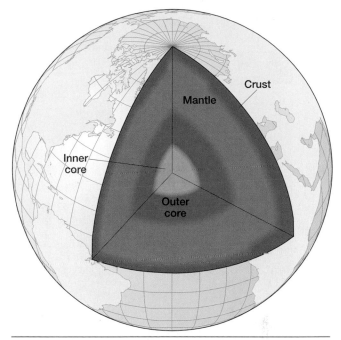

Figure 13-1   The presumed vertical structure of Earth's interior.

1. The **crust**, the outermost shell, consists of a broad mixture of rock types. Beneath the oceans the crust has an average thickness of only 5 or 6 miles (8 to 10 kilometers), whereas beneath the continents the thickness averages more than three times as much. There is a gradual increase in density with depth in this rigid outer shell. Altogether, the crust makes up less than 1 percent of Earth's volume and about 0.4 percent of Earth's weight.

    At the base of the crust there is thought to be a significant change in mineral composition; this relatively narrow zone of change is called the **Mohorovičić discontinuity** (named for the Yugoslavian seismologist, A. Mohorovičić, who discovered it), or **Moho** for short.

2. Beneath the Moho is the **mantle**, which extends downward to a depth of approximately 1800 miles (2900 kilometers). Volumetrically, the mantle is by far the largest of the four shells. Although its depth is only about one-half the distance from the surface to the center of the Earth, its location on the periphery of the sphere gives it a vast breadth; it makes up 84 percent of the total volume of Earth and about two-thirds of Earth's total weight.

    Earth scientists now believe there are three zones in the mantle, as Figure 13-2 shows. The uppermost zone is relatively thin but hard and rigid extending down to about 25 miles (40 kilometers) beneath the

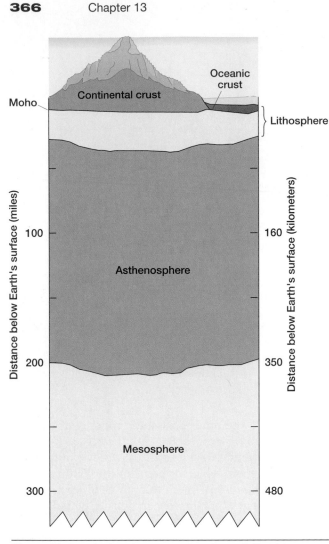

Figure 13-2 Idealized cross section through Earth's crust and part of the mantle. The crust and uppermost mantle, both rigid zones, are together called the lithosphere. The asthenosphere part of the mantle is hot and therefore weak and easily deformed. In the mesosphere part, the mantle is rigid again. The mesosphere is much deeper than shown here, extending to about 1800 miles (2900 kilometers) below Earth's surface.

oceanic crust and perhaps 50 miles (80 kilometers) below the continental crust. To geologists, this uppermost mantle zone plus the crust are together called the **lithosphere.** (This use of the word is different from what we learned in Chapter 1. There, in describing the four spheres that interact to make Earth what it is, we used "lithosphere" to mean all the rock that makes up the solid part of Earth's surface. These two meanings should not be confused and are usually distinguishable from context. Except in the present discussion, "lithosphere" in this book always has the geographer's meaning—in other words, the solid part of Earth's surface.) Beneath this rigid zone, and extending downward to a depth of possibly 200 miles (350 kilometers), is a zone in which the rocks are so hot that they lose

much of their strength and are easily deformed, like tar. This is called the **asthenosphere** (weak sphere). Below the asthenosphere is the deep mantle, the **mesosphere**, where the rocks are believed to be rigid again.

3. Beneath the mantle is the **outer core**, thought to be molten (liquid) and extending to a depth of about 3100 miles (5000 kilometers).

4. The innermost portion of Earth is the **inner core**, a supposedly solid and very dense mass having a radius of about 900 miles (1450 kilometers). Both the inner and outer cores are thought to be made of iron/nickel or iron/silicate. These two zones together make up about 15 percent of Earth's volume and 32 percent of its weight.

This generalized model of Earth's interior is inexact and probably inadequate. The fragility of our supposed understanding of the depths was dramatically revealed in the 1960s when the notion of **continental drift**, propounded in the early 1900s but held in disdain by most scientists for half a century, was revived and expanded. Recent seismic and magnetic evidence makes it clear that the drift concept is valid, and its elaboration as the theory of **plate tectonics** is now almost universally accepted by Earth scientists. The mechanics of plate tectonics are discussed in Chapter 14; the point here is that although our understanding of the crust and upper mantle has fundamentally changed in the last three decades, it is still by no means complete, which only emphasizes the vastness of our ignorance about the nature of the deeper interior of our planet.

GEOGRAPHERS FOCUS ON THE SURFACE

Even if we possessed relatively precise knowledge of the Earth's interior, however, it would not hold our attention as geographers for long. The focus of geographic study is in the zone of human habitation, that part of the environment having some human interaction. In the same way that the geographer's concern with the upper atmosphere is limited largely to its effects on weather and climate, so the geographer's concern with the deeper portion of Earth is restricted primarily to its influence on topography and other surface features. As geographers we want to know about Earth's interior only as it helps us to comprehend the nature and characteristics of the surface.

COMPOSITION OF THE CRUST

About 90 chemical elements are found in Earth's crust, occasionally as discrete elements but usually bonded with one or more other elements to form compounds. These naturally formed compounds and elements of the lithosphere are called **minerals**, which are solid substances having a specific chemical composition and a characteristic crystal structure (Table 13-1). About 3500 minerals are known so far, with about 50 new types discovered each year. Most have been found in the crust, as that is the only part of Earth that humans have been able to investigate. A few known minerals are extraterrestrial, having been either identified in meteorites or brought back from the moon.

Inside Earth is an unknown amount of molten mineral matter called **magma**. At or near the surface, however, almost all the lithosphere is a solid, generally known as **rock**, composed of aggregated mineral particles that occur in bewildering variety and complexity. In coarse-grained rocks the crystalline mineral component can easily be seen with the naked eye (Figure 13-3), whereas in many fine-grained rocks the minerals can be distinguished only under magnification. Fewer than 20 minerals account for more than 95 percent of the composition of all continental and oceanic crust.

Solid rock sometimes is found right at the surface, in which case it is called an **outcrop** (Figure 13-4). Over most of Earth's land area, though, solid rock exists as a buried layer called **bedrock** and covered by a layer of regolith (Chapter 12). Soil, when present, is above the regolith.

The enormous variety of rocks can be systematically classified. A detailed knowledge of petrology (the characteristics of different kinds of rocks) is unnecessary for our purposes here, however, and so we restrict coverage to a survey of the three major rock classes—igneous, sedimentary, and metamorphic—and their basic attributes (Figure 13-5 and Table 13-2).

IGNEOUS ROCKS

In the beginning, all rocks on Earth were igneous. The initial formation of the planet involved the cooling and solidification of magma, which produced the first solid material—**igneous rocks**. The word *igneous* signifies a fiery inception; all igneous rocks were formed by magmatic cooling, which has continued to occur throughout the history of our planet. There are within Earth

(a)

(b)

Figure 13-3 Some rocks, such as this piece of granite **(a)**, are coarse-grained. Others, such as this sample of rhyolite **(b)**, are very fine-grained. (Breck P. Kent/Earth Sciences)

TABLE 13-1
Necessary Features of a Mineral
By definition, all minerals must—
1. be found in nature
2. be made up totally of inorganic substances
3. have the same chemical composition wherever found
4. contain atoms arranged in a regular pattern and forming solid units called *crystals*

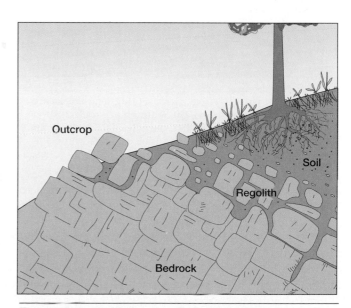

Figure 13-4 Bedrock is usually buried under a layer of soil and regolith but occasionally appears as an outcrop.

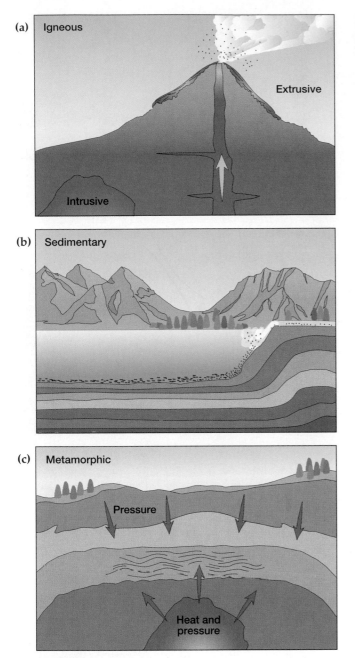

Figure 13-5 The three rock classes. **(a)** Igneous rocks are formed when magma cools. **(b)** Sedimentary rocks result from consolidation of deposited particles. **(c)** Metamorphic rocks are produced when heat and/or pressure act on existing igneous and sedimentary rocks.

T A B L E 1 3 - 2 **Rock Classification**		
Class	*Subclass*	*Type*
Igneous	Intrusive	Granite Syenite Diorite Gabbro Peridotite Pyroxenite Hornblendite
	Extrusive	Obsidian Pumice Tuff Rhyolite Trachyte Andesite Basalt Diabase
Sedimentary	Calcareous	Limestone Dolomite
	Siliceous	Shale Sandstone Chert Conglomerate Breccia
Metamorphic	Foliated	Gneiss Schist Amphibolite Slate
	Nonfoliated	Quartzite Marble Serpentinite

forces of tremendous strength that cause magma to rise toward the surface, where it cools and solidifies, becoming part of the solid crust.

There are a great many kinds of igneous rocks, and their characteristics are quite variable. Their principal shared trait is that they have a crystalline structure. If the magma from which a particular igneous rock formed cooled slowly, as happens far beneath Earth's surface where thousands of years may be required for full cooling,

the crystals can be large, giving the rock a very coarse-grained appearance, as shown in Figure 13-3. When the magma cools rapidly, as on Earth's surface, and solidification may be complete within hours, the crystals may be so small as to be invisible without microscopic inspection.

As Figure 13-5(a) shows, an important distinction is made among igneous rocks on the basis of the conditions under which they solidified. **Extrusive rocks** were spewed out onto Earth's surface while still molten, solidifying quickly in the open air. Their most familiar form is the spectacular ejection of lava from volcanoes, but quieter instances of extrusion, such as oozing of magma from fissures, have also been frequent in Earth's history. During extrusion, much of the volatile matter in the magma escapes as gases prior to crystallization, and the rapid cooling may produce a glassy texture in all or part of the resulting rock. Of the many kinds of extrusive rocks, by far the most common is a dark, fine-grained (often speckled) rock called *basalt* (Figure 13-6).

Figure 13-6 Several horizontal layers of flood basalt are exposed on the wall of the Snake River Canyon near Twin Falls, Idaho. (TLM photo)

Intrusive rocks (also called plutonic rocks) cool and solidify beneath Earth's surface, where surrounding nonmagmatic material serves as insulation that greatly retards the rate of cooling. Although originally buried, intrusive rocks may subsequently become important to topographic development by being pushed upward to the surface or by being exposed by erosion. The most common and well-known intrusive rock is *granite*, a light-colored, coarse-grained igneous rock (Figure 13-7).

The original crust that formed as Earth cooled consisted of igneous rocks; thus all subsequently formed rocks are derived from this original igneous supply.

SEDIMENTARY ROCKS

External processes, mechanical and chemical, operating on rocks cause them to disintegrate. This disintegration produces fragmented mineral material, some

Figure 13-7 Massive outcrops of granite at Sylvan Lake in the Black Hills of South Dakota. Granite is an intrusive igneous rock that solidified beneath Earth's surface and was subsequently exposed by erosion. (TLM photo)

of which is removed by water, wind, ice, gravity, or a combination of these agents. Much of this material is transported by water moving in rivers or streams as **sediment**. Eventually the sediment is deposited somewhere in a quiet body of water, particularly on the floor of an ocean [Figure 13-5(b)]. Over a long period of time, sedimentary deposits can build to a remarkable thickness—many thousands of feet. The sheer weight of this massive overburden exerts an enormous pressure, which causes individual particles in the sediment to adhese to each other and to interlock. In addition, chemical cementation normally takes place. Various cementing agents—especially silica, calcium carbonate, and iron oxide—precipitate from the water into the pore spaces in the sediment (Figure 13-8). This combination of pressure and cementation consolidates and transforms the sediments to **sedimentary rock**.

During sedimentation, materials are sorted roughly by size; the finer particles are carried farther than the heavier particles. Other variations in the composition of the sediments are due to processes and rates of deposition, changes in climatic conditions, ocean current movements, and other factors. Consequently, most

sedimentary deposits are built up in more or less distinct horizontal layers called **strata**, which vary in thickness and composition (Figure 13-9). The resulting parallel structure, or *stratification*, is a characteristic feature of most sedimentary rocks. Although originally deposited and formed in horizontal orientation, the strata may later be uplifted, tilted, and deformed by pressures from within Earth (Figure 13-10).

Sedimentary rocks can be classified on the basis of how they were formed: mechanically, chemically, or organically . The mechanically accumulated sedimentary rocks are composed of fragments of preexisting rocks, in the form of boulders, gravel, sand, silt, or clay; by far the most common are shale (composed of silt and clay particles) and sandstone (made up of compacted sand grains). Chemically accumulated sedimentary rocks are usually formed by the precipitation of soluble materials or sometimes by more complicated chemical reactions. Calcium carbonate is a common component of such rocks, and limestone is the most widespread result. Organically accumulated sedimentary rocks, including coal, are formed from the accumulated remains of dead plants or animals. Limestone can also be formed in this fashion from skeletal remains of coral and other lime-secreting sea animals.

There is considerable overlap among these three formation methods, with the result that, in addition to there being many different kinds of sedimentary rocks, there also are many gradations among them. Taken together, however, the vast majority of all sedimentary rocks are either limestone, sandstone, or shale (Figure 13-11).

METAMORPHIC ROCKS

Originally either igneous or sedimentary, **metamorphic rocks** are those that have been drastically changed by heat and/or pressure. The effects of heat and pressure on rocks are complex, being strongly influenced by such things as the presence or absence of fluids in the rocks and the length of time the rocks are heated and/or subjected to high pressure. The metamorphic result is often quite different from the original rock; the rocks are changed in structure, texture, composition, and appearance (Figure 13-12). Metamorphism is virtually a cooking process that partially melts the rock, causing its mineral components to be recrystallized and rearranged. The usual result is a banding, or *foliation*, that gives a wavy-layered appearance, although if the original rock was dominated by a single mineral (as sandstone or limestone) such foliation does not normally develop.

Some rocks, when metamorphosed, change in predictable fashion. Thus, limestone usually becomes marble, sandstone normally is changed to quartzite, and shale to slate. In many cases, however, the metamor-

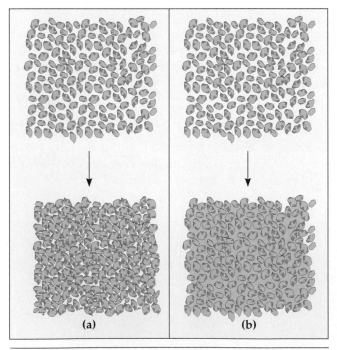

Figure 13-8 Sedimentary rocks are composed of small particles of matter deposited by water or wind in layers and then consolidated by compaction and/or cementation. **(a)** Compaction consists of the packing of the particles as a result of the weight of overlying material. **(b)** Cementation involves the infilling of pore spaces among the particles by a cementing agent, such as silica, calcium carbonate, or iron oxide.

Figure 13-9   Nearly horizontal strata of limestone and shale in a road cut near Lyons, Colorado. (TLM photo)

phosis is so great that it is difficult to ascertain the nature of the original rock. By far the most common metamorphic rocks are *schist* (in which the foliations are very narrow) and *gneiss* (broad foliations), both of which represent a high degree of metamorphism (Figure 13-13).

Figure 13-14 diagrams the relationships among the three classes of rocks.

RELATIVE FREQUENCY OF ROCK CLASSES

The lithosphere has a very uneven distribution of the three principal rock classes. Sedimentary rocks compose the most common bedrock (perhaps as much as 75 percent) of the continents (Figure 13-15), and sediments cover nearly all the sea floor, except in areas where volcanoes are active. The sedimenta-

Figure 13-10   Sedimentary strata that have been folded and tilted into an almost vertical orientation on Mount Angeles in the Olympic Mountains of Washington. (TLM photo)

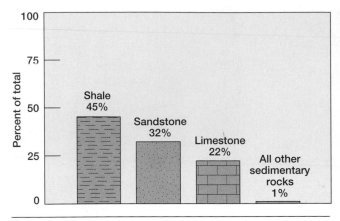

Figure 13-11 The estimated relative abundance of various types of sedimentary rocks.

(a)

ry cover is not thick, however, averaging less than 1.5 miles (2.4 kilometers), and sedimentary rocks accordingly constitute only a very small proportion (perhaps 5 percent) of the total volume of the crust (Figure 13-16). Igneous rocks apparently make up the bulk of the crust, but the volume of metamorphic rocks (which are relatively minor at Earth's surface) is conceivably even greater because an enormous amount of metamorphosis has taken place beneath the crustal surface.

Figure 13-12 This bedrock exposure in northeastern California, near Alturas, shows a light brown basalt (a common extrusive igneous rock) overlying a colorful layer of tuff (an extrusive igneous rock formed by consolidation of volcanic ash). The basalt was extruded onto the tuff in molten form, and its great heat "cooked," or metamorphosed, the upper portion of the tuff. Visual evidence of the metamorphosis is seen in the difference in color between the metamorphosed and unmetamorphosed part of the tuff stratum. (TLM photo)

(b)

Figure 13-13 (a) An outcrop of banded gneiss in Greenland. This particular formation is one of the oldest on Earth; its age is 3.8 billion years. (Kevin Schafer/Peter Arnold, Inc.) (b) An outcropping of schist bedrock in the inner gorge of the Colorado River in Grand Canyon National Park. (C. C. Lockwood/Earth Scenes)

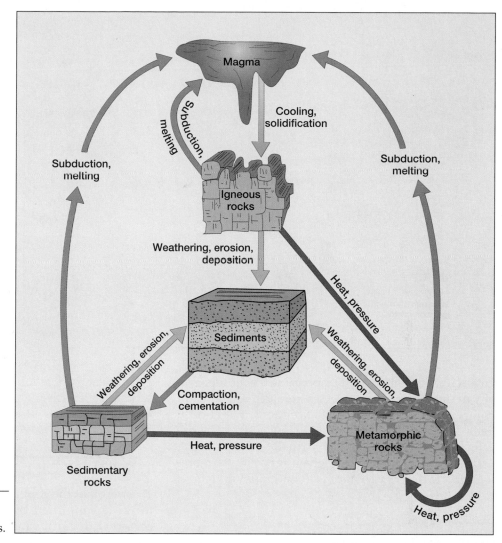

Figure 13-14 The rock cycle, showing the relationships among the three classes of rocks.

SOME CRITICAL CONCEPTS

Before we begin a geographic analysis of the lithosphere, we need to learn about a few concepts and terms essential to the subsequent discussion.

BASIC TERMS

Our attention is directed primarily to **topography**, which is the surface configuration of Earth. A **landform** is an individual topographic feature, of whatever size; thus the term could refer to something as minor as a cliff or a sand dune, as well as to something as major as a peninsula or a mountain range. The plural—landforms—is less restrictive and is generally considered synonymous with *topography*. Our focus in this section of the book is **geomorphology**, the study of the characteristics, origin, and development of landforms.

One other term used frequently in the following pages is **relief**, which refers to the difference in elevation between the highest and lowest points in an area. The term can be used at any scale. Thus, as we saw in Figure 1-6, the maximum world relief is approximately 13 miles (21 kilometers), which is the difference in elevation between the top of Mount Everest and the bottom of the Mariana Trench. At the other extreme, a *local relief* in some place like Florida can be a matter of merely a few feet.

UNIFORMITARIANISM

Fundamental to understanding topographic development is familiarity with the doctrine called **uniformitarianism,** which holds that "the present is the key to the past." This means that the processes that formed the topography of the past are the same ones that have shaped contemporary topography; these processes are still functioning in the same fashion and, barring un-

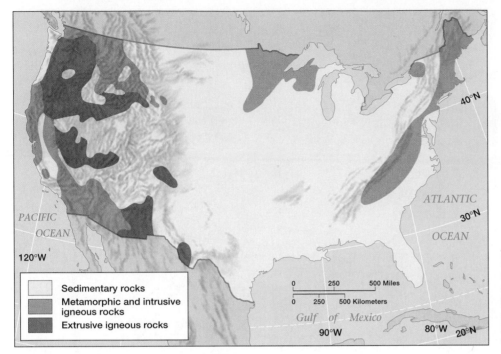
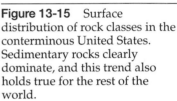

Figure 13-15 Surface distribution of rock classes in the conterminous United States. Sedimentary rocks clearly dominate, and this trend also holds true for the rest of the world.

foreseen cataclysm, will be responsible for the topography of the future. The processes involved are not temporary and, with only a few exceptions, not abrupt. They are mostly permanent and slow acting. The development of landforms is a virtually endless event, with the topography at any given time simply representing a temporary balance in a continuum of change.

GEOLOGIC TIME

Probably the most mind-boggling concept in all physical geography is the vastness of geologic time (Figure 13-17). In our puny human scale of time, we deal with such brief intervals as hours, months, and centuries, which does nothing to prepare us for the scale of

Earth's history. The colossal sweep of geologic time encompasses epochs of millions and hundreds of millions of years.

THE STUDY OF LANDFORMS

To assert that the geographer's task in studying landforms is simplified by being only marginally concerned with Earth's interior does not, by any stretch of the imagination, mean that the task is simple. Although our focus is on Earth's surface, that surface is vast, complex, and often obscured. Even without considering the 70 percent of our planet covered with oceanic waters, we must realize that more than 58

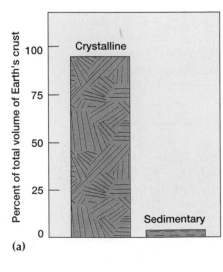

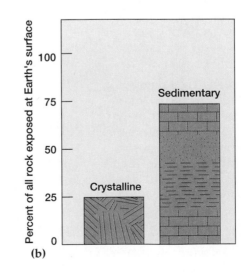

Figure 13-16 Relative abundances of sedimentary and crystalline (igneous and metamorphic) rocks **(a)** in Earth's crust and **(b)** exposed at Earth's surface.

PEOPLE AND THE ENVIRONMENT

THE MAGNITUDE OF GEOLOGIC TIME

Geologic time is the time span during which events commonly termed *geologic* have taken (and still are taking) place. In order for uniformitarianism to work as a basic premise for interpreting Earth history, that history must have occurred over a span of time vast enough to allow feats of considerable magnitude to be accomplished by means of processes infinitesimally slow when measured on a human timescale.

How can an ordinary mortal grasp the grandiose sweep of geologic time? To state that Earth is thought to have existed for about 4.6 billion years, or that the Age of Dinosaurs persisted for some 160 million years, or that the Rocky Mountains were initially uplifted approximately 65 million years ago is to enumerate temporal expanses of almost unfathomable scope.

Analogy offers at least the momentary illusion of comprehension. Let us consider both a long and a short analogy in an effort to reduce the almost measureless numbers of geologic time into a manageable frame of reference.

If we could envision the entire 4.6-billion-year history of Earth compressed into a single calendar year, each day would be equivalent to 12.6 million years, each hour to 525,000 years, each minute to 8750 years, and each second to 146 years (see the accompanying tabulation). On such a scale, the planet was lifeless for the first 4 months, with primal forms of one-celled life not appearing until early May. These primitive algae and bacteria had the world to themselves until early November, when the first multicelled organisms began to evolve. The first vertebrate animals,

Geologic time, years	Its equivalent on 1-year scale
4.6 billion	1 year
12.6 million	1 day
525,000	1 hour
8750	1 minute
146	1 second

antediluvian fishes, appeared about November 21st, and before the end of the month, amphibians began establishing themselves as the first terrestrial vertebrates. Vascular plants, mostly tree ferns, club mosses, and horsetails, appeared about November 27th, and reptiles began their era of dominance about December 7th. Mammals arrived about December 14th, with birds on the following day. Flowering plants would first bloom about December 21st, and on Christmas Eve would appear both the first grasses and the first primates. The first hominids walked upright in mid-afternoon on New Year's Eve, and *Homo sapiens* appeared on the scene about an hour before midnight. The age of written history encompassed the last minute of the year—just the last 60 ticks of the clock!!

For our short analogy, we turn to the out-of-doors. If all earthly time could be represented as a cliff 1 mile (1.6 kilometers) high, then all historic time would be found in the uppermost tenth of an inch (0.25 centimeter), and an individual's existence would be encompassed within less than the thickness of the finest hair.

Only with such an extraordinary timescale as this can one give credence to the doctrine of uniformitarianism or accept that the Grand Canyon is a youthful feature carved by that relatively small river seen deep in its inner gorge or believe that Africa and South America were once joined together and have drifted 2000 miles (3200 kilometers) apart. Indeed, the remainder of this book can be considered only as fanciful fiction unless one can rely on the concept of geologic time. The geomorphic processes generally operate with excruciating slowness, but the vastness of geologic history provides a suitable time frame for their accomplishments.

If daunted by the noxious stench
Exhaled from Time's abyss,
Retreat into some lesser trench
Where ignorance is bliss.

—Claude C. Albritton Jr.,
The Abyss of Time

million square miles (150 million square kilometers) of land are scattered over seven continents and innumerable islands. This area encompasses the widest possible latitudinal range and the full diversity of environmental conditions. Moreover, much of the surface is obscured from view by the presence of vegetation, soil, or the works of humankind. We must try to penetrate those obstructions, observe the characteristics of the lithospheric surface, and encompass the immensity and diversity of a worldwide landscape. This is far from a simple task.

To organize our thinking for such a complex endeavor, we can isolate certain basic elements for an analytic approach:

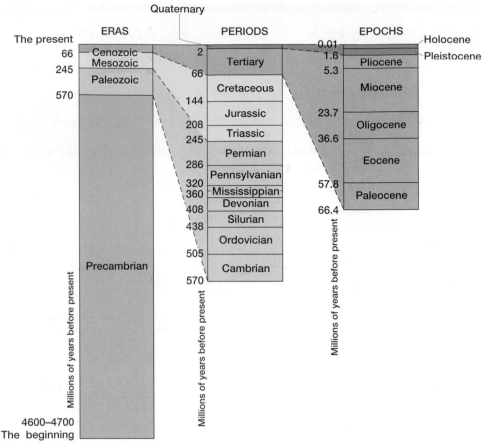

Figure 13-17 The relative length of geologic time intervals. Geologic history is divided into four major units, called *eras*. The eras are subdivided into *periods*, and the more recent periods are further subdivided into *epochs*. The Precambrian era was seven or eight times longer than the other three eras combined.

1. **Structure** refers to the nature, arrangement, and orientation of the materials making up the feature being studied. Structure is essentially the geologic underpinning of the landform. Is it composed of bedrock or not? If so, what kind of bedrock and in what configuration? If not, what are the nature and orientation of the material? With a structure as clearly visible as that shown in Figure 13-18, these questions are easily answered, but such is not always the case, of course.

2. **Process** considers the actions that have combined to produce the landform. A variety of forces—usually geologic, hydrologic, atmospheric, and biotic—are always at work shaping the features of the lithospheric surface, and their interaction is critical to the formation of the feature(s). A landscape resulting from the process known as glaciation is shown in Figure 13-19.

3. **Slope** is the fundamental aspect of shape for any landform. The angular relationship of the surface is essentially a reflection of the contemporary balance among the various components of structure and process (Figure 13-20). The inclinations and lengths of the slopes provide details that are im-

portant both in describing and in analyzing the feature.

4. **Drainage** refers to the movement of water (from rainfall and snowmelt) either over Earth's surface or down into the soil and bedrock. Although moving water is an outstanding force under the "process" heading, the ramifications of slope wash, streamflow, stream patterns, and other aspects of drainage are so significant that the general topic of drainage is considered a basic element in landform analysis.

Once these basic elements have been recognized and identified, the geographer is prepared to analyze the topography by answering the fundamental questions at the heart of any geographic inquiry:

What? The *form* of the feature(s)
Where? The *distribution* and *pattern* of the landform assemblage
Why? An explanation of *origin* and *development*
So What? The *significance* of the topography in relationship to other elements of the environment and to human life and activities

Figure 13-18 The structure of these abrupt sandstone cliffs in southeastern Utah is easy to see. (TLM photo)

INTERNAL AND EXTERNAL GEOMORPHIC PROCESSES

The topography of Earth has infinite variety, apparently being much more diverse than on any other known planet. This variety reflects the complexity of interactions between process and structure—the multiplicity of shapes and forms that result as the geomorphic processes exert their inexorable effects.

These processes are relatively few in number but extremely varied in nature and operation. Basically they are either internal or external, as Figure 13-21 shows. The *internal* processes operate from within Earth, energized by internal heat that generates extremely strong

Figure 13-19 The spectacular terrain of Montana's Glacier National Park was produced by a variety of interacting processes, but the contemporary topography is particularly due to glaciation. This view is across St. Mary Lake toward Mount Jackson. (TLM photo)

Figure 13-20   Slope is a conspicuous visual element of some landscapes. The abrupt cliffs of the volcanic backbone of the island of Bora-Bora (in French Polynesia) dominate this scene. (TLM photo)

forces that apparently operate outside of any surface or atmospheric influences. These forces result in crustal movements of various kinds. In general, they are constructive, uplifting, building forces that tend to increase the relief of the land surface.

In contrast, the *external* processes are largely subaerial (which means they operate at the base of the atmosphere) and draw their energy mostly from sources above the lithosphere, either in the atmosphere or in the oceans. Unlike internal processes, external processes are well understood and their behavior is often predictable. Moreover, their behavior may be significantly influenced by the existing topography, particularly its shape and the nature of the surface materials. The external processes may be thought of generally as wearing-down or destructive forces that eventually tend to diminish topographic irregularities and decrease the relief of Earth's surface.

Internal and external processes thus work in more or less direct opposition to one another. Their battleground is Earth's surface, the interface between lithosphere and atmosphere, where this remarkable struggle has persisted for billions of years and may continue endlessly into the future.

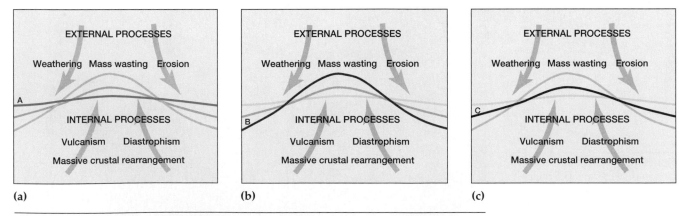

(a)   (b)   (c)

Figure 13-21   Schematic relationship between external and internal geomorphic processes. **(a)** A surface that has been worn down by external forces but not uplifted by internal forces. **(b)** A surface that has been uplifted by internal forces but not worn away by external forces. **(c)** A "normal" surface that has been both uplifted by internal forces and worn away by external forces.

In succeeding chapters we consider these various processes—their nature, dynamics, and effects—in detail, but it is useful here to summarize them so that they can be glimpsed in totality before we treat them one at a time. Table 13-3 presents such an overview. Note, however, that our classification scheme is imperfect; some items are clearly separate and discrete, whereas others overlap with each other. The table, then, represents a simple, logical way to begin a study of the processes but is not necessarily the only or ultimate framework.

THE QUESTION OF SCALE

In any systematic study of geomorphic processes, two general topics should be kept in mind—*scale* and *pattern*. The question of scale is fundamental in geography. Regardless of the subject of geographic inquiry, recognizable features and associations are likely to vary considerably depending on the scale of observation. This simply means that the aspects of the landscape one observes in a close-up view are different from those observed from a more distant view.

At least five orders of **relief** can be recognized on the surface of the lithosphere (Figure 13-22):

1. *First-order relief* represents the small-scale end of the spectrum, which means that the features are the largest that can be recognized: *continental platforms* and *ocean basins*. Although the shoreline at sea lev-

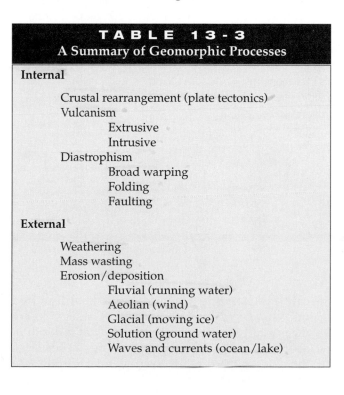

TABLE 13-3
A Summary of Geomorphic Processes

Internal

      Crustal rearrangement (plate tectonics)
      Vulcanism
            Extrusive
            Intrusive
      Diastrophism
            Broad warping
            Folding
            Faulting

External

      Weathering
      Mass wasting
      Erosion/deposition
            Fluvial (running water)
            Aeolian (wind)
            Glacial (moving ice)
            Solution (ground water)
            Waves and currents (ocean/lake)

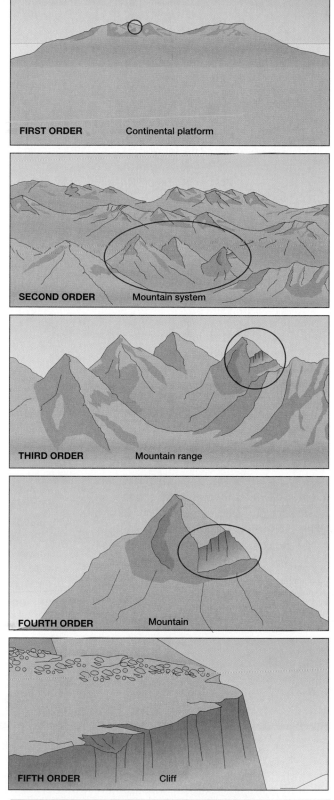

FIRST ORDER   Continental platform

SECOND ORDER   Mountain system

THIRD ORDER   Mountain range

FOURTH ORDER   Mountain

FIFTH ORDER   Cliff

Figure 13-22   Five orders of relief. The circle represents a focal point for transition from one order to the next.

FOCUS

AN EXAMPLE OF SCALE

As an example of the complexity and significance of scale, let us focus our attention on a particular place on Earth's surface and view it from different perspectives. The location is north-central Colorado within the boundaries of Rocky Mountain National Park, some 8 miles (13 kilometers) due west of the town of Estes Park. It encompasses a small valley called Horseshoe Park, through which flows a clear mountain stream named Fall River and adjacent to which is the steep slope of Bighorn Mountain (Figure 13-2-A).

1. To illustrate the largest scale of ordinary human experience, we will hike northward from the center of Horseshoe Park up the side of Bighorn Mountain. At this level of observation, the first topographic feature of note is a smooth stretch of Fall River we must cross. We walk over a small sand bar at the south edge of the river, wade for a few steps in the river, and step up an 18-inch (46-centimeter) bank onto the mountainside, noting the dry bed of a small intermittent pond on our left. After 20 minutes or so of steep uphill scrambling, we reach a rugged granite outcrop (locally called Hazel Cone or Poop Point),

which presents us with an almost vertical cliff face to climb.

2. At a significantly different scale of observation, we might travel for 20 minutes by car in this same area. The road through Horseshoe Park is part of U.S. Highway 34, which here is called the Trail Ridge Road. After 20 minutes, we reach a magnificent viewpoint high on the mountain to the southwest of Horseshoe Park. From this vantage point, our view of the country through which we have hiked is significantly enlarged. We can no longer recognize the sand bar, the bank, or the dry pond, and even the rugged cliff of Hazel Cone appears as little more than a pimple on the vast slope of Bighorn Mountain. Instead, we see that Fall River is a broadly meandering stream in a flat valley and that Bighorn Mountain is an impressive peak rising high above.

3. Our third observation of this area might take place from a plane flying at 39,000 feet (12 kilometers) on a run between Omaha and Salt Lake City. From this elevation, Fall River is nearly invisible, and only careful observation reveals Horseshoe Park. Bighorn Moun-

tain is now merged indistinguishably as part of the Mummy Range, which is seen to be a minor offshoot of a much larger and more impressive mountain system called the Front Range.

4. A fourth level of observation would be available to us if we could hitch a ride on a satellite orbiting 100 miles (160 kilometers) above Earth. Our brief glimpse of northern Colorado would probably be inadequate to distinguish the Mummy Range, and even the 250-mile (400-kilometer) long Front Range would appear only as a component of the mighty Rocky Mountain cordillera, which extends from New Mexico to northern Canada.

5. At the smallest scale, the final viewpoint possible to humans could come from a spacecraft rocketing toward some distant heavenly body. Looking back in the direction of Horseshoe Park from a near-space position, one might possibly recognize the Rocky Mountains, but the only conspicuous feature in this small-scale view would be the North American continent.

el appears as a conspicuous demarcation between land and water, it is not the accepted boundary between platforms and basins. Each continent has a margin that is submerged, called the **continental shelf**. This shelf is a terrace of varying width, in some places only a few miles wide and in other localities extending seaward for several hundred miles; its average width is about 44 miles (74 kilometers). At its outer edge, generally at a depth of about 450 feet (135 meters) below present sea level, the slope pitches more steeply and abruptly into the ocean basins.

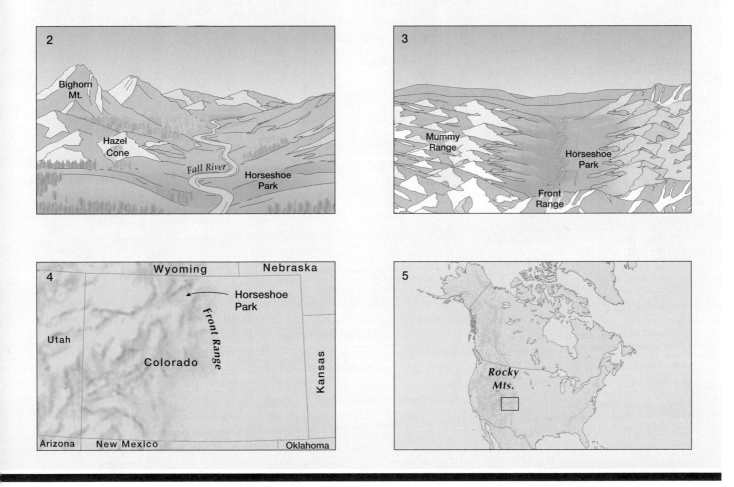

Figure 13-2-A An experience with change of scale: **(1)** A close view of Horseshoe Park, Colorado from the west. The dashed line shows our hiking route. **(2)** Looking down on Horseshoe Park from Trail Ridge Road. **(3)** An aerial view of the Mummy Range and part of the Front Range. **(4)** A high-altitude look at Colorado. **(5)** North America as it might be seen from a distant spacecraft.

2. *Second-order relief* consists of major mountain systems and other extensive surface formations of subcontinental extent (such as the Mississippi lowland or the Amazon basin). Second-order relief features (like those of all other orders) may be found in ocean basins as well as on continental platforms, most conspicuously in the form of the great undersea mountain ranges usually referred to as *ridges*.

3. *Third-order relief* encompasses specific landform complexes of lesser extent and generally of smaller size than those of the second order, with no precise

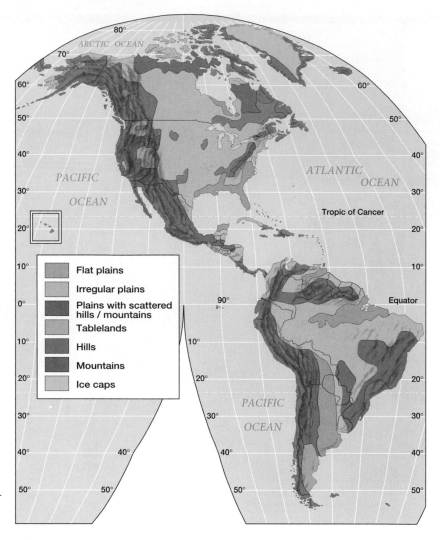

Figure 13-23 Major landform assemblages of the world.

separation between the two. Typical third-order features include discrete mountain ranges, groups of hills, and large river valleys.

4. *Fourth-order relief* comprises the sculptural details of the third-order features, including such individual landforms as a mountain, mesa, or hill.
5. *Fifth-order relief* consists of small individual features that may be part of the fourth-order relief, such as a sandbar, cliff, or waterfall.

An important correlation with this classification scheme is the degree of permanency represented by the various orders. Although no features of Earth's crust are permanent, the lower-numbered orders include features that are usually more long-lived than those of the higher numbered orders.

As an analogy, one might compare the orders of relief with a setting for theatrical presentations in which geomorphic dramas are featured. The first-order features represent the foundation for construction, a stable and long-lasting undergirding that may outlast a sequence of buildings erected on it. The second-order features represent a theater built on the foundation, which will be in use for a long time and in which will be presented many different dramas. The third-order features represent the stage installed in the theater, which will be used for a number of presentations but will be changed several times during the life of the theater. The fourth-order features represent the scenery, which will be reconstructed with each succeeding drama. The fifth-order features represent the props, which will be revised several times during the course of a single play.

THE PURSUIT OF PATTERN

A prime goal of any geographic study is to detect patterns in the areal distribution of phenomena. If features have a disordered and apparently haphaz-

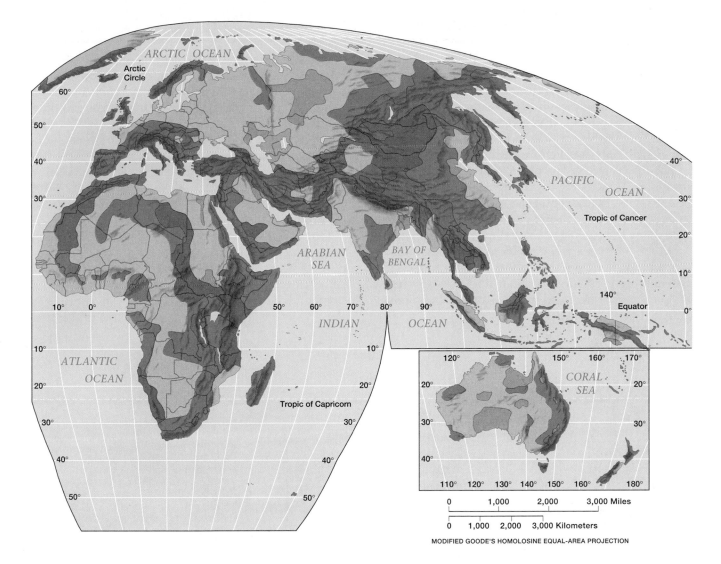

MODIFIED GOODE'S HOMOLOSINE EQUAL-AREA PROJECTION

ard distribution, it is more difficult to comprehend the processes that formed them and the relationships that exist among them. If there is some perceptible pattern to their distribution, however—some more or less predictable spatial arrangement—it becomes simpler for us to understand both the reasons for the distribution pattern and the interrelationships that pertain.

In previous portions of this book we were concerned with geographic elements and complexes having some predictability to their distribution patterns. For example, one can say with some assurance that, at about 50° latitude on the western coasts of continents, there can be found a certain type of climate, a particular association of natural vegetation and native animal life, and a set of roughly predictable soil-forming processes. These relatively orderly arrangements have been helpful, and perhaps even comforting, in our studies thus far.

Now, alas, we enter into a part of physical geogra-

phy in which orderly patterns of distribution are much more difficult to discern, as the landform distribution pattern in Figure 13-23 so readily shows. There are a few aspects of predictability; for example, one can anticipate that in desert areas, certain geomorphic processes are more conspicuous than others and certain landform features are likely to be found. Overall, however, the global distribution of topography is very disordered and irregular. The pursuit of a predictable pattern is, for the most part, a thankless endeavor, and to comprehend the world distribution of landforms requires an unfortunate degree of sheer memorization.

Largely for this reason, the geomorphology portion of this book concentrates less on distribution and more on process. Comprehending the dynamics of topographic development is more important to an understanding of systematic physical geography than any amount of detailed study of landform distribution.

CHAPTER SUMMARY

Although our knowledge of Earth's interior is very incomplete, it is generally accepted that our planet has a heavy, solid inner core surrounded by three concentric shells—outer core, mantle, and crust—of various densities and compositions. As geographers, our interest is focused on Earth's surface, and we are concerned with the interior only as it helps us understand the surface.

Earth's crust consists of a variety of minerals that form many kinds of rocks. These rocks can be separated into three fundamental classes—igneous, sedimentary, and metamorphic. Igneous rocks are those formed by the cooling of molten magma. Sedimentary rocks are produced when layers of fragmented mineral material (sediment) solidify as a result of cementation and high pressure. Metamorphic rocks are created by the action of heat and pressure on existing igneous and sedimentary rocks.

The doctrine of uniformitarianism holds that the processes that formed the topography of the past are the same as those functioning today. This is a fundamental concept in the understanding of contemporary landforms.

The study of landforms is facilitated by considering four basic elements—structure, process, slope, and drainage. The following are questions that constitute the basis of geographic analysis of landforms: *what?* (the form of the feature); *where?* (its distribution); *why?* (its origin); and *so what?* (its significance).

Our understanding of geomorphic processes and features depends on the scale of our study. Five orders of relief are generally recognized, ranging from very large features (such as a continental platform) to very small ones (such as a sand bar).

KEY TERMS

asthenosphere
bedrock
continental drift
continental shelf
crust
drainage
extrusive rock
geomorphology
inner core
intrusive rock

landform
lithosphere
magma
mantle
mesosphere
metamorphic rock
mineral
Mohoroviăc discontinuity (Moho)
outcrop
outer core
plate tectonics

process
relief
rock
sediment
sedimentary rock
slope
strata (*plural; stratum, singular*)
structure
topography
uniformitarianism

REVIEW QUESTIONS

1. Why are geographers more interested in Earth's surface than its interior?
2. Distinguish among sedimentary, igneous, and metamorphic rocks.
3. Which of the major rock classes is most widespread across Earth's surface?
4. Distinguish between intrusive and extrusive rocks.
5. What is the difference between topography and relief?
6. Distinguish between internal and external processes that shape Earth's surface.

SOME USEFUL REFERENCES

ALBRITTON, CLAUDE C., JR., *The Abyss of Time*. San Francisco: Freeman, Cooper & Company, 1980.

BRIDGES, E.M., *World Geomorphology*. New York: Cambridge University Press, 1990.

CHORLEY, RICHARD J., STANLEY A. SCHUMM, AND DAVID E. SUGDEN, *Geomorphology*. London: Methuen, 1984.

COOKE, R. U., AND J. C. DOORNKAMP, *Geomorphology in Environmental Management*, 2d ed. York, England: Oxford University Press, 1990.

CURRAN, H. ALLEN, ET AL, *Atlas of Landforms*. New York: John Wiley & Sons, 1984.

GERRARD, J.A., *Rocks and Landforms*. Winchester, MA: Allen & Unwin, 1988.

NEWMAN, WILLIAM L., *Geologic Time*. Washington, DC: U.S. Geological Survey, 1988.

RICE, R.I., *Fundamentals of Geomorphology*. New York: John Wiley & Sons, 1988.

STANLEY, STEVEN M., *Earth and Life Through Time*. San Francisco: W. H. Freeman, 1989.

14

THE INTERNAL PROCESSES

THE FORCES OPERATING WITHIN EARTH ARE COMPLEX, of unimaginable strength, and largely mysterious. Even though year by year we understand more and more as a result of geophysical and geologic research, the energetics and dynamics of Earth's internal processes are as yet only partially comprehended.

THE IMPACT OF INTERNAL PROCESSES ON THE LANDSCAPE

In our endeavor to understand the development of the earthly landscape, no pursuit is more rewarding than a consideration of the internal processes, for they are the supreme builders of terrain. Energized by awesome forces within Earth, the internal processes actively reshape the crustal surface. The crust is buckled and bent, land is raised and lowered, rocks are fractured and folded, solid material is melted, and molten material is solidified. These actions have been going on for billions of years and are fundamentally responsible for the gross shape of the lithospheric landscape at any given time. They do not always act independently and separately, but in this chapter we isolate them in order to simplify our analysis.

CRUSTAL REARRANGEMENT

Until recently, most Earth scientists assumed that the planet's crust was rigid, with continents and ocean basins fixed in position and significantly modified only by changes in sea level and periods of mountain building. The uneven shapes and irregular distribution of the continents were a puzzlement, but it was generally accepted that the present arrangement was emplaced in some ancient age when Earth's crust cooled from its original molten state.

PLASTICITY

The rigid-Earth theory was seriously called into question in recent years by a variety of discoveries and hypotheses. Prominent among these was the recognition that the igneous rocks of the upper crust occur in two layers that are well differentiated in several characteristics, especially density (Figure 14-1). The lower layer is thought to be continuous, underlying both ocean basins and continents, at or near the surface under basins and deeper beneath continents. This layer is called the **sima**, named for its two most prominent

mineral compounds, *si*lica and *ma*gnesium. It is alkaline, dark, relatively young, basaltic, and relatively dense [about 3.0 g/cm^3 (grams per cubic centimeter)].

The upper layer is believed to underlie only the continents, where it sits as immense bodies of rock embedded in the sima. This uppper layer is called the **sial**, named for its common constituents *si*lica and *al*uminum. Lighter in color and more acidic than the sima, the sial is generally granitic and notably less dense (averaging about 2.8 g/cm^3) because the heavier elements iron and magnesium are less abundant in the sial than in the sima.

The general crustal structure, then, appears to be continents of lightweight sial floating on a foundation of denser sima. Such a structure casts serious doubts on the theory of rigidity and introduces the concept of **plasticity** and a model in which the continents are free to move rather than being cemented in place.

ISOSTASY

Related to plasticity is the principle of **isostasy**, which is defined as the maintenance of hydrostatic equilibrium in the crust. In simplest terms, isostasy means that the addition of a significant amount of mass onto a portion of the crust causes the crust to sink, whereas the removal of a large mass allows the crust to rise.

The details of isostasy are not clear. For example, the depth of the sinking is unknown; it may involve only the surface section in contact with the added mass, or it may involve the entire sialic layer, or it may extend clear through the sima to the base of the lithosphere. Another thing we have not figured out yet is what determines the areal extent of an isostatic adjustment: some adjustments cover only a few hundred square miles, while others extend over half a continent. Also unclear is the immediacy of the isostatic response: in some cases, the raising or lowering seems to take place just as soon as a mass is added or removed; in others, there is a lengthy lag before the adjustment begins.

These points of confusion notwithstanding, the eventual topographic results are obvious—the raising or lowering of some portion of the surface.

Isostatic reactions have a variety of causes. The crust may be depressed, for example, by deposition of a large amount of sediment on a continental shelf, or by the accumulation of a great mass of glacial ice on a landmass, as illustrated in Figure 14-2, or even by the weight of water trapped behind a large dam. Depressed crust may rebound to a higher elevation as material is eroded away, as an ice sheet melts, or as a

An eruption of the Tolbachik Volcano in Kamchatka, eastern Russia. *(Photri/The Stock Market)*

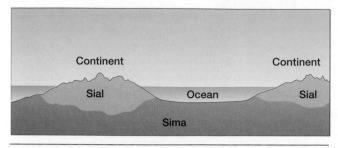

Figure 14-1 Igneous rocks in Earth's upper crust are either sial or sima. Sial (silica and aluminum) is a topmost layer associated with the continents, whereas sima (silica and magnesium) underlies both ocean basins and continental sial.

large body of water drains. Florida, for example, has experienced recent isostatic uplift because of the mass removed as groundwater dissolved the extensive limestone bedrock underlying the state.

CONTINENTAL DRIFT

Plasticity and isostasy provide a basis for understanding a much more dramatic restructuring of the crust. The shapes and positions of the continents may seem fixed at the timescale of human experience, but at the geologic timescale, measured in millions or tens of millions of years, continents are quite mobile. They have moved, collided and merged, and then been torn apart again. The theory of continental drift proposes that the present continents were originally connected as one enormous landmass that has broken up and drifted apart over the last few hundred million years. The drifting continues today, so that the contemporary position of the continents is by no means their ultimate one.

The idea of a single supercontinent from which large fragments separated has been around for a long time. Various naturalists, physicists, astronomers, geologists, botanists, and geographers from a number of countries have been putting forth this idea since the days of geographer Abraham Ortelius in the 1590s and philosopher Francis Bacon in 1620. Until fairly recently, however, the idea was generally unacceptable to the scientific community at large.

During the second and third decades of the twentieth century, the notion of continental drift was revived, most notably by the German meteorologist Alfred Wegener, who put together the first comprehensive theory to describe and partially explain the phenomenon. Wegener postulated a massive supercontinent, which he called **Pangaea** (Greek for "whole land"), as having existed about 250 million years ago and then breaking up into several large sections that have continued to move away from one another to this day (Figure 14-3). Wegener accumulated much evidence to support his hypothesis, most notably the remarkable number of close affinities of geologic features on both sides of the Atlantic Ocean. The continental margins of the subequatorial portions of Africa and South America fit together with jigsaw-puzzle-like precision, as Figure 14-4 shows, and there are other important elements of congruity in the coastline shapes on both sides of the North Atlantic. Moreover, the petrologic and paleontologic records on both sides of the Atlantic show many distributions that would be continuous if the ocean did not intervene, as Figure 14-4 shows.

The dramatic evidence of this former transatlantic connection, along with data Wegener collected from other areas, attracted much attention in the 1920s and generated much controversy. Some Southern Hemisphere geologists, particularly in South Africa, responded with enthusiasm. They gathered supportive evidence in the form of ancient glacial deposits and fossilized plant remains that exhibited similar distributions in South America, South Africa, Madagascar, India, Australia, and—more recently—Antarctica, evidence that could be most logically explained by continental drift.

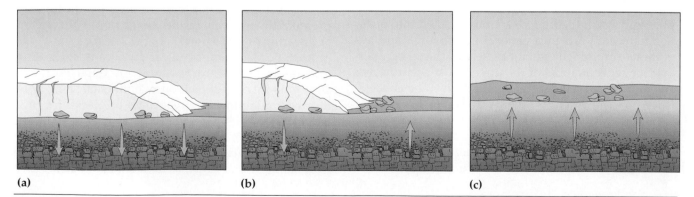

(a) (b) (c)

Figure 14-2 Isostatic adjustment. **(a)** During a glacial epoch, the heavy weight of accumulated ice depresses the crust. **(b)** Deglaciation removes the weighty overburden as the ice melts and the water flows away. **(c)** The crust rises, or "rebounds," after the weight is removed.

Figure 14-3 The presumed arrangement of the supercontinent Pangaea about 250 million years ago. The initial separation produced Laurasia, containing what is now North America and Eurasia (excluding India), and Gondwanaland, composed of all the present-day Southern Hemisphere continents plus India.

The general response to the Wegenerian hypothesis, however, was disbelief. Despite the vast number of distributional coincidences, most scientists felt that two difficulties made the theory improbable if not impossible: (1) Earth's crust was believed to be too rigid to permit such large-scale motions, and (2) no suitable mechanism could be conceived of that would provide enough energy to displace such large masses for a long journey. For these reasons, most Earth scientists ignored or even debunked the idea of continental drift for the better part of half a century after Wegener's theory was presented.

PLATE TECTONICS

The questionable validity of continental drift notwithstanding, continuing research revealed more and more about crustal mechanics. Geologists, geophysicists, seismologists, oceanographers, and physicists accumulated a large body of data about the ocean floor and the underlying simatic crust. It became apparent that heat convection (the process of convection was discussed in Chapter 4) is at work in Earth's interior. A very sluggish thermal convection system appears to operate within the planet, bringing deep-seated magma slowly to the surface and pulling remelted crustal rocks into the depths.

The Evidence Depth soundings have proven that running across the floors of all the oceans for 40,000 miles (64,000 kilometers) is a continuous system of large ridges located some distance from the continents, often in midocean (Figure 14-5). They remind one of the stitching on a baseball, both stitches and ridges being arranged not in simple lines but in oscillating offset segments. In addition, deep trenches occur at many places in the ocean floors, often around the margins of the ocean basins.

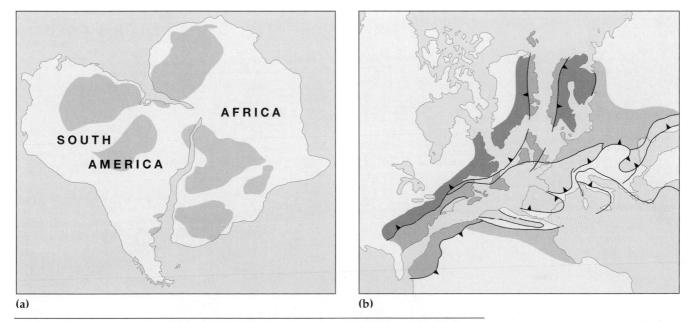

(a)

(b)

Figure 14-4 Some of the most decisive evidence for continental drift comes from the transatlantic connection. **(a)** The matching of the coasts of Africa and South America is persuasive. The ancient continental blocks of the two continents are shown in darker tone, with the younger areas (mostly regions of sedimentary deposition and volcanic activity) shown in lighter tone. **(b)** Across the North Atlantic are zones of similar geological activity, which have a complex relationship but are very difficult to correlate unless Europe, North America, and North Africa were once together.

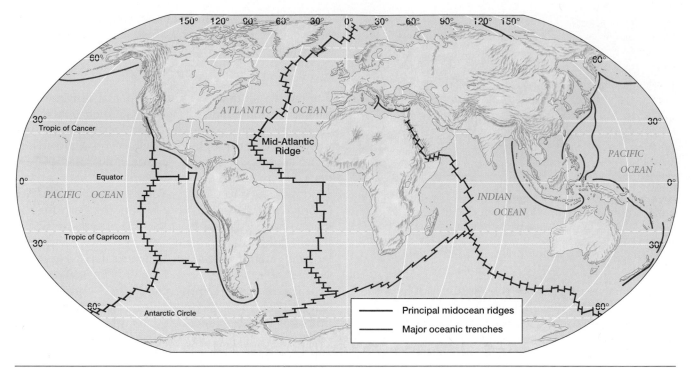

Figure 14-5   A continuous system of ridges runs across the floor of the world ocean. In addition to the ridges, this schematic map also shows the major oceanic trenches, which lie along many of the world's coasts.

In the early 1960s, a new theory was propounded, most notably by the American oceanographer Harry Hess, to explain these ridges and trenches. Known as **seafloor spreading,** this theory says that oceanic ridges are formed by currents of magma rising up from the mantle, often during volcanic eruptions, and spreading laterally (Figure 14-6). Thus the oceanic ridges contain the newest crust formed on the planet.

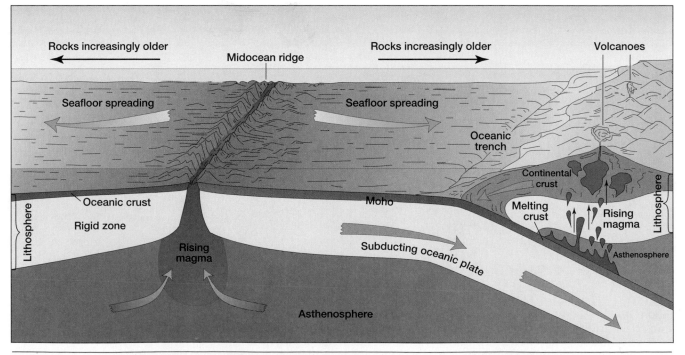

Figure 14-6   Seafloor spreading. Convection currents bring magma from the asthenosphere up through fissures in the oceanic crust. The cooled and therefore solidified magma becomes a new portion of ridge along the ocean floor, and the two sides of the ridge drift away from each other, as indicated by the arrows. Where denser oceanic crust meets lighter continental crust, the former slides under the latter in a process called subduction. As this subducted crust melts in the heat of the asthenosphere, it again becomes magma.

At other places in the ocean basins, usually near the trenches, older crust descends into the interior, in a process called **subduction**, where it is presumably melted and recycled into the convective system.

Oceanic crust has a relatively short life at Earth's surface. New crust is formed at the oceanic ridges, as just described, and within 200 million years is returned to the mantle by subduction. Because lower density continental crust cannot be subducted, once it forms it is virtually permanent. The continual recycling of oceanic crust means that its average age is only about 100 million years, whereas the average age of continental crust is 20 times that. Indeed, some fragments of continental crust have been discovered that are 4 billion years old—about four-fifths of the age of Earth!

The validity of the theory of seafloor spreading has been confirmed by two sets of evidence: paleomagnetism and core sampling. When any rock containing iron is formed, it is magnetized so that the iron grains become oriented toward Earth's magnetic pole. This orientation then becomes a permanent record of the polarity of Earth's magnetic field at the time the rock solidified. During the last 100 million years, Earth's magnetic field is known to have reversed itself, with the north and south magnetic poles changing places more than 170 times. Thus if the seafloor has spread laterally by the addition of new crust at the oceanic ridges, there should be a relatively symmetrical pattern of magnetic orientation on both sides of the ridges. Such is the case, as Figure 14-7 shows.

Final confirmation of seafloor spreading was obtained from holes drilled into the floor by a research ship, the *Glomar Challenger*. Several thousand *core samples* from sea-bottom sediments have been analyzed, and it is evident from this work that, almost invariably,

sediment thickness and age increase with increasing distance from the oceanic ridges, indicating that sediments farthest from the ridges are oldest. Sediments near ridges are thinner and younger, and right at ridges the material is almost all igneous, with little accumulation of sediment.

Thus the seafloors can be likened to gigantic conveyor belts, moving ever outward from the oceanic ridges and toward the trenches. As geologic movements go, seafloor spreading is quite rapid, expansion taking place about as fast as fingernails grow.

On the basis of these details and a variety of other evidence, the theory of plate tectonics is now generally accepted. The lithosphere is believed to be a mosaic of rigid plates embedded in an underlying plastic asthenosphere (Figure 14-8). The plates vary considerably in areal extent: some are almost hemispheric in size, whereas others are much smaller. The number of plates and some of their boundaries are as yet unclear. Thirteen major plates, and perhaps an equal number of smaller ones, are postulated. Many of the smaller plates are remnants of once larger plates that are now being subducted. All plates are thought to be about 60 miles (100 kilometers) thick, and most consist of both sialic and simatic crust.

These plates are coherent masses of lithospheric material that move ever so slowly over the asthenosphere, the movements sometimes bringing two plates together on a collision course. The plates are rigid and therefore deformed significantly only at the edges and only where one plate impinges upon another. The oceanic portions of the plates are more stable and coherent than the continental portions. The latter are continuously breaking up, rafting about, colliding, and amalgamating, whereas oceanic areas are less likely to break into smaller pieces.

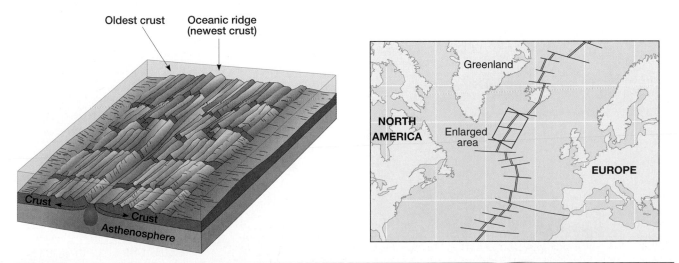

Figure 14-7   Sea floor spreading involves the rise of rock material from within Earth and its lateral movement away from the zone of upwelling. This extremely gradual process moves the older material further away as it is replaced by newer material from below.

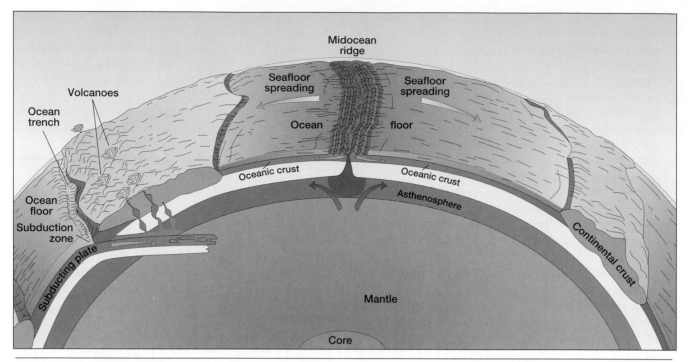

Figure 14-8 The lithosphere consists of huge rigid plates floating around on a sea of asthenosphere.

The Boundaries Only three types of contact between plates are possible: two plates may diverge from one another, converge toward one another, or slide laterally past one another (Figure 14-9).

At a **divergent boundary**, magma from the asthenosphere wells up in the opening between plates. This upward flow of molten material produces a continuous line of active volcanoes that spill out basalt onto the ocean floor. A divergent boundary is usually represented by an oceanic ridge. Most of the midocean ridges of the world are either active or extinct spreading ridges. Such locations often are associated with shallow earthquakes and volcanic activity. Divergent boundaries are said to be constructive because material is being added to the crustal surface at such locations.

Because new oceanic crust is being created at divergent boundaries, there must be a counterbalancing mechanism for destroying crust: otherwise the planet

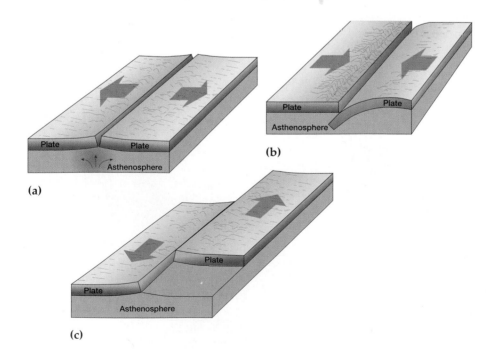

Figure 14-9 The three kinds of plate boundaries: **(a)** divergent; **(b)** convergent; **(c)** transform.

would be growing larger. Worldwide, old oceanic crust must be consumed at the same rate that new crust is formed. At a **convergent boundary**, plates moving in opposite directions meet, and the result of the collision normally is a vast crumpling of the edges as one plate subducts under the other (Figure 14-10). Subduction is a slow but nevertheless catastrophic event in which the crust of the submerging plate is gradually incorporated into the superheated depths. Convergent boundaries are destructive because they result in removal of part of the surface crust.

Convergent plate boundaries are responsible for some of the most massive and spectacular of earthly landforms: major mountain ranges, volcanoes, and deep oceanic trenches. Oceanic crust is denser than continental crust, and so the former always underrides the latter when the two collide [Figure 14-10(a)]. The oceanic slab slowly and inexorably sinks into the mantle. This submerging slab exerts a pull on the rest of the plate and may be the ultimate cause of plate movement, pulling the plate in after itself, as it were. Wherever such an oceanic/continental convergent boundary exists, a mountain range is formed on land (the Andes range is one example) and a parallel trench develops beneath the sea. In such a situation both volcanoes and earthquakes are common.

If the convergent boundary is between two oceanic plates, as in Figure 14-10(b), an oceanic trench is formed and volcanic activity is initiated, with the volcanoes being on the ocean floor. With time, a volcanic island arc (such as the Aleutians) develops, and such an arc eventually may become a more mature island arc system (such as Japan is today).

Where there is a convergent boundary between two continental plates, huge mountain ranges, such as the Alps and Himalayas, are built up [Figure 14-10(c)]. Under these conditions, volcanoes are rare or nonexistent but shallow-focus earthquakes are common.

Early researchers thought that a subducted plate would melt when pushed down into the hot mantle. However, more recent calculations indicate that such a result is unlikely. Oceanic crust is relatively cold when it approaches a subduction zone and would take a long time to become hot enough to melt. A more likely explanation is that water is driven off from the oceanic crust as it is subducted and this water lowers the melting point of the mantle below, causing it to melt and producing both extrusive and intrusive igneous rocks.

At a **transform boundary**, two plates slip past one another laterally [Figure 14-9(c)]. The slippage edge is a great vertical fracture called a *strike-slip fault* (discussed in more detail when we consider faulting later in this chapter). A transform boundary is classified as *conservative* because the plate movements are basically

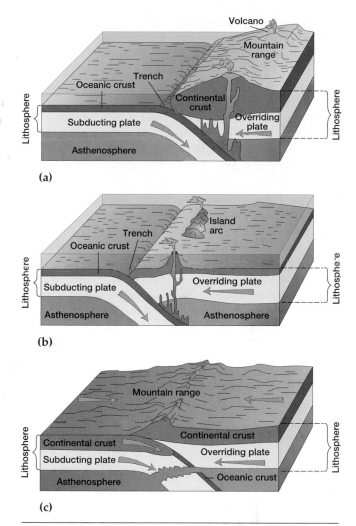

Figure 14-10   Idealized portrayals of the boundary action when plates converge: **(a)** Where continental plate meets oceanic plate, the latter is subducted and an oceanic trench and coastal mountains usually are created. **(b)** Where oceanic plate meets oceanic plate, an oceanic trench usually results. **(c)** Where continental plate meets continental plate, mountains generally are thrust upward.

parallel to the boundary, a situation that neither creates new crust nor destroys old. Such boundaries generally form huge faults, commonly hundreds of miles long, and result in a great deal of seismic (earthquake) activity. For example, the most famous earthquake fault system in the United States, the San Andreas in California, is on a transform boundary between the Pacific and North American plates (Figure 14-11).

The Rearrangement   Plate tectonics provides us with a grand framework for understanding the extensive crustal rearrangement that apparently has taken place during the relatively recent history of Earth. Wegener's Pangaea is now generally accepted as having existed. There is substantial evidence to indicate that, 450 million years ago, five continents existed; these continents sutured ("to suture" means to be crushed together) to form Pangaea. For the next 200 million years or so, Earth had but a single major continent and a single world ocean. About 250 million years ago, Pangaea began to break up, first into two massive pieces—Laurasia in the Northern Hemisphere and Gondwanaland in the Southern Hemisphere (Figure 14-3)—and then into a number of smaller bits (Figure 14-12). The absolute original locations of the plates are not accurately known, but their relative positions in the past can be determined by matching such things as rocks, fossils, and magnetic patterns. The various plates, along with their attached continents and parts of continents, became separated and drifted in various directions, their divergence often associated with seafloor spreading and their convergence frequently involving collision, subduction, and mountain building (Figure 14-13).

One of the great triumphs of the theory of plate tectonics is that it explains topographic patterns. It can account for the formation of many mountains, midocean ridges, oceanic trenches, island arcs, and the associated earthquake and volcanic zones. Where these features appear, there are usually plates either colliding or separating.

Despite the correlation of mountain building with plate convergence, the contemporary pattern of plates does not explain all mountain belts. Many of the major mountain ranges of North America and Europe are in the middle of plates rather than in boundary zones. The genesis of such midplate ranges is not fully understood but presumably is related to changing tectonic conditions in the past. There is convincing evidence that during some past eras there were fewer plates than there are today and during other eras there were more plates than there are today. In addition, the sizes and shapes of past plates differed from the current sizes and shapes. For example, there was no Atlantic Ocean 250 million years ago, and yet today it is a major feature of the planet, widening at a rate of about 2 inches (5 centimeters) per year. Seafloor spreading is proceeding even faster in parts of the Pacific. Indeed, some geophysicists have postulated that oceans are being created and removed by crustal rearrangement on about a 100-million-year cycle.

Thus crust continues to be shifted across the planet on top of plates that, over eons of time, may change in size, shape, and direction of movement. The evidence for these ancient developments is difficult to find, and it is probable that Earth's tectonic history will never be completely deciphered. In any case, the events associated with the fragmentation of Pangaea represent only the most recent of an unknown number of plate tectonic developments in the history of our planet. Indeed, in the early history of Earth, plate tectonics may have been much more active than it is today.

Modifications to the Original Theory With each passing year, we learn more about plate tectonics. Two examples of recently acquired knowledge are accreted terranes and mantle plumes.

Accreted Terranes A **terrane** is a small-to-medium mass of crust carried a long distance by a drifting plate that eventually converges with another plate. The terrane is too buoyant to be subducted in the collision and instead is fused ("accreted") to the other plate, often

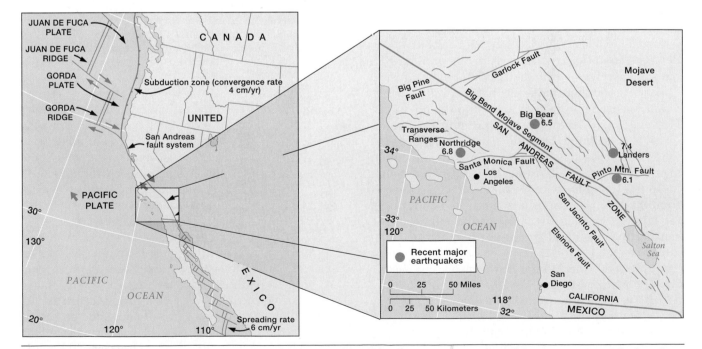

Figure 14-11 The San Andreas fault system of southern California represents a zone of much tectonic activity. The locations of major recent earthquakes, along with their Richter numbers, are shown.

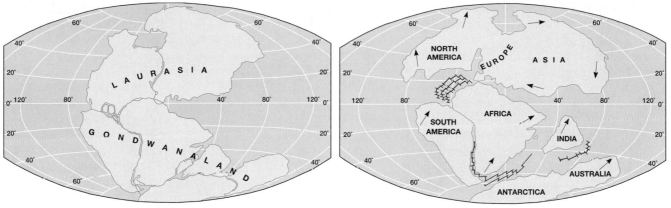

(a) 225 million years before present

(b) 135 million years before present

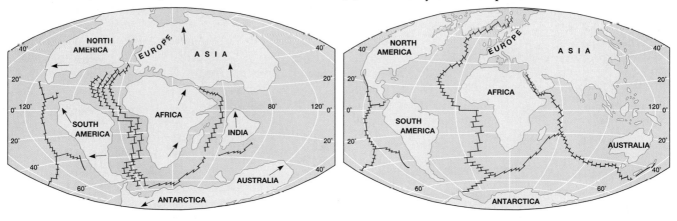

(c) 65 million years before present

(d) Today

Figure 14-12   How Pangaea might have broken into separate continental masses. Part (d) is essentially the contemporary arrangement.

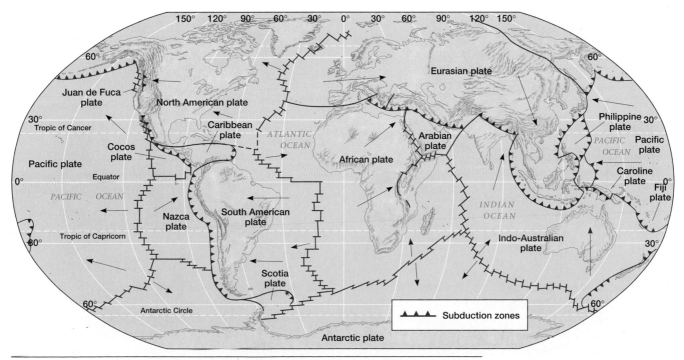

Figure 14-13   The major contempory tectonic plates and their generalized direction of drift.

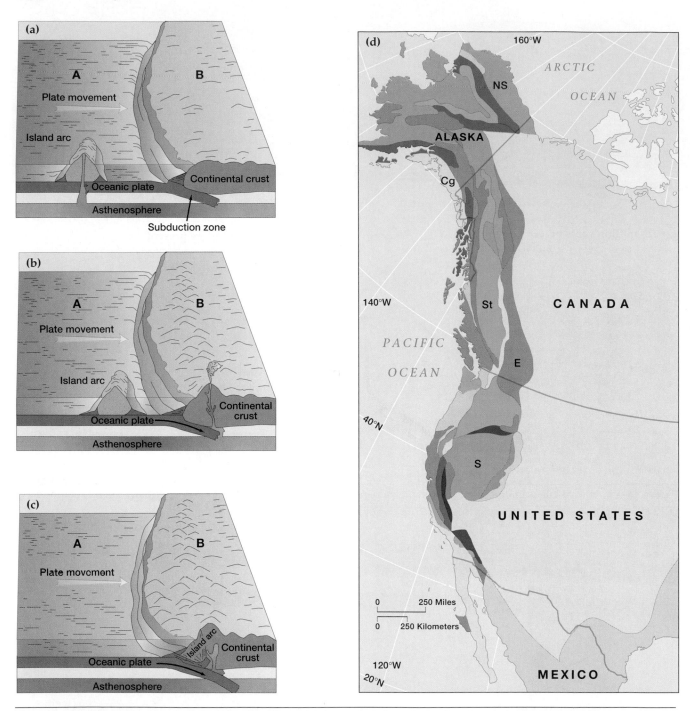

Figure 14-14 One way an accreted terrane can form. **(a)** A moving plate, A, carries along an old island arc. Plate A converges with another plate, B. **(b)** Plate A begins to be subducted under plate B, but the island arc is too buoyant for subduction. **(c)** Plate A is completely subducted, and the island arc is now accreted to plate B. **(d)** The western part of North America consists of a complicated mixture of terranes that have been accreted to the North American plate.

being fragmented in the process (Figure 14-14). Terranes are distinctive geologically because their lithologic complement (types of rock) is generally quite different from that of the plate to which they are accreted. It is generally believed that every continent has grown outward by the accumulation of accreted ter-

ranes on one or more of its margins. North America is a prominent example, as Figure 14-14(d) shows: most of Alaska and much of western Canada and the western United States consist of a mosaic of several dozen accreted terranes, some of which have been traced to origins south of the equator.

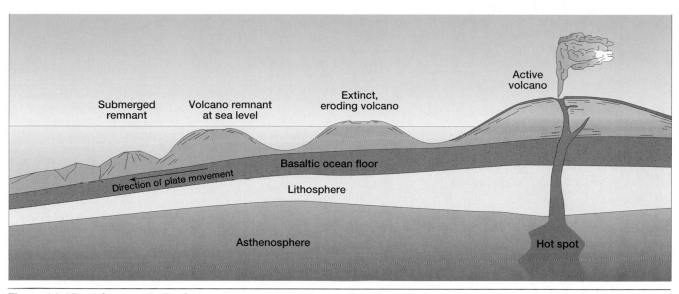

Figure 14-15   A hot spot in Earth's crust spews out magma, and a volcano is formed. The plate on which the volcano is being formed is moving leftward in this drawing, and so what is now a submerged volcano remnant at the extreme left was once a newly formed, active volcano at the extreme right.

Mantle Plumes   There are many places on Earth where magma from deep in the mantle comes either to or almost to the surface at locations that are not anywhere near a plate boundary; these leaky spots in the interior of a plate are referred to as **mantle plumes** or **hot spots**. We do not yet know the cause of these plumes, but more than 100 of them have thus far been identified. As the magma rises through the crust, it creates volcanoes and/or hydrothermal (hot water) features on the crust surface (Figure 14-15). The plate containing the hot spot is on the move, however, and so the volcanoes or other plume features appear to drift away from the plume. From a surface perspective, the hot spot appears to migrate in the direction opposite the direction of plate movement.

One striking result of a mantle plume is the Hawaiian Islands, where the ancient volcanic remnants of Midway Island are now 1600 miles (2500 kilometers) northwest of the presently active volcanoes on the big island of Hawaii, although both developed over the same mantle plume, separated in time by several million years. The volcanoes of the Hawaiian chain are progressively younger from west to east, as the Pacific plate drifts northwestward while new volcanoes are produced on an "assembly line" moving over one persistent hot spot (Figure 14-16). Other famous hot spots are in Yellowstone Park and in Iceland.

Mantle plumes help explain both topographic development and the rate and direction of plate movement. Because the plumes are effectively fixed in position, the "trails" they produce indicate absolute plate motions.

The Questions   Despite the intellectual and scientific enrichment we have received from plate tectonics the-ory, we still have many unanswered questions. For example: Why are some plates so much larger than others? What determines the zones of crustal weakness where plate boundaries occur? What is the ultimate cause of plate movement? We know that the causal forces must be in the interior because it is inconceivable that an external action could create such persistent movements in opposite directions on a spinning and orbiting planet. Yet, what is the nature of these interior forces? These and a host of other puzzles await solutions.

Most basic is this question: What is the source of the incredible amount of energy required? The engine that drives the motion of these massive plates continues to defy easy analysis because it is so utterly hidden from view. It is known that deep within Earth there is much dynamism. Thermal energy left over from when Earth formed, augmented by energy released in the radioactive decay of such elements as uranium and thorium deep inside Earth, may be sufficient to set in motion the convective circulations that appear to be in effect (Figure 14-6). We do not know if this is a sufficient answer to the basic question, and we certainly do not completely understand the mechanisms involved.

The present state of our knowledge about plate tectonics, however, is ample to provide a firm basis for understanding the gross patterns of most of the world's first-order and second-order relief features—the size, shape, and distribution of the continents and ocean basins and many of the **cordilleras** (a cordillera is a chain of mountains, often encompassing many ranges). In order to understand more detailed topographic features, we must turn to less spectacular, but no less fundamental, internal processes, which are often directly associated with tectonic movement.

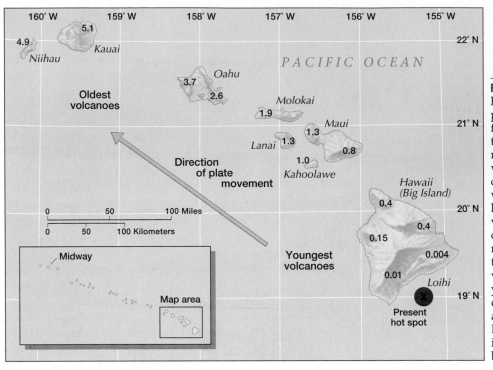

Figure 14-16   The Hawaiian hot spot. A mantle plume has persisted at the point marked X for many millions of years. As the Pacific plate moved northwesterly, a progression of volcanoes was created and then died as their source of magma was shut off. Among the earliest known was at Midway. Later volcanoes developed down the chain. The numbers on the eight main islands indicate dates of the basalt that formed the volcanic cones, in millions of years before the present. The only recently active volcanoes are on the Big Island of Hawaii. Loihi in the southeastern corner is the next volcanic island being built.

VULCANISM

Vulcanism is a general term that refers to all the phenomena connected with the origin and movement of molten rock. These phenomena include the well-known explosive volcanic eruptions that are among the most spectacular and terrifying events in all nature, along with much more quiescent events, such as the slow solidification of molten material below the surface.

We noted in Chapter 13 the distinction between extrusive and intrusive igneous rocks; a similar differentiation is made between extrusive and intrusive vulcanism. When magma is expelled onto Earth's surface while still molten, the activity is *extrusive* and is called **volcanism;** when magma solidifies in the shallow crust near the surface, it is an *intrusive* process; when magma solidifies deep inside Earth far below the surface, the process is called **plutonic activity**. Thus vulcanism is the general category, and volcanism (extrusive), intrusive vulcanism (shallow), and plutonic activity (intrusive, very deep) are three types of vulcanism.

VOLCANISM

Magma extruded onto Earth's surface is called **lava** (Figure 14-17). The ejection of lava into the open air is sometimes volatile and explosive, devastating the area for miles around; in other cases, it is gentle and quiet, affecting the landscape more gradually. All eruptions, however, alter the landscape because the fiery lava is an inexorable force until it cools, even if it is expelled only slowly and in small quantities.

The explosive eruption of a volcano is an awesome spectacle. In addition to an outward flow of lava, such solid matter as rock fragments, solidified lava blobs, ashes, and dust (collectively called **pyroclastic material**), as well as gas and steam, may be hurled upward in prodigious quantities. In some cases, the volcano literally explodes, disintegrating in an enormous self-destructive blast. The supreme example of such self-destruction within historic times was the final eruption of Krakatau, a volcano which occupied a small island in the Netherlands East Indies (now Indonesia) between Sumatra and Java. When it exploded in 1883, the noise was heard 1500 miles (2400 kilometers) away in Australia, and 6 cubic miles (20 cubic kilometers) of material was blasted into the air. The island disappeared, leaving only open sea where it had been. The tsunamis (great seismic sea waves) it generated drowned more than 30,000 people, and sunsets in various parts of the world were colored by fine volcanic dust for many months afterward.

The nature of a volcanic eruption apparently is determined largely by the chemistry of the magma that feeds it, although the relative strength of the surface crust and the degree of confining pressure to which the magma is subjected may also be important. The chemical relationships are complex, but the critical component seems to be the relative amount of silica (in the form of SiO_2) in the magma. A high silica content usually indicates acidic, cooler magma in which some of the heavier minerals have already crystallized and a considerable amount of gas has already separated. Some of this gas is trapped in pockets in the magma under great pressure.

Figure 14-17   An explosive eruption of Anak Krakatay Volcano in Indonesia. (Photo Researchers, Inc.)

As the magma approaches the surface, the confining pressure is diminished and the pent-up gases are released explosively. The initial explosion eases the pressure within the magma, but it may also trigger a lengthy chain reaction of further gas formation and explosions lasting for days. If, on the other hand, the magma is basic and has a low silica content, it is likely to be considerably hotter, and consequently most of the magmatic gases stay in solution, producing a much more fluid mixture. The resulting eruption yields a great outpouring of lava, quietly and without explosions.

The eruptive style of a volcano can change with time, but it usually does not. If the initial eruptions from a particular crater are explosive, future eruptions will probably also be explosive. Most other aspects of volcanic activity are much less predictable, however, particularly the timing of eruptions. In the vast majority of cases, an eruption begins with little or no warning.

A volcano is considered *active* if it has erupted at least once within recorded history. There are about 550 active volcanoes in the world; about 10 percent of them are in the United States, mostly in Alaska. On average, about 15 of them will erupt this week, 55 this year, and perhaps 160 this decade. Moreover, there will be one or two eruptions per year from volcanoes with no historic activity. In addition to surface eruptions, there is a great deal of underwater volcanic activity; indeed, it is estimated that more than three-fourths of all volcanic activity is undersea activity.

Within the conterminous 48 United States prior to the 1980 eruption of Mount St. Helens, there was only one volcano classified as active—Lassen Peak in California, which last erupted in 1917 but still occasionally produces gas and steam. A few old volcanoes, notably California's Mount Shasta and Washington's Mount Baker and Mount Rainier, show signs of potential activity but have not erupted in recorded time, and there are hundreds of extinct volcanoes, primarily in the west coast states. Both Alaska and Hawaii have many volcanoes, both active and inactive.

All volcanoes are temporary features of the landscape. Some may have an active life of only a few years, whereas others are sporadically active for thousands of centuries. At the other end of the scale, new volcanoes are spawned from time to time. One of the more spectacular recent developments was the birth of Surtsey, which rose out of the sea as a new island above a mantle plume off the coast of Iceland in 1964.

In spite of the destruction they cause, volcanoes do provide vital services to the planet. Just as our blood carries nutrients that nourish our bodies, so volcanoes do the same for the "skin" of Earth. Magma contains the major elements—phosphorus, potassium, calcium, magnesium, sulfur—required for plant growth. When this magma is extruded as lava that hardens into rock, the weathering that releases the nutrients into soil may require decades or centuries. When the magma is blasted out as ash, however, nutrients can be leached into the soil within months. It is no coincidence that Java, one of the most volcanically active parts of the planet, is also one of the world's most fertile areas.

Many of the world's most extensive lava flows were not extruded from volcanoes but rather issued quietly from midocean ridges and mantle plumes. The lava that flows out of these fissures is nearly always basaltic and frequently comes forth in great volume. The term **flood basalt** is applied to the vast accumulations of lava that build up, layer upon layer, sometimes covering tens of thousands of square miles to depths of many hundreds of feet. A prominent example of flood basalt in the United States is the Columbia Plateau, which covers 50,000 square miles (130,000 square kilometers) in Washington, Oregon, and Idaho [Figure 14-18(b)]. Larger outpourings are evidenced on other continents, most notably the Deccan Plateau of India (200,000 square miles; 520,000 square kilometers) [Figure 14-18(a)], and even more extensive flood basalts probably

PEOPLE AND THE ENVIRONMENT

THE LESSONS OF MOUNT ST. HELENS

In the spring and summer of 1980, Mount St. Helens, a splendid peak in south-central Washington (Figure 14-1-A), suddenly began extrusive activity. Starting on March 27, the long-dormant volcano entered a period of sporadic eruption that devastated the surrounding area, temporarily paralyzing much human activity within hundreds of miles and becoming the principal topic of conversation across the country for weeks.

After 123 years of dormancy, the unstable crust of Mount St. Helens abruptly turned turbulent. An initial earthquake was recorded on March 20, 1980, followed in the succeeding week by several hundred more, generally between 2.0 and 4.5 on the Richter scale. The eruptions began on March 27 and continued sporadically for the next 51 days, consisting almost entirely of steam, smoke, and ash and creating a prominent crater. Then came the first and (to date) most devastating of the explosive blasts—on May 18, 1980.

Without warning, an earthquake unhinged the entire north slope of the

mountain, uncapping bottled-up gases and magma with the force of 500 Hiroshima-sized atomic bombs. A major vertical eruption was within the range of expectations, but this explosion extended laterally as well. A column of steam and ash was projected 70,000 feet (21,300 meters) above the mountain, and 150 square miles (390 square kilometers) on the northern side of the volcano was devastated by the sideways blast. A cubic mile of mountain, 12 percent of its total bulk, was pulverized and blown away, the elevation of the peak being instantly reduced from 9677 feet (2950 meters) to 8400 feet (2560 meters).

Spirit Lake, one of the most beautiful bodies of water in America, disappeared in an avalanche of boiling mud and rock, as did 25 smaller lakes. Millions of trees were scattered like matchsticks over an 8-by-15-mile (13-by-24-kilometer) area (Figure 14-1-B). A churning mudflow, moving at 200 miles (320 kilometers) per hour, carried millions of tons of debris into the Toutle River, where a raging mud flood knocked out every bridge for 30 miles (48 kilometers), overwhelmed the Cowlitz River valley and clogged the Columbia River shipping channel to less than half its normal depth of 40 feet (12 meters). The toll was 70 dead and missing people and the loss of an estimated 11 million fish and 1.5 million larger animals and birds (Figure 14-1-B).

Later in 1980, there were four other major eruptions, several minor ones, and more than 5000 earthquakes. The mountaintop has been significantly lowered and totally reshaped (Figure 14-1-C). The upper slopes remain devastated; yet lower down, life has reju-

venated. Spiders were spinning webs in the devastated area within a week of the greatest eruption. Ferns, skunk cabbages, and even trees were sprouting before the end of the summer. Larger animals returned more slowly, but even elk were exploring parts of the debris avalanche within weeks of the May 18 eruption. The resilience of the biosphere has been rousingly confirmed. It will take decades for the evergreen forest and alpine tundra to regenerate, but subclimax plant formations and a vast horde of faunal species are increasingly apparent.

Perhaps the most remarkable response was demonstrated by the salmon of the Toutle River. Returning from the Pacific Ocean up the Columbia and Cowlitz rivers, the salmon apparently sensed the change in their natal stream and returned downstream to the Columbia. Finding the next stream hospitable, they made their spawning run on the Kalama River, thus adopting that river as their "new" ancestral stream, a process unprecedented among anadromous fish.

Mount St. Helens presents an unparalleled opportunity for volcano watching. More scientists have more instruments perched at more vantage points than ever before in history. Yet clearly the most important lesson is for the public. The saga of Mount St. Helens in 1980 may appear to be an unusual, brief, and isolated sequence of events, but such is not the case.

Mount St. Helens is a young and not particularly notable volcanic peak in the Cascade Range, which includes 14 other volcanoes of equal or greater significance (Figure 14-1-D). Eight of those peaks have erupted in the past two centuries, the most recent (prior to Mount St. Helens) being Lassen Peak, which blew more than 170 times during 1914/15.

Figure 14-1-A   Mount St. Helens prior to the 1980 eruption. (Pat and Tom Leeson/Photo Researchers, Inc.)

Apart from Lassen Peak, at its extreme southern end, the Cascade Range during the twentieth century has been noted for its serene and majestic beauty rather than its potential turbulence. It is, however, clearly a part of the Pacific Ring of Fire, and the evidence of past volcanic activity is ubiquitous. In 1975, for the first time since Lassen Peak became quiescent in 1921, sulfurous activity appeared in the Cascades: Mount Baker began spouting steam, its first fumings in 125 years. Much scientific interest was focused on this phenomenon, but it was thoroughly upstaged by Mount St. Helens some 5 years later.

What predictive conclusions can be gleaned from all this? Will Mount St. Helens subside into dormancy again? Possible, but not probable. Once an eruptive period begins, volcanoes tend to be active sporadically for years, if not decades. The citizens of Yakima, Washington, and even Portland, Oregon, have no reason to dispose of their ash-filtering face masks yet.

Will a different volcano become

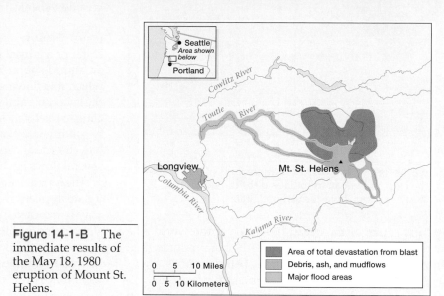

Figure 14-1-B The immediate results of the May 18, 1980 eruption of Mount St. Helens.

0   5   10 Miles
0   5   10 Kilometers

Area of total devastation from blast
Debris, ash, and mudflows
Major flood areas

important? Don't bet against it. Mount Baker has already given some cause for anxiety with its recent steam emissions and might be the next to erupt; beware, Vancouver!! Or, since Mount Rainier is situated halfway between Mount Baker and Mount St. Helens, should Seattle perhaps be the wary city?

The most meaningful conclusion from all this is that our predictive knowledge of volcanoes is still extraordinarily limited. Mount St. Helens was undoubtedly a good teacher, but we have a great deal more to learn.

Figure 14-1-C Mount St. Helens after the 1980 eruption. The posteruption elevation is 1300 feet (400 meters) less than the preeruption elevation. A new bulging lava dome has formed in the blown-out crater. (TLM photo)

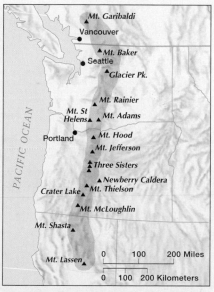

Figure 14-1-D Prominent volcanoes (active, dormant, and extinct) of the Cascade Range.

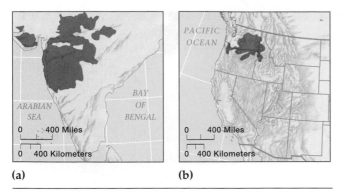

(a)   **(b)**

Figure 14-18   Two extensive outpourings of flood basalt: **(a)** the Deccan Plateau in India and **(b)** the Columbia Plateau in the northwestern United States.

occupy much of the ocean floors. Over the world as a whole, more lava has issued quietly from fissures than from the combined outpourings of all volcanoes.

As is the case with all other internal processes, volcanism can produce an almost infinite variety of terrain features (Figure 14-19). Four particular landforms are

distinctively associated with volcanic activity, however: lava flows, volcanic peaks, calderas, and volcanic necks.

Lava Flows   Whether originating from a volcanic crater or a crustal fissure, a lava flow spreads outward at an attitude approximately parallel with the surface over which it is flowing, and this parallelism is maintained as the lava cools and solidifies. Although some viscous flows cling to relatively steep slopes, the vast majority eventually solidify in a horizontal orientation that grossly resembles the stratification of sedimentary rock, particularly if several flows have accumulated on top of one another.

The topographic expression of a lava flow, then, is often a flattish plain or plateau. The strata of sequential flows may be exposed by erosion, as streams usually incise very steep-sided gullies into lava flows (Figure 14-20). The character of the flow surface varies with the nature of the lava and with the extent of erosion, but as a general rule the surface of relatively recent lava flows tends to be extremely irregular and fragmented.

Figure 14-19   Basalt cliffs in Devils Postpile National Monument, California. (John Gerlach/Earth Scenes)

Figure 14-20   Lava flows displayed on the walls of the narrow gorge of the Rio Grande, near Taos, New Mexico. (C. W. Biedel, M. D./Photri)

Volcanic Peaks Volcanoes are surface expressions of subsurface igneous activity. Often starting small, a volcano may grow into a conspicuous hill or a massive mountain. Most volcanic peaks take the form of a cone that has a symmetrical profile (Table 14-1). Magma differences produce different types of cones. Basaltic ("basic") lava tends to flow quite easily over the surface, forming low-lying shield volcanoes, whereas high silica lavas ("acidic") such as rhyolite often are thick and pasty, forming lava domes. Magmas of inter-

TABLE 14-1					
Principal Types of Volcanoes					
Type	Composition	Relative size	Typical eruption pattern	Characteristics	Example
Cinder cone	Loose rock fragments ejected from central vent; pyroclastic material is dust, ash, cinders, larger pieces; chemistry of magma varies, often basaltic	Small, maximum height 1500 ft (500 m)		Short life span; steep slopes (up to) 30°	Paricutín Sunset Crater
Composite (stratovolcano)	Alternating layers of pyroclastics and solified lava; magma usually intermediate in chemistry, often andesitic	Large	100s or 1000s of years of inactivity separating few years of violent activity	Steep slopes, symmetrical; long life span	Mt. Fuji Mt. Rainier Mt. Shasta Mt. Vesuvius Mt. St. Helens
Shield	Solified lava that was nonviscous when molten and flowed quietly from central vent; magma usually basaltic	Large	Quiet, nonviolent because lava is very fluid	Broad, gently sloping cone, much broader than high; slope is 10° at summit, 2° at base	Hawaiian Islands Tahiti
Lava dome	Solidified lava that was thick and viscous when molten; magma usually high in silica, often rhyolitic	Small	Steep slope; dome grows by expansion from within; frequently occurs within crater of composite volcano		Lassen Peak

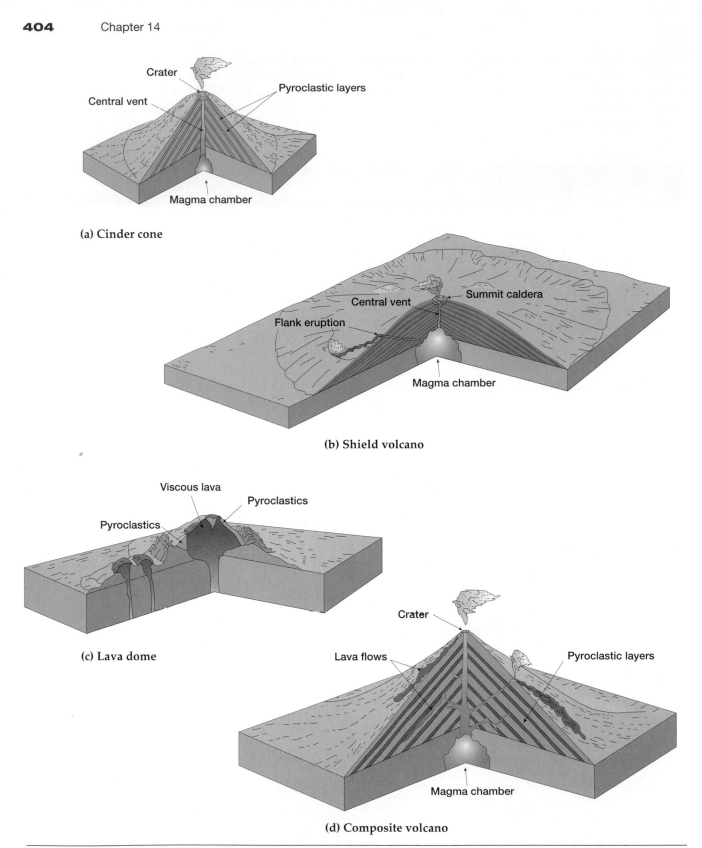

(a) Cinder cone

(b) Shield volcano

(c) Lava dome

(d) Composite volcano

Figure 14-21   The four principal types of volcanic cones.

mediate composition, such as those producing andesitic lava, typically produce composite volcanic cones (Figure 14-21). A common denominator of nearly all volcanic peaks is a crater, normally set conspicuously at the apex of the cone. Frequently, smaller subsidiary cones develop around the base or on the side of a principal peak, or even in the crater.

Calderas   Uncommon in occurrence but spectacular in result is the formation of a **caldera**, which is produced when a volcano explodes, collapses, or does both. The result is an immense basin-shaped depression, generally circular, that has a diameter many times larger than that of the original volcanic vent or vents. Some calderas are tens of miles in diameter. North America's most famous caldera is Oregon's misnamed Crater Lake (Figure 14-22). Less than 7000 years ago, a mountain on this site, Mount Mazama, was a volcanic cone that reached an estimated altitude of 12,000 feet (3660 meters) above sea level [Figure 14-23(a)]. Once magma ceased to flow out from the top of this volcano [Figure 14-23(b)], pressure built up and the walls

weakened because they were no longer supported by the upflowing magma. The final, cataclysmic eruption [Figure 14-23(c)] removed—by explosion and collapse—the upper 4000 feet (1220 meters) of the peak and produced a caldera whose bottom is 4000 feet (1220 meters) below the crest of the remaining rim. Lat-

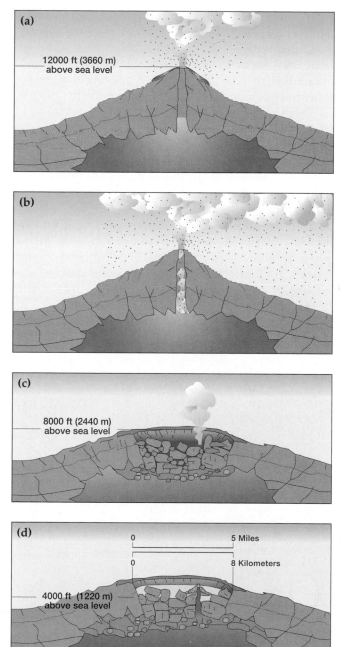

Figure 14-23   Formation of Crater Lake. **(a)** Mount Mazama 7000 years ago. **(b)** The magma flow is cut off, and the pressure builds. **(c)** An explosion blows off the top 4000 feet (1220 meters) of the mountain. Collapse of the crater walls forms a caldera 4000 feet (1220 meters) deep. **(d)** A new fissure allows magma to flow again, and the volcano known as Wizard Island is created.

(a)

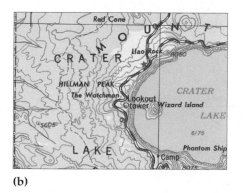

(b)

Figure 14-22   Oregon's Crater Lake occupies an immense caldera. Wizard Island represents a more recent subsidiary volcanic cone. (TLM Photo)

er, half this depth filled with water, creating one of the deepest lakes in North America. A subsidiary volcanic cone has subsequently built up from the bottom of the caldera and now breaks the surface of the lake as Wizard Island [Figure 14-23(d)].

Volcanic Necks   More limited still but very prominent where it does occur is a **volcanic neck**, a small, sharp spire that rises abruptly above the surrounding land. It represents the pipe, or throat, of an old volcano that filled with solidified lava after its final eruption. Millennia of erosion have removed the less resistant material that made up the cone, leaving the harder, lava-choked neck as a conspicuous remnant.

Volcano Distribution   Areas of volcanism are widespread over the world, but, as Figure 14-24 shows, their distribution is highly irregular, with many volcanoes in some regions and none in others. Volcanic activity is primarily associated with plate boundaries. At a divergent boundary, magma wells up from the interior both by eruption from active volcanoes and by flooding out of fissures. At convergent boundaries, where subduction is taking place, volcanoes are often formed in association with the turbulent descent and melting of crust and the crust crumpling that results from plate collision.

It is apparent from Figure 14-24 that the most notable area of vulcanism is around the margin of the Pacific Ocean, often referred to as the Pacific Ring of Fire or the Andesite Line (because the lava from the volcanoes consists primarily of andesite, a distinctive sialic mineral association). Some 80 percent of the world's volcanoes, both active and inactive, are associated with the Pacific ring.

INTRUSIVE VULCANISM

When magma solidifies below Earth's surface, it produces igneous rock. If this rock is pushed upward into the crust either before or after solidification, it is called an igneous intrusion. Most such intrusions have no effect on the surface landscape, but sometimes the igneous mass is raised high enough to deform the overlying material and change the shape of the surface. In many cases, the intrusion is exposed at the surface. When intrusions are thus exposed to the external processes, they often become conspicuous because they are usually resistant to erosion and with the passage of time stand up relatively higher than the surrounding land.

Intrusions come in all shapes, sizes, and compositions. Moreover, their relationship to overlying or surrounding rock is also quite variable. The intrusive process is usually a disturbing one for preexisting rock. Rising magma makes room for itself by a process called *stoping* (a mining term for ore removal by working upward). The molten invading magma can assimilate the rock being invaded or heat it enough to make it flow out of the way or either split or bow it upward. Adjacent to

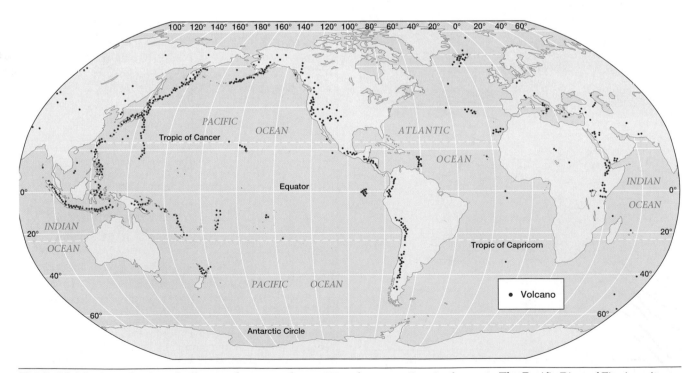

Figure 14-24   Distribution of volcanoes known to have erupted at some time in the past. The Pacific Ring of Fire is quite conspicuous.

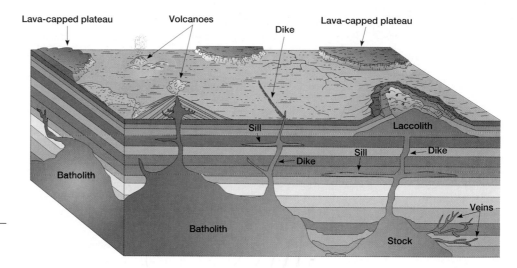

Lava-capped plateau   Volcanoes   Dike   Lava-capped plateau

Sill   Laccolith

Batholith   Dike   Sill   Dike

Batholith   Veins

Batholith   Stock

Figure 14-25   The six typical forms of igneous intrusions.

the area the magma has invaded, the invaded rock usually experiences contact metamorphism from being exposed to the heat and pressure of the rising intrusion.

In rare cases, there may be little or no disturbance, as with a small intrusion that inserts itself between preexisting sedimentary beds with little deformation or metamorphism.

Although there is almost infinite variety in the forms assumed by igneous intrusions, most can be broadly classified according to a scheme that contains only a half-dozen types (Figure 14-25).

Batholiths

By far the largest and most amorphous intrusion is the **batholith**, which is a subterranean igneous body of indefinite depth and enormous size [it must have a surface area of at least 40 square miles (100 square kilometers) to be a batholith]. Batholiths often form the core of major mountain ranges, their intrusive uplift being a fundamental part of the mountain-building process. Such notable ranges as the Sierra Nevada in California, Idaho's Sawtooth Mountains, and Colorado's Front Range were created at least partially by the uplift of massive batholiths; almost all the bedrock exposed in the high country of these ranges consists of batholithic granite, which was originally covered by an extensive overburden of other rocks that has since been eroded away (Figure 14-26).

Stocks

Similar to a batholith but much smaller is a **stock**. It is also amorphous and of indefinite depth, but it has a surface area of only a few square miles at most. Many stocks apparently are offshoots of batholiths.

Laccoliths

A specialized form of intrusion is the **laccolith**, which is produced when slow-flowing, viscous magma is forced between horizontal layers of preexisting rock. Although continuing to be fed from some subterranean source, the magma resists flowing and instead builds up into a mushroom-shaped mass that domes the overlying strata. If this dome is near enough to Earth's surface, a rounded hill will rise, like a blister, above the surrounding area. Many laccoliths are small, but some are so large as to form the cores of hills or mountains in much the same fashion as batholiths. The Black Hills of South Dakota, for example, have a laccolithic core, as do several of the geologically famous mountain groups (Henry, Abajo, LaSal) in southern Utah.

Batholiths, stocks, and laccoliths are sometimes grouped under the general heading *plutons*, which simply means any large intrusive igneous body.

Dikes

Probably the most widespread of all intrusive forms is the **dike**, a vertical or nearly vertical sheet of magma thrust upward into preexisting rock, sometimes forcing its way into vertical fractures and sometimes melting its way upward. Dikes are notable because they are vertical, narrow (a few inches to a few yards wide), and usually quite resistant to erosion. As with most other igneous intrusions, their depth is indeterminate; they often serve as conduits through which deep-seated magma can reach the surface. In some cases, dikes are quite long, extending for miles or even tens of miles in one direction. When exposed at the surface by erosion, they commonly form sheer-sided walls that rise above the surrounding terrain, as Figure 14-27 shows. Dikes often are found in association with volcanoes, occurring as radial walls extending outward from the volcano like spokes of a wheel. Notable examples of radial dike development can be seen around Shiprock in northwestern New Mexico and around the Spanish Peaks in south-central Colorado.

Sills

A **sill** is also a long, thin intrusive body, but its orientation is determined by the structure of the preexisting rocks. It is formed when magma is forced between strata that are already in place; the result is often

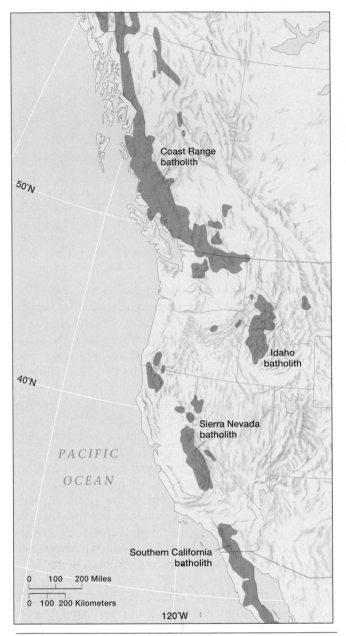

Figure 14-26 The major batholiths of western North America. These extensive intrusions are mostly associated with the subduction boundary along the western edge of the North American plate.

a horizontal igneous sheet between horizontal sedimentary layers. Sills are much less common than dikes, and their landscape expression is usually inconspicuous except where a sill serves as a caprock protecting softer rocks underneath, which can produce a steep-walled mesa, butte, or cliff.

Veins Least prominent among igneous intrusions but widespread in occurrence are thin **veins** of magma that may occur individually or in profusion. They are commonly formed when molten material forces itself into small fractures in the preexisting rocks, but they can also result from melting by an upward surge of

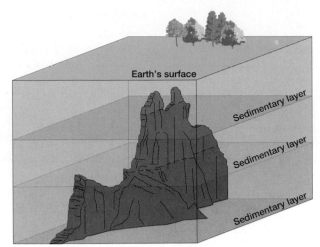

(a)

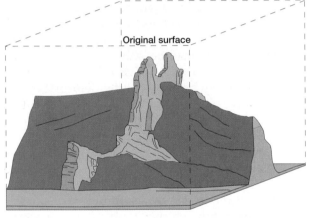

(b)

(c)

Figure 14-27 (a) An igneous dike was intruded into softer bedrock. (b&c) Subsequent erosion has left the resistant dike standing as an abrupt natural wall. This scene is in southern Colorado, near LaVeta. (TLM photo)

magma. Intrusive veins may take very irregular shapes, but normally they have a generally vertical orientation.

DIASTROPHISM

Diastrophism is a general term that refers to the deformation of Earth's crust. The term covers various kinds of crust movement and implies that the material is solid and not molten. The rocks may be bent or broken in a variety of ways, in response to great pressures exerted either from in the mantle or from in the crust. Some of these vast pressures clearly result from plate movement, but others appear to be unrelated to plate tectonics. Some are caused by the rise of molten material from below, either intrusively or extrusively, but most seem to have no causal relationship with magma.

Whatever the causes of diastrophic pressures, the results are often conspicuous in the landscape. Diastrophic movements are particularly obvious in sedimentary rocks because all sedimentary strata are initially deposited in a horizontal or near horizontal attitude—so if they now are bent, broken, or nonhorizontal, they have clearly experienced some sort of diastrophic deformation.

For purposes of description and analysis, diastrophic movements can be separated into three types—broad warping, folding, and faulting—although the separation in nature is not always so discrete and clear-cut.

BROAD WARPING

Any time the crust is deformed gently over a large area, we say that **broad warping** has taken place. Earth's crust has been subjected to broad warping—both uplift and depression—innumerable times. Areas that are now well above sea level are often deeply covered with marine sediments, indicating that they were once in a submarine environment and consequently at a lower elevation (Figure 14-28). Conversely, many ar-

Figure 14-28 Fossil seashells 10,000 feet (3050 meters) above present sea level in the Bighorn Mountains of Wyoming. (Earl Scott/Photo Researchers, Inc.)

eas that are now covered by ocean (for example, the North Sea) were once well above sea level.

There are many causes of broad crustal warping. In many cases, isostasy is involved—downwarping resulting from some addition of mass on the surface or upwarping due to a removal of mass. Frequently, however, there is no apparent relationship to an isostatic adjustment, and some other explanation must be sought.

When broad warping takes place in an ocean basin or in the interior of a continent, the resulting landform may be relatively inconspicuous and recognizable only through instrumented measurements. The warping of a coastal area causes an immediate change in sea level, however, and the resulting topographic development is readily apparent.

The effects of broad warping can be seen by comparing the east and west coasts of the United States (Figure 14-29).

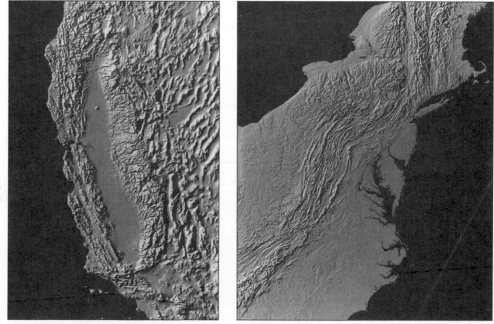

Figure 14-29 The Pacific coast is primarily a coastline of emergence, manifesting a smooth shoreline with steep slopes immediately adjacent. The Atlantic coast, on the other hand, is a coastline of submergence, as shown by extreme irregularities and numerous estuaries.

Although the geomorphic history of these two coast-lines is fairly complex and involves more than simple warping, the gross effects of warping are both obvious and strongly contrasting. The Atlantic coast is highly irregular, indented, embayed, and fretted by long estuaries. It is a classic example of a coastline of submergence, in which the land has been sinking in relation to sea level. The flat coastal plain is seamed by broad, shallow valleys, through which flow a number of rivers on their way from the Appalachian Mountains to the ocean. As the coastal plain was warped downward in the recent geologic past, the sea has invaded these valleys, and the lower reaches of each one are now drowned and have become an estuary (a valley flooded by the sea).

Opposite conditions prevail on the Pacific coast, which has been upwarped and is clearly a coastline of emergence. No coastal plain is present, the topography being hilly or mountainous right down to the coast. The present coastline is regular and smooth, with few embayments and no estuaries. Topographic evidence of the emergent upwarping is manifested in a series of old beach lines, wave-cut benches, and wave-eroded cliffs that now occur well above the present sea level.

FOLDING

When crustal rocks are subjected to certain forces, particularly lateral compression, they are often deformed by being bent, in a process called **folding** (Figure 14-30). The notion of folding sometimes is difficult to conceptualize. Our common experience is that rocks are hard and brittle, and if subjected to stresses, they might be expected to break; bending is harder to visualize. In nature, however, when great pressure is applied for long periods, particularly in an enclosed, buried, subterranean environment, the result is often a slow plastic deformation that can produce folded structures of incredible complexity. Folding can occur in any kind of rock, but it is obviously most recognizable in sedimentary strata.

Folding can take place at almost any scale. Some folds can be measured in no more than inches or centimeters, whereas others can develop over such broad areas that crest and trough are tens of miles apart.

The configuration of the folds can be equally variable. In some cases, the folding is simple and symmetrical; elsewhere it may be of extraordinary complexity and totally without symmetry. Moreover, the structure may become even more complex by breakage of the rock (faulting), which the stresses may engender in addition to the complicated folding.

Structural geologists recognize many kinds of folds, with a lengthy nomenclature to classify them. For introductory physical geography, however, only a few terms are necessary (Figure 14-31). A **monocline** is a one-sided slope connecting two horizontal or gently inclined strata. A simple symmetrical upfold is an **anticline**, and a simple downfold is a **syncline**. Also relatively common is an upfold that has been pushed so vigorously from one side that it becomes oversteepened enough to have a reverse orientation on the other side; such a structure is referred to as an **overturned fold**. If the pressure is enough to break the oversteepened limb and cause a shearing movement, the result is an **overthrust fold**, which causes older rock to ride above younger rock.

Folds frequently occur multiply rather than individually, as enormous forces crumple the land surface. When sedimentary strata are subjected to lateral compression, a series of simple parallel folds is often formed, with an alternating sequence of anticlines and synclines. If the folding takes place near enough to the surface, or if it is subsequently exposed by uplift and/or erosion, the topographic result is likely to be a series of long, narrow, parallel ridges and valleys.

The simplest relationship between structure and topography, and one that often occurs in nature, finds the upfolded anticlines producing ridges and the downfolded synclines forming valleys. The converse relationship is also possible, however, with valleys developing on the anticlines and ridges on the synclines [Figure 14-32(a)]. This inverted topography is most easily explained by the effects of tension (pulling apart) and compression (pushing together) on the folded strata. Where a layer is arched over an upfold, tension cracks can form and provide easy footholds for erosional forces to remove materials and incise downward into the underlying strata. Conversely, the compression that acts upon the downfolded beds increases their density and therefore their resistance to erosion. Thus over a long period of time, the upfolds may be eroded away faster than the downfolds, producing anticlinal valleys and synclinal ridges.

All these types of folding, as well as many variations, are found in what is called the ridge-and-valley section of the Appalachian Mountains, a world-famous area noted for its remarkably parallel sequence of mountains and valleys developed on folds [Figure 14-32(b)]. The section extends for about 1000 miles (1600 kilometers) in a northeast-southwest direction across parts of nine states, with a width that varies from 25 to 75 miles (40 to 120 kilometers).

The geometry of complex folding is of infinite variability, with extremes of crumpling and over-

(a)

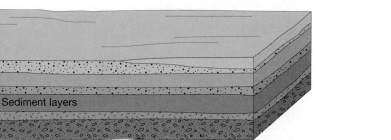

Sediment layers

Figure 14-30 **(a)** Compressive lateral pressures cause sedimentary strata that are initially horizontal to fold, much as pushing on a tablecloth causes it to bunch up into folds. **(b)** Tightly folded sedimentary strata in a road cut near Los Angeles. (TLM photo)

(b)

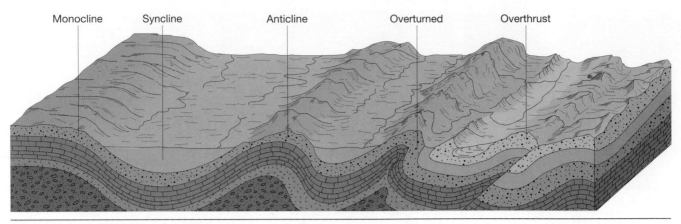

Monocline   Syncline   Anticline   Overturned   Overthrust

Figure 14-31   The basic types of folds.

turning. The resultant topographic features can be equally variable, depending, as always, on the actions of the external processes. One example of this complexity is portrayed in Figure 14-33, which displays some of the remarkable compressional folding that has taken place in the Swiss Alps and the generalized topographic profile that has resulted.

The landforms that result from folding are even more varied than the folds themselves because other processes are involved in the final shaping of the surface. Most folding takes place in a constricted subterranean environment; when the structures are exposed at the surface, erosion modifies them in a variety of ways.

FAULTING

Another prominent result of crustal stresses is the breaking apart of rock material. When rock is broken with accompanying displacement (that is, actual movement of the crust on one or both sides of the break), the action is called **faulting** (Figures 14-34 and 14-35 on page 418). The movement can be vertical or horizontal or a combination of both. Faulting usually takes place along zones of weakness in the crust; such an area is referred to as a fault zone, and the intersection of that zone with Earth's surface is called a fault line.

Movement of crust along a fault zone is sometimes very slow, but it may also occur as a sudden slippage. A single slippage may result in a displacement of only a centimeter or so, as in Figure 14-36, but in some cases the movement may be as much as 20 or 30 feet (6 or 9 meters). Successive slippages may be years or even centuries apart, but the cumulative displacement over millions of years could conceivably amount to hundreds of miles horizontally and tens of miles vertically.

Usually, but not exclusively, associated with

faults are the abrupt movements of the crust known as earthquakes. These movements vary from mild, imperceptible tremors to wild shakings that persist for many seconds. Although usually of very limited importance in the shaping of landforms, earthquakes sometimes cause much destruction and suffering for humans.

The depth of fault actions is unknown, but major faults seem to penetrate many miles into Earth's crust. Indeed, the deeper fault zones apparently serve as conduits to allow both water and heat from inside Earth to approach the surface. Frequently springs are found along fault lines, sometimes with hot water gushing forth. Volcanic activity is also associated with some fault zones, as magma forces its way upward in the zone of weakness.

Fault lines are often marked by other prominent topographic features. Most notable are **fault scarps** (Figure 14-35), which are steep cliffs that represent the edge of a vertically displaced block. Some fault scarps are as much as 2 miles (3.2 kilometers) high and extend for more than 100 miles (160 kilometers) in virtually a straight line. The abruptness of their rise, the steepness of their slope, and the linearity of their orientation combine to make some fault scarps extremely spectacular features in the landscape, as exemplified by the eastern face of California's Sierra Nevada (Figure 14-37) and the western face of Utah's Wasatch Range.

Other prominent fault-line topographic features include linear erosional valleys, carved in the weakened rock along the fault zone; displaced stream courses, associated with lateral movement of the fault blocks; and sag ponds, caused by spring and runoff water that collects in sunken ground resulting from fault movement.

Types of Faults   Although structural geologists recognize more than two dozen kinds of faults, they

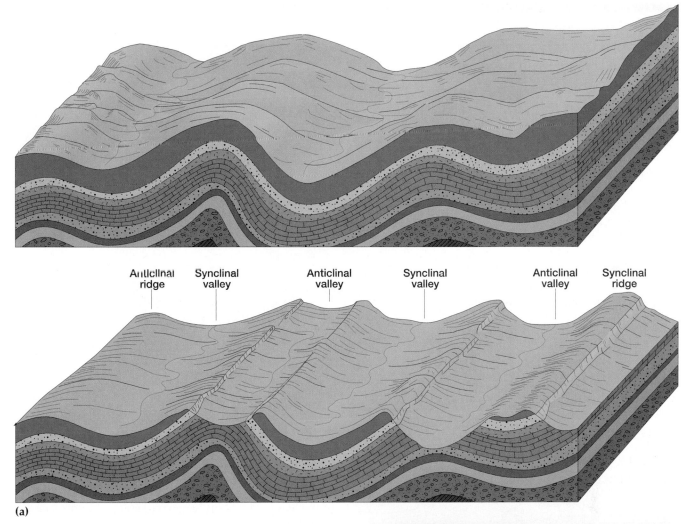

Anticlinal ridge | Synclinal valley | Anticlinal valley | Synclinal valley | Anticlinal valley | Synclinal ridge

(a)

Figure 14-32 **(a)** Formation of anticlinal valleys and synclinal ridges. **(b)** Satellite image of the intensely folded Appalachian topography of the eastern United States. (Worldsat International Inc./Science Photo Library/Photo Researchers, Inc.)

can be generalized into four principal types on the basis of direction and angle of movement (Figure 14-38). Two types involve displacement that is mostly vertical, a third encompasses only horizontal movement, and the fourth includes both horizontal and vertical offsets.

1. A **normal fault** results from tension stresses in the crust. It produces a very steeply inclined fault zone, with the block of land on one side being pushed up, or *upthrown*, relative to the *downthrown* block on the other side. A prominent fault scarp usually is formed.

2. A **reverse fault** is produced from compression stresses, with the upthrown block rising steeply above the downthrown block, so that the fault

scarp would be severely oversteepened if erosion did not act to smooth the slope somewhat. Landslides normally accompany reverse faulting.

FOCUS

EARTHQUAKES

Although most natural processes are either benign or inoffensive to humankind, a few are abrupt and capable of an enormous amount of destruction. As shapers of terrain, earthquakes are of minor significance, but as producers of instantaneous havoc they are overwhelming.

An earthquake is essentially a vibration in the crust produced by shock waves resulting from a sudden displacement along a fault. The fault movement allows an abrupt release of energy after a long, slow accumulation of strain. The faulting

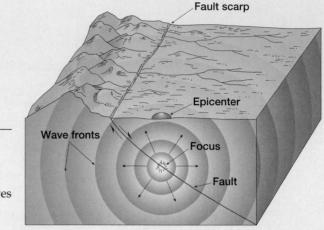

Figure 14-2-A
Relationship among focus, epicenter, and seismic waves of an earthquake. The waves are indicated by the concentric circles.

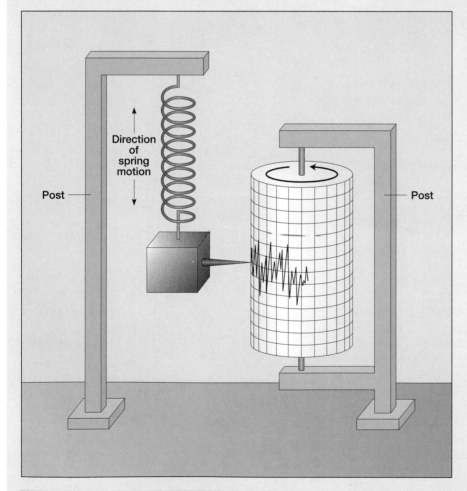

Figure 14-2-B A simple seismograph. The two posts, anchored in bedrock, pick up vibrations in the crust. The pendulum, suspended by a spring from one of the posts, traces the movement of the other post on the rotating drum.

may take place right at the surface, but it usually originates at considerable depth, as much as 400 miles (640 kilometers) beneath the surface. The pent-up energy that is released moves through the lithosphere in several kinds of seismic waves that originate at the center of motion (called the focus of the quake). These waves travel outward in widening circles, exactly like the ripples produced when a rock is thrown into a pond, gradually losing momentum with increasing distance from the focus. The strongest shocks and greatest crustal vibration felt at the ground surface are usually directly above the focus, and this ground-surface point is referred to as the **epicenter** of the earthquake (Figure 14-2-A).

Seismographs around the world record the arrival times and forces of seismic waves (Figure 14-2-B). Comparing records from different stations allows seismologists to pinpoint a focus with great precision and so to determine the strength of the quake.

Several scales are used to indicate quake strength, but by far the most widely quoted is the Richter magnitude scale, devised by Cali-

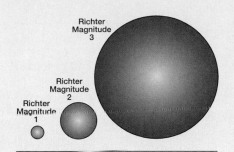

Figure 14-2-C Relationship between earthquake magnitude and energy. The volumes of the spheres are roughly proportional to the amount of energy released by earthquakes of the magnitude indicated. Using the same scale to represent the energy released by the San Francisco earthquake of 1906 (Richter magnitude 8.3) would require a sphere with a diameter of 220 feet (66 meters). (From Stephen L. Harris, *Agents of Chaos*. Missoula, MT: Mountain Press Publishing Co., 1990, p. 14.)

fornia seismologist Charles F. Richter in 1935 to describe the amount of energy released in a quake. The scale is logarithmic, and so each successively higher number represents an amount of ground vibration ten times greater than that represented by the preceding number (Figure 14-2-C). The scale ranges from 0 to 9, but theoretically there is no upper limit. Any earthquake with a Richter number of 8 or more is considered catastrophic, and shocks of lesser magnitude can cause immense damage under certain conditions.

The most violent known earthquakes have recorded 8.9 on the Richter scale. In comparison, the famous San Francisco quake of 1906 reached an estimated 8.3, the southern Alaska earthquake of 1964 recorded 8.5, and the Northridge, California, quake of 1994 attained only 6.7.

Another measure of earthquake violence is intensity. Whereas magnitude depends only on the energy released at the focus, intensity involves the amount of damage caused, particularly to human-built structures. The most widely used intensity scale was designed in 1902 by Italian geologist Giuseppe Mercalli (after whom it is named) and has been modified by the U.S. Coast and Geodetic Survey to its present form. Intensity is determined not only by energy released but also by the distance from the epicenter, nature of the land surface, structural characteristics of the buildings, and degree to which ground motions are felt.

The most notable earthquake-caused terrain modifications are usually landslides, which may be triggered in hilly terrain. The land-

(continued)

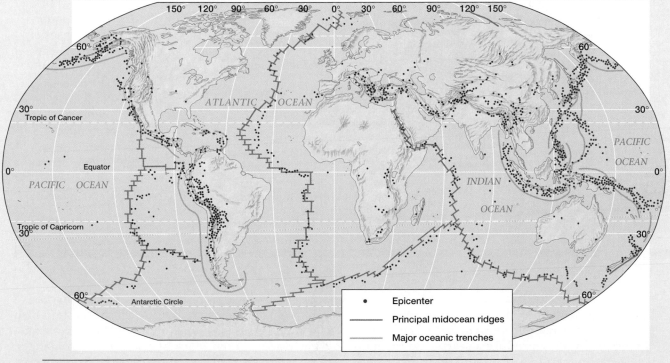

Figure 14-2-D The distribution of epicenters for all earthquakes of at least 5.5 magnitude from 1963 through 1977. Their relationship to midocean ridges and oceanic trenches is striking.

EARTHQUAKES

(continued)

slides sometimes produce significant secondary effects such as blocking streams and thereby creating instant new lakes.

Another kind of hazard associated with earthquakes involves water movements in lakes and oceans. The abrupt crustal vibrations can set great waves in motion in lakes and reservoirs, causing them to overflow shorelines or dams in the same fashion that water can be sloshed out of a dishpan. Much more significant, however, are great seismic sea waves called tsunamis, which are

sometimes generated by undersea earthquakes. These waves, sometimes occurring in a sequential train, move quickly across the ocean. They are all but imperceptible in deep water, but when they reach shallow coastal waters they sometimes build up to tens of feet in height and crash onto the shoreline with devastating effect (where they are often incorrectly called tidal waves).

In any given year, tens of thousands of earthquakes occur somewhere in the crust, most of them

followed by multiple aftershocks. "Significant" earthquakes (to be classed as significant, the quake must have a magnitude of at least 6.5 on the Richter scale, cause casualties, or create considerable damage) occur on an average of between 60 and 70 times a year throughout the world, and the long term worldwide average is 10,000 human deaths caused by quakes annnually. (See the accompanying table.)

Earthquakes may occur anywhere, even in the middle of apparently very stable continental areas. Most, however, take place in the boundary zones of the great crustal plates, particularly along the midocean ridges and in the subduction areas of ocean margins. The greatest concentrations of earthquake epicenters are found around the rim of the Pacific Ocean (Figure 14-2-D).

Despite our expanding understanding of the nature and causes of earthquakes and the increasing sophistication of quake-detecting instruments, it is still quite impossible to predict a quake with any assurance. Earthquake prediction has been characterized as a "conjectural science." As one famous contemporary geophysicist states it, "At present we can't predict earthquakes any better than the ancients did."

Worldwide Earthquake Frequency and Damage

Richter magnitude	Number per year	Modified Mercalli intensity scale	Characteristic effects
<3.4	800,000	I	Recorded only by seismographs
3.5–4.2	30,000	II, III	Felt by some people
4.3–4.8	4800	IV	Felt by many people
4.9–5.4	1400	V	Felt by everyone
5.5–6.1	500	VI, VII	Slight building damage
6.2–6.9	100	VIII, IX	Much building damage
7.0–7.3	15	X	Serious damage, bridges twisted, walls fractured
7.4–7.9	4	XI	Great damage, buildings collapse
>8.0	1 every 5–10 years	XII	Total damage, waves on ground surface, objects thrown into air

3. In a **strike-slip fault**, the movement is entirely horizontal, with the adjacent blocks being displaced laterally relative to each other. This action, of course, does not produce a scarp, and the fault line is topographically inconspicuous except where stream courses are displaced or sag ponds develop.

4. More complicated in structure and more impressive in their dynamics are **overthrust faults**, in which compression forces the upthrown block to override the downthrown block at a relatively low angle, sometimes for many miles. Overthrusting occurs frequently in mountain building, resulting in unusual geologic relationships such as older strata being piled on top of younger rocks.

Figure 14-33 Cross section through the Swiss Alps, showing the enormous complexity of fold structures.

Prominent Faulted Landforms   Certain conspicuous fault associations recur frequently in various parts of the world, causing a very notable gross configuration of the terrain.

Fault-Block Mountains   Under certain stresses, a surface block may be severely faulted and upthrown on one side without any faulting or uplift on the other. When this happens, the block is tilted asymmetrically, producing a steep slope along the fault scarp and a relatively gentle slope on the other side of the block. The classic example of a **fault-block mountain range** is California's Sierra Nevada (Figure 14-37), an immense

(a)

(b)

Figure 14-34 Horizontal sedimentary strata that have been slightly folded and mildly faulted. This is a road cut near Newman in Western Australia. (TLM Photo)

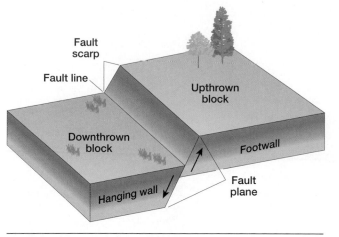

Figure 14-35   A simple fault structure.

Figure 14-36   A prominent fault line runs through the town of Hollister, California. Dramatic evidence of its activity is shown by this offset wall and sidewalk. (TLM photo)

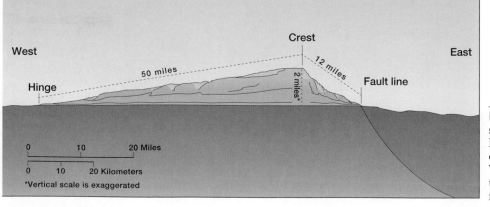

Figure 14-37   The western slope of the Sierra Nevada is long and gentle, whereas the eastern slope is short and steep. This gentle/steep combination is the result of enormous block faulting on the eastern side.

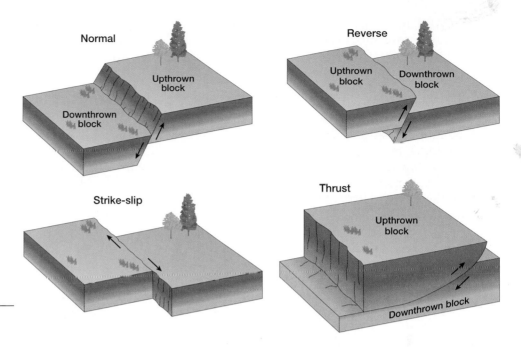

Figure 14-38 The principal types of faults.

block nearly 400 miles (640 kilometers) north-south and about 60 miles (96 kilometers) east-west. The spectacular eastern face is a fault scarp that has a vertical relief of about 2 miles (3.2 kilometers) in a horizontal distance of only about 12 miles (19 kilometers). In contrast, the general slope of the western flank of the range, from crest to hinge line (the line along which the gentle side begins to rise), has a vertical dimension of about 2 miles (3.2 kilometers) spread over a horizontal distance of nearly 50 miles (80 kilometers). The shape of the range has of course been modified by other processes, but its general configuration was determined by block faulting.

Horst   Another frequent occurrence in nature is the uplift of a block of land between two parallel faults, an action that produces a structure called a **horst** (Figure 14-39). The same result can be achieved if the land on both sides has been downthrown. In either case, the horst is a block elevated above the surrounding land. It may take the form of a plateau or a mountain mass with two steep, straight sides.

Graben   At the other extreme is a **graben**, which is a block of land bounded by parallel faults in which the block has been downthrown, producing a distinctive structural valley with a straight, steep-sided fault scarp on either side (Figure 14-39). A world-famous graben is Death Valley in California.

Grabens and horsts often occur side by side. The basin-and-range country of the western interior of the United States is a vast sequence of horsts and grabens, mixed with fault-block ranges and some more complex structures, encompassing most of Nevada and portions of surrounding states.

Rift Valleys   Downfaulted grabens occasionally extend for extraordinary distances as linear valleys enclosed between steep fault scarps. Such lengthy trenches are called **rift valleys**, and they comprise some of Earth's most notable structural lineaments, particularly the Great Rift Valley in East Africa.

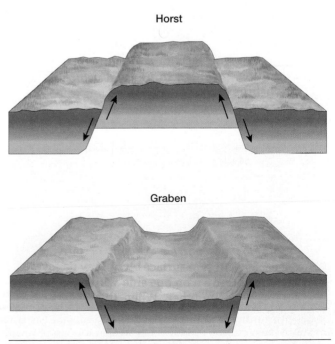

Figure 14-39 Horst and graben.

Wherever Earth's crust experiences stresses sufficient to cause faulting, other kinds of diastrophic movement (folding, warping) may also occur. Moreover, such stresses are frequently associated with zones of weakness in the crust, and molten magma sometimes finds its way to the surface through these zones. Thus a fault may occur in isolation, but much more common are multiple faults, or a mixture of faulting and folding, or an even more complex association of diastrophism and vulcanism and ultimately the forces of plate tectonics.

Earth's surface is greatly stressed by crust movement, by the associated rise of magma into the divergent zones, and particularly by the crumpling and subduction that occur when plates collide or slide over one another. The resulting tension and compression cause many of the lesser diastrophic movements we have just studied. Thus an appreciation of plate tectonics helps us to understand the formation of not only first-order relief features but also many of the second-, third-, and fourth-order features as well.

THE COMPLEXITIES OF CRUSTAL CONFIGURATION

Considering each internal process in turn as we have just done is a helpful way to systematize knowledge. Doing so presents an artificial and misleading picture, however, because in nature these processes are interre-

lated. To attempt a more balanced assessment, let us consider a highly simplified statement of the origin of the gross contemporary topographic features of a small part of Earth's surface—a mountainous section of northwestern Montana that encompasses the spectacular scenery of Glacier National Park.

This area, now part of the northern Rocky Mountains, was below sea level for many millions of years. Most of the rocks in the region were formed in the Precambrian era, when much of the area now occupied by the Rocky Mountain cordillera consisted of a large, shallow, seawater-filled trough. Muds and sands were washed into this Precambrian sea for millennia, and a vast thickness of sedimentary strata built up. These limestones, shales, and sandstones accumulated as six distinct formations, each with a conspicuous color variation, known collectively as the Belt Series.

Occasionally during this lengthy epoch of sedimentary accumulation, igneous activity added variety. Most notable was a vast outpouring of flood basalt that issued from fissures in the ocean floor and was extruded in the form of a submarine lava flow. Further sediments were then deposited on top of the lava flow.

Igneous intrusions were also injected from time to time, including one large sill and a number of dikes. The igneous rocks, both the flood basalt and the various intrusions, initiated contact metamorphism, whereby the tremendous heat of the igneous material converted some of the adjacent sedimentaries into metamorphic rocks (mostly changing limestone to

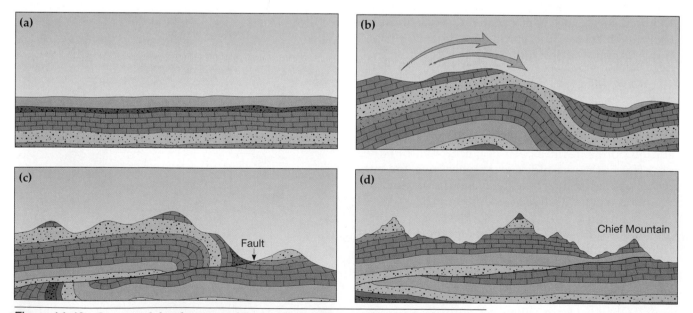

Figure 14-40 Sequential development of the Lewis Overthrust: **(a)** initial sea-bottom deposition of sediments; **(b)** uplift and folding; **(c)** continued pressure from the west causes overturning of the fold and faulting along the eastern limb of the anticline; **(d)** subsequent erosion produces the present topography, with Chief Mountain as a residual outlier to the east of the range.

Figure 14-41 Chief Mountain sits in splendid isolation because of the massive movement associated with the Lewis Overthrust and subsequent erosion. (TLM photo)

marble and sandstone to quartzite).

After a long gap in the geologic record, during which the Rocky Mountain region was mostly above sea level, the land once again sank below the ocean during the Cretaceous period and another thick series of sediments was deposited. This was followed by a period of mountain building so significant that it has been named the Rocky Mountain Revolution. In the Glacier National Park area, the rocks were compressed and uplifted, converting the site of the former sea into a mountainous region.

Along with uplift came extreme lateral pressure from the west, convoluting the gently downfolded strata into a prominent anticline. Continuing pressure then overturned the anticline toward the east, with a lengthening western limb and a truncating eastern limb. This additional strain on the rock and the persistent crustal pressure eventually caused a vast rupture and faulting. The entire block was then pushed eastward by one of the greatest thrust faults known, the Lewis Overthrust (Figure 14-40). This remarkable fault forced the Precambrian sedimentaries out over the Cretaceous strata that underlie the plains to the east, by as much as 20 miles (32 kilometers). The plane of the thrust fault was only slightly above the horizontal, nowhere exceeding a dip of 10°. This had the peculiar effect of placing older rock layers on top of much younger strata. The terrain thus produced is referred to as "mountains without roots." Chief Mountain is world famous as a rootless mountain because of its conspicuous location as an erosional outlier east of the main range (Figure 14-41).

CHAPTER SUMMARY

Strong forces inside Earth are instrumental in reshaping the crustal surface. These internal processes are extraordinarily powerful, infinitely ongoing, and only incompletely understood. The big picture of crustal shaping and reshaping has become much clearer in the last quarter century because scientists have learned more and more about the dynamics of plate tectonics.

Earth's upper crust consists of a lower layer of dense sima that underlies both continents and ocean basins and a discontinuous layer of sial found only beneath the continental masses.

The generally derided theory of continental drift, notably expounded by Alfred Wegener in the early 1900s, was revived in the second half of the twentieth century and has become almost universally accepted as the concept of plate tectonics. It postulates a massive supercontinent called Pangaea, which formed from smaller continents about 450 million years ago and then, about 250 million years ago, began to break up again into large sections (plates) that have continuous-ly moved away from one another and now comprise the contemporary continents.

Plates move apart at divergent boundaries, usually represented by an undersea mountain range (called an oceanic ridge), with new seafloor material being created by the upwelling of hot magma from below. At a convergent boundary, plates moving in opposite directions meet, resulting in mountain building at the edges as one plate slides (subducts) under the other. At a transform boundary, two plates slip past one another laterally; this usually does not produce major landforms but may result in fault movements and earthquakes.

Vulcanism is a general term that covers all processes in which magma moves from Earth's interior to or near the surface: volcanism (extrusive vulcanism), intrusive vulcanism, and deep plutonic activity.

Diastrophism refers to the deformation of Earth's crust by bending or breaking in response to great pressures exerted either from below or from within the crust. The major diastrophic movements are broad warping, folding, and faulting.

KEY TERMS

anticline	horst	rift valley
batholith	hot spot	seafloor spreading
broad warping	isostasy	sial
caldera	laccolith	sill
convergent boundary	lava	sima
cordillera	mantle plume	stock
diastrophism	monocline	strike-slip fault
dike	normal fault	vein
divergent boundary	overthrust fault	subduction
epicenter	overthrust fold	syncline
fault-block mountain range	overturned fold	terrane
faulting	Pangaea	transform boundary
fault scarp	plasticity	vein
flood basalt	plutonic activity	volcanic neck
folding	pyroclastic material	volcanism
graben	reverse fault	vulcanism

REVIEW QUESTIONS

1. Explain why the rigid Earth theory is no longer considered correct.
2. What is the relationship between plasticity and isostasy?
3. Why was Alfred Wegener's continental drift theory rejected for so long?
4. Explain how the presence of the Mid-Atlantic Ridge supports the theory of plate tectonics.
5. Explain how seafloor spreading supports the theory of plate tectonics.
6. Distinguish among the three kinds of plate boundaries.

7. Explain the Hawaiian Islands in terms of mantle plumes.
8. Is there any relationship between tectonic plates and mantle plumes? Explain.
9. Explain the difference between folding and faulting.
10. Explain how it is possible for a syncline to produce a ridge.
11. Why is there such a concentration of earthquakes and volcanoes around the margin of the Pacific Ocean?
12. How important are earthquakes in topographic development?
13. Describe the major forms of igneous intrusions.
14. Distinguish between a dike and a sill; a batholith and a laccolith.

S O M E   U S E F U L   R E F E R E N C E S

BOLT, BRUCE A., *Earthquakes*, rev. ed. New York: W. H. Freeman, 1993.

BONATTI, ENRICO, "The Earth's Mantle Below the Oceans," *Scientific American*, Vol. 270, March 1994, pp. 44-51.

DECKER, R., AND B. DECKER, *Volcanoes*, rev. ed. San Francisco: W. H. Freeman, 1989.

HARRIS, STEPHEN L., *Agents of Chaos: Earthquakes, Volcanoes, and Other Natural Disasters*. Missoula, MT: Mountain Press, 1991.

KEAREY, PHILIP, AND FREDERICK J. VINE, *Global Tectonics*. Oxford, England: Blackwell Scientific Publications, 1990.

MOORES, ELDRIDGE M., (ed.), *Shaping the Earth: Tectonics of Continents and Oceans*. New York: W. H. Freeman, 1990.

OLLIER, CLIFF, *Volcanoes*. New York: Basil Blackwell, 1988.

RAYMO, CHET, *The Crust of Our Earth: An Armchair Traveler's Guide to the New Geology*. Englewood Cliffs, NJ: Prentice Hall, 1983.

15

PRELIMINARIES TO EROSION: WEATHERING AND MASS WASTING

I F THE INTERNAL PROCESSES OF LANDSCAPE FORMATION are overwhelming, the external processes are inexorable. During all the time that continents are drifting, that the crust is bending and breaking, that volcanoes are erupting and intrusions forming, another suite of natural forces is simultaneously at work. These are the external forces, often mundane and even minuscule in contrast to those described in the preceding chapter. Yet the cumulative effect of the external processes is awesome; they are capable of wearing down anything the internal forces can erect. No rock is too resistant and no mountain too massive to withstand their unrelenting power. Ultimately, the detailed configuration of peaks, slopes, valleys, and plains is molded by the work of gravity, water, wind, and ice.

The overall effect of the disintegration, wearing away, and removal of rock material is generally referred to as **denudation**, a term that implies a lowering of continental surfaces. Denudation is accomplished by the interaction of three types of activities :

- Weathering is the breaking down of rock into smaller components by atmospheric and biotic agencies.
- Mass wasting involves the downslope movement of broken rock material due to gravity.
- Erosion consists of more extensive and generally more distant removal of fragmented rock material.

All three processes are illustrated in Figure 15-1.

THE IMPACT OF WEATHERING AND MASS WASTING ON THE LANDSCAPE

The most readily observable landscape effect of weathering is the fragmentation of bedrock—the reduction of large rock units into more numerous and less cohesive smaller units. This activity occurs in place, without conspicuous displacement of the rocks.

Mass wasting always involves the downslope movement of rock material. Its impact on the landscape normally is twofold: an open scar is left on the surface that was vacated, and an accumulation of debris is deposited somewhere downslope from the scar.

Although weathering and mass wasting often are deemphasized as processes of landform modification and shaping, the role they play is critical. Take the Grand Canyon of the Colorado River, for example. One of the most impressive landscape features of the world, the canyon is usually attributed only to the ero-

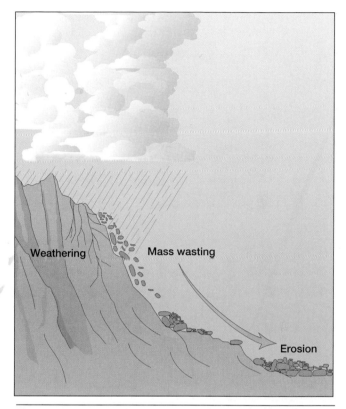

Figure 15-1 Denudation, the lowering of continental surfaces, is accomplished by a combination of weathering, mass wasting, and erosion.

sive powers of the river. However, most of what we see in the Grand Canyon are forms produced by weathering and mass wasting (Figure 15-2). Besides deepening its channel, the river's main role is to transport the sediment loosened by weathering and pulled into the stream by gravity and mass wasting.

WEATHERING

The first step in the shaping of Earth's surface by external processes is **weathering**, the mechanical disintegration and/or chemical decomposition that destroys the coherence of bedrock and begins to fragment rock masses into progressively smaller components. Fast or slow, mechanical and chemical weathering occur wherever the lithosphere and atmosphere meet. It occurs with great subtlety, however, breaking chemical bonds, separating grain from grain, pitting the smooth, fracturing the solid. It is the aging process of rock surfaces, the process that must precede the other, more active, forms of denudation.

Whenever bedrock is exposed, it weathers. It often

The eroded sandstone layers of Antelope Canyon in northeastern Arizona. *(Mark Newman/Photo Researchers, Inc.)*

Figure 15-2 The Colorado River, shown here in Arizona's Grand Canyon, is the principal erosive force, but weathering and mass wasting have also contributed significantly to the carving of the massive canyon. (Sylvain Grandadam/Photo Researchers, Inc.)

has a different color or texture than neighboring unexposed bedrock, for instance. Most significant from a topographic standpoint, exposed bedrock is likely to be looser and less coherent than the underlying rock. Blocks or chips may be so loose that they can be detached with little effort. Sometimes pieces are so "rotten" that they can be crumbled by finger pressure. Slightly deeper in the bedrock, there is firmer, more solid rock, although along cracks or crevices, weathering may extend to considerable depths. In some cases, the weathering may reach as much as several thousand feet beneath the surface. This penetration is made possible by open spaces in the rock bodies and even in the mineral grains. Subsurface weathering is initiated along these openings, which can be penetrated by such weathering agents as water, air, and plant roots. As time passes, the weathering effects spread from the immediate vicinity of the openings into the denser rock beyond (Figure 15-3).

Openings in the surface and near-surface bedrock are frequently submicroscopic, but they may also be large enough to be conspicuous and are sometimes huge. In any case, they occur in vast numbers and provide avenues along which weathering agents can attack the bedrock and break it apart. Broadly speaking, five types of openings are common:

1. *Microscopic spaces* occur in profusion. Although tiny, they are so numerous that they can be responsible for extensive weathering. They may consist of spaces between crystals of igneous or metamorphic rocks, pores between grains of sedimentary rocks,

or minute fractures within or alongside mineral grains in any kind of rock.

2. *Joints*, the most common structural features of the rocks of the lithosphere, are cracks that develop as a result of stress, but the rocks show no appreciable movement parallel to the joint walls. Joints are innumerable in all rock masses, dividing them into blocks of various sizes. Because of their ubiquity, joints are the most important of all rock openings in facilitating weathering.

3. *Faults* are cracks in bedrock along which there is relative movement of the walls making up the crack (Figure 15-4). Faults generally are individual or occur only in small numbers, whereas joints normally are multitudinous. A further difference between faults and joints is that the former sometimes appear as major landscape features, extending for tens or even hundreds of miles, whereas the latter are normally minor structures extending only a few inches or feet. Faults allow easy penetration of weathering agents into subsurface areas because not only fracturing but also displacement is involved.

4. *Lava vesicles* are holes of various size, usually small, that develop in cooling lava when gas is unable to escape as the lava solidifies.

5. *Solution cavities* are holes formed in calcareous rocks (particularly limestone) as the soluble minerals are dissolved and carried away by percolating water. Most solution cavities are small, but sometimes huge holes and even massive caverns are created when large amounts of solubles are removed.

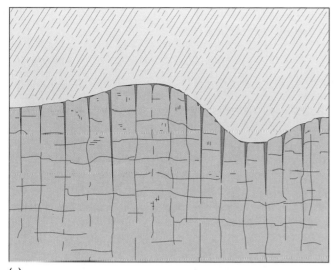

(a)

(b)

Figure 15-3 How deep weathering develops in bedrock containing many cracks: **(a)** before weathering; **(b)** after weathering.

JOINTING

Almost all lithospheric bedrock is jointed, resulting sometimes from contractive cooling of molten material, sometimes from contraction of sedimentary strata as they dry, and sometimes from diastrophic tension. At Earth's surface, the separation between blocks on either side of a joint may be conspicuous because weathering emphasizes the fracture. Below the surface, however, the visible separation is minimal.

Joints are relatively common in most rock, but they are clearly more abundant in some places than in others. Where numerous, they are usually arranged in *sets*, each set comprising a series of approximately parallel fractures. Frequently, two prominent sets intersect almost at right angles; such a combination constitutes a *joint system* (Figure 15-5). A well-developed joint system, particularly in sedimentary rock having prominent natural bedding planes, can divide stratified rock into a remarkably regular series of close-fitting blocks. Generally speaking, jointing is more regularly patterned, and the resulting blocks are more sharply defined in fine-grained rocks than in coarse grained ones.

In some places, large joints or joint sets extend for long distances and through a considerable thickness of rocks; these are termed **master joints** (Figure 15-6). Master joints play a role in topographic development by functioning as a plane of weakness, a plane more susceptible to weathering and erosion than the rock around it. Thus the location of large features of the landscape, such as valleys and cliffs, may be influenced by the position of master joints.

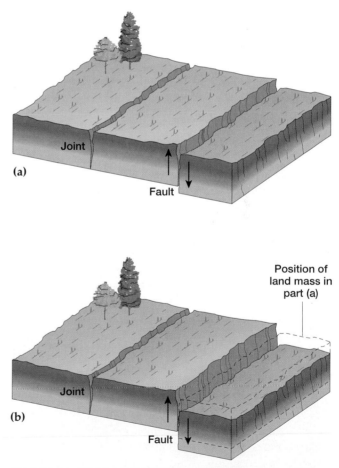

(a)

Joint

Fault

Position of land mass in part (a)

Joint

(b)

Fault

Figure 15-4 The essential difference between joints and faults is that the former involve no relative displacement on either side of the crack. Here we see a hypothetical joint and fault in the same piece of bedrock at two moments in time. At the later time **(b)**, the rock masses on the two sides of the fault have moved relative to each other but those on the two sides of the joint are still in the same relative positions as they were in part **(a)**.

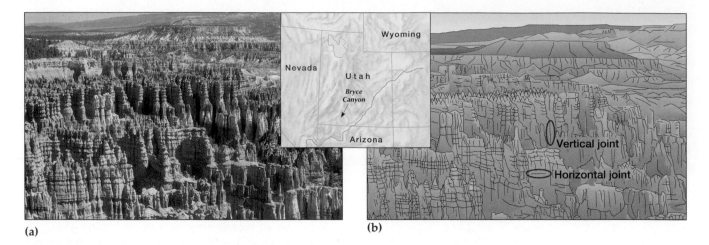

(a)

(b)

Figure 15-5 **(a)** A view of the badlands topography of Bryce Canyon in southern Utah from Inspiration Point. (TLM photo) **(b)** The closely-spaced joint systems contribute to intricate sculpturing by weathering and erosion.

WEATHERING AGENTS

Most weathering agents are atmospheric. Because it is gaseous, the atmosphere is able to penetrate readily into all cracks and crevices in bedrock. From a chemical standpoint, oxygen, carbon dioxide, and water vapor are the three atmospheric components of greatest importance in rock weathering.

Temperature changes are a second important weathering agent. Most notable, however, is water, which can penetrate downward effectively into openings in the bedrock. Biotic agents also contribute to weathering, in part through the burrowing activities of animals and the rooting effects of plants, but especially through the production of chemical substances that attack the rock.

The total effect of these agents is complicated and is influenced by a variety of factors; the nature and struc-ture of the bedrock, the abundance and size of openings in it, the surface configuration, prevailing climatic conditions, the vegetative cover, and the variety and abundance of digging animals. For analytical purposes, however, it is convenient to recognize three principal categories of weathering—mechanical, chemical, and biotic. While we now consider each of them in turn, we should bear in mind that they usually act in concert.

Mechanical Weathering
Mechanical weathering is the physical disintegration of rock material without any change in its chemical composition. In essence, big rocks are mechanically weathered into little ones by various stresses that cause the rock to fracture into smaller fragments. Most mechanical weathering occurs at or very near the surface, but under certain conditions it may occur at considerable depth.

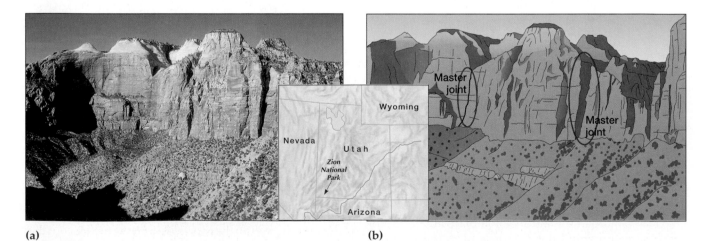

(a)

(b)

Figure 15-6 **(a)** An array of mighty monoliths in Utah's Zion National Park. (TLM photo) **(b)** Master joints are widely spaced in this area, allowing for the development of massive blocks and precipitous cliffs.

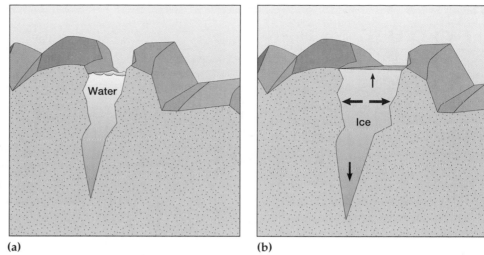

Figure 15-7 Frost wedging. When water freezes, its volume increases. When this water is in a rock crack, the expanded ice exerts a force that can deepen and widen the crack, especially if the process is repeated many times.

(a)

(b)

Probably the most important single agent of mechanical weathering is the freeze/thaw action of water. When water freezes, it expands by almost 10 percent. Moreover, the upper surface of the water freezes first, which means that the principal force of expansion is exerted against the wall of the confining rock rather than upward. This expanding wedge of ice splits the rock, as shown in Figure 15-7.

Even the strongest rocks cannot withstand frequent alternation of freezing and thawing. Repetition is the key to understanding the inexorable force of **frost shattering** or **frost wedging** (Figure 15-8). Regardless of its

size, if an opening in rock contains water, when the temperature falls below 32°F (0°C), ice forms, wedging its way downward. When the temperature rises above 32°F, the ice melts and the water sinks farther into the slightly enlarged cavity. With renewed freezing, the wedging is repeated. Such a freeze/thaw pattern may be repeated millions of times through the eons of Earth history, providing what is literally an irresistible force.

Frost wedging in large openings may produce large boulders, whereas that occurring in small openings may granulate the rock into sand and dust particles, with every size gradation in between. A common form

Figure 15-8 An example of the powerful force of freeze/thaw action on a granite boulder in Colorado's Forest Canyon. (TLM photo)

FOCUS

EXFOLIATION

One of the most striking of all weathering processes is **exfoliation**, in which curved layers peel off bedrock. (We saw in Chapter 13 that the term *foliation* refers to the parallel alignment of textural and structural features of a rock; *exfoliation* involves the stripping away of roughly parallel, concentric rock slabs.) Curved and concentric sets of joints, usually minor and inconspicuous, develop in the bedrock, and parallel shells of rock break away in succession, somewhat analogous to the way layers of an onion separate. The sheets that split off are sometimes only a millimeter or so thick; in other cases, however, they may be several feet thick.

Exfoliation occurs mainly in granite and related intrusive rocks, but under certain circumstances it is also common in sandstone and other sedimentary strata.

The dynamics of exfoliation are not fully understood. Both chemical and mechanical weathering probably are involved, but the latter appears to be much more significant than the former. Fracturing resulting from temperature variations may be an important component of the mechanism, although exfoliation is also found in areas where temperature fluctuations

Figure 15-1-A Exfoliation on a grand scale. This is Half Dome in Yosemite National Park, California. (Breck Kent/Earth Scenes)

are not marked. Volumetric changes in minerals, which set up strains in the rock, may also be involved. Such changes are most notably produced by *hydration*, in which water molecules become attached to other substances and then the added water causes the original substance to swell without any change in its chem-

ical composition. This swelling weakens the rock mass and is usually sufficient to produce some fracturing.

Another exfoliation mechanism that has been hypothesized and widely accepted is that the rock cracks after an overlying weight has been removed, a process called both *unloading* and *pressure release*. The intrusive bedrock originally may have been deeply buried beneath a heavy overburden. When the overlying material is stripped away by erosion, the release of pressure allows expansion in the rock. The outer layers are unable to contain the expanding mass, and the expansion can be absorbed only by cracking along the sets of joints.

Whatever the causes of exfoliation, the results are conspicuous. If the rock mass is a large one, such as Half Dome or one of the other granite monoliths overlooking Yosemite Valley, California, its surface consists of imperfect curves punctuated by several partially fractured shells of the surface layers, and the mass is referred to as an *exfoliation dome*. Exfoliation may also occur on detached boulders in which case the result is usually a rounder shape, with each layer of shelling revealing a smaller spherical mass.

---

of breakup in coarse-grained crystalline rocks is a shattering caused by frost wedging between grains. This type of shattering produces gravel or coarse sand in a process termed *granular disintegration.*

The physical characteristics of the rock are important determinants of the rate and magnitude of mechanical weathering, as are temperature and moisture variations. The process is most effective where freezing is prolonged and intense—in high latitudes, in midlatitudes during winters, and at high altitudes. It is most conspicuous above the treeline of mountainous areas (Figure 15-9), where broken blocks of rock are likely to be found in profusion everywhere except on slopes that are too steep to allow them to lie without sliding downhill.

Related to frost wedging but much less significant is **salt wedging**, which happens when salts crystallize out of solution as water evaporates. In areas of dry climate, water is often drawn upward in rock openings by capillary action, as discussed in Chapter 12. This water nearly always carries dissolved salts. When the water evaporates, as it commonly does, the salts are left behind as tiny crystals. With time, the crystals grow, prying apart the rock and weakening its internal structure, much in the fashion previously described for freezing water, although less intensely. In humid areas, salt wedging is inconsequential because the soluble salts are flushed away by percolating groundwater.

Figure 15-9 Frost wedging is an especially pervasive force on mountaintops above the treeline, as with these granite boulders on Australia's highest peak, Mount Kosciusko, in New South Wales. (TLM photo)

Temperature changes not accompanied by freeze/thaw cycles also weather rock mechanically, but they do so much more gradually than the processes just described. The fluctuation of temperature from day to night and from summer to winter can cause minute changes in the volume of most mineral particles, forcing expansion when heated and contraction when cooled. This volumetric variation weakens the coherence of the mineral grains and tends to break them apart. Millions of repetitions are normally required for much weakening or fracturing to occur, although the intense heat of forest fires or brushfires can speed up the process. This factor is most significant in arid areas and near mountain summits, where direct solar radiation is intense during the day and radiational cooling is prominent at night.

Chemical changes (discussed in more detail in the next subsection) also contribute to mechanical weathering. Various chemical actions can cause an increase in volume of the affected mineral grains. This swelling sets up strains that weaken the coherence of the rock and cause fractures.

Some biotic activities also contribute to mechanical weathering. Most notable is the penetration of growing plant roots into cracks and crevices, which exerts an expansive force that widens the openings. This factor is especially conspicuous where trees grow out of joint or fault planes, with their large roots showing amazing tenacity and persistence as wedging devices. Additionally, burrowing animals sometimes are factors in rock disintegration. The total effect of these biotic actions is probably significant, but it is difficult to assess because it is obscured by subsequent chemical weathering

When acting alone, mechanical weathering breaks up rock masses into ever smaller pieces, producing boulders, cobbles, pebbles, sand, silt, and dust (Figure 15-10).

Chemical Weathering   Mechanical weathering is usually accompanied by **chemical weathering**, which is the decomposition of rock by the alteration of its minerals. Almost all minerals are subject to chemical alteration when exposed to atmospheric and biotic agents. Some minerals, such as quartz, are extremely resistant to chemical change, but many others are very susceptible. There are very few rocks that cannot be significantly affected by chemical weathering because the alteration of even a single significant mineral constituent can lead to the eventual disintegration of an entire rock mass.

One important effect of mechanical weathering is to expose bedrock to the forces of chemical weathering. The greater the surface area exposed, the more effective the chemical weathering. Thus finer grained materials decompose more rapidly than coarser grained materials of identical composition because in the former there is more exposed surface area.

Virtually all chemical weathering requires moisture. Thus an abundance of water enhances the effectiveness of chemical weathering, and chemical processes operate more rapidly in humid climates than in arid areas. Moreover, chemical reactions are more rapid under high-temperature conditions than in cooler regions. Consequently, chemical weathering is most efficient and conspicuous in warm climates. In cold lands, there is less chemical weathering (although the removal of dissolved salts is sometimes notable) and the influence of physical weathering is dominant.

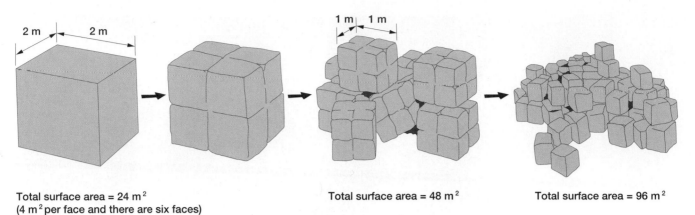

Total surface area = 24 m^2
(4 m^2 per face and there are six faces)

Total surface area = 48 m^2

Total surface area = 96 m^2

Figure 15-10 As mechanical weathering fragments rock, the amount of surface area exposed to further weathering is increased. Each successive step shown here doubles the surface area of the preceding step. (After William M. Marsh and Jeff Dozier, *Landscape: An Introduction to Physical Geography*. Reading, MA: Addison-Wesley, 1981. By permission of John Wiley & Sons.)

Some of the chemical reactions that affect rocks are very complex, but others are simple and predictable. The principal reacting agents are oxygen, water, and carbon dioxide, and the most significant processes are oxidation, hydrolysis, and carbonation. These processes often take place more or less simultaneously, largely because they all involve water that contains dissolved atmospheric gases. Water percolating into the ground acts as a weak acid because of the presence of these gases and of decay products from the local vegetation; the presence of these impurities increases the water's capacity to drive chemical reactions.

When the oxygen dissolved in water comes into contact with certain rock minerals, the minerals undergo **oxidation**, in which the oxygen atoms combine with atoms of various metallic elements making up the minerals in the rock and form new products. The new substances are usually more voluminous, softer, and more easily eroded than the original compounds.

When iron-bearing minerals react with oxygen (that is, become oxidized), iron oxide is produced. This reaction, probably the most common oxidation in the lithosphere, is called *rusting*, and the prevalence of rusty red stains on the surface of many rocks attests to its widespread occurrence (Figure 15-11). Similar effects are produced by the oxidation of aluminum. Since iron and aluminum are very common in Earth's crust, a reddish brown color is seen in many rocks and soils, particularly in tropical areas because there oxidation is the most notable chemical weathering process. Rusting contributes significantly to weathering because oxides usually are softer and more easily removed than the original iron and aluminum compounds from which the oxides were formed.

Hydrolysis is the chemical union of water with another substance to produce a new compound that is nearly always softer and weaker than the original. Igneous rocks are particularly susceptible to hydrolysis because their silicate minerals combine readily with water. Hydrolysis invariably increases the volume of the mineral, and this expansion can contribute to mechanical disintegration. In tropical areas, where water frequently percolates to considerable depth, hydrolysis often occurs far below the surface.

Carbonation is the reaction between the carbon dioxide in water and carbonate rocks to produce a very soluble product (calcium bicarbonate) that can readily be removed by runoff or percolation and can also be deposited in crystalline form if the water is evaporated.

These and other less common chemical weathering processes are continuously at work at and beneath Earth's surface. Most chemically weathered rocks are changed physically, too. Their coherence is weakened, and the loose particles produced at Earth's surface are quite unlike the parent material. Beneath the surface, the rock holds together—but in a chemically altered condition. The major eventual products of chemical weathering are clays.

Biological Weathering   As already mentioned, plants frequently and animals occasionally contribute to weathering, and such processes involving living organisms are called **biological weathering**. Most notable is the penetration of growing plant roots into cracks and crevices, as in Figure 15-12, with the result that the opening is expanded. Roots are especially effective as weathering agents where trees grow out of joint or fault planes, with their large roots showing amazing tenacity and persistence as wedging devices.

Lichens are primitive organisms that consist of algae and fungi living as a single unit. Typically they live on bare rock, bare soil, or tree bark (Figure 15-13). They draw minerals from the rock by ion exchange, and this leaching can weaken the rock. Moreover, expansion

Figure 15-11 Iron oxide (rust) stains on a cliff in Capitol Reef National Park in Utah. (Charlie Ott/Photo Researchers, Inc.)

and contraction of lichens as they get alternately wet and dry flake off tiny particles of rock.

Burrowing by animals mixes soil effectively and is sometimes a factor in rock disintegration.

The total effect of all these biotic actions is probably significant but difficult to evaluate because it is obscured by subsequent chemical weathering.

CLIMATE AND WEATHERING

Geomorphic research tells us more and more every year about how climate influences weathering. The basic generalization is that weathering, particularly chemical weathering, is enhanced by a combination of high temperatures and abundant precipitation. Of these two factors, the latter is usually more important than the former. There are many variations on this theme; Figure 15-14 represents a generalized model of relevant relationships.

MASS WASTING

As we learned above, the denudation of Earth's surface is accomplished by weathering followed by mass wasting followed by erosion. The ultimate destiny of all weathered material is to be carried away by erosion, a topic we cover in the remaining chapters of this book. The remainder of this chapter is concerned with **mass wasting**, the process whereby weathered material is moved a relatively short distance downslope under the direct influence of gravity (Figure 15-1, p. 425). Although it is sometimes circumvented when erosive agents act on weathered material directly, mass wasting is normally the second step in a three-step denudation process.

Throughout our planet, gravity is inescapable; everywhere, it pulls objects toward the center of Earth. Where the land is flat, the influence of gravity on topographic development is minimal. Even on gentle slopes, however, minute effects are likely to be significant in the long run, and on steep slopes, the results are often immediate and conspicuous. Any loosened material is impelled downslope by gravity—in some cases falling abruptly or rolling rapidly, in others flowing or creeping with imperceptible gradualness.

The materials involved in these movements are all the varied products of weathering. Gigantic boulders respond to the pull of gravity in much the same fashion as do particles of dust, although the larger the object, the more immediate and pronounced the effect. Of particular importance, however, is the implication of "mass" in mass wasting; the accumulations of material moved—fragmented rock, regolith, soil—are often extremely large and contain enormous amounts of mass.

All rock materials, from individual fragments to cohesive layers of soil, lie at rest on a slope if undisturbed unless the slope has a certain steepness. The steepest angle that can be assumed by loose fragments on a slope without downslope movement is called the **angle of repose**. This angle, which varies with the nature and internal cohesion of the material, represents a fine balance between the pull of gravity and the cohesion and friction of the rock material. If additional material accumulates on a debris pile lying on a slope that is near the angle of repose, the newly added material may upset the balance (because the added weight overcomes the friction force that is keeping the pile

Figure 15-12 Roots can serve as formidable natural tools to enlarge cracks and crevices in bedrock. **(a)** Grass growing out of a vertical cliff of basalt near Cairns, Queensland, Australia. **(b)** Juniper trees that have managed to find footholds in bare sandstone surfaces near Tuba City in northern Arizona. (TLM photos)

from sliding) and may cause all or part of the material to slide downward (Figure 15-15).

If water is added to the rock material through rainfall, snowmelt, or subsurface flow, the mobility is usually increased, particularly if the rock fragments are small. Water is a lubricating medium, and it diminishes friction between particles so that they can slide past one another more readily. Water also adds to the buoyancy and weight of the mass, which makes for a lower angle of repose and adds momentum once movement is under way. For this reason, mass wasting is particularly likely during and after heavy rains.

Another facilitator of mass wasting is clay. As noted in Chapter 12, clays readily absorb water. This absorbed water combined with the fine-grained texture of the material makes clay a very slippery and mobile substance. Any material resting on clay can often be set in motion by rainfall or an earthquake shock, even on very gentle slopes. Indeed, some clay formations are called **quick clays** because they spontaneously change from a relatively solid mass to a near-liquid condition as the result of a sudden disturbance or shock.

In subarctic regions and and at high latitudes, mass wasting is often initiated by the heaving action of frozen groundwater. The presence of thawed, water-saturated ground in summer overlying permanently frozen subsoil (*permafrost*) also contributes to mass wasting in such regions. Some geomorphologists assert that, in the subarctic, mass wasting is the single most important means of transport of weathered material.

Although some types of mass wasting are rapid and conspicuous, others are slow and gradual. The principles involved are generally similar, but the extent of the activity and particularly the rate of movement are quite variable. In our consideration here, we proceed from the most rapid to the slowest, discussing the characteristics of each as if they were discrete movements, although in nature the various types often overlap (Figure 15-16).

FALL

The simplest and most obvious form of mass wasting is **fall**, the falling of pieces of rock downslope. When loosened by weathering on a very steep slope, a rock fragment may simply be dislodged and fall, roll, or bounce down to the bottom of that segment of slope. This is a very characteristic event in mountainous areas, particularly as a result of frost. Normally the fragments do not travel far before they become lodged, although the lodging may be unstable and temporary.

Pieces of rock that fall in this fashion are referred to collectively as **talus** or **scree**. Sometimes the fragments accumulate relatively uniformly along the base of the slope, in which case the resultant landform is called a **talus slope** or **talus apron** [Figure 15-17(a)]. More characteristically, however, the dislodged rocks collect in sloping, cone-shaped heaps called **talus cones** [Figure 15-17(b)]. This cone pattern is commonplace because most steep bedrock slopes and cliffs are seamed by vertical ravines and gullies that funnel the falling rock fragments into piles directly beneath the ravines, usually producing a series of talus cones side by side along the base of the slope or cliff. Some falling fragments, especially larger ones with their greater momentum, tumble and roll to the base of the cone. Most of the new talus, however, comes to rest at the upper end of the cone. The cone thereby grows up the mountainside.

The angle of repose for talus is very high, generally about 35° and sometimes as great as 40°. The slope of a talus accumulation is gently concave upward, with the steepest angle near the apex of the cone. New material is frequently added to the top of the cone, where the fragments invariably are in delicate equilibrium, and each new piece that bombards down from above may cause a readjustment, with considerable downhill sliding. The freshest blocks, then, tend to be at the upper end of the cone. There is also a rough sorting of fragments according to size, however, with the larger pieces rolling farther downslope and the smaller bits lodging higher up. With the passage of time, all talus slowly migrates downslope, encouraged by the freeze/thaw action of water in the many open spaces of the cones and aprons. All may be further reduced in size by more weathering.

Figure 15-13 Rocks covered with multi-colored lichens in the Lakes District of England. (Martin Bond/Science Photo Library/Photo Researchers, Inc.)

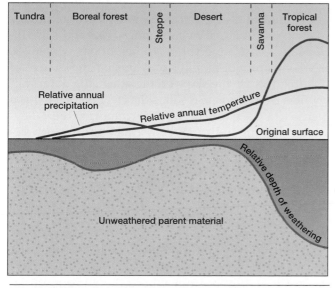

Figure 15-14 Relationship between important climatic elements and depth of weathering. The extreme effectiveness of weathering in the humid tropics is clearly shown. Humid midlatitude regions are omitted from this diagram because the relationships are more complex there.

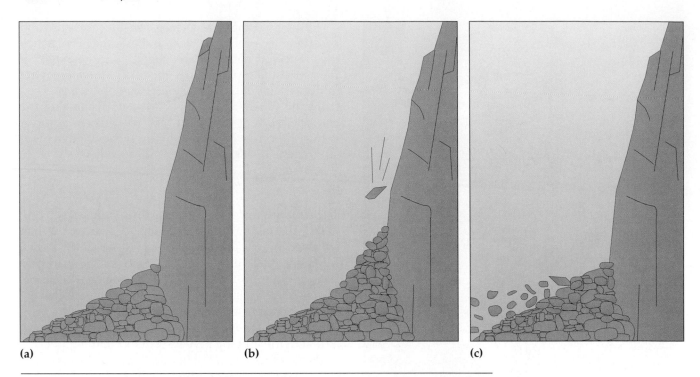

(a) (b) (c)

Figure 15-15 Loose material lies quietly on a slope when the angle of slope is smaller than the angle of repose. On a slope where the angle of slope is larger than the angle of repose, the loose material slides downslope under the pull of gravity. Every type of material has its own angle of repose. Here, for instance, if the material were a grouping of large boulders rather than the small pebbles and cobbles shown, the boulders would be lying still even on the steeper slope.

SLIDE

In mountainous terrain, landslides carry large masses of rock and soil downslope abruptly and often catastrophically. A **landslide** is essentially an instantaneous collapse of a slope and does not necessarily involve the lubricating effects of water or clay. In other words, the sliding material represents a rigid mass that is suddenly displaced without any plastic flow. The presence of water may contribute to the action, however; many slides are triggered by rains that add weight to already overloaded slopes. Landslides may be activated by other stimuli as well, most notably by earth tremors. Slides are also sometimes initiated simply by lateral erosion of a stream that undercuts its bank and thus oversteepens the slope above.

Some slides move only regolith, but many large slides also involve masses of bedrock detached along joint planes and fracture zones. The rapid downslope surge invariably involves the violent disintegration of much of the material, regardless of the size of the blocks originally dislodged.

Landslide action is not only abrupt but also rapid. Precise measurement of the rate of movement is obviously impossible, but eyewitness accounts affirm speeds of 100 miles (160 kilometers) per hour and oc-

casionally more than twice that rate. Thunderous noise accompanies the slide, and the blasts of air that the slide creates can strip leaves, twigs, and even branches from nearby trees.

As most landslides occur in steep, mountainous terrain, the great mass of material (displaced volume is sometimes measured in cubic miles) that roars downslope may choke the valley at the bottom. Moreover, the momentum of the slide may push material several hundred feet up the slope on the other side of the valley. One characteristic result is the creation of a natural dam across the width of the valley, blocking the valley-bottom stream and producing a new lake, which becomes larger and larger until it either overtops the dam or cuts a path through it.

The immediate topographic result of a landslide is threefold:

1. On the hill where the slide originated, there is a deep and extensive scar, usually exposing a mixture of bedrock and scattered debris.
2. In the valley bottom where the slide material comes to rest, there is a massive pile of highly irregular debris, usually in the form of either a broad ridge or low-lying cone. Its surface consists of a

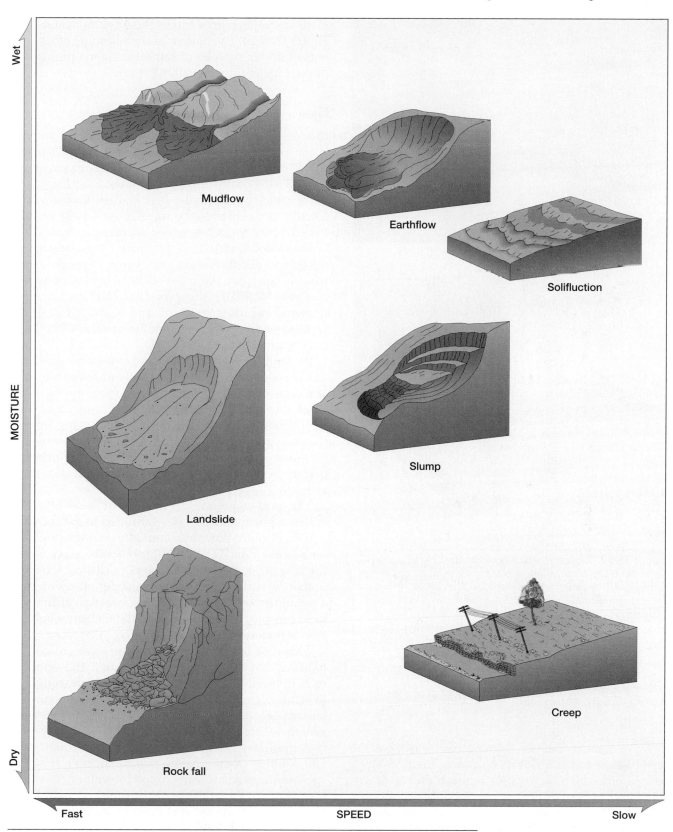

Figure 15-16   Speed and moisture relationships for the various types of mass wasting.

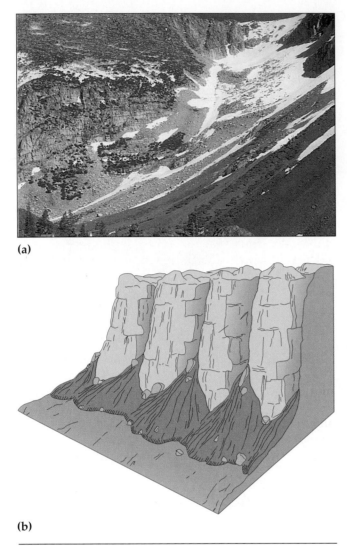

(a)

(b)

Figure 15-17 Talus accumulations at the base of a steep slope: **(a)** If the accumulation is fairly even all along the slope base, it is called a talus slope or talus apron. **(b)** Often the accumulation is uneven, taking the form of individual talus cones.

jumble of unsorted material, ranging from immense boulders to fine dust.

3. On the up-valley side of the debris, a lake may form.

An extremely common form of mass wasting is the type of slide called a **slump**. Slumping involves slope collapse in which the rock or regolith moves downward and at the same time rotates outward along a curved surface that has its concave side facing upward. The upper portion of the moving material tilts down and back, and the lower portion moves up and out. The top of the slump is usually marked by a crescent-shaped scarp face, sometimes with a steplike arrange-

ment of smaller scarps and terraces below, as shown in Figure 15-18(a). The bottom of the slumping block consists of a bulging lobe of saturated debris protruding downslope or into the valley bottom.

FLOW

In a less spectacular form of mass wasting, but one that is conspicuous where it occurs, a sector of a slope becomes unstable, normally owing to the addition of water, and slips gently downhill. In some cases the flow is fairly rapid, but normally it is gradual and sluggish. Usually the center of the mass moves more rapidly than the base and sides, which are retarded by friction.

Many flows are relatively small, often encompassing an area of only a few square yards. More characteristically, however, they cover several tens or hundreds of acres. Normally, they are relatively shallow phenomena, including only soil and regolith, but under certain conditions a considerable amount of bedrock may be involved.

As with other forms of mass movement, gravity is the impelling force. Water is nearly always an important catalyst to the movement; the surface materials become unstable with the added weight of water, and their cohesion is diminished by waterlogging, so that they are more responsive to the pull of gravity. The presence of clay also promotes flow, as clay minerals become very slippery when lubricated with any sort of moisture.

The most common flow movement is **earthflow**, in which a portion of a water-saturated slope moves a limited distance downhill, normally during or right after a heavy rain [Figure 15-18(b)]. At the place where the flow originates, there is usually a distinct interruption in the surface of the preflow slope, either cracks or a prominent oversteepened scarp face. An earthflow is most conspicuous in its lower portion, where a bulging lobe of material pushes out onto the valley floor.

This type of slope failure is relatively common on hillsides that are not densely vegetated and often results in blocked transportation lines (roads, railways) in valley bottoms. Property damage is occasionally extensive, but the rate of movement is usually so sluggish that there is no threat to life.

A **mudflow** originates on slopes in arid and semiarid country when a heavy rain following a long dry spell produces a cascading runoff too voluminous to be absorbed into the soil. Fine debris is picked up from the hillsides by the runoff and concentrated in the valley bottoms, where it flows down-valley as a viscous mass. The leading edge of the mudflow continues to accumulate load, becoming increasingly stiff and retarding the flow of the more liquid upstream portions, so that the

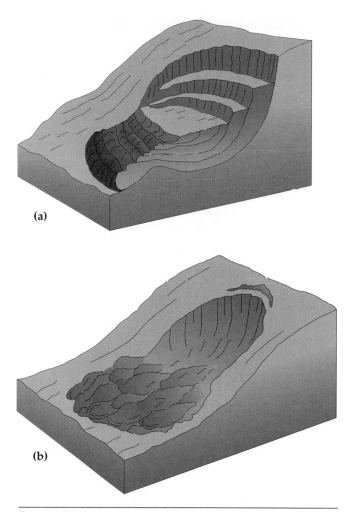

(a)

(b)

Figure 15-18 Comparison of **(a)** slump and **(b)** earthflow.

entire mudflow moves haltingly down the valley. When a mudflow reaches the mouth of the valley and abruptly leaves its confining walls for the more open slopes of the piedmont zone, the pent-up liquid behind the glutinous leading edge breaks through with a rush, spreading muddy debris into a wide sheet. Mudflows often pick up large rocks, including huge boulders, and carry them along as part of their load. In some cases, the large pieces are so numerous that the term **debris flow** is used in preference to mudflow.

An important distinction between earthflow and mudflow is that the latter moves along the surface of a slope and down established drainage channels, whereas the former involves a slope collapse and has no relationship to the drainage network. Moreover, mudflows are normally much more rapid, with a rate of movement intermediate between the sluggish surge of an earthflow and the rapid flow of a stream of water. Mudflows are potentially (and actually) more dangerous to humans than earthflows because of the more rapid movement, the larger quantity of debris in-

volved, and the fact that the mudflow often discharges abruptly across a piedmont zone, which is likely to be a favored area for human settlement and intensive agriculture.

In some rugged mountain areas, talus accumulates in great masses, and these masses may move slowly downslope under their own weight. As we shall see in Chapter 20, glaciers move in this same way, and for this reason these extremely slow flows of talus are called **rock glaciers.** The flow in rock glaciers is caused primarily by the pull of gravity, aided by freeze/thaw temperature changes, and may be largely independent of any lubricating effect of water. Rock glaciers occur primarily in glacial environments; many are found in midlatitude mountains that are relicts of periods of glaciation. They are normally found on relatively steep slopes but sometimes extend far down-valley and even out onto an adjacent plain.

CREEP

The slowest and least perceptible form of mass wasting, **creep,** consists of a very gradual downhill movement of soil and regolith so unobtrusive that normally it can be recognized only by indirect evidence. Generally the entire slope is involved. Creep is such a pervasive phenomenon that it occurs all over the world on sloping land. Although most notable on steep, lightly vegetated slopes, it also occurs on gentle slopes that have a dense plant cover. Wherever weathered materials are available for movement on land that is not flat, creep is a persistent form of mass wasting.

Creep is universal. Infinite numbers of tiny bits of lithospheric material, as well as many larger pieces, march slowly and sporadically downslope from the places where they were produced by weathering or deposited after erosion. When free water is present in the surface material, creep is usually accelerated because the lubricating effect allows individual particles to move more easily and because water adds to the weight of the mass.

Creep is caused by the interaction of various factors, the most significant being alternation of freeze/thaw and wet/dry conditions. When water in the soil freezes, soil particles tend to be displaced upward and in the direction perpendicular to the ground surface, as Figure 15-19 shows. After thawing, however, the particles settle downward, not directly into their original position but rather pulled slightly downslope by gravity. With countless repetitions, this process can result in downhill movement of the entire slope.

Many other agents also contribute to creep. Indeed, any activity that disturbs soil and regolith on a sloping surface is a contributor because gravity affects every

PEOPLE AND THE ENVIRONMENT

LANDSLIDE CATASTROPHES

The spectacular and destructive nature of large landslides makes them a matter of great concern, on a par with hurricanes, tornadoes, tsunamis, volcanic eruptions, and earthquakes. The danger to human life and property is mostly concentrated in the immediate path of the slide, although subsequent flooding from naturally dammed streams may wreak more extensive havoc.

Countless examples of catastrophic landslides come from many parts of the world, particularly in Asia, along the Andes Mountains in South America, and in various parts of Europe. Presented here are descriptions of three that have occurred in North America in the twentieth century (Figure 15-2-A).

Frank, Alberta

In the spring of 1903, an enormous mass of limestone abruptly broke loose from the side of Turtle Mountain, which towers 3000 feet (910 meters) above the valley of the Crow's Nest River in the eastern foothills of the Canadian Rockies in southwestern Alberta. The land is structurally unstable because, during the moun-

Figure 15-2-A The location of three disastrous landslides.

tain-building epoch, massive limestone blocks were thrust-faulted over weaker rocks and the lower part of the limestone was shattered by the movements along the fault zone, producing an area of weakness beneath more cohesive rock. Moreover, a set of large joints was inclined downward toward the valley.

Although earthquakes in the area in 1901 had created no obvious problems, they may have contributed to slope instability. The landslide occurred without warning, and it moved 40 million cubic yards (30 mil-

lion cubic meters) of bedrock and overburden down the side of Turtle Mountain. The leading edge advanced more than 2 miles (3.2 kilometers) across the valley and climbed 400 feet (120 meters) up the opposite slope (Figure 15-2-B). The small coal-mining town of Frank, in the valley bottom at the northern edge of the slide, was partly buried by the debris, with the loss of 70 lives in approximately 2 minutes.

Gros Ventre, Wyoming

In the foothills of the Gros Ventre Mountains south of Yellowstone Park in northwestern Wyoming, a large landslide broke loose in June 1925. It roared across the valley of the Gros Ventre River and climbed more than 300 feet (90 meters) up the opposite slope. No people were in the area at the time, and thus the only known fatalities were six head of cattle. The debris formed a natural dam that blocked the river and created a lake 5 miles (8 kilometers) long. Two years later, the lake waters suddenly breached the dam and a wall of water and debris flooded down the valley, washing

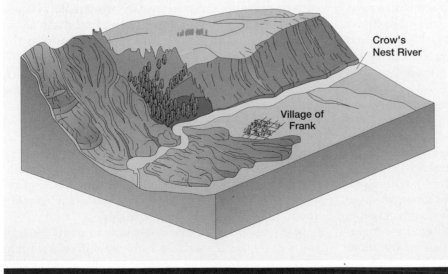

Figure 15-2-B The Frank (Alberta) slide. The mountain slope at the upper left broke loose and roared down into the valley, partly burying the town of Frank. The river was temporarily blocked by the slide, although it soon reestablished its course.

Figure 15-2-C The slope collapse and landslide that dammed the valley of the Madison River. This three-stage sketch looks down-valley.

away the village of Kelly and drowning seven people.

Madison Valley, Montana

Just before midnight in the late summer of 1959, the geologically unstable area of Yellowstone Park was rocked by a major earthquake (7.8 on the Richter scale). Of the many striking geomorphic results of the shock, by far the most conspicuous and tragic took place in the narrow valley of the Madison River a few miles west of the park in Montana. Hebgen Lake, a reservoir in the valley 7 miles (11 kilometers) long, experienced a series of *seiches* (abrupt wave actions in which sudden oscillation of the surface causes a surge of water to flow across the lake). An enormous quantity of water was hurled over the dam as if sloshed out of a dishpan, and it swept down the valley as a flash flood (Figure 15-2-C).

This wall of water sped 7 miles (11 kilometers) downstream, almost to the point where the constricted canyon opened out into rolling wheat and grazing land. Just as the surge approached the canyon mouth, a resistant buttress of sedimentary rock on the steep south slope gave way, releasing an enormous mass of unstable rock that had been retained above and behind it. Some 43 million cubic yards (33 million cubic meters) of debris cascaded down the slope, filled the valley bottom, spurted up the opposite hillside, and sprawled up and down the canyon for 1.5 miles (2.4 kilometers). A U.S. Forest Service campground, crowded with sleeping vacationers, was partly buried by the slide.

The slide dammed the river and forced the surging water back into the unburied portion of the campground. Thus the campers who had escaped being crushed were further endangered by the force of the ricocheting water and then the inexorable flooding as a new lake formed behind the natural dam. The final death toll from this tragic episode will never be known. Nine bodies were found, and indirect evidence (mostly reports from families and friends of people thought to be in the area at the time) indicated that at least 19 other people were buried in the slide.

An immediate result of the cataclysm was the creation of a continu-

(Continued)

LANDSLIDE CATASTROPHES

Continued

ously enlarging lake, called Quake Lake, upstream from the blocking debris (Figure 15-2-D). Not only did this impoundment produce pressures that might rupture the natural dam, but its upstream end would eventually lap against the foundations of Hebgen Dam, quite possibly undermining it. The down-valley area, which includes three towns, thus faced the threat of abrupt flooding from two broken dams, and memories of the postlandslide Gros Ventre tragedy were much in the minds of officials. Consequently, nearly $2 million was spent in the next 10 weeks to construct a deep channel across the top of the slide, allowing the Madison River to resume its normal flow regime.

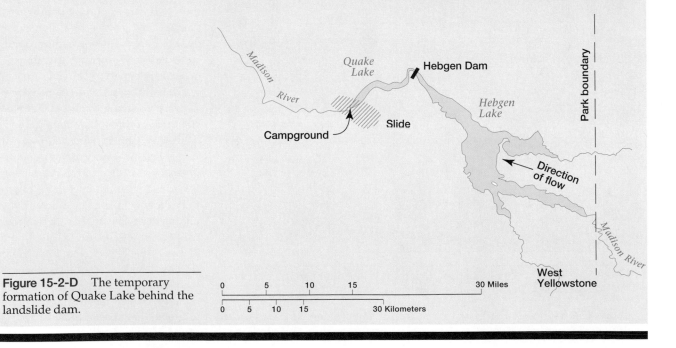

Figure 15-2-D The temporary formation of Quake Lake behind the landslide dam.

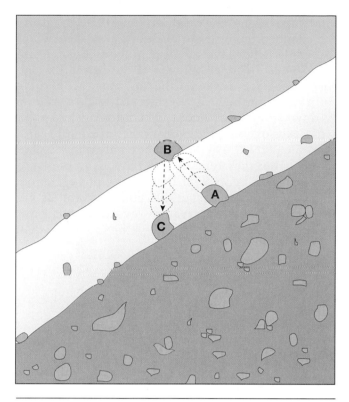

Figure 15-19 The movement of a typical rock particle in a freeze/thaw situation. Freezing lifts the particle perpendicular to the slope (from A to B); upon thawing, the particle settles slightly downslope (from B to C).

rearrangement of particles, attracting them downslope. For example, burrowing animals pile most of their excavated material downslope, and subsequent infilling of burrows is mostly by material from the upslope side. As plant roots grow, they also tend to displace particles downslope. Animals that walk on the surface exert a downslope movement as well. Even the shaking of earthquakes or thunder produces disturbances that stimulate creep.

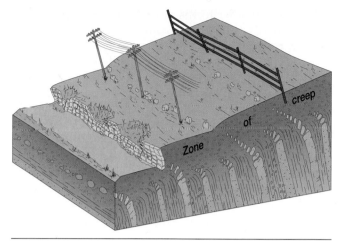

Figure 15-20 Visual evidence of soil creep: displacement and/or bending of fences, utility poles, and retaining walls.

Whenever it occurs, creep is a very slow process, but its rate of movement is faster under some circumstances than others. The principal variables are slope angle, vegetative cover, and moisture supply. Creep operates faster on steep slopes than on gentle ones, for obvious reasons. Deep-rooted and dense-growing vegetation inhibits creep because the roots bind the soil. Also, as mentioned previously, creep is generally faster on water-saturated slopes than on dry ones. Although extreme creep rates of up to 2 inches (5 centimeters) per year have been reported, much more common are rates of just a fraction of a centimeter per year.

Whatever the creep rate, it is much too slow for the eye to perceive and the results are all but invisible. Creep is usually recognized only through the displacement of human-built structures—most commonly when fence posts and utility poles are tilted downhill. Retaining walls may be broken or displaced, and even roadbeds may be disturbed (Figure 15-20).

Figure 15-21 A hillside laced with terracettes near Palmerston North, in New Island, New Zealand. Heavy use of the slope by sheep accentuates the ridges. (TLM photo)

Unlike the other forms of mass wasting, creep produces few distinctive landforms. Rather it induces an imperceptible diminishing of slope angles and gradual lowering of hilltops—in other words, a widespread but minute smoothing of the land surface.

Under certain conditions, and usually on steep, grassy slopes, creep can produce a complicated terracing effect that resembles a network of faint trails. The individual ridges are called **terracettes**. Grazing animals tend to walk along the terracettes, accentuating their outlines until the entire hillside is covered with a maze of them (Figure 15-21).

A special form of creep that produces a distinctive surface appearance is **solifluction** (meaning "soil flowage"), a process largely restricted to tundra landscape beyond the tree line (Figure 15-22). During the summer, the near-surface portion of the ground (called the active layer) thaws but the meltwater cannot percolate deeper because of the permafrost below. The spaces between the soil particles become saturated, and the heavy surface material sags slowly downslope. Movement is erratic and irregular, with lobes overlapping one another in a haphazard, fishscale pattern. The lobes move only a few inches per year, but they remain very obvious in the landscape, in part because of the scarcity of vegetation. Where solifluction occurs, drainage channels are usually scarce because water flow during the short summer is mostly lateral through the soil rather than across the surface.

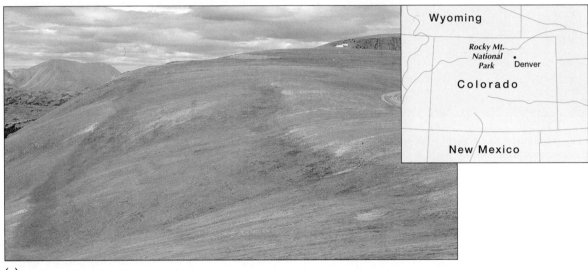

(a)

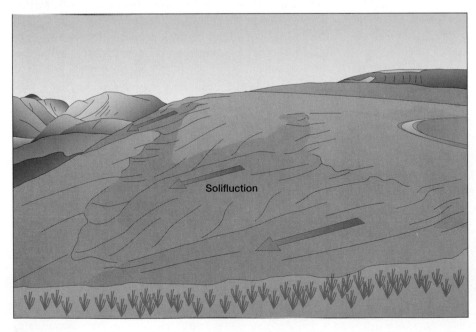

(b)

Figure 15-22   Solifluction in the high country. The surface has slipped gradually downslope about 200 feet (60 meters) in some places here on Trail Ridge in north-central Colorado. (TLM photo)

CHAPTER SUMMARY

The roles of weathering and mass wasting in the denudation of continental surfaces are fundamental but preliminary. The effects of these two processes are sometimes distinctive and obvious but more often unobtrusive and inconspicuous.

Weathering loosens surface and near-surface material in bedrock and makes it less coherent. Mechanical weathering is the physical disintegration of rock material—making little rocks out of big ones. It is accomplished particularly by frost wedging resulting from temperature fluctuations. Chemical weathering

is the decomposition of rock by alteration of its minerals, usually in the presence of moisture. The most significant chemical weathering processes are oxidation, hydrolysis, and carbonation.

Mass wasting is the movement of weathered material—soil, regolith, and rock particles—a relatively short distance downslope under the influence of gravity. It includes rapid movements such as fall, slow movements such as creep, and movements of intermediate speed such as slide and flow.

KEY TERMS

angle of repose
biological weathering
carbonation
chemical weathering
creep
debris flow
denudation
earthflow
exfoliation
fall
fault

frost shattering
frost wedging
hydrolysis
joint
landslide
mass wasting
master joint
mechanical weathering
mudflow
oxidation
quick clay

rock glacier
salt wedging
scree
slump
solifluction
talus
talus apron
talus cone
talus slope
terracette
weathering

REVIEW QUESTIONS

1. Distinguish among weathering, mass wasting, and erosion.
2. Explain how it is possible for weathering to take place beneath the surface of bedrock.
3. What is the difference between a joint and a fault?
4. Explain the mechanics of frost wedging.
5. Why is chemical weathering more effective in humid than in arid climates?
6. What is the relationship between oxidation and rusting?

7. What is the relationship between gravity and mass wasting?
8. How does the angle of repose affect mass wasting?
9. What is the role of clay in mass wasting?
10. What is the difference between earthflow and mudflow?
11. Explain soil creep.
12. In what ways does moisture expedite mass wasting?

SOME USEFUL REFERENCES

BRABB, E.E., AND B. L. HARROD (eds.), *Landslides: Extent and Economic Significance.* Rotterdam: Balkema, 1989.

CROZIER, M.J., *Landslides: Causes, Consequences, and Environment.* London: Croom Helm, 1986.

OLLIER, CLIFF, *Weathering,* 2nd ed. London: Longman, 1984.

SELBY, M.J., *Hillslope Materials and Processes.* New York: Oxford University Press, 1983.

16

THE FLUVIAL
PROCESSES

LAND IS SHAPED BY A VARIETY OF AGENTS FUNCTIONING in concert. In Chapter 14 we considered internal land-shaping agents, and in Chapter 15 we studied the external processes called weathering and mass wasting. In Chapters 16 through 20 we concentrate on the external land-shaping agents that work by erosion and deposition: running water, wind, coastal waters, subsurface waters, and moving ice. By far the most important of these external agents is running water moving over the land, the topic of this chapter.

Running water probably contributes more to shaping landforms than all other external agents combined. This is true not because running water is more forceful than other agents (moving ice and pounding waves often apply much greater amounts of energy per unit area) but rather because running water is ubiquitous: it exists everywhere except in Antarctica. (Wind is ubiquitous, too, but trivial in its power as a terrain sculptor.)

Fluvial processes, defined as those that involve running water, encompass both the unchanneled downslope movement of surface water, called **overland flow,** and the channeled movement of water along a valley bottom, or **streamflow**.

THE IMPACT OF FLUVIAL PROCESSES ON THE LANDSCAPE

Moving water is so widespread and so effective as an agent of erosion and deposition that its impact on the landscape is usually prominent, if not dominant. With few exceptions, the shapes of valleys are either determined or significantly modified by the water that runs over them (Figure 16-1). Areas above the valleys are less affected by running water, but even there overland flow may significantly influence the shape of the land.

The basic landscape-sculpturing effect of running water is to smooth irregularities—in simplest terms, to wear down the hills by erosion and fill up the valleys by deposition. Before achieving these results, however, fluvial action often creates deep canyons and steep slopes that persist for very long periods of time.

SOME FUNDAMENTAL CONCEPTS

Before we get into the specifics of fluvial processes and their topographic results, we need to clarify a few basic concepts.

VALLEYS AND INTERFLUVES

All the surfaces of the continents can be considered to consist of two topographic elements—valleys and interfluves. The distinction between the two is not always obvious in nature, but the conceptual demarcation is clear. A **valley** is that portion of the terrain in which a drainage system is clearly established (Figure 16-2). It includes the valley bottom, which is the lower, flatter area that is partially or totally occupied by the channel of a stream, as well as the valley sides, the steeper slopes that rise above the valley bottom on either side. Some valley bottoms are narrow, elongated, and limited in area, while others are broad and extensive. Valley

Figure 16-1 The erosive work of some rivers is very conspicuous. This gorge carved out by the Green River in northeastern Utah is called Whirlpool Canyon. (TLM photo)

The upper reaches of the Kobuk River in Alaska's Brooks Range. *(Michael Male/Photo Researchers, Inc.)*

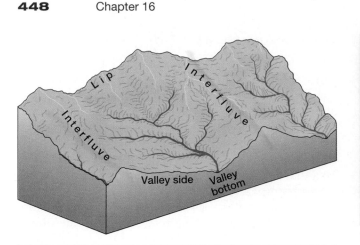

Figure 16-2   Valleys and interfluves. Valleys normally have clear-cut drainage systems; interfluves do not.

Figure 16-3   The awesome impact of a raindrop produces splash erosion. (U. S. Department of Agriculture)

sides are also variable, with slopes that are steep or gentle and an extent that is limited or expansive. The upper limit of a valley is not always readily apparent, but it can be clearly conceptualized as a lip or rim at the top of the valley sides above which drainage channels are either absent or indistinct.

An **interfluve** (from Latin: *inter*, between and *fluvia*, rivers) is the higher land above the valley sides that separates adjacent valleys. Some interfluves consist of ridgetops or mountain crests with precipitous slopes, but others are simply broad and flattish divides between drainage systems. Conceptually, all parts of the terrain not in a valley are part of an interfluve.

These simplistic definitions are not always applicable in nature, as some terrain elements defy classification. Swamps and marshes, for example, may be on interfluves but are usually in valleys, although with no clearly established drainage system. Such exceptional cases, however, should not inhibit our acceptance of the valley/interfluve concept.

EROSION AND DEPOSITION

All external forces remove fragments of bedrock, regolith, and soil from their original positions (*erosion*), transport them to another location, and set them down (*deposition*). The fluvial processes, our concern in this chapter, produce one set of landforms by erosion and another quite different set by deposition.

Erosion by Overland Flow   On the interfluve, fluvial erosion begins when rain starts to fall (Figure 16-3). Unless the impact of rain is absorbed by vegetation or some other protective covering, the collision of raindrops with the ground is strong enough to blast fine soil particles upward and outward, shifting them a few millimeters laterally [Figure 16-4(a)]. On sloping ground, most particles move downhill by this *splash erosion* [Figure 16-4(b)].

Figure 16-4   Comparison of the effect of splash erosion on flat and sloping ground. **(a)** On level ground, all particles are displaced equal distances, with the result that there is no net (overall) movement. **(b)** On sloping ground, the general movement of soil particles dislodged by splash erosion is downslope because the downslope splash moves particles a greater distance than does the upslope splash.

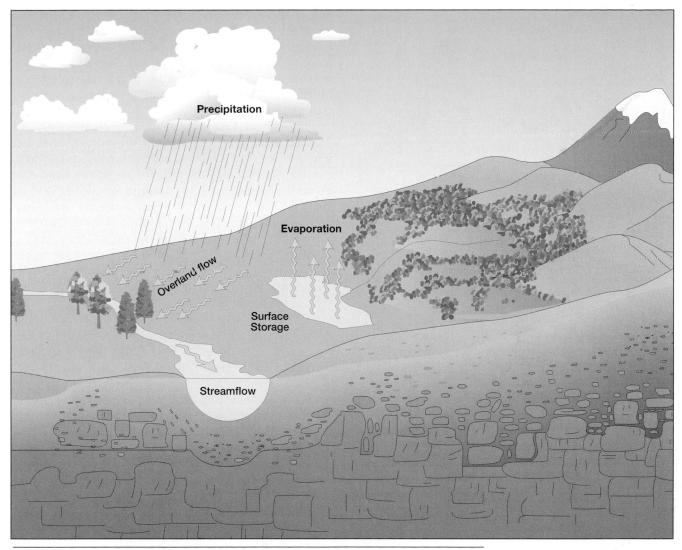

Figure 16-5   The difference between overland flow and streamflow.

In the first few minutes of a rain, much of the water infiltrates the soil and consequently there is little runoff. During heavy or continued rain, however, particularly if the land is sloping and has only a sparse vegetative cover, infiltration is greatly diminished and most of the water moves downslope as overland flow (Figure 16-5). The water flows across the surface as a thin sheet, transporting material already loosened by splash erosion, in a process termed *sheet erosion* (Figure 16-6). As overland flow moves downslope and its volume increases, the resulting turbulence tends to break up the sheet flow into multitudinous tiny channels called rills. This more-concentrated flow loosens additional material and scores the slope with numerous parallel seams; this sequence of events is termed *rill erosion*. If the process continues, the rills begin to coalesce into fewer and larger channels called gullies, and *gully erosion* becomes recognizable. As the gullies get larger and larger, they tend to become incorporated

into the drainage system of the adjacent valley, and the flow changes from overland flow to streamflow.

Erosion by Streamflow   Once surface flow is channeled, its ability to erode is greatly increased by the increased volume of water. Erosion is accomplished in part by the direct hydraulic power of the moving water, which can excavate and transport material at the bottom and sides of the stream. Banks can also be undermined by streamflow, particularly at times of high water, dumping more loose material into the water to be swept downstream.

The erosive capability of streamflow is also significantly enhanced by the abrasive tools it picks up and carries along with it. All sizes of rock fragments, from silt to boulders, chip and grind the stream bed as they travel downstream in the moving water. These rock fragments break off more fragments from the bottom and sides of the channel, and they collide with one an-

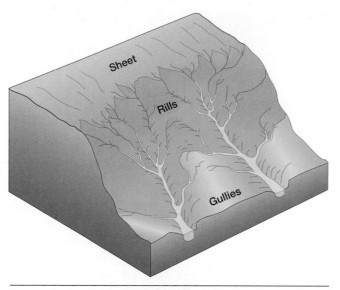

Figure 16-6 Sheet, rill, and gully erosion.

other, becoming both smaller and rounder from the wear and tear of the frequent collisions (Figure 16-7). The eventual result of this *abrasion* is to reduce almost all stream-carried debris to very small silt particles.

A certain amount of chemical action also accompanies streamflow, and some chemical weathering processes—particularly solution action and hydrolysis—also help erode the stream channel by *corrosion*.

The erosive effectiveness of streamflow varies enormously from one situation to another, determined primarily by the speed and turbulence of the flow, on the one hand, and by the resistance of the bedrock, on the other. Flow speed is governed by the gradient (slope angle) of the streambed (the steeper the gradient, the faster the flow) and by the volume of flow (more water normally means higher speed). The degree of turbulence is determined in part by the flow speed (faster flows are more turbulent) and in part by the roughness of the stream channel (an irregular channel surface increases turbulence).

Transportation   Any water moving downslope, whether moving as overland flow or as streamflow, can transport rock material. At any given time and place, the load carried by overland flow is likely to be small in comparison with what a stream can transport. In total, however, the amount of material transported by overland flow is incredibly large. Eventually, most of this great mass of material reaches the streams in the valley bottoms, where it is added to the stream-eroded debris to constitute the **stream load**.

Streams effectively sort the debris they transport because finer, lighter material moves more rapidly than coarser, heavier material. Essentially, the stream load contains three factions (Figure 16-8):

1. Some minerals, mostly salts, are dissolved in the water and carried in solution as the **dissolved load**.
2. Very fine particles of clay and silt are carried in suspension, moving along with the water without ever touching the streambed. These tiny particles, called the **suspended load,** have a very slow settling speed, even in still water [fine clay may require as much as a year to sink 100 feet (30 meters) in perfectly quiet water].

Figure 16-7 Rock particles transported by streams are inevitably rounded by frequent collisions. These stream-worn rocks are in the bed of the Aluna River in the Southern Highlands of Papua New Guinea. (TLM photo)

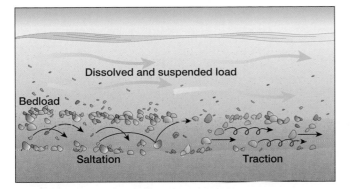

Figure 16-8   A stream moves its load in three ways. The dissolved and suspended loads are carried in the general water flow. The bedload is moved by traction (dragging) and saltation (bouncing).

3. Sand, gravel, and larger rock fragments constitute the **bedload**. The smaller particles are moved along with the general streamflow in a series of jumps or bounces collectively referred to as **saltation**. Coarser pieces are moved by **traction**, which is defined as rolling or sliding along the streambed.

As a general rule, most of the material transported by a stream is in the suspended load and the least amount is in the bedload. The load is normally moved spasmodically: debris is transported some distance, dropped, then picked up later and carried farther.

Geomorphologists employ two concepts—competence and capacity—in describing the load a stream can transport. **Competence** is a measure of the particle size a stream can transport, expressed by the diameter of the largest particles that can be moved. Competence depends mainly on flow speed, with the size of the largest particle that can be moved varying with the sixth power of the water speed. In other words, if the flow speed is doubled, the size of the largest movable particle is increased 64-fold (2^6). Thus a stream that normally can transport only sand grains might easily be able to move large boulders during a flood.

Capacity is a measure of the amount of solid material a stream has the potential to transport, normally expressed as the volume of material passing a given point in the stream channel during a given time interval. Capacity may vary tremendously over time, depending mostly on fluctuations in volume and flow speed but also on the characteristics of the load (particularly the mix of coarse and fine sediments). It is difficult to overemphasize the significance of the greatly expanded capacity of a stream to transport material at flood time.

Deposition   Whatever is picked up must eventually be set down, which means that erosion is inevitably followed by deposition. Moving water, whether moving as overland flow or as streamflow, carries its load downslope or down-valley toward an ultimate destination—either ocean, lake, or interior drainage basin. A large volume of fast-moving water can carry its debris a great distance, but sooner or later deposition will take place as either flow speed or water volume decreases. Diminished flow is often the result of a change in gradient, but it may also occur when a channel either widens or changes direction. Therefore stream deposits are found at the mouths of canyons, on floodplains, and at riverbends. Eventually, however, most waterborne debris is dumped by moving water (stream or river) into quiet water (ocean or lake).

The general term for stream-deposited debris is **alluvium**; it is characterized by a sorting of particles on the basis of size.

TIME AND THE RIVER

Most of us have stood on a bridge or on the brink of a canyon looking at a small stream far below and wondered how that meager flow could have carved the enormous gorge we see spread beneath us. On such occasions, we might be tempted to discount uniformitarianism and consider some brief, catastrophic origin as a more logical linking of what we can see with what we can imagine. Such reasoning, however, is erroneous, for uniformitarianism easily accounts for the existence of tiny streams in extensive valleys by invoking two geomorphic principles: (1) the extraordinary length of geologic time and (2) the remarkable effectiveness of floods.

We have already discussed the incredible expanse of time involved in the evolution of topography (Chapter 13). This vast temporal sweep provides the opportunity for countless repetitions of an action, and the repetitive movement of even a tiny flow of water can wear away the strongest rock.

Of perhaps equal importance is that the amount of water flowing is not always tiny. Most streams have very erratic regimes, with great fluctuation in **discharge**, or volume of flow. Most of the world's streams carry a relatively small amount of water during most of the year and a relatively large volume during floodtime. In some streams, flood flow occurs for several weeks or even months; but in the vast majority, the duration of high-water flow is much more restricted. In many cases, the "wet season" may be only 1 or 2 days a year. Yet the amount of denudation that can be accomplished during high-water flow is supremely greater than that done during periods of normal discharge. The epic work of streams—the carving of great valleys, the forming of vast floodplains—is primarily accomplished by flood flows.

STREAM CHANNELS

Overland flow is a relatively simple process. It is affected by such factors as rainfall intensity and duration, vegetative cover, surface characteristics, and slope shape, but its general characteristics are straightforward and easily understood. Streamflow is much more complicated, in part because streams represent not only a process of denudation but also an element of the landscape—an active force as well as an object of study.

CHANNEL FLOW

A basic characteristic of streamflow, and one that distinguishes it from the randomness of overland flow, is that it is normally confined to channels, which gives it a three-dimensional nature with scope for considerable complexity. In any channel having even a slight gradient, gravitational pull overcomes friction forces to move the water down-channel. Except under unusual circumstances, however, this movement is not straight, smooth, and regular. Rather it tends to be unsystematic and irregular, with many directional components and with different speeds in different parts of the channel.

A principal cause of flow irregularity is the retarding effect of friction along the bottom and sides of the channel, which causes the water to move most slowly there and fastest in the center of the stream (Figure 16-9). The amount of friction is determined by the width and depth of the channel and the roughness of its surface. A narrow, shallow channel with a rough bottom has a much greater retarding effect on streamflow than a wide, deep, smooth-bottomed one. One effect of friction is that it uses up much of the stream's energy, decreasing the amount available for erosion and transportation.

TURBULENCE

In turbulent flow, the general downstream movement is interrupted by irregularities in the direction and speed of the water. Such irregularities produce momentary currents that can move in any direction, including upward (Figure 16-10). Turbulence in streamflow is caused partly by friction, partly by internal shearing stresses between currents within the flow, and partly by surface irregularities in the channel. Stream speed also contributes to the development of turbulence, with faster streams more turbulent than slow-moving ones.

Eddies and whirlpools are conspicuous results of turbulence, as is the roiling whitewater of rapids. Even streams that appear very placid and smooth on the surface, however, are often turbulent at lower levels.

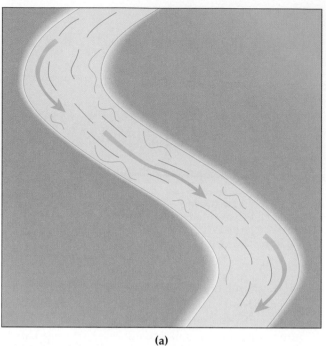

(a)

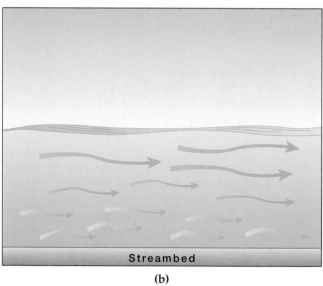

Streambed

(b)

Figure 16-9 Friction forces between the solid and liquid surfaces cause a stream to flow fastest in the middle and at the surface. **(a)** Overhead view. The water speed is proportional to arrow length, fastest in the center of the stream and slowest near the two banks. **(b)** Cross section through center of stream. The arrow lengths indicate that water speed is greatest at the surface, where the water is farthest from any solid surfaces, and least at the bottom, where the water drags along the streambed.

Turbulent flow creates a great deal of frictional stress as the numerous internal currents interfere with one another. This stress dissipates much of the stream's energy, decreasing the amount available to erode the channel and transport sediment. On the other hand, turbulence contributes to erosion by creating flow patterns that pry and lift rock materials from the streambed.

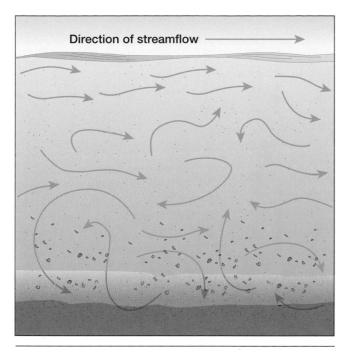

Figure 16-10 Turbulent flow is characterized by movement in all directions, a condition that allows sediment to be picked up and carried within the flow.

CHANNEL CHANGES

Nearly every stream continuously rearranges its sediment in response to variations in flow speed and volume. During high-water periods, when flow is fast and voluminous, the stream scours its bed by detaching particles from it and shifting most or all sediment downstream. During low-water periods, the flow is slowed and sediment is more likely to settle to the bottom, which results in filling of the channel (Figure 16-11).

Irregularities in streamflow are manifested in various ways, but perhaps most conspicuously by variations in channel patterns. If streamflow were smooth and regular, one might expect stream channels to be straight and direct. Few natural stream channels are straight and uniform for any appreciable distance, however. Instead, they wind about to a greater or lesser extent, sometimes developing remarkable sinuosity. In some instances, this winding is a response to the underlying geologic structure; but even in areas of perfectly uniform structure, stream channels are normally not straight. The mechanics of this winding habit are not fully understood and are beyond our concern here, although we discuss some aspects later in this chapter.

Stream channel patterns are generally grouped into four categories: straight, sinuous, meandering, and braided, each with variations. **Straight channels** are short, uncommon, and usually indicative of strong control by the underlying geologic structure. A straight channel does not necessarily mean straight flow, however. A line running in the direction of the water and through the deepest parts of the channel, called the **thalweg** (German: *Thal*, valley, and *Weg*, way), rarely follows a straight path midway between the stream banks; rather, it wanders back and forth across the channel, as Figure 16-12(a) shows. Opposite the place where the thalweg approaches one bank, a deposit of alluvium is likely to be found. Thus straight channels are likely to have many of the characteristics of sinuous channels.

Sinuous channels are much more common than straight ones. They are winding channels and are found in almost every type of topographic setting. Their curvature is usually gentle and irregular.

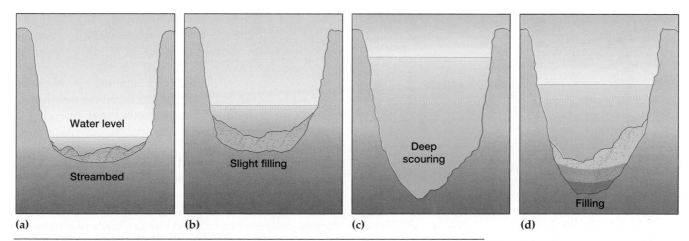

(a) (b) (c) (d)

Figure 16-11 Changing channel depth and shape during a flood: **(a)** Water flow is low prior to flooding. **(b)** As the volume of streamflow increases, the streambed is raised slightly by filling. **(c)** Flood flow significantly deepens the channel by scouring. **(d)** As the flood recedes, considerable filling raises the channel bed again.

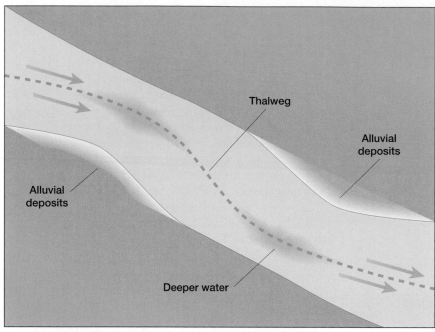

(a)

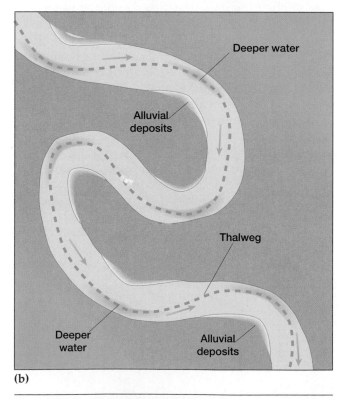

(b)

Figure 16-12 **(a)** Straight and **(b)** meandering channels. Pools of deeper water and alluvial deposits often occur on opposite sides of the channel from one another. Where the channel curves in part (b), there is usually a deeper water pool on the outside of the curve and an alluvial deposit on the inside.

Meandering channels exhibit an extraordinarily intricate pattern of smooth curves in which the stream follows a serpentine course, twisting and contorting and turning back on itself, forming tightly curved loops and then abandoning them, cutting a new and different and equally tortuous course [Figure 16-12(b)]. Meandering generally occurs on flat land and in a stream that has an unhurried flow but is moving with enough force to erode its banks and carry sediment. A meander shifts its location almost continuously. This is accomplished by erosion on the outside of curves and deposition on the inside (Figure 16-13). In this fashion, meanders migrate across the floodplain and also tend to shift down-valley, producing rapid and sometimes abrupt changes in the channel.

The length of a stream's meanders and the radius of its curves are related to flow volume, so that the pattern of meandering rivers is relatively constant, no matter what their size. A meandering river swings back and forth in ever-expanding loops. Eventually, when the radius of a loop reaches about 2.5 times the stream's width, the loop stops growing. Often a loop is short-circuited as the stream cuts a new channel across its neck and starts meandering again. This short-circuiting is shown in Figure 16-13(c), and the cutoff portion of the channel is called an **oxbow lake** because it rounded shape resembles the bow part of yokes used on teams of oxen.

Braided streams consist of a multiplicity of interwoven and interconnected shallow channels separated by low bars or islands of sand, gravel, and other loose debris (Figures 16-14 and 16-15). At any given time, the active channels of a braided stream may cover no more than one-tenth of the width of the entire channel system, but in a single year most or all of the surface sediments may be reworked by the flow of the laterally shifting channels.

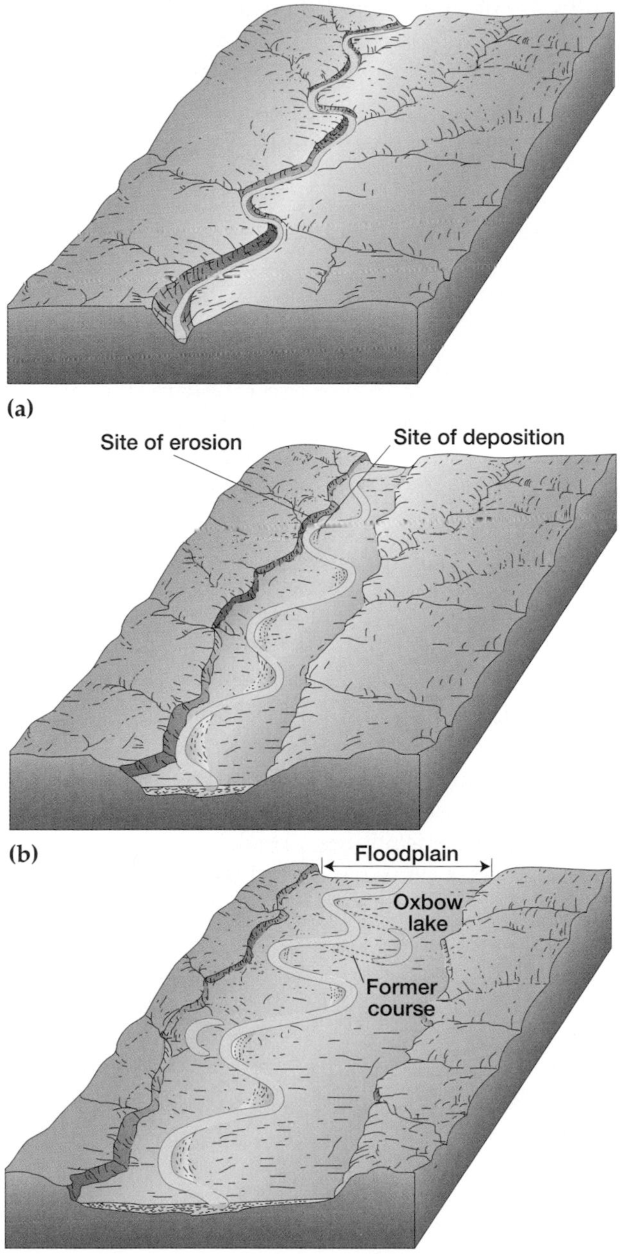

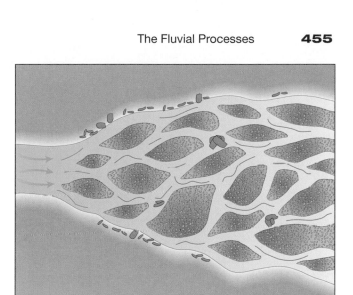

(a)

Site of erosion   Site of deposition

(b)

Floodplain

Oxbow lake

Former course

(c)

Figure 16-13   A meandering stream slowly widens its valley and may eventually create a broad floodplain.

STREAM SYSTEMS

Streams and their valleys dissect the land in a myriad of patterns. However, there are some important threads of similarity, and the study of individual streams can produce additional generalities to aid in our understanding of topographic development.

Figure 16-14   A braided channel develops when bars split the main channel into many smaller channels and thereby greatly widens the stream.

DRAINAGE BASINS

The **drainage basin**, or **watershed**, of a particular stream is all the area that contributes overland flow and groundwater to that stream. In other words, the drainage basin consists of a stream's valley bottom, valley sides, and those portions of the surrounding interfluves that drain toward the valley. Conceptually, the drainage basin is bounded in all directions except down-valley by a **drainage divide**, which is the line of separation between runoff that descends in the direc-

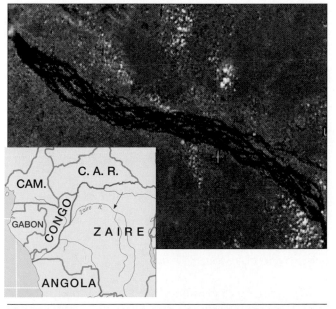

Figure 16-15   The complexly braided channel of the middle course of the Zaire River in equatorial Africa. (EROS Data Center)

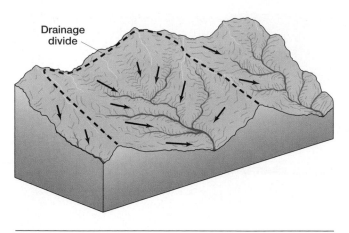

Drainage divide

Figure 16-16   The drainage basins of adjacent streams are separated by drainage divides, which are located on the crests of the surrounding interfluves.

tion of the drainage basin in question and runoff that goes toward an adjacent basin (Figure 16-16).

Every stream of any size has its own drainage basin, but for practical purposes the term is often reserved for major streams. The drainage basin of a principal river encompasses the smaller drainage basins of all its tributaries; consequently, larger basins include a hierarchy of smaller tributary basins (Figure 16-17).

STREAM ORDERS

In every drainage basin, small streams come together to form successively larger ones and small valleys join more extensive ones. This relationship, although vari-

able in detail, holds true for drainage basins of any size. This systematic characteristic makes it possible to recognize a natural organization within a watershed, and the concept of **stream order** has been devised to describe the arrangement (Figure 16-18).

A first-order stream, the smallest unit in the system, is one without tributaries. Where two first-order streams unite, a second-order stream is formed. At the confluence of two second-order streams, a third-order stream begins, and this uniting principle applies through successively higher orders.

Note from Figure 16-18 that the joining of a lower order stream with a higher order stream does not increase the order below that junction. For example, the confluence of a first-order stream and a second-order one does not produce a third-order stream. A third-order stream is formed only by the joining of two second-order streams.

The concept of stream order is more than simply a numbers game, as several significant relationships are involved (Figure 16-19). In a well-developed drainage system, for example, one can predict with some certainty that first-order streams and valleys are more numerous than all others combined and that each succeeding higher order is represented by fewer and fewer streams [Figure 16-19(a)]. Some other predictable relationships are that average stream length increases regularly with increasing order [Figure 16-19(b)], that average watershed area increases regularly with increasing order [Figure 16-19(c)], and that average stream gradient decreases with increasing order.

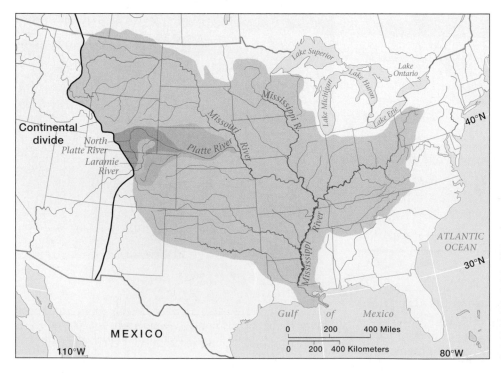

Figure 16-17   A nested hierarchy of drainage basins: the Laramie River flows into the North Platte; the North Platte drains into the Platte; the Platte is a tributary of the Missouri; the Missouri flows into the Mississippi; finally, the Mississippi empties into the Gulf of Mexico.

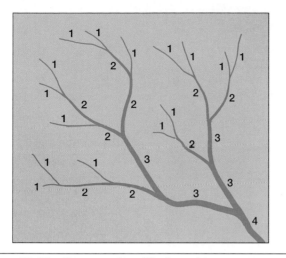

Figure 16-18   Stream orders. The branching component of a stream and its tributaries can be classified into a hierarchy of segments, ranging from smallest to largest (in this case, from 1 to 4).

STRUCTURAL RELATIONSHIPS

Many factors affect stream development, and perhaps the most important is the geologic/topographic structure over or through which the stream must make its way and carve its valley. Each stream faces particular structural obstacles as it seeks the path of least resistance in its descending course to the sea. Most streams respond directly and conspicuously to structural controls, which is to say that their courses are guided and shaped by the nature and arrangement of the underlying bedrock.

PERMANENCE OF FLOW

We tend to think of rivers as permanent, but in fact many of the world's streams do not flow year-round. In humid regions, the large rivers and most tributaries are **perennial streams**—that is, permanent—but in less-well-watered parts of the world, many of the major streams and most tributaries carry water only part of the time, either during the wet season or during and immediately after rains. These impermanent flows are called **ephemeral streams** if they carry water only during and immediately after a rain and **intermittent streams** if they flow for only part of the year, although the term *intermittent* is sometimes applied to both cases (Figure 16-20).

Even in humid regions, many first-order and second-order streams are only intermittent. These are generally short streams with relatively steep gradients and small watersheds. If rain is not frequent, or snowmelt not continuously available, these low-order streams simply run out of water. Higher order streams in the same regions are likely to have per-

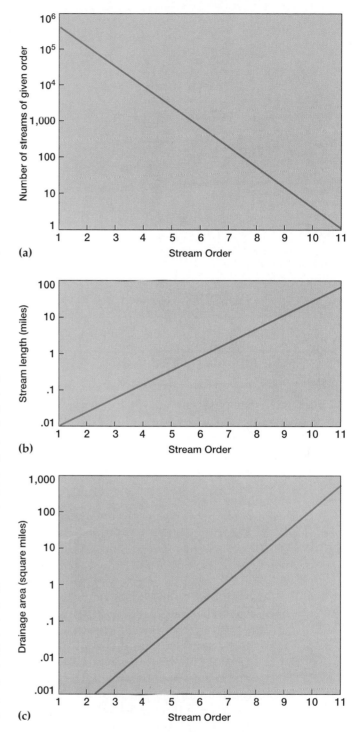

Figure 16-19   Some typical relationships between stream order and other factors: **(a)** first-order streams are the most numerous; **(b)** first-order streams are the shortest; **(c)** first-order streams have the smallest drainage areas.

manent flow because of their larger drainage areas and because previous rainfall or snowmelt that sank into the ground can emerge in the valleys as groundwater runoff long after the rains have ceased.

AUSTRALIA

Wiluna

Figure 16-20   Most desert streams are either intermittent or ephemeral, carrying water only rarely. This dry bed is near Wiluna in Western Australia. (TLM photo)

THE SHAPING AND RESHAPING OF VALLEYS

Running water shapes terrain partly by overland flow on interfluves but mostly by streamflow in the valleys. The shaping of valleys and their almost continuous modification through time produce a changing balance between valleys and interfluves and consequently an ongoing dynamism in the configuration of most parts of the continental surfaces.

A stream excavates its valley by eroding the channel bed. If only downcutting were involved, the resulting valley would be a narrow, steep-sided gorge. Such gorges sometimes occur, but usually other factors are at work also and the result is a wide valley. In either case, there is a lower limit to how much downcutting a stream can do, and this limit is called the base level of the stream. **Base level** is an imaginary surface extending underneath the continents from sea level at the coasts (Figure 16-21). This imaginary surface is not simply a horizontal extension of sea level, however; inland it is gently inclined at a gradient that allows streams to maintain some flow. Sea level, then, is the *ultimate base level*, or lower limit of downcutting.

As Figure 16-21 shows, there are also local, or temporary, base levels, which are lower limits of downcutting imposed on particular streams or sections of streams by structural or drainage conditions. For example, no tributary can cut deeper than its level of confluence with the higher order stream it joins, and so the level of their point of junction is a local base level for the tributary. Similarly, a lake normally serves as the temporary base level for all streams that flow into it.

Some valleys have been downfaulted to elevations below sea level (Death Valley in California is one example), a situation producing a temporary base level lower than the ultimate base level.

The longitudinal profile of a stream (a diagram showing streambed elevation from source to mouth) is ultimately restricted by base level, but the profile at any given time depends upon a variety of factors. The long-term tendency is toward a smooth profile in which all factors are balanced. This hypothetical condition is called a **graded stream**, defined as one in which the gradient just allows the stream to transport its load. A graded stream is more theoretical than actual because equilibrium is so difficult to achieve and so easy to upset.

VALLEY DEEPENING

Wherever it has either a relatively rapid speed or a relatively large volume, a stream expends most of its energy in downcutting. This lowering of the streambed

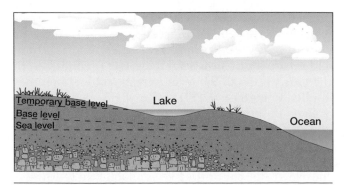

Temporary base level   Lake

Base level
Sea level   Ocean

Figure 16-21   Comparison of sea level, base level, and temporary base level.

involves the hydraulic power of the moving water, the prying and lifting capabilities of turbulent flow, and the abrasive effect of the stream's bedload as it rolls, slides, and bounces along the channel. Downcutting is most frequent in the upper reaches of a stream, where the gradient is usually steep and the valley narrow. The general effect of downcutting is to produce a deep valley with steep sides and a V-shaped cross section.

Waterfalls and rapids are often found in valleys where downcutting is prominent. They occur in steeper sections of the channel, and their faster, more turbulent flow intensifies erosion. These irregularities in the channel are collectively termed **knickpoints**. They may originate in various ways but commonly are the result of abrupt changes in bedrock resistance. The more resistant material inhibits downcutting, and as the water plunges over the waterfall or rapids with accelerated vigor it tends to scour the channel above and along the knickpoint and fill the channel immediately downstream. This intensified action eventually wears away the harder material, so that the knickpoint migrates upstream with a successively lower profile until it finally disappears and the channel gradient is smoothed (Figure 16-22).

The principle of upstream knickpoint migration is important to an understanding of fluvial erosion because it illustrates dramatically the manner in which valley shape often develops first in the lower reaches and then proceeds progressively upstream, even though the water obviously flows in the opposite direction.

VALLEY WIDENING

Where a stream gradient is steep and the channel well above the local base level, downcutting is the dominant activity, and as a result valley widening is likely to be slow. Even at this stage, however, some widening takes place as the combined action of weathering, mass wasting, and overland flow removes material from the valley sides. In the valley bottom, downcutting diminishes with time and eventually ceases as the stream develops a gentle profile. The stream's energy is then increasingly diverted into a meandering flow pattern, the reasons for which are not yet fully understood. As the stream waffles from side to side, *lateral erosion* begins. In essence, this means that the principal current swings from one bank to the other, eroding where the water speed is greatest and depositing where it is least. The water moves fastest on the outside of curves, and there it undercuts the bank; on the inside of a curve, alluvium is likely to accumulate, as was shown in Figure 16-12.

The current often shifts position, so that undercutting is not concentrated in just a few locations. Rather, over a long period of time, most or all parts of the valley sides

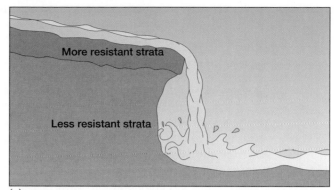

(a)

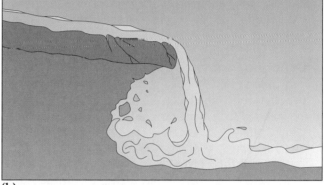

(b)

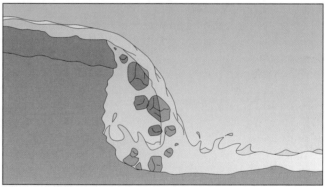

(c)

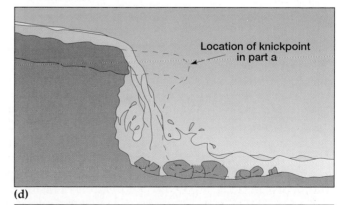

(d)

Figure 16-22 (a) Knickpoint formed where a stream flows over a resistant layer of rock. **(b)** The water flow undercuts the lip and **(c)** causes it to collapse. **(d)** Position of knickpoint has migrated upstream.

FOCUS

DRAINAGE PATTERNS

Stream systems are frequently orderly, forming conspicuous drainage patterns in the landscape. These patterns develop largely in response to the underlying structure and slope of the land surface. Geologic/topographic structure can often be deduced from a drainage pattern, and, conversely, drainage pattern often can be predicted from structure.

Dendritic Pattern

The most common drainage pattern over the world is a treelike, branching one called a **dendritic pattern** (Figure 16-1-A). It consists of a random merging of streams, with tributaries joining larger streams irregularly but always at an angle smaller than 90°. The pattern resembles branches on a tree or veins on a leaf. The relationship between drainage pattern and land structure is negative, which is to say that the underlying structure does not control the evolution of the drainage pattern. Dendritic patterns are more numerous than all others combined and can be found almost anywhere.

Trellis Pattern

A **trellis pattern** usually develops on alternating bands of tilted hard and soft strata, with long, parallel streams linked by short, right-angled segments. Two regions of the United States are particularly noted for their trellis drainage patterns: the ridge-and-valley section of the Appalachian Mountains and the Ouachita Mountains of western Arkansas and southeastern Oklahoma.

In the ridge-and-valley section, which extends northeast-southwest for more than 800 miles (1280 kilometers) from New York to Alabama, the drainage pattern developed in response to tightly folded Paleozoic sedimentary strata forming a world-famous series of parallel ridges and valleys. Parallel streams flow in the valleys between the ridges, with short, right-angled connections here and there cutting through the ridges (Figure 16-1-B).

The marked contrast between trellis and dendritic patterns is shown dramatically by the principal streams of West Virginia (Figure 16-1-C). The folded structures of the eastern part of the state produce trellising, whereas the nearly horizontal strata of the rest of the state are characterized by dendritic patterns.

Radial Pattern

A **radial pattern** usually is found when streams descend from some sort of concentric uplift, such as an

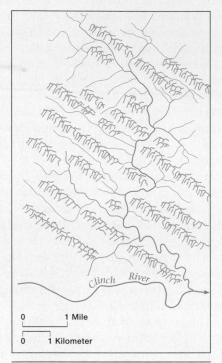

Figure 16-1-B   A trellis drainage pattern in the Norris (Tennessee) topographic quadrangle.

isolated volcano. Figure 16-1-D shows one example: Mount Egmont on the North Island of New Zealand.

Centripetal Pattern

A **centripetal pattern**, essentially the opposite of a radial one, is usually associated with streams converging in a basin. Occasionally, however, centripetal drainage develops on a much grander scale. Shown in Figure 16-1-E is the northeastern part of Australia, where rivers from hundreds of miles away converge toward the Gulf of Carpentaria, a basin partially inundated by the sea.

Annular Pattern

More complex is an **annular pattern**, which can develop either on a

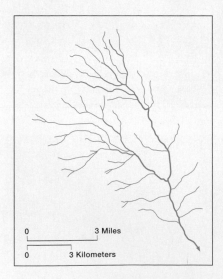

Figure 16-1-A   A dendritic drainage pattern in the Pat O'Hara Mountain (Wyoming) topographic quadrangle.

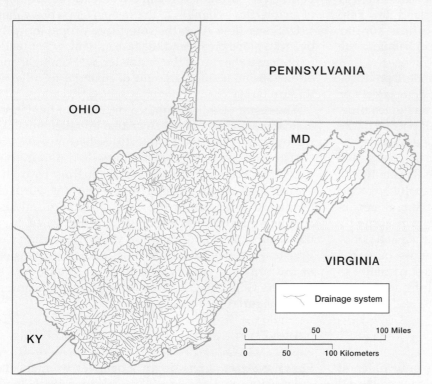

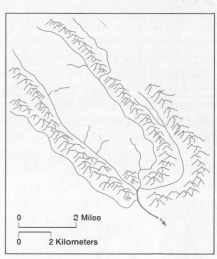

Figure 16-1-F   An annular drainage pattern in the Maverick Spring Dome (Wyoming) topographic quadrangle.

dome or in a basin where dissection has exposed alternating concentric bands of tilted hard and soft rock. The principal streams follow curving courses on the softer material, occasionally breaking through the harder layers in short, right-angled segments. The Maverick Spring Dome of Wyoming portrays a prominent example of annnular drainage (Figure 16-1-F). This dome of ancient crystalline rocks was pushed up through a sedimentary overlay and has been deeply eroded, thus exposing crystallines in the higher part of the hills, with upturned concentric sedimentary ridges (called hogbacks) around the margin. The streams are mostly incised into the softer layers.

Figure 16-1-C   Drainage pattern contrasts in West Virginia. The trellis systems in the east are a response to parallel folding; the dendritic drainages in the west have developed because there are no prominent structural controls.

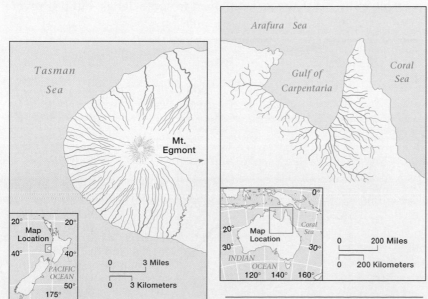

Figure 16-1-D   The extraordinary radial drainage pattern of Mount Egmont in New Zealand.

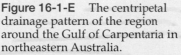

Figure 16-1-E   The centripetal drainage pattern of the region around the Gulf of Carpentaria in northeastern Australia.

are undercut. The undercutting allows material to slump into the stream. All this time the valley floor is being widened, mass wasting and overland flow help wear down the valley sides. In addition, similar processes along tributary streams also contribute to the general widening of the main valley.

The frequent shifting of stream meanders produces an increasingly broader, flattish valley floor largely or completely covered with alluvium. At any given time, a stream is likely to occupy only a small portion of the flatland, although during periods of flood flow, the entire floor may be flooded. For this reason, the valley bottom is properly termed a **floodplain** (Figure 16-23). The outer edges of the floodplain are usually bounded by a slope, marking the outer limit of lateral erosion and undercutting where the flat terrain abruptly changes to a line of **bluffs**. Valley widening and floodplain development can extend for great distances; the floodplains of many of the world's largest rivers are so broad that a person standing on the bluffs at one side cannot see the bluffs on the other.

Valley Lengthening

A stream may lengthen its valley in two quite different ways: (1) by headward erosion at the upper end or (2) by delta formation at the lower end.

Headward Erosion   No concept is more fundamental to an understanding of fluvial processes than headward erosion because it is the basis of rill, gully,

and valley formation and extension. The upper perimeter of a valley is the line where the gentle slope of an interfluve changes to the steeper slope of a valley side. Sheet flow from the interfluve drops abruptly over this slope break, and the fast-moving water tends to undercut the rim of the perimeter, weakening it and often causing a small amount of material to collapse (Figure 16-24).

The result of this action is a decrease in interfluve area and a commensurate increase in valley area. As the overland flow of the interfluve becomes part of the streamflow of the valley, there is a minute but distinct extension of rills and gullies into the drainage divide of the interfluve—in other words, a headward extension of the valley. Although minuscule as an individual event, when multiplied by a thousand gullies and a million years, this action can lengthen a valley by tens of miles and the expansion of a drainage basin by hundreds of square miles. Thus the valley lengthens at the expense of the interfluve (Figure 16-25 on page 466).

Delta Formation   A valley can also be lengthened at its seaward end, in this case by deposition. Flowing water slows down whenever it enters the quiet water of a lake or ocean and deposits its load. Most of this debris is dropped right at the mouth of the river in a landform called a **delta**, after a fancied resemblance to the Greek capital letter delta, Δ (Figure 16-26 on page 467). The classic triangular shape is maintained in some deltas, but it is severely

Figure 16-23   Looking east across the floodplain of the Illinois River near the town of Hardin, Illinois. Our viewpoint is from the bluff on the western margin of the floodplain, which rises 200 feet (60 meters) above the river. In the distance we can see the bluff that marks the eastern edge of the floodplain, 4 miles (6 kilometers) away. The river has overflowed its bank in this springtime flood scene and has inundated the trees growing along the edge of the channel. Hidden in the trees is an artificial levee that keeps the floodwaters out of the low-lying floodplain lands in the distance. (TLM photo)

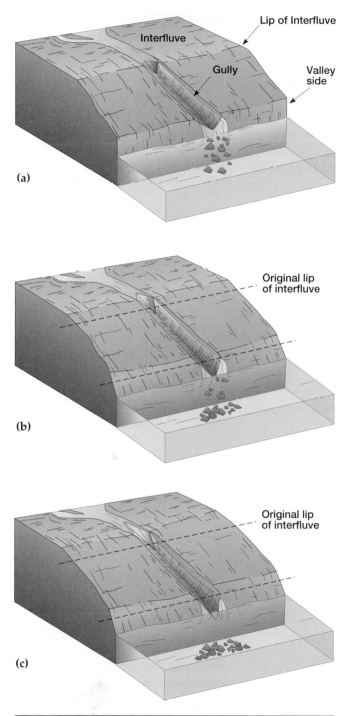

Figure 16-24 **(a)** Headward erosion occurs at the upper end of a stream where overland flow pours off the lip of the interfluve into the valley. **(b)** The channeled streamflow wears back the lip of the interfluve. **(c)** Over time, this erosion extends the valley headward at the expense of the interfluve.

modified in others because of imbalances between the amount of sediment deposited by rivers and the removal of those sediments by ocean waves and currents. At some river mouths, the seawater moves so vigorously that no delta is formed (Table 16-1).

TABLE 16-1	
The World's Delta-less Rivers	
Rank[a]	River (Country)
1	Amazon (Brazil)
2	Zaire (Zaire)
6	Yenisey (Russia)
10	Paraná (Argentina)
11	St. Lawrence (Canada)
15	Tocantins (Brazil)
20	Columbia (United States)
21	Zambezi (Mozambique)

[a]In terms of average discharge.

Generally, however, a prominent and clear-cut delta does form (Table 16-2 and Figure 16-27 on page 467). The stream slows down, losing both competence and capacity, and drops much of its load, which partially blocks the channel and forces the stream to seek another path. Later this new path is likely to become clogged, and the pattern is repeated. As a result, deltas usually consist of a maze of roughly parallel channels called **distributaries** through which the water flows

TABLE 16-2		
The World's Largest Deltas		
Rank	River (Country)	Area $(km^2 \times 10^3)$
1	Indus (Pakistan)	163.0
2	Nile (Egypt)	160.0
3	Hwang Ho (China)	127.0
4	Yangtze (China)	124.0
5	Ganges/Brahmaputra (Bangladesh)	91.0
6	Orinoco (Venezuela)	57.0
7	Yukon (Alaska)	54.0
8	Mekong (Vietnam)	52.0
9	Irrawaddy (Myanmar)	31.0
10	Lena (Russia)	28.5
11	Mississippi (United States)	28.0
12	Chao Phraya (Thailand)	24.6
13	Rhine (Netherlands)	22.0
14	Colorado (Mexico)	19.8
15	Niger (Nigeria)	19.4

FOCUS

STREAM CAPTURE

Headward erosion is illustrated dramatically when a portion of the drainage basin of one stream is diverted into the basin of another stream by natural processes. This event, called **stream capture** or **stream piracy**, is relatively uncommon in nature, but evidence that it does sometimes occur is found in many places

As a hypothetical example, let us consider two streams flowing across a coastal plain, as shown in Figure 16-2-A. Their valleys are separated by an interfluve, which for this example can be thought of as an undulating area of low relief. Stream A is shorter than stream B but is also more powerful, and A's valley is aligned so that headward extension will project it in the direction of B's valley.

In the normal course of events, rainfall and snowmelt accumulate on the interfluve and move down as overland flow into both valleys. The overland flow becomes concentrated into rills and then gullies, and the gullies eventually become tiny distributaries to the valleys.

Owing to its power and orientation, stream A lengthens its valley headward at the expense of the interfluve. As the valley becomes larger and the interfluve smaller, the drainage divide between the two valleys is shifted toward stream B. As the process continues, the headwaters of stream A eventually extend completely into the valley bottom of stream B and the flow from the upper reaches of B is diverted into A. In the parlance of the geographer, A has captured part of B. Stream A is called the *captor stream*, the lower part of B is the *beheaded stream*, the upper part of B is the *captured stream*, and the abrupt bend in the stream channel where the capture took place is called an *elbow of capture*.

An example from the real world is shown in Figure 16-2-B: capture of the upper portion of Beaverdam Creek by the Shenandoah River in Virginia. Beaverdam Creek had cut a narrow gap through the Blue Ridge Mountains as this linear belt of hills was uplifted. Such a narrow notch in which a stream flows through a ridge is called a **water gap**. Stream capture then beheaded Beaverdam Creek as the young Shenandoah River extended its valley by headward erosion. Because of the way Beaverdam Creek was positioned relative to the Shenandoah valley, no elbow of capture was produced by this stream piracy. The evidence that tells us this piracy took place is the now-dry water gap; the diversion of streamflow away from the gap as Beaverdam Creek was beheaded left the gap high and dry as the adjacent streams continued their downcutting. Such an abandoned water gap is referred to as a **wind gap**.

Stream capture on a grand scale can be detected on a map of West Africa. The mighty Niger River has its headwaters relatively near the Atlantic Ocean, but it flows inward rather than seaward. After flowing northeast for nearly 1000 miles (1600 kilometers), it makes an abrupt turn to the southeast and then continues in that direction for another 1000 miles before finally emptying into the Atlantic. At some time in the past, the upper reaches of what is now the Niger was a separate river, one that did not change course but rather flowed northeast until it reached a great inland lake in what is now central Sahara [Figure 16-2-C(a)]. This river was beheaded by the ancestral Niger, producing a great elbow of capture and leaving the beheaded stream

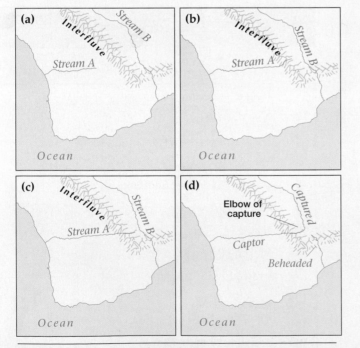

Figure 16-2-A   A hypothetical stream-capture sequence. Stream A is stronger than stream B. The valley of A is extended by headward erosion until A captures and beheads B.

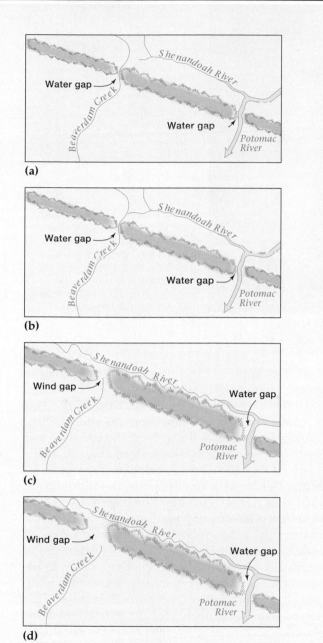

(a)

(b)

(c)

(d)

Figure 16-2-B Stream capture in Virginia. Headward erosion by the Shenandoah River beheaded Beaverdam Creek and captured the flow from its upper reaches.

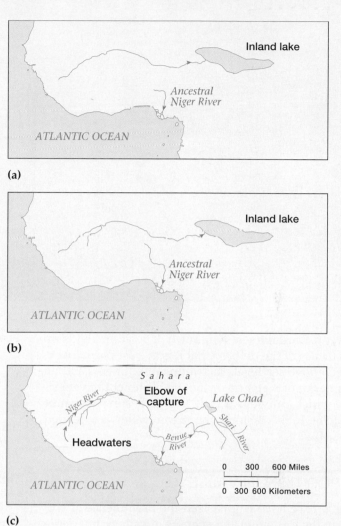

(a)

(b)

(c)

Figure 16-2-C Stream capture, actual and anticipated. **(a)** The upper course of the Niger River was once part of an unnamed stream that flowed into a large lake in what is now the Sahara. **(b)** This ancient stream was captured by headward erosion of the ancestral Niger. **(c)** At present, headward erosion on the Benue River gives promise of capturing the Shari River just a few tens or hundreds of centuries from now.

to wither and dry up as the climate became more arid [Figure 6-2-C(b)].

The map of Africa provides us with still another major point of interest concerning stream capture, but in this case the capture has not yet taken place. The Shari River of central Africa flows northwesterly into Lake Chad. Because its lower course has a very flat gradient, the stream flows sluggishly there without any power for downcutting. West of the Shari is an active and powerfully downcutting river in Nigeria, the Benue, a major tributary of the Niger. Some of the tributary headwaters of the Benue originate in a flat, swampy interfluve only a short distance from the flood-plain of the Shari. Since the Benue is more active than the Shari and its alignment is such that headward erosion cuts directly into the Shari drainage, the Benue is likely to behead the Shari before our very eyes, so to speak, provided we can wait a few thousand years.

Figure 16-25 This Wisconsin scene illustrates headward erosion. The grassy area being enjoyed by the horses is an interfluve that is being cut into by headward erosion of the irregular gorgelike valley in the foreground. (TLM photo)

slowly toward the sea. Continued deposition builds up the surface of the delta so that it is at least partially exposed above sea level. Rich alluvial sediments and an abundance of water favor the establishment of vegetation, which provides a base for further expansion of the delta. In this fashion, the stream valley is extended downstream.

DEPOSITION IN VALLEYS

Thus far we have emphasized the prominence of erosion in the formation and shaping of valleys, but deposition, too, has a role in these processes.

Fluvial deposits may include all sizes of rock debris, but smaller particles constitute by far the bulk of the total, primarily because most soil and surface sediment are either silt or clay. In addition, all particles get smaller as they are transported downstream because the constant battering and buffeting eventually reduces boulders, cobbles, and pebbles to sand, silt, and clay. Moreover, all rock fragments are chemically weathered as they are transported in the water and temporarily stored in stream deposits.

Because of fluctuations in flow and turbulence, alluvium can be deposited almost anywhere in a valley bottom: on the stream bottom, on the sides, in the center, at the base of knickpoints, in overflow areas, and in a variety of other locations. Under some circumstances, particularly after a period of flood flow, alluvium may accumulate on the streambed to such an extent that the bed's elevation is raised in a process called **aggradation**.

FLOODPLAINS

The most prominent depositional landscape is the *floodplain*, formed when a meandering stream erodes laterally to produce a broad, flattish valley floor. This floor is inundated by overflow from the stream channel during floodtime, and the floods leave broad and sometimes deep deposits of alluvium over the entire floor, which then is called a floodplain. The most conspicuous feature of a floodplain is the meandering channel of its river (Figure 16-28 on page 471).

Meanders often develop narrow necks that are easily cut through by the stream, leaving abandoned cutoff meanders, which initially hold water as oxbow lakes but gradually fill with sediment and vegetation to become oxbow swamps and eventually retain their identity only as meander scars (Figure 16-29 on page 472).

A floodplain is slightly higher along the edges of the stream channel for the following reason. As the stream overflows at floodtime, the current, as it leaves the normal channel, is abruptly slowed by friction with the floodplain surface. This slowdown causes the principal deposition to take place along the margins of the main channel, producing **natural levees** (from the Old French word *levée*, act of raising, derived from the Latin *levare*, to raise) on each side of the stream (Figure 16-30 on page 473). The natural levees merge outwardly and almost imperceptibly with the less-well-drained and lower portions of the floodplain, generally referred to as **backswamps** (Figure 16-31 on page 473).

Sometimes a tributary stream entering a floodplain that has prominent natural levees cannot flow directly into the main channel and so flows down valley in the

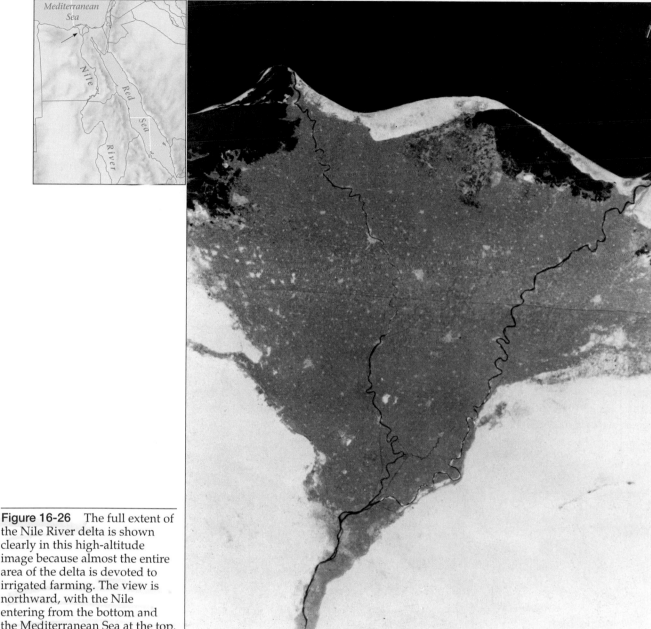

Figure 16-26   The full extent of the Nile River delta is shown clearly in this high-altitude image because almost the entire area of the delta is devoted to irrigated farming. The view is northward, with the Nile entering from the bottom and the Mediterranean Sea at the top. (Photri)

backswamp zone, running parallel to the main stream for some distance before finding an entrance. A tributary stream with such a pattern is referred to as a **yazoo stream,** after Mississippi's Yazoo River, which flows parallel to the Mississippi River for about 175 miles (280 kilometers) before joining it.

Some floodplains have no natural levees. The reasons for this lack are not clear but apparently are related to the nature of the sediment load, especially the relative proportions of bedload and suspended load.

STREAM REJUVENATION

All parts of the continental surfaces experience episodes, sometimes frequent and sometimes rare, when their elevation relative to sea level changes. This change is occasionally caused by a drop in sea level (as occurred throughout the world during the various ice ages when frozen water accumulated on the land, diminishing the amount of water in the oceans), but it is much more commonly the result of tectonic uplift of the land surface. When such uplift occurs, it rejuvenates

PEOPLE AND THE ENVIRONMENT

MODIFYING THE MISSISSIPPI

There are obvious attributes—flat land, abundant water, productive soils—that attract humans to valley bottoms, and therefore such areas are often places of intensive agriculture, transportation routes, and urban development. However, fluvial processes are always ongoing in valley bottoms, with the result that nature and humans coexist in an uneasy juxtaposition shrouded by the specter of flood. We have seen that every river is subject to at least occasional flooding, and the existence of a floodplain, so very attractive for human settlement, is incontrovertible evidence that floods do occur from time to time.

Accordingly, wherever humans have settled in considerable concentrations in river valleys, they have gone to extraordinary lengths to mitigate potential flood damage. The principal means for averting disaster are sizable earthwork or concrete water-containment and diversion structures in the form of dams, levees, and overflow floodways. As an example of the remarkable efforts that go into such endeavors, we might consider the major river system of North America: the Mississippi.

The Mississippi originates in Minnesota and flows more or less directly southward to the Gulf of Mexico below New Orleans (Figure 16-3-A). It is joined by a number of right- and left-bank tributaries along the way, of which the Missouri is by far the most important. There are also many left-bank tributaries, of which the Ohio, with its tributary the Tennessee, is the most notable. (Tribuary streams are designated as "right bank" or "left bank" from the perspective of an observer looking downstream.)

All four of these rivers have been thoroughly dammed, largely for flood control but also for such other benefits as hydroelectricity production, navigation stabilization, and recreation. On the Mississippi, there are 27 low dams between St. Louis and the head of navigation at Minneapolis, most equipped with hydroelectricity facilities and each with locks to allow barges and other shallow-draft boats to pass. The Missouri has fewer but larger dams, six in all, widely scattered from Montana to Nebraska. These are primarily flood-control dams, but they also have ancillary purposes. The Ohio is punctuated by more than three dozen low dams, whose primary purpose is maintenance of pools deep enough for barge navigation, with flood control as a sec-

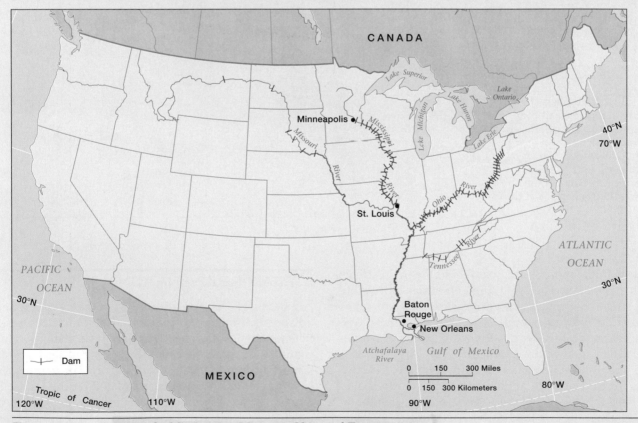

Figure 16-3-A   Dams on the Missisipipi, Missouri, Ohio, and Tennessee rivers.

Figure 16-3-B   Artificial levees are built to prevent rivers from spreading out over their floodplains. Often they have to be raised and/or repaired to constrain the river at flood time. Shown here is an emergency effort in August, 1993, to protect the town of Prairie du Rocher, Illinois, from the flooding Mississippi River by adding sandbags to the levee. (Reuters/Bettmann)

ondary consideration. Most thoroughly dammed is the Tennessee, whose nine mainstream dams have reduced it to a series of quiet reservoirs for its entire length, apart from the upper headwaters. These are Tennessee Valley Authority (TVA) dams built during the 1930s particularly for flood control but with various subsidiary benefits.

These four river valleys contain an extensive series of levees designed to protect the local floodplain and move floodwaters downstream (Figure 16-3-B). There is a vicious-circle aspect to levee building: anytime a levee is raised in an upstream area most downstream locales require higher levees to pass the floodwaters on without overflow. These levees are usually the highest parts of the landscape, particularly in the ever-flatter and more extensive floodplains downstream.

At the lower end of the Mississippi system, in southern Louisiana, river control is extremely complicated and the results of human efforts are ambiguous. This region, which receives the full flow of the continent's mightiest river system, is exceedingly flat and thus has poor natural drainage. During the past 20 centuries, the lower course of the river has shifted several times, producing at least seven subdeltas that are the principal elements of the present complicated "bird's foot" delta of the Mississippi. The main flow of the river during the past 600 years or so has been along its present course, southeast from New Orleans. This portion of the delta was built out into the Gulf of Mexico at a rate of more than 6 miles (9.6 kilometers) per century in that period.

Under the normal pattern of deltaic fluctuation, the main flow of the river would now be shifting to the shorter and slightly steeper channel of the Atchafalaya River, a prominent distributary of the ancestral delta a few miles west of the present main channel of the Mississippi. However, an enormous flood-control structure was erected above Baton Rouge in an effort (thus far successful) to prevent the river from abandoning its present channel and delta.

Another major disruption of the natural fluvial pattern has produced particularly serious effects on the lower delta. The herculean artificial-drainage and river-channeling efforts to provide a dry surface for human settlement in southeastern Louisiana have restricted the Mississippi and its distributaries to relatively narrow channels, thus keeping both silt and fresh water from getting to the extensive surrounding marshlands. In addition, a maze of canals dredged to drain potential farmland and provide for increased boat traffic accelerates erosion

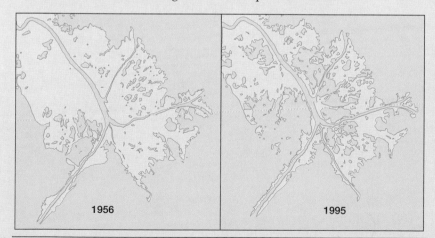

1956   1995

Figure 16-3-C   Land loss in the lower Mississippi delta. Much of the marshland shown in the left diagram (1956) has sunk or been washed away by the intruding ocean (1995).

(Continued)

MODIFYING THE MISSISSIPPI

(Continued)

and allows salt water to encroach. The result is a continuing diminution of land at a rate that is unprecedented on the North American coastline. The delta is both washing away and sinking. The amount of marsh continues to decrease, as Figure 16-3-C shows, and the amount of open water in the delta continues to expand. Over the last three decades, the land loss has averaged about 40 square miles (104 square kilometers) annually. The freshwater and brackish ecosystems suffer, and some of the higher land containing human settlements is sinking.

The eternal battle between humans and rivers continues. Can people live on a floodplain without being flooded? Immense efforts are necessary to "control" rivers and mitigate flood damage. In every case the economic costs are exorbitant, and the ecological costs are sometimes equally high (Figure 16-3-D).

Figure 16-3-D   A sample of the Midwest floods of 1993. Here the Des Moines River inundates the area around Sec Taylor Stadium in Des Moines, Iowa. (Skyview/The Stock Market)

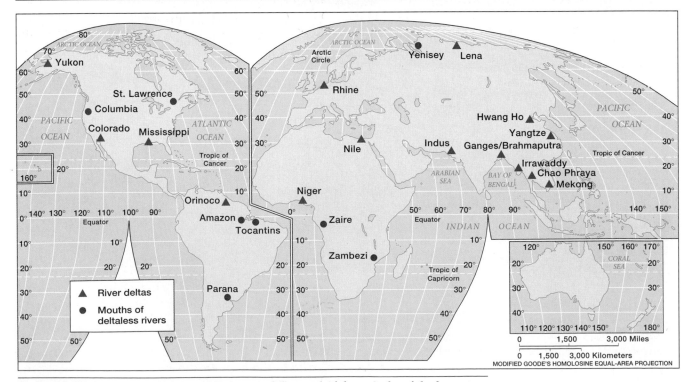

Figure 16-27   Locations of the world's largest deltas and of the mouths of the largest deltaless rivers.

Figure 16-28 The floodplain of the Bear River near Brigham City, Utah. As with all other floodplains, the meandering river channel is the most conspicuous feature. In this case, snow cover highlights the pattern. (TLM photo)

the streams in the area. The increased gradient causes the streams to flow faster, which provides renewed energy for downcutting. Vertical incision, which may have been long dormant, is initiated or intensified.

If a stream occupied a broad floodplain prior to rejuvenation, rejuvenation cuts a steep-walled gorge (Figure 16-32 on page 471). This means that the floodplain can no longer function as an overflow area and instead becomes an abandoned stretch of flat land overlooking the new gorge. This remnant of the previous valley floor is called a **stream terrace**. Terraces often, but not always, occur in pairs, one on either side of the newly incised stream channel.

Under certain circumstances, rejuvenative uplift has a different and even more conspicuous topographic effect: **entrenched meanders** (Figure 16-33 on page 475). These topographic features are formed when an area containing a meandering stream is uplifted slowly and the stream incises downward while still retaining the meandering course. In some cases, such meanders may become entrenched in narrow gorges hundreds of feet deep.

THEORIES OF LANDFORM DEVELOPMENT

Slope is the basic element of landforms. In the final analysis, all topographic study is the study of how slopes change through time. In the long run, slopes are everchanging, and our understanding of topography is predicated largely on how well we can comprehend the nature, extent, and causes of these changes.

The various internal and external processes operating on Earth's surface produce an infinite variety of landscapes. Systematizing this vast array of facts and relationships into a coherent body of knowledge has been the goal of many students of geomorphology. If principles can be recognized, theories and models can be formulated. The theories and models can then be tested, and out of this procedure we can hope to gain a broader understanding of the features of the terrain and of the processes that produced them. Many scholars have contributed to this organizational task, and several comprehensive theories of terrain evolution have been devised, three of which we discuss here.

THE GEOMORPHIC CYCLE

The first, and in many ways most important, model of landscape development was propounded by William Morris Davis, an American geographer/geomorphologist active in the 1890s and early 1900s. He called it the *geographical cycle*, but many of his students considered the adjective too generalized, and so the theory came to be known as the *cycle of erosion*. This term in turn was considered too restrictive because weathering and deposition are also involved, and so the Davisian theory is now usually referred to as the **geomorphic cycle**.

Davis envisioned a circular sequence of terrain evolution in which a relatively flat surface is uplifted, then incised by fluvial erosion into a landscape of slopes and valleys, and finally denuded until it is once again a flat surface at low elevation (Figure 16-34 on page 476). He likened this sequence to the life cycle of an organism, recognizing stages of development he called youth, maturity, and old age.

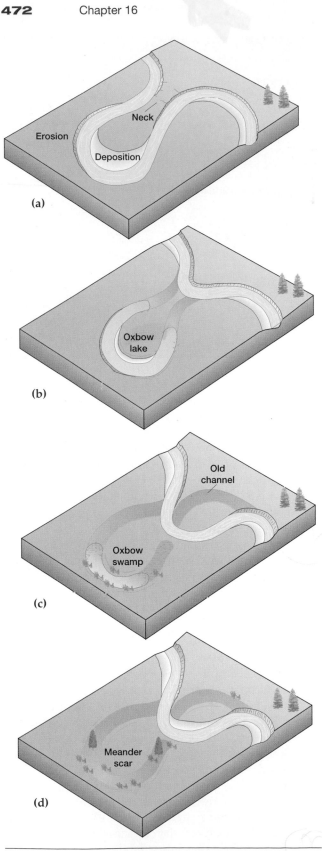

Figure 16-29 The evolution of features in a river floodplain. As the river cuts across the narrow neck of a meander, the river bend becomes an oxbow lake which becomes an oxbow swamp which becomes a meander scar.

Davis stressed that any landscape can be comprehended by analyzing structure, process, and stage. *Structure* refers to the type and arrangement of the underlying rocks and surface materials; *process* is concerned with the internal and external forces that shape the landforms; and *stage* is the length of time during which the processes have been at work.

Initial Surface   Davis postulated that the initial surface is uplifted relatively rapidly, so that erosion has little time to act until the uplift is complete. Thus the initial surface consists of relatively flat land far above sea level. He further assumed no significant subsequent crustal movement or deformation for the duration of the cycle as well as a stable base level during all stages.

Youth   During the youthful stage of development, streams become established and a drainage pattern begins to take shape. These streams, which incise deep, narrow, steep-sided, V-shaped valleys, flow rapidly and have irregular gradients marked by waterfalls and rapids. During this stage, most of the initial surface is broad, flattish interfluves, largely unaffected by stream erosion, and encompassing shallow lakes and swamps because of the incomplete drainage system.

Maturity   In the mature stage, the main streams approach an equilibrium condition, having worn away the falls and rapids and developed smooth profiles. Vertical erosion ceases in the main valleys, the streams begin to meander, floodplains are formed, and the drainage system is more extensive than in the youth stage. The interfluves are thoroughly dissected during this stage, their lakes and swamps drained by headward erosion and their remnants existing only as narrow drainage divides between valleys.

Whereas youth is characterized by the presence of a vast area of initial surface, maturity is marked by the absence of initial surface.

Old Age   With the passage of a vast amount of time, erosion reduces the entire landscape to near base level. Sloping land is virtually absent, and the entire region is dominated by extensive floodplains over which a few major streams meander broadly and slowly.

Peneplain   The end product of the geomorphic cycle is a flat, featureless landscape with minimal relief. Davis called this a **peneplain** (*paene* is Latin for almost; hence, "almost a plain"). He envisioned occa-

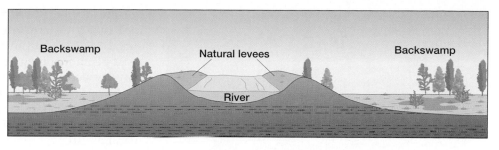

Figure 16-30   Cross section of a stream channel bordered by natural levees.

sional remnants of exceptionally resistant rock rising slightly above the peneplain surface; such erosional remnants are dubbed *monadnocks*, after a mountain of this name in New Hampshire.

Rejuvenation   Because Davis recognized that the extraordinarily long time without crustal deformation required by his model is unlikely, his theory also covered *rejuvenation*, whereby regional uplift could raise the land and interrupt the cycle at any stage. This tectonic activity would reenergize the system, initiate a new period of downcutting, and restart the cycle.

CRUSTAL CHANGE AND SLOPE DEVELOPMENT

The second theory of terrain evolution we consider grew out of critical analysis of Davis's geomorphic cycle. Davis was both a prolific writer and a persuasive teacher, and his theory had a profound influence for many decades, especially in the United States. Even in the early days, however, there were strong dissenters. Other geomorphologists recognized imperfections in some of his assumptions and questioned some of his conclusions. For example, apparently no intact peneplains exist; remnants of peneplain surfaces are recognized in some areas, but nowhere does an actual peneplain occur. A more im-

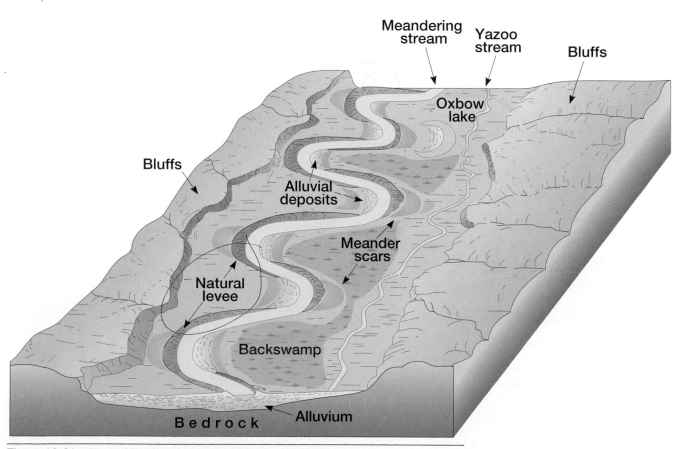

Figure 16-31   Typical landforms in a floodplain.

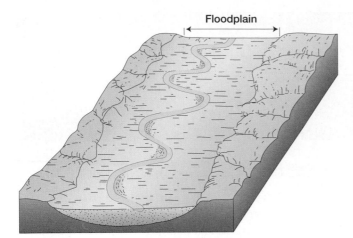

(a) Before uplift

(b) Uplift

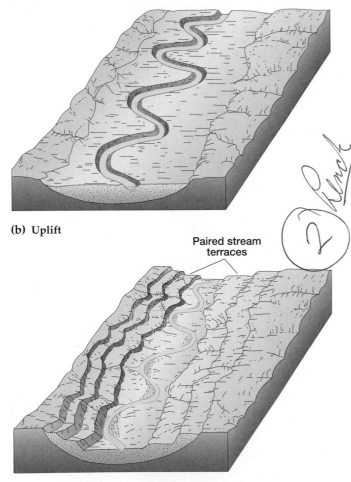

Paired stream
terraces

(c) After uplift

Figure 16-32 Stream terraces normally represent sequences of uplift and rejuvenation.

portant difficulty with Davis's model concerns his idea that little erosion takes place while the initial surface is being uplifted, a notion unacceptable to most geomorphologists. Moreover, the causal interplay between uplift and erosion is open to varying interpretations. Finally, there are serious doubts about sequential development, and some people feel that the biological analogy may be more misleading than helpful.

The sequential aspect of the geomorphic cycle is very appealing because it provides an orderly, evolutionary, and predictable train of development. Moreover, many areas in nature have the appearance of youthful, mature, or old-age topography. However, no proof has ever been found that one stage commonly precedes another in regular fashion, and even in a single valley the terrain characteristic of the various stages is often jumbled. For landform analysis, therefore, it is probably better to use the terms youth, maturity, and old age as descriptive summaries of regional topography rather than as distinct implications of sequential development.

In the Davisian cycle, drainage divides waste away in a steady and predictable pattern. As a slope retreats, it becomes less steep and more rounded, always maintaining a convex form. In nature, however, not all slopes are convex; some are straight and others are concave. Walther Penck, a young German geomorphologist and a prominent early critic of Davis, pointed out in the 1920s that slopes assume various shapes as they erode. Penck stressed that uplift stimulates erosion immediately and that slope form is significantly influenced by the rate of uplift or other crustal deformation. He argued that steep slopes, particularly, maintain a constant angle as they erode, retaining their steepness as they diminish in a sort of "parallel retreat" rather than being worn down at a continually lower slope angle. He viewed eroding slopes as retreating rather than being reduced overall, which means that some initial surface is retained long after it would have been worn away in the Davisian concept (Figure 16-35). Many, but not all, of Penck's ideas have been substantiated by subsequent workers, and his ideas have come to be called the **theory of crustal change and slope development.**

EQUILIBRIUM THEORY

The third model of landform development is called equilibrium theory. In the last three decades or so, many geomorphologists have been studying the mechanics of landform development. This approach

Figure 16-33 **(a)** Deeply entrenched meanders of the Green River in southeastern Utah. (TLM photo) **(b)** Flood-plain meanders some times become rejuvenated by uplift, causing the stream to renew downcutting and producing entrenched meanders.

(a)

(b) Floodplain meanders

(c) Entrenched meanders

emphasizes the delicate balance between form and process in the landscape. It is believed that the influence of crustal movement and the resistance of the underlying rock vary significantly from place to place and that these variations are as significant as differences in process in determining terrain. Thus, **equilibrium theory** suggests that slope forms are adjusted to geomorphic processes so that there is a balance of energy—the energy provided is just adequate for the work to be done. For example, harder rock develops steeper slopes and higher relief, and softer rock has gentler slopes and lower relief. The uniformity inherent in both the Davis and Penck theories is thus called into question.

A prime example of the application of equilibrium theory can be seen in any hilly area where the land is being simultaneously uplifted tectonically and eroded fluvially, as is happening in the Alps and the Hi-

malayas today (Figure 16-36 on page 478). If the slopes are in equilibrium, they are being wasted away at the same rate as they are being regenerated by uplift. Thus the rocks are being changed through erosion from above and uplift from below, but the form of the surface remains the same; the landscape is in dynamic equilibrium. A change in either the rate of erosion or the rate of uplift forces the landscape through a period of adjustment until the slopes again reach a gradient at which the rate of erosion equals the rate of uplift.

Equilibrium theory has serious shortcomings in areas that are tectonically stable or have limited streamflow (deserts, for example). It does, however, focus more precisely than our other two models on the relationship between geomorphic processes and surface forms and for this reason has dominated fluvial geomorphology since the 1960s.

(a)

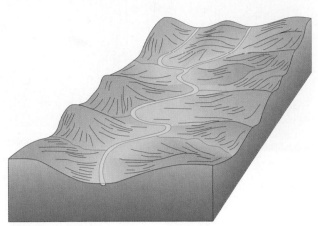

(d)

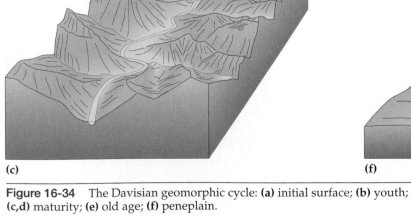

(b)

(e)

(c)

(f)

Figure 16-34   The Davisian geomorphic cycle: **(a)** initial surface; **(b)** youth; **(c,d)** maturity; **(e)** old age; **(f)** peneplain.

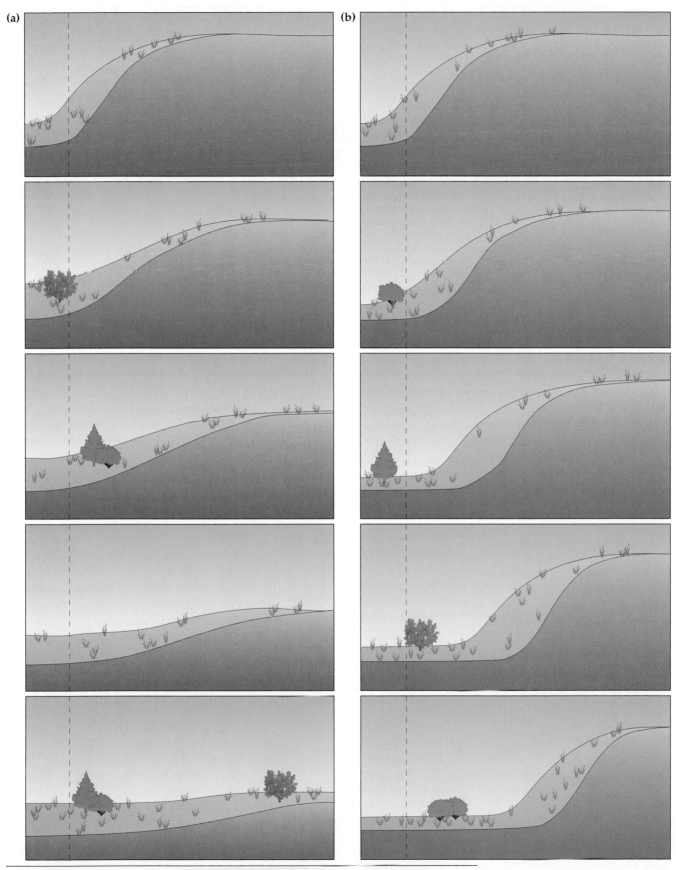

(a)

(b)

Figure 16-35 Slope retreat in **(a)** the Davis model and **(b)** the Penck model. The Davis concept proposes a continually diminishing angle of slope, whereas Penck theorized parallel retreat in which the slope angle remains approximately the same over time.

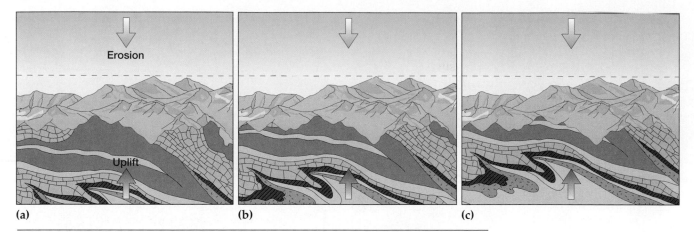

(a)   (b)   (c)

Figure 16-36   The dynamic equilibrium concept. This vertical cross section through an area of the Swiss Alps shows that erosion reduces relief just about as rapidly as uplift raises the land, with the result that the elevation of the mountains remains essentially constant over time. The surface rocks are continually changing through uplift and erosion, but the shape of the surface remains approximately the same because removal and replacement are in balance.

CHAPTER SUMMARY

Continental surfaces consists of valleys and interfluves, the former cut by water moving either as overland flow or as streamflow.

Stream channels can be changed rapidly by scouring and filling, as well as by turbulent flow. The principal channel patterns are straight, meandering, and braided. Stream systems develop in drainage basins. The branching components of a stream and its tributaries can be classified into a logical series of stream orders. Often there is a direct relationship between the location and pattern of a stream, on one hand, and the underlying geologic structure, on the other.

The shaping of stream valleys over time produces a changing sequence of landforms and a changing balance between streams and interfluves. Valley

deepening is simple and straightforward. Valley widening is accomplished primarily by lateral erosion. Valley lengthening is achieved either by headward erosion at the upper end or by delta formation at the lower end.

Fluvial deposition takes place wherever a stream's speed is inadequate to transport the load. The most prominent depositional landscape is the floodplain, which is formed by a combination of erosion and deposition.

Three theories of terrain evolution are the geomorphic cycle concept of William Morris Davis, Walther Penck's theory of crustal change and slope development, and equilibrium theory.

KEY TERMS

aggradation	delta	fluvial processes
alluvium	dendritic pattern	geomorphic cycle
annular pattern	discharge	graded stream
backswamp	dissolved load	interfluve
base level	distributary	intermittent stream
bedload	drainage basin	knickpoint
bluff	drainage divide	meandering channel
braided stream	entrenched meander	natural levee
capacity	ephemeral stream	oxbow lake
centripetal pattern	equilibrium theory	overland flow
competence	floodplain	peneplain

perennial stream
radial pattern
saltation
sinuous channel
straight channel
stream capture
streamflow
stream load

stream order
stream piracy
stream terrace
suspended load
thalweg
theory of crustal change and slope
    development
traction

trellis pattern
valley
water gap
watershed
wind gap
yazoo stream

REVIEW QUESTIONS

1. What is the distinction between overland flow and streamflow?
2. What determines the erosive effectiveness of streamflow?
3. What are the abrasive tools used by a stream in its erosive activities?
4. What are the components of a stream's load?
5. How does a stream sort alluvial material?
6. Why is streamflow so unsystematic and irregular?
7. Discuss some of the relationships between stream order and other characteristics of streams.
8. Is it ever possible for a stream to erode below sea level? Explain.

9. Why do most knickpoints tend to migrate upstream?
10. Explain how a stream valley is lengthened.
11. What is the difference between stream capacity and competence?
12. How does the equilibrium theory differ from earlier theories of topographic development?
13. How is it possible for stream meanders to become deeply entrenched?
14. Why don't all rivers form deltas?
15. What does the presence of stream terraces tell you about the erosional history of the stream valley?

SOME USEFUL REFERENCES

KNIGHTON, DAVID, *Fluvial Forms and Processes*. Baltimore: Edward Arnold, 1984.

MORISAWA, MARIE, *Rivers: Form and Process*. New York: John Wiley & Sons, 1986.

PARSONS, A. J., *Hillslope Form*. New York: Routledge, 1989.

PETTS, GEOFFREY E., *Rivers and Landscape*. Baltimore: Edward Arnold, 1985.

RICHARDS, KEITH, *Rivers: Form and Process in Alluvial Channels*. London: Methuen, 1982.

SCHUMM, S. A., M. PAUL MOSLEY, AND WILLIAM E. WEAVER, *Experimental Fluvial Geomorphology*. New York: John Wiley & Sons, 1987.

17

THE TOPOGRAPHY OF ARID LANDS

A RID LANDS ARE IN MANY WAYS DISTINCTIVE FROM humid ones, but there are no obvious boundaries to separate the two. In this chapter we focus on the dry lands of the world, without attempting to establish precise definitions or borders. We are concerned not with where such borders lie but rather with the processes that shape desert landscapes. It should be understood, however, that both the processes and the landforms of desert landscapes occur more widely than the term *desert* might imply. Thus, while this chapter is about deserts, much of what is discussed is also applicable to semiarid, subhumid, and even humid regions.

A SPECIALIZED ENVIRONMENT

As we learned in Chapter 11, desert terrain is usually stark and abrupt, unsoftened by regolith, soil, or vegetation. Despite the great difference in appearance between arid lands and humid, however, most of the terrain-forming processes active in humid areas are also at work in desert areas. The desert landforms that result are often conspicuously different from those found in wetter locations, however, as a result of a variety of factors, the most important of which are as follows:

1. *Weathering.* In dry lands, mechanical weathering is dominant (whereas chemical weathering dominates in humid regions). This predominance of mechanical weathering results not only in a generally slower rate of total weathering in deserts but also in the production of more angular particles of weathered rock.
2. *Creep.* Soil creep is a relatively minor phenomenon on most desert slopes. This is due partly to the lack of soil but primarily to the lack of the lubricating effects of water. Creep is a smoothing phenomenon in more humid climates, and its lack in deserts accounts in part for the angularity of desert slopes.
3. *Soil and regolith.* In deserts, the covering of soil and regolith is either thin or absent in most places, a condition that exposes the bedrock to erosion and contributes to the stark, rugged, rocky terrain.
4. *Impermeable surfaces.* A relatively large proportion of the desert surface is impermeable to percolating water, permitting little moisture to seep into the ground. *Caprocks* (resistant bedrock surfaces) and hardpans (hardened and generally water-impermeable subsurface soil layers) of various types are

widespread, and what soil has formed is usually thoroughly compacted.

5. *Sand.* Deserts have an abundance of sand in comparison with other parts of the world (Figure 17-1). This is not to say that deserts are mostly sand-covered; indeed, the notion that deserts consist of great seas of sand is quite erroneous. Nevertheless, the relatively high proportion of sand in deserts has three important influences on topographic development: (1) a sandy cover allows water to infiltrate the ground and inhibits drainage via streams and overland flow, (2) sand is readily moved by heavy rains, and (3) it can be shifted and shaped by the wind.
6. *Rainfall.* Much of the rainfall in desert areas is intense, which means that runoff is usually rapid. Floods, although often brief and covering only a limited area, are the rule rather than the exception. Thus fluvial erosion and deposition, however sporadic and rare, are remarkably effective and conspicuous.
7. *Drainage.* Almost all streams in desert areas are ephemeral, flowing only during and immediately after a rain. Such streams are effective agents of erosion, shifting enormous amounts of material in a short time. This is mostly short-distance trans-

Figure 17-1 Most deserts contain considerable accumulations of sand, which sometimes builds up to form conspicuous dunes. Here, tourists explore dunes in the extreme southeastern corner of California. (TLM photo)

Taghit Oasis in Algeria is dwarfed by the immense sand dunes of the Sahara Desert's Grand Erg Oriental. *(RAGA/The Stock Market)*

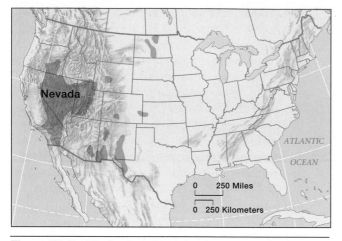

Figure 17-2   The basins of interior drainage in the United States are all in the western part of the nation. By far the largest area of interior drainage is in Nevada and the adjoining states.

portation, however. A large volume of unconsolidated debris is moved to a nearby location, and as the stream dries up, the debris is dumped on slopes or in valleys, where it is readily available for the next rain. As a consequence, depositional features are unusually common in desert areas.

8. *Wind.* Another fallacy associated with deserts is that their landforms are produced largely by wind action. This is not true, even though high winds are characteristic of most deserts and even though sand and dust particles are easily shifted.

9. *Basins of interior drainage.* Desert areas contain many watersheds that do not drain ultimately into any ocean. For most continental surfaces, rainfall has the potential of flowing all the way to the sea. In dry lands, however, drainage networks are frequently underdeveloped, and the terminus of a drainage system is often a basin or valley with no external outlet, as Figure 17-2 shows for the United States. Any rain that falls in Nevada, for example, except in the extreme southeastern and northeastern corners of the state, has no hope of reaching the sea.

10. *Vegetation.* All the environmental factors listed above have important effects on topographic development, but the single most significant feature of dry lands is the lack of vegetation. The absence of a continuous vegetative cover is the key factor that makes deserts what they are (Figure 17-3). The plant cover consists mostly of widely spaced shrubs or sparse grass, which provide little protection from the force of raindrops and function inadequately to bind the surface material with roots.

ERGS, REGS, AND HAMADAS

There are three types of landscape found only in desert areas: the erg, the reg, and the hamada.

Most notable is the **erg**, the classic "sea of sand" often associated in the public mind with the term desert (Figure 17-1). An *erg* (Arabic for sand) is a large area covered with loose sand, generally arranged in some sort of dune formation by the wind. The accumula-

Figure 17-3   In most arid regions, the vegetation is sparse and spindly. This is a Death Valley (California) scene. (TLM photo)

tion of the vast amount of sand necessary to produce an erg is not easily explained. Since desert weathering processes are very slow, it is probable that ergs have developed only where a more humid climate originally formed the weathering products. After being formed, these products were carried by streams into an area of accumulation. Then the climate became drier and consequently wind, rather than water, became the principal agent of transportation and deposition.

Several large ergs occur in the Sahara and Arabian deserts, and smaller ergs are found in most other deserts. The Australian deserts are dominated by large accumulations of sand, including extensive dunefields, but these are not true ergs because most of the sand is anchored by vegetation and therefore not free to move with the wind. Relict ergs (usually in the form of sand dunes covered with vegetation) are sometimes found in nonarid areas, indicative of a drier climate in the past. Much of western Nebraska has such relicts.

A second type of desert landscape is the **reg**, a surface covering of coarse gravel, pebbles, and/or boulders from which all sand and dust have been removed by wind and water. A *reg* (Arabic for stone), then, is a stony desert, although the surface covering of stones may be very thin (in some cases, it is literally one pebble deep). The finer material having been removed, the surface pebbles often fit closely together, sealing whatever material is below from further erosion. For this reason, a reg is often referred to as *desert pavement* or

desert armor. In Australia, where regs are widespread, they are called *gibber plains* (Figure 17-4).

A striking feature of some deserts, one particularly but not exclusively associated with regs, is **desert varnish**. This is a dark, shiny coating, consisting mostly of iron and manganese oxides, that forms on the surface of pebbles, stones, and larger outcrops after long exposure to the desert air. Desert varnish is an important dating tool because the longer a rock surface has been exposed to weathering, the greater is the concentration of the oxide coating and thus the darker the color.

A third desert landscape is the **hamada** (Arabic for rock), a barren surface of consolidated material. A hamada surface usually consists of exposed bedrock, but it is sometimes composed of sedimentary material that has been cemented together by salts evaporated from groundwater. In either case, fragments formed by weathering are quickly swept away by the wind, so that little loose material remains.

Ergs, regs, and hamadas are all limited to plains areas. Regs and hamadas are exceedingly flat, whereas ergs are as high as the sand dunes built by the wind. The boundaries of these landscapes are often sharp because of the abrupt change in friction-layer speed as the wind moves from a sandy surface to a nonsandy surface. For example, wind passing from an erg to a reg or hamada is no longer slowed by the drag of loose sand and so can speed up and sweep the hard surface clean. On the other hand, wind moving from a barren

Figure 17-4 A gibber plain near William Creek, South Australia. An endless expanse of pebbles forms a desert pavement in this typical reg scene. (TLM photo)

PEOPLE AND THE ENVIRONMENT

DESERTIFICATION

Desertification is sure to appear on almost any short list of major world environmental problems. Although coined much earlier, this term has come into prominent use in the last three decades to refer to the expanding of desert conditions into areas previously not desert. Desertification is occurring today all over the world: southern Africa, the Middle East, India, western China, southern Australia, Chile and Peru, northeastern Brazil, the southwestern United States, Mexico, and other places. The United Nations Environment Program estimates that 8100 square miles (21 million hectares) of land is desertified each year.

Deserts can enlarge through entirely natural causes, primarily recurrent drought. However, they can also expand as a result of human activity: overgrazing of livestock, imprudent agricultural practices, deforestation, and improvident use of water resources.

Desertification is normally associated with the margins of existing deserts and implies an expansion of an already-existing desert, but this is not always the case. For example, "dust bowl" conditions in the North American Great Plains in the 1930s were not peripheral to any existing desert, yet they represented a clear and spectacular case of desertification long before the term was in common use.

The specter of desertification was brought to worldwide attention in the late 1960s by the onset of a cruel six-year drought in the African Sahel. This subhumid to semiarid region on the southern margin of the Sahara occupies parts of ten countries, from the Atlantic Ocean on the west to the Ethiopian highlands on the east (Figure 17-1-A). It has an east-west expanse of more than 3000 miles (4800 kilometers), and a north-south extent that varies from 300 to 500 miles (480 to 800 kilometers).

As the Sahelian drought intensified from 1968 to 1974, vegetation disappeared, millions of livestock

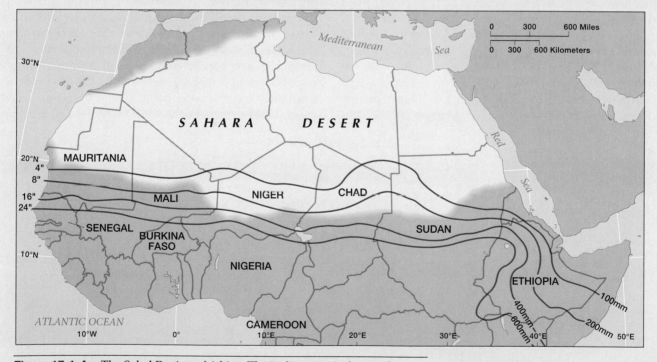

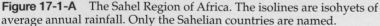

Figure 17-1-A The Sahel Region of Africa. The isolines are isohyets of average annual rainfall. Only the Sahelian countries are named.

perished, tens of thousands of people died, and there was a prominent outflow of surviving humans to an uncertain future in already overcrowded lands, particularly cities to the south of the drought-stricken region. Since 1974 the grip of the drought has been less intense, but almost every year the Sahel has had below-average rainfall, and the cumulative effect has been awesome (Figure 17-1-B).

Traditional life in the Sahel was dominated by nomads herding goats, cattle, and camels. Herdspeople and their livestock followed the ITC (intertropical convergence zone) rainbelt northward and southward in its seasonal migration. This generally kept human use of the land in balance with environmental conditions. Crop-growing was mostly limited to favored river valleys. During the 1950s and 1960s, however, significant changes ensued. The population of both humans and livestock soared, due to improved medical conditions, above-average rainfall, and economic and political changes. Newly independent nations, however, closed international borders to traditional pastoral migration routes. Dry-land farming and some irrigation agriculture spread into traditionally nomadic lands, tilling soil that should never have been broken. An enormous demand for firewood (cooking and heating fuel for 80 percent of the Sahelian population) resulted in the elimination of most trees and large shrubs. Overgrazing removed most grasses and small shrubs. Then came year after year of searing drought.

In the last two decades, $10 billion has come from outside sources to ease the Sahelian situation. Much of this money has been for immediate famine relief and has had no positive long-term effect other than to prevent or postpone starvation for some people. There are no major success stories in the continuing, exhausting, complex problem of combating Sahelian desertification.

What of the future? We know that the desert environment is fragile, but it is also resilient. The North American Dust Bowl recovered from its devastation of the 1930s. If human and livestock pressure on the Sahelian land could be relieved and if rainfall levels were normal for a few years, one could expect striking improvement in the landscape. However, the exploding population makes this an unlikely scenario. Indeed, cynical drought-relief workers are known to be placing bets as to whether the first nation in the world to become literally uninhabitable because of ecological catastrophe will be Mali or Mauritania.

Figure 17-1-B   Cattle raising clouds of dust on desertified land in the Sudan. (Y. Artus/Peter Arnold, Inc.)

reg or hamada to a sandy erg is slowed down perceptibly, and deposition results.

Although the extent of ergs, regs, and hamadas is significant in some desert areas, the majority of the arid land in the world contains only a limited number of these surfaces. For example, only one-third of the Arabian Desert, the sandiest desert of all, is covered with sand, and much of that is not in the form of an erg.

RUNNING WATER IN WATERLESS REGIONS

Probably the most fundamental fact of desert geomorphology is that running water is by far the most important external agent of landform development. The erosional and depositional work of running water accounts for the shape of the terrain surface almost everywhere outside ergs and regs. The lightly vegetated ground is defenseless to whatever rainfall may occur, and erosion by rain splash, sheetwash, rilling, and streamflood is enormously effective. Despite the rarity of precipitation, its intensity produces abrupt runoff, and great volumes of sediment can be moved in a very short time.

The steeper gradients of mountain streams increase the capacity of these streams for carrying large loads, of course, but the sporadic flow of mountain streams in arid lands results in an unpredictable imbalance between erosion and deposition. At any given time, therefore, much transportable rock debris and alluvium sit at rest in the dry stream bed of a desert mountain, awaiting the next flow. Loose surface material is either thin or absent on the slopes, and bedrock is often clearly exposed, with the more resistant strata standing out as caprocks and cliff faces.

Where slopes are gentle in an arid land, the streams rapidly become choked with sediment as the brief floodtime subsides. Here stream channels are readily subdivided by braiding, and main channels often break up into distributaries in the basins. Much silt and sand are thus left on the surface for the next flood to move, unless wind moves them first.

SURFACE WATER IN THE DESERT

Surface water in deserts is conspicuous by its absence. These are lands of sandy streams and dusty lakes, in which the presence of surface water is usually episodic and brief. Permanent streams in the dry lands are few, far between, and, with scarce exceptions, *exotic*, meaning they are sustained by water that originates outside the desert. This water that feeds exotic streams

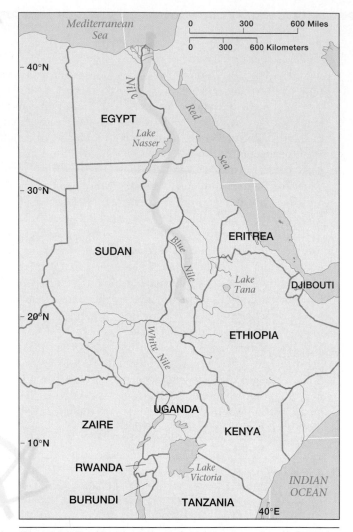

Figure 17-5 The world's preeminent example of an exotic stream, the Nile River, flows for many hundreds of miles without being joined by a tributary.

comes from an adjacent wetter area or a higher mountain area in the desert and has sufficient volume to survive passage across the dry lands. The Nile River is the classic example of an exotic stream (Figure 17-5). Its water comes from the mountains and lakes of central Africa and Ethiopia in sufficient quantity to survive a 2000-mile (3200-kilometer) journey across the desert without benefit of tributaries.

In humid regions, a river becomes larger as it flows downstream, nourished by tributaries and groundwater inflow. In dry lands, however, the flow of exotic rivers diminishes downstream because the water seeps into the riverbed, evaporates, and is diverted for irrigation.

Although almost every desert has a few prominent exotic rivers, more than 99 percent of all desert streams are ephemeral. The brief periods during which these

Figure 17-6 An ephemeral streambed in central New Mexico. It carries water only a few days each year. (Dr. E. R. Degginger)

streams flow are times of intense erosion, transportation, and deposition, however. Most ephemeral desert streamflow eventually dissipates through seepage and evaporation, although sometimes such a stream is able to reach the sea, a lake, or an exotic river. The normally dry beds of ephemeral streams typically have flat floors, sandy bottoms, and steep sides (Figure 17-6). In the United States they are variously referred to as *arroyos*, *gullies*, *washes*, or *coulees*. In North Africa and Arabia the name *wadi* is common; in South Africa, *donga*; in India, *nullah*.

Although lakes are uncommon in desert areas, dry lake beds are not (Figure 17-7). We have already noted the prevalence of basins of interior drainage in dry lands; most of them have a lake bed occupying their area of lowest elevation, which functions as the local base level for that basin. These dry lake beds are called **playas**, although the term **salina** may be used if there is an unusually heavy concentration of salt in the lake-bed sediments. If a playa surface is heavily impregnated with clay, the formation is called a **claypan.** On rare occasions, the intermittent streams may have sufficient flow to bring water to the playa, forming a temporary **playa lake**.

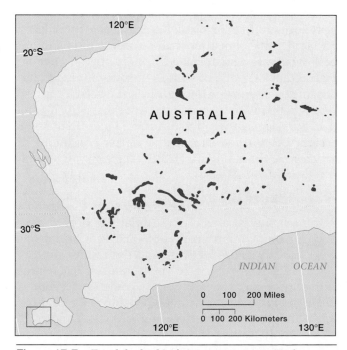

Figure 17-7 Dry lake beds often are numerous in desert areas. This map shows the principal playas and salinas in Western Australia.

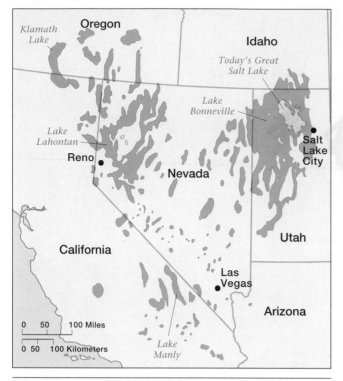

Figure 17-8 Pleistocene lakes of the intermontane region of the United States. Today's Great Salt Lake (Utah) is shown outlined in blue inside the boundaries of the ancestral Lake Bonneville.

A few desert lakes are permanent. The smaller ones are nearly always the product of either subsurface structural conditions that provide water from a permanent spring or exotic streams from nearby mountains. The larger ones are almost all remnants of still larger bodies of water that formed in a previously wetter climate. Utah's Great Salt Lake is the outstanding example in this country. Although it is the second largest lake wholly in the United States (after Lake Michigan), it is a mere shadow of the former Lake Bonneville, which was formed during the wetter conditions of Pleistocene times (Figure 17-8).

FLUVIAL EROSION IN ARID LANDS

Although fluvial erosion takes place in desert areas only during a small portion of each year, it does its work rapidly and effectively and the results are conspicuous. In desert areas of any significant relief, large expanses of exposed bedrock are common because of the lack of soil and vegetation. During the rare rains, this bedrock is both mechanically weathered and eroded by running water, and the result of the latter process is steep, rugged, rocky surfaces.

Whenever a land surface erodes, variations in rock type and structure produce differences in the slope and shape of the resulting landform, a process called **differential erosion** (Figure 17-9). Rocks resistant to erosion form cliffs, pinnacles, spires, and other sharp crests, while softer rocks wear away more rapidly to produce gentler slopes. Differential erosion is very common in sedimentary landscapes because there are significant differences in resistance from one stratum to the next; such areas often have vertical escarpments (precipitous clifflike slopes) and abrupt changes in slope angle. In areas dominated by igneous or metamorphic bedrock, however, there is not much differential erosion because there is not much difference in resistance from one part of the bedrock to another.

The steep rock faces that are evidence of differential erosion are more noticeable in dry lands than in humid lands. The reason for this difference is that in humid areas the lush vegetation overgrows the faces and obscures their steepness, as Figure 17-10 shows.

Scattered throughout the arid and semiarid lands of the world are isolated landforms that rise abruptly from the surrounding plains. Such steep-sided mountains, hills, or ridges are referred to as **inselbergs** (island mountain) because they resemble rocky islands standing above the surface of a broad sea. A notable type of inselberg is the **bornhardt**, which is composed of highly resistant rock and has a rounded form. Differential weathering and erosion lower the surrounding terrain, leaving the resistant bornhardt standing high (Figures 17-11 and 17-12). Bornhardts are very stable and may persist for tens of millions of years.

Along the lower slopes of desert mountains and hills, a distinctive kind of surface often develops. This gently inclined bedrock platform, called a **pediment,** is a *residual surface* (that is, a surface created by erosion rather than deposition) extending outward from the mountain front (Figure 17-13). Pediments have a complicated origin that is still incompletely understood. They develop as desert mountains are worn down by weathering and erosion. Many geomorphologists believe that pediments were formed in previous periods of wetter climate and are only marginally related to present-day erosion. In any case, they seem to have been produced by a complex of fluvial processes that wear away the lower slopes, although we do know that stream dissection is not a critical agent in pediment formation because the pediment surface is relatively smooth and has few lines of concentrated drainage. As the slopes of the upland erode, the pediment becomes more extensive.

Pediments are found in all deserts and are sometimes the dominant terrain feature. They are not easily recognizable, however, because almost invariably they

Figure 17-9 The effects of differential erosion are conspicuous on the Red Cliffs in western Colorado. The more resistant layers near the top form an abrupt escarpment, whereas the softer layers below are weathered and eroded into gentler slopes. (TLM photo)

are covered with a veneer of debris deposited by water and wind.

Stream channels in desert areas tend to be deep. Both mountain canyons and flatland arroyos typically

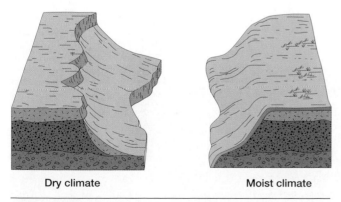

Dry climate   Moist climate

Figure 17-10 Slope comparisons in dry and moist climates. The steep relief is more easily seen in a dry climate because there is very little obscuring vegetation cover. In a humid climate, the steep faces are to some degree obscured by vegetation.

have flat, narrow bottoms and steep, often near-vertical sides. Deep accumulations of sand and other loose debris usually cover the channel bed, although flash floods sometimes scour away all alluvial fill right down to the bedrock.

FLUVIAL DEPOSITION IN ARID LANDS

Except in hills and mountains, depositional features are more notable than erosional ones in a desert landscape. Depositional features consist mostly of talus accumulations at the foot of steep slopes and deposits of alluvium and other fragmented debris in ephemeral stream channels, the latter representing bedload left behind with the subsidence of the last flood.

Piedmont is a generic term meaning any zone at the foot of a mountain range. (Do not confuse it with *pediment*, which comes from the same Latin root but refers to a specific landform.) The piedmont of a desert mountain range is one of the most prominent areas of fluvial deposition. There is normally a pronounced

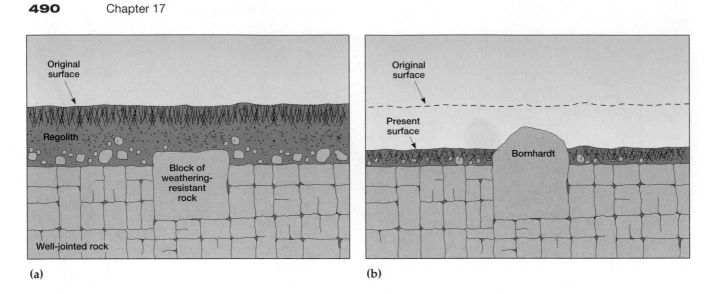

(a)

(b)

Figure 17-11 The development of a bornhardt. **(a)** The well-jointed rock is more susceptible to weathering and erosion than the resistant block in the center. **(b)** As a result, the bornhardt emerges as a conspicuous result of erosion.

Figure 17-12 Kata Tjuta (The Olgas) is a massive bornhardt in the desert of central Australia. (TLM photo)

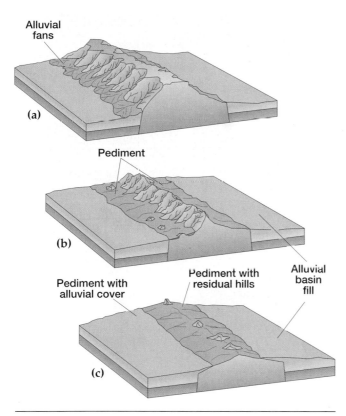

Figure 17-13   The development of a pediment as a desert mountain range is worn down by erosion.

change in the angle of slope at the mountain base (the **piedmont angle**), with a steep slope giving way abruptly to a gentle one (Figure 17-14). This is a logical area for significant accumulation of rock debris because the break in slope greatly reduces the speed of any sheetwash or streamflow that crosses the piedmont angle. Moreover, the streams issue more or less

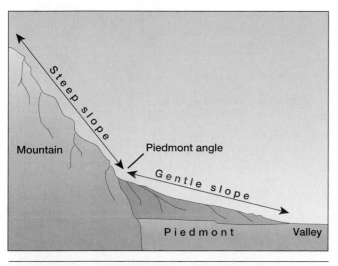

Figure 17-14   A piedmont is any gently sloping land surface that extends outward from the base of mountains or hills.

abruptly from canyons onto the more open piedmont and so are freed from lateral constraint. The resulting fluvial deposition in the piedmont often reaches depths of hundreds of feet.

The flatter portions of desert areas, particularly in basins of interior drainage, also hold a prominent accumulation of water-deposited material (Figure 17-15). Any sheetwash or streamflow that reaches into such low-lying flatlands usually had to travel a considerable distance over low-angle slopes, which means that both flow volume and flow speed are likely to be limited. Consequently, larger rock fragments are rarely transported into the basins; instead, they are covered with fine particles of sand, silt, and clay, sometimes to a considerable depth.

THE WORK OF THE WIND

The irrepressible winds of the desert create spectacular sand and dust storms and continuously reshape minor details of the landscape. However, the effect of wind as a sculptor of terrain is very limited, with the important exception of such relatively impermanent features as sand dunes.

The term *wind* is theoretically restricted to horizontal air movement. Some turbulence is nearly always involved, however, and so there is usually a vertical component of flow as well. In general, the motion of air passing over the ground is similar to that of water flowing over a stream bed, and that similarity is the cause of the turbulence. In a thin layer right at the ground surface, wind speed is zero, just as the speed of the water layer touching the banks and bed of a stream is zero, but wind speed increases with distance above the ground. The shear developed between different layers of air moving at different speeds causes turbulence similar to that in a stream of water. Wind turbulence can also be caused by heating from below, which causes the air to expand and move upward (Figure 17-16).

Aeolian processes are those related to wind action (Aeolus was the Greek god of the winds). They are most pronounced, widespread, and effective wherever fine-grained unconsolidated sedimentary material is exposed to the atmosphere, without benefit of vegetation, moisture, or some other form of protection—in other words, in deserts and along sandy beaches. Our focus here is wind action in desert regions.

AEOLIAN EROSION

The erosive effect of wind can be divided into two categories—deflation and abrasion. **Deflation** is the shifting of loose particles as a result of their being blown

Figure 17-15 Sediments and salts often accumulate in desert basins, with the salts imparting a whitish color to the landscape. This is a portion of the vast Etosha Pan in Namibia, with its typical wildlife of gemsbok, wildebeest, and ostrich. (TLM photo)

Figure 17-16 The wind sometimes is a prominent force in rearranging loose particles. This scene is near Barrow Creek in the Northern Territory of Australia. (TLM photo)

either through the air or along the ground. Except under extraordinary circumstances, the wind is not strong or buoyant enough to move anything more than dust and small sand grains, and therefore no significant landforms are created by deflation. Sometimes a **blowout,** or *deflation hollow,* may be formed; this is a shallow depression from which an abundance of fine material has been deflated. Most blowouts are small, but some exceed a mile (1.6 kilometers) in diameter. Along with fluvial erosion, deflation is also a factor in the formation of a reg surface.

Aeolian abrasion is analogous to fluvial abrasion, except that the aeolian variety is much less effective. Whereas deflation is accomplished entirely by air currents, abrasion requires "tools" in the form of airborne sand and dust particles. The wind drives these particles against rock and soil surfaces in a form of natural sandblasting. Wind abrasion does not construct or even significantly shape a landform; it merely sculptures those already in existence. The principal results of aeolian abrasion are the pitting, etching, faceting, and polishing of exposed rock surfaces and the further fragmenting of rock fragments.

Aeolian Transportation

Rock materials are transported by wind in much the same fashion as they are moved by water, but less effectively. The finest particles are carried in suspension as dust. Strong, turbulent winds can lift and carry thousands of tons of suspended dust. Some dust storms extend for thousands of feet above Earth's surface and may move material through more than 1000 miles (1600 kilometers) of horizontal distance.

Particles larger than dust are moved by wind through saltation and traction, just as in streamflow (Figure 17-17). Wind is unable to lift particles larger than the size of sand grains, and even these are likely to be carried only a foot or two above the surface. Indeed, most sand, even when propelled by a strong wind, leaps along in the low, curved trajectory typical of saltation, striking the ground at a low angle and bouncing onward. Larger particles move by traction, being rolled or pushed along the ground by the wind. It is estimated that three-fourths of the total volume of all wind-moved material in dry lands is shifted by saltation and traction, particularly the former. At the same time, the entire surface layer of sand moves slowly downwind as a result of the frequent impact of the saltating grains; this process is called *creep,* but it should not be confused with soil creep.

Because the wind can lift particles only so high, a true sandstorm is a cloud of horizontally moving sand

Figure 17-17   Wind carries tiny dust particles in suspension, larger particles are moved by saltation and traction.

that extends for only a few inches or feet above the ground surface. Persons standing in its path have their legs peppered by sand grains, but their heads are most likely above the sand cloud. The abrasive impact of a sandstorm, while having little erosive effect on the terrain, may be quite significant for the works of humans near ground level; unprotected wooden poles and posts can be rapidly cut down by the sandblasting, and cars traveling through a sandstorm are likely to suffer etched windshields and chipped paint.

Aeolian Deposition

Sand and dust moved by the wind are eventually deposited when the wind dies down. The finer material, which may be carried long distances, is usually laid down as a thin coating of silt and has little or no landform significance. The coarser sand, however, is normally deposited locally. Sometimes it is spread across the landscape as an amorphous sheet called a **sandplain**. The most notable of all aeolian deposits, however, is the **sand dune**, in which loose windblown sand is heaped into a mound or low hill.

Desert Sand Dunes   In some instances, dunefields are composed entirely of unanchored sand that is mostly uniform grains of quartz (occasionally gypsum, rarely some other minerals) and usually duncolored, although sometimes a brilliant white. Unanchored dunes are deformable obstructions to air flow. Because they are unanchored, they can move, divide, grow, or shrink. They develop sheltered air pockets on their leeward sides that slow down and baffle the wind, so that deposition is promoted there.

Unanchored dunes are normally moved by local winds. The wind erodes the windward slope of the dune, forcing the sand grains up and over the crest to be deposited on the steeper leeward side, or **slip face** (Figure 17-18). If the wind prevails from one direction for many hours, the dune may migrate downwind without changing shape. Such migration is usually

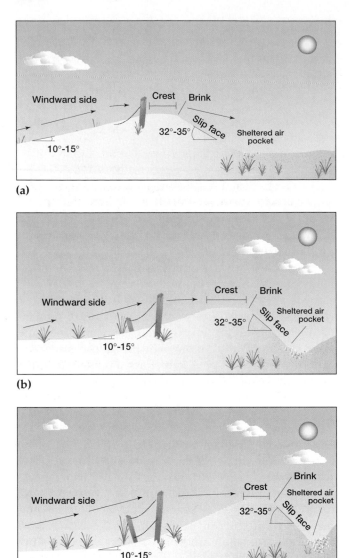

(a)

(b)

(c)

Figure 17-18   Components and slope angles of an idealized sand dune.

slow, but in some cases, dunes can move several hundred feet in a year.

Not all dunes are unanchored, however, and another characteristic dunefield arrangement is one in which the dunes are mostly or entirely anchored and therefore no longer shifting with the wind. Various agents can anchor dunes, the primary one being vegetation. Dunes provide little nourishment or moisture for plant growth, but desert vegetation is remarkably hardy and persistent and often able to survive in a dune environment. Where vegetation manages to gain a foothold, it may proliferate and anchor the dunes.

Dune patterns are almost infinite in their variety. Several characteristic dune forms are widespread in the world's deserts, however, and here we consider three of the most common.

1. Best known of all dune forms is the **barchan**, which usually occurs as an individual dune migrating across a nonsandy surface, although barchans may also be found in groups. A barchan is crescent-shaped, with the horns of the crescent pointing downwind (Figures 17-19a and 17-20). Sand movement in a barchan is not only over the crest, from windward side to slip face, but also around the edges of the crescent to extend the horns. Barchans form where strong winds blow consistently from one direction. They tend to be the fastest moving of all dunes and are found in all deserts except those of Australia. They are most widespread in the deserts of central Asia (Thar and Takla Makan) and in parts of the Sahara.

2. **Transverse dunes,** which are also crescent-shaped but less uniformly so than barchans (Figure 17-19b), occur where the supply of sand is much greater than that found in locations that have barchans; normally the entire landscape leading to transverse-dune formation is sand-covered. As with a barchan, the convex side of a transverse dune faces the prevailing wind direction. In a formation of transverse dunes, all the crests are perpendicular to the wind direction, and the dunes are aligned in parallel waves across the land. They migrate downwind just as barchans do, and if the sand supply decreases, they are likely to break up into barchans.

3. **Seifs** are long, narrow dunes that usually occur in multiplicity and in a generally parallel arrangement [Figures 17-19(c) and 17-21]. Typically they are a few dozen to a few hundred feet high, a few tens of yards wide, and miles or even tens of miles long. Their origin is still not well understood, although their lengthy, parallel orientation apparently represents an intermediate direction between two dominant wind directions. Seifs are rare in American deserts but may be the most common dune forms in other parts of the world.

Coastal Dunes   Winds are also active in dune formation along many stretches of ocean and lake coasts, whether the climate is dry or otherwise. On almost all flattish coastlines, ocean waves deposit sand along the beach. A prominent onshore wind can blow some of the sand inland, often forming dunes. In some areas, particularly if vegetative growth is inhibited, the dunes slowly migrate inland, occasionally inundating forests, fields, roads, and buildings. Most coastal dune aggregations are small, but they sometimes cover extensive areas. The largest area is probably along the Atlantic coastline of southern France, where dunes extend for 150 miles (240 kilometers) along the shore and reach inland for 2 to 6 miles (3 to 10 kilometers).

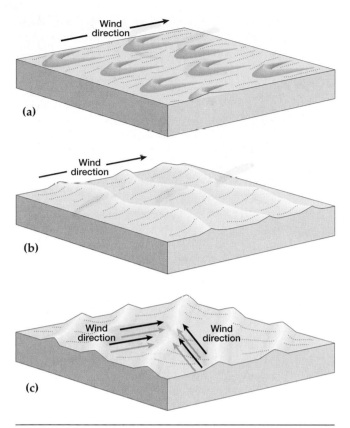

(a)

(b)

(c)

Figure 17-19   Common dune types. **(a)** Barchans **(b)** Transverse dunes **(c)** Seifs.

Loess   A form of aeolian deposit *not* associated with dry lands is **loess,** a wind-deposited silt that is fine-grained, calcareous, and usually buff-colored. (Pronounced *luhss* to rhyme with "hearse" if one left out the *r*, this is a German word derived from the name of a village in Alsace.) Despite its depositional origin, loess lacks horizontal stratification. Perhaps its most distinctive characteristic is its great vertical durability, which results from its fine grain size, high porosity, and vertical jointlike cleavage planes. The tiny grains have great molecular attraction for one another, making the particles very cohesive. Moreover, the particles are angular, which increases porosity. Thus loess accepts and holds large amounts of water. Although relatively soft and unconsolidated, when exposed to erosion, loess maintains almost vertical slopes because of its structural characteristics, as though it were firmly cemented rock (Figure 17-22). Prominent bluffs are often produced as erosional surfaces in loess deposits.

The formational history of loess is varying, complicated, and not completely understood. Its immediate origin is clearly aeolian deposition, but the materials apparently can be derived from a variety of sources. Much of the silt was produced in association with Pleistocene glaciation. During both glacial and interglacial periods, rivers carried large amounts of debris-laden meltwater from the glaciers, producing many broad floodplains. During periods of low water, winds whipped the smaller, dust-sized particles from the floodplains and dropped them in some places in very thick deposits. Much loess also seems to have been generated by deflation of dust from desert areas, especially in central Asia.

Most deposits of loess are in the midlatitudes, where some are very extensive, particularly in the United States, Russia, China, and Argentina (Figure 17-23). Indeed, some 10 percent of Earth's land surface is covered with loess, and in the conterminous United

Figure 17-20   The characteristic crescent shape of a barchan dune in White Sands National Monument, New Mexico. (B. G. Murray, Jr./Earth Scenes)

Figure 17-21 The Namib Desert extends for 2000 miles (3200 kilometers) along the coast of southwestern Africa. This Landsat photo shows the driest part of the desert, whose surface is dominated by long linear sand ridges. The river crossing the top of the picture from east to west is the Kuiseb. (NASA Headquarters)

Figure 17-22 Loess has a remarkable capability for standing in vertical cliffs, as seen in this road cut near Maryville, Missouri. (TLM photo)

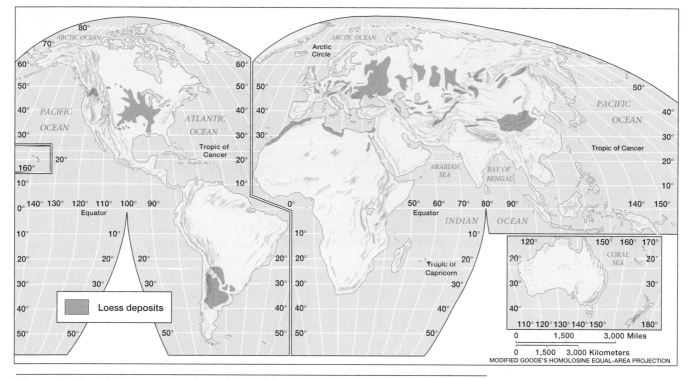

Figure 17-23   Major loess deposits of the world.

States the total approaches 30 percent. Loess deposits provide fertile possibilities for agriculture, for they serve as the parent material for some of the world's most productive soils, especially for growing grain.

The loess areas have been particularly significant in China because of their agricultural productivity. The Yellow River (Hwang Ho) received its name from the vast amount of buff-colored sediment it carries, as did its destination, the Yellow Sea. Also, numerous cave dwellings have been excavated in the Chinese loess because of its remarkable capability for standing in vertical walls. Unfortunately, however, this region is also prone to earthquakes, and the cave homes collapse readily when tremors occur; thus some of the world's greatest earthquake disasters in terms of loss of life have taken place there.

TWO CHARACTERISTIC DESERT LANDFORM ASSEMBLAGES

Surface features vary considerably from one desert to the next. Overall, however, the most common desert landform is a mountain or mountain range, and the most frequently recurring desert profile is a mountain flanked by plains.

Two particular assemblages of landforms are the most common in the deserts of the United States:

basin-and-range terrain and mesa-and-scarp terrain. Their pattern of development is repeated time and again over thousands of square miles of the American Southwest (Figure 17-24).

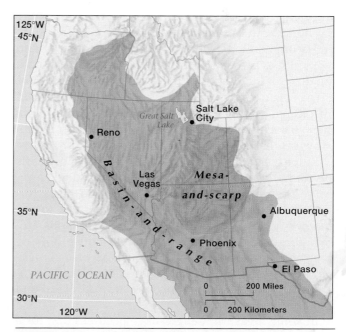

Figure 17-24   The southwestern interior of the United States contains two principal assemblages of landforms: basin-and-range and mesa-and-scarp.

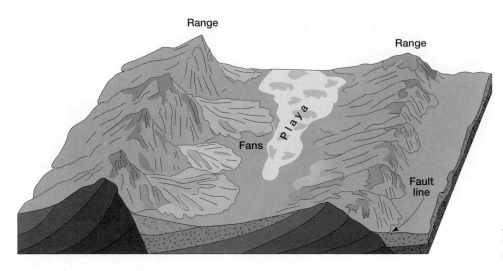

Range

Range

Playa

Fans

Fault line

Figure 17-25 A typical basin-and-range landscape.

BASIN-AND-RANGE TERRAIN

As Figure 17-24 shows, most of the southwestern interior of the United States is characterized by basin-and-range topography. This is a land largely without external drainage, with only a few exotic rivers (notably the Colorado and Rio Grande) flowing through or out of the region. The landscape consists of numerous ranges of mountains and hills scattered around a series of interior drainage basins (Figure 17-25).

Basin-and-range terrain has three principal features: ranges, piedmont zones, and basins.

The Ranges If we stand in the basin of any basin-and-range landscape, the mountain ranges dominate the horizon in all directions. Some of the ranges are high, some quite low, but the prevalence of steep and rocky slopes presents an aura of ruggedness. Although the diastrophic origins of these mountains vary (most have been uplifted by faulting, but others were formed by folding, by vulcanism, or in more complex fashion), their surface features have been shaped almost entirely by weathering, mass wasting, and fluvial processes.

Ridge crests and peaks are usually sharp, steep cliffs are common, and rocky outcrops protrude at all elevations. The mountain ranges of a basin-and-range formation are usually long, narrow, and parallel to one another. Most of them are seamed by numerous gullies, gorges, and canyons that rarely house flowing streams. These dry drainage channels are usually narrow and steep-sided and have a V-shaped profile. Typically the channel bottoms are well supplied with sand and other loose debris.

In some areas, the ranges have been eroding for a long time and consequently have been worn down to a very low relief. Where this has happened, slopes are gentler, summits more rounded, and gorges less in-

cised. Bare rock outcrops are still prominent, however. If the range stands in isolation and the alluvial plains and basins roundabout are extensive, the term *inselberg* is applied to the mountain remnant.

The Piedmont At the base of the ranges, there is usually a sharp break in slope (the piedmont angle) that marks the change from range to piedmont. The piedmont is a transition area from the steep slopes of the ranges to the near-flatness of the basins. Slope angles vary in any given piedmont but are generally gentle, with a somewhat steeper upper margin (Figure 17-26).

Much of the piedmont is underlain by an erosional pediment, although the pediment is rarely visible. It is normally covered with several feet of unconsolidated debris because the piedmont is particularly well suited for deposition. During the occasional rainfall, streams come roaring out of the gullies and gorges of the surrounding ranges, heavily laden with sedimentary material. As they burst out of the mouths of the confining canyons onto the piedmont, their speed and load capacity drop abruptly. Significant deposition is the inevitable result.

One of the most prominent and widespread topographic features to be found in any desert area is the **alluvial fan,** particularly characteristic of a basin-and-range piedmont. As a stream leaves the narrow confines of a mountain gorge and debouches into the open piedmont, it abruptly loses both capacity and competence, and breaks into distributaries that wend their way down the piedmont slope, sometimes cutting shallow new channels in the loose alluvium but frequently depositing more debris atop the old (Figure 17-27). Channels become choked and overflow, developing new ones. In this fashion a fan-shaped deposit is constructed at the mouth of the canyon. When one part of the fan is built up, the channeled flow shifts to an-

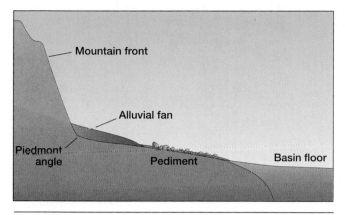

Figure 17-26   An idealized cross section of a desert piedmont zone.

other section and builds that up. This means that the entire fan is eventually covered more or less symmetrically with sediment. As deposition continues, the fan is extended outward across the piedmont and onto the basin floor.

The material in a alluvial fan is not well sorted. Instead, the fan may have a heterogeneous mixture of particle sizes because volumes of water flow vary, often considerably, from year to year. Thus there is not the neat sorting that occurs in a delta. In general, however, large boulders are dropped near the apex of the fan and finer material around the margins, with a considerable mixture of particle sizes throughout. The dry drainage channels across the fan surface frequently shift their positions as well as their balance between erosion and deposition.

As alluvial fans become larger, neighboring ones often overlap and are called *coalescing alluvial fans* (Figure 17-28). Continued growth and more complete overlap may eventually result in a continuous alluvial surface all across the piedmont, in which case it is difficult to distinguish between individual fans. This feature is known as either a piedmont alluvial plain or a **bajada**. Near the mountain front, a bajada surface is undulating, with convex sections near the canyon mouths and concave sections in the overlap areas between the canyons. In the portion of the bajada away from the range and out on the basin floor, however, no undulations occur because the component fans have coalesced so thoroughly.

The Basin   Beyond the piedmont is the flattish floor of the basin, which has a very gentle slope from all sides toward some low point. This low point is usually in a playa (or claypan or salina, depending on the nature of the surface). Drainage channels across the basin floor are sometimes clear-cut but more often shallow and ill-defined, frequently disappearing be-

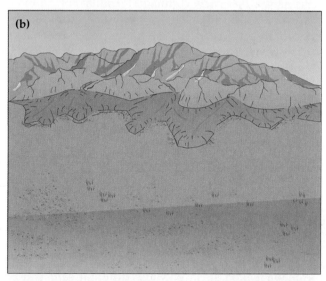

Figure 17-27   Alluvial fans are deposited in the piedmont zone at the base of mountain ranges.

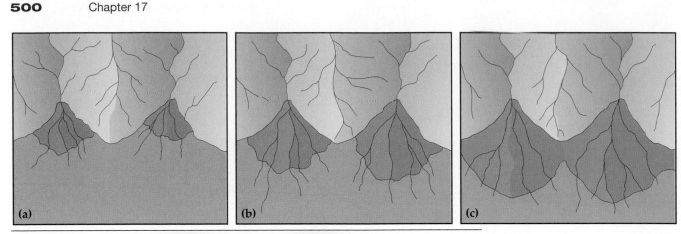

Figure 17-28 The development and coalescence of alluvial fans.

fore reaching the low point. This low point thus functions as the *theoretical* drainage terminal for all overland and streamflow from the near sides of the surrounding ranges, but only sometimes does much water reach it. Most is lost by evaporation and seepage long before.

Salt accumulations are commonplace on the playa surrounding the low point of a desert basin because of all the water-soluble minerals washed out of the surrounding watershed. Once out of the mountains and in the basin, the water evaporates or seeps away, as described above, but the salts cannot evaporate and are only marginally involved in seepage. There is usually sufficient water to allow flow across the outer rim of the basin floor and into the playa area, though, and so salts become increasingly concentrated in the playa, which is then more properly called a salina. The presence of the salt usually gives a brilliant whitish color to the playa/salina surface. Many different salts can be involved, and their accumulations are sometimes large enough to support a prosperous mining enterprise.

In the rare occasions when water does flow into a playa, the formation becomes a *playa lake*. Such lakes may be extensive, but they are usually very shallow and normally persist for only a few days or weeks. Saline lakes are marked by clear water and a salty froth around the edges, whereas shallow freshwater lakes have muddy water because they lack salt to make the silt settle.

The basin floor is covered with very-fine-grained material because the contributory streams are too weak to transport large particles. Silt and sand predominate and sometimes accumulate to remarkable depths. Indeed, the normal denudation processes in basin-and-range country tend to raise the floor of the basins. Debris from the surrounding ranges has nowhere to go but the basin of interior drainage. Thus as the mountains are being worn down, the basin floor is gradually rising.

The fine material of basin floors is very susceptible to the wind, with the result that small concentrations of sand dunes are often found in some corner of the basin. Free-moving sand is rarely found in the center of the basin because winds push it to one side or another.

Mesa-and-Scarp Terrain

The other major landform assemblage of the American Southwest is mesa-and-scarp terrain. It is most prominent in Four Corners country, the place where Colorado, Utah, Arizona, and New Mexico come together. **Mesa** is Spanish for table and implies a flat-topped surface. **Scarp** is short for escarpment and pertains to steep, more or less vertical cliffs.

Mesa-and-scarp terrain is normally associated with horizontal sedimentary strata. Such strata invariably offer different degrees of resistance to erosion, and so abrupt changes in slope angle are characteristic of this terrain. The most resistant layers, typically limestone or sandstone, often play a dual role in the development of a mesa-and-scarp terrain: they form an extensive caprock, which becomes the mesa, and at the eroded edge of the caprock, the hard layer protects underlying strata and produces an escarpment. Thus it is the resistant layers that are responsible for both elements of slope (*mesa* and *scarp*) that describe this terrain type.

Often mesa-and-scarp topography has a broad and irregular stair-step pattern (Figure 17-29). Figure 17-30 shows a cross section throught a typical formation. An extensive, flat erosional platform (the mesa) in the topmost resistant layer of a sedimentary accumulation terminates in an escarpment (the scarp) that extends downward to the bottom of the resistant layer(s). From here, another slope, steep but not as steep as the escarpment, continues down through softer strata. This inclined slope extends downward as far as the next re-

Nevada
Utah
California
Grand
Canyon
Arizona

Figure 17-29 The stair-step pattern of mesa-and-scarp terrain is shown on an imposing scale in Arizona's Grand Canyon. (TLM photo)

sistant layer, which forms either another escarpment or another mesa ending in an escarpment.

The erosional platforms are properly referred to as **plateaus** if they are bounded on one or more sides by a prominent escarpment. If a scarp edge is absent or relatively inconspicuous, the platform is called a **stripped plain**.

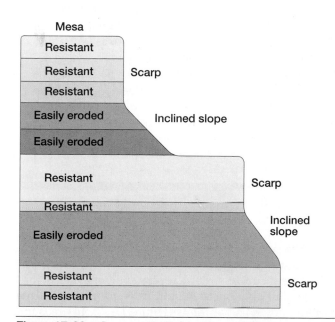

Mesa

| Resistant |
| Resistant | Scarp |
| Resistant |
| Easily eroded | Inclined slope |
| Easily eroded |
| Resistant | Scarp |
| Resistant |
| Easily eroded | Inclined slope |
| Resistant | Scarp |
| Resistant |

Figure 17-30 Cross section of a mesa-and-scarp formation. Differential erosion shows up prominently: the resistant strata weather and erode into mesas or scarps, whereas the more easily eroded strata yield gentler inclined slopes.

The escarpment edge is worn back by fluvial erosion. The cliffs retreat, maintaining their perpendicular faces, as they are undermined by the more rapid erosion of the less resistant strata (often shale) beneath the caprock. Much of the undermining is accomplished by a process called **sapping**, in which groundwater seeps and trickles out of the scarp face, eroding fine particles and weakening the cohesion of the face. When thus undermined, blocks of the caprock break off, usually along vertical joint lines. Throughout this process, the harder rocks are the cliff-formers and the less resistant beds develop more gently inclined slopes. Talus often accumulates at the base of the slope.

Although the term *mesa* is applied generally to many flat surfaces in dry environments, it properly refers to a particular landform: a flat-topped, steep-sided hill with a limited summit area. It is a remnant of a formerly more extensive surface, most of which has been worn away by erosion. Sometimes it stands in splendid isolation as a final remnant in an area where most of the previous surface has been removed, but more commonly it occurs as an outlying mass not very distant from the retreating escarpment face to which it was once connected. A mesa is invariably capped by some sort of resistant material that helps keep the summit flat even as the bulk of the rock mass is reduced by continuing erosion of its rimming cliffs (Figure 17-31 on page 504).

A related but smaller topographic feature is the **butte**, an erosional remnant having a very small surface area and cliffs that rise conspicuously

FOCUS

DEATH VALLEY: A PRIMER OF BASIN-AND-RANGE TERRAIN

California's Death Valley is a vast topographic museum, a veritable primer of basin-and-range terrain. Located in east-central California, close to the Nevada border, the valley is a classic graben, with extensive and complex fault zones both east and west of the valley floor (Figure 17-2-A). The trough is about 140 miles (225 kilometers) long, in a general northwest-southeast orientation; its width ranges from 4 to 16 miles (6 to 26 kilometers). The downfaulting has been so pronounced that nearly 550 square miles (1425 square kilometers) of the valley floor is below sea level, reaching a depth of –282 feet (–86 meters). Lengthy, upfaulted mountain ranges border the valley on either side. The Panamint Mountains on the west are the most prominent; their high point at Telescope Peak (11,049 feet or 3368 meters above sea level) is only 18 miles (29 kilometers) due west from the low point of the valley. The Amargosa Range on the east is equally steep and rugged but a bit lower overall.

The most conspicuous topography associated with Death Valley is the surrounding mountains (Figure 17-2-B). The ranges have all the characteristic features of desert mountains, being rugged, rocky, and generally barren. Their erosional slopes are steep and their escarpments steeper. The canyons that seam the ranges are invariably deep, narrow, V-shaped gorges. Many of those in the Amargosa Range and some in the Panamints are wine-glass canyons: the cup of the glass is the open area of dispersed headwater tributaries high in the range, the stem is the narrow gorge cut through the mountain front, and the base is the fan that opens out onto the piedmont.

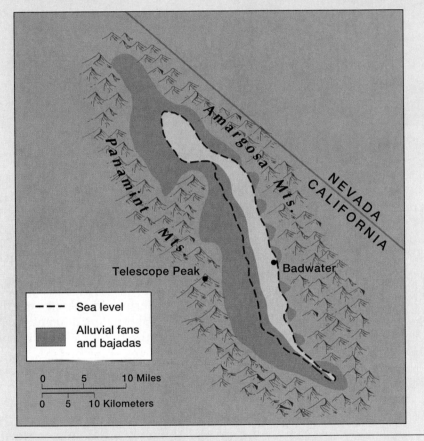

Figure 17-2-A   The setting of Death Valley.

above their surroundings. Some buttes have other origins, but most are formed by the mass wasting of mesas.

With further erosion, a still smaller residual feature, usually referred to as a pinnacle or **pillar**, may be all that is left—a final spire of resistant caprock protecting weaker underlying beds. Buttes, mesas, and pinnacles typically are found not far from some retreating escarpment face (Figure 17-32).

One of the most striking topographic features of arid and semiarid regions is the intricately rilled and barren terrain known as **badlands** (Figure 17-33). In areas underlain by horizontal strata of shale and other clay formations that are poorly consolidated, overland flow

The piedmont at the foot of the Panamints and the Amargosas is almost completely alluviated in one of the most extensive fan complexes imaginable. Every canyon mouth is the apex either of a fan or of a fan-shaped debris flow, and most of the fans overlap with neighbors to the north and south. The fans on the western side of the valley (those formed by debris from the Panamints) are much more extensive than those on the eastern side. For the most part, the Panamint fans are thoroughly coalesced into a conspicuous bajada that averages about 5 miles (8 kilometers) wide, with the outer margin of the bajada as much as 2000 feet (610 meters) lower than the canyon mouths.

The Amargosa fans are much smaller, primarily because the fault pattern of the graben has tilted the valley eastward so that its lowest portions are nestled close to the base of the Amargosa Range. Thus the west-side fans have been able to extend outward onto the valley floor, whereas tilting has reduced the size of the east-side fans by creating shorter slopes and by facilitating their partial burial by valley-floor deposits. The Amargosa fans, then, are mostly short, steep, and discrete, so there is no bajada on the eastern side of Death Valley, although some of the fans do overlap with one another.

The floor of Death Valley is also of great topographic interest, although its flatness makes the features less easy to see and understand. The valley is filled with an incredible depth of alluvium, most of which has been washed down from the surrounding mountains; in places the fill is estimated to consist of 3000 feet (915 meters) of young alluvium resting atop another 6000 feet (1830 meters) of Tertiary sediment. The surface of the valley floor has little relief and slopes gently toward the low point near Badwater, a permanent salt pond. Drainage channels appear irregularly on the valley floor, trending toward Badwater. In some places, distinct braided channels appear; in other locations, the channels disappear in sand or playa.

There are several extensive crusty-white salt pans, particularly in the middle of the valley, along with several sand accumulations, with one area of mobile dunes covering 14 square miles (36 square kilometers).

During the most recent Ice Age, Death Valley was occupied by an immense lake. Lake Manly was more than 100 miles (160 kilometers) long and 600 feet (180 meters) deep. It was fed by three rivers that flowed into the valley from the west, carrying meltwater from Sierra Nevada glaciers. As the climate became drier and warmer, the lake eventually disappeared through evaporation and seepage, but traces of its various shoreline levels can still be seen at several places on the lower slopes. Much of the salt accumulated in the valley is due to the evaporating waters of the lake.

Figure 17-2-B A small alluvial fan in California's Death Valley. (Dan Suzio/Photo Researchers, Inc.)

after the occasional rains is an extremely effective erosive agent. Innumerable tiny rills that develop over the surface rapidly evolve into ravines and gullies that dissect the land in an extraordinarily detailed manner. A maze of short but very steep slopes is etched in a filigree of rills, gullies, and gorges, with a great many ridges, ledges, and other erosional remnants scattered throughout. Erosion is too rapid to permit soil to form or plants to grow, and so badlands are barren, lifeless wastelands of almost impassable terrain. They are found in scattered locations (most of them mercifully small) in every western state, the most famous areas being in Bryce Canyon National Park in southern Utah and in Badlands National Park in western South Dakota.

Figure 17-31 Spectacular mesa-and-scarp terrain in the Monument Valley on the Arizona–Utah border. (Harvey Lloyd/The Stock Market)

Mesa-and-scarp terrain is also famous for numerous minor erosional features, most produced by a combination of weathering and fluvial erosion, with probably the most spectacular being a *natural bridge*. Such a bridge can form anytime the rock over which water flows changes from an erosion-resistant type to a less resistant type. One place a natural bridge frequently forms is where an entrenched meander wears away the rock in a narrow neck between meander loops.

Pedestals (Figure 17-34) and *pillars*, sometimes larger at the top than at the bottom, rise abruptly above their surroundings, their caps resistant material but their narrow bases continuously weathered by rainwater trickling down the surface. This water dissolves the cementing material that holds the sand grains together, and the loosened grains are easily blown or washed away.

One other notable characteristic of mesa-and-scarp terrain is vivid colors. The sedimentary outcrops and sandy debris of these regions often are resplendent in various shades of red, brown, yellow, and gray, due mostly to iron compounds.

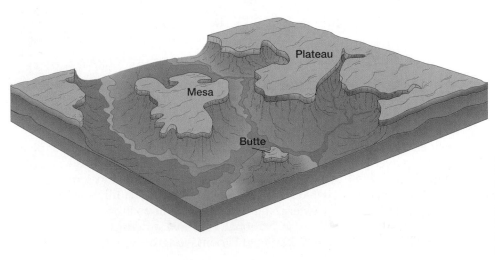

Figure 17-32 Typical development of residual land forms in horizontal sedimentary strata with a hard caprock. With the passage of time, larger features are eroded into smaller features.

Figure 17-33 Badlands are characterized by innumerable ravines and gullies dissecting the land and forming a maze of low but very steep slopes. This scene is in western Wyoming, near Dubois. (TLM photo)

Figure 17-34 The Goblet of Venus was a notable pedestal rock in southern Utah until it was pushed over by vandals. (TLM photo)

CHAPTER SUMMARY

The terrain in arid lands is either stark and angular (where bedrock outcrops are the main feature) or rounded and sandy (where depositional features predominate). Unique desert landscapes include vast sandy areas (ergs), extensive sections surfaced with gravel (regs), and barren bedrock surfaces (hamadas).

Despite the scarcity of rain, fluvial processes are by far the principal agents of landform sculpting in the desert. Most desert streams are either ephemeral or exotic, but they can erode the land surface rapidly and effectively when water is flowing. Notable erosional features include pediments and inselbergs.

The restless winds of desert areas represent a minor force overall, although they continuously reshape depositional landforms. Most prominent of all desert aeolian deposits are sand dunes, which produce one of the world's most distinctive landscapes. Common dune forms include barchans, transverse dunes, and seifs. Notable aeolian deposits that are not necessarily associated with arid regions include coastal dunes and loess.

In the dry lands of the United States, two particular landform assemblages—basin-and-range and mesa-and-scarp—are dominant.

KEY TERMS

aeolian processes	differential erosion	playa lake
alluvial fan	erg	reg
badlands	hamada	salina
bajada	inselberg	sand dune
barchan	loess	sandplain
blowout	mesa	sapping
bornhardt	pediment	scarp
butte	piedmont	seif
claypan	piedmont angle	slip face
deflation	pillar	stripped plain
desertification	plateau	transverse dune
desert varnish	playa	

REVIEW QUESTIONS

1. List several ways in which topographic development in arid lands is different from that in humid regions.
2. Why are basins of interior drainage particularly prevalent in desert areas?
3. Distinguish between erg and reg.
4. Characterize the surface water found in a typical desert.
5. What is the difference between a playa and a salina?
6. What is the difference between pediment and piedmont?
7. Distinguish between deflation and abrasion.
8. How does an alluvial fan differ from a delta?
9. How important is wind in the sculpturing of desert landforms?
10. Most loess is not found in arid regions; why is it discussed in this chapter?
11. Describe some typical landforms of basin-and-range country and some typical of mesa-and-scarp country.
12. How does a mesa differ from a butte?
13. What is distinctive about badlands terrain? How did it come to be that way?

SOME USEFUL REFERENCES

BRYAN, RORKE, AND AARON YAIR, (eds.), *Badland Geomorphology and Piping*. Norwich, England: Geo Books, 1982.

COOKE, RON, ANDREW WARREN, AND ANDREW GOUDIE, *Desert Geomorphology*. London: University College London Press, 1993.

DOEHRING, D.O., *Geomorphology in Arid Regions*. London: George Allen & Unwin Ltd., 1981.

MABBUTT, J.A., *Desert Landforms*. Cambridge, MA: The M.I.T. Press, 1977.

THOMAS, DAVID S.G., *Arid-Zone Geomorphology*. New York: Halstead Press, 1989.

THOMAS, DAVID S.G., AND NICHOLAS J. MIDDLETON, *World Atlas of Desertification*. London: Edward Arnold, 1992.

18

COASTAL PROCESSES AND TERRAIN

I N ADDITION TO RUNNING WATER, TWO OTHER FORMS OF moving water—waves and ocean currents—shape the land. Coastlines are shaped by the agitated waters crashing or lapping against them, with the result that coastal terrain is often quite different from the terrain just a short distance from shore.

THE IMPACT OF WAVES AND CURRENTS ON THE LANDSCAPE

Coastal processes affect only a tiny fraction of the total area of Earth's surface, but they create a landscape that is almost totally different from any other on the planet. Although there is some overlap, in essence waves are agents of erosion and currents are agents of deposition. The most notable land features created by waves are rocky cliffs and **headlands** (promontories of sloping land projecting into the sea). Depositional features are diverse in form, but by far the most common are beaches and sandbars.

Beaches along the shorelines of both oceans and lakes are the most distinctive aspect of coastal landscapes. They provide a transition from land to water and usually are impermanent features of the landscape, built up during times of "normal" weather and eroded during storms.

COASTAL PROCESSES

The coastlines of the world's oceans and lakes extend for hundreds of thousands of miles. Every conceivable variety of structure, relief, and topography can be found somewhere along these coasts. The distinctiveness of the coastal milieu, however, is that it is at the interface of three major components of Earth's environment—hydrosphere, lithosphere, and atmosphere. This interface is dynamic and highly energetic, primarily because of the restless motions of the waters (Figure 18-1).

We saw in Chapter 17 that wind is sometimes an important shaper of landforms on the continents. Along coastlines, the wind has an even greater influence on topography because the surface of a large body of water can be deformed abruptly and rapidly by wind action. This deformation of the water surface is what creates waves and ocean currents, both of which shape coastlines, producing topographic features found throughout the world in all latitudes and in a variety of climates.

The wind is not the only force causing water to move, of course, but from the standpoint of geomorphic effects it is the most important. Oceanic coastlines also experience daily tidal fluctuations, which often move enormous quantities of water. Diastrophic events, particularly earthquakes, contribute to water motion, as does volcanic activity upon occasion. Even more fundamental are long-term changes in sea or lake level caused by tectonic forces and **eustatic forces** (the latter producing sea-level change entirely by increase or decrease in the amount of water in the world ocean) and to a lesser extent by the actions of continental ice sheets.

The forces that shape the topography of oceanic coastlines are similar to the forces acting on lakeshores, with three important exceptions:

1. Along lakeshores, the range of tides is so small that they are insignificant to landform development.
2. The causes of sea-level fluctuations are quite different from the causes of lake-level fluctuations.
3. Reefs are built only in tropical and subtropical oceans, not in lakes.

With these exceptions, the topographic forms produced on seacoasts and lakeshores are generally similar. Even so, the larger the body of water, the greater the effects of the coastal processes. Thus topographic features developed along seacoasts are normally larger, more conspicuous, and more distinctive than those found along lakeshores.

SHORE-SHAPING FORCES

There are seven principal processes that contribute to the shaping of coastal features. These are discussed in the following subsections.

Changes in Water Level Sea-level changes can result either from the uplift/sinking of a landmass (tectonic cause) or from an increase/decrease in the amount of water in the oceans (eustatic cause). During Earth's recent history, there have been many changes in sea level, sometimes worldwide and sometimes only around one or a few continents/islands. The eustatic changes of greatest magnitude and most extensive effect are those associated with seawater volume before, during, and after the Pleistocene glaciations. As a result of all these changes in sea level over the centuries, some present-day coastlines show emergent characteristics, in which shoreline topography of the past is now situated well above the contemporary sea level, while others show

Some of the world's most spectacular landforms are found along rugged coastlines. This is The Pinnacle near Point Lobos on the central coast of California. (*TLM photo*)

Figure 18-1 A coastline is the place where hydrosphere, lithosphere, and atmosphere meet. It is often an interface of ceaseless movement and energy transfer. Here a series of waves roll into the beach at Cape Byron, on the eastern coast of Australia. (TLM photo)

submergence characteristics, with a portion of a previous landscape now under water (Figure 18-2).

Most water-level changes in lakes have been less extensive and less notable. These changes are usually the result of the total or partial drainage of a lake, and their principal topographic expression is exposed ancestral beach lines and wave-cut cliffs above present lake levels.

Tides As we learned in Chapter 9, the waters of the world ocean oscillate in a regular and predictable pattern called tides, resulting from the gravitational influence of the sun and moon. The tides rise and fall in a cycle that takes about 12 hours, producing two high tides and two low tides a day on most (but not all) seacoasts.

Despite the enormous amount of water moved by tides and despite the frequency of this movement, the topographic effects are surprisingly small. Tides are significant agents of erosion only in narrow bays, around the margin of shallow seas, and in passages between islands, where they produce currents strong enough to scour the bottom and erode cliffs and shorelines (Figure 18-3).

(a)

(b)

Figure 18-2 **(a)** The deeply embayed coastline of the western shore of Chesapeake Bay in Maryland, a classic example of a submerging coast where the ocean has invaded river valleys, turning them into estuaries. (NASA/Peter Arnold, Inc.) **(b)** An uplifted coastal terrace well above the present shoreline. This is the central coast of California. (John Buitenkani/Photo Researchers, Inc.)

Figure 18-3   Under certain circumstances, tidal flow can be strong enough to shape landforms. These gigantic pedestal rocks on the edge of the Bay of Fundy in New Brunswick, Canada, were carved by the highest tides in the world. For scale, the spruce trees on top of the rocks are about 30 feet (9 meters) tall. (TLM photo)

Waves   Waves are generated largely by wind, and they are the most important wind-driven agents that shape landforms. As mentioned in Chapter 9, waves in the open ocean and those in the interior of large lakes are no more than shapes. As a wave passes a given point on the water surface, the water at that point rises up with the wave. After the wave has passed, the water drops back down to its original position. You can observe this phenomenon with a leaf floating on the surface of a still pond (Figure 18-4). Throw a stone into the water a few feet from the leaf. As the waves the stone creates spread out to where the leaf is, the leaf bobs up and down as the waves pass but never moves

horizontally. Once the pond surface is smooth again, you will find the leaf still at its original location. As waves reach shallow water, however, the bottom interferes with their motion. This interference causes the wave height to increase and the slope to steepen, a change that quickly causes the wave to collapse, or "break," sending a cascading mass of water shoreward (Figure 18-5 on page 514). This surge can carry sand and rock particles onto the beach, and it can pound onto rocky headlands and sea cliffs with considerable force (Figure 18-6 on page 514). When the wave is spent, the return flow sweeps seaward, carrying loose material with it.

(a)

(b)

(c)

(d)

Figure 18-4   Water waves are often no more than shapes. Their passage does not displace floating objects. In these drawings, the leaf is not moved horizontally at all.

FOCUS

WAVES

Waves are rising and falling motions transmitted from particle to particle in a substance, carrying energy from one place to another in the process. Our interest here is in water waves, which are simply undulations in the surface layers of a water body. Although water waves appear to move water horizontally, this appearance is misleading. Except where waves crest and break, the water through which a wave passes is shifted only very slightly.

Most water waves are wind generated, set in motion largely by the friction of air blowing across the water. This transfer of energy from wind to water initiates wave motion. Some water waves (called *forced waves*) are generated directly by wind stress on the water surface; they can develop to considerable size if the wind is strong and turbulent but usually last only for a limited time and do not travel far. Water waves become **swell** when they escape the influence of the generating wind, and swell can travel enormous distances. A small number of all water waves are generated by something other than the wind, such as a tidal surge, volcanic activity, or undersea diastrophic movement.

As a wave passes a given point on the water surface, the water particles at that point make a small circular or oscillatory movement, with very little forward motion (to oscillate means to move back and forth over the same space time and again). Waves that cause water to move this way are called *waves of oscillation*. As the wave passes, the water moves upward, producing a **wave crest**, as shown in Figure 18-1-A. Then crest formation is followed by a sinking of the surface that creates a **wave trough**. The horizontal distance from crest to crest or from trough to trough is called the **wavelength**. The vertical dimension of wave development is determined by the circular orbit of the surface water particles as the wave form passes; the vertical distance from crest to trough is equivalent to the diameter of this orbit and is called the **wave height**. The height of any wave depends on wind speed, wind duration, water depth, and fetch (the area of open water). **Wave amplitude** is one-half the height; in other words, the vertical distance from the still-water level either upward to the crest or downward to the trough.

The passage of a wave of oscillation normally moves water only very slightly in the direction of flow. Thus an object floating on the surface simply bobs up and down without advancing, except as it may be pushed by the wind. The influence of wave movement diminishes rapidly with depth; even very high waves stir the subsurface water to a depth of only a few tens of yards.

Waves often travel great distances across deep water with relatively little change in speed or shape. As they roll into shallow water, however, a significant metamorphosis occurs. When the water depth becomes equal to about half the wavelength, the wave motion begins to be affected by frictional drag on the sea bottom. The waves of oscillation then rapidly become changed into *waves of translation*, and the result is significant horizontal movement of the surface water. Friction retards the progress of the waves so that they are slowed and bunch together, marking a decrease in wavelength, while at the same time their height is increased. As the wave becomes higher and steeper, frictional drag becomes even greater, which causes the wave to tilt forward and become more and more unstable. Soon and abruptly the wave breaks (Figure 18-1-B), collapsing into whitewater surf or plunging forward as a breaker, or, if the height is small, perhaps simply surging up the beach without cresting. When a wave breaks, the motion of the water instantly becomes turbulent, like that of a swift river. The breaking wave rushes toward shore or up the beach as **swash**. The momentum of the surging swash is soon overcome by fric-

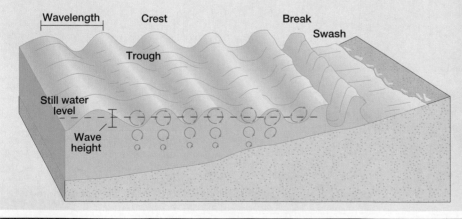

Wavelength   Crest   Break   Swash   Trough   Still water level   Wave height

Figure 18-1-A In deep water, the passage of a wave involves almost circular movements of individual water particles. Agitation diminishes rapidly with depth, as shown here by decreasing orbital diameter downward. As the wave moves into shallow water, the orbits of the revolving particles become more elliptical, the wavelength becomes shorter, and the wave becomes steeper. Eventually the wave "breaks" and dissipates its remaining energy as it washes up onto the beach.

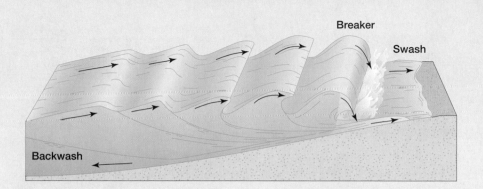

Figure 18-1-B   A breaking wave. In shallow water the ocean bottom impedes oscillation, causing the wave to become increasingly steeper until it is so oversteepened that it collapses and tumbles forward as a breaker. The surging water then rushes up the beach as swash.

tion, and a reverse flow, called **backwash**, drains much of the water seaward again, usually to meet the oncoming swash of the next wave.

Waves often change direction as they approach the shore, a phenomenon known as **wave refraction**. It occurs because a line of waves does not approach exactly parallel to the shore, either because the coastline is uneven or because of irregularities in water depth in the nearshore zone. For one or more of these reasons, one portion of a wave reaches shallow water sooner than other portions and is thus slowed down. This slowing down causes the wave line to bend (the physicist's term for this bending is *refraction*) as it pivots toward the obstructing area. The wave energy then becomes concentrated in the vicinity of the obstruction and is diminished in other areas (Figure 18-1-C).

The most conspicuous geomorphic result of wave refraction is the focusing of wave action on headlands, subjecting them to the direct onslaught of pounding waves, whereas an adjacent bay experiences much gentler, low-energy wave action. Other things being equal, the differential effect of wave refraction tends to smooth the coastal outline by wearing back the headlands and increasing sediment accumulation in the bays.

Occasionally, major oceanic wave systems are triggered by undersea tectonic or volcanic events. These waves, called **tsunamis** or **seismic sea waves** (improperly called *tidal waves*), are inconspicuous in the open sea because they are low and have enormous wavelengths (usually several tens of miles), although they can move several hundred miles per hour. As they approach a coast, they

build to considerable height—on rare occasions as much as 100 feet (30 meters)—and can cause great devastation as they engulf the coastal zone. They occur rarely but can be disastrous. Happily, tsunamis can usually be anticipated long enough in advance to allow time for evacuation of the impact area, as they usually originate in deep-ocean trenches from earthquake shocks that are readily detectable by seismographs.

Whether they are awesome tsunamis or mild swells, the peculiar contradiction of water waves is that they normally pass harmlessly under such fragile things as boats or swimmers in open water but can wreak devastation on even the hardest rocks of a shoreline. In other words, a wave of oscillation is a relatively gentle phenomenon, but a wave of translation can be a powerful force of destruction.

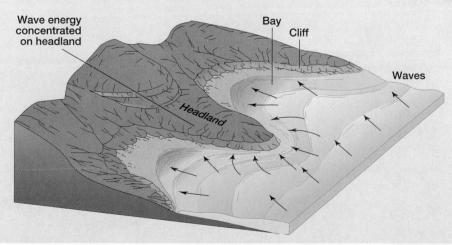

Figure 18-1-C   Refraction of waves on an irregular coastline. The waves approach the headland first and then pivot toward it. Thus wave energy is concentrated on the headlands and is diminished in the bays.

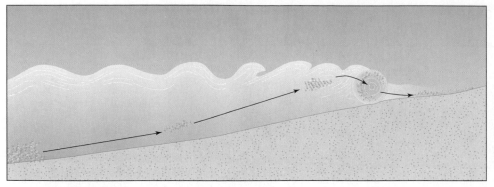

Figure 18-5   As a wave approaches a shoreline, the water becomes shallow, the wave form drags on the bottom, and the oversteepened wave then "breaks" or cascades forward toward the shore.

Currents   Currents consist of large volumes of water moving horizontally. Many kinds of currents flow in the oceans and lakes of the world, usually confined to surface areas but sometimes deeper. Coastal topography is affected most by **longshore currents**, in which the water moves roughly parallel to the shoreline in a generally downwind direction. (Think of longshore as a contraction for "along the shore.") Such currents can exert a significant erosive force on weak coastal rocks. Moreover, they can transport and deposit vast amounts of sand and other sedimentary material.

Stream Outflow   The outflow of streams and rivers into oceans and lakes is included here in our discussion of shore-shaping forces because these outflows are important feeders of sediments to oceans and lakes. They provide much of the sand and other sedimentary material that is moved around and deposited by coastal waters (Figure 18-7).

Ice Push   The shores of bodies of water that freeze over in winter are sometimes significantly affected by ice push, which is usually the result of the contraction and expansion that occur when the water freezes and thaws as the weather changes. As more and more water turns to ice and therefore expands in volume (recall the discussion of frost wedging in Chapter 15), nearshore ice is shoved onto the land, where it can deform the shoreline by pushing against it, more or less in the fashion of a small glacial advance.

Ice push is usually unimportant on seashores outside the Arctic and Antarctic, but it can be responsible for numerous minor alterations of the shorelines of high-latitude or high-altitude lakes.

Organic Secretions   Several primitive aquatic animals and plants produce solid masses of rocklike material by secreting lime (which is the lay term for calcium carbonate). By far the most significant of these organisms

Figure 18-6   The continuous pounding of waves can erode even the most resistant coastal rocks. This scene is near Port Edward on the Natal coast of South Africa. (TLM photo)

Figure 18-7 The sediment plume of the Betsiboka River of Madagascar spreads broadly into the blue waters of the Mozambique Channel, which separates the island of Madagascar from the African coast. (NASA Headquarters)

is the coral polyp, a tiny animal that builds a hard external skeleton of calcium carbonate and then lives inside it. Coral polyps are of many species, and they cluster together in social colonies of uncounted billions of individuals. Under favorable conditions (clear, shallow, salty warm water), the coral can accumulate into enormous masses, forming reefs, platforms, and atolls, all commonplace features in tropical and subtropical oceans.

EROSION

The most notable erosion along coastlines is accomplished by wave action. The incessant pounding of even small waves is a potent force in wearing away the shore, and the enormous power of storm waves almost defies comprehension. Waves break with abrupt and dramatic impact, hurling water, debris, and air in a thunderous crash onto the shore. Spray from breaking waves commonly moves as fast as 70 miles (113 kilometers) per hour, and small jets have been measured at more than twice that speed. This speed coupled with the sheer mass of the water involved in such hydraulic pounding is responsible for much coastal erosion, which is made much more effective by the abrasive rock particles carried by the waves (Figures 18-8 and 18-9).

Wherever the land along a shore is rocks or cliffs rather than sand, there is another dimension to wave erosion: the air forced into cracks in the rock as the wave hits the shore. The resulting compression is abruptly released as the water recedes, allowing instant expansion of the air. This pneumatic action is often very effective in loosening rock particles of various sizes.

Chemical action also plays a part in the erosion of rocks and cliffs because most rocks are to some extent soluble in seawater. In another form of chemical action, salts from seawater crystallize in the crevices and pores of onshore rocks and cliffs, and this deposition is a further mechanism for weakening and breaking up the rock.

On shores made up of cliffs, the most effective erosion takes place just at or slightly above sea level, so that a notch is cut in the base of the cliff. The cliff face then retreats as the slope above the undercutting collapses (Figure 18-10). The resulting debris is broken, smoothed, and made smaller by further wave action, and eventually most of it is carried seaward.

Where a shoreline is composed of sand or other unconsolidated material, currents and tides also cause rapid erosion. Storms greatly accelerate the erosion of sandy shores; a violent storm can remove an entire beach in just a few hours, cutting it right down to bedrock.

Figure 18-8 The incessant pounding of waves on this soft-rock headland on the southern coast of the state of Victoria, Australia produced a double arch that eventually eroded into a single arch. The double-arch photograph was taken in 1985 and the single-arch one in 1992. The view is in Port Campbell National Park. (TLM photos)

COASTAL SEDIMENT TRANSPORT

Nearly all movement of rock debris along coastlines is accomplished by wave action. The most obvious, but not always the most significant, movements involve the short-distance shifting of sand directly onshore by breaking waves and directly offshore by the retreating water. The debris (mostly sand) that advances and retreats with the repeated surging and ebbing of the waves abrades and smooths the beach, while the continuous abrasion reduces the size of the particles.

A movement generally of greater consequence is **beach drifting** along a coastline, a zigzag movement of particles that results in downwind displacement parallel to the coast (Figure 18-11). Nearly all waves approach the coast obliquely rather than at a right angle, as Figure 18-1 shows, and therefore the sand and other debris carried onshore by the breaking wave move up the beach at an oblique angle. Some of the water soaks into the beach, but much of it returns seaward directly downslope, which is normally at a right angle to the shoreline.

Figure 18-9 Pounding waves do not erode evenly and smoothly. Stacks and cliffs testify to the principle of differential erosion in this scene from Port Campbell National Park, Victoria, Australia. (TLM photo)

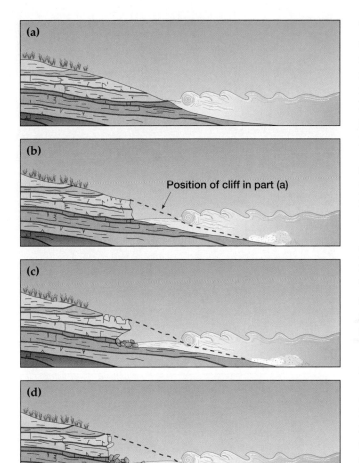

Figure 18-10 **(a)** Waves pounding an exposed rocky shoreline erode the rock most effectively at water level, with the result that a notch may be cut in the face of the headland. **(b)** The presence of this notch undermines the higher portion of the headland, which may subsequently collapse **(c)**, producing a steep cliff. **(d)** The notching/undercutting/collapse sequence may be repeated many times, causing the cliff face to retreat.

This return flow takes some of the debris with it, much of which is picked up by the next surging wave and carried shoreward again along an oblique path. This infinitely repetitious pattern of movement shifts the debris farther and farther along the coastline longitudinally. Because wind is the driving force for wave motion, the strength, direction, and duration of the wind are the principal determinants of beach drifting.

Longshore currents are prominent transporters of sand and other sediment, moving material generally parallel to the coast. Tides are not very effective in moving coarser debris, but where channels are narrow and relatively shallow tides can move fine material.

Some debris transport along shorelines is accomplished directly by the wind. Wherever waves have carried or hurled sand and finer grained particles to positions above the water level, these particles can be picked up by a breeze and moved overland. This type of movement frequently results in dune formation and sometimes moves sand a considerable distance inland (Figure 18-12).

COASTAL DEPOSITION

Although the restless waters of coastal areas accomplish notable erosion and transportation, in many cases the most conspicuous topographic features of a shoreline are formed by deposition. Just as in streamflow over the surface of a continent, marine deposition occurs wherever the energy of moving water is diminished.

Maritime deposits along coastlines tend to be more ephemeral than noncoastal deposits. This is due primarily to the composition of marine deposits, which typically consist of relatively small particles (sand and gravel), and to the fact that the sand is not stabilized by a vegetation cover. Most coastal deposits are under constant onslaught by agitated waters, which can

Figure 18-11 Beach drifting involves a zigzag movement in a general downwind direction along the coast. Sand is brought obliquely onto the beach by the wave and then is returned seaward by the backwash.

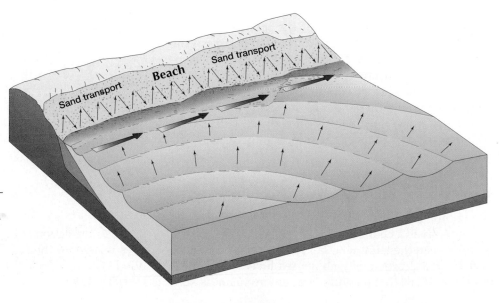

Figure 18-12 Sand sometimes is heaped into dunes that cover extensive areas. One of the largest sand accumulations in North America is along the central coast of Oregon, near Florence. (Farrell Grehan/Photo Researchers, Inc.)

rapidly wash away portions of the sediment. Consequently, the sediment budget must be in balance if the deposit is to persist; for the budget to be in balance, removal of sand must be offset by addition of sand. Most marine deposits have a continuing sediment flux, with debris arriving at one end and departing at the other. During storms, the balance is often upset, with the result that the deposit is either significantly reshaped or totally removed.

SIGNIFICANCE OF RECENT SEA-LEVEL FLUCTUATIONS

One of the most prominent factors influencing coastal topography is a change in sea level. Where the sea rises relative to the land, whether the cause be eustatic or tectonic, land previously above sea level is drowned, with the result that a continental environment is sub-

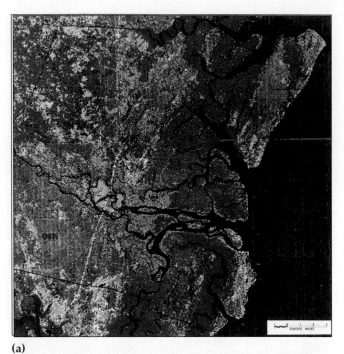

(a)

(b)

(c)

Figure 18-13 (a) A ria shoreline, showing numerous estuaries, along the Atlantic coast of Georgia, near Brunswick. (NASA/Peter Arnold, Inc.) (b,c) These drawings show a coastal area being flooded by the ocean.

jected to the actions of coastal waters. Where the land rises relative to the sea, land previously at sea level is lifted high and dry, with the result that previously submarine landscape is exposed to coastal processes. The topographic result of such fluctuations is usually more pronounced in areas of low relief, but even where coastal mountains are involved, terrain modification may be significant.

Submergence Almost all the world's oceanic coastlines show evidence of submergence during the last 15,000 years or so, a result of the melting of Pleistocene ice. As water from melting glaciers returned to the oceans, rising sea level caused widespread submergence of coastal zones.

The most prominent result of submergence is the drowning of previous river valleys, which produces **estuaries**, or long fingers of seawater projecting inland. A coast along which there are numerous estuaries is called a **ria shoreline** (Figure 18-13); a *ria* (from the Spanish *ría*, river) is a long, narrow inlet of a river that gradually decreases in depth from mouth to head. If a hilly or mountainous coastal area is submerged, numerous offshore islands may indicate the previous location of hilltops and ridge crests.

Emergence Evidence of previously higher sea levels sometimes is related to ice melting during past interglacial ages but more often is associated with tectonic uplift. The clearest topographic result of coastal emergence is shoreline features raised well above the present water level. Often the emerged portion of a continental shelf appears as a broad, flat coastal plain.

The Global Warming Phenomenon In Chapter 4 we discussed the possibility of global warming in the near future. If the scenario of a generally continuing increase in worldwide temperatures is indeed correct, we can anticipate another period of deglaciation, with the ice sheets of Antarctica and Greenland slowly melting. Such a situation would cause a global eustatic rise in sea level that would inundate many islands and coastal plains of the world and have a profound effect on all humankind.

COASTAL LANDFORMS

Certain coastal landforms are widespread on Earth. How they develop illustrates many of the processes we have just discussed.

BEACHES

The most widespread marine depositional feature is the **beach**, which is an exposed deposit of loose sediment adjacent to a body of water. Although the sediment can range in size from fine sand to large cobbles, it is usually relatively homogeneous in size on a given section of beach. Beaches composed of smaller particles (which is to say sand because silt and clay get carried away in suspension and do not form beaches) are normally broad and slope gently seaward, whereas those formed of larger particles (gravel, cobbles) generally slope more steeply.

Beaches occupy the transition zone between land and water, sometimes extending well above the normal sea level into elevations reached only by the highest storm waves. On the seaward side, they generally extend down to the level of the lowest tides and can often be found at still lower levels, where they merge with muddy bottom deposits. Figure 18-14 portrays an idealized beach profile. The **backshore** is the upper part of the beach, landward of the high-water line. It is usually dry, being covered by waves only during severe storms. It contains one or more **berms,** which are flattish wave-deposited sediment platforms. The **foreshore** is the zone that is regularly covered and uncoverd by the rise and fall of tides. The **offshore** is the zone that is permanently under water; it is the place where waves break and where surf action is greatest.

Beaches sometimes extend for dozens of miles along straight coastlines, particularly if the relief of the land is slight and the bedrock unresistant. Along irregular shorelines, beach development may be restricted largely or entirely to bays, with the bays frequently alternating with rocky headlands.

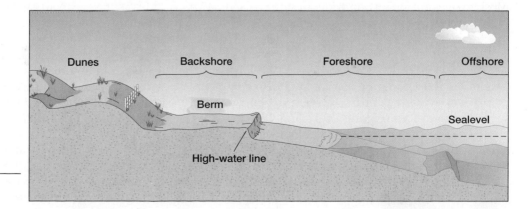

Figure 18-14 An idealized beach profile.

PEOPLE AND THE ENVIRONMENT

HUMAN INFLUENCES ON COASTAL TOPOGRAPHY

Worldwide, humans have made very few changes in the terrain of most ocean and lake coastlines. In some areas, however, natural coastal processes have been significantly influenced by human enterprise, resulting in major topographic modification.

The most prominent terrain reshaping is associated with port activities in urban areas, where dredging, draining, blasting, digging, and other construction endeavors have completely rearranged the landscape so that the reshaped port bears little resemblance to the original shoreline.

Beyond commercial harbors, human interference with coastal processes is mostly associated with beaches. In part this interference is due to the great recreational attractiveness of beaches, but it also reflects the fact that beaches are ephemeral and easily modified by both natural and artificial processes. A typical beach may be nourished during the calm season (usually summer) by sand deposition from low-energy swell and then stripped of most or all of its sand by the erosive action of high-energy waves during the storm season (normally winter).

These fluctuations generally are not well received by people who own beachfront property or municipalities that need beach recreational areas for their citizens. To remedy the situation, artificial structures are often built to interrupt the movement of sand. This interruption must be done with care, however, as any modification of the sediment balance in one place is likely to affect other locales along the shore as well.

Structures designed to help control coastal sand movement are usually built either perpendicular or parallel to the shoreline. Two major perpendicular structures are groins and jetties, with the former particularly favored for retaining beach sand. A **groin** is a short wall built out from the beach to impede the longshore current and force sand deposition on the upshore side of the groin as the current speed is diminished. It is inevitable that deposition along a groin robs downshore beaches of incoming sediment. Thus the current often removes sand from the beach downshore of the groin. To offset this effect, another groin may be built downshore, and then another and another. Such famous recreational shorelines as Florida's Miami Beach and much of coastal New Jersey are adorned with hundreds of groins.

A **jetty** is a groin built along one side of a river or harbor entrance, and most often the entrance is lined with two jetties, as Figure 18-2-A shows. The idea is to confine the flow of water to a narrow zone, thereby keeping the sand carried by the water in motion and inhibiting its deposition in the navigation channel. Jetties tend to interfere with longshore currents in the same way groins do.

The two main parallel structures built to modify a coastline are breakwaters and seawalls. A **breakwater** is an offshore wall designed to absorb the force of breaking waves and thus create a quiet water area inshore. The reduced wave activity afforded by the breakwater often facilitates sand deposition around the breakwater, thereby starving downshore beach areas.

Beach shape may change greatly from day to day and even from hour to hour. Normally beaches are built up during quiet weather and removed rapidly during storms. Most beaches are longer and wider in summer and greatly worn away by the storminess of winter.

CLIFFS/BENCHES/TERRACES

One of the most common coastal landform complexes comprises wave-cut cliffs, wave-cut benches, and wave-built terraces (Figures 18-15 and 18-16). As discussed earlier, as waves eat away at a rocky headland, steep **wave-cut cliffs** are formed, and these cliffs receive the greatest pounding at their base, where the power of the waves is concentrated. A combination of hydraulic pounding, abrasion, pneumatic push, and chemical solution at the cliff base frequently cuts a notch at the high-water level. As the notch is enlarged, the overhang sporadically collapses, and the cliff recedes as the ocean advances.

Seaward of the cliff face, the pounding and abrasion of the waves create a broad erosional platform called a **wave-cut bench**, usually slightly below water level. The combination of wave-cut cliff and wave-cut bench produces a profile that resembles a letter L, with the steep vertical cliff descending to a notched base and the flat horizontal bench extending seaward.

The debris eroded from cliff and bench is mostly removed by the swirling waters. The larger fragments are battered into smaller and smaller pieces until they are small enough to be transportable. Some of the sand

Seawalls, often called *revetments*, are structures built along the shore or along the backshore slope of a beach to intercept breaking waves, particularly storm waves, and diminish their erosive effect on the shoreline.

Upstream flood-control projects (particularly dams) on rivers and streams, sometimes well inland, can also influence coastal processes, primarily by diminishing the supply of sand and silt transported from stream to ocean. The stream delivers less sediment to the ocean, but the work of waves and currents along the shore continues. The result is increased beach erosion without a full sand supply for replenishment; hence the beach eventually starves unless some other provision is made for nourishing it.

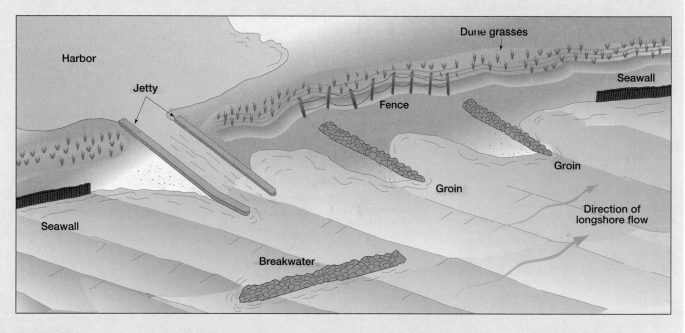

Figure 18-2-A Structures built to modify sand erosion, transportation, and deposition.

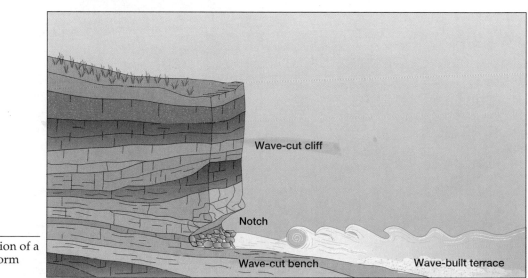

Figure 18-15 Cross section of a cliff/bench/terrace landform complex.

Figure 18-16   A conspicuous wave-cut notch in resistant coralline rock on the Fiji island of Vanua Levu. (TLM photo)

and gravel produced in this fashion may be washed into an adjacent bay to become, at least temporarily, a part of the beach. Much of the debris, however, is shifted directly seaward, where a great deal of it is deposited just beyond the wave-cut bench as a **wave-built terrace**. With the passage of time and the wearing away of the cliff by weathering and erosion, the terrace may become so large that it buries the bench entirely. Here the result is formation of a beach that extends to the base of the cliff.

BARRIER ISLANDS/LAGOONS

Another prominent coastal deposition is the **barrier island** (Figure 18-17), a long, narrow sandbar built up in shallow offshore waters, sometimes only a few hundred yards from the coast but often several miles at sea. Barrier islands are always oriented approximately parallel to the shore. They are believed to result from the heaping up of debris where large waves (particularly storm waves) begin to break in the shallow waters of continental shelves. However, many larger barrier islands may

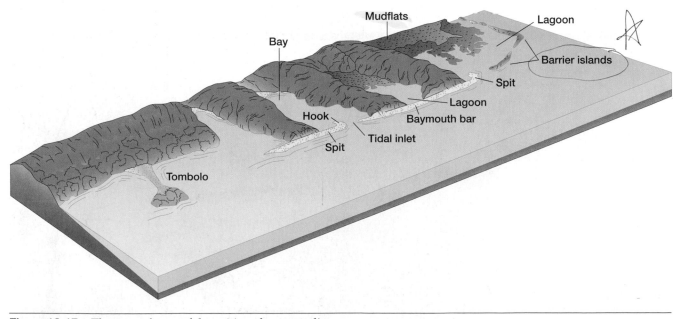

Figure 18-17   The many forms of deposition along coastlines.

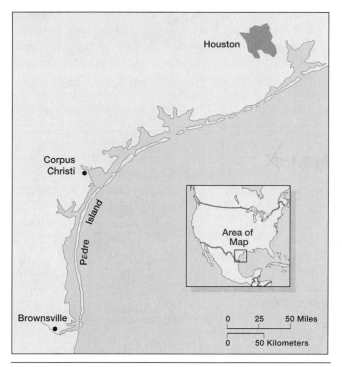

Figure 18-18 Most of the Atlantic and Gulf coasts of the United States are fringed by barrier islands and lagoons. This map depicts the situation along the Texas coast.

have more complicated histories linked to the lowered sea level that resulted when a large amount of seawater was locked up in glaciers during Pleistocene times.

Barrier islands often become the dominant element of a coastal terrain. Although they usually rise at most only a few feet above sea level and are typically only a few hundred yards wide, they may extend many miles in length. Most of the Atlantic and Gulf of Mexico coastline of the United States, for instance, is paralleled by lengthy barrier islands, several more than 30 miles (48 kilometers) long (Figure 18-18).

An extensive barrier island isolates the water between itself and the mainland, forming a body of quiet salt or brackish water called a **lagoon** (Figure 18-19). Over time, a lagoon becomes increasingly filled with water deposited sediment from coastal streams, wind-deposited sand from the barrier island, and tidal deposits if the lagoon has an opening to the sea. All three of these sources contribute to the build up of **mudflats** on the edges of the lagoon. Unless tidal inlets across the barrier island permit vigorous tides or currents to carry lagoon debris seaward, therefore, the ultimate destiny of most lagoons is to slowly be transformed first to marshes and then to meadows (Figure 18-20).

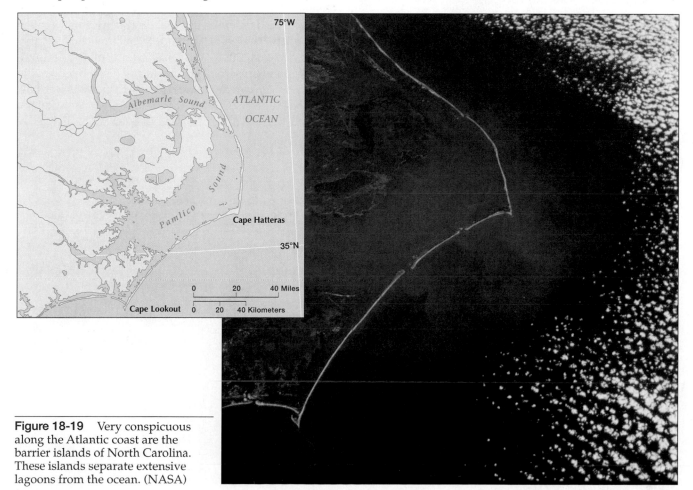

Figure 18-19 Very conspicuous along the Atlantic coast are the barrier islands of North Carolina. These islands separate extensive lagoons from the ocean. (NASA)

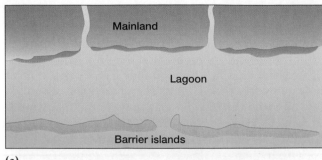

Mainland

Lagoon

Barrier islands

(a)

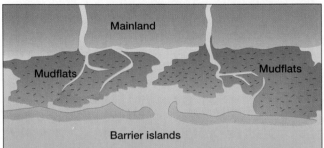

Mainland

Mudflats   Mudflats

Barrier islands

(b)

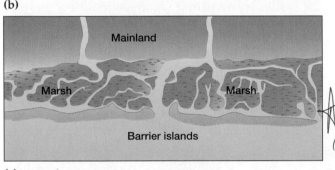

Mainland

Marsh   Marsh

Barrier islands

(c)

Figure 18-20   Barriers are separated from the mainland by lagoons. With the passage of time, the lagoons often become choked with sediment, and become converted to marsh, mudflat, or meadow.

In addition to infilling by sediment, another factor contributes to a lagoon's disappearance. After a barrier island becomes a certain size, it often begins to migrate slowly shoreward as waves wear away its seaward shore and sediments accumulate to build up its landward shore. Eventually, if the pattern is not interrupted by such things as changing sea level, the island and the mainland shore will merge.

SPITS

Nearshore ocean waters normally contain a considerable amount of fine-grained sediment that is shifted about by waves, currents, and tides. At the mouth of a bay, sediment may be moved by longshore drift into deeper water. There the flow speed is slowed and the sediment is deposited. The growing bank of land guides the current farther into the deep water, where still more material is dropped. Any such linear deposit attached to the land at one end and extending into open water is called a **spit** (Figure 18-17).

Although most spits are straight, sandy peninsulas projecting out into a bay or other coastal indentation, the vicissitudes of currents, winds, and waves often give them other configurations. In some cases, the spit becomes extended clear across the mouth of a bay to connect with land on the other side, producing what is called either a **bay barrier** or a **baymouth bar** and transforming the bay to a lagoon. Another common modification of spit shape is caused by conflicting water movements in the bay, which can cause the deposits to curve toward the mainland, forming a **hook** at the outer end of the spit (Figure 18-21). A less common but even more distinctive

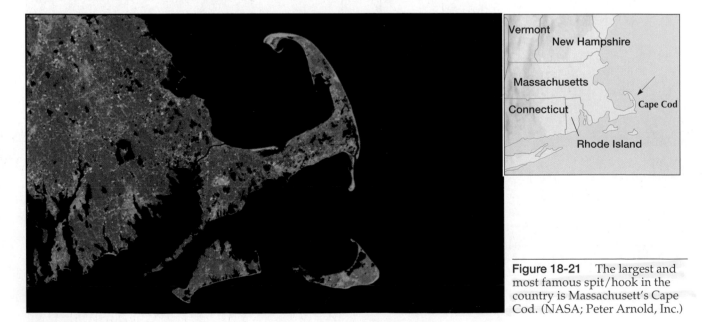

Vermont

New Hampshire

Massachusetts

Connecticut   Cape Cod

Rhode Island

Figure 18-21   The largest and most famous spit/hook in the country is Massachusett's Cape Cod. (NASA; Peter Arnold, Inc.)

development is a **tombolo**. This is a spit formed when waves converging in two directions on the landward side of a nearshore island deposit their sand, so that the bar connects the island to the land (Figure 18-22).

FJORDED COASTS

The most spectacular coastlines occur where high-relief coastal terrain has undergone extensive glaciation. Troughs once gouged out either by glaciers or by ice sheets have been cut so deep that their bottoms are presently far below sea level. Gradually, therefore, as the ice has melted, the troughs have filled up with seawater. In some localities these deep, sheer-walled coastal indentations—called **fjords**—are so numerous that they create an extraordinarily irregular coastline, often with long, narrow fingers of saltwater reaching more than 100 miles (160 kilometers) inland.

The most extensive and spectacular fjorded coasts are in Norway, western Canada, Alaska, southern Chile, the South Island of New Zealand, Greenland, and Antarctica (Figure 18-23).

CORAL COASTS

In tropical oceans, nearly all continents and islands are fringed with either **coral reefs** or some other type of coralline formation (Figure 18-24). As mentioned earlier, coral polyps are tiny marine animals that secrete external skeletons of calcium carbonate, and this material is commonly known as coral (Figure 18-25). The polyps live in shallow tropical waters and can build coralline formations anywhere a coastline provides a stable foundation. Coral reefs in the shallows off the coasts of Florida, for instance, are built on such stable bases. The famous Great Barrier Reef off the northeastern coast of Australia is an immense shallow-water

Figure 18-22 A famous tombolo is found on the northwestern coast of France. A large rock, called Mont Saint Michel, is tied to the mainland by a narrow sandbar. The Village of Mont Saint Michel has been built on the rock.

Figure 18-23 A typical fjord. This is Olden Fjord in Norway. (Harvey Lloyd/The Stak Market)

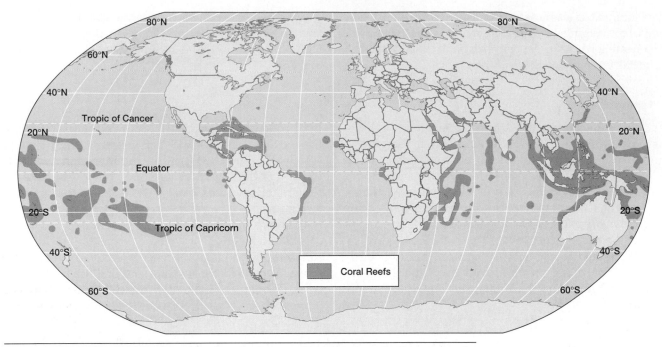

Figure 18-24   Distribution of coral reefs and other coralline structures in the oceans of the world.

platform of bedrock largely but not entirely covered with coral. Its enormously complex structure includes many individual reefs, irregular coral masses, and a number of islands (Figure 18-26).

One favored location for coral reefs is around a volcanic island in tropical waters; as the volcano forms and then subsides, the following different types of reefs grow upward. When the volcano first forms, as, for example, over the Hawaiian hot spot described in Chapter 14, coral accumulates on the part of the mountain flank just below sea level because it is in these shallow waters that polyps live. The result is a reef built right onto the volcano, as shown in Figure 18-27; such an attached reef is called a **fringing reef** (Figure 18-28).

Figure 18-25   A small coral head in the Caribbean Sea. (Andrew J. Martinez/Photo Researchers, Inc.)

Figure 18-26 A view of Australia's Great Barrier Reef. (Tony Roberts/Sportshot/Photo Researchers, Inc.)

(a) Active volcano

(b) Dormant volcano

(c) Submerged volcano

Figure 18-27 Coral reef formation around a sinking volcano. **(a)** Around a newly formed volcano rising above the water of a tropical ocean, secretions from coral polyps living along the shallow-water flanks of the volcano accumulate into a fringing reef attached to the mountain. **(b)** As the volcano becomes dormant and begins to sink, the coral continues to grow upward over the original base, essentially a cylinder surrounding the mountain. Breaks in the coral ring allow water in, and a lagoon forms between coral and mountain. Such a reef separated from its mainland by a lagoon is a barrier reef. **(c)** Once the volcano is completely submerged, the coral surrounding a landless lagoon is called an atoll reef.

As new layers are laid down over old, the coral builds upward around the volcano as a cylinder of irregular height. At the same time, the volcano is sinking and pulling the original reef base downward. When the coral has been built up enough and the volcano has sunk enough, the result at the water surface is a coral ring separated by a lagoon from the part of the volcano still above water, as in Figure 18-27(b). This ring of coral, which may be a broken circle because of the varying thickness of the coral, appearing to float around a central volcanic peak (but in actuality attached to the flanks of the sinking mountain far below the water surface), is called a **barrier reef.** The surface of a barrier reef is usually right at sea level, with some portions projecting upward into the air.

Coral polyps continue to live in the upper, shallow-water portions of a barrier reef, and so the reef continues to grow upward. Once the top of the volcano sinks below the water surface, the reef sur-

rounding a now landless lagoon is called an **atoll**. The term *atoll* implies a ring-shaped structure. In actuality, however, the ring is rarely unbroken; rather it consists of a string of closely spaced coral islets separated by narrow channels of water. Each individual islet is called a **motu**.

Figure 18-28 A part of the fringing reef on the island of Moorea in French Polynesia. (Jack Fields/Photo Researchers, Inc.)

CHAPTER SUMMARY

The principal forces shaping coastlines are changes in sea level, tides, waves, currents, stream outflow, ice push, and organic secretions.

Waves cause erosion as they constantly batter the rocks, cliffs, and/or sand of a coastline. Waves and longshore currents transport sediments along a coast, usually in the downwind direction. Because of the constant movement of waves, coastal deposition results in built-up land formations whenever the sediment budget is in balance, which happens when the sand-addition rate equals the sand-removal rate. Most marine deposits have a continuing sediment flux, with debris arriving at one end and departing at the other.

Landforms along coastlines include beaches, wave-cut cliffs, wave-cut benches, wave-built terraces, barrier islands and their lagoons, spits, fjords, and coral reefs.

KEY TERMS

atoll
backshore
backwash
barrier island
barrier reef
bay barrier
baymouth bar
beach
beach drifting
berm
breakwater
coral reef
estuary
eustatic forces
fjord

foreshore
fringing reef
groin
headland
hook
jetty
lagoon
longshore current
motu
mudflat
offshore
ria shoreline
seawall
seismic sea wave
spit

swash
swell
tombolo
tsunami
wave amplitude
wave-built terrace
wave crest
wave-cut bench
wave-cut cliff
wave height
wavelength
wave refraction
wave trough

REVIEW QUESTIONS

1. What effect did Pleistocene glaciation have on sea level?
2. How does air serve as a tool of erosion in wave action?
3. Explain the roles of beach drifting, tides, and currents in sand transportation.
4. Explain the concept of the sediment budget of a coastal depositional landform.
5. What causes beaches to change shape and size?
6. Distinguish between the topography of shorelines of submergence and that of shorelines of emergence.
7. Why are coastlines of submergence so common today?
8. Explain how a wave-cut cliff forms.
9. What happens to most coastal lagoons with the passage of time?
10. Explain how a fringing coral reef forms and how it can subsequently become transformed first to a barrier reef and then to an atoll.

SOME USEFUL REFERENCES

BIRD, ERIC C. F., *Coasts: An Introduction to Coastal Geomorphology*, 3rd ed. New York: Basil Blackwell, 1984.

BIRD, ERIC C. F., *Submerging Coasts: The Effects of a Rising Sea Level on Coastal Environments*. New York: John Wiley & Sons, 1993.

DOLAN, R. AND H. LINS, "Beaches and Barrier Islands," *Scientific American*, Vol. 257, July 1987, pp. 68–77.

HARDISTY, J., *Beaches: Form and Process*. New York: Harper Collins Academic, 1990.

LEATHERMAN, STEPHEN, *Barrier Island Handbook (Coastal Publications Series)*. College Park: Laboratory for Coastal Research, University of Maryland, 1988.

SUNAMURA, TSUGUO, *Geomorphology of Rocky Coasts*. New York: John Wiley & Sons, 1992.

TOOLEY, MICHAEL J., AND IAN SHENNAN (eds.), *Sea-Level Changes*. New York: Basil Blackwell, 1987.

TRENHAILE, ALAN S., *The Geomorphology of Rock Coasts*. New York: Oxford University Press, 1987.

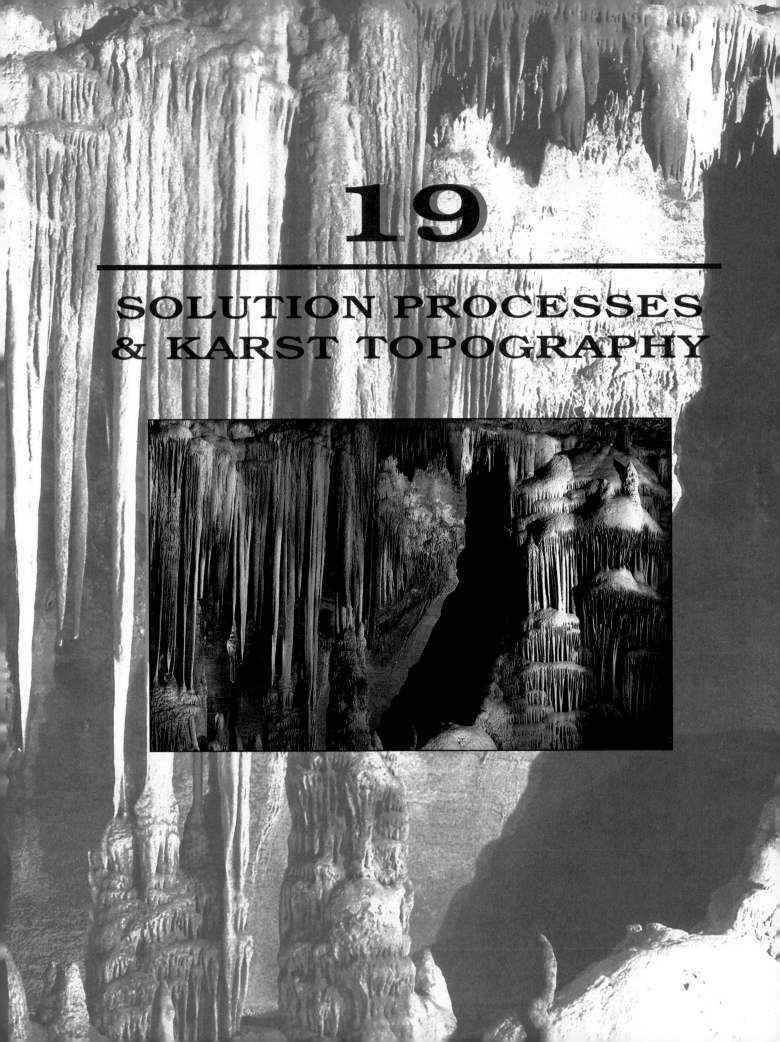

19

SOLUTION PROCESSES & KARST TOPOGRAPHY

IN STUDYING TOPOGRAPHIC DEVELOPMENT, WE PAID A great deal of attention to the role of water. We noted that water running across the ground is the most significant external shaper of terrain and that coastal waters produce distinctive landforms around the margins of oceans and lakes. In both cases, the water moves rapidly and much energy is expended in erosion, transportation, and deposition.

In this brief chapter we focus on underground water, which, because it is confined, functions in a much more restricted fashion than surface water. Underground water is largely unchanneled and therefore generally diffused, and it moves very slowly for the most part. Consequently, it is almost totally ineffective in terms of hydraulic power, corrosion, and other kinds of mechanical erosion.

THE IMPACT OF SOLUTION PROCESSES ON THE LANDSCAPE

The mechanical effects of underground water have only a limited influence on topographic development. Some subsurface mechanical weathering does take place, but the surface landscape rarely is directly affected by it, although certain forms of mass wasting (such as earthflows and slumps) are facilitated when loose materials are lubricated by underground water.

Through its chemical action, however, underground water is an effective shaper of the topographic landscape. Water is a solvent for certain rock-forming chemicals, dissolving them from rock and then carrying them away in solution and depositing them elsewhere. Under particular circumstances, the aboveground results of this dissolution are widespread and distinctive.

Underground water also affects surface topography via the creation of such hydrothermal features as hot springs and geysers, formed when hot water from underground is discharged at the ground surface.

SOLUTION AND PRECIPITATION

The chemical reactions involving underground water are relatively simple. Although pure water is a poor solvent, almost all underground water is laced with enough chemical impurities to make it a good solvent for the compounds that make up a few common minerals (Figure 19-1). Basically, underground water is a weak solution of carbonic acid because it contains dissolved carbon dioxide gas.

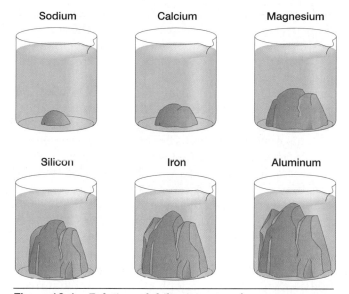

Figure 19-1 Relative solubility in water of some common rock-forming elements. The darker portion of each diagram indicates the proportion of the element that is *insoluble* in water. Sodium and calcium can be almost completely dissolved, for instance, whereas iron and aluminum are essentially insoluble.

Dissolution is an important weathering/erosion process for all rocks, but it is particularly effective on carbonate sedimentary rocks, particularly limestone. A common sedimentary rock, limestone is composed largely of calcium carbonate, which reacts strongly with carbonic acid solution to yield calcium bicarbonate, a compound that is very soluble in (and thus easily removed by) water. Other limy rocks, such as gypsum and chalk, undergo similar reactions. Dolomite is a calcium magnesium carbonate rock that dissolves almost as quickly as limestone. The pertinent chemical equations for the dissolving of limestone and dolomite are as follows:

$$\underset{\text{lime}}{CaCO_3} + \underset{\text{water}}{H_2O} + \underset{\substack{\text{carbon}\\\text{dioxide}}}{CO_2} \rightarrow \underset{\text{calcium bicarbonate}}{Ca(HCO_3)_2}$$

$$\underset{\text{dolomite}}{CaMg(CO_3)_2} + \underset{\text{water}}{2H_2O} + \underset{\substack{\text{carbon}\\\text{dioxide}}}{2CO_2} \rightarrow \underset{\substack{\text{calcium}\\\text{bicarbonate}}}{Ca(HCO_3)_2} + \underset{\substack{\text{magnesium}\\\text{bicarbonate}}}{Mg(HCO_3)_2}$$

These reactions are the most notable dissolution processes. Water percolating down into a limy bedrock dissolves and carries away a part of the rock mass. Since limestone and related rocks are composed largely of soluble minerals, great volumes of rock are sometimes dissolved and removed, leaving conspicuous voids in the bedrock. This action occurs more rapidly

Multiple speleothems in Blanchard Springs Cave in the Ozark section of Arkansas.
(Mark E. Gibson/The Stock Market)

and on a vaster scale in a humid climate, where abundant precipitation provides plenty of the aqueous medium necessary for dissolution. In arid regions, dissolution action is unusual except for relict features dating from a more humid past.

Bedrock structure is also a factor in dissolution. A profusion of joints and bedding planes permits groundwater to penetrate the rock readily. That the water is moving also helps because, as a given volume of water becomes saturated with dissolved calcium bicarbonate, it can drain away and be replaced by fresh unsaturated water that can dissolve more rock. Such drainage is enhanced by some outlet at a lower level, such as a deep subsurface stream.

Most limestone is resistant to mechanical erosion and often produces rugged topography. Thus its ready solubility contrasts notably with its mechanical durability—a vulnerable interior beneath a durable surface.

Complementing the removal of lime is its precipitation from solution. Mineralized water may trickle in along a cavern roof or wall. The reduced air pressure in the open cavern induces precipitation of whatever minerals the water is carrying.

One other type of precipitation is worth mentioning, despite its scarcity, because of its dramatic distinctiveness. Hot springs and geysers nearly always provide an accumulation of precipitated minerals, frequently brilliant white but sometimes orange, green, or some other color due to associated algae. Wherever it comes in contact with magma, underground water becomes heated, and this water sometimes finds its way back to the surface through a natural opening so rapidly that it is still hot when it reaches the open air. Hot water generally is a much better solvent than cold, and so a hot spring or geyser usually contains a significant quantity of dissolved minerals. When exposed to the open air, the hot water precipitates much of its mineral content as its temperature and the pressure on it decrease, as the dissolved gases that helped keep the minerals in solution dissipate, and as algae and other organisms living in it secrete mineral matter. These deposits contain a variety of calcareous minerals (such as travertine, tufa, and sinter) and take the form of mounds, terraces, walls, and peripheral rims.

It should be noted, however, that the solubility of carbon dioxide increases as water temperature declines. Thus cool water often is more potent than hot water as a solvent for calcium carbonate.

CAVERNS AND RELATED FEATURES

Some of the most spectacular landforms produced by dissolution are not visible at Earth's surface. Solution along joints and bedding planes in limestone beneath the surface often creates large open areas called **caverns**.

The largest of these openings are usually more expansive horizontally than vertically, indicating a development along bedding planes. In many cases, however, the cavern pattern has a rectangularity that demonstrates a relationship to the joint system (Figure 19-2).

Caverns are found anywhere there is a massive limestone deposit at or near the surface. The state of Missouri, for example, has more than 6000. Caverns often are difficult to find because their connection to the surface may be extremely small and obscure. Beneath the surface, however, some caverns are very extensive, with an elaborate system of galleries and passageways, usually very irregular in shape and sometimes including massive openings ("rooms") scattered here and there along the galleries. A stream may flow along the floor of a large cavern, adding another dimension to erosion/deposition.

There are two principal stages in cavern formation. First there is the initial excavation, wherein percolating water dissolves the limy bedrock and leaves voids. This dissolution is followed by a "decoration stage" in which ceilings, walls, and floors are decorated with a wondrous variety of **speleothems** (Figure 19-3). These forms are deposited when water leaves behind the compounds (principally carbon dioxide and calcite) it was carrying in solution. Once out of solution, the carbon dioxide gas diffuses into the cave atmosphere, and calcite is deposited. Much of the deposition occurs on the sides of the cavern, but the most striking features are formed on the roof and floor. Where water drips from the roof, a pendant structure grows slowly downward like an icicle (**stalactite**). Where the drip hits the floor, a companion feature (**stalagmite**) grows upward. Stalactites and stalagmites may extend until they meet to form a pillar (Figure 19-4).

KARST TOPOGRAPHY

In many areas where the bedrock is limestone or similarly soluble rock, dissolution has been so widespread and effective that a distinctive landform assemblage has developed at the surface, in addition to whatever caves may exist underground. The term **karst** (a Germanized form of an ancient Slavic word) is applied to this topography. The name derives from the Karst (meaning barren land) region of the former Yugoslavia, a rugged hilly area that has been shaped almost entirely by solution action in limestone formations (Figure 19-5).

The term *karst* connotes both a set of processes and an assemblage of landforms. The word is used as the catchall name of a cornerstone concept that describes the special landforms that develop on exceptionally soluble rocks, although there is a broad international vocabulary to refer to specific features in specific regions.

(a)

(b)

Figure 19-2 Caverns are formed by solution action of underground water as it trickles along bedding planes and joint systems.

Figure 19-3 A multitude of stalactites hang from the ceiling of this room in Lehman Cave in Nevada's Great Basin National Park. Some of them have joined with stalagmites to form pillars. (Tom Bean/Tony Stone Images)

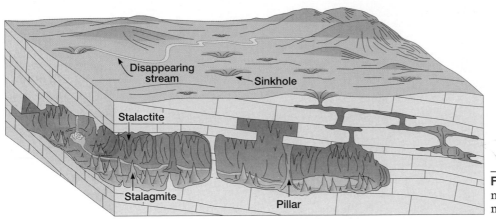

Figure 19-4 Extensive caverns may be found in regions of massive limestone bedrock.

Karst landscapes usually evolve where there is massive limestone bedrock. However, karstic features may also occur where other highly soluble rocks—dolomite, gypsum, or halite—predominate. Karst landforms are worthy of study not only because of their dramatic appearance but also because of their abundance. It is estimated that about 10 percent of Earth's land area has soluble carbonate rocks at or near the surface; in the conterminous United States, this total rises to 15 percent.

The most common surface features of karst landscapes are **sinkholes** (also called *dolines*), which occur by the hundreds and sometimes by the thousands. Sinkholes are rounded depressions formed by the dissolution of surface carbonate rocks, typically (but by no means always) at joint intersections (Figure 19-6). The sinkholes erode more rapidly than the surrounding area, forming closed depressions. Their sides generally slope inward at the angle of repose (usually 20° to 30°) of the adjacent material, although some have more gentle side slopes. A sinkhole that results from the collapse of the roof of a subsurface cavern is called a **collapse doline**; these may have vertical walls or even overhanging cliffs.

Sinkholes range in size from shallow depressions a few yards in diameter and a few feet deep to major features miles in diameter and hundreds of feet deep. The largest are associated with tropical regions, where they develop rapidly and where adjacent holes often intersect.

The bottom of a sinkhole may lead into a subterranean passage, down which water pours during a rainstorm. More commonly, however, the subsurface entrance is blocked by rock rubble, soil, or vegetation, and rains form temporary lakes until the water percolates away. Indeed, sinkholes are the karst analog of river valleys, in that they are the fundamental unit of both erosion and weathering.

Figure 19-5 Irregular topography is prominent in the namesake karst region of the former Yugoslavia. (Alex S. MacLean/Peter Arnold, Inc.)

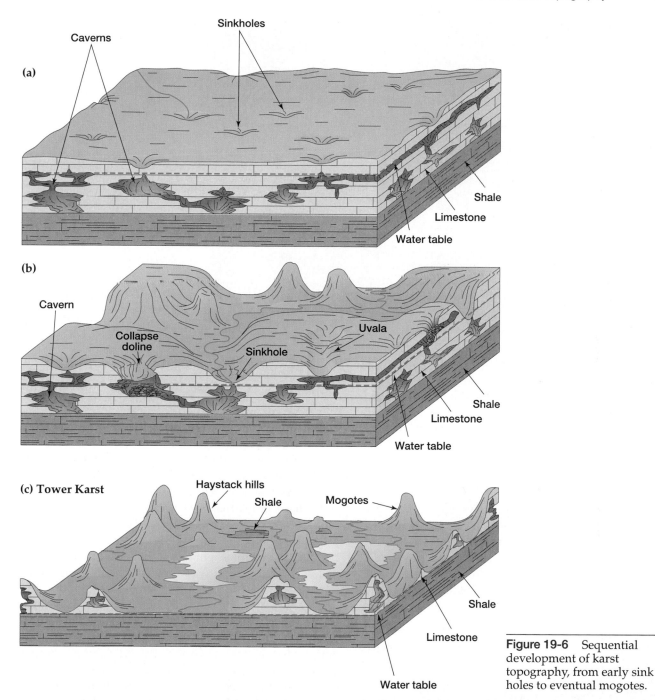

Figure 19-6 Sequential development of karst topography, from early sink holes to eventual mogotes.

Where sinkholes occur in profusion, they often channel surface runoff into the groundwater circulation, leaving networks of dry valleys as relict surface forms. The Serbo-Croatian term **uvala** refers to such a chain of intersecting sinkholes. In many cases, sinkholes evolve into uvala over time.

In some karst regions, depressions are so common that they dominate the landscape. Most of central Florida, for example, is underlain by massive limestone bedrock, which is particularly susceptible to the formation of collapse dolines, and the process is accelerated when the water table drops. Sinkholes have

been forming in Florida for a long time, with several thousand of them having appeared in the present century. Indeed, most of central Florida's scenic lakes began as sinkholes.

The frantic population growth that Florida has experienced in recent decades has put a heavy drain on its underground water supply. This depletion was exacerbated by several years of below-normal rainfall in the 1970s, which caused accelerated drawdown of the water table. As a result, the number and size of Florida sinkholes increased at a disturbing pace (Figure 19-7). New sinkholes materialized somewhere in the state at

Figure 19-7 The sinkhole that swallowed a house; central Florida. (St. Petersburg Times/Gamma-Liaison, Inc.)

a rate of about one per day during the early 1980s, although the tempo has slowed a bit since then.

Sinkholes are also commonplace in much of the Midwest, particularly Kentucky, Illinois, and Missouri (Figure 19-8). For example, at the University of Missouri in Columbia, both the football stadium and the basketball fieldhouse are built in sinkholes, and parking lots around them occasionally sink.

Apart from the ubiquity of sinkholes, karst areas show considerable topographic diversity. Where the relief is slight, as in central Florida, sinkholes are the dominant features. Where the relief is greater, however, cliffs and steep slopes alternate with flat-floored, streamless valleys. Limestone bedrock exposed at the surface tends to be pitted, grooved, etched, and fluted with a great intricacy of erosive detail.

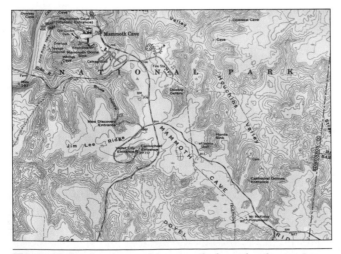

Figure 19-8 A topographic map of a karst landscape in central Kentucky.

Residual karst features, in the form of very steep-sided hills, dominate some parts of the world (Figure 19-9). These formations are sometimes referred to as **tower karst** because of their almost vertical sides and conical or hemispheric shapes, sometimes riddled with caves. The tower karst of southeastern China and adjacent parts of northern Vietnam is world-famous for its spectacular scenery, as are the **mogotes** (haystack hills) of western Cuba.

In many ways the most notable feature of karst regions is what is missing—surface drainage. Most rainfall and snowmelt seep downward along joints and bedding planes, enlarging them by dissolution. Surface runoff that does become channeled usually does not go far before it disappears into a sinkhole or joint crack. The water that collects in sinkholes generally percolates downward, but some sinkholes have distinct openings at their bottom (called **swallow holes**) through which surface drainage can pour directly into an underground channel, often to reappear at the surface through another hole some distance away. Where dissolution has been effective for a long time, there may be a complex underground drainage system that has superseded any sort of surface drainage net. An appropriate generalization concerning surface drainage in karst regions is that valleys are relatively scarce and mostly dry.

HYDROTHERMAL FEATURES

In many parts of the world, there are small areas where hot water comes to the surface through natural openings. Such outpouring of hot water, often accompanied by steam, is a **hydrothermal activity** and usually takes the form of either a hot spring or a geyser.

Figure 19-9 **(a)** Spectacular tower karst hills in Guilin, China. (Tom Nebbia/The Stock Market) **(b)** Mogotes in limestone country in western Cuba. (Ray Pfortner)

HOT SPRINGS

The appearance of hot water at Earth's surface usually indicates that the underground water has come in contact with heated rocks or magma and has been forced upward through a fissure by the pressures that develop when water is heated anywhere (Figure 19-10). The usual result at the surface is a **hot spring**, with water bubbling out either continuously or intermittently. The hot water invariably contains a large amount of dissolved mineral matter, and a considerable proportion of this load is precipitated out as soon as the water reaches the surface and its temperature and the pressure on it both decrease.

The deposits around and downslope from hot springs can take many forms. If the opening is on sloping land, terraces are usually formed. Where the springs emerge onto flat land, there may be cones, domes, or irregularly concentric deposits. Since lime is so readily soluble in water containing carbonic acid, the deposits of most springs are composed largely of massive (*travertine*) or porous (*tufa* or *sinter*) accumulations of calcium carbonate. Various other minerals are also contained in the deposits on occasion, especially silica, but are much less common than calcium compounds.

Sometimes the water bubbling out of a hot spring builds a continually enlarging mound or terrace. As the structure is built higher, the opening through which the hot water comes to the surface also rises, so that the water is always emerging above the highest point. As the water flows down the sides of the structure, more deposition takes place there, thus broadening the structure as well, often with brilliantly colored algae, which add to the striking appearance as well as contribute mineral secretions to the deposit (Figure 19-11).

GEYSERS

A special form of intermittent hot spring is the **geyser**. Hot water usually issues from a geyser only sporadically, and most or all of the flow is a temporary ejection (called an *eruption*) in which hot water and steam spout upward. Then the geyser subsides into apparent inactivity until the next eruption.

The basic principle of geyser activity involves the building up of pressure within a restricted subter-

Figure 19-10 Cross section through a hot spring.

ranean tube until that pressure is relieved by an eruption. The process begins when underground water seeps into subterranean openings that are connected in a series of narrow caverns and shafts. Heated rocks and/or magma are close enough to these storage reservoirs to provide a constant source of heat. As the water accumulates in the reservoirs, it is heated to 400°F (204°C) or higher, which is much above the boiling point at sea level and normal pressure. (Such superheating is possible because of the high underground water pressure.) At these high temperatures, much of the water becomes steam. The accumulation of steam deep in the tube and of boiling water higher in the tube eventually causes a great upward surge that sends water and steam showering out of the geyser vent. This eruption releases the pressure, and when the eruption subsides, underground water again begins to collect in the reservoirs in preparation for a repetition of the process.

A tremendous supply of heat is essential for geyser activity. Recent studies in Yellowstone Park's Upper Geyser Basin indicate that the heat emanating from that basin is at least 800 times greater than the heat flowing from a nongeyser area of the same size. This extraordinary flow of extra heat has been going on for at least 40,000 years.

Some geysers erupt continuously, indicating that they are really hot springs that have a constant supply of water through which steam is escaping. Most geysers are only sporadically active, however, apparently depending on the accumulation of sufficient water to force an eruption. Some eruptions are very brief, whereas others continue for many minutes. The interval between eruptions for most geysers is variable. Most erupt at intervals of a few hours or a few days, but some wait years or even decades between eruptions. The temperature of the erupting water generally is near the boiling point for pure water (212°F or 100°C at sea level). In some geysers the erupting water column goes up only a few inches in the air, whereas in others the column rises to more than 150 feet (45 meters).

Geyser comes from the Icelandic word *geysir* (to gush or to rage), the Great Geysir in southern Iceland being the namesake origin for this term. The most famous of all geysers is Old Faithful in Yellowstone Park (Figure 19-12). Its reputation is based partly on the force of its eruptions (the column goes more than 100 feet or 30 meters high) but primarily on its regularity. Since first timed by scientists more than a century ago, Old Faithful had maintained an average interval of 65 minutes between eruptions, day and night, winter and summer, year after year. In the early 1980s, however, several consecutive earthquakes on the Yellowstone plateau apparently upset the geyser's internal plumbing, and Old Faithful became more erratic, which is to say, more like other geysers.

The deposits resulting from geyser activity are usually much less notable than those associated with hot springs. Some geysers erupt from open pools of hot water, throwing tremendous sheets of water and steam into the air but usually producing relatively minor depositional features. Other geysers are of the "nozzle" type and consequently build up a depositional cone and erupt through a small opening in it (Figure 19-13). Most deposits resulting from geyser activity are simply sheets of precipitated mineral matter spread irregularly over the ground.

FUMAROLES

A third hydrothermal feature is the **fumarole,** a surface crack directly connected to a deep-seated heat source (Figure 19-14 on page 542). For some reason, very little water drains into the tube of a fumarole. The water that does drain in is instantly converted to steam by the heat, and a cloud of steam is then expelled from the opening. Thus a fumarole is marked by steam issuing either continuously or sporadically from a surface vent; in essence, a fumarole is simply a hot spring that lacks water.

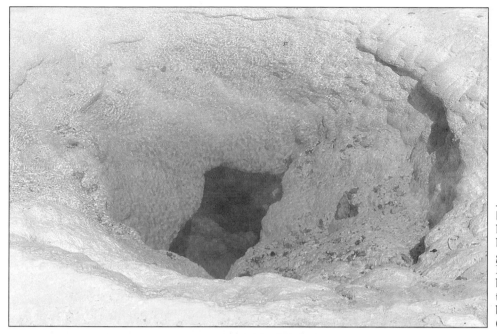

Figure 19-11 The sides and bottoms of hot springs often are brilliantly colored by algal growth. This is Morning Glory Pool in Yellowstone Park. The black rectangle in the center is the opening to the fissure that brings hot water to the surface. (TLM photo)

Figure 19-12 Old Faithful in all its eruptive grandeur. (TLM photo)

Figure 19-13 Most geysers erupt from vents or hot pools, but some build up prominent depositional "nozzles" through which water is expelled. This is Lone Star Geyser in Yellowstone Park. (TLM photo)

FOCUS

HYDROTHERMAL FEATURES IN YELLOWSTONE

Hydrothermal features are found in many volcanic areas, being particularly notable in Iceland, New Zealand, Chile, and Siberia's Kamchatka Peninsula. By far the largest concentration, however, occurs in Yellowstone National Park in northwestern Wyoming, which contains about 225 of the world's 425 geysers as well as more than half of the world's other hydrothermal phenomena. The area consists of a broad, flattish plateau bordered by extensive mountains (the Absaroka Range) on the east and by more limited highlands (particularly the Gallatin Mountains) on the west (Figure 19-1-A). The bedrock surface of the plateau is almost entirely volcanic materials, and although no volcanic cones are in evidence, we know that this clearly is a region of extensive recent volcanism.

The uniqueness of Yellowstone's geologic setting stems from the pres-

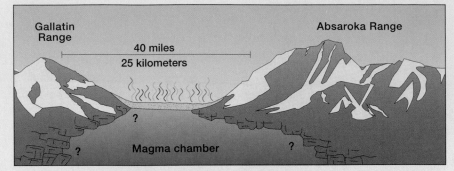

Figure 19-1-A Schematic west-east cross section through the Yellowstone Plateau.

ence of a large, shallow magma chamber beneath the plateau. Test boreholes reveal an abnormally high thermal gradient, in which the temperature increases with depth at a rate of about 36 F° per 100 feet (20 C° per 30 meters), indicating molten material less than 1 mile (1.6 kilometers) below the surface. This shallow magma pool

provides a remarkable heat source—the most important of the three conditions necessary for the development of hydrothermal features.

The second requisite is an abundance of water that can seep downward and become heated. Yellowstone receives copious summer rain and a deep winter snowpack (averaging more than 100 inches or 254 centimeters).

The third necessity for hydrothermal development is a weak or broken ground surface that allows water to move up and down easily. Here, too, Yellowstone fits the bill, for the ground surface there is very unstable and subject to frequent earthquakes, faulting, and volcanic activity. Consequently, many fractures and weak zones provide easy avenues for vertical water movement.

The park contains about 225 geysers, more than 3000 hot springs, and 7000 other thermal features (fumaroles, steam vents, hot-water terraces, hot-mud cauldrons, and so forth). There are five major geyser basins, a half dozen minor ones, and an extensive scattering of individual or small groups of thermal features.

The principal geyser basins are all in the same watershed on the western side of the park (Figure 19-1-B). The Gibbon River from the north and the Firehole River from

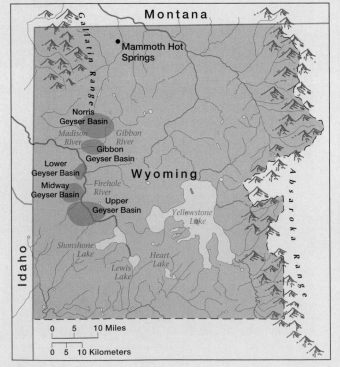

Figure 19-1-B Yellowstone Park and its major geyser basins.

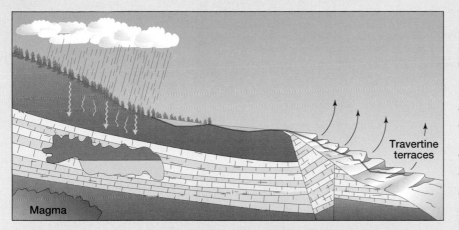

Figure 19-1-C Jupiter Terrace at Mammoth Hot Springs, Yellowstone Park. Water from rainfall and snowmelt percolates down into the underlying limestone, where it is heated from below and seeps downslope. Some of the hot water issues onto the surface and precipitates travertine deposits when exposed to the air.

the south unite to form the Madison River, which flows westerly into Montana, eventually to join two other rivers in forming the Missouri. The Gibbon River drains the Norris and Gibbon geyser basins, whereas the Firehole drains the Upper, Midway, and Lower basins. The Firehole River derives its name from the great quantity of hot water fed into it from the hot springs and geysers along its way. Approximately two-thirds of the hydrothermal features of Yellowstone are in the drainage area of the Firehole River.

All the major geyser basins consist of gently undulating plains or valleys covered mostly with glacial sediments and large expanses of whitish siliceous material called *geyserite*. Each basin contains from a few to several dozen geysers, some of which are inconspicuous holes in the geyserite but others of which are built-up cones that rise a few feet above the basin. In addition, each basin contains a number of hot springs and fumaroles.

Yellowstone's geysers exhibit an extraordinary range of behavior. Some erupt continually; others have experienced only a single eruption in all history. Most, however, erupt irregularly several times a day or week. Some shoot their hot water only a few inches into the air, but the largest (Steamboat, Excelsior) erupts to heights of 300 feet (100 meters), with clouds of steam rising much higher.

In the northeastern portion of the park is the most remarkable aggregation of hot-water terraces in the world, the Mammoth Hot Springs Terraces (Figures 19-1-C and 19-1-D). There, groundwater percolates down from surrounding hills into thick layers of limestone. Hot water, carbon dioxide, and other gases rise from the heated magma to mingle with the groundwater and produce a mild carbonic acid solution that rapidly dissolves great quantities of the limestone. Saturated with lime, the temporarily carbonated water seeps downslope until it gushes forth near the base of the hills as the Mammoth Hot Springs. The carbon dioxide escapes into the air, and the lime is precipitated as massive deposits of travertine in the form of flat-topped, steep-sided terraces.

Figure 19-1-D This is neither snow nor ice, but rather a small portion of the travertine terraces at Mammoth Hot Springs in Yellowstone Park. (TLM photo)

Figure 19-14 A fumarole is like a geyser except that it erupts no water; it sends out only steam. This scene is from Iceland. (Torleif Svensson/The Stock Market)

CHAPTER SUMMARY

Water percolating underground produces both erosional and depositional landforms. The water acts as a solvent, dissolving certain chemicals (especially calcium and magnesium carbonates), carrying them away in solution, and depositing them elsewhere.

This dissolution produces numerous subsurface caverns in limestone bedrock and is sometimes so widespread and effective that such prominent surface features as sinkholes are created. Some landscapes are thoroughly dominated by solution-produced forms, and the terrain is referred to as karst topography.

A hot spring forms when underground water comes in contact with heated rocks or magma and bubbles to the surface through a fissure. A geyser is a special form of intermittent hot spring that erupts when pressure builds in a restricted subterranean tube. A fumarole is a hot spring that lacks water and spouts only steam from its vent.

KEY TERMS

cavern	hydrothermal activity	stalactite
collapse doline	karst	stalagmite
fumarole	mogote	swallow hole
geyser	sinkhole	tower karst
hot spring	speleothem	uvala

REVIEW QUESTIONS

1. How is it possible for percolating groundwater to both erode and deposit?
2. Explain how a sinkhole is formed.
3. Why is there a scarcity of surface drainage in karst areas?
4. What is the difference between a hot spring, a geyser, and a fumarole?
5. What causes this difference?

S O M E   U S E F U L   R E F E R E N C E S

JENNINGS, J. N., *Karst Geomorphology*. New York: Blackwell, 1985.

LA FLEUR, R. R., *Groundwater as a Geomorphic Agent*. Boston: Allen & Unwin, 1984.

TRUDGILL S. A., *Limestone Geomorphology*. New York: Longman, 1986.

WHITE, W. B., *Geomorphology and Hydrology of Karst Terrains*. New York: Oxford University Press, 1988.

20

GLACIAL MODIFICATION
OF TERRAIN

IN THE LONG HISTORY OF OUR PLANET, ICE AGES HAVE occurred an unknown number of times. (The cause or causes of the climate changes that lead to these ice ages are still unknown to us, a mystery we discuss further at the end of the chapter.) With one outstanding exception, however, nearly all evidence of past glacial periods has been eradicated by subsequent geomorphic events, with the result that only the most recent ice age has influenced contemporary topography. Consequently, when referring to the Ice Age, capitalized, we usually mean this most recent ice age, which is the main feature of the geologic epoch known as Pleistocene, a period that began about 2 million years ago and ended less than 10,000 years ago (if indeed it has ended).

In this chapter we are concerned with Pleistocene events both because they significantly modified pre-Pleistocene topography and because their results are so thoroughly imprinted on many parts of the continental terrain today. Glacial processes are still at work, to be sure, but their importance is much less now than it was just a few thousand years ago simply because so much less glacial ice is present today.

THE IMPACT OF GLACIERS ON THE LANDSCAPE

When the amount of snow that falls in a winter is greater than the amount that melts the following summer, a glacier forms. The snow that falls the next winter weighs down on the old snow and turns it to ice. After many years of such accumulation, the ice mass begins to move under the pull of gravity. Wherever glaciers have developed, they have had an overwhelming impact on the landscape simply because moving ice grinds away almost anything in its path. Human-built structures and preexisting vegetation are destroyed, most soil is bulldozed away, and bedrock is polished, scraped, gouged, plucked, and abraded. In short, preglacial topography is significantly reshaped. Moreover, when the glacier ceases its advance, and even before, under certain circumstances, the debris that it picked up is deposited in a new location, further changing the shape of the terrain.

About 7 percent of all contemporary erosion is accomplished by glaciers. This is a paltry total in comparison with fluvial erosion, to be sure, but considering how small a land area is covered by glacial ice today, it is clear that glaciers make a respectable contribution to continental denudation. It has been cal-

culated that glaciation increases the erosion rate on a mountain by at least ten times over the rate on a comparable unglaciated mountain.

Glaciation modifies flat landscapes greatly, with the result that postglacial slope, drainage, and surficial material are likely to be totally different from what they were before the glacier passed by. In mountainous areas, the metamorphosis of the landscape may be less complete, but the topography is deepened, steepened, and sharpened throughout.

GLACIATIONS PAST AND PRESENT

The amount of glacial ice on Earth's surface has varied remarkably over the last few million years, with periods of accumulation interspersed with periods of melting and times of ice advance alternating with times of ice retreat. A great deal of secondary evidence was left behind by the moving and melting ice, and scientists have been remarkably perspicacious in piecing together the chronology of past glaciations. Nevertheless, the record is incomplete and often approximate. As is to be expected, the more recent events are best documented; the farther one delves into the past, the murkier the evidence becomes.

PLEISTOCENE GLACIATION

The precise boundaries of the Pleistocene epoch (Figure 13-17) are unknown. It began at least 1.5 million years B.P. (before the present), but evidence of an earlier beginning is discovered almost every year. Geochronologists keep pushing the starting date farther and farther back, therefore, and the most recent findings tell us that some parts of the Northern Hemisphere were covered by glaciers as much as 2.5 million years B.P.

New evidence has also changed the date of the close of the Pleistocene, with glaciologists now believing that the last major ice retreat took place more recently than 9000 years ago. Even this most recent estimate for the close date cannot be cast in stone, however, because the Ice Age may not yet have ended at all, a possibility we consider later in the chapter. For now, let us just say that, to the best of present knowledge, the Pleistocene epoch occupied almost all of the most recent couple of million years of Earth's history.

The dominant environmental characteristic of the Pleistocene was the refrigeration of high-latitude and high-altitude areas, so that a vast amount of ice accu-

Sawyer Glacier debouches into the fjord called Tracy Arm in southeastern Alaska.
(*Tom Bean/The Stock Market*)

Figure 20-13 Unsorted glacial till in a roadcut near Bridgeport, California. (TLM photo)

CONTINENTAL ICE SHEETS

Apart from the oceans and continents, continental ice sheets are the most extensive features ever to appear on the face of the planet. Their actions during Pleistocene time significantly reshaped both the terrain and the drainage of nearly one-fifth of the total surface area of the continents.

DEVELOPMENT AND FLOW

Pleistocene ice sheets, with the exception of the one covering Antarctica, did not originate in the polar regions. Rather, they developed in subpolar and midlatitude locations and then spread outward in all directions, including poleward. Several (perhaps several dozen) centers of original ice accumulation have been identi-

Figure 20-14 A glacial erratic carried many miles before being deposited by a Pleistocene glacier in what is now Yellowstone National Park. The ice has long since melted, and a forest has grown up around the erratic. (TLM photo)

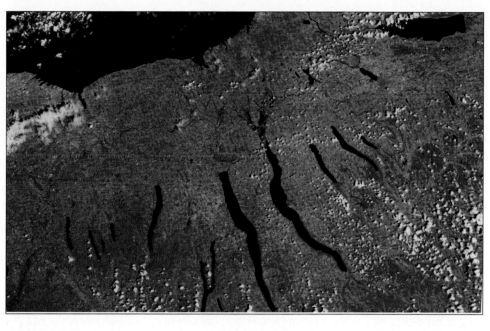

Figure 20-15 The Finger Lakes of upstate New York occupy glacial valleys that were gouged out by ice sheets moving from north-northwest to south-southeast. The large body of water at upper left is Lake Ontario. This false color composite image was taken from an altitude of about 570 miles (920 kilometers) by the ERTS-1 satellite. (NASA/Photri)

fied. The accumulated snow/névé/ice eventually produced such a heavy weight that the ice began to flow outward from each center of accumulation.

The initial flow was channeled by the preexisting terrain along valleys and other low-lying areas, but in time the ice developed to such depths that it overrode almost all preglacial topography. In many places, it submerged even the highest points under thousands of feet of ice. Eventually the various ice sheets coalesced into only one, two, or three massive sheets on each continent. These vast ice sheets flowed and ebbed as the climate changed, always modifying the landscape with their enormous erosive power and the great masses of debris they deposited. The elaborate result was nothing less than a total reshaping of the land surface and a total rearrangement of the drainage pattern.

EROSION BY ICE SHEETS

Except in mountainous areas of great initial relief, the principal topography resulting from the erosion caused by an ice sheet is a gently undulating surface. The most conspicuous features are valley bottoms gouged and deepened by the moving ice. Such U-shaped troughs are deepest where the preglacial valleys were oriented parallel to the direction of ice movement, particularly in areas of softer bedrock. A prime example of such development is the Finger Lakes District of central New York, where a set of parallel stream valleys was reshaped by glaciation into a group of long, narrow, deep lakes (Figure 20-15). Even where the preglacial valley was not oriented parallel to the direction of ice flow, however, glacial gouging and scooping normally produced a large number of shallow excavations that became lakes after the ice disappeared. Indeed, the postglacial landscape in areas of ice sheet erosion is notable for its profusion of lakes.

Hills are generally sheared off and rounded by the moving ice. A characteristic shape is the **roche moutonnée,** which is often produced when a bedrock hill is overridden by moving ice (Figure 20-16). (The origin of this French term is unclear. It is often translated as

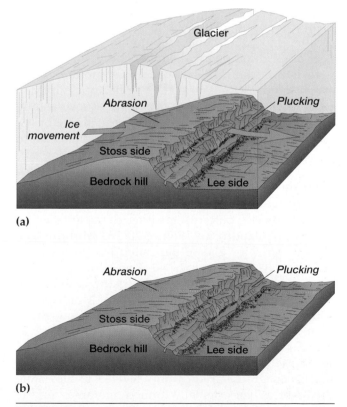

(a)

(b)

Figure 20-16 The formation of a roche moutonneé. **(a)** The glacier rides over a resistant bedrock surface, smoothing the stoss side by abrasion and steepening the lee side by plucking. **(b)** When the ice has melted, an asymmetrical hill is the result.

"sheep's back," but some authorities believe it is based on a fancied resemblance to wavy wigs that were fashionable in France in the late 1700s and were known as *moutonnées* because they were pomaded with mutton tallow.) The stoss side (facing in the direction from which the ice came) of a roche moutonnée is smoothly rounded and streamlined by grinding abrasion as the ice rides up the slope, but the lee side (facing away from the direction from which the ice came) is shaped largely by plucking, which produces a steeper and more irregular slope.

The postglacial landscape produced by ice sheets is one of relatively low relief but not absolute flatness. The principal terrain elements are ice-scoured rocky knobs and scooped-out depressions. Soil and weathered materials are largely absent, with bare rock and lakes dominating the surface. Stream patterns are erratic and inadequately developed because the preglacial drainage net was deranged by ice erosion. Once eroded by the passing ice sheet, however, most of this landscape was subjected to further modification by glacial deposition. Thus the starkness of the erosional landscape is modified by depositional debris.

DEPOSITION BY ICE SHEETS

In some cases, the till transported by ice sheets is deposited heterogeneously and extensively, without forming any identifiable topographic features; a veneer of unsorted debris is simply laid down over the preexisting terrain. This veneer is sometimes quite shallow and does not mask the original topography. In other cases, till is deposited to a depth of several hundred feet, completely obliterating the shape of the preglacial landscape. In either case, deposition tends to be uneven, producing an irregularly undulating surface of broad, low rises and shallow depressions. Such a surface is referred to as a **till plain.**

In many instances glacial sediments are laid down in more precise patterns, creating characteristic and identifiable landforms (Figure 20-17). **Moraine** is a general term for glacier-deposited landforms composed entirely or largely of till. They typically consist of irregular rolling topography rising some small height above the surrounding terrain. Moraines are usually much longer than they are wide, although the width can vary from a few tens of feet to as much as several miles. Some moraines are distinct ridges, whereas others are much more irregular in shape. Their relief is not great, varying from a few feet to a few hundred feet. When originally formed, moraines tend to have relatively smooth and gentle slopes, which become more uneven with the passage of time, as the blocks of stagnant ice, both large and small, included within the till eventually melt, creating irregular depressions, known as **kettles,** in the morainal surface.

Three types of moraines are particularly associated with deposition from continental ice sheets, although all three may be produced by alpine glaciers as well. A *terminal moraine* is a ridge of till that marks the outermost limit of glacial advance. It can vary in size from a conspicuous rampart tens of feet high to a low, discontinuous wall of debris. A terminal moraine is formed when a glacier reaches its equilibrium point and so is wasting at the same rate that it is being nourished. Although the toe of the glacier is not advancing, the interior continues to flow forward, delivering a supply of till. As the ice melts around the margin, the till is deposited, and the moraine grows (Figure 20-18).

Behind the terminal moraine (in other words, facing the direction from which the ice advanced), *recessional moraines* may develop. These are ridges that mark positions where the ice front was temporarily stabilized during the final retreat of the glacier. Both terminal and recessional moraines normally occur in the form of concave arcs that bulge outward in the direction of ice movement, indicating that the ice sheets advanced not along an even line but rather as a connecting series of great tongues of ice, each with a curved front (Figure 20-19).

The third type of moraine is the *ground moraine*, formed when large quantities of till are laid down from out of the body of the glacier rather than from its edge. A ground moraine usually means gently rolling plains across the landscape. It may be shallow or deep and often consists of low knolls and shallow kettles.

Another prominent feature deposited by ice sheets is a low, elongated hill called a **drumlin,** a term that comes from *druim*, an old Irish word for ridge (Figure 20-20). Drumlins are much smaller than moraines but composed of similarly unsorted till. The long axis of the drumlin is aligned parallel with the direction of ice movement. The end of the drumlin facing the direction from which the ice came is blunt and slightly steeper than the opposite end. Thus the configuration is the reverse of that of a roche moutonnée. The origin of drumlins is not completely understood, but most of them apparently are the result of ice readvance into an area of previous glacial deposition. In other words, they are depositional features subsequently shaped by erosion. Drumlins usually occur in groups, sometimes numbering in the hundreds, with all drumlins in a group oriented parallel to each other (Figure 20-21). The greatest concentrations of drumlins in the United States are found in central New York and eastern Wisconsin.

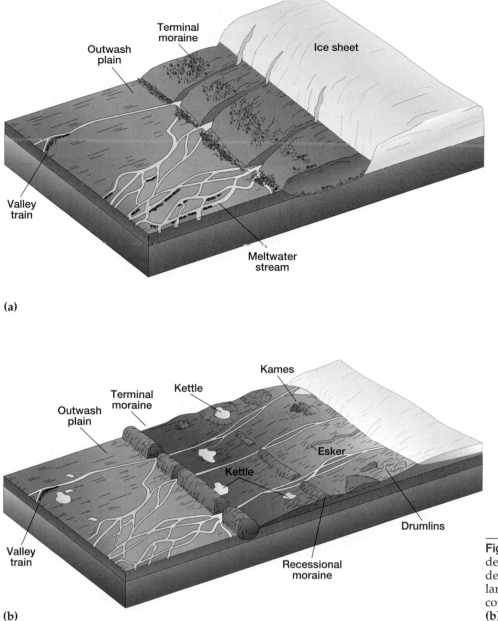

Figure 20-17 Glacier-deposited (G) and glaciofluvially deposited (GF) features of a landscape **(a)** covered by a continental ice sheet and **(b)** after the sheet has retreated.

GLACIOFLUVIAL FEATURES

The deposition or redeposition of debris by ice-sheet meltwater produces certain features both where the sheet covered the ground and in the periglacial region. These features are composed of **stratified drift,** which means that there has been some sorting of the debris as it was carried along by the meltwater. Glaciofluvial features, then, are composed largely or entirely of gravel, sand, and silt because meltwater is incapable of moving larger material.

The most extensive glaciofluvial features are **outwash plains,** which are smooth, flat alluvial aprons deposited beyond recessional or terminal moraines by streams issuing from the ice (Figure 20-17). Streams of water, heavily loaded with reworked till or with debris washed directly from the ice, issue from the melting glacier to form a braided pattern of channels across the area beyond the glacial front. As they flow away from the ice, these braided streams, choked with debris, rapidly lose their speed and deposit their load. Such outwash deposits sometimes cover many hundreds of square miles. They are occasionally pitted by kettles that often become ponds or small lakes. Beyond the outwash plain, there is sometimes a lengthy deposit of glaciofluvial alluvium confined to a valley bottom; such a deposit is termed a **valley train** (Figure 20-17).

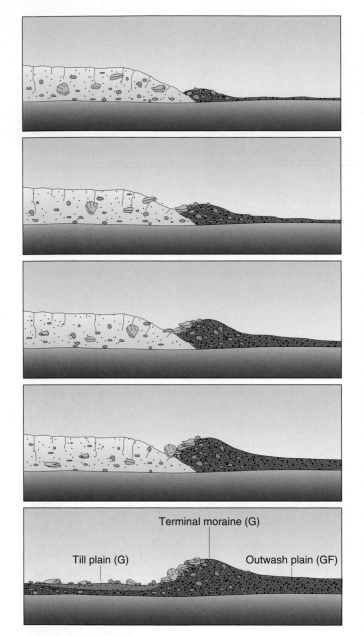

Terminal moraine (G)

Till plain (G)   Outwash plain (GF)

Less common than outwash plains but more conspicuous are long sinuous ridges of stratified drift called **eskers,** named from *eiscir,* another Irish word for ridge. These landforms are composed largely of glaciofluvial gravel and are thought to have originated when streams flowing through tunnels in the interior of the ice sheet became choked off during a time in which the ice was neither flowing nor advancing. These streams beneath the stagnating sheet often carry a great deal of debris, and as the ice melts, the streams deposit much of their load in the tunnel. Eskers are this debris exposed once the ice melts away. They are usually a few dozen feet high, a few dozen feet wide, and may be a few dozen miles [or up to a hundred miles (160 kilometers)] long.

Small, steep mounds or conical hills of stratified drift are found sporadically in areas of ice-sheet deposition. These **kames** (the word derives from *comb,* an old Scottish word referring to an elongated steep ridge) appear to be of diverse origin, but they are clearly associated with meltwater deposition in stagnant ice. They are mounds of poorly sorted sand and gravel that probably formed within glacial fissures or between the glacier and the land surface (Figure 20-22). Many seem to have been built as steep fans or deltas against the edge of the ice that later collapsed partially when the ice melted.

Morainal surfaces containing a number of mounds and depressions is called kame-and-kettle topography (Figure 20-23).

Figure 20-18   Terminal moraine growth at the toe of a stable glacier. The movement of a large boulder is shown from the time it is plucked from the bedrock until it is deposited as part of the moraine. The final diagram represents the situation after the ice has melted. See Figure 20-17 for an explanation of the symbols G and GF. (After Sheldon Judson, Marvin E. Kauffman, and L. Don Leet, *Physical Geology.* Englewood Cliffs, NJ: Prentice Hall, 1987.)

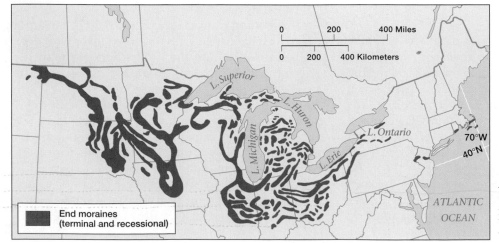

Figure 20-19   Terminal and recessional moraines in the United States resulting from the Wisconsin glaciation.

Figure 20-20 A small drumlin on the South Island of New Zealand. (Dr. E. R. Degginger)

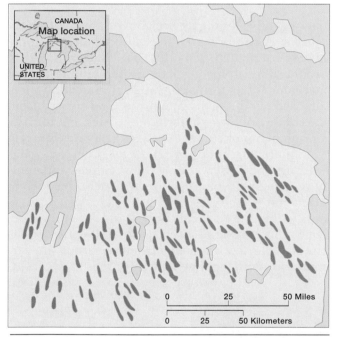

Figure 20-21 A drumlin field in western Michigan.

Figure 20-22 A kame in southeastern Wisconsin, near Dundee. (TLM photo)

Figure 20-23 Kame-and-kettle topography in an early stage of development. This is the Tasman Valley on the South Island of New Zealand. (Dr. E. R. Degginger)

MOUNTAIN GLACIERS

Most of the world's high mountain regions experienced extensive Pleistocene glaciation, and many mountain glaciers exist today. Mountain glaciers usually do not reshape the terrain as completely as the ice sheets did, partly because some portions of the mountains protrude above the ice and partly because the movement of mountain glaciers is channeled by the mountains. However, the effect of glacial action, particularly erosion, on mountainous topography is to create slopes that are steeper and relief that is greater than the preglacial slope and relief. This action, of course, is in contrast to ice sheet-action, which smooths and rounds the terrain.

DEVELOPMENT AND FLOW

A highland icefield can extend broadly across the high country, submerging all but the uppermost peaks and finding its outlet in a series of lobes that issue from the icefield and move down adjacent drainage channels (Figure 20-24). Individual alpine glaciers usually form in sheltered depressions near the heads of stream valleys (often far below the level of the peaks). Glaciers from either source advance downslope, pulled by gravity and normally finding the path of least resistance along a preexisting stream valley (Figure 20-25). A system of merging glaciers usually develops, with a trunk glacier in the main valley joined by tributary glaciers from smaller valleys.

EROSION BY MOUNTAIN GLACIERS

Erosion by highland icefields and alpine glaciers re-shapes the topography in dramatic fashion, as Figure 20-26 shows. It remodels the peaks and ridges and thoroughly transforms the valleys leading down from the high country.

In the High Country   The basic landform feature in glaciated mountains is the **cirque,** a broad amphitheater hollowed out at the head of a glacial valley (Figure 20-27). It has very steep, often perpendicular, head and side walls and a floor that is either

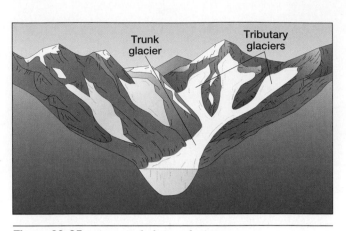

Figure 20-25   A typical alpine glacier scene.

flat or gently sloping or else gouged enough to form a basin. A cirque marks the place where an alpine glacier originated. It is the first landform feature produced by alpine glaciation, although the precise mechanics of its formation are hazy. Essentially, it is quarried out of the mountainside. One theory is that the shifting of the equilibrium line back and forth as a result of minor climatic changes generates much of this quarrying action, abetted by plucking, mass wasting, and frost wedging. Most cirques apparently owe their development to repeated episodes of glaciation. As the glacier grows, its erosive effectiveness within the cirque increases, and when the glacier begins to extend itself down-valley out of the cirque, quarried fragments from the cirque are carried away with the flowing ice. Cirques vary consid-

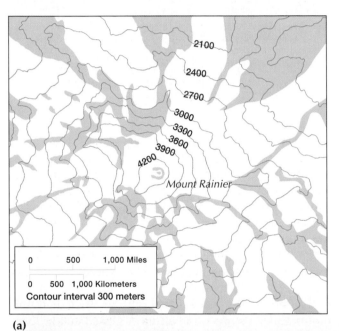

(a)

(b)

Figure 20-24   **(a)** Topographic map of the icefield at the top of Mt. Rainier in the Cascade Range in Washington. The green indicates areas not covered by the ice. Note the tongues of valley glaciers radiating from the central field. **(b)** The icefield at Mt. Rainier. (NASA/Photri)

Figure 20-26 The Grand Teton Mountains of Wyoming have been spectacularly modified by the action of mountain glaciers. (TLM Photo)

Figure 20-27 Only a small remnant glacier remains in this cirque on the north side of Wheeler Peak in Nevada's Great Basin National Park. (TLM photo)

erably in size, ranging from a few acres to a few square miles in extent (Figure 20-28)..

A cirque grows steeper as its glacier plucks rock from the head and side walls. Where cirques are close together, the upland interfluve between neighboring cirques is reduced to little more than a steep rock wall. Where several cirques have been cut back into an interfluve from opposite sides of a divide, a narrow, jagged, serrated spine of rock may be all that is left of the ridge crest; this is called an **arête** (French for fishbone; derived from the Latin *arista*, spine). If two adjacent cirques on opposite sides of a divide are being cut back enough to remove part of the arête between them, the sharp-edged pass or saddle through the ridge is referred to as a **col** (collum is Latin for neck) (Figure 20-28). An even more

prominent feature of glaciated highland summits is a **horn,** a steep-sided, pyramidal rock pinnacle formed by expansive quarrying of the headwalls where three or more cirques intersect (Figure 20-29). The name is derived from Switzerland's Matterhorn, the most famous example of such a glaciated spire.

When the glacial ice in a cirque has melted away, there is often enough of a depression formed to hold water. Such a cirque lake is called a **tarn.**

In the Valleys   Some alpine glaciers, such as those shown in Figure 20-30, never leave their cirques, presumably because of insufficient accumulation of ice to force a down-valley movement, and as a result the valleys below are unmodified by glacial erosion. Most alpine glaciers, however, as well as those issuing from

Figure 20-28 As alpine glaciers escape the confines of their original cirques and move across the mountain, the cirques are enlarged and gorges, horns, arêtes, and cols are formed. (After William Morris Davis, "The Colorado Front Range," *Annals of the Association of American Geographers*, 1911. By permission of the Association of American Geographers.)

Figure 20-29 Mount Aspiring is a prominent horn in the southern Alps of New Zealand. (TLM photo)

highland icefields, flow down preexisting valleys and reshape them (Figure 20-31).

A glacier moves down a mountain valley with much greater erosive effectiveness than a stream. The glacier is denser, carries more abrasive tools, and has an enormously greater volume. It erodes by both abrasion and plucking. The lower layers of the ice can even flow uphill for some distance if blocked by resistant rock on the valley floor, permitting rock fragments to be dragged out of depressions in the valley floor.

The principal erosive work of a valley glacier is to deepen, steepen, and widen its valley. Abrasion and plucking take place not only on the valley floor but along the sides as well. The cross-sectional profile is changed from its stream-cut V-shape to an ice-eroded

U-shape flared at the top (Figure 20-32). Moreover, the general course of the valley is straightened because the ice does not meander like a stream; rather it tends to grind away the protruding spurs that separate side canyons, creating what are called truncated spurs, and thereby replacing the sinuous course of the stream with a straight **glacial trough.**

As might be expected, a glacier grinding along the floor of a glacial trough does not produce a very smooth surface. Valley glaciers do not erode a continuously sloping channel because differential erosion works with ice as well as with water. Therefore resistant rock on the valley floor is gouged less deeply than weaker or more fractured rock. As a result, the long profile of the glaciated valley floor is ir-

Figure 20-30 Three cirques at about the 11,000-foot (3300-meter) level on Mount Nebo in central Utah. From the shape of the valleys below the cirques, it appears that the original cirque glaciers never moved out to become valley glaciers. (TLM photo)

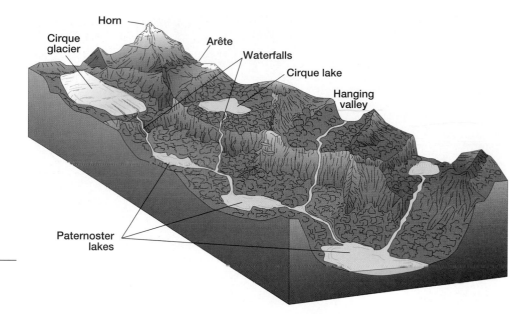

Figure 20-31 Some features associated with mountain glaciation.

regular, with parts that are gently sloping, flat, or steep, and with some excavated depressions alternating in erratic sequence.

The resulting landscape of the glacial trough, after the ice has melted away, usually shows an irregular series of rock steps or benches, with steep (though usually short) cliffs on the down-valley side and small lakes in the shallow excavated depressions of the benches. The postglacial stream that flows down-valley out of the cirque has a relatively straight course but a fluctuating gradient. Rapids and waterfalls are common, particularly on the cliffs below the benches. The various shallow lakes occur in a sequence called **paternoster lakes,** after a fancied resemblance to beads on a rosary.

As discussed in Chapter 18, some of the most spectacular glacial troughs occur along coastlines where valleys have been partly drowned by the sea to create fjords.

Our focus in the last few paragraphs has been on major valleys occupied by trunk glaciers. The same processes are at work and the same features produced in tributary valleys, although usually on a lesser scale. However, one important distinction between main valleys and tributary valleys is the amount of glacial erosion. Erosive effectiveness is determined largely by the amount of ice that passes through the valley; thus the smaller ice streams of tributary valleys cannot widen and deepen as much as the main valley glaciers.

Figure 20-32 The magnificent glacial landscape of Yosemite, looking past Half Dome up the U-shaped Tenaya Valley. (TLM photo)

Figure 20-33 The waters of Upper Yosemite Falls plummet from the hanging valley of Yosemite Creek into the glacial trough of the Merced River in Yosemite National Park. (TLM photo)

T A B L E 2 0 - 4
Landforms Produced by Glacial and Glaciofluvial Processes

Landform	Process	Agent[a]	Composition
Arête	Glacial erosion	MG	Bedrock
Cirque	Glacial erosion	MG	Bedrock
Col	Glacial erosion	MG	Bedrock
Drumlin	Glacial deposition	CIS	Till
Esker	Glaciofluvial deposition	CIS	Stratified drift
Fjord	Glacial erosion	MG	Bedrock
Glacial trough	Glacial erosion	MG	Bedrock
Ground moraine	Glacial deposition	MG, CIS	Till
Hanging valley	Glacial erosion	MG	Bedrock
Horn	Glacial erosion	MG	Bedrock
Kame	Glaciofluvial deposition	MG, CIS	Stratified drift
Kettle	Glaciofluvial deposition	MG, CIS	Stratified drift
Lateral moraine	Glacial deposition	MG	Till
Medial moraine	Glacial deposition	MG	Till
Nunatak	Glacial erosion	MG	Bedrock
Outwash plain	Glaciofluvial deposition	MG, CIS	Stratified drift
Recessional moraine	Glacial deposition	MG, CIS	Till
Roche moutonnée	Glacial erosion	MG, CIS	Bedrock
Terminal moraine	Glacial deposition	MG, CIS	Till
Till plain	Glacial deposition	MG, CIS	Till
Valley train	Glaciofluvial deposition	MG, CIS	Stratified drift

[a] MG = mountain glacier; CIS = continental ice sheet.

When occupied by glaciers and thus covered by a relatively level field of ice, main and tributary valleys may appear equally deep. When the ice melts, however, that valleys are of different depths becomes obvious; the mouths of the tributary valleys are characteristically perched high along the sides of the major troughs, forming **hanging valleys.** Typically, streams that drain the tributary valleys must plunge over waterfalls to reach the floor of the main trough. Several of the world-famous falls of Yosemite National Park are of this type (Figure 20-33).

DEPOSITION BY MOUNTAIN GLACIERS

Depositional features are less significant in areas of mountain glaciation than in areas where continental ice sheets have been at work. The high country is almost totally devoid of drift; only in the middle and lower courses of glacial valleys can much deposition be found.

The principal depositional landforms associated with mountain glaciation are moraines. Terminal and recessional moraines form just as they do with ice sheets. Moraines resulting from mountain glaciation are much smaller and less conspicuous, however, because they are restricted to glacial troughs.

The largest depositional features produced by mountain glaciation are lateral moraines; these are well-defined ridges of unsorted debris built up along the sides of valley glaciers (Figure 20-34). The debris is partly material deposited by the glacier and partly rock that falls or is washed down the valley walls.

Figure 20-34 The Athabaska valley glacier issuing from a highland icefield in the Canadian Rockies. The distance from the highest icefall to the toe is about 7 miles (11 kilometers). The glacier has heaped up conspicuous lateral moraines along both sides. (TLM photo)

FOCUS

PROGLACIAL LAKES

By far the largest proglacial lake in North America was Lake Agassiz, named after the Swiss-American scientist, Louis Agassiz, who first developed the theory of an Ice Age. Formed from meltwater impounded between the great ice sheet on the north and various ridges and moraines on the south, this lake first appeared about 13,000 years B.P. and existed for 5000 years before its final drainage. During its lifetime, its various sizes added up to a combined coverage of 180,000 square miles (466,000 square kilometers). Only about half of this area was occupied by the lake at its maximum extent, but this 90,000 square miles (233,000 square kilometers) was a more extensive water surface than that of all five Great Lakes. Lake Agassiz initially drained southward into the Mississippi River system [Figure 20-1-A(a)], but as the ice sheet retreated the lake found a new drainage outlet into Lake Superior and eventually into Hudson Bay [Figure 20-1-A(b,c)].

One significant result of proglacial lakes was the development of lacustrine (lake-built) topographic features. Terraces and beaches built by erosion and deposition along the lakeshores are still

found on the land today, well removed from any present-day lakes. More important from an economic standpoint was the accumulation of fine sediments in the old lake beds. The plain of Lake Agassiz, for example, is a remarkably flat surface in Minnesota, North Dakota, Manitoba, and Saskatchewan, underlain by very productive soils derived from the lacustrine sediments.

The Great Lakes are the largest glacially formed lakes in the world and constitute the most conspicuous result of Pleistocene glaciation in North America. The lake basins probably originated as broad lowlands developed on weak sedimentary rocks by fluvial processes and presumably drained into the Mississippi. The early advances of the ice sheet over this region deepened the basins by glacial erosion.

Subsequent retreat of the ice produced the first large proglacial lakes—Chicago and Maumee—both of which drained southward [Figure 20-1-B(a)]. Further minor advances and retreats of ice lobes broke up Lake Maumee and formed the smaller Lake Saginaw to the west and the huge Lake Whittlesey to the east, as well as established a new drainage outlet eastward along

(a)

(b)

(c)

Figure 20-1-A The presumed development of Lake Agassiz as the ice sheet melted.

the ice front into what is now the Mohawk Valley of New York (Figure 20-1-B(b)]. The final retreating phase of the ice sheet opened up Lake Superior for the first time and established a drainage outlet down the valley of the Ottawa River into the St. Lawrence estuary [Figure 20-1-B(c)].

The immediate postglacial stage of the Great Lakes was the maximum extent of their development, with the three upper lakes being joined into a single enormous Lake Nipissing, which had three drainage outlets [Figure 20-1-B(d)]. The final form of the lakes came when isostatic rebound (due to the unloading of the great weight of ice) uplifted the area to the north, draining Lake Nipissing sufficiently to create the three separate upper lakes and shutting off the Ottawa River drainage outlet. The Great Lakes were then in their contemporary form, but they still had two drainage outlets, one via the Illinois River into the Mississippi and the other down the lakes into the St. Lawrence River. A final lowering of the lake levels caused the abandonment of the Illinois River outlet, and the entire Great Lakes system now drains into the Atlantic via the St. Lawrence River.

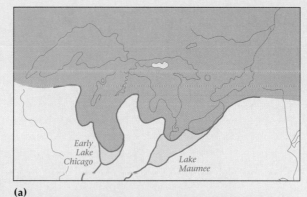

(a)

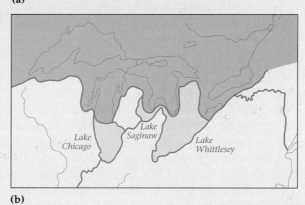

(b)

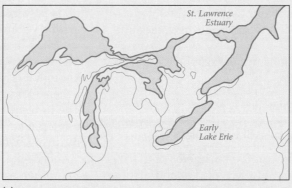

(c)

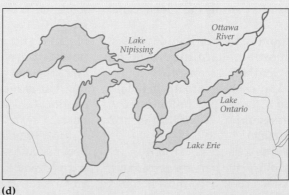

(d)

Figure 20-1-B Probable sequential development of the Great Lakes.

Figure 20-35   Prominent medial moraines on McBride Glacier, near Glacier Bay in southeastern Alaska. (TLM photo)

Where a tributary glacier joins a trunk glacier, their lateral moraines become united at the intersection and often continue together down the middle of the combined glacier as a dark band of rocky debris known as a medial moraine. Medial moraines are sometimes found in groups of three or four running together, indicating that several glaciers have joined to produce a candy-cane effect of black (moraine) and white (ice) bands extending down the valley (Figure 20-35).

The debris left by meltwater below mountain glaciers is similar to that bordering ice sheets because similar outwash is produced.

Table 20-4 (on page 566) summarizes the landforms associated with glaciation.

THE PERIGLACIAL ENVIRONMENT

The term *periglacial* means on the perimeter of glaciation. More than 20 percent of the world's land area is presently periglacial, but most of this was covered by ice on one or more occasions during the Pleistocene epoch.

Periglacial lands are found either in high latitudes or in high altitudes. Almost all are in the Northern Hemisphere because the Southern Hemisphere continents either do not extend far enough into the high latitudes to be significantly affected by contemporary glaciation (Africa, South America, Australia) or are mostly ice-covered (Antarctica).

Nonglacial land-shaping processes function in periglacial areas, but in addition the pervasive cold imparts some distinctive characteristics, the most notable of which is permafrost (Chapter 9). Either continuous

or discontinuous permafrost occurs over most of Alaska and more than half of Canada and Russia. There are also extensive high-altitude areas of permafrost in Asia, Scandinavia, and the western United States. In some cases the frozen ground extends to extraordinary depths; a thickness of 3000 feet (1000 meters) has been found in Canada's Northwest Territories and 4500 feet (1500 meters) in north-central Siberia.

The most unique and eye-catching periglacial terrain is **patterned ground,** the generic name applied to various geometric patterns that repeatedly appear over large areas in the Arctic (Figure 20-36). The patterns are of apparently varied but still unknown origin, although it is accepted that frost action is instrumental in their formation. The principal significance of patterned ground is that it demonstrates the mobility of periglacial surfaces, emphasizing the role of soil ice in producing geomorphic activities largely unknown in warmer regions.

Another sometimes conspicuous development in periglacial regions is **proglacial lakes** (*pro* here means marginal to or in advance of). Where ice flows across a land surface, the natural drainage is either impeded or blocked, and meltwater from the ice can become impounded against the ice front, forming a proglacial lake. Such an event sometimes occurs in alpine glaciation but is much more common along the margin of continental ice sheets, particularly when the ice stagnates.

Most proglacial lakes are small and quite temporary because subsequent ice movements cause drainage changes and because normal fluvial processes, accelerated by the growing accumulation of meltwater in the lake, cut spillways or channels to drain the impounded

waters. Sometimes, however, proglacial lakes are large and relatively long-lived. Such major lakes are characterized by considerable fluctuations in size, due to the changing location of the receding or advancing ice front. Several huge proglacial lakes were impounded along the margins of the ice sheets as they advanced and retreated during the Pleistocene epoch in North America, Europe, and Siberia.

ARE WE STILL IN AN ICE AGE?

Ice ages are fascinating not only because of the landscape changes they bring about but also because of their mystery. What initiates massive accumulations of ice on the continental surfaces, stimulates their advances and retreats, and finally causes them to disappear? Scientists and other scholars have pondered these questions for decades. They have developed many scenarios, postulated many theories, and constructed many models in attempts to explain the sporadic glaciation and deglaciation of our planet.

Any satisfactory theory must be able to account for four main glacial characteristics:

1. The accumulation of ice masses more or less simultaneously at various latitudes in both hemispheres but without uniformity (for example, much less in Siberia and Alaska than in similar latitudes in Canada and Scandinavia).
2. The apparently concurrent development of pluvial conditions in dry-land areas.

3. Multiple cycles of ice advance and retreat, including both minor fluctuations over decades and centuries, and major glaciations and deglaciations over tens of thousands of years.
4. Eventual total deglaciation, either actual (in past geologic eras) or potential (for the Pleistocene epoch).

It is easy enough to state that glaciers develop when there is a net accumulation of snow over a period of time and that glaciers waste away when summer melting exceeds winter snowfall. Beyond that simplistic statement, however, theorists cannot even agree whether a colder climate would be more conducive to glaciation than a warmer one! Although colder conditions would inhibit summer wastage and thus enhance the longevity of the winter accumulation, cold air cannot hold much water vapor. Hence, warmer winters would favor increased snowfall, whereas cooler summers are needed for decreased melting. Even a theory that accommodated either significantly increased snowfall or significantly decreased melting, or a combination of both, would still have to take into account the advance and retreat of the ice.

Various indeed are the explanations of Pleistocene climatic changes (and, by extension, of climatic fluctuations during previous ice ages). Some theories are based on variations in the intensity of solar radiation received by the Earth; others are founded on such astronomical cycles as the shifting of the Earth's axis or variations in the eccentricity of the Earth's orbit around the sun, and still others are focused on changes in the amount of carbon dioxide in the atmosphere. Some are founded on changes in the position of the continents and ocean circulation patterns. Some are rooted in the increased altitude of continental masses after a period of tectonic upheaval, and some incorporate elements of more than one of the above. We make no attempt here to summarize this multitude of theories, primarily because none of them is widely accepted; the search for a convincing explanation continues.

One significant outgrowth of our inadequate understanding of what causes a glacial age is the further puzzle of whether or not the Pleistocene, the beginning and ending dates of which are defined as being the beginning and ending dates of the most recent Ice Age, has ended. Are we now, in the twentieth century A.D., living in a postglacial period or in an interglacial period? Has the Pleistocene epoch closed? Are we now in the early centuries of a new geologic era or have the glaciers merely "gone back to get some more rocks"? Can the Ice Age be over when 7 percent of the ocean surface is sheathed by ice in winter, 10 percent of the land surface remains covered by glaciers, an additional 22 percent is underlain by permafrost, seasonal snow covers the continents throughout the midlatitudes, and higher mountain summits are ice-capped even at the equator? We are, of course, much too close to the event (in a temporal sense) to obtain a definitive answer.

However, some hints may be straws in the wind, or they may be only minor aberrations in a long-term pattern. For most of the last 10,000 years, there is evidence of a general deglaciation of Earth—a general retreat of almost all glaciers. Over the last three decades, however, this trend has been partly but not entirely either slowed or reversed. Many glaciers with a history of nothing but retreat as long as they have been known by humankind began to readvance in the 1960s and 1970s. The world has lately experienced several years in which the weather was favorable for glacier growth. Are these temporary conditions, or does this mark the onset of another glacial stage—and the continuation of the Pleistocene?

CHAPTER SUMMARY

Earth has experienced an unknown number of ice ages, but only the most recent—the Pleistocene epoch—is significant to our understanding of the contemporary landscape. About one-third of Earth's land area was ice-covered during the Pleistocene; glaciers today occupy only about one-tenth of the surface, and most of that is in Antarctica and Greenland.

The two basic kinds of glaciers—ice sheets and mountain glaciers—are both inexorable forces of erosion and prominent agents of deposition. Glaciers form wherever more snow falls in winter than melts in summer, over a period of time. As snow accumulates deeper and deeper, its weight compresses the lower portion into ice. If the compression continues, the ice moves outward or downslope from the area of accumulation. The inexorable force of the moving ice severely erodes the preglacial terrain by abrasion and plucking, and fragmented rock debris is subsequently deposited in various forms by the ice and its meltwater.

The most intriguing aspect of Pleistocene glaciation is the set of mysteries it presents. What factors initiate glacial advance? What causes their retreat? And most enigmatic of all, is the Ice Age over?

KEY TERMS

ablation	glacial trough	periglacial zone
accumulation	hanging valley	piedmont glacier
alpine glacier	horn	plucking
arête	icefield	proglacial lake
cirque	ice sheet	roche moutonnée
cirque glacier	kame	stratified drift
col	kettle	tarn
drift	moraine	till
drumlin	névé	till plain
equilibrium line	outlet glacier	valley glacier
esker	outwash plain	valley train
glacial erratic	paternoster lake	
glacial flour	patterned ground	

REVIEW QUESTIONS

1. Why is the Pleistocene epoch so important to physical geography, whereas other ice ages are not?
2. Describe the extent of glaciation during the Pleistocene epoch.
3. Discuss the periglacial and pluvial developments associated with Pleistocene glaciation.
4. Describe the three types of alpine glaciers.
5. Describe the metamorphosis from snow to glacial ice.
6. Discuss the mechanics of glacial movement.
7. Compare and contrast the roles of abrasion and plucking in glacial erosion.
8. Why are mountainous areas that have experienced glaciation usually quite rugged?
9. Why are regions once covered by a continental ice sheet usually poorly drained?
10. How is it possible for glacial ice to be advancing while the margin of the glacier is retreating?
11. How does glaciofluvial deposition differ from glacial deposition?
12. Why are recessional moraines less prominent than terminal moraines?
13. Describe the modifications a glacier can make to a stream-cut valley in mountainous terrain.
14. Are we still in an ice age?

SOME USEFUL REFERENCES

DAWSON, ALASTAIR G., *Ice Age Earth: Late Quaternary Geology and Climate.* New York: Routledge, Chapman & Hall, 1992.

DREWRY, DAVID, *Glacial Geologic Processes.* London: Edward Arnold, 1986.

FRENCH, H. M., *The Periglacial Environment.* London: Longman, 1976.

GERRARD, A. J., *Mountain Environments: An Examination of the Physical Geography of Mountains.* Cambridge, MA: M.I.T. Press, 1990.

GROVE, JEAN M., *The Little Ice Age.* New York: Routledge, 1988.

HARRIS, STUART A., *The Permafrost Environment.* Totowa, NJ: Rowman & Littlefield, 1986.

KRANTZ, E. B., et al., "Patterned Ground," *Scientific American,* Vol. 198, December 1988, p. 68.

SHARP, ROBERT P., *Living Ice: Understanding Glaciers and Glaciation.* New York: Cambridge University Press, 1991.

THEAKSTONE, W. H., J.A. MATTHEWS, AND C. HARRIS. *Glaciers and Environmental Change.* London: Edward Arnold, 1992.

APPENDIX I
FAMILIES OF MAP PROJECTIONS

Although there are dozens of map types, the great majority of them have very arcane uses, and so here we discuss only the four types used most frequently. These types are summarized in table I-1.

CYLINDRICAL PROJECTIONS

As figure 2-9 shows, a **cylindrical projection** is obtained when the markings on a center-lit globe are projected onto a cylinder wrapped tangent to the globe, the line of tangency usually being the equator. Having the equator as the tangency line produces a right-angled grid (meridians and parallels meet at right angles) on a rectangular map. There is no size distortion at the circle of tangency, but size distortion increases progressively with increasing distance from this circle, a characteristic clearly exemplified by the Mercator projection.

Cylindrical projections, with all their size distortions, generally are used for maps of the whole world. They are popular for this purpose apparently because a rectangular grid is psychologically comfortable and the rectangular format fits nicely on walls and pages. Such rationale, regrettably, has little intellectual virtue.

ELLIPTICAL PROJECTIONS

An **elliptical projection** (also called an *oval projection*) is a roughly football-shaped map, usually of the entire world, although sometimes only the central section of an elliptical projection is used for maps of lesser areas.

In most elliptical projections, a central parallel (usu-
ally the equator) and a central meridian (generally the prime meridian) cross at right angles in the middle of the map, which is a point of no distortion. (Distortion in size and shape normally increases progressively as one moves away from this point in any direction.) All the other parallels are then drawn parallel to the central one. All the other meridians are drawn so that all of them begin at a common point at the top of the oval and end at a common point at the bottom; the result, of course, is that all meridians except the central one must be drawn as curved lines.

AZIMUTHAL PROJECTIONS

An **azimuthal projection** (also called either a *planar projection* or a *zenithal projection*) is obtained by projecting the markings of a center-lit globe onto a flat piece of paper that is tangent to the globe at some point. Such a projection is shown in figure 2-7. The point of tangency can be any spot on the globe, but it usually is either the North or South Pole or some point on the equator. There is no distortion immediately around the point of tangency, but distortion increases progressively away from this point.

No more than one hemisphere can be displayed with any success on an azimuthal projection. Thus these projections show the same view as one gets when looking at a globe in the classroom and the same view an astronaut sees from distant space. This half-view-only characteristic can be a drawback, of course, just as it is with a globe, although azimuthal projections can be useful for focusing attention on a specific region.

CONIC PROJECTIONS

As figure 2-8 shows, a **conic projection** is obtained by projecting the markings of a center-lit globe onto a cone wrapped tangent to, or intersecting, a portion of the globe. Normally the apex of the cone is positioned above a pole, which means that the circle of tangency coincides with a parallel. This parallel then becomes the principal parallel of the projection; distortion is least in its vicinity and increases progressively as one moves away from it. Consequently, conic projections are best suited for regions of east-west orientation in the middle latitudes, being particularly useful for maps of the United States, Europe, or China.

It is impractical to use conic projections for more than one-fourth of Earth's surface (a semihemisphere), and they are particularly well adapted for mapping relatively small areas, such as a state or county.

TABLE I-1
The Main Families of Map Projections

Family	Some examples
Cylindrical	Central perspective cylindrical, Equirectangular, Gall's stereographic, Gnomonic cylindrical, Lambert's cylindrical equal-area, Mercator
Elliptical	Aitoff's, Denoyer semielliptical, Goode's homolosine, Mollweide's, Robinson, Sinusoidal
Azimuthal	Azimuthal equidistant, Gnomonic, Lambert's equal-area, Orthographic, Stereographic
Conic	Alber's conic equal-area, Bonne, Lambert's conformal conic, Polyconic, Simple conic

APPENDIX II
TOPOGRAPHIC MAP SYMBOLS

The United States Geological Survey (USGS) has an ongoing national Mapping Program, an important component of which is several series of topographic maps at a variety of scales from 1:24,000 or 1:1,000,000. These maps are designed particularly to show the shape of the land surface by means of contour lines, but they also portray a variety of other features.

Six colors and shades are in standard use to aid in distinguishing various kinds of map features:

Brown—contour lines
Blue—hydrographic (water) features
Black—features constructed or designated by humans, such as buildings, roads, boundary lines, and names
Green—woodlands, forests, orchards, vineyards
Red—important roads and lines of the public land survey system
Red tint—urban areas

The principal standard symbols used on USGS topographic maps are shown on the accompanying pages.

Provisional edition maps - metric or conventional units
Metric unit maps
Conventional unit maps

CONTROL DATA AND MONUMENTS

	Metric/Conv	Metric	Conventional
Aerial photograph roll and frame number	Not Shown	Not Shown	3-20
Horizontal control:			
Third order or better, permanent mark	Neace △	Neace △	Neace ⟁
With third order or better elevation	BM △ 148	BM △ 45.1	Pike BM 45.1
Checked spot elevation	△64	△19.5	Not Shown
Coincident with section corner	Cactus △	Cactus △	Cactus △
Unmonumented	Not Shown	Not Shown	+
Vertical control:			
Third order or better, with tablet	BM × 53	BM × 16.3	BM × 53.4
Third order or better, recoverable mark	× 394	× 120.0	× 393.6
Bench mark at found section corner	BM + 61	BM + 18.6	BM + 60.9
Spot elevation	× 17	× 5.3	× 17
Boundary monument:			
With tablet	BM □ 71	BM □ 21.6	BM 71
Without tablet	□ 562	□ 171.3	□ 562
With number and elevation	67 □ 988	67 □ 301.1	67 □ 988
U.S. mineral or location monument	▲	▲	USMM ▲

BOUNDARIES

National			
State or territorial			
County or equivalent			
Civil township or equivalent			
Incorporated-city or equivalent			
Park, reservation, or monument			
Small park			

LAND SURVEY SYSTEMS

U.S. Public Land Survey System:
Township or range line
Location doubtful
Section line
Location doubtful
Found section corner; found closing corner
Witness corner; meander corner

Provisional edition maps - metric or conventional units
Metric unit maps
Conventional unit maps

Other land surveys:
Township or range line
Section line
Land grant or mining claim; monument — □   — □   — □
Fence line

ROADS AND RELATED FEATURES

Primary highway
Secondary highway
Light duty road
Unimproved road
Trail
Dual highway
Dual highway with median strip
Road under construction   U. C.
Underpass; overpass
Bridge
Drawbridge
Tunnel

BUILDINGS AND RELATED FEATURES

Dwelling or place of employment: small; large
School; church
Barn, warehouse, etc.: small; large
House omission tint
Racetrack
Airport
Landing strip
Well (other than water); windmill
Water tank: small; large
Other tank: small; large
Covered reservoir
Gaging station
Landmark object
Campground; picnic area
Cemetery: small; large   Cem   Cem   Cem

RAILROADS AND RELATED FEATURES

Standard gauge single track; station
Standard gauge multiple track
Abandoned
Under construction
Narrow gauge single track
Narrow gauge multiple track
Railroad in street
Juxtaposition
Roundhouse and turntable

TRANSMISSION LINES AND PIPELINES

Power transmission line: pole; tower
Telephone or telegraph line
Aboveground oil or gas pipeline
Underground oil or gas pipeline

CONTOURS

Topographic:
Intermediate
Index
Supplementary
Depression
Cut; fill
Bathymetric:
Intermediate
Index
Primary
Index Primary
Supplementary

MINES AND CAVES

Quarry or open pit mine
Gravel, sand, clay, or borrow pit
Mine tunnel or cave entrance
Prospect; mine shaft
Mine dump
Tailings

SURFACE FEATURES

Levee
Sand or mud area, dunes, or shifting sand
Intricate surface area
Gravel beach or glacial moraine
Tailings pond

VEGETATION

Woods
Scrub
Orchard
Vineyard
Mangrove

MARINE SHORELINE

Topographic maps:
Approximate mean high water
Indefinite or unsurveyed
Topographic-bathymetric maps:
Mean high water
Apparent (edge of vegetation)

COASTAL FEATURES

Foreshore flat
Rock or coral reef
Rock bare or awash
Group of rocks bare or awash
Exposed wreck
Depth curve; sounding
Breakwater, pier, jetty, or wharf
Seawall

BATHYMETRIC FEATURES

Area exposed at mean low tide; sounding datum
Channel
Offshore oil or gas: well; platform
Sunken rock

RIVERS, LAKES, AND CANALS

Intermittent stream
Intermittent river
Disappearing stream
Perennial stream
Perennial river
Small falls; small rapids
Large falls; large rapids
Masonry dam
Dam with lock
Dam carrying road
Intermittent lake or pond
Dry lake
Narrow wash
Wide wash
Canal, flume, or aqueduct with lock
Elevated aqueduct, flume, or conduit
Aqueduct tunnel
Water well; spring or seep

GLACIERS AND PERMANENT SNOWFIELDS

Contours and limits
Form lines

SUBMERGED AREAS AND BOGS

Marsh or swamp
Submerged marsh or swamp
Wooded marsh or swamp
Submerged wooded marsh or swamp
Rice field
Land subject to inundation

APPENDIX III
THE WEATHER-STATION MODEL

Weather data are recorded at regular intervals for a great many locations on Earth, each location being called a **weather station**. These data are then plotted on weather maps according to a standard format and code. The format for a standard station model is shown in figure III-1, along with an explanation of the code. Figure III-2 shows the same model but with the codes replaced by data from a particular station, and figures III-3 and III-4 list some of the codes and symbols used by meteorologists. Tables III-1 through III-3 give addition codes and symbols, and figure III-5 is a sample weather map.

The wind symbol (Table III-3) gives two pieces of information. Wind direction is indicated by an arrow shaft entering the station circle from the direction in which the wind is blowing, as at the one-o'clock position in figure III-2, and wind speed is indicated by the number of "feathers" and half-feathers protruding from the shaft; each half feather represents a five-knot increase in speed. (A knot is one nautical mile per hour, which is the same as 1.15 statute miles per hour.)

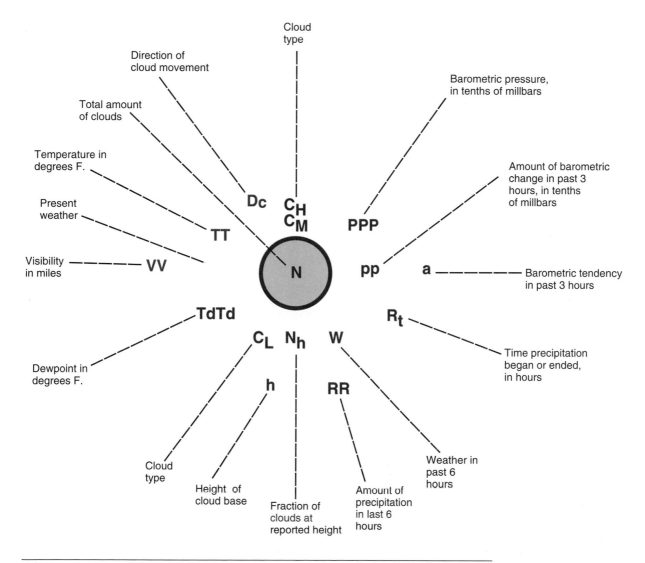

Figure III-1   A standard weather-station model. (No symbol for wind speed and direction is shown here because there is no one assigned place for this symbol; instead, its position on the model depends on wind direction.)

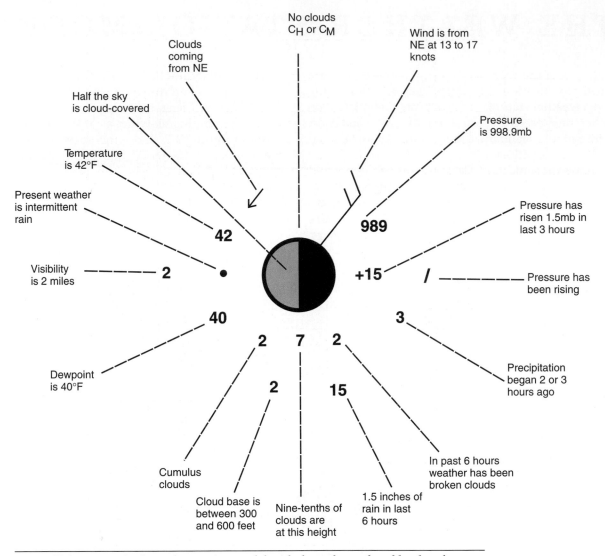

Figure III-2 A standard weather-station model with the codes replaced by data from a particular station.

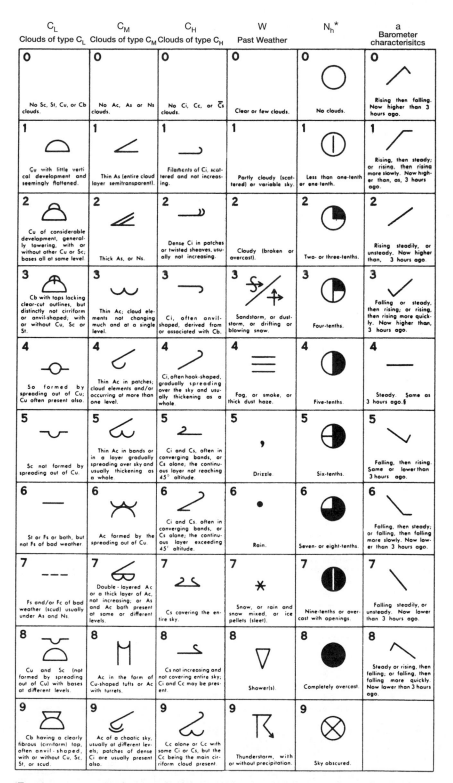

C_L Clouds of type C_L	C_M Clouds of type C_M	C_H Clouds of type C_H	W Past Weather	N_h* 	a Barometer characterisitcs
0 No Sc, St, Cu, or Cb clouds.	**0** No Ac, As or Ns clouds.	**0** No Ci, Cc, or C̄s clouds.	**0** Clear or few clouds.	**0** No clouds.	**0** Rising then falling. Now higher than 3 hours ago.
1 Cu with little vertical development and seemingly flattened.	**1** Thin As (entire cloud layer semitransparent).	**1** Filaments of Ci, scattered and not increasing.	**1** Partly cloudy (scattered) or variable sky.	**1** Less than one-tenth or one-tenth.	**1** Rising, then steady; or rising, then rising more slowly. Now higher than, as, 3 hours ago.
2 Cu of considerable development, generally towering, with or without other Cu or Sc; bases all at same level.	**2** Thick As, or Ns.	**2** Dense Ci in patches or twisted sheaves, usually not increasing.	**2** Cloudy (broken or overcast).	**2** Two- or three-tenths.	**2** Rising steadily, or unsteady. Now higher than, 3 hours ago.
3 Cb with tops lacking clear-cut outlines, but distinctly not cirriform or anvil-shaped; with or without Cu, Sc or St.	**3** Thin Ac; cloud elements not changing much and at a single level.	**3** Ci, often anvil-shaped, derived from or associated with Cb.	**3** Sandstorm, or dust-storm, or drifting or blowing snow.	**3** Four-tenths.	**3** Falling or steady, then rising; or rising, then rising more quickly. Now higher than, 3 hours ago.
4 So formed by spreading out of Cu; Cu often present also.	**4** Thin Ac in patches; cloud elements and/or occurring at more than one level.	**4** Ci, often hook-shaped, gradually spreading over the sky and usually thickening as a whole.	**4** Fog, or smoke, or thick dust haze.	**4** Five-tenths.	**4** Steady. Same as 3 hours ago.§
5 Sc not formed by spreading out of Cu.	**5** Thin Ac in bands or in a layer gradually spreading over sky and usually thickening as a whole.	**5** Ci and Cs, often in converging bands, or Cs alone; the continuous layer not reaching 45° altitude.	**5** Drizzle.	**5** Six-tenths.	**5** Falling, then rising. Same or lower than 3 hours ago.
6 St or Fs or both, but not Fs of bad weather	**6** Ac formed by the spreading out of Cu.	**6** Ci and Cs, often in converging bands, or Cs alone; the continuous layer exceeding 45° altitude.	**6** Rain.	**6** Seven- or eight-tenths.	**6** Falling, then steady; or falling, then falling more slowly. Now lower than 3 hours ago.
7 Fs and/or Fc of bad weather (scud) usually under As and Ns.	**7** Double-layered Ac or a thick layer of Ac, not increasing; or As and Ac both present at same or different levels.	**7** Cs covering the entire sky.	**7** Snow, or rain and snow mixed, or ice pellets (sleet).	**7** Nine-tenths or overcast with openings.	**7** Falling steadily, or unsteady. Now lower than 3 hours ago.
8 Cu and Sc (not formed by spreading out of Cu) with bases at different levels.	**8** Ac in the form of Cu-shaped tufts or Ac with turrets.	**8** Cs not increasing and not covering entire sky; Ci and Cc may be present.	**8** Shower(s).	**8** Completely overcast.	**8** Steady or rising, then falling; or falling, then falling more quickly. Now lower than 3 hours ago.
9 Cb having a clearly fibrous (cirriform) top, often anvil-shaped, with or without Cu, Sc, St, or scud.	**9** Ac of a chaotic sky, usually at different levels; patches of dense Ci are usually present also.	**9** Cc alone or Cc with some Ci or Cs, but the Cc being the main cirriform cloud present.	**9** Thunderstorm, with or without precipitation.	**9** Sky obscured.	

*Fraction representing how much of the total cloud cover is at the reported base height.

Figure III-3   Standard symbols used to indicate cloud conditions, past weather, and barometer characteristics. The numbers in the upper left corners of the cells are used in a standard model, and the icons are used on weather maps.

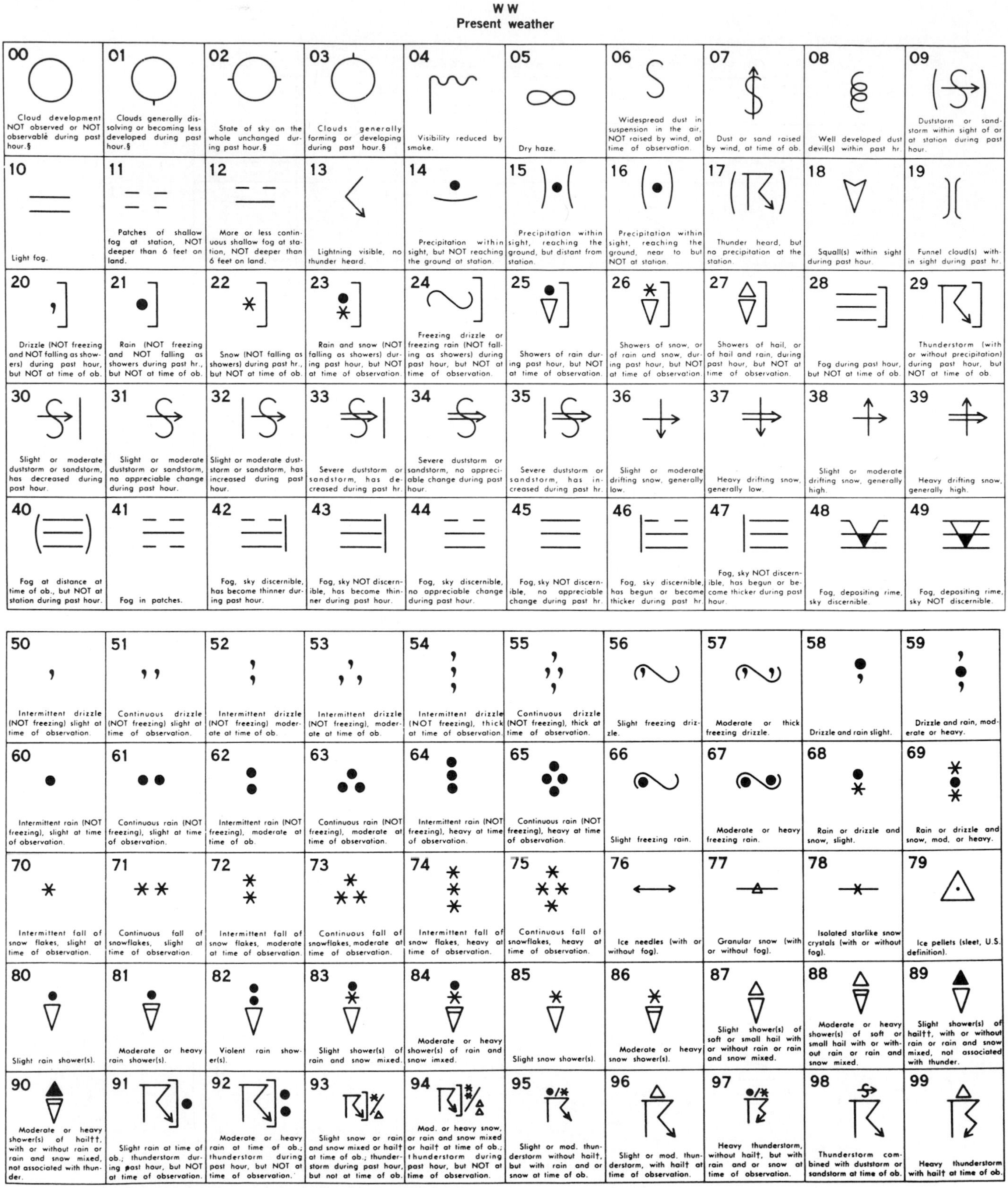

W W
Present weather

00 Cloud development NOT observed or NOT observable during past hour.§	**01** Clouds generally dissolving or becoming less developed during past hour.§	**02** State of sky on the whole unchanged during past hour.§	**03** Clouds generally forming or developing during past hour.§	**04** Visibility reduced by smoke.	**05** Dry haze.	**06** Widespread dust in suspension in the air, NOT raised by wind, at time of observation.	**07** Dust or sand raised by wind, at time of ob.	**08** Well developed dust devil(s) within past hr.	**09** Duststorm or sandstorm within sight of or at station during past hour.
10 Light fog.	**11** Patches of shallow fog at station, NOT deeper than 6 feet on land.	**12** More or less continuous shallow fog at station, NOT deeper than 6 feet on land.	**13** Lightning visible, no thunder heard.	**14** Precipitation within sight, but NOT reaching the ground at station.	**15** Precipitation within sight, reaching the ground, but distant from station.	**16** Precipitation within sight, reaching the ground, near to but NOT at station.	**17** Thunder heard, but no precipitation at the station.	**18** Squall(s) within sight during past hour.	**19** Funnel cloud(s) within sight during past hr.
20 Drizzle (NOT freezing and NOT falling as showers) during past hour, but NOT at time of ob.	**21** Rain (NOT freezing and NOT falling as showers during past hr., but NOT at time of ob.	**22** Snow (NOT falling as showers) during past hr., but NOT at time of ob.	**23** Rain and snow (NOT falling as showers) during past hr., but NOT at time of observation.	**24** Freezing drizzle or freezing rain (NOT falling as showers) during past hour, but NOT at time of observation.	**25** Showers of rain during past hour, but NOT at time of observation.	**26** Showers of snow, or of rain and snow, during past hour, but NOT at time of observation.	**27** Showers of hail, or of hail and rain, during past hour, but NOT at time of observation.	**28** Fog during past hour, but NOT at time of ob.	**29** Thunderstorm (with or without precipitation) during past hour, but NOT at time of ob.
30 Slight or moderate duststorm or sandstorm, has decreased during past hour.	**31** Slight or moderate duststorm or sandstorm, no appreciable change during past hour.	**32** Slight or moderate duststorm or sandstorm, has increased during past hour.	**33** Severe duststorm or sandstorm, has decreased during past hr.	**34** Severe duststorm or sandstorm, no appreciable change during past hour.	**35** Severe duststorm or sandstorm, has increased during past hour.	**36** Slight or moderate drifting snow, generally low.	**37** Heavy drifting snow, generally low.	**38** Slight or moderate drifting snow, generally high.	**39** Heavy drifting snow, generally high.
40 Fog at distance at time of ob., but NOT at station during past hour.	**41** Fog in patches.	**42** Fog, sky discernible, has become thinner during past hour.	**43** Fog, sky NOT discernible, has become thinner during past hour.	**44** Fog, sky discernible, no appreciable change during past hour.	**45** Fog, sky NOT discernible, no appreciable change during past hr.	**46** Fog, sky discernible, has begun or become thicker during past hr.	**47** Fog, sky NOT discernible, has begun or become thicker during past hour.	**48** Fog, depositing rime, sky discernible.	**49** Fog, depositing rime, sky NOT discernible.
50 Intermittent drizzle (NOT freezing) slight at time of observation.	**51** Continuous drizzle (NOT freezing) slight at time of observation.	**52** Intermittent drizzle (NOT freezing) moderate at time of ob.	**53** Continuous drizzle (NOT freezing), moderate at time of ob.	**54** Intermittent drizzle (NOT freezing), thick at time of observation.	**55** Continuous drizzle (NOT freezing), thick at time of observation.	**56** Slight freezing drizzle.	**57** Moderate or thick freezing drizzle.	**58** Drizzle and rain slight.	**59** Drizzle and rain, moderate or heavy.
60 Intermittent rain (NOT freezing), slight at time of observation.	**61** Continuous rain (NOT freezing), slight at time of observation.	**62** Intermittent rain (NOT freezing), moderate at time of ob.	**63** Continuous rain (NOT freezing), moderate at time of observation.	**64** Intermittent rain (NOT freezing), heavy at time of observation.	**65** Continuous rain (NOT freezing), heavy at time of observation.	**66** Slight freezing rain.	**67** Moderate or heavy freezing rain.	**68** Rain or drizzle and snow, slight.	**69** Rain or drizzle and snow, mod. or heavy.
70 Intermittent fall of snow flakes, slight at time of observation.	**71** Continuous fall of snowflakes, slight at time of observation.	**72** Intermittent fall of snow flakes, moderate at time of observation.	**73** Continuous fall of snowflakes, moderate at time of observation.	**74** Intermittent fall of snow flakes, heavy at time of observation.	**75** Continuous fall of snowflakes, heavy at time of observation.	**76** Ice needles (with or without fog).	**77** Granular snow (with or without fog).	**78** Isolated starlike snow crystals (with or without fog).	**79** Ice pellets (sleet, U.S. definition).
80 Slight rain shower(s).	**81** Moderate or heavy rain shower(s).	**82** Violent rain shower(s).	**83** Slight shower(s) of rain and snow mixed.	**84** Moderate or heavy shower(s) of rain and snow imxed.	**85** Slight snow shower(s).	**86** Moderate or heavy snow shower(s).	**87** Slight shower(s) of soft or small hail with or without rain or rain and snow mixed.	**88** Moderate or heavy shower(s) of soft or small hail with or without rain or rain and snow mixed.	**89** Slight shower(s) of hail††, with or without rain or rain and snow mixed, not associated with thunder.
90 Moderate or heavy shower(s) of hail††, with or without rain or rain and snow mixed, not associated with thunder.	**91** Slight rain at time of ob.; thunderstorm during past hour, but NOT at time of observation.	**92** Moderate or heavy rain at time of ob.; thunderstorm during past hour, but NOT at time of observation.	**93** Slight snow or rain and snow mixed or hail† at time of ob.; thunderstorm during past hour, but not at time of ob.	**94** Mod. or heavy snow, or rain and snow mixed or hail† at time of ob.; thunderstorm during past hour, but NOT at time of observation.	**95** Slight or mod. thunderstorm without hail†, but with rain and or snow at time of ob.	**96** Slight or mod. thunderstorm, with hail† at time of observation.	**97** Heavy thunderstorm, without hail†, but with rain and or snow at time of observation.	**98** Thunderstorm combined with duststorm or sandstorm at time of ob.	**99** Heavy thunderstorm with hail† at time of ob.

§ The symbol is not plotted for "ww" when "00" is reported. When "01, 02, or 03" is reported for "ww," the symbol is plotted on the station circle. Symbols are not plotted for "a" when "3 or 8" is reported.

† Refers to "hail" only.

†† Refers to "soft hail," "small hail," and "hail."

Figure III-4 Standard weather-map symbols used to indicate present weather conditions.

TABLE III-1
Standard Cloud Height Codes

h (height of cloud base)	Approximate cloud height	
	(feet)	(meters)
0	0–149	0–49
1	150–299	50–99
2	300–599	100–199
3	600–999	200–299
4	1000–1999	300–599
5	2000–3499	600–999
6	3500–4999	1000–1499
7	5000–6499	1500–1999
8	6500–7999	2000–2499
9	>8000 or no clouds	>2500 or no clouds

TABLE III-2
Standard Precipitation Codes

R_t code	Time of precipitation
0	No precipitation
1	Less than one hour ago
2	1 to 2 hours ago
3	2 to 3 hours ago
4	3 to 4 hours ago
5	4 to 5 hours ago
6	5 to 6 hours ago
7	6 to 12 hours ago
8	More than 12 hours ago
9	Unknown

TABLE III-3
Wind Speed/Direction Symbols

Symbol	Wind speed (knots)
	Calm
	1-2
	3-7
	8-12
	13-17
	18-22
	23-27
	28-32
	33-37
	38-42
	43-47
	48-52
	53-57
	13-17
	63-67
	68-72
	73-77

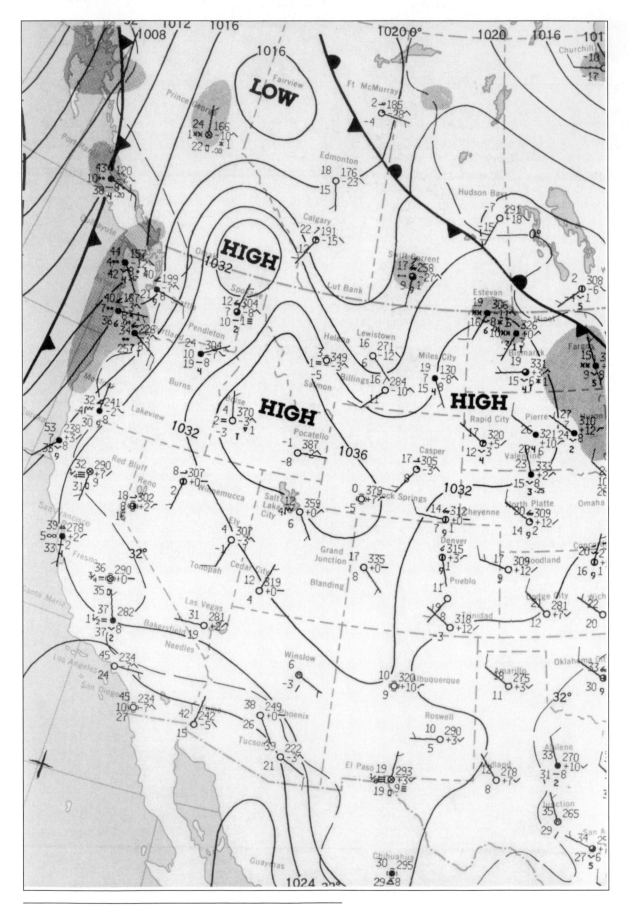

A sample weather map.

APPENDIX IV
METEOROLOGICAL TABLES

DETERMINING RELATIVE HUMIDITY

Measure the wet-bulb and dry-bulb temperatures, preferably using a sling psychrometer. Then use these two values to read the relative humidity from table IV-1.

For example, if the dry-bulb temperature is 70°F and the wet-bulb temperature is 65°F, the wet-bulb thermometer is showing a "depression" of 5°. Find 70°F in the air-temperature column of the table. Then find 5° in the top row. Move to the right from 70° and down from 5° and at the point of intersection, read the relative humidity: 77 percent.

THE BEAUFORT SCALE OF WIND SPEED

Admiral Beaufort of the British Navy developed, early in the nineteenth century, a scale of wind speed widely used in the English-speaking world. It has been modified through the years, but the essentials have not changed. The scale is shown in Table IV-2.

TABLE IV-1
Relative Humidity Values

Depression of wet-bulb thermometer, F°

Air temp., °F	1	2	3	4	5	6	7	8	9	10	11	12	13	14	15	16	17	18	19	20	21	22	23	24	25	26	27	28	29	30	31	32	33	34	35
0	67	33	1																																
5	73	46	20																																
10	78	56	34	13	15																														
15	82	64	46	29	11																														
20	85	70	55	40	26	12																													
25	87	74	62	49	37	25	13	1																											
30	89	78	67	56	46	36	26	16	6																										
35	91	81	72	63	54	45	36	27	19	10	2																								
40	92	83	75	68	60	52	45	37	29	22	15	7																							
45	93	86	78	71	64	57	51	44	38	31	25	18	12	6																					
50	93	87	74	67	61	55	49	43	38	32	27	21	16	10	5																				
55	94	88	82	76	70	65	59	54	49	43	38	33	28	23	19	11	9	5																	
60	94	89	83	78	73	68	63	58	53	48	43	39	34	30	26	21	17	13	9	5	1														
65	95	90	85	80	75	70	66	61	56	52	48	44	39	35	31	27	24	20	16	12	9	5	2												
70	95	90	86	81	77	72	68	64	59	55	51	48	44	40	36	33	29	25	22	19	15	12	9	6	3										
75	96	91	86	82	78	74	70	66	62	58	54	51	47	44	40	37	34	30	27	24	21	18	15	12	9	7	4	1							
80	96	91	87	83	79	75	72	68	64	61	57	54	50	47	44	41	38	35	32	29	26	23	20	18	15	12	10	7	5	3					
85	96	92	88	84	81	77	73	70	66	63	59	57	53	50	47	44	41	38	36	33	30	27	25	22	20	17	15	13	10	8	6	4	2		
90	96	92	89	85	81	78	74	71	68	65	61	58	55	52	49	47	44	41	39	36	34	31	29	26	24	22	19	17	15	13	11	9	7	5	3
95	96	93	89	86	82	79	76	73	69	66	63	61	58	55	52	50	47	44	42	39	37	34	32	30	28	25	23	21	19	17	15	13	11	10	8
100	96	93	89	86	83	80	77	73	70	68	65	62	59	56	54	51	49	46	44	41	39	37	35	33	30	28	26	24	22	21	19	17	15	13	12
105	97	93	90	87	84	81	78	75	72	69	66	64	61	58	56	53	51	49	46	44	42	40	38	36	34	32	30	28	26	24	22	21	19	17	15
110	97	93	90	87	84	81	78	75	73	70	67	65	62	60	57	55	52	50	48	46	44	42	40	38	36	34	32	30	28	26	25	23	21	20	18
115	97	94	91	88	85	82	79	76	74	71	69	66	64	61	59	57	54	52	50	48	46	44	42	40	38	36	34	33	31	29	28	26	25	23	21
120	97	94	91	88	85	82	80	77	74	72	69	67	65	62	60	58	55	53	51	49	47	45	43	41	40	38	36	34	33	31	29	28	26	25	23
125	97	94	91	88	86	83	80	78	75	73	70	68	66	64	61	59	57	55	53	51	49	47	45	44	42	40	38	37	35	33	32	30	29	27	26
130	97	94	91	89	86	83	81	78	76	73	71	69	67	64	62	60	58	56	54	52	50	48	47	45	43	41	40	38	37	35	33	32	30	29	28

TABLE IV-2
Beaufort Scale

Beaufort force	Speed (miles per hour)	(knots)	Description
0	1	1	Calm
1	1–3	1–3	Light air
2	4–7	4–6	Light breeze
3	8–12	7–10	Gentle breeze
4	13–18	11–16	Moderate breeze
5	19–24	17–21	Fresh breeze
6	25–31	22–27	Strong breeze
7	32–38	28–33	Near gale
8	39–46	34–40	Gale
9	47–54	41–47	Strong gale
10	55–63	48–55	Storm
11	64–72	56–63	Violent storm
12	73–82	64–71	Hurricane
13	83–92	72–80	Hurricane
14	93–103	81–89	Hurricane
15	104–114	90–99	Hurricane
16	115–125	100–108	Hurricane
17	126–136	109–118	Hurricane

APPENDIX V
THE INTERNATIONAL SYSTEM OF UNITS (SI)

With the major exception of the United States, the system of weights and measures used worldwide both in scientific work and in everyday life is the International System, usually abbreviated SI from its French name, *Systeme Internationale*. The system has seven base units and supplementary units for angles (Table V-1). The beauty of the system lies in its reliance on multiples of the number 10, with the prefixes shown in Table V-2 being used to cover a magnitude range from the astronomically large to the infinitesimally small.

Table V-3 lists the most frequently used conversion factors.

TABLE V-1
SI Units

Quantity	Unit	Symbol
Base units		
Length	Meter	m
Mass	Kilogram	kg
Time	Second	s
Electric current	Ampere	A
Temperature	Kelvin	K
Amount of substance	Mole	mol
Luminous intensity	Candela	cd
Supplementary units		
Plane angle	Radian	rad
Solid angle	Steradian	sr

TABLE V-2
Common Multiples and SI Prefixes

Multiple	Value	Prefix	Symbol
1 000 000 000 000	10^{12}	tera	T
1 000 000 000	10^{9}	giga	G
1 000 000	10^{6}	mega	M
1000	10^{3}	kilo	k
100	10^{2}	hecto	h
10	10^{1}	deka	da
0.1	10^{-1}	deci	d
0.01	10^{-2}	centi	c
0.001	10^{-3}	milli	m
0.000001	10^{-6}	micro	μ
0.000000001	10^{-9}	nano	n
0.000000000001	10^{-12}	pico	p

TABLE V-3
SI—British Conversion Units

Multiply	By	To Get
Inches	2.54	Centimeters
Feet	0.3048	Meters
Yards	0.9144	Meters
Miles	1.6093	Kilometers
Millimeters	0.039	Inches
Centimeters	0.39	Inches
Meters	3.281	Feet
Kilometers	0.621	Miles
Area		
Square inches	6.452	Square centimeters
Square feet	0.0929	Square meters
Square yards	0.836	Square meters
Square miles	2.59	Square kilometers
Acres	0.4	Hectares
Square centimeters	0.155	Square inches
Square meters	10.764	Square feet
Square meters	1.196	Square yards
Square kilometers	0.386	Square miles
Hectares	2.471	Acres
Volume		
Cubic inches	16.387	Cubic centimeters
Cubic feet	0.028	Cubic meters
Cubic yards	0.7646	Cubic meters
Fluid ounces	29.57	Milliliters
Pints	0.47	Liters
Quarts	0.946	Liters
Gallons	3.785	Liters
Cubic centimeters	0.061	Cubic inches
Cubic meters	35.3	Cubic feet
Cubic meters	1.3	Cubic yards
Milliliters	0.034	Fluid ounces
Liters	1.0567	Quarts
Liters	0.264	Gallons
Mass (weight)		
Ounces	28.35	Grams
Pounds	0.4536	Kilograms
Tons (2000 lb)	907.18	Kilograms
Tons (2000 lb)	0.90718	Tonnes
Grams	0.035	Ounces
Kilograms	2.2046	Pounds
Kilograms	0.0011	Tons (2000 lb)
Tonnes	1.1023	Tons (2000 lb)

APPENDIX VI
THE SOIL TAXONOMY

The Soil Taxonomy, described in chapter 12, utilizes a nomenclature "invented" for the purpose. This nomenclature consists of "synthetic" names, which means that syllables from existing words are rearranged and combined to produce new words for the names for the various soil types. The beauty of the system is that each newly coined name is highly descriptive of the soil it represents.

The awkwardness of the nomenclature is threefold:

1. Most of the terms are new and have never before appeared in print or in conversation. Thus they look strange, and the words do not easily roll off the tongue. It is almost a new language.
2. Many of the words are difficult to write and great care must be taken in the spelling of seemingly bizarre combinations of letters.
3. Many of the syllables sound so much alike that differences in pronunciation are often slight, although enunciating these slight differences is essential if the new nomenclature is to serve its purpose.

Nevertheless, the Soil Taxonomy has a sound theoretical base, and once the user is familiar with the vocabulary, every syllable of every word gives important information about a soil. Almost all the syllables are derived from Greek or Latin roots, in contrast to the English and Russian terms used in previous systems. In some cases, the appropriateness of the classical derivatives may be open to question, but the uniformity and logic of the terminology extend throughout the system.

THE HIERARCHY

The top level in the Soil Taxonomy hierarchy is soil *order*. The names of orders are made up of three or four syllables, the last of which is always *sol* (from the Latin word *solum*, "soil"). The next-to-last syllable consists of a single linking vowel, either i or o. The syllable or two that begin the word contain the formative element of the name and give information regarding some distinctive characteristic of the order. Thus the names of the eleven soil orders contain eleven syllables that (1) serve as formative elements for the order names and (2) appear in all names in the next three levels of the hierarchy.

Suborder is the second level in the hierarchy. All suborder names contain two syllables: the first indicates some distinctive characteristic of the suborder, the second identifies the order to which the suborder belongs. For example, *aqu* is derived from the Latin word for water, and so *Aquent* is the suborder of wet soils of the order Entisols. The names of the four dozen suborders are constructed from the two dozen root elements shown in table VI-1.

The third level of the hierarchy contains great groups, the names of which are constructed by grafting one or more syllables to the beginning of a suborder name. Hence, a Cryaquent is a cold soil that is a member of the Aquent suborder of the Entisol order (*cry* comes from the Greek word for coldness). The formative prefixes for the great group names are derived from about 50 root words, samples of which are presented in table VI-2.

The next level is called *subgroup*, and there are more than 1000 subgroups recognized in the United States. Each subgroup name consists of two words: the first derived from a formative element higher up in the hierarchy (with a few exceptions), the second the same as that of the relevant great group. Thus a *Sphagnic Cryaquent* is of the order Entisol, the suborder Aquent, the great group Cryaquent, and it contains sphagnum moss (derived from the Greek word *sphagnos*, "bog.")

The *family* is the penultimate level in the hierarchy. It is not given a proper name but is simply described by one or more lower-case adjectives, as "a skeletal, mixed, acidic family."

Finally, the lowest level of the hierarchy is the *series*, named for geographic location. Table VI-3 gives a typical naming sequence.

SUMMARY OF SOIL SUBORDERS

Alfisols   Alfisols have five suborders. *Aqualls* have characteristics associated with wetness. *Boralfs* are associated with cold boreal forests. *Udalfs* are brownish or reddish soils of moist midlatitude regions. *Ustalfs* are similar to Udalfs in color but subtropical in location and usually have a hard surface layer in the dry season. *Xeralfs* are found in mediterranean climates and are characterized by a thick, hard surface horizon in the dry season.

Andisols   Seven suborders are recognized in the Andisols, five of which are distinguished by moisture content. *Aquands* have abundant moisture, often with poor drainage. *Torrands* are associated with a hot, dry regime. *Udands* are found in humid climates, *Ustands* in dry climates with hot summers. *Xerands* have a pronounced annual dry season. In addition, *Cryands* are found in cold climates, and *Vitrands* are distinguished by the presence of glass.

Aridisols Two suborders generally are recognized on the basis of degree of weathering. *Argids* have a distinctive subsurface horizon with clay accumulation, whereas *Orthids* do not.

Entisols There are five suborders of Entisols. *Aquents* occupy wet environments where the soil is more or less continuously saturated with water; they may be found in any temperature regime. *Arents* lack horizons because of human interference, particularly that involving large agricultural or engineering machinery. *Fluvents* form on recent water-deposited sediments that have satisfactory drainage. *Orthents* develop on recent erosional surfaces. *Psamments* occur in sandy situations, where the sand is either shifting or stabilized by vegetation.

Histosols The four suborders of Histosols—*Fibrists*, *Folists*, *Hemists*, and *Saprists*—are differentiated on the basis of degree of plant-material decomposition.

Inceptisols The six suborders of Inceptisols—*Andepts*, *Aquepts*, *Ochrepts*, *Plaggepts*, *Troperts*, and *Um-*

T A B L E V I - 1 **Name Derivations of Soil Suborders**			
Root	*Derivation*	*Connotation*	*Example of suborder name*
alb	Latin *albus*, "white"	Presence of a bleached eluvial horizon	Alboll
and	Japanese *ando*, a volcanic soil	Derived from pyroclastic material	Andept
aqu	Latin *aqua*, "water"	Associated with wetness	Aquent
ar	Latin *arare*, "to plow"	Horizons are mixed	Arent
arg	Latin *argilla*, "white clay"	Presence of a horizon containing illuvial clay	Argid
bor	Greek *boreas*, "northern"	Associated with cool conditions	Boroll
ferr	Latin *ferrum*, "iron"	Presence of iron	Ferrod
fibr	Latin *fibra*, "fiber"	Presence of undecomposed organic matter	Fibrist
fluv	Latin *fluvius*, "river"	Associated with floodplains	Fluvent
fol	Latin *folia*, "leaf"	Mass of leaves	Folist
hem	Greek *hemi*, "half"	Intermediate stage of decomposition	Hemist
hum	Latin *humus*, "earth"	Presence of organic matter	Humult
ochr	Greek *ochros*, "pale"	Presence of a light-colored surface horizon	Ochrept
orth	Greek *orthos*, "true"	Most common or typical group	Orthent
plag	German *plaggen*, "sod"	Presence of a human-induced surface horizon	Plaggept
psamm	Greek *psammos*, "sand"	Sandy texture	Psamment
rend	Polish *rendzino*, a type of soil	Significant calcareous content	Rendoll
sapr	Greek *sapros*, "rotten"	Most decomposed stage	Saprist
torr	Latin *torridus*, "hot and dry"	Usually dry	Torrox
trop	Greek *tropikos*, "of the solstice"	Continuously warm	Tropert
ud	Latin *udud*, "humid"	Of humid climates	Udoll
umbr	Latin *umbro*, "shade"	Presence of a dark surface horizon	Umbrept
ust	Latain *ustus*, "burnt"	Of dry climates	Ustert
xer	Greek *xeros*, "dry"	Annual dry season	Xeralf

brepts—have relatively complicated distinguishing characteristics.

Mollisols   The seven suborders of Mollisols—*Albolls, Aquolls, Borolls, Rendolls, Udolls, Ustolls,* and *Xerolls*—are distinguished largely, but not entirely, on the basis of relative wetness/dryness.

Oxisols   The five suborders of Oxisols—*Aquox, Humox, Orthos, Torros,* and *Ustox*—are distinguished from one another primarily by what effect varying amounts and seasonality of rainfall have on the profile.

Spodosols   Of the four suborders of Spodosols, most widespread are the *Orthods*, which represent the typical Spodosols. *Aquods, Ferrods,* and *Humods* are differentiated on the basis of the amount of iron in the spodic horizon.

Ultisols   Five suborders of Ultisols—*Aquults, Humults, Udults, Ustults,* and *Xerults*—are recognized. The distinction among them is largely on the basis of temperature and moisture conditions and on how these parameters influence the epipedon.

Vertisols   The four principal suborders of Vertisols are distinguished largely on the frequency of "cracking," which is a function of climate. *Torrerts* are found in arid regions, and the cracks in these soils remain open most of the time. *Uderts* are found in humid areas, and in these soils cracking is irregular. *Usterts* are associated with monsoonal climates and have a relatively complicated cracking pattern. *Xererts* occur in mediterranean climates and have cracks that open and close regularly once each year.

TABLE VI-2 Name Derivations for Great Groups			
Root	*Derivation*	*Connotation*	*Example of great group name*
calc	Latin *calcis*, "lime"	Presence of calcic horizon	Calciorthid
ferr	Latin *ferrum*, "iron"	Presence of iron	Ferrudalf
natr	latin *natrium*, "sodium"	Presence of a natric horizon	Natraboll
pale	Greek *paleos*, "old"	An old development	Paleargid
plinth	Greek *plinthos*, "brick"	Presence of plinthite	Plenthoxeralf
quartz	the German name	High quartz content	Quartzipsamment
verm	Latin *vermes*, "worm"	Notable presence of worms	Vermudoll

TABLE VI-3 A Typical Soil Taxonomy Naming Sequence	
Order	*Entisol*
Suborder	Aquent
Great group	Cryaquent
Subgroup	Sphagnic Cryaquent
Family	skeletal, mixed, acidic, Sphagnic Cryaquent
Series	Aberdeen

GLOSSARY

Ablation Wastage of glacial ice through melting and sublimation.

Ablation zone The lower portion of a glacier where there is a net annual loss of ice due to melting and sublimation.

Abrasion The chipping and grinding effect of rock fragments as they are swirled or bounced or rolled downstream by moving water.

Absolute humidity A direct measure of the water vapor content of air, expressed as the weight of water vapor in a given volume of air, usually as grams of water per cubic meter of air.

Absorption The ability of an object to assimilate energy from electromagnetic waves that strike it.

Accelerated erosion Soil erosion occurring at a rate faster than soil horizons can be formed from the parent regolith.

Accrete Grow together, or add by growth.

Accumulation Addition of ice into a glacier by incorporation of snow.

Accumulation zone The upper portion of a glacier where there is a greater annual increment of ice than there is wastage.

Acid rain Precipitation with a pH less than 5.6. It may involve dry deposition without moisture.

Adiabatic cooling Cooling by expansion in rising air.

Adiabatic warming Warming by contraction in descending air.

Adret slope A slope oriented so that the sun's rays arrive at a relatively direct angle. Such a slope is relatively hot and dry, and its vegetation will not only be sparser and smaller but is also likely to have a different species composition from adjacent slopes with different exposures.

Advection Horizontal movement of air across the Earth's surface.

Advectional inversion An inverted temperature gradient caused by a horizontal inflow of colder air into an areausually colder air "draining" downslope into a valleyor by cool maritime air blowing into a coastal locale. Advectional inversions are usually short-lived and shallow and are more common in winter than summer.

Advection fog The condensation that results when warm, moist air moves horizontally over a cold surface.

Aeolian processes Processes related to wind action that are most pronounced, widespread, and effective in dry lands.

Aerial photograph A photograph taken from an elevated "platform," such as a balloon, airplane, rocket, or satellite.

Aggradation The process in which a stream bed is raised as a result of the deposition of sediment.

Agonic line The isogonic line on which there is no magnetic declination because of the exact alignment of true north and magnetic north.

A horizon Upper soil layer in which humus and other organic materials are mixed with mineral particles.

Air drainage The sliding of cold air downslope to collect in the lowest spots, usually at night.

Air mass An extensive body of air that has relatively uniform properties in the horizontal dimension and moves as an entity.

Albedo The fraction of total solar radiation that is reflected back, unchanged, into space.

Alcove arch A naturally excavated chamber that formed when trickling surface water and percolating groundwater softened and dissolved the underlying rock, cementing the material of this softer rock under more resistant overhanging caprock. The loosened grains were removed by water and wind erosion.

Alfisol A widely distributed soil order distinguished by a subsurface clay horizon and a medium-to-generous supply of plant nutrients and water.

Alluvial fan A fan-shaped deposition feature laid down by a stream issuing from a mountain canyon.

Alluvium Any stream-deposited sedimentary material.

Alpine glacier Individual glacier that develops near a mountain crest line and normally moves down valley for some distance.

Altocumulus Middle-level clouds, between about 6500 and 20,000 feet (2 and 6 km), which are puffy in form and are composed of liquid water.

Altostratus Middle-level clouds, between 6500 and 20,000 feet (2 and 6 km), which are layered and are composed of liquid water.

Amphibians Semiaquatic vertebrate animals. In the larval stage they are fully aquatic and breathe through gills; as adults they are air-breathers by means of lungs and through glandular skin.

Anadromous fish Fish that spawn in freshwater streams but live most of their lives in the ocean.

Andesite A volcanic rock composed largely of a distinctive mineral association of plagioclase feldspar.

Andisol The newest of the eleven orders in The Soil Taxonomy. Andisols are derived from volcanic ash.

Anemometer An instrument that measures wind speed.

Angiosperms Plants that have seeds encased in some sort of protective body, such as a fruit, a nut, or a seedpod.

Angle of incidence The angle at which the sun's rays strike Earth's surface.

Angle of repose Steepest angle that can be assumed by loose fragments on a slope without downslope movement.

Animalia The kingdom of organisms that consists of all multicellular animals.

Annual plants Plants that perish during times of climatic stress but leave behind a reservoir of seeds to germinate during the next favorable period.

Annular drainage pattern A network in which the major streams are arranged in a ringlike, concentric pattern in response to a structural dome or basin.

Antarctic Circle The parallel of 66 1/2° south latitude.

Antelope A collective term that refers to several dozen species of hoofed animals, nearly all of which have permanent, unbranched horns.

Anthropogenic Human-induced.

Anticline A simple symmetrical upfold.

Anticyclone A high-pressure center.

Antitrade winds Tropical upper air winds that blow toward the northeast in the Northern Hemisphere and toward the southeast in the Southern Hemisphere.

Aphelion The point in Earth's elliptical orbit at which the Earth is farthest from the sun (94,555,000 miles or 152,171,500 km).

Apogee The point at which the moon is farthest from Earth in its elliptical orbit.

Aquale A suborder of Alfisol soils that has characteristics that are associated with wetness.

Aquent One of the suborders of Entisol soils that occupies wet environments where the soil is more or less continuously saturated with water.

Aquiclude An impermeable rock layer that is so dense as to exclude water.

Aquifer A permeable subsurface rock layer that can store, transmit, and supply water.

Arboreal Tree-dwelling.

Arctic Circle The parallel of 66 1/2° north latitude.

Arcuate delta A delta with its shoreline curved convexly outward from the land.

Arent A suborder of Entisol soils that lacks horizons because of human interference.

Arete A narrow, jagged, serrated spine of rock; remainder of a ridge crest after several cirques have been cut back into an interfluve from opposite sides of a divide.

Argid A suborder of Aridisol soils that has a distinctive subsurface horizon with clay accumulation.

Arid Dry; receiving limited precipitation (generally less than10 inches [25 cm] annually).

Aridisol A soil order occupying dry environments that do not have enough water to remove soluble minerals from the soil; typified by a thin profile that is greatly lacking in organic matter and a sandy texture.

Arroyo The term used in the United States for the normally dry bed of an intermittent stream.

Artesian well The free flow that results when a well is drilled from the surface down into the aquifer and the confining pressure is sufficient to force the water to the surface without artificial pumping.

Association (plant) An assemblage of plants living in close inter-dependence, with similar habitat requirements, and with one or more dominant species that may be used to denote it.

Asthenosphere   Plastic layer of the upper mantle that underlies the lithosphere. Its rock is very hot and therefore weak and easily deformed.

Atmosphere   The gaseous envelope surrounding Earth.

Atmospheric pressure   The weight of the air.

Atoll   Coral reef in the general shape of a ring or partial ring that encloses a lagoon.

Autumnal equinox   Equinox that occurs about September 23rd in the Northern Hemisphere and March 21st in the Southern Hemisphere.

Axis (Earth's axis)   The diameter line that connects the points of maximum flattening on Earth's surface.

Azimuthal projection   A family of maps derived by the perspective extension of the geographic grid from a globe to a plane that is tangent to the globe at some point.

Backshore   The upper part of the shore, beyond the reach of ordinary waves.

Backswamp   Area of low, swampy ground on a floodplain between the natural levee and the bluffs.

Backwash   Water moving seaward after the momentum of the wave swash is overcome by gravity and friction.

Badlands   Intricately rilled and barren terrain of arid and semiarid regions, characterized by a multiplicity of short, steep slopes.

Baguio   The term used for a tropical cyclone affecting the Philippines.

Bajada (piedmont alluvial plain)   A continual alluvial surface that extends across the piedmont zone, slanting from the range toward the basin, in which it is difficult to distinguish between individual fans.

Bar   Ridge of sand or gravel deposited offshore or in a river.

Barchan   A crescent-shaped sand dune with cusps of the crescent pointing downwind.

Barometer   An instrument that measures atmospheric pressure.

Barrier reef   A prominent ridge of coral that roughly parallels the coastline, but lies offshore, with a shallow lagoon between the reefs and the coast.

Basal slip   The term used to describe the sliding of the entire mass at the bottom of a glacier over its bed on a lubricating film of water.

Basalt   Fine-grained speckled extrusive rock.

Base element   A chemical element that is an important plant nutrient, such as calcium and potassium.

Base level   An imaginary surface extending underneath the continents from sea level at the coasts, and indicating the lowest level to which land can be eroded.

Batholith   The largest and most amorphous of igneous intrusions.

Bayhead   The portion of the bay that extends furthest inland.

Baymouth bar   A spit that has become extended across the mouth of a bay to connect with a headland on the other side, transforming the bay into a lagoon.

Beach   An exposed deposit of loose sediment, normally composed of sand and!/or gravel, and occupying the coastal transition zone between land and water.

Bedload   Sand, gravel, and larger rock fragments moving in a stream by saltation and traction.

Bedrock   Residual rock that has not experienced erosion.

Benthos   Animals and plants that live at the ocean bottom.

Berm   The relatively flat part of a backshore beach, composed of wave-deposited material.

B horizon   Mineral soil horizon located beneath the A horizon.

Binomial classification scheme   A taxonomic classification scheme that consists of a generic and a specific name.

Biomass   The total weight of all living organisms.

Biome   A large, recognizable assemblage of plants and animals in functional interaction with its environment.

Biosphere   The living organisms of Earth.

Biota   The total complex of plant and animal life.

Bird's-foot delta   A delta that has long, projecting distributary channels branching outward like the toes or claws of a bird.

Blackbody   A body that emits the maximum amount of radiation possible, at every wavelength, for its temperature.

Blowout (deflation hollow)   A shallow depression from which an abundance of fine material has been deflated.

Bluff   A relatively steep slope at the outer edge of a flood plain.

Boralf   A suborder of Alfisol soils associated with wetness.

Boreal forest (taiga)   An extensive needle-leaf forest in the subarctic regions of North America and Eurasia.

Bornhardt   A rounded or domal inselberg composed of very resistant rock that stands above the surrounding terrain because of differential erosion and weathering.

Brackish   Slightly saline, with a salt content less than that of sea water.

Braided stream   A stream that consists of a multiplicity of interwoven and interconnected shallow channels separated by low islands of sand, gravel, and other loose debris.

Breakwater   A structure protecting a nearshore area from breaking waves.

Broadleaf trees   Trees that have flat and expansive leaves.

Bryophytes   Mosses and liverworts.

Butte   An erosional remnant of very small surface area and clif-flike sides that rises conspicuously above the surroundings.

Buys-Ballot's law   A physical principle stating that if one stands with one's back to the wind in the Northern Hemisphere, low pressure is on the left and high pressure is on the right. Directions are reversed in the Southern Hemisphere.

Calcification   One of the dominant pedogenic regimes in areas where the principal soil moisture movement is upward because of a moisture deficit. This regime is characterized by a concentration of calcium carbonate ($CaCO_3$) in the B horizon forming a hardpan, an upward movement of $CaCO_3$ by capillary water by grass roots, and a return of $CaCO_3$ when grass dies.

Caldera   Large, steep-sided, roughly circular depression resulting from the explosion and subsidence of a large volcano.

Capacity   The maximum load that a stream can transport under given conditions.

Capillarity   The action by which water can climb upward in restricted confinement as a result of its high surface tension, and thus the ability of its molecules to stick closely together.

Capillary water   Moisture held at the surface of soil particles by surface tension.

Caprock   A flattish erosional surface formed of the most resistant layers of horizontal sedimentary or volcanic strata.

Carbonate   A category of minerals composed of carbon and oxygen combined with another element or elements.

Carbonates   Minerals that are carbonate compounds of calcium or magnesium.

Carbonation   A process in which carbon dioxide in water reacts with carbonate rocks to produce a very soluble product (calcium bicarbonate), which can readily be removed by runoff or percolation, and which can also be deposited in crystalline form if the water is evaporated.

Carbon cycle   The change from carbon dioxide to living matter and back to carbon dioxide.

Carnivore   A flesh-eating animal.

Cartographer   A map-maker.

Cartography   The construction and production of maps.

Cation   An atom or group of atoms with a positive electrical charge.

Cation exchange capacity (CEC)   Capability of soil to attract and exchange cations.

Centripetal drainage pattern   A basin structure in which the streams converge toward the center.

Cetaceans   Members of an order of aquatic, largely marine mammals, such as whales and dolphins.

Channel precipitation   Rain or snow that falls directly onto a water surface.

Chaparral   Shrubby vegetation of the mediterranean climatic region of North America.

Chemical weathering   The decomposition of rock by the alteration of rock-forming minerals.

Chinook   A localized downslope wind of relatively dry and warm air, which is further warmed adiabatically as it moves down the leeward slope of the Rocky Mountains.

Chlorofluorocarbons (CFCs)   Synthetic chemicals that destroy ozone in the upper atmosphere.

Chordata   The phylum within the animal kingdom that includes all animals with backbones.

C horizon   Lower soil layer composed of weathered parent material that has not been significantly affected by translocation or leaching.

Cinder cone   Small, common volcano that is composed primarily of pyroclastic material blasted out from a vent in small but intense explosions. The structure of the volcano is usually a conical hill of loose material.

Circle of illumination   The edge of the sunlit hemisphere that is a great circle separating Earth into a light half and a dark half.

Cirque A broad amphitheater hollowed out at the head of a glacial valley by ice erosion.

Cirque glacier A small glacier confined to its cirque and not moving down-valley.

Cirriform cloud A cloud that is thin and wispy, composed of ice crystals rather than water particles, and found at high elevations.

Cirrocumulus High cirriform clouds arranged in a patchy pattern.

Cirrostratus High cirriform clouds that appear as whitish, translucent veils.

Cirrus High cirriform clouds of feathery appearance.

Cladistics A biological taxonomy based on shared characteristics of organisms rather than on shared ancestry.

Clay Very small inorganic particles produced by chemical alteration of silicate minerals.

Claypan A playa surface that is heavily impregnated with clay.

Cliff Sheer residual rock face.

Climate An aggregate of day-to-day weather conditions over a long period of time.

Climax vegetation A stable plant association of relatively constant composition that develops at the end of a long succession of changes.

Cloud seeding The introduction of particles (usually dry ice or silver iodide) into clouds for the purpose of altering the cloud's normal development, usually in an effort to enhance precipitation.

Coalesce To come together into one body.

Col A pass or saddle through a ridge produced when two adjacent cirques on opposite sides of a divide are cut back enough to remove part of the arete between them.

Cold front The leading edge of a cool air mass actively displacing warm air.

Collapse doline A sinkhole produced by the collapse of the roof of a subsurface cavern.

Colloids Organic and inorganic microscopic particles of soil that represent the chemically active portion of particles in the soil.

Combined water Water held in chemical combination with various soil minerals.

Competence The size of the largest particle that can be transported by a stream.

Composite volcanoes Volcanoes with the classic symmetrical cone-shaped peak, produced by a mixture of lava outpouring and pyroclastic explosion.

Computer rectification Adjustment by means of computer.

Condensation Process by which water vapor is converted to liquid water.

Condensation nuclei Tiny atmospheric particles of dust, smoke, and salt that serve as collection centers for water molecules.

Conditional instability A lapse rate somewhere between the dry and wet adiabatic rates.

Conduction The movement of energy from one molecule to another without changing the relative positions of the molecules. It enables the transfer of heat between different parts of a stationary body.

Cone of depression The phenomenon whereby the water table has sunk into the approximate shape of an inverted cone in the immediate vicinity of a well as the result of the removal of a considerable amount of the water.

Conformality The property of a map projection that maintains proper shapes of surface features.

Conglomerate Sedimentary rock composed of pebbles or larger particles in a matrix of finer material.

Conic projection A family of maps in which one or more cones is set tangent to, or intersecting, a portion of the globe and the geographic grid is projected onto the cone(s).

Continental drift Theory that proposes that the present continents were originally connected as one or two large landmasses that have broken up and literally drifted apart over the last several hundred million years.

Continentality Tendency of areas remote from the ocean, especially in high and mid-latitudes, to have large annual and daily temperature ranges.

Continental shelf Submerged margin of continents.

Contour line A line joining points of equal elevation.

Convection Vertical movements of parcels of air due to density differences.

Convergent precipitation Showery precipitation that occurs as a result of the forced uplift of air due to crowding in areas of air convergence.

Coral reef A coralline formation that fringes continents and islands in warm-water tropical oceans.

Core Spherical central portion of Earth; divided into a molten outer layer and a solid inner layer.

Coriolis effect The apparent deflection of free-moving objects to the right in the Northern Hemisphere and to the left in the Southern Hemisphere, in response to the rotation of Earth.

Corrosion Chemical reactions (especially hydrolysis and solution action) associated with streamflow that assist in erosion.

Crater The depression at the summit of a volcano, or a depression caused by the impact of a meteorite.

Creep The slowest and least perceptible form of mass wasting, which consists of a very gradual downhill movement of soil and regolith.

Crust The outermost solid layer of Earth.

Cuesta A low asymmetrical ridge formed by a resistant, gently dipping sedimentary layer.

Cumuliform cloud A cloud that is massive and rounded, usually with a flat base and limited horizontal extent, but often billowing upward to great heights.

Cumulonimbus Cumuliform cloud of great vertical development, often associated with a thunderstorm.

Cumulus Puffy white cloud that forms from rising columns of air.

Cumulus stage The early stage of thunderstorm formation in which updrafts prevail and the cloud continues to grow.

Cutoff meander A sweeping stream channel curve that is isolated from streamflow because the narrow meander neck has been cut through by stream erosion.

Cycle of erosion See Geomorphic cycle.

Cyclone Low-pressure center.

Cylindrical projection A family of maps derived from the concept of projection onto a paper cylinder that is tangential to, or intersecting with, a globe.

Debris flow Streamlike flow of muddy water heavily laden with sediments of various sizes.

Deciduous tree A tree that experiences an annual period in which all leaves die and usually fall from the tree, due either to a cold season or a dry season.

Deflation The shifting of loose particles by wind blowing them into the air or rolling them along the ground.

Delta A landform at the mouth of a river produced by the sudden dissipation of a stream's velocity and the resulting deposition of the stream's load.

Dendritic drainage pattern A treelike, branching pattern that consists of a random merging of streams, with tributaries joining larger streams irregularly but always at acute angles.

Denitrification The conversion of nitrates into free nitrogen in the air.

Denudation The total effect of all actions that lower the surface of the continents.

Desertification Degradation of the soil and biota due to a combination of drought and improper land use.

Desert varnish A dark shiny coating that forms on rock surfaces that are exposed to desert air for a long time.

Dew The condensation of beads of water on relatively cold surfaces.

Dew point The critical air temperature at which saturation is reached.

Diastrophism A general term that refers to the deformation of Earth's crust.

Differential erosion The process whereby different rocks or parts of the same rock erode at different rates.

Dike A vertical or nearly vertical sheet of magma that is thrust upward into preexisting rock.

Discharge Volume of flow of a stream.

Dissolved load The minerals, largely salts, that are dissolved in water and carried invisibly in solution.

Distributaries Branching stream channels that cross a delta.

Diurnal Daily.

Doldrums Belt of calm air associated with the region between the trade winds of the Northern and Southern hemispheres; generally in the vicinity of the equator.

Drainage basin An area that contributes overland flow and groundwater to a specific stream. (Also called a watershed or catchment.)

Drainage density The ratio of the total length of all stream segments in a drainage basin to the area of the basin.

Drainage divide The line of separation between runoff that descends into two different drainage basins.

Drainage net   The complex of streams within a drainage basin.

Drift   All material carried and deposited by glaciers.

Dripstone   Features formed by precipitated deposits of minerals that decorate the walls, floor, and roof of a cavern.

Drumlin   A low, elongated hill formed by ice sheet deposition. The long axis is aligned parallel with the direction of ice movements, and the end of the drumlin that faces the direction from which the ice came is somewhat blunt and slightly steeper than the narrower and more gently sloping end that faces in the opposite direction.

Dry adiabatic rate (adiabat)   The relatively steady rate at which a parcel of air cools as it rises (5 ½° F per 1000 feet) (10°C/km).

Dyne   The force needed to accelerate 1 gram of mass 1 centimeter per second.

Earthquake   Abrupt movement of Earth's crust.

Easterly wave   A long but weak migratory low-pressure trough in the tropics.

Ebb tide   A periodic falling of sea level.

Ecosystem   The totality of interactions among organisms and the environment in the area of consideration.

Ecotone   The transition zone between biotic communities in which the typical species of one community intermingle or inter-digitate with those of another.

Edaphic   Having to do with soil.

E horizon   A light-colored, eluvial layer that usually occurs between the A and B horizons.

Electromagnetic spectrum   Electromagnetic radiation, arranged according to wavelength.

Elliptical projection   A family of map projections in which the entire world is displayed in an oval shape.

Eluviation   The process by which gravitational water picks up fine particles of soil from the upper layers and carries them downward.

Embayment   An open rounded bay in a coastline.

Endemic   Found only in a particular area.

Energy   The capacity to do work.

Entisol   The least developed of all soil orders, with little mineral alteration and no pedogenic horizons. These soils are commonly thin and/or sandy and have limited productivity, although those developed on recent alluvial deposits tend to be quite fertile.

Entrenched meanders   A winding, sinuous stream valley with abrupt sides.

Ephemeral stream   A stream that carries water only during the "wet season" or during and immediately after rains.

Epiphytes   Plants that live above ground level out of contact with the soil, usually growing on trees or shrubs.

Equator   The parallel of 0 latitude.

Equatorial countercurrent   An east-moving ocean current, between the two equatorial currents.

Equilibrium line   A theoretical line separating the ablation zone and accumulation zone of a glacier along which accumulation exactly balances ablation.

Equilibrium theory   Idea that slope forms are adjusted to geomorphic processes so that there is a balance of energythe energy provided is just adequate for the work to be done.

Equinox   The time of the year when the perpendicular rays of the sun strike the equator, the circle of illumination just touches both poles, and the periods of daylight and darkness are each 12 hours long all over Earth.

Equivalence   The property of a map projection that maintains equal areal relationships in all parts of the map.

Erg   "Sea of sand." A large area covered with loose sand, generally arranged in some sort of dune formation by the wind.

Erosion   Detachment and removal of fragmented rock material.

Escarpment   An abrupt slope that breaks up the general continuity of the terrain.

Eskers   Long, sinuous ridges of stratified drift composed largely of glaciofluvial gravel and formed by the choking of subglacial streams during a time of glacial stagnation.

Estivation   The act of spending a dry/hot period in a torpid state.

Estuary   A finger of the sea projecting inland along drowned river valleys.

Evaporation   Process by which liquid water is converted to gaseous water vapor.

Evaporation fog   The condensation that results from the addition of water vapor to cold air that is already near saturation.

Evapotranspiration   The transfer of moisture to the atmosphere by transpiration from plants and evaporation from soil and plants.

Evergreen   A tree or shrub that sheds its leaves on a sporadic or successive basis, but at any given time appears to be fully leaved.

Exfoliation   Weathering process in which curved layers peel off bedrock in sheets. This process apparently occurs only in granite and related intrusive rocks.

Exfoliation dome   A large rock mass with a surface configuration that consists of imperfect curves punctuated by several partially fractured shells of the surface layers.

Exosphere   The highest zone of Earth's atmosphere.

Exotics   Organisms that are introduced into "new" habitats in which they did not naturally occur.

Exotic stream   A stream that flows into a dry region, bringing its water from somewhere else.

Extinction   The dying out of the entire population of a taxa.

Extirpation   Extermination.

Extratropical anticyclone   An extensive migratory high-pressure cell of the midlatitudes that moves generally with the westerlies.

Extratropical cyclone   Large migratory low-pressure system that occurs within the middle latitudes and moves generally with the westerlies.

Extrusive rock   Molten rock ejected onto Earth's surface, solidifying quickly in the open air.

Eye   The nonstormy center of a tropical cyclone, which has a diameter of 10 to 25 miles (16 to 40 km) and is a singular area of calmness in the maelstrom that whirls around it.

Eye wall   Peripheral zone at the edge of the eye where winds reach their highest speed.

Fault   A fracture or zone of fracture where the rock is forcefully broken with an accompanying displacement; i.e., an actual movement of the crust on one or both sides of the break. The movement can be horizontal or vertical, or a combination of both.

Fault-block mountain   A mountain formed under certain conditions of crustal stress, whereby a surface block may be severely faulted and upthrown on one side without any faulting or uplift on the other side. The block is tilted asymmetrically, producing a steep slope along the fault scarp and a relatively gentle slope on the other side of the block.

Fault line   The intersection of a fault zone with Earth's surface.

Fault plane   Interface along which movement occurs in faulting.

Fault scarp   Cliff formed by faulting.

Fault zone   Zone of weakness in the crust where faulting may take place.

Fauna   Animals.

Feral   A domesticated creature that has reverted to a wild existence, or the progeny thereof.

Ferromagnesian   Containing iron and/or magnesium as a major constituent.

Field capacity   The maximum amount of water that can be retained in the soil after the gravitational water has drained away.

Firn (névé)   Snow granules that have become packed and begin to coalesce due to compression, achieving a density about half as great as that of water.

First-order relief   The largest scale of relief features, consisting of continental platforms and ocean basins.

Fix (the verb)   To make firm or stable.

Fjord   A glacial trough that has been partly drowned by the sea.

Flood basalt   A large-scale outpouring of basaltic lava that may cover an extensive area of Earth's surface.

Floodplain   A flattish valley floor covered with stream-deposited sediments and subject to periodic or episodic inundation by overflow from the stream.

Flood tide   The movement of ocean water toward the coastfrom the ocean's lowest surface level the water rises gradually for about 6 hours and 13 minutes.

Flora   Plants.

Flow   A gentle, smooth motion as when a sector of a slope becomes unstable, normally due to an addition of water, and slips gently downhill.

Fluvent   A suborder of Entisol soils that form on recent water-deposited sediments that have satisfactory drainage.

Fluvial   Running water including both overland flow and streamflow.

Foehn   See Chinook. The word foehn is used particularly in Europe.

Fog   A cloud whose base is at or very near ground level.

Folding   The bending of crustal rocks by compression and/or uplift.

Foliation   Bending that gives a wavy layered appearance to metamorphic rock.

Food chain   Sequential predation in which organisms feed upon one another, with organisms at one level providing food for organisms at the next level, etc. Energy is thus transferred through the ecosystem.

Food pyramid   See also Food chain. Another conceptualization of energy transfer through the ecosystem from large numbers of "lower" forms of life through succeedingly smaller numbers of "higher" forms, as the organisms at one level are eaten by the organisms at the next higher level.

Forbs   Broadleaf herbaceous plants.

Foreshore   The lower shore zone of a beach, generally between the levels of high and low tide.

Forest   An assemblage of trees growing closely together so that their individual leaf canopies generally overlap.

Fractional scale   Ratio of distance between points on a map and the same points on Earth's surface; expressed as a ratio or fraction.

Freezing   Change from liquid to solid state.

Friction layer   Zone of the atmosphere, between Earth's surface and about 3300 feet (1000 meters) where most frictional resistance is found.

Fringing reef   A coral reef built out laterally from the shore, forming a broad bench that is only slightly below sea level, often with the tops of individual coral "heads" exposed to the open air at low tide.

Front   A zone of discontinuity between unlike air masses.

Frost wedging   Fragmentation of rock due to expansion of water that freezes in rock openings.

Fumarole   A hydrothermal feature consisting of a surface crack that is directly connected with a deep-seated source of heat. The little water that drains into this tube is instantly converted to steam by heat and gases, and a cloud of steam is then expelled from the opening.

Genesis   Origin or beginning.

Genus   The usual major subdivision of a family in the taxonomy of plants and animals, normally consisting of more than one species.

Geomorphic cycle (cycle of erosion)   A conceptual model of landscape development that was propounded by William Morris Davis. Davis envisioned a circular sequence of terrain evolution in which a relatively flat surface was uplifted from a lower to a higher elevation, where it was incised by fluvial erosion into a landscape of slopes and valleys and then thoroughly denuded until it became a flat surface at low elevation. He recognized three stages"youth," "maturity," and "old age."

Geomorphology   The study of the characteristics, origin, and development of landforms.

Geostrophic wind   A wind that moves parallel to the isobars as a result of the balance between the pressure gradient force and the Coriolis effect.

Geyser   A specialized form of intermittent hot spring with water issuing only sporadically as a temporary ejection, in which hot water and steam are spouted upward for some distance.

Geyserite   Whitish-colored siliceous sinter that often covers a geyser basin.

Gibber plain   A gravel-covered desert in Australia.

Glacial erratic   Outsize boulder included in the glacial till, which may be very different from the local bedrock.

Glacial flour   Rock material that has been ground to the texture of very fine talcum powder by glacial action.

Glacial trough   A valley reshaped by an alpine glacier.

Glacier   A large natural accumulation of land ice that flows either downslope or outward from its center of accumulation.

Glaciofluvial deposition   The action whereby much of the debris that is carried along by glaciers is eventually deposited or redeposited by glacial meltwater.

Gleization   The dominant pedogenic regime in areas where the soil is saturated most of the time due to poor drainage.

Gley soils   The general term for soils produced by gleization.

Gneiss   One of the most common metamorphic rocks; it is characterized by broad foliations.

Gondwanaland   Presumed Southern Hemisphere supercontinent resulting from the initial breakup of Pangaea.

Graben   A glock of land bounded by parallel faults in which the block has been downthrown, producing a distinctive structural valley with a straight, steep-sided fault scarp on either side.

Grade   A smooth longitudinal profile of a stream; a profile of equilibrium.

Graded stream   A stream that is at equilibrium with its load; the stream is just able to transport the load.

Gradient   Horizontal rate of variation of some quantity, such as atmospheric pressure or elevation of stream bed.

Granite   The most common and well-known intrusive rock.

Graphic scale   The use of a line marked off in graduated distances as a map scale.

Grassland   Plant association dominated by grasses and forbs.

Graticule   The network of parallels and meridians on a map.

Gravitational water   Soil water that is temporary in occurrence in that it results from prolonged infiltration from above (usually due to prolonged precipitation) and is pulled downward through the interstices toward the groundwater zone below by gravitational attraction.

Great Artesian Basin   The most notable of the world's artesian basins, which underlies some 670,000 square miles (1,750,000 km2) of east-central Australia.

Greenhouse effect   The trapping of heat in the lower troposphere because of differential transmissivity for short and long wavesthe atmosphere lets shortwave radiation in but doesn't let long-wave radiation out.

Groin   A short wall built perpendicularly from the beach into the shore zone to interrupt the longshore current and trap sand.

Ground fog   See radiation fog.

Ground ice   Ice located below the surface of the land.

Ground moraine   A moraine consisting of till deposited widely over a land surface beneath an ice sheet.

Groundwater   Water found in the vadose zone.

Groundwater flow   Water that moves through the groundwater system below the water table.

Gully erosion   Overland flow that erodes conspicuous channels in the soil.

Gymnosperms ("naked seeds")   Seed-reproducing plants that carry their seeds in cones.

Hadley cells   Two complete vertical circulation cells between the equator, where warm air rises, and 25° to 30° of latitude, where much of the air subsides.

Hail   Rounded or irregular pellets or lumps of ice produced in cumulonimbus clouds as a result of active turbulence and vertical air currents. Small ice particles grow by collecting moisture from supercooled cloud droplets.

Halide   A category of minerals that is notably salty.

Halley, Edmund An English astronomer and cartographer who, in his published map of 1700, was the first person to use isolines that appeared in print. His map showed isogonic lines in the Atlantic Ocean.

Hamada   A barren desert surface of consolidated material that usually consists of exposed bedrock but is sometimes composed of sedimentary material that has been cemented together by salts evaporated from groundwater.

Hanging valley   A tributary glacial trough, the bottom of which is considerably higher than the bottom of the principal trough that it joins.

Hardpan   A thin hard stratum within or beneath the soil surface.

Hardwoods   Angiosperm trees that are usually broad-leaved and deciduous. Their wood has a relatively complicated structure, but it is not always hard.

Headland   A promontory with a steep cliff face projecting into the sea.

Headward erosion   Erosion that eats into the interfluve at the upper end of a gully or valley.

Heat   A form of energy associated with the random motion of molecules. Things are made hotter by the collision of the moving molecules.

Heat index   A calculation of apparent temperature (how hot the air feels to one's skin) from temperature and relative humidity data.

Herbaceous plants   Plants that have soft stems, mostly grasses, forbs, and lichens.

Herbivore   An animal that feeds on vegetation.

Heterosphere   That portion of the atmosphere above the homosphere where there is an irregular layering of gases in accordance with their molecular weights.

Hibernation   The act of spending winter in a dormant state.

Histosol   A soil order characterized by organic, rather than mineral, soils, which is invariably saturated with water all or most of the time.

Homosphere   A zone of homogenous composition comprising the lowest 50 miles (80 km) of the atmosphere, where the principal gases have a uniform pattern of vertical distribution.

Hook   A curving sand bar at the outer end of a spit, produced by conflicting water movements in a bay that guide deposition in a curved fashion toward the coast.

Horizon   The more or less distinctly recognizable layer of soil, distinguished from one another by differing characteristics and forming a vertical zonation of the soil.

Horn   A steep-sided, pyramidal, rock pinnacle formed by expansive quarrying of the headwalls where three or more cirques intersect.

Horse latitudes   Areas in the Subtropical Highs characterized by warm, tropical sunshine and an absence of wind.

Horst   An uplifted block of land between two parallel faults.

Hot spot. See mantle plume.

Hot spring   Hot water at Earth's surface that has been forced upward through fissures or cracks by the pressures that develop when underground water has come in contact with heated rocks or magma beneath the surface.

Humidity   Water vapor in the air.

Humus   A dark-colored, gelatinous, chemically stable fraction of organic matter on or in the soil.

Hurricane   A tropical cyclone affecting North or Central America.

Hurricane warning   Notification issued by the Weather Service when hurricane conditions are expected in a specified area within 24 hours or less.

Hurricane watch   Notification issued by the Weather Service when hurricane conditions are expected in a specified area within 24 to 36 hours.

Hydration   The process whereby water molecules become attached to other substances and cause a wetting and swelling of the original substance without any change in its chemical composition.

Hydrography   The study of the waters of Earth.

Hydrologic cycle   A series of storage areas interconnected by various transfer processes, in which there is a ceaseless interchange of moisture in terms of its geographical location and its physical state.

Hydrology   Study of water on or below the surface of Earth.

Hydrolysis   A chemical union of water with another substance to produce a new compound that is nearly always softer and weaker than the original.

Hydrophyte   A "water-loving" plant that is adapted to living more-or-less permanently immersed in water.

Hydrosphere   Total water realm of Earth, including the oceans, surface waters of the lands, groundwater, and water held in the atmosphere.

Hydrothermal activity   The outpouring or ejection of hot water, often accompanied by steam, which usually takes the form of either a hot spring or a geyser.

Hygrometer   Any instrument for measuring humidity.

Hygrophyte   A water-tolerant plant that requires a saturated or semi- saturated environment.

Hygroscopic water   A microscopically thin film of moisture that is bound rigidly to soil particles by adhesion.

Iceberg   A great chunk of floating ice that breaks off an ice shelf or the end of an outlet glacier.

Ice cap   A small ice sheet, normally found in the summit area of high mountains.

Ice floe   A mass of ice that breaks off from larger ice bodies (ice sheets, glaciers, ice packs, and ice shelves) and floats independently in the sea. This term is generally used with large, flattish, tabular masses.

Ice pack   The extensive and cohesive mass of floating ice that is found in the Arctic and Antarctic oceans.

Ice sheet   A vast blanket of ice that completely inundates the underlying terrain to depths of hundreds or thousands of feet.

Ice shelf   A massive portion of an ice sheet that projects out over the sea.

Igneous rock   Rock formed by solidification of molten magma.

Illuviation   The process by which fine particles of soil from the upper layers are deposited at a lower level.

Inceptisol   An immature order of soils that has relatively faint characteristics; not yet prominent enough to produce diagnostic horizons.

Infiltration   Downward movement of water into the soil and regolith.

Infrared radiation   Electromagnetic radiation in the wavelength range of about 0.7 to 1000 micrometers.

Inertia   Inactivity; remaining at rest.

Inner core   The solid, dense, innermost portion of Earth, believed to consist largely of iron and nickel.

Inselberg ("island mountain")   Isolated summit rising abruptly from a low-relief surface.

Insolation   Incoming solar radiation.

Interflow   Water that infiltrates from the land surface above and then moves laterally in the soil or rock below.

Interfluve   The higher land above the valley sides that separates adjacent valleys.

Intermittent stream   A stream that carries water only part of the time, during the "wet season" or during and immediately after rains.

International date line   The line marking a time difference of an entire day from one side of the line to the other. Generally, this line falls on the 180th meridian except where it deviates to avoid separating an island group.

Interpolate   To insert something additional between other things, particularly by estimation.

Interstices   The pore spaces; a labyrinth of interconnecting passageways among the soil particles that makes up nearly half the volume of an average soil.

Intertropical convergence zone   The region where the northeast trades and the southeast trades converge.

Interval   The numerical difference between one isoline and the next.

Intrusive igneous rock   Rocks that cool and solidify beneath Earth's surface.

Invertebrates   Animals without backbones.

Ionosphere   A deep atmospheric layer containing electrically charged molecules and atoms in the upper mesosphere and lower thermosphere.

Isobar   A line joining points of equal atmospheric pressure.

Isogonic line   A line joining points of equal magnetic declination.

Isohyet   A line joining points of equal numerical value of precipitation.

Isoline   A line on a map connecting points that have the same quality or intensity of a given phenomenon.

Isostasy   Maintenance of the hydrostatic equilibrium of Earth's crust.

Isotherm   A line joining points of equal temperature.

Jet stream   A rapidly moving current of air concentrated along a quasi-horizontal axis in the upper troposphere or in the stratosphere, characterized by strong vertical and lateral wind shears and featuring one or more velocity maxima.

Jetty   A wall built into the ocean at the entrance of a river or harbor to protect against sediment deposition, storm waves, and currents.

Joints   Cracks that develop in bedrock due to stress, but in which there is no appreciable movement parallel to the walls of the joint.

Jungle   A type of forest that is overgrown with dense vegetation, usually found in tropical or subtropical regions.

Kame   A relatively steep-sided mound or conical hill composed of stratified drift found in areas of ice sheet deposition and associated with meltwater deposition in close association with stagnant ice.

Karst   Topography developed as a consequence of subsurface solution.

Katabatic wind   A wind that originates in cold upland areas and cascades toward lower elevations under the influence of gravity.

Kettle   An irregular depression in a morainal surface created when blocks of stagnant ice eventually melt.

Kilopascal   Equivalent to 10 millibars.

Knickpoint   A sharp irregularity (such as a waterfall, rapid, or cascade) in a stream-channel profile.

Knot   A unit of speed equal to 1 nautical mile, or 1.15 statute miles, per hour.

Köppen system   A climatic classification of the world devised by Wladimir Köppen.

Laccolith   An igneous intrusion produced when slow-moving viscous magma is forced between horizontal layers of preexisting rock. The magma resists flowing and builds up into a mushroom-shaped mass that domes the overlying strata. If near enough to Earth's surface, a rounded hill will rise above the surrounding area.

Lagomorphs   Rabbits and hares.

Lagoon   A body of quiet salt or brackish water in an area between a barrier island or a barrier reef and the mainland.

Lake   A body of water surrounded by land.

Land breeze   Local wind blowing from land to water, usually at night.

Landform   An individual topographic feature, of whatever size.

Landforms   Topography.

Landsat   A series of unmanned satellites that orbit Earth at an altitude of 570 miles (915 km) and are capable of imaging all parts of Earth, except the polar regions, every nine days.

Landslide   An abrupt and often catastrophic event in which a large mass of rock and earth slides bodily downslope in only a few seconds or minutes. An instantaneous collapse of a slope.

Langley (ly)   The unit of measure of radiation intensity that is 1 calorie per square centimeter.

Lapse rate   The rate of temperature decrease with height in the troposphere. The average lapse rate has been calculated to be about 3.6°F per 1000 feet (6.5°C/km).

Large-scale map   A map with a scale that is a relatively large representative fraction and therefore portrays only a small portion of Earth's surface, but in considerable detail.

Latent heat   Energy stored or released when a substance changes state.

Latent heat of vaporization   Stored energy absorbed by escaping molecules during evaporation.

Lateral erosion   Erosion that occurs when the principal current of a stream swings laterally from one bank to the other, eroding where the velocity is greatest and depositing where it is least.

Lateral moraine   Well-defined ridge of unsorted debris built up along the sides of valley glaciers, parallel to the valley walls.

Laterization   The dominant pedogenic regime in areas where temperatures are relatively high throughout the year and which is characterized by rapid weathering of parent material, dissolution of nearly all minerals, and the speedy decomposition of organic matter.

Latitude   Distance measured north and south of the equator.

Latosol   The general term applied to soils produced by laterization.

Laurasia   Presumed Northern Hemisphere supercontinent resulting from the initial breakup of Pangaea.

Lava   Molten magma that is extruded onto the surface of Earth, where it cools and solidifies.

Lava vesicles   Holes of various sizes, usually small, that develop in cooling lava when gas is unable to escape during solidification.

Leaching   The process in which gravitational water dissolves soluble materials and carries them downward in solution to be redeposited at lower levels.

Liana   A tropical rainforest vine usually rooted in the ground, with leaves and flowers mostly in the tree canopy above.

Lifting condensation level   The altitude at which rising air cools sufficiently to reach 100 percent relative humidity at the dew point temperature, and condensation begins.

Lightning   A luminous electric discharge in the atmosphere caused by the separation of positive and negative charges associated with cumulonimbus clouds.

Limestone   A common, chemically accumulated sedimentary rock composed largely of calcium carbonate.

Linnaean system   A biological classification that focuses on the morphology of the organisms and groups them on the basis of structural similarity.

Lithosphere   The solid, inorganic portion of the earthly fundament comprised of the rocks of Earth's crust as well as of the broken and unconsolidated particles of mineral matter that overlie the unfragmented bedrock. Sometimes used as a general term for the entire solid Earth.

Litter   The collection of dead plant parts that accumulate at the surface of the soil.

Loam   A soil texture in which none of the three principal soil separates—sand, silt, and clay—dominates the other two.

Loess   A fine-grained, wind-deposited silt. Loess lacks horizontal stratification, and its most distinctive characteristic is its ability to stand in vertical cliffs.

Longitude   Distance measured in degrees, minutes, and seconds, east and west from the prime meridian on Earth's surface.

Longshore current (littoral current)   A current in which water moves roughly parallel to the shoreline in a generally downwind direction.

Longshore drift   The zigzag movement of particles in which the net result is a displacement parallel to the coast in a downwind direction.

Loxodrome (rhumb line)   A true compass heading; a line of constant compass direction.

Magma   Molten material in Earth's interior.

Magnetic declination   The angular difference between a magnetic north line and a true north line, expressed as degrees east or west of the meridian in question.

Magnetic north   The direction in which a magnetic compass needle points.

Mammals   The highest form of animal life, distinguished from all other animals by several internal characteristics and two prominent external features—the production of milk to feed their young and the possession of true hair.

Mantle   That portion of Earth beneath the crust and surrounding the core.

Mantle plume   A location where molten mantle magma rises to, or almost to, Earth's surface.

Map projection   A systematic representation of all or part of the three-dimensional Earth surface on a two-dimensional flat surface.

Marble   Metamorphosed limestone.

Marine terrace   A platform of marine erosion that has been uplifted above sea level.

Maritime   Of or pertaining to the sea.

Marsh   Flattish surface area that is submerged in water at least part of the time, but is shallow enough to permit the growth of water-tolerant plants—primarily grasses and sedges.

Marsupials   A small group of mammals whose females have pouches in which the young, which are born in a very undeveloped condition, live for several weeks or months after birth.

Mass wasting   The downslope movement of broken rock material by gravity, sometimes lubricated by the presence of water.

Maturity (mature stage)   The second stage in Davis's Geomorphic Cycle, which is dominated by slope land and marked by the absence of the vast area of initial surface.

Meandering stream   A stream with a sinuous course of elaborate, smooth curves.

Meander scar   A former stream meander through which the stream no longer flows.

Mechanical weathering   The physical disintegration of rock material without any change in its chemical composition.

Medial moraine   A dark band of rocky debris down the middle of a glacier created by the union of the lateral moraines of two adjacent glaciers.

Mercalli intensity scale   Scale to evaluate the intensity of earthquake action on the basis of damage caused.

Mercator projection   A cylindrical projection mathematically adjusted to attain complete conformality, and which has a rapidly increasing scale with increasing latitude.

Meridian   An imaginary line of longitude extending from pole to pole, crossing all parallels at right angles, and being aligned in true north-south directions.

Mesa   A flat-topped, steep-sided hill with a limited summit area.

Mesopause   Transition zone at the top of the mesosphere.

Mesosphere   Atmospheric layer above the stratosphere, where temperature decreases with height; also refers to the rigid part of the deep mantle, below the asthenosphere.

Mesozoic era   The second youngest geologic age; the name refers to "middle life."

Metamorphic rock   Rock that was originally something else but has been drastically changed by massive forces of heat and/or pressure working on it from within Earth.

Midocean ridge   A lengthy system of deep-sea mountain ranges, generally located at some distance from any continent.

Millibar   An "absolute" measure of pressure, consisting of one-thousandth part of a bar, or 1000 dynes per square centimeter.

Mineral   A naturally formed inorganic substance that has an unvarying chemical composition.

Mistral   A cold, high-velocity wind that sometimes surges down France's Rhone Valley, from the Alps to the Mediterranean Sea.

Mogote   A steep-sided hill of residual limestone bedrock formed largely by solution action.

Mohorovičić discontinuity   The boundary between Earth's crust and mantle. Also known simply as the Moho.

Mollisol   A soil order characterized by the presence of a mollic epipedon, which is a mineral surface horizon that is dark, thick, contains abundant humus and base nutrients, and retains a soft character when it dries out.

Monadnock Erosional remnant of resistant rock rising slightly above a landscape of limited relief.

Monera The kingdom of organisms that comprises the simplest known organismsone-celled bacteria and blue-green algae.

Monocline A one-limbed fold connecting horizontal or gently inclined strata.

Monotremes Egg-laying mammals.

Monsoon A seasonal reversal of winds; a general onshore movement in summer and a general offshore flow in winter, with a very distinctive seasonal precipitation regime.

Moraine The largest and generally most conspicuous landform feature produced by glacial deposition, which consists of irregular rolling topography that rises somewhat above the level of the surrounding terrain.

Morphology Form or shape or structure.

Motus Coral islets that are closely spaced and separated by narrow channels of water and that together form a ring-shaped atoll.

Mudflow Downslope movement of a thick mixture of soil and water.

Multispectral scanning system (MSS) A remote sensing instrument that collects multiple digital images simultaneously in different bands.

Muskeg A poorly drained bog in northern North America or Eurasia, commonly vegetated with mosses, sedges, and stunted trees.

Natural levee An embankment of slightly higher ground fringing a stream channel in a floodplain; formed by deposition during floodtime.

Natural selection The Darwinian theory of "the survival of the fittest," which explains the origin of any species as a normal process of descent, with variation, from parent forms.

Nautical mile A unit of distance in sea and air navigation equal to 6076 feet or 1.852 kilometers.

Neap tides The lower-than-normal tidal variations that occur twice a month as the result of the alignment of the sun and moon at a right angle to one another.

Nebula A cloud of interstellar gas and dust, perhaps one light-year in diameter.

Needle-leaf trees Trees adorned with thin slivers of tough, leathery, waxy needles rather than typical leaves.

Nekton The term applied to animals that swim freely in the oceans.

Névé See Firn.

Newton The force that must be exerted on a mass of 1 kilogram in order to accelerate it at a rate of 1 meter per second per second.

Nimbostratus A low, dark cloud, often occurring as widespread overcast and normally producing precipitation.

Nitrogen cycle An endless series of processes in which nitrogen moves through the environment.

Nitrogen fixation Conversion of gaseous nitrogen into forms that can be used by plant life.

Nitrogenous Containing nitrogen.

Nocturnal Active at night.

Nonferromagnesian Lacking iron and magnesium in its chemical composition.

Normal fault The result of tension producing a steeply inclined fault plain, with the block of land on one side being pushed up, or upthrown, in relation to the block on the other side, which is downthrown.

Nunatak A rocky pinnacle protruding above an ice field.

Obsequent streams Streams that commonly flow in the opposite direction of consequents and are often short tributaries of subsequents.

Occluded front A complex front formed when a cold front overtakes a warm front.

Oceanic trench Deep linear depression in the ocean floor where subduction is taking place.

Offshore That portion of the shore zone seaward from the low tide line and extending to the area where wave erosion and deposition do not occur.

Offshore bar (barrier bar) Long, narrow sand bar built up in shallow offshore waters.

Offshore flow Wind movement from land to water.

O horizon The immediate surface layer of a soil profile, consisting mostly of organic material.

Old age Third stage in Davis's Geomorphic Cycle in which the entire landscape has been eroded down to a plain of low relief.

Onshore flow Wind movement from water to land.

Orographic precipitation Precipitation that occurs when air, forced to ascend over topographic barriers, cools to the dew point.

Orthent A suborder of Entisol soils that develops on recent erosional surfaces.

Orthid A suborder of Aridisol soils that does not have a distinctive subsurface horizon.

Orthophotomap A map produced through computerized rectification of aerial imagery.

Oscillation A swinging movement to and fro.

Outcrop Surface exposure of bedrock.

Outer core The liquid shell beneath the mantle that encloses Earth's inner core.

Outlet glacier A tongue of ice around the margin of an ice sheet that extends between rimming hills to the sea.

Outwash plain Extensive glaciofluvial feature that is a relatively smooth, flattish alluvial apron deposited beyond recessional or terminal moraines by streams issuing from ice.

Overland flow The general movement of surface water down the slope of the land surface.

Overthrust fault A fault created by compression forcing the upthrown block to override the downthrown block at a relatively low angle.

Overthrust fold A fold in which the pressure was great enough to break the oversteepened limb and cause a shearing movement.

Overturned fold An upfold that has been pushed so vigorously from one side that it becomes oversteepened enough to have a reverse orientation on the other side.

Oxbow lake A cutoff meander that initially holds water.

Oxbow swamp An oxbow lake that has been at least partly filled with sediment and vegetation.

Oxidation The chemical union of oxygen atoms with atoms from various metallic elements to form new products, which are usually more voluminous, softer, and more easily eroded than the original compounds.

Oxide A category of minerals composed of oxygen combined with another element.

Oxisol The most thoroughly weathered and leached of all soils. This soil order invariably displays a high degree of mineral alteration and profile development.

Oxygen cycle The movement of oxygen by various processes through the environment.

Ozone A gas composed of molecules consisting of three atoms of oxygen, O3.

Ozone layer A layer of ozone between 10 and 25 miles (16 and 40 km) high, which absorbs ultraviolet solar radiation.

Ozonosphere Zone of relatively rich concentration of ozone in the atmosphere.

Paleomagnetism Past magnetic orientation.

Pangaea The massive supercontinent that Alfred Wegener postulated to have existed about 200 million years ago. He visualized Pangaea as breaking up into several large sections that have continually moved away from one another and that now comprise the present continents.

Parallel A circle resulting from an isoline connecting all points of equal latitude.

Parallel drainage pattern A drainage pattern that can emerge in areas of pronounced regional slope, particularly if the gradient is gentle, with long consequent streams flowing parallel to one another.

Parasites Organisms of one species that infest the body of a creature of another species, obtaining their nutriment from the host, which is almost invariably weakened and sometimes killed by the actions of the parasite.

Parent material The source of the weathered fragments of rock from which soil is madesolid bedrock or loose sediments that have been transported from elsewhere by the action of water, wind, or ice.

Particulate Composed of distinct particles or small pieces.

Pascal (Pa) A pressure of 1 newton per square meter.

Paternoster lakes A sequence of small lakes found in the shallow excavated depressions of a glacial trough.

Pediment A gently inclined bedrock platform that extends outward from a mountain front, usually in an arid region.

Pedogenic regimes Soil-forming regimes that can be thought of as environmental settings in which certain physical/chemical/biological processes prevail.

Ped A larger mass or clump that individual soil particles tend to aggregate into and that determines the structure of the soil.

Peneplain A flat and relatively featureless landscape with minimal relief; considered to be the end product of the geomorphic cycle. Peneplain means "almost a plain."

Percolate To filter through a porous substance.

Perennials Plants that can live more than a single year despite seasonal climatic variations.

Perennial stream A permanent stream that contains water the year-round.

Perigee The point at which the moon is closest to Earth in its elliptical orbit—231,200 miles (370,000 km).

Periglacial zone An area of indefinite size beyond the outermost extent of ice advance that was indirectly influenced by glaciation.

Perihelion The point in its orbit at which a planet is nearest the sun.

Permafrost Permanent ground ice or permanently frozen subsoil.

Permeability A soil or rock characteristic in which there are interconnected pore spaces through which water can move.

Phases of the moon The recurring appearances of the moon or a planet with regard to the form of its illuminated disk.

Photogrammetry The science of obtaining reliable measurements from photographs and, by extension, mapping from aerial photos.

Photoperiodism The response of an organism to the length of exposure to light in a 24-hour period.

Photosynthesis The basic process whereby plants produce stored chemical energy from water and carbon dioxide and which is activated by sunlight.

Phreatic zone (zone of saturation) The second hydrologic zone below the surface of the ground, whose uppermost boundary is the water table. The pore spaces and cracks in the bedrock and regolith of this zone are fully saturated.

Piedmont Zone at the "foot of the mountains."

Piedmont angle The pronounced change in the angle of slope at a mountain base, with a steep slope giving way abruptly to a gentle one.

Piedmont glacier A valley glacier that extends to the mouth of the valley and spreads out broadly over the flat land beyond.

Piezometric surface The elevation to which water will rise under natural confining pressure in a well.

Pillar An erosional remnant in the form of a steep-sided spire that has a resistant caprock; normally found in an arid or semiarid environment.

Pinnacle See pillar.

Pixel An individual picture element of a remote sensing image.

Placental mammals Mammals whose young grow and develop in the mother's body, nourished by an organ known as the placenta, which forms a vital connecting link with the mother's bloodstream.

Plane of the ecliptic The imaginary plane that passes through the sun and through Earth at every position in its orbit around the sun.

Plankton Plants and animals that float about, drifting with the currents and tidesmostly microscopic in size.

Plantae The kingdom of organisms that includes the green plants and higher algae.

Plant succession The process whereby one type of vegetation is replaced naturally by another.

Plateau Flattish erosional platform bounded on at least one side by a prominent escarpment.

Plate suturing The crushing together of two continental plates when they collide.

Plate tectonics A coherent theory of massive crustal rearrangement based on the movement of continent-sized crustal plates.

Playa Dry lake bed in a basin of interior drainage.

Playa lake Shallow and short-lived lake formed when water flows into a playa.

Pleistocene Epoch An epoch of the Cenozoic Era between the Pliocene and the Holocene.

Plucking (quarrying) Action in which rock particles beneath the ice are grasped by the freezing of meltwater in joints and fractures and pried out and dragged along in the general flow of a glacier.

Pluton A large intrusive igneous body.

Pluvial Pertaining to rain; often used in connection with a past rainy period.

Podzol General term for a soil formed by podzolization.

Podzolization The dominant pedogenic regime in areas where winters are long and cold, and which is characterized by slow chemical weathering of soils and rapid mechanical weathering from frost action, resulting in soils that are shallow, acidic, and with a fairly distinctive profile.

Polar easterlies A global wind system that occupies most of the area between the Polar Highs and about 60 of latitude. The winds move generally from east to west and are typically cold and dry.

Polar front The contact between unlike air masses in the subpolar low-pressure zone.

Polar high A high-pressure cell situated over either polar region.

Polarity A characteristic of Earth's axis wherein it always points toward Polaris (the North Star) at every position in Earth's orbit around the sun.

Pond A lake of very small size.

Porosity The amount of pore space between the soil particles and between the peds, which is a measure of the capacity of the soil to hold water and air.

Potential evapotranspiration The maximum amount of moisture that could be lost from soil and vegetation if the water were available.

Prairie A tall grassland in the midlatitudes.

Pressure release See Unloading.

Primary consumer Animals that eat plants, as the first stage in a food pyramid or chain.

Primates Humans and such humanlike mammals as apes and monkeys.

Prime meridian The meridian passing through the Royal Observatory at Greenwich (England), just east of London, and from which longitude is measured.

Progeny Offspring.

Proglacial lake A lake formed when ice flows across or against the general slope of the land and the natural drainage is impeded or completely blocked so that meltwater from the ice becomes impounded against the ice front.

Protista The kingdom of organisms that consists of one-celled organisms outside the Monera Kingdom, and some simple multicelled algae.

Psamment A suborder of Entisol soils that occurs in sandy situations.

Pteridophytes Spore-bearing plants such as ferns, horsetails, and clubmosses.

Pyroclastic material Rock fragments thrown into the air by volcanic explosions.

Quartz A mineral composed of silicon dioxide.

Quartzite A metamorphosed rock derived from sandstone or another rock composed largely of quartz.

Quick clays Clay formations that spontaneously change from a relatively solid mass into a near-liquid condition as the result of a sudden disturbance or shock.

Radial drainage pattern Drainage pattern in which consequent streams flow outward in all directions from a central dome or peak.

Radiation The process by which energy is emitted from a body.

Radiational inversion Surface inversion that results from rapid radiational cooling of lower air, typically on cold winter nights.

Radiation fog A fog produced by condensation near the ground, where air is cooled to the dew point by contact with the colder ground.

Rain The most common and widespread form of precipitation, consisting of drops of liquid water.

Rain gauge An instrument used to measure the amount of rain that has fallen.

Rain shadow Area of low rainfall on the leeward side of a topographic barrier.

Recessional moraine A glacial deposit formed during a pause in the retreat of the ice margin.

Recharge rate The speed at which a substance is replenished.

Rectangular drainage pattern Pattern where streams are essentially all subsequents that follow sets of faults and/or joints, with prominent right-angled relationships.

Rectification Adjustment.

Reef A mass of rock with its surface at or just below the low-tide line.

Reflection The ability of an object to repel waves without altering either the object or the waves.

Refraction Change of direction

Reg   A desert surface of coarse material from which all sand and dust have been removed by wind and water erosion. Often referred to as desert pavement or desert armor.

Regime   Seasonal pattern or sequential process.

Regolith   A layer of broken and partly decomposed rock particles that covers bedrock.

Rejuvenation   Concept that regional uplift could raise the land and interrupt the geomorphic cycle at any stage by reenergizing the degradational processes.

Relative humidity   An expression of the amount of water vapor in the air in comparison with the total amount that could be there if the air were saturated. This is a ratio that is expressed as a percentage.

Relief   The difference in elevation between the highest and lowest points in an area; the vertical variation from mountain top to valley bottom.

Remote sensing   Study of an object or surface from a distance by using various instruments.

Representative fraction (r.f.)   The ratio that is an expression of a fractional scale that compares map distance with ground distance.

Reptiles   Cold-blooded vertebrates of which most are land-based.

Resequent streams   Streams that flow in the same direction as consequents but develop on slopes that were formed later. They are found typically on the upslope sides of subsequent valleys.

Residual landform   A topographic feature caused by the erosion of bedrock.

Return beam vidicon (RBV)   A remote sensing system consisting of three cameras that image the same section of Earth's surface simultaneously.

Reverse fault   A fault produced from compression, with the upthrown block rising steeply above the downthrown block, so that the fault scarp would be severely oversteepened if erosion did not act to smooth the slope.

Revetment   See seawall.

R horizon   The consolidated bedrock at the base of a soil profile.

Rhumb line   See Loxodrome.

Ria shoreline   An embayed coast with numerous estuaries.

Richter scale   A scale of earthquake magnitudes, devised by California seismologist Charles F. Richter, in 1935, to describe the amount of energy released in a single earthquake.

Rift valley   A downfaulted graben structure extended for extraordinary distances as linear structural valleys enclosed between typically steep fault scarps.

Rill erosion   A more concentrated flow than that of sheet erosion, which loosens additional material and scores the slope with numerous parallel seams.

Rills   Tiny drainage channels.

Riparian vegetation   Anomalous stream-side growth, particularly prominent in relatively dry regions, where stream courses may be lined with trees, although no other trees are to be found in the landscape.

Roche moutonnée   A characteristic landform produced when a bedrock hill or knob is overridden by moving ice. The stoss side is smoothly rounded and streamlined by grinding abrasion as the ice rides up the slope, but the lee side is shaped largely by plucking, which produces a steeper and more irregular slope.

Rock   Solid material composed of aggregated mineral particles.

Rock glacier   An accumulated talus mass that moves slowly but distinctly downslope under its own weight.

Rodents   Gnawing mammals.

Rossby wave   A very large north-south undulation of the upper-air westerlies.

Runoff   Flow of water from land to oceans by overland flow, streamflow, and groundwater flow.

Rusting   Production of reddish/yellowish/brownish iron oxide minerals by oxidation.

Sag pond   A pond caused by the collection of water from springs and/or runoff into sunken ground, resulting from the jostling of Earth in the area of fault movement.

Salina   Dry lake bed that contains an unusually heavy concentration of salt in the lake-bed sediment.

Salinity   A measure of the concentration of dissolved salts.

Salinization   One of the dominant pedogenic regimes in areas where principal soil moisture movement is upward because of a moisture deficit.

Saltation   Process in which small particles are moved along by streamflow or wind in a series of jumps or bounces.

Salt wedging   Rock disintegration caused by the crystallization of salts from evaporating water.

Sand dune   A mound, ridge, or low hill of loose wind-blown sand.

Sandplain   An amorphous sheet of coarse sand spread across the landscape with no particular surface shape or significant relief.

Sandstone   A common mechanically accumulated sedimentary rock composed largely of sand grains.

Sand storm   A cloud of generally horizontally moving sand that extends for only a few inches or feet above the surface.

Sapping   An erosional process in which soil particles are removed from a slope by the seepage or trickling out of underground water.

Savanna   A low-latitude grassland characterized by tall forms.

Scattering   A change in direction, but not in wavelength, of light waves.

Schist   One of the most common metamorphic rocks in which the foliations are very narrow.

Scree   See Talus.

Sea breeze   A wind that blows from the sea toward the land, usually during the day.

Sea-floor spreading   The pulling apart of crustal plates to permit the rise of deep-seated magma to Earth's surface in midocean areas.

Seawall   A strong wall or embankment built along the shore to act as a breakwater.

Secondary consumer   Animals that eat other animals, as the second and further stages in a food pyramid or chain.

Second-order relief   Major mountain systems and other extensive surface formations of subcontinental extent.

Sedges   Rushlike plants with solid stems growing in wet places.

Sediment   Small particles of rock debris or organic material deposited by water, wind, or ice.

Sedimentary rock   Rock formed of sediment that is consolidated by the combination of pressure and cementation.

Seep (the noun)   A small spring or pool where liquid from the ground has oozed to the surface.

Seep (the verb)   To ooze gradually.

Seif (longitudinal dune)   Long, narrow desert dunes that usually occur in multiplicity and in parallel arrangement.

Selva (tropical rainforest)   A distinctive assemblage of tropical vegetation that is dominated by a great variety of tall, high-crowned trees.

Semiarid   Characterized by a small amount of annual precipitation (generally between 10 and 20 inches [25 and 50 cm]).

Sensible temperature   A concept of the relative temperature that is sensed by a person's body.

Separates   The size groups within the standard classification of soil particle sizes.

Seral   A stage that is not the final stage in an ecological succession.

Seventh   Approximation See Soil Taxonomy.

Shale   A common mechanically accumulated sedimentary rock composed of silt and clay.

Sheet erosion   The transportation by water flowing across the surface as a thin sheet, of material already loosened by splash erosion.

Shield volcanoes   Volcanoes built up in a lengthy outpouring of very fluid basaltic lava.

Shrub   Woody, low-growing perennial plant.

Shrubland   Plant association dominated by relatively short woody plants.

Sial   The upper layer of two igneous rock layers, which is discontinuous, apparently underlying only the continental masses, where it sits as immense bodies of rock embedded in the sima beneath. "Sial" is named for its common constituents of silica and aluminum.

Sidereal day   A complete rotation of Earth with respect to the stars (23 hours, 56 minutes, and 4.099 seconds).

Silica   Silicon dioxide in any of several mineral forms.

Silicate   A category of minerals composed of silicon and oxygen combined with another element or elements.

Sill   A long, thin intrusive body that is formed when magma is forced between parallel layers of preexisting rock to solidify eventually in a sheet.

Sima   The lower of two igneous rock layers that is continuous and underlies both the ocean basins and the continents. "Sima" is named for its two most prominent mineral compounds, silica and magnesium.

Sinkhole (doline) A small, rounded depression that is formed by the dissolution of surface limestone, typically at joint intersections.

Slate Metamorphosed shale.

Sleet Small raindrops that freeze during descent.

Slip face Steeper leeward side of a sand dune.

Slump A slope collapse with a backward rotation.

Small-scale map A map whose scale is a relatively small representative fraction and therefore shows a large portion of Earth's surface in limited detail.

Snow Solid precipitation in the form of ice crystals, small pellets, or flakes, which is formed by the direct conversion of water vapor into ice.

Snowline The elevation above which some winter snow is able to persist throughout the year.

Sod A ground surface covered with grass.

Softwoods Gymnosperm treesnearly all are needle-leaved evergreenswith wood of simple cellular structure but not always soft.

Soil An infinitely varying mixture of weathered mineral particles, decaying organic matter, living organisms, gases, and liquid solutions. Soil is that part of the outer skin of Earth occupied by plant roots.

Soil profile A vertical cross section from Earth's surface down through the soil layers into the parent material beneath.

Soil Taxonomy The system of soil classification currently in use in the United States. It is genetic in nature and focuses on the existing properties of the soil rather than on environment, genesis, or the properties it would possess under virgin conditions.

Soil water Water found in the phreatic zone.

Soil-water balance The relationship between gain, loss, and storage of soil water.

Soil-water budget An accounting that demonstrates the variation of the soil-water balance over a period of time.

Solar constant The fairly constant amount of solar insolation received at the top of the atmosphereslightly less than 2 calories per square centimeter per minute or 2 langleys per minute.

Solar day A complete rotation of Earth with respect to the sun (24 hours).

Solifluction A special form of creep in tundra areas that produces a distinctive surface appearance. During the summer the near-surface portion of the ground thaws, but the meltwater cannot percolate deeper because of the permafrost below. The spaces between the soil particles become saturated, and the heavy surface material sags slowly downslope.

Solstice One of those two times of the year in which the sun's perpendicular rays hit the northernmost or southernmost latitudes (23°) reached during Earth's cycle of revolution.

Solum The true soil that includes only the top four horizons: O, the organic surface layer; A, the topsoil; E, the eluvial layer; and B, the subsoil.

Source region A part of Earth's surface that is particularly suited to generate air masses.

Species The major subdivisioin of a genus, regarded as the basic category of biological taxonomic classification.

Specific heat The amount of energy required to raise the temperature of a unit mass of a substance by 1C.

Specific humidity A direct measure of water vapor content expressed as the mass of water vapor in a given mass of air (grams of vapor/kilograms of air).

Speleothem A feature formed by precipitated deposits of minerals on the wall, floor, or roof of a cave.

Spit A linear deposit of marine sediment that is attached to the land at one or both ends.

Splash erosion The direct collision of a raindrop with the ground, which blasts fine particles upward and outward, shifting them a few millimeters laterally.

Spodosol A soil order characterized by the occurrence of a spodic subsurface horizon, which is an illuvial layer where organic matter and aluminum accumulate, and which has a dark, sometimes reddish, color.

Spring A stream of surface water that emerges from the ground.

Spring tide A time of maximum tide that occurs as a result of the alignment of sun, moon, and Earth.

Squall line A line of intense thunderstorms.

Stalactite A pendant structure hanging downward from a cavern's roof.

Stalagmite A projecting structure growing upward from a cavern's floor.

Stationary front The common "boundary" between two air masses in a situation in which neither air mass displaces the other.

Statute mile A unit of distance on land equal to 5280 feet or 1760 yards (1693 meters).

Steppe A plant association dominated by short grasses and bunchgrasses of the midlatitudes.

Stock A small body of igneous rock intruded into older rock, amorphous in shape and indefinite in depth.

Storm surge A surge of wind-driven water as much as 25 feet (7.5 meters) above normal tide level, which occurs when a hurricane pounds into a shoreline.

Strata Distinct layers of sediment.

Stratified drift Drift that was sorted as it was carried along by the flowing glacial meltwater.

Stratiform cloud A cloud form characterized by clouds that appear as grayish sheets or layers that cover most or all of the sky, rarely being broken into individual cloud units.

Stratocumulus Low clouds, usually below 6500 feet (2 km), which sometimes occur as individual clouds, but more often appear as a general overcast.

Stratopause The top of the stratosphere elevation about 30 miles (48 km) where maximum temperature is reached.

Stratosphere Atmospheric layer directly above the troposphere.

Stratus Low clouds, usually below 6500 feet (2 km), which sometimes occur as individual clouds, but more often appear as a general overcast.

Stream capture (stream piracy) An event where a portion of the flow of one stream is diverted into that of another by natural pro-cesses.

Streamflow Channeled movement of water along a valley bottom.

Stream load Solid matter carried by a stream.

Stream order Concept that describes the hierarchy of a drainage net.

Stream terrace Remnant of a previous valley floodplain of a rejuvenated stream.

Striations Marks in a bedrock surface produced by the direct impact of glacial ice and its load of abrasive debris, which parallel the direction of ice movement.

Stripped plain A flattish erosional platform where the scarp edge is absent or relatively inconspicuous.

Structure Nature, arrangement, and orientation of the materials.

Subaerial Occurring or forming on Earth's surface.

Subartesian well The free flow that results when a well is drilled from the surface down into a confined aquifer and which requires artificial pumping to raise the water to the surface because the confining pressure forces the water only partway up the well shaft.

Subduction Descent of the edge of a crustal plate under the edge of an adjoining plate, presumably involving melting of the subducted material.

Sublimation The process by which water vapor is converted directly to ice, or vice versa.

Subpolar low A zone of low pressure that is situated at about 50° to 60° of latitude in both Northern and Southern hemispheres (also referred to as the polar front).

Subside To sink to a lower level.

Subsidence inversion A temperature inversion that occurs well above Earth's surface as a result of air sinking from above.

Subtropical Bordering on the tropics; between tropical and temperate regions.

Subtropical high Large semipermanent high-pressure cells centered at about 30° latitude over the oceans, which have average diameters of 2000 miles (3200 km) and are usually elongated east-west.

Succulents Plants that have fleshy stems that store water.

Sulfate A category of minerals composed of sulfur and oxygen combined with another element or elements.

Sulfide A category of minerals composed of sulfur combined with another element or elements.

Summer solstice The dates that represent the most poleward extent of the perpendicular rays of the sun. (Northern Hemisphere—23.5° N on about June 21; Southern Hemisphere—23.5° S on about December 21).

Supercooled water Water that persists in liquid form at temperatures below freezing.

Surf The swell of the sea which breaks upon a reef or a shore.

Surrogate Substitute

Suspended load   The very fine particles of clay and silt that are in suspension and move along with the flow of water without ever touching the stream bed.

Swallow hole   The distinct opening at the bottom of some sinks through which surface drainage can pour directly into an underground channel.

Swamp   A flattish surface area that is submerged in water at least part of the time, but is shallow enough to permit the growth of water-tolerant plants predominantly trees.

Swash   The cascading forward motion of a breaking wave that rushes up the beach.

Swell   A water wave, usually produced by stormy conditions, that can travel enormous distances away from the source of the disturbance.

Symbiosis   A mutually beneficial relationship between two organisms.

Syncline   A simple downfold.

Taiga (boreal forest)   The great northern coniferous forest.

Talus (scree)   Pieces of rock, of whatever size, that fall directly downslope.

Talus cone   Sloping, cone-shaped heaps of dislodged talus.

Talus slope (talus apron)   The fragments of rocks (talus) that accumulate relatively uniformly along the base of the slope.

Tarn   Small lake in the shallow excavated depression of rock benches of a glacial trough or cirque.

Taxon/taxa   Any taxonomic category, as species, genus, etc.

Taxonomy   The science of classification.

Tectonic activity   Crustal movements of various kinds.

Temperature   A measure of the degree of hotness or coldness of a substance.

Temperature inversion   A situation in which temperature increases upward, and the normal condition is inverted.

Terminal moraine   A glacial deposit that builds up at the outermost extent of ice advance.

Terracettes   A complicated terracing effect, resembling a network of faint trails, which is produced by a creep, usually on steep grassy slopes.

Terrane   A mass of continental crust that has become accreted to a tectonic plate margin that has different lithologic characteristics from those of the terrane.

Terrestrial   Growing or living on the ground.

Thalweg   A line connecting the deepest points of a stream channel.

Thematic mapper   A sophisticated multispectral scanning system that senses on seven narrowly defined spectral bands.

Thermometer   An instrument designed for the measurement of temperature.

Thermosphere   The highest recognized thermal layer in the atmosphere, above the mesopause, where temperature remains relatively uniform for several miles and then increases continually with height.

Third-order relief   Specific landform complexes of lesser extent and generally of smaller size than those of the second order.

Thunder   The sound wave that results from the shock wave produced by the instantaneous expansion of air that is abruptly heated by a lightning bolt.

Thunderstorm   A relatively violent convective storm accompanied by thunder and lightning.

Tidal bore   A wall of seawater several inches to several feet in height that rushes up a river as the result of enormous tidal inflow.

Tidal range   The vertical difference in elevation between high and low tide.

Tides   The rise and fall of the coastal water levels caused by the alternate increasing and decreasing gravitational pull of the moon and the sun on varying parts of Earth's surface.

Till   Rock debris that is deposited directly by moving or melting ice, with no meltwater flow or redeposition involved.

Till plain   An irregularly undulating surface of broad, low rises and shallow depressions produced by the uneven deposition of glacial till.

Tombolo   A spit formed by sand deposition of waves converging in two directions on the landward side of a nearshore island, so that a spit connects the island to the land.

Topography   Surface configuration of Earth.

Tornado   A localized cyclonic low-pressure cell surrounded by a whirling cylinder of wind spinning so violently that centrifugal force creates partial vacuum within the funnel.

Torret   A suborder of Vertisol soils found in arid regions where cracks remain open most of the time.

Tower karst   Tall, steep-sided hills in an area of karst topography.

Tracheophyta   A division in the plant kingdom that consists of vascular plants that have efficient internal systems for transporting water and sugars and a complex differentiation of organs into leaves, stem, and roots.

Traction   Process in which coarse particles are rolled or slid along the stream bed.

Trade winds   The major wind system of the tropics, issuing from the equatorward sides of the Subtropical Highs and diverging toward the west and toward the equator.

Trajectory   The path followed by a moving body.

Transcurrent boundary   Two plates slipping past one another laterally in a typical fault structure.

Transferscope   An instrument that allows the operator to bring a map and photo, or photo and other image, to a common scale for comparison.

Transform fault   A fault produced by shearing, with adjacent blocks being displaced laterally with respect to one another. The movement is entirely horizontal.

Transmission   The ability of a medium to allow rays to pass through it.

Transpiration   The transfer of moisture from plant leaves to the atmosphere.

Transverse dune   A crescent-shaped dune that has convex sides facing the prevailing direction of wind and which occurs where the supply of sand is great. The crest is perpendicular to the wind vector, and aligned in parallel waves across the land.

Travertine   Massive accumulation of calcium carbonate.

Treeline   The altitude above sea level above (or below) which trees do not grow.

Trellis drainage pattern   A drainage pattern that is usually developed on alternating bands of hard and soft strata, with long parallel subsequent streams linked by short right-angled segments and joined by short obsequent and resequent tributaries.

Tropical cyclone   A storm most significantly affecting the tropics and subtropics, which is intense, revolving, rain-drenched, migratory, destructive, and erratic. Such a storm system consists of a prominent low-pressure center that is essentially circular in shape and has a steep pressure gradient outward from the center.

Tropical depression   By international agreement, an incipient tropical cyclone with winds not exceeding 33 knots.

Tropical disturbance   A term used by the Weather Service for a cyclonic wind system in the tropics that is in its formative stage.

Tropical rainforest   See Selva.

Tropical storm   By international agreement, an incipient tropical cyclone with winds between 34 and 63 knots.

Tropical year   The amount of time it takes for Earth to revolve completely around the sun (365.25 days).

Tropic of Cancer   The parallel of 23.5° north latitude, which marks the northernmost location reached by the vertical rays of the sun in the annual cycle of Earth's revolution.

Tropic of Capricorn   The parallel of 23.5° south latitude, which marks the southernmost location reached by the vertical rays of the sun in the annual cycle of Earth's revolution.

Tropopause   A transition zone at the top of the troposphere, where temperature ceases to decrease with height.

Troposphere   The lowest thermal layer of the atmosphere, in which temperature decreases with height.

True (geographic) north   The actual direction toward the North Pole from any point, measured along the meridian that passes through that point.

Tsunami   Very long sea wave generated by submarine earthquake or volcanic eruption.

Tufa (sinter)   Porous accumulations of calcium carbonate.

Tundra   A complex mix of very low growing plants, including grasses, forbs, dwarf shrubs, mosses, and lichens, but no trees. Tundra occurs only in the perennially cold climates of high latitudes or high altitudes.

Turbulent flow   The general downstream movement is interrupted by continuous irregularities in direction and speed, producing momentary currents that can move in any direction, including upward.

Typhoons   The term used for tropical cyclones affecting the western North Pacific region.

Ubac slope   A slope oriented so that sunlight strikes it at a low angle and hence is much less effective in heating and evaporating than on the adret slope, thus producing more luxuriant vegetation of a richer diversity.

Udalf   A suborder of Alfisol soils, characterized by brownish or reddish soils of moist midlatitude regions.

Udert   A suborder of Vertisol soils, found in humid areas where cracking is irregular.

Ultisol   A soil order similar to Alfisols, but more thoroughly weathered and more completely leached of bases.

Ultraviolet radiation   Electromagnetic radiation in the wavelength range of 0.2 to 0.4 micron.

Ultraviolet waves   Waves in the electromagnetic spectrum between 0.4 and 0.1 micrometer in length.

Undulation   A wavelike motion.

Ungulates   Hoofed mammals.

Uniformitarianism   The concept that the present is the key to the past in geomorphic processes. The processes now operating have also operated in the same way in the past.

Upslope fog   Condensation that occurs when humid air is caused to ascend a topographic slope and consequently cools adiabatically.

Urban heat island   Higher temperatures in the air over a city than over adjacent rural areas.

Ustalf   A suborder of Alfisol soils, characterized by brownish or reddish soils of subtropical regions and a hard surface layer in the dry season.

Ustert   A suborder of Vertisol soils, associated with monsoonal climates and with a complicated cracking pattern.

Uvala   A compound doline or chain of intersecting dolines.

Vadose zone (zone of aeration)   The topmost hydrologic zone within the ground, which contains a fluctuating amount of moisture (soil water) in the pore spaces of the soil (or soil and rock).

Valley   That portion of the total terrain in which a drainage system is clearly established.

Valley glacier   A long, narrow feature resembling a river of ice, which spills out of its originating basins and flows down-valley.

Valley train   A lengthy deposit of glaciofluvial alluvium confined to a valley bottom beyond the outwash plain.

Vector (of a disease)   An insect or other organism that transmits a disease.

Vein   Small igneous intrusions, usually with vertical orientation.

Vernal equinox   The equinox that occurs about March 20 in the Northern Hemisphere and September 22 in the Southern Hemisphere.

Vertebrates   Animals that have a backbone that protects their spinal cordfishes, amphibians, reptiles, birds, and mammals.

Vertical zonation   The horizontal layering of different plant associations on a mountainside or hillside.

Vertisol   A soil order comprising a specialized type of soil that contains a large quantity of clay and has an exceptional capacity for absorbing water. An alternation of wetting and drying, expansion and contraction, produces a churning effect that mixes the soil constituents, inhibits the development of horizons, and may even cause minor irregularities in the surface of the land.

Visible light   Waves in the electromagnetic spectrum in the narrow band between about 0.4 and 0.7 micrometer in length.

Volcanic ash   Fine particles of extrusive igneous rock blown out of a volcanic vent.

Volcanic neck   Small, sharp spire that rises abruptly above the surrounding land. It represents the pipe or throat of an old volcano, filled with solidified lava after its final eruption. The less resistant material that made up the cone is eroded, leaving the harder, lava-choked neck as a remnant.

Volcano   A conical mountain or hill from which extrusive material is ejected.

Vulcanism   General term that refers to movement of magma from the interior of Earth to or near the surface.

Wadi   The normally dry beds of an intermittent stream. This term is used in the Sahara Desert region.

Warm front   The leading edge of an advancing warm air mass.

Waterfall   Abrupt descent of a stream over a prominent knickpoint.

Water gap   A narrow notch in which a stream flows through a ridge.

Waterless zone   The fifth and lowermost hydrologic zone that generally begins several miles or kilometers beneath the land surface and is characterized by the lack of water in pore spaces due to the great pressure and density of the rock.

Water table   The top of the saturated zone within the ground.

Water vapor   The gaseous state of moisture.

Wave amplitude   One-half the wave height; i.e., the vertical distance from still-water level, either upward to the crest or downward to the trough.

Wave-built terrace   Submarine deposit of sand at the outer margin of an erosional platform or bench.

Wave crest   Highest point of a wave.

Wave-cut notch   An indentation in a sea cliff cut at water level by the combined effects of hydraulic pounding, abrasion, pneumatic push, and solution action.

Wave height   The vertical distance from crest to trough.

Wavelength   The horizontal distance from crest to crest or from trough to trough.

Wave refraction   Phenomenon whereby waves change their directional trend as they approach a shoreline.

Wave trough   Lowest part of a wave.

Weather   The short-term atmospheric conditions for a given time and a specific area.

Weathering   The physical and chemical disintegration of rock that is exposed to the weather.

Westerlies   The great wind system of the midlatitudes that flows basically from west to east around the world in the latitudinal zone between about 30° and 60° both north and south of the equator.

Wet adiabatic rate (pseudoadiabat)   The diminished rate of cooling, averaging about 2°F per 1000 feet (6.5°C/km), of rising air above the lifting condensation level.

Wetland   Landscape characterized by shallow standing water all or most of the year, with vegetation rising above the water level.

Wilting point   The point at which plants are no longer able to extract moisture from the soil because the capillary water is all used up or evaporated.

Wind gap   An abandoned water gap.

Winter solstice   The dates at which the most poleward extent of the perpendicular rays of the sun occur in the opposite hemisphere.

Woodland   Tree-dominated plant association in which the trees are spaced more widely apart than those of forests and do not have interlacing canopies.

Woody plants   Plants that have stems composed of hard fibrous materialmostly trees and shrubs.

Xeralf   A suborder of Alfisol soils found in regions of mediterranean climate and characterized by a massive hard surface horizon in the dry season.

Xerert   A suborder of Vertisol soils found in mediterranean climates and characterized by cracks that open and close regularly once each year.

Xerophytes   Plants that are structurally adapted to withstand protracted dry conditions.

Yazoo stream   A tributary unable to enter the main stream because of natural levees along the main stream.

Youth (youthful stage)   The initial, down-cutting stage in Davis's Geomorphic Cycle, in which the consequent streams are established and a drainage net begins to take shape.

Zone of aeration (vadose zone)   The topmost hydrologic zone within the ground, which contains a fluctuating amount of moisture (soil water) in the pore spaces of the soil (or soil and rock).

Zone of confined water   The third hydrologic zone below the surface of the ground, which contains one or more permeable rock layers (aquifers) into which water can infiltrate and is separated from the zone of saturation by impermeable layers.

Zone of saturation (phreatic zone)   The second hydrologic zone below the surface of the ground, whose uppermost boundary is the water table. The pore spaces and cracks in the bedrock and the regolith of this zone are fully saturated.

INDEX